Chiral Ana

MW00716853

Chiral Analysis

Edited by

KENNETH W. BUSCH
AND
MARIANNA A. BUSCH
Baylor University, Waco
Texas, USA

ELSEVIER

AMSTERDAM • BOSTON • HEIDELBERG • LONDON • NEW YORK • OXFORD • PARIS
SAN DIEGO • SAN FRANCISCO • SINGAPORE • SYDNEY • TOKYO

Elsevier
Radarweg 29, PO Box 211, 1000 AE Amsterdam, The Netherlands
The Boulevard, Langford Lane, Kidlington, Oxford OX5 1GB, UK

First edition 2006

Copyright © 2006 Elsevier B.V. All rights reserved

No part of this publication may be reproduced, stored in a retrieval system or transmitted in any form or by any means electronic, mechanical, photocopying, recording or otherwise without the prior written permission of the publisher

Permissions may be sought directly from Elsevier's Science & Technology Rights Department in Oxford, UK: phone (+44) (0) 1865 843830; fax (+44) (0) 1865 853333; email: permissions@elsevier.com. Alternatively you can submit your request online by visiting the Elsevier web site at http://elsevier.com/locate/permissions, and selecting *Obtaining permission to use Elsevier material*

Notice
No responsibility is assumed by the publisher for any injury and/or damage to persons or property as a matter of products liability, negligence or otherwise, or from any use or operation of any methods, products, instructions or ideas contained in the material herein. Because of rapid advances in the medical sciences, in particular, independent verification of diagnoses and drug dosages should be made

Library of Congress Cataloging-in-Publication Data
A catalog record for this book is available from the Library of Congress

British Library Cataloguing in Publication Data
A catalogue record for this book is available from the British Library

ISBN-13: 978-0-444-51669-5
ISBN-10: 0-444-51669-7

For information on all Elsevier publications
visit our website at books.elsevier.com

Transferred to Digital Printing in 2010

Working together to grow
libraries in developing countries

www.elsevier.com | www.bookaid.org | www.sabre.org

ELSEVIER BOOK AID International Sabre Foundation

Preface

This book is intended to serve as a resource for scientists interested in the general area of chirality and chiral analysis, in particular. The term chiral analysis, as used here, refers to the broad range of techniques that can be used not only for the determination of enantiomeric composition, but also includes methods for the determination of absolute configuration.

Asymmetric synthesis by means of enantioselective catalysis continues to be a major thrust in modern organic chemistry. In the area of chiral catalyst discovery, for example, catalyst screening to pick out the most likely candidates from a large library of potential enantioselective catalysts is often the limiting step. In the pharmaceutical industry, the need for improved strategies for the assessment of enantiomeric purity arises from the increased demand by governmental agencies for documentation on the pharmacological effects of individual enantiomers along with the simultaneous need for high-throughput screening methods to determine enantiomeric excess in large combinatorial libraries. In addition, rapid methods for the determination of the absolute configuration of newly synthesized molecules of pharmaceutical interest are important in the characterization of new drug molecules.

In spite of the important role of chiral analysis in drug discovery, there is currently no monograph available that covers this rapidly developing topic in its entirety. While some books are available on chromatographic and electrophoretic methods of chiral analysis, they do not include emerging chiroptical methods, mass spectrometric methods, or methods based on NMR. As a result, a major goal in compiling the chapters in this book has been to provide a monograph that covers both spectroscopic and separation techniques.

With this goal in mind, the eighteen chapters that make up the book have been divided into three parts. Part I consists of Chapters 1–4, and provides introductory material on chirality and the need for chiral analysis. Part II (Chapters 5–9) discusses chromatographic and electrophoretic methods for chiral separation, and Part III (Chapters 10–18) discusses spectroscopic techniques. While each chapter has been written to be self-contained, their arrangement in the book is intended to provide a unified, structured, logical sequence of topics. The book is written for science professionals with an undergraduate background in chemistry, and is targeted for a broad audience, including chemistry graduate students, analytical chemists, organic chemists, professionals in the pharmaceutical industry, and others with an interest in chirality and chiral analysis. One of us (KWB) would like to thank Baylor University for a summer sabbatical to work on this project.

Kenneth W. Busch
Marianna A. Busch
Waco, Texas

Contents

PART I

© 2006 Elsevier B.V. All rights reserved.

CHAPTER 1

A history of chirality

Seymour Mauskopf

Department of History, Duke University, Durham NC 27708-0719, USA

In the August 25, 2003 issue of Chemical and Engineering News, the results of a survey of the journal's readers on "the most beautiful experiments in the history of chemistry" were reported. First place was given to Louis Pasteur's manual separation of tartrate enantiomers (1848) [1]. One prominent chemist (and historian of chemistry) characterized its scientific significance:

> Pasteur's separation of optical isomers opened up an area of chemical structure particularly important to organic chemistry and biochemistry [2].

In particular, Pasteur's achievement constituted the basis for unfolding the concept of chirality.

In this chapter, I will explore the background to Pasteur's first major experimental work, explicate what he did in studying tartrates and other organic crystalline compounds, and, finally, trace the development of the chemical concept of chirality from Pasteur's discoveries.

1.1 THE CONCEPT OF CHIRALITY

The word "chirality" did not exist in Pasteur's day. Pasteur used the word "dissymétrie," or "dissymétrie moléculaire." The words "chiral" and "chirality" were introduced by William Thompson, Lord Kelvin, in 1884:

> I call any geometrical figure, or group of points, *chiral*, and say that it has chirality, if its image in a plane mirror, ideally realized, cannot be brought to coincide with itself [3].

"Chiral" and "chirality" have a classical derivation from the Greek word, "cheir," which means "hand." Pairs of human hands (and feet) are the most common and prominent examples of Kelvin's geometrical type; they are "identical opposites," mirror-images of each other [4]. The word "enantiomorph," closely associated with chirality, captures in its Greek-derived meaning of "opposite form" (*enantios morphe*) just this essence [5].

This is what Pasteur meant by "*dissymétrie moléculaire*." In a lecture that he delivered to the *Société chimique de Paris* in 1860, titled, "*Recherches sur la dissymétrie*

References pp. 21–24

moléculaire des produits organiques naturels," he speculated (almost uniquely for him) on the atomic arrangements in dextro- and levo-organic compounds that are optically active [6]:

> Are the atoms of the right acid grouped on the spirals of a dextrogyrate helix, or placed at the summits of an irregular tetrahedron, or disposed according to some particular dissymmetric grouping or other? We cannot answer these questions. But it cannot be doubted that there exists an arrangement of the atoms in a dissymmetric order, having a non-superposable image, and it is no less certain that the atoms of the *levo*-acid realize precisely the inverse dissymmetric grouping to this [7].

1.2 THE BACKGROUND TO PASTEUR

The basis for Pasteur's speculations on molecular dissymmetry lay in his discoveries regarding tartrate crystals that so entranced and impressed the readers of Chemical and Engineering News. Here is the traditional version of the discovery, originating from Pasteur's own accounts. In 1844, the German chemist, Eilhard Mitscherlich announced the discovery of a major anomaly to the conception of chemical isomers that was generally accepted at the time. Namely, in the case of the isomer pair, sodium-ammonium tartrate, and sodium-ammonium racemate (or paratartrate), Mitscherlich had found no differences in chemical compositions, specific weights, optical structures, or crystal forms of these isomers. Yet the tartrate isomer was optically active, the racemate inactive.

In the traditional version, Pasteur studied the crystal forms of the two isomers four years later with more finesse than had Mitscherlich, and discovered something that Mitscherlich had missed. The crystal forms were not identical at all but had subtle yet very important differences. They were all "hemihedral," i.e. they had tiny asymmetrically placed modifications on some of the edges. Sodium-ammonium tartrate crystals had the modifications on the right edges but the corresponding racemate crystals were more complicated. Half were like the tartrate crystals but other half had the modifications on the left edges, and were mirror images of the tartrate crystals. If the racemate crystals were separated, segregated, and then re-dissolved, both the solutions were optically active. Solutions of the tartrate-form crystals produced optical activity like that of the corresponding tartrate salt, but solutions of the mirror-image form were optically active in the opposite direction from that of the tartrate. It was the identification of these crystallographical differences by Pasteur, his meticulous and laborious separation of the racemate crystals into their two mirror-image forms, and his correlation of crystal form with optical activity, that constituted Pasteur's "most beautiful experiment in the history of chemistry."

The *bare* facts of this narrative are incontrovertible, and it is a good story with a nice moral for the aspiring scientist: pay close attention to what you are investigating! But, from the perspective of history of science, this narrative reads like an unproblematic case of empirical success in scientific research. Pasteur had simply observed his crystals more carefully than Mitscherlich.

I do not mean to deny the importance of Pasteur's observational skill and finesse in making his achievement possible. He *did* discover features of his crystals that Mitscherlich had earlier overlooked. However, important scientific discovery is almost never without a theoretical context. And, in this case, my own research and that of the late Pasteur scholar, Gerald Geison, have demonstrated that Pasteur's discovery had its own complex theoretical context in developments particularly associated with French science in the half century before his discovery [8]. I shall now proceed to my own narrative, based on our research.

The term, "isomer," meaning "composed of equal parts," had been coined by the Swedish chemist, Jöns Jacob Berzelius in 1830 to denominate substances possessing the same chemical composition but different chemical properties [9]. A focus of Berzelius' article had been the two organic acid isomers and their salts: tartaric and racemic (paratartaric) acid. Berzelius was one of the most important advocates of John Dalton's chemical atomic theory, and it followed from that theory that substances with identical chemical composition should have the same atomic makeup. With isomers, Berzelius assumed that the differences in chemical properties must arise from differences in the spatial arrangements of the constituent atoms. Another assumption was that, if the isomers existed in crystalline form, the differences in atomic arrangements ought to be manifested in different crystalline forms. Berzelius put Mitscherlich, his former pupil and a chemist who had done important work in crystallography, onto the study of the crystalline forms of the tartaric–racemic isomers to see if this assumption was borne out. It appeared to be for all the crystalline isomeric salts Mitscherlich studied except one pair: sodium-ammonium tartrate, and sodium-ammonium racemate, which appeared to have identical crystalline forms. But despite the interest in this anomaly to both Berzelius and Mitscherlich, Mitscherlich did not publish anything at the time, and further investigation was discontinued for over a decade.

In that interval, the French physicist, Jean-Baptiste Biot carried out extensive studies of the "optical activity" of a large number of organic compounds, including the tartaric-racemic isomers. Optical activity, or optical rotatory power, was the ability of substances, in these cases, in solution, to rotate the plane of polarized light passed through them [10]. Because optical activity was manifested in solutions of these substances, Biot believed it to be a *molecular* property of the substances, correlated with their chemical composition and arrangement.

Biot had found that tartaric acid and tartrates were optically active; racemic acid and racemates were not. In the early 1840s, Mitscherlich returned to his work on tartrates and racemates; the anomaly of sodium-ammonium tartrate, and sodium-ammonium racemate had been enhanced by Biot's discoveries regarding their optical activity, and he communicated his thoughts directly to Biot. The identities that Mitscherlich had found with these isomers, particularly the crystallographical one, would seem to indicate that "the nature, and the number of atoms, their arrangement and their distances, are the same in the two substances under comparison." But there was the dissimilarity in optical activity that Biot had discovered. Biot communicated

Mitscherlich's note to the Paris *Académie des Sciences* and underscored the anomaly in his own comments:

> But it would be difficult to conceive mechanically how *dissimilar* molecules, placed in the same number and at equal distances and arranged in the same manner, could produce material systems of similar types, whose crystal form and physical properties are as exactly parallel as the two substances under comparison here; at least nothing gives assurances of it and the contrary would be much more presumable [11].

Pasteur himself employed the note in the construction of the narrative of his great discovery:

> I meditated for a long time on that note; it troubled all my student ideas; I could not understand how two substances could be so similar as Mitscherlich said without being completely identical. To be capable of surprise is the first movement of the spirit towards discovery [12].

When Mitscherlich's note was read to the *Académie des Sciences* and published in its *Comptes Rendus*, the twenty-year-old Pasteur was embarking upon his scientific studies at the prestigious *École normale supérieure* in Paris. He left no record of when he actually read (and meditated on) Mitscherlich's note. But what we now know on the basis of my and Geison's research is that the path from meditating on the note, whenever that was, to the discovery was a complex one. Moreover, it involved Pasteur deeply in a particular French crystallographical/chemical tradition, which originated in the theory of crystallography devised in the late eighteenth century by the abbé-scientist, René-Just Haüy, and whose greatest mid-nineteenth-century exemplar was his own mentor in his postgraduate years at the *École normale*, the chemist Auguste Laurent. It is necessary to digress and give some of the background of this tradition.

In 1784, Haüy published an *Essai d'une théorie sur la structure des cristaux*, in which he presented, virtually full blown, the first comprehensive mathematical theory of crystal structure [13]. Appearing twenty-five years before Dalton's A New System of Chemical Philosophy [14], this work anticipated Dalton's in employing theoretical molecular units to construct a quantitative description of the macroscopic material world—in this case, of crystal structure.

Both Haüy and Dalton built their respective theories on what I have called the "molecular assumption:" all material bodies are composed of discrete material units—molecules—which have the same basic properties as the macroscopic bodies that they compose. Molecules are formed by chemical synthesis and then aggregate together to form the macroscopic body without losing their discreteness. Moreover, molecules of each substance are invariant in their properties, which are specific to that substance.

For Haüy, the essential property was molecular *shape* (for Dalton, it was weight). Haüy posited that crystal were built up out of small polyhedral molecules (*molécules intégrantes*) of shapes and dimensions specific to each crystalline substance. Envisioned as stacked together like bricks, the geometry of the molecular shapes and their aggregations generated the particular forms that crystals of a given substance could assume. As Haüy put it, the task of the crystallographer was:

> Given a crystal, to determine the precise form of its constituent molecules [later termed integrant molecule], their respective arrangements, and the laws which the variation of the [molecular] layers of which it composed, follow [15].

Crystals themselves were envisioned by Haüy as having a two-part structure: an inner core, "primitive form" or "nucleus" (*noyau*), often revealed when a crystal was cleaved, and external secondary form—the actual shape of the un-cleaved crystal. All parts of the crystal were composed of *molécules intégrantes* but the secondary forms were generated from the nucleus by means of recessed depositions of molecular layers to build up a kind of molecular pyramid on each face (Fig. 1.1). It was through his "determination" of the shapes of the molecules, and of the "laws of decrement," as he called them, that a quantitative, molecular description of crystal structure could be generated.

Although Haüy never adopted Dalton's chemical atomic theory, he did believe that his *molécules intégrantes* were chemical molecules, as invariant in their chemical composition as they were in their crystalline forms. The specifics of Haüy's molecular geometry underwent modification after Haüy, particularly under the impact of Daltonian atomism, but his general precept that crystalline form reflected atomic arrangement and molecular shape remained, as Berzelius' and Mitscherlich's concern with the anomaly of the sodium-ammonium tartrate–racemate isomers testifies.

During his education at the *École normale*, Haüy's theories reached Pasteur through transformations by two intermediaries: the mineralogist and crystallographer, Gabriel Delafosse, and the chemist, Auguste Laurent. Both of these scientists combined Haüy's crystal structure theory with Daltonian chemical atomism.

Delafosse, who had been Haüy's own student, was Pasteur's lecturer in mineralogy. Pasteur's notes on Delafosse's lectures survive, and they give a good idea of Pasteur's

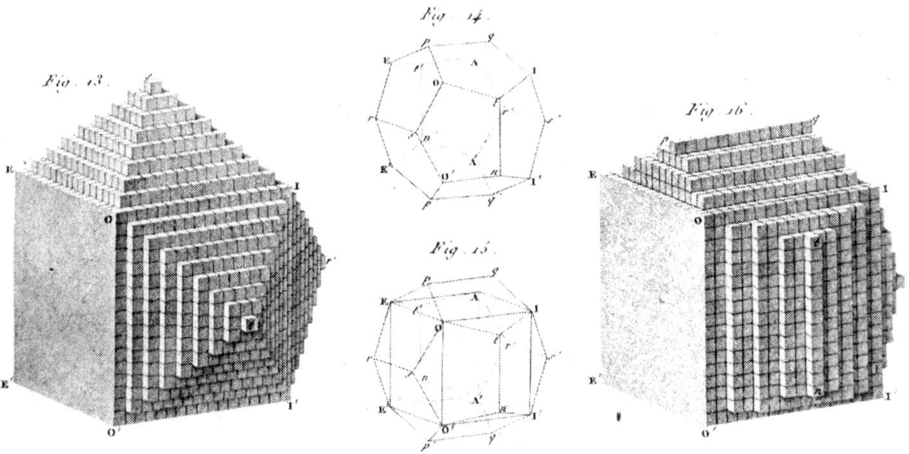

Fig. 1.1. Haüy's molecular structural model of iron pyrite crystals. Source: Figs. 15 and 16, René-Just Haüy, Traité de minérologie, 4 Vols. + Atlas, Chez Louis, Paris, 1801.

References pp. 21–24

introduction to contemporary French crystal structure theory and some of its research problems.

Delafosse had made two major modifications in his teacher's crystal structure theory. One was to distinguish between Haüy's *molécules intégrantes*, which he regarded as mathematical descriptions of the reticulated intermolecular spaces in a crystal, and what he regarded as the true physical/chemical crystalline molecule, comprised of atoms arranged in space. However, like Haüy's *molécules intégrantes*, Delafosse's physical/chemical molecules were polyhedral.

His theoretical program was to try to get at some of the actual shapes of these molecules and, to do this, he focused on crystals with hemihedral characteristics and with certain peculiar physical properties like surface striations, electrical polarity, and optical activity. Hemihedral crystals possess incomplete symmetry; the requirement that any modification of an angle or edge be reproduced on all other symmetrically placed angles and edges, was not fulfilled.

It is possible that Pasteur learned about Mitscherlich's note through the lectures of Delafosse. He certainly did learn about the striking case of optically active hemihedral quartz crystals. Normally, quartz crystallizes as hexagonal prisms with six-faced pyramids at each terminus [16]. But, in Pasteur's words:

> Haüy was the first to observe, in certain specimens, a face very different from these, which he designated by the letter x, inclined more towards one side than the other, without being double, as the law of symmetry would require in this case. Another strange peculiarity of these crystals did not escape crystallographers, namely, that this face x was inclined sometimes in one sense, sometimes in the other. Haüy, who liked to give suitable names to each variety of a species called the variety of quartz bearing the face x *plagihedral.* The crystals in which the face x was inclined to the right, when the crystal was oriented in a given manner, were called right plagihedra; those in which x was inclined in the opposite sense, left plagihedra [17].

In fact, quartz was the first substance in which optical activity had been found [18], and in 1820, the English scientist, J.W.F. Herschel made the correlation between the left-handed and right-handed hemihedry of quartz crystals that were optically active, and the directions in which they deviated the plane of polarized light [19].

Delafosse attempted to devise a model of a quartz physical molecule that would account for these properties. He accepted the traditional view that the basic shape was rhombohedral, like the primitive form of quartz crystals generally. But to account for the hemihedry of the optically active varieties (and the optical activity), Delafosse proposed that these molecules were themselves asymmetrical [20].

In his retrospective construction of the path leading to his discovery, Pasteur gave considerable credit to Delafosse for establishing a proper mindset in him. He claimed that he was "guided" by Biot's studies of "molecular rotatory polarization" in the tartrates, by Herschel's correlation between optical activity and hemihedry in quartz crystals, and finally, by the "sagacious views" of Delafosse:

> With whom hemihedry has always been a law of structure and not an accident of crystallization, I believed that there might be a relation between the

hemihedry of the tartrates and their property of deviating the plane of polarized light [21].

In this retrospect account (and others), Pasteur omitted any mention of the other scientific intermediary, Auguste Laurent. In my view and Geison's, Laurent's influence on Pasteur was the most significant of any of his teachers or mentors. Space constraints preclude any discussion of Laurent's fascinating and tragic career and his important place in the development of structural organic chemistry. Suffice it to say that he, like Delafosse, profoundly influenced by Haüy's crystallography, believed that there was an intimate relationship between crystal form and atomic–molecular arrangement within the crystal. Moreover, he extended, by analogy, Haüy's two-part crystal structure model to the explication of organic chemistry taxonomy. Laurent believed that chemical properties and relations depended ultimately on chemical structure. Taking as his point of departure, organic substitution reactions, he suggested that families of similar chemical substances all shared a common nuclear "radical," modified among the members of a family (e.g. naphthalene compounds) by substitutions by atoms (or atomic groups) of other elements of the hydrogen atoms in the outer layers of the molecule:

> I shall imagine a given hydrocarbon $C^{12}H^{12}$, if you wish, and I shall represent it by a prism of six sides, having at the twelve solid angles twelve atoms of carbon and in the middle of the edges of its bases twelve atoms of hydrogen. This prism could be surmounted by two pyramids as modifications, one on each base. If the pyramid is of water, the formula will be $C^{12}H^{12} + 2HO$½ or $2H^2O$ and it will represent an ether or an alcohol. If these pyramids are replaced by others of sulfuric acid, hydrochloric acid, etc.,... salts will be formed which ought to be represented by an hexagonal prism plus the modifications of the bases....
>
> Thus one sees that, just as in crystallography, one could remove the modifications, or a part thereof, without replacing them, or replacing them wholly or partially.
>
> But it will not be the same in the central frame, or in the radical; one cannot remove from it a sole piece – that is to say – a sole atom without destroying it, at least if one does not replace the removed atom by another equivalent atom which will continue to maintain the equilibrium of the frame [22].

Pasteur graduated from the *École normale*, in 1846 but continued there as an *agregé* to the chemist, Antoine Jerome Balard for two years until the fall of 1848. During this period, Pasteur wrote theses in chemistry and physics for his doctorate, and, sometime in the spring of 1848, made his momentous discovery. During just these years, Laurent was also working in Balard's laboratory, and served, in effect, as Pasteur's mentor for the theses and the subsequent work leading to the discovery.

In the early 1840s, Laurent developed his crystallographical–chemical reasoning in a number of directions. One was to extend the concept of "isomorphism," devised by Mitscherlich some twenty-years earlier. This concept linked chemistry and crystallography by positing a close correlation between crystal form and chemical combination:

> The same number of atoms combined in the same manner produced the same crystalline form, and [the same crystalline form] is independent of the chemical nature of the atoms and it is only determined by the number and relative position of the atoms [23].

Laurent proposed to subsume under this concept, cases where chemically substances appeared to share a common radical but crystallized in incompatible crystalline forms:

> From the chemical point of view, I think that the barrier raised between the different crystalline systems ought to be removed. Let us suppose that, in a hydrocarbon, the atoms are grouped cubically, and that each face carries an atom of hydrogen. If two opposite atoms of hydrogen are replaced by two atoms of chlorine, these latter can elevate or depress slightly the cube in one direction, and one will obtain a prism with a square base, only slightly different from the cube [24].

By the same reasoning, the crystallographical and chemical distinctions in the concepts of dimorphism [25] and isomerism might also be blurred.

In fact, Pasteur was exploring just these kinds of interrelations when he made his discovery: whether crystals of eight tartrate salts could be characterized as isomorphic (in Laurent's wider sense) even though five crystallized in a form incompatible with that assumed by the other three. This project led him to inquire closely as to their chemical formulas (particularly their water of crystallization) and their crystallography. While closely examining whether two of these (sodium-ammonium tartrate and sodium-potassium tartrate) (Seignette salt) salts were truly isomorphic, Pasteur noticed tiny hemihedral facets on their crystals (Figs. 1.2 and 1.3). Although at first insensitive to their implications for asymmetry, he soon picked up on the fact that they were always oriented in one direction. He then turned to the case of sodium-ammonium racemate (or paratartrate), whose crystallization he expected to be different from the corresponding tartrate because of the differences in the water of crystallization. His anticipation was that the racemate crystals would be symmetrical but he soon found a confusing situation: some were hemihedral in the same way as the tartrate but others were hemihedral in the opposite sense [26].

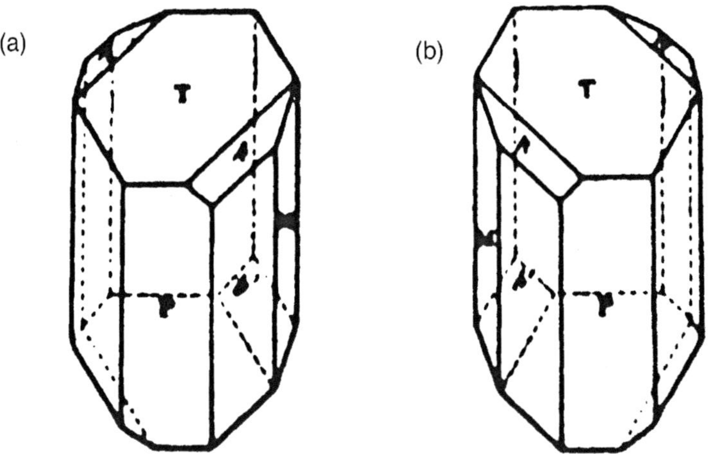

Fig. 1.2. Hemihedral crystals of sodium-ammonium tartrate. Source: G.B. Kauffman and R.D. Myers, The resolution of racemic acid: a classic stereochemical experiment for the undergraduate laboratory. J. Chem. Educ., 52 (1975) 778.

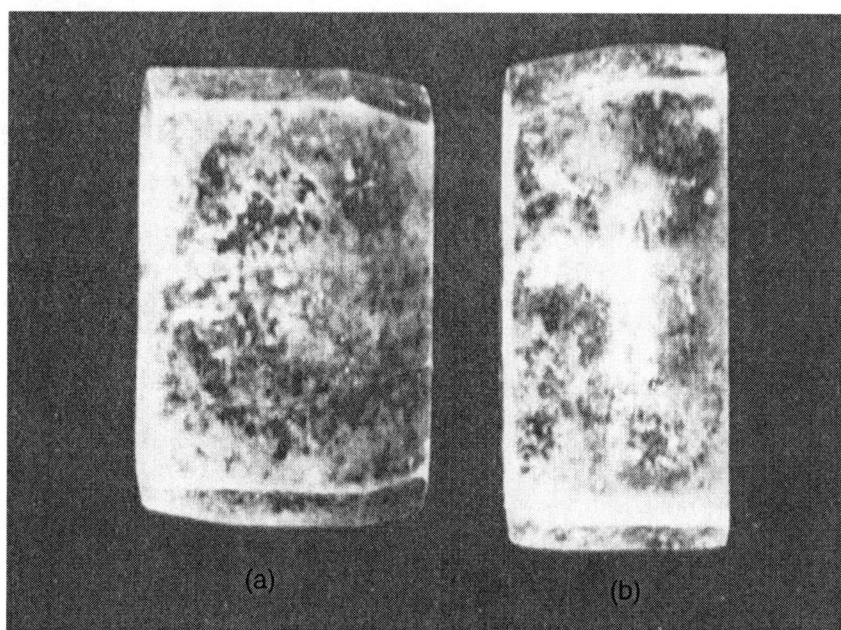

Fig. 1.3. Sodium-ammonium tartrate crystals, L- and D-forms. Source: G.B. Kauffman and R.D. Myers, The resolution of racemic acid: a classic stereochemical experiment for the undergraduate laboratory. J. Chem. Educ., 52 (1975) 780.

Exactly what his sequence of thoughts and actions were during these experiments is hard to determine from the laboratory notes but Geison has argued persuasively that it was only at the point when Pasteur accepted that he had indeed discovered two opposite forms of hemihedral crystals in the racemate that he thought to correlate this with their optical activity. Certainly at this point, if not earlier, the thoughts and discoveries Mitscherlich, Biot, and Delafosse must have intersected with those of Laurent, which had been guiding him heretofore.

There is a danger in my emphasis of the theoretical context in this reconstruction of Pasteur's discovery in losing sight of the discovery itself—the sighting of the hemihedral facets. My point is that the significance of such sightings might well have been reduced or even lost on a scientist who was not so deeply immersed in the kind of crystallographical/chemical research and reasoning that characterized Pasteur's Laurent-inspired research at this time.

Yet the sightings themselves do testify to a very high degree of laboratory artistry on Pasteur's part. There was also a considerable degree of contingency associated with them. As George Kauffman has noted, Pasteur was very fortunate both in his choice of which tartrate/racemate he was studying, and in the ambient atmospheric conditions in which he carried out his observations and separations. The sodium-ammonium racemate was one of only two in this class of crystals that are resolvable into mirror-image hemihedral forms. Moreover, even this only would occur when the temperature is below 26°C. Finally, it is even possible that Pasteur's sighting was aided by the presence in the laboratory air of "proper" dust, containing seed crystals of both hemihedral forms [27].

Pasteur first framed his discoveries regarding the tartrate and racemate in Laurentian terms. In a draft written after he had recognized the hemihedry of his tartrate and racemate crystals (possibly!), Pasteur used language that could have been Laurent's:

> All the tartrates are hemihedral. Thus, the molecular group common to all these salts, and which the introduction of water of crystallization and of oxides comes to modify at the extremities, does not receive the same element at each extremity, or, at least, they are distributed in a dissymmetrical manner. On the contrary, the extremities of the prism of the paratartrates are all symmetrical. Behold all the difference. It suffices to explain the isomerism of these two types of salts, and, at the same time, we see that this isomerism is not very profound. There is a difference in the molecular arrangement of tartrates and paratartrates, but this difference seems only to stem from a regular distribution of the oxide or water of crystallization molecules in the one case and a dissymmetrical distribution in the other [28].

The exception to this racemate symmetry, the case of his discovery, "could be accommodated to this Laurentian scheme by recognizing that it could be separated into two 'veritable' tartrates, one hemihedral to the left, the other to the right [29]."

But by the time Pasteur published his discovery, this kind of imagery had been excised, and the discovery of agent of Laurent would soon follow. The latter may, as Geison suggested, have been due to career opportunism on Pasteur's part [30]; it may also have been due to the ascendant positivism that was coming to characterize French scientific and chemical thought at mid-century—the two explanations may, in fact, be reinforcing.

Positivism—in which speculations about atomic and molecular arrangements or even about atoms and molecules were eschewed—became ascendant among French chemists in the second half of the nineteenth century, with the significant exception of Adolphe Wurtz [31].

This having been said, it is difficult for me to say how Pasteur might have made any further conceptual advance on the atomic and molecular nature of chirality had he continued to work under the Laurentian models that had guided him in his work and his initial interpretation of his discovery. As Geison has put it:

> Laurent's models were flexible enough to accommodate any consistent crystallographic difference between the sodium-ammonium tartrate and its corresponding paratartrate [32].

There was nothing in Laurent's molecular approach that I can see would have helped to elucidate chirality. In any case, although Pasteur continued to carry out research on optical activity and crystalline asymmetry in organic substances for a full decade after his discovery, he never attempted to speculate about causes at the atomic and molecular level again, with the exception of the highly suggestive, if very general, suggestion in his 1860 lecture, "*Recherches sur la dissymétrie moléculaire* [33]."

1.3 TOWARDS A "CHEMISTRY" IN SPACE

If French chemistry moved towards a positivist, even anti-atomistic stance after mid-nineteenth century, chemists, elsewhere, particularly in Germanic countries, were developing a wealth of ideas on chemical structure. A number of advances in chemistry in the 1850s, associated with what Alan Rocke has termed "the quiet revolution [34]," enhanced the ability of chemist, so inclined, to envision structural formulas for organic compounds. These included the clarification of (and agreement on) atomic and molecular weights and molecular formulas and the development of the concept of valence [35]. August Kekulé, Aleksandr Butlerov, and Alexander Crum Brown all developed ideas and graphic representations of chemical structure. But, as Peter Ramberg put it, these graphics were "symbolic forms of representation" and not "iconic images;" they were not meant to be "'windows' into the physical reality of the molecule [36]." None dealt with three-dimensional mapping of atomic arrangement, and certainly not with the issues of *dissymétrie moléculaire* and optical activity that had exercised Pasteur, if only for a while.

These issues reappeared in the 1860s in the research of the German chemist, Johannes Wislicenus. Receiving his doctorate in 1860 for a thesis on structural chemistry, Wislicenus took up the study of lactic acid, particularly challenging to a structuralist because of its hybrid behavior as an acid and an alcohol. The issues were sharpened for him by his discovery of isomers that were optically active and optically inactive. Throughout the 1860s, Wislicenus made tentative conceptual moves beyond the chemical structural formulas and the graphic representations of them towards something more physically real and three-dimensional. A representative example dates from his lectures in the Zúrich Polytechnical Institute, from the notes taken by Robert Gnehm:

> Our formulas (structural formulas) are pictures in a plane, but molecules are nevertheless bodies, for example [Gnehm gives the structure for ethanol with Crum Brown notation] gives us the sequence of atoms, but by no means the three-dimensional arrangement of the molecule's atoms in space. We still lack clues about the mathematical form of the molecule. But certain essential properties of the molecule depend on these spatial properties, at least, for example, the effect of light. [and] the crystal form [37].

But Wislicenus went no further at this point.

1.4 VAN'T HOFF, LE BEL, AND THE TETRAHEDRAL CARBON ATOM

In 1874, two scientists who had recently been fellow researchers in Wurtz's laboratory at the *École de Médicine* in Paris, fulfilled Wislicenus' quest for models of "the three-dimensional arrangement of the molecule's atoms in space." They were the Dutchman, Jacobus Henricus Van't Hoff and the Frenchman, Joseph-Achilles Le Bel.

Van't Hoff published first and was directly inspired by Wislicenus' work. He had studied chemistry at the technical school in Delft and then worked briefly in a sugar factory. He then turned to mathematics and physics at the University of Leiden but returned to chemistry with a research stay with Kekulé at Bonn (apparently disappointing) and a doctoral degree at the University of Utrecht (December, 1873). Finally, he worked in Wurtz's laboratory in Paris. Van't Hoff thus had a combined background in chemistry, in structural chemistry (with Kekulé and Wurtz) and in mathematics and physics. To Ramberg, this last was very important, giving Van't Hoff a "Galilean" conception of nature (and its molecules) as being inherently geometrical [38]. But his geometry was not abstract mathematics; Van't Hoff's objective was to provide "a more definite statement about the actual positions of the atoms [39]."

It is more difficult to correlate Le Bel's educational trajectory with his interest in elucidating atomic and molecular arrangements. A graduate of the *École polytechnique*, Le Bel then studied at the *Collège de France* and the *École de Médicine*, where he worked with Wurtz at the same time as Van't Hoff. It is most likely that Wurtz's influence was strong in directing Le Bel towards issues of atomic structure, as it was in Van't Hoff's case.

Le Bel never held an academic post and, unlike Van't Hoff, who went on to a very distinguished career in chemistry (both as a developer of stereochemistry and as a founder of physical chemistry), Le Bel never followed up his original 1874 article on stereochemistry. He published in a variety of fields in and outside of chemistry, in none of which did he achieve the distinction of the 1874 publication. Of that publication, Ramberg wrote:

> His paper might well have drifted off into obscurity...if in the later versions of his own theory Van't Hoff had not repeatedly given credit to Le Bel for the same idea [40].

A comparison of the titles of the two articles gives a clue into their somewhat different points of departure and foci. The lengthy title of Van't Hoff's original pamphlet clearly places it in the tradition of structural chemistry: "Proposal for Extending the Currently Employed Structural Formulae in Chemistry into Space, Together With a Related Remark on the Relationship Between Optical Activating Power and the Chemical Constitution of Organic Compounds [41]." The conception of Le Bel was more restricted—or focused—to this issue of optical activity, as its title illustrates: "On the Relationships that Exist Between the Atomic Formulas of Organic Bodies and the Rotatory Power of their Solution [42]."

And the research problem each identified early in his article illustrated this difference. For Van't Hoff, the primary problem was the inability of structural formulas to explain isomerism adequately. For Le Bel, the problem was the lack of a "certain rule" for predicting optical activity. Pasteur's correlation between "molecular asymmetry" and optical activity was invoked almost immediately by him [43]; Van't Hoff only raised this issue after he had outlined his approach for dealing with isomers [44]. Moreover, there was a subtle but important difference in the two writers' attitude towards elucidating atomic arrangement. Van't Hoff posited a realistic (or, in Ramberg's terminology, "iconic") approach to the issue of atomic arrangement at the outset [45]; By contrast, Le Bel was more agnostic and abstract [46].

Van't Hoff's realism was vividly displayed (in a literal graphical sense) in the manner with which he dealt with the challenge that he claimed had stimulated his thinking in the first place: Wislicenus' studies of lactic acid and the problem of the lactic acid isomers and optical activity. His response was remarkably simple. Assuming that the four valences of a carbon atom were satisfied by bonds that were fixed and rigid, he imagined them directed to the four corners of a tetrahedron. To deal with optically active isomers, he made the following additional assumption:

> In cases where the four affinities of the carbon atom are saturated with four mutually different univalent groups, two and not more than two different tetrahedral can be formed, which are each other's mirror images, but which cannot ever be imagined as covering each other, that is, we are faced with two isomeric structural formulas in space (Figs. 1.4 and 1.5) [47].

Van't Hoff suggested that the arrangement of his atoms in optically active molecules was analogous to the arrangement of molecules in the enantiomorphic crystals of optically active substances discovered by Pasteur [48].

Although it dealt brilliantly with the case of optical activity and *dissymétrie moléculaire*, Van't Hoff's model of the tetrahedral bonding of carbon was intended as a general geometrical structural model for all carbon bonding. Thus, Van't Hoff readily

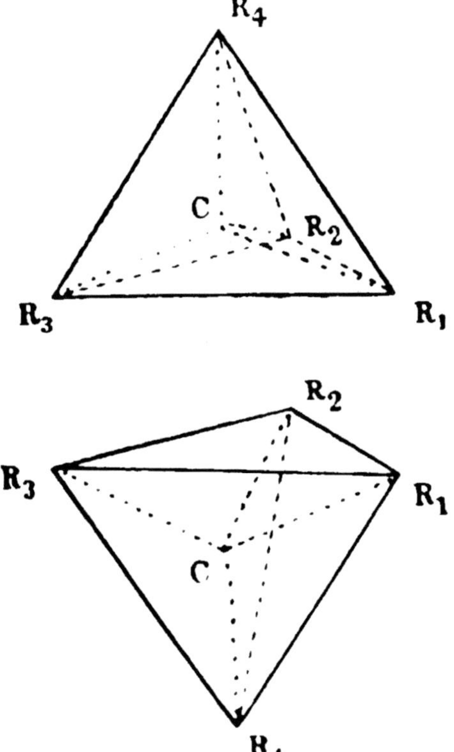

Fig. 1.4. Van't Hoff's illustration of enantiomers $CR_1R_2R_3R_4$. Source: O.B. Ramsay, Stereochemistry, Heyden, London, 1981, p. 83.

Fig. 1.5. Van't Hoff's cardboard tetrahedral models. Source: O.B. Ramsay, Stereochemistry, Heyden, London, 1981, p. 83.

extended the tetrahedral model to the case of unsaturated carbon bonding, where there existed double bonding (or "double linking," as Van't Hoff termed it) or triple bonding (acetylene). Double bonding was graphically represented as two tetrahedra with one edge in common (Fig. 1.6) and triple bonding by two tetrahedra with three summits or a face in common (Fig. 1.7) [49].

Le Bel's article was not as chemically inclusive as Van't Hoff's. Nor, because of his greater agnosticism regarding atomic structure, did he postulate a tetrahedral model of carbon bonding (although he assumed an equivalence to the tetravalency of the carbon in his argument) or employ the kind of perspective-based graphics that Van't Hoff used to visualize this bonding. His argument was couched in more abstract terms of symmetry:

> Let us consider a molecule of a chemical compound having the formula MA_4; M being a single or complex radical combined with four univalent atoms A, capable of being replaced by substitution. Let us replace three of them by simple or complex univalent radicals differing from one another and from M; the body obtained will be asymmetric.
>
> Indeed, the group of radicals R, R′, R″, A, when considered as material points differing among themselves form a structure which is enantiomorphous with its reflected image, and the residue M cannot re-establish the symmetry [50].

His second principle was that substitution of only one or two groups on MA_4 would result in either a symmetric or asymmetric substance depending on whether symmetry was present originally. If it were, then the substitution of two groups would produce

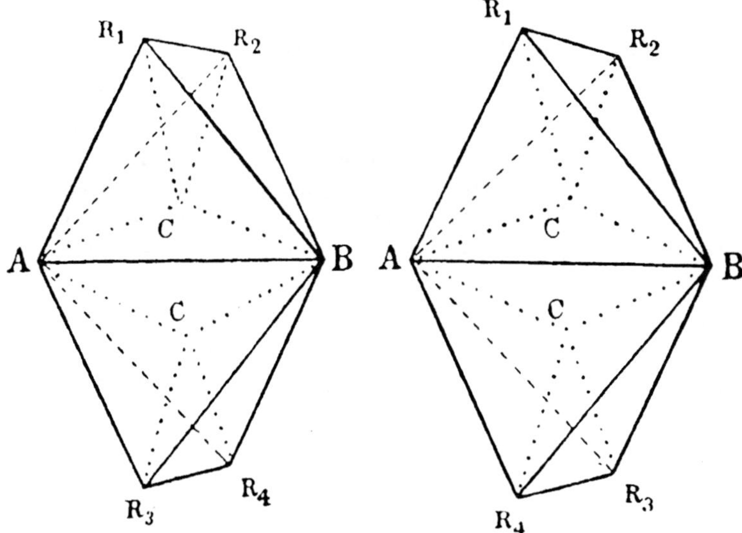

Fig. 1.6. Van't Hoff's model of double bond isomers. Source: O.B. Ramsay, Stereochemistry, Heyden, London, 1981, p. 86.

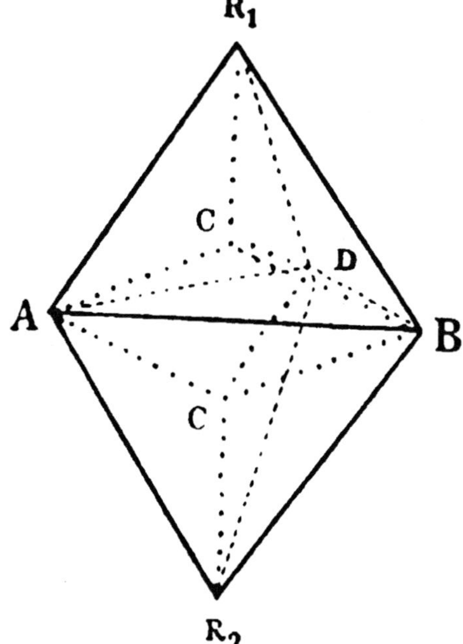

Fig. 1.7. Van't Hoff's model of an alkyne (triple bond). Source: O.B. Ramsay, Stereochemistry, Heyden, London, 1981, p. 87.

a symmetrical—i.e. an optically inactive—molecule. It was only here that he introduced the geometry of the tetrahedron, but rather obscurely and as a kind of addendum, rather than as a fundamental datum about carbon bonding, and certainly not as a graphic for illustrating asymmetry:

> Again, if it happens not only that a single substitution furnishes but one derivative, but also that two and even three substitutions give only one and the same chemical isomer, we are obliged to admit that the four atoms A occupy the angles of a regular tetrahedron, whose planes of symmetry are identical with those of the whole molecule MA_4; in this case, no bisubstitution product can have rotatory power [51].

Although Le Bel did not refer to tetrahedra again in the article, and offered no graphics, what he did could be interpreted (and has been) as utilizing, in effect, something close to Van't Hoff's spatial models. Namely, he used as his basis of analysis for "saturated bodies of the fatty series" the methane or "marsh gas" type of a carbon atom bonded to four hydrogen atoms or to their radical substitutes. In considering whether a compound derived from this "type" would be optically active, Le Bel applied his "general principles." Thus, those cases where three of the hydrogen atoms were substituted by three different radicals (first principle) yielded optically active compounds [52].

For unsaturated isomers (ethylene derivatives), Le Bel postulated that the hydrogen atoms lie "at the angles of a hemihedral quadratic pyramid superposable upon its image P/2, and we should obtain by two substitutions two isomers, one of which would be symmetrical and the other unsymmetrical [53]." Le Bel went beyond Van't Hoff (at this time) to try to account for optical activity in the aromatic series [54].

1.5 RECEPTION OF THE TETRAHEDRAL CARBON MODEL

As already noted, Van't Hoff's and not Le Bel's theory prevailed. This was as much due to Van't Hoff's development and marketing of his ideas as to any intrinsic superiority of them over Le Bel's. Van't Hoff's original Dutch pamphlet, necessarily, limited in accessibility, received a translation into French and publication in the *Bulletin de la société chimique de France* [55]. Of far greater moment, Van't Hoff expanded his ideas in a book form with the title, *La chimie dans l'espace* [56], and sent copies along with cardboard molecular models to many prominent chemists, one of whom was Wislicenus.

The chemist who had inspired Van't Hoff in the first place was, in turn, so enthusiastic about *La chimie dans l'espace* that he organized a translation into German by his assistant, Felix Hermann as *Die Lagerung der Atome in Raume* [57]. It was through this version that Van't Hoff's ideas became known to the German chemical community. Moreover, Wislicenus himself contributed to both the further development of Van't Hoff's ideas and advocated them among his German colleagues. It would go far beyond the scope of this chapter to discuss either of these activities, which are detailed by Ramberg [58]. By the time of the second German edition (1894), Wislicenus could write:

> For that matter, the previous categorical resistance has almost died out; where it does still rise, it turns to the last foundation: against the atomistic view itself, but it does not deny that the doctrine of spatial atomic arrangements is a stage, perhaps the last,

> in the consistent and necessary development of chemical atomism. The opposition is primarily directed—indeed often justified—towards particular applications that explain individual, real relationships and processes, but that do not actually question the theory's basic content [59].

In the early twentieth century, chirality was extended to non-organic compounds. The development of coordination chemistry by Alfred Werner for metal amines led to the prediction (and identification) of optically active, asymmetrical compounds, whose core was a metallic rather than a carbon atom. Although the first one identified in 1911 by Werner's American student, Victor King, were carbon containing compounds, by 1914, an optically active cobalt complex containing no carbon was identified [60].

1.6 CHIRALITY AND LIFE

Pasteur carried on his research into optical activity and "dissymétrie moléculaire" of organic substances for ten years. In 1857, he made a seemingly abrupt change in research orientation: abandoning the basic research of crystallography and chemistry for practical research on fermentation. This, in turn, led him further and further from his original orientation to those biological and biomedical achievements for which he is best remembered: disproof of spontaneous generation and the germ theory of disease.

Although much of the biographical literature on Pasteur has traced this change of research to the agricultural and industrial context to his life in Lille, a center for the manufacture of beetroot alcohol, Geison has shown that the cause has deeper roots in a "preconceived idea" that may have gone back to the start of Pasteur's earlier scientific research: "the power to rotate polarized light [i.e. optical activity] has never been found in a compound [*corps*] artificially prepared from other compounds not possessing this power [61]."

This view that optical activity (and molecular dissymmetry) could only arise from organic materials produced by vital processes and/or through the agency of life remained with him throughout his life. It led him to envisage a fundamental molecular difference between "living nature" (dissymmetric) and "dead nature" (symmetric). Pasteur was not, however, a traditional vitalist; he believed that molecular dissymmetry ultimately must have some physical cause, which he enveloped in rather grandiose terms:

> Can these asymmetric actions be connected with cosmic influences? Do they reside in light, electricity, magnetism, heat? Could they be related to the movement of the earth, the electrical currents by which physicists explain the terrestrial poles [62]?

Pasteur had actually tried to test whether he could invoke experimentally some of these "cosmic influences." In his first position, at Strasbourg, he had asked Heinrich Daniel Ruhmkorff to build him some powerful magnets; subsequently, in Lille, he turned to other mechanisms to alter the polarity of solar radiation [63].

But these stratagems *per se* proved to be ineffective in producing optical activity. Paulo Palladino has laid out the vigorous debates over both the origin of life and of the possibility of creating optically active organic substances "artificially" that ensued in the late nineteenth and early twentieth century; the issue of this possibility is still unresolved [64].

1.7 OPTICAL ACTIVITY AND DRUG ACTION

Not only did Pasteur elucidate the correlations between optical activity, molecular dissymmetry, and vital agent; he also discovered and studied the abilities of optically active compounds or organisms to affect selectively the direction of chemical reactions to produce only one of the enantiomers. For instance, he had found that a microbe that fermented ammonium paratartrate (racemate), selectively metabolized the right-handed enantiomer while leaving the left-handed one alone. Studying the course of the reaction with a polarimeter revealed an increasing optical activity to the left [65].

This discovery was extended to physiological correlates and, by the early twentieth century, to pharmacological ones by Arthur Cushny (1866–1926) [66]. Born and trained in Scotland (to where he eventually returned) and Germany in pharmacology, he was appointed to a chair in material medica and therapeutics at the University of Michigan in 1893. It was here that he carried out his first work on the differential pharmacological action of optical isomers, L-hyoscyamine and atropine, a racemic mixture of the L- and D-hyoscyamine [67]. He suggested that the differential effect (the L-hyoscyamine had twice the activity of atropine) could be explained by the assumption that D-isomer was inactive and the atropine contained equal amounts of each isomer. Over the next five years, Cushny published two more examples [68].

Although he used the term "chemical affinity" as an explanation of the differential effects, Parascandola argues against ascribing to Cushny a chemical interpretation. For Cushny, difference in physical properties between isomers like solubility seemed always to have primacy over chemical properties. Nor, although he used the term, "receptive substance," did he mean anything like the Ehrlich–Langley theory of receptors, or Fischers "lock and key" analogy for enzyme bonding. Nor, finally, did Cushny attribute much importance to the spatial configuration of the isomers to account for their differential action.

However, Cushny knew that he could not completely ignore the chemical nature and configuration of the molecular reactants. He devised an explanation of the differential pharmacological action of optical isomers that involved consideration of the compounds the two isomers formed with its optically active receptive substance: he posited that they would form two diastereomers (stereoisomers that are dissymmetric but not mirror images) that would have different properties, particularly physical ones.

There were contemporaries of Cushny who did try to account for the differential pharmacological behavior of optical isomers in terms of their molecular configurations but it was only in 1933 that Leslie Easson and Edgar Stedman of the Department of Medicinal Chemistry at the University of Edinburgh came up with a detailed theoretical and experimental scenario. They argued that the pharmacological effect of a drug depended on the attachment or "fit" of the groups bonded to its asymmetrical carbon atom to the receptor. When three of the four groups were aligned to corresponding groups on the receptor molecule, there was maximal effect. By the 1950s, this explanation approach had become ascendant [69].

REFERENCES

1 M. Freemantle, Chemistry at its most beautiful. Chem. Eng. News, August 25, 2003, pp. 27–30. Lavoisier's work on the oxidation of metals only came in second.

2 C.J. Giunta, Chem. Eng. News, August 25, 2003, p. 27.

3 Lord Kelvin, Baltimore Lectures, 1884. http://www.chem.gla.ac.uk/~laurence/Chirality.htm W.H. Thomson, Lord Kelvin, Baltimore Lectures on Molecular Dynamics and the Wave Theory of Light, C.J. Clay & Sons, London 1904, pp. 436 and 619. Reference from: http://www.flack.ch/howard/cristallo/cacs.pdf, Footnote 29.

4 See http://www.chiraltech.com/new/introduction.htm for a discussion of this term. The term does not seem to have figured in stereochemical accounts until the mid-1960s. See E.L. Eliel, S.H. Wilen and M.P. Doyle, Basic Organic Stereochemistry, Wiley Interscience, New York, 2001, p. 3. Eliel did not use the terms "chiral" or "chirality" in his earlier book, Stereochemistry of Carbon Compounds, McGraw-Hill, New York, 1962, but instead suggested Pasteur's term, "dissymmetric" (p. 12).

5 See http://www.madsci.org/posts/archives/oct99/941203265.Sh.r.html for suggestion that "enantiomorph" was coined by the German crystallographer, Karl Friedrich Naumann in his Handbuch der theoretische Krystallographie (1856). Chemically, "enantiomorph" is often used interchangeably with "enantiomer."

6 Optical activity, involving the rotation of plane-polarized light, will be discussed later.

7 Quoted from O.B. Ramsay, Stereochemistry, Heyden, London, 1981, pp. 77–78.

8 S.H. Mauskopf, Crystals and Compounds: Molecular Structure and Composition in Nineteenth-Century French Science, Chapter 6, Transactions of the American Philosophical Society, NS, Vol. 66, Part 3, The American Philosophical Society, Philadelphia, 1966; G.L. Geison, The Private Science of Louis Pasteur, Chapter 3, Princeton University Press, Princeton, 1995.

9 J.J. Berzelius, Composition de l'acide tartrique et de l'acide racémique (traubensáure). Annales de chimie et de physique, 46 (1831), 113–147.

10 The first case discovered, and the major inorganic example of an optically active substance, known at this time, was quartz.

11 J.B. Biot, Communication d'une note de M. Mitscherlich, Paris. Comptes rendus hebdomadaires des séances de l'Académie des sciences, 19 (1844) 723.

12 L. Pasteur, La dissymétrie moléculaire (Conférence faite à la Société chimique de Paris le 22 Decembre 1883), in: P. Vallery-Radot (Ed.), Oeuvres de Pasteur, 7 Vols., Masson et Cie, Paris, 1922–1939, 1, p. 370.

13 Gogue & Neé de la Rochelle, Paris. Some of his idea had been adumbrated earlier by Torbern Bergman and, more generally, J.-B.L de Romé de l'Isle.

14 2 Vols., R. Bickerstaff, London. The first volume containing the first comprehensive published account of the chemical atomic theory by Dalton appeared in 1808.

15 R.-J. Haúy, Essai d'une théorie, p. 25, quoted in Mauskopf, Crystals and Compounds, p. 12.

16 And a rhomohedral primitive form.

17 Pasteur, La dissymétrie moléculaire, quoted from http://webserver.lemoyne.edu/faculty/giunta/pasteur60.html.

18 In 1811 by François Arago, and studied by Biot in the next decade.

19 Mauskopf, Crystals and Compounds, pp. 61ff.

20 Either the atoms themselves were helically arranged, or the atoms at the apices of the lateral solid angles of the molecular rhombohedra were somehow inclined to one side or another, or the rhombohedral molecule had suffered distortion through a lateral twist of one half relative to the other. Delafosse, "Recherches sur la cristallisation," pp. 684–688. Biot did not share Delafosse's view that optical activity was a molecular phenomenon in the case of quartz because the property disappeared when the crystals were fused.

21 Pasteur, La dissymétrie moléculaire, http://webserver.lemoyne.edu/faculty/giunta/pasteur60.html. Geison disagreed with me about the importance of Delafosse for Pasteur but I saw nothing in his argument that compelled me to change mine.

22 A. Laurent, Recherches diverses de chimie organique. Sur la densité des argiles cuites à diverses temperatures (Thèse de chimie et de physique présentée à la faculté des sciences de Paris le 20 décembre 1837), translated and quoted in Mauskopf, Crystals and Compounds, p. 47.

23 E. Mitscherlich, Sur la relation qui existe entre la forme crystalline et les proportions chimiques: IIme mémoire sur les arsenates et les phosphates. Annales de chimie et de physique, 19 (1821) 419. Quoted and translated in Mauskopf, Crystals and Compounds, p. 31.

24 A. Laurent, Nouvelles recherches sur les rapports qui existent entre la constitution des corps et leurs formes cristallines, sur l'isomérimorphisme et sur l'hémimorphisme. Comptes rendus… de l'Académie des sciences, 15 (1842) 351. Translated and quoted in Mauskopf, Crystals and Compounds, p. 49.

25 Also devised by Mitscherlich. Dimorphism, complementary to isomorphism, recognized that substances of the same chemical composition could, under special conditions, crystallize in incompatible forms.

26 See Geison, Private Science, pp. 62–85, for a detailed exposition of his reconstruction of Pasteur's train of experiments.

27 G.B. Kauffman and R.D. Myers, The resolution of racemic acid: a classic stereochemical experiment for the undergraduate laboratory. Journal of Chemical Education, 52 (1975) 778–779. The other type of racemate that would be resolvable into mirror-image hemihedral crystals is the sodium-potassium salt (Rochelle Salt).

28 Quoted in Mauskopf, Crystals and Compounds, p. 78.

29 Geison, Private Science, p. 86.

30 Ibid., pp. 88–89.

31 A.J. Rocke, Nationalizing Science: Adolphe Wurtz and the Battle for French Chemistry, MIT Press, Cambridge, MA, 2001.

32 Geison, Private Science, p. 79.

33 See above, p. 4 and footnote 7.

34 A.J. Rocke, The Quiet Revolution: Hermann Kolbe and the Science of Organic Chemistry, University of California Press, Berkeley, 1993.

35 P.J. Ramberg, Chemical Structure, Spatial Arrangement, Ashgate, Aldershot, Hampshire, 2003, p. 21. Much of my subsequent account will rely on Chemical Structure.

36 Ibid., p. 51.

37 Lecture, November 8, 1871. Quoted in ibid, pp. 49–50. See pp. 42–50 for an account of his work on lactic acid.

38 Ramberg, Chemical Structure, p. 56.

39 Jacobus Henricus van't Hoff, A Suggestion Looking to the Extension into Space of the Structural Formulas at Present used in Chemistry. And a Note upon the Relation between the Optical Activity and the Chemical Constitution of Organic Compounds (English translation of the French translation of the original Dutch pamphlet, G.M. Richardson, in Memoirs by Pasteur, van't Hoff, le Bel and Wislicenus), O. Theodor Benfey (Ed.), Classics in the Theory of Chemical Combination, Dover Publications, New York 1963, p. 151.

40 Ibid., p. 53.

41 Published in Utrecht and dated September 5, 1874, Title cited in Ramberg, Chemical Structure, pp. 54–55. The title given in Benfey (Ed.), Classics in the Theory of Chemical Combination, is from the French translation of the pamphlet: A Suggestion Looking to the Extension into Space of the Structural Formulas at Present used in Chemistry. And a Note upon the Relation between the Optical Activity and the Chemical Constitution of Organic Compounds. This latter title will be used when citing quotations from Benfey.

42 Published in the Bulletin de la société chimique de France, 22 (1874) 337–347. Title translated and quoted in ibid., p. 60. The title given in Benfey (Ed.), Classics in the Theory of Chemical Combination is almost exactly the same.

43 Le Bel, On the relationships, in: Benfey (Ed.), Classics, p. 161.
44 Van't Hoff, A suggestion, in: Benfey (Ed.), Classics, p. 153ff.
45 See above, Ref. [38] and related text.
46 "In the reasoning which follows, we shall ignore the asymmetries which might arise from the arrangement in space possessed by the atoms and univalent radicals; but shall consider them as spheres or material points, which will be equal if the atoms or radicals are equal, and different if they are different." Le Bel, On the relationships, Benfey (Ed.), Classics, pp. 161–162.
47 Translated and quoted in Ramberg, Chemical Structure, p. 57. Carbon double bonds could be envisioned as two tetrahedral sharing an edge, resulting in isomers with the same structure.
48 Ibid., p. 59.
49 Van't Hoff, A suggestion, Benfey (Ed.), Classics, pp. 156–158.
50 Le Bel, On the relationships, Benfey (Ed.), Classics, p. 162. Ramberg denominates this as "one of the more cryptic statements in Rotatory Power, for the production of an asymmetric object by this method is not at all obvious from the text—it is made as a declarative statement without diagrams, justification or clarification." Ramberg, Chemical Structure, p. 60. Le Bel saw two exceptions: (1) When plane of symmetry of the molecule contains the four atoms, A; (2) When the fourth radical is composed of all of the atoms of the first three radicals. Le Bel, On the relationships, Benfey (Ed.), Classics, p. 162
51 Ibid., p. 163. Benfey notes that "chemical isomer" does not refer to optical activity.
52 E.g. the lactic, malic, tartaric, amyl, and (to some degree) sugar groups. Ibid., pp. 163–167. In later years, Le Bel was quite adamant about eschewing the tetrahedral model of the carbon atom, e.g.: "I use the greatest effort in all my explanations to abstain fully from basing my ideas on the preliminary hypothesis that compounds of carbon of the formula CR_4 have the shape of a regular tetrahedron." Cited and dated 1890 in Ramsay, Stereochemistry, p. 90.
53 Ibid., p. 168. Le Bel's lack of diagrams has caused even modern commentators difficulty, e.g. Ramsay, who wrote that "the best guess [for a hemihedral quadratic pyramid] is that he was talking about a square pyramid." Ramsay, Stereochemistry p. 90.
54 Le Bel, On the relationships, Benfey (Ed.), Classics, pp. 169–170. None were known at the time. See Ramberg, Chemical Structure, pp. 62–63 for a discussion and criticism of Le Bel's attempt. But Le Bel's ideas on aromatic compounds probably influenced Van't Hoff when he subsequently took up this subject. Ramberg, Chemical Structure, pp. 72–73.
55 'Sur les formulas de structure dans l'espace,' Bull. Soc. Chim., 23 (1875) 295–301.
56 Bazendijk, Rotterdam, 1875, The book was published at Van't Hoff's own expense.
57 Braunschweig, Viewig, 1877. Wislicenus wrote the Foreword.
58 Ramberg, Chemical Structure, Chapters 4–5.
59 Translated and quoted in ibid., p. 155.
60 Chemical Structure, Chapter 9, especially pp. 313ff. For one of the numerous works on Werner published G.B. Kauffman, his principal historical investigator, see The first resolution of a coordination compound, in: O.B. Ramsay (Ed.), van't Hoff–Le Bel Centennial [ACS Symposium Series 12] American Chemical Society, Washington, DC, 1975, pp. 126–142.
61 Pasteur laboratory notebook, February, 1851. Translated and quoted in Geison, Private Science, p. 135. One of the products of fermentation, amyl alcohol, had recently been the subject of Pasteur's research. The prevalent theory for the formation of this product was that it involved the decomposition of sugar. To Pasteur, the dissimilarities between sugar and amyl alcohol were too great to permit the optical activity of the former to carry over to its product. But, then, why was amyl alcohol optically active? It could only be due to the agency of a living ferment.
62 1860., quoted in ibid., p. 104.
63 P. Palladino, Stereochemistry and the nature of life. Mechanist, vitalist, and evolutionary perspectives. Isis, 81 (1990) 47–48.
64 The whole article is ibid., pp. 44–67.
65 Geison, Private Science, p. 103. The microbe was identified in 1860 as *Penicillium glaucumer*. Other cases were discovered later in the century, e.g. with microorganisms and ferments such as zymase, and with tissues of high organism, which displayed differential oxidation of optical isomers.

J. Parascandola, Arthur cushny, optical isomerism, and the mechanism of drug action. J. His. Biol., 8 (1975) 147.

66 My remarks on Cushny are based on ibid., pp. 145–165.

67 Published in 1903. In particular, he found that L-hyoscyamine had twice the activity of atropine on the motor nerve endings, the salivary glands, and the eye pupils of frogs and various mammals. Ibid., pp. 148–149.

68 L-hyoscine and racemic hyoscine; L-adrenaline and racemic adrenaline.

69 J. Parascandola, The evolution of stereochemical concepts in pharmacology, in: Ramsay (Ed.), van't Hoff–Le Bel Centennial, pp. 148–155.

© 2006 Elsevier B.V. All rights reserved.

CHAPTER 2

Chiral asymmetry in nature

Dilip Kondepudi[1] and Kouichi Asakura[2]

[1] *Wake Forest Professor, Department of Chemistry, Wake Forest University, Winston-Salem, NC 27109, USA*
[2] *Department of Applied Chemistry, Faculty of Science and Technology, Keio University, 3-14-1, Hioshi, Kohoku, Yokohama 223-8522, Japan*

2.1 INTRODUCTION

The wide-ranging chiral asymmetry in nature has given rise to a multi-disciplinary interest in chirality. It is remarkable that nature does not exhibit left–right symmetry at any level, morphological, molecular and even at the most fundamental level of elementary particle interactions [1–3]. At the astronomical level, circularly polarized light has been found in the Orion Nebula [4], indicating the existence of large regions of space with asymmetry.

Chiral asymmetry in biomolecules—the predominance of L-amino acids and D-sugars—has profound consequences for pharmaceutical and agricultural chemistry. It necessitates us to develop a theoretical framework in which we can quantitatively study chirality and the processes that generate chiral asymmetry. In this chapter, we discuss the basic definitions and nomenclature used to describe chiral systems. This is followed by a brief survey of known chiral asymmetries in nature and questions they raise. We then present a general theory of spontaneous chiral symmetry breaking and discuss examples of spontaneous generation of asymmetry.

2.2 CHIRALITY: TERMINOLOGY AND QUANTIFICATION ISSUES

When an object is not identical to its mirror image, it is said to be "chiral". Lord Kelvin, who introduced this terminology, defined it thus: "I call any geometrical figure, or any group of points, chiral, and say it has chirality, if its image in a plane mirror, ideally realized, cannot be brought to coincide with itself" [5]. A chiral object or a geometrical figure and its mirror image are thus distinguishable and they could be simply

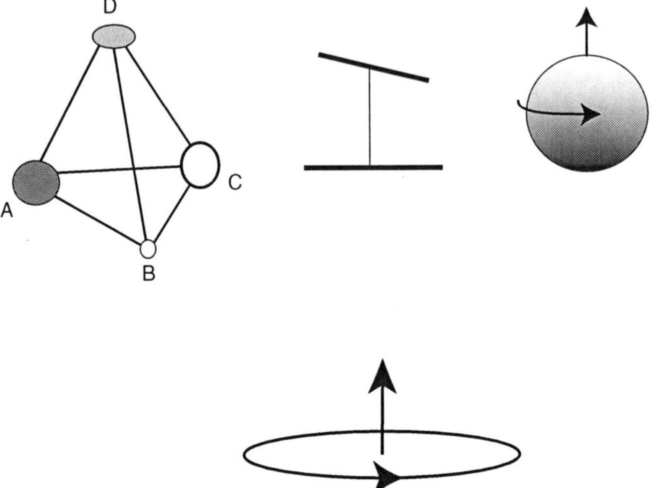

Fig. 2.1. Examples of basic chiral units. All complex objects can be reduced to a set of basic chiral units. As shown in the lower part, every basic chiral unit can be associated with a directed circle and a vector which together define a right- or left-handed helix.

identified as "left-handed" or "right-handed", a terminology that can readily be understood because we are asymmetric in the use of our hands. But there is a subtle issue here—how do you specify what is meant by a "left-handed object" to someone who does not know what it is, say to a being on a distant planet? Could we send this information as a radio signal coded as a sequence of digits ("0" and "1")? The answer to this question turns out to be negative; it is not possible to specify what one means by "left" or "right" in a linear sequence of digits. A three-dimensional object or a chirally asymmetric phenomenon in nature is needed to specify this information [6–8].

While Kelvin's definition clearly identifies a chiral object, we often have a need to distinguish between a simple chiral object, such as a tetrahedral molecule with four different atoms, and a complex chiral object, such as a protein. In order to describe chirality of assemblies of simple achiral building blocks, we may define a basic chiral unit as an object that cannot be divided into two or more chiral objects; any disassembly of such a unit will result in achiral subunits. Some examples of basic chiral units are shown in Fig. 2.1. According to this definition, a helix made of a continuous line segment is not a basic chiral unit; however, a helical assembly of a sphere consisting of four spheres is a basic chiral unit. All basic chiral units can be associated with a direction of rotation and an arrow that together define a helix as shown in Fig. 2.1. (In certain situations, one might identify the direction of rotation with an axial vector, while the arrow is a vector. The combination of a vector and an axial vector defines chirality.) Through such association, we can specify the absolute configuration of all basic chiral units. The chirality of large-assembly of simple achiral units has to be described in terms of chirality of such basic chiral units. We will discuss more about this point later in this chapter.

2.2.1 Cahn–Ingold–Prelog classification of chiral molecules

In the case of chiral molecules, a system of rules that identifies the absolute structure of the two enantiomers as *R* (Rectus, Latin word for "right") and *S* (Sinister, Latin word for "left") was proposed by Cahn, Ingold and Prelog [9]. These rules have been adopted by International Union of Pure and Applied Chemistry (IUPAC). The two molecular structures are called enantiomers of a chiral molecule. This designation is particularly suited for chiral organic molecules in which a carbon is bonded to four different atoms or groups, R1, R2, R3 and R4 as shown in Fig. 2.2. Each of the four groups bonded to the carbon are ranked by the atomic number of the atom directly bonded to the carbon, higher atomic number corresponding to higher rank. If two or more atoms have the same atomic number, then other atoms in the chain of each group are compared until a difference is found. If all the atomic numbers are the same, then atomic masses are used instead for ranking the groups. Thus, a CH_3 has a lower rank than CH_2OH or CH_2CH_3; similarly, the isotope D has a higher ranking than H. In Fig. 2.2, the ranking is assumed to be R1 > R2 > R3 > R4. For such a molecule, the "arrow" is defined as that pointing from the central carbon to the group with the lowest rank, R4. The rotation direction is obtained from the three highest ranking groups by moving from the highest to the lowest ranking group, i.e. R1 → R2 → R3. If this rotation and the arrow define a right-handed helix, the molecule is classified as *R*; if the helix is left-handed, the molecule is classified as *S*. It is possible that two of the four groups, say R1 and R2, are stereoisomers. In that case the Cahn–Ingold–Prelog rules specify a precise way of ranking the two groups. These rules may be found in [9].

Before the Cahn–Ingold–Prelog rules were formulated, the terminology for identifying the two enantiomers of a chiral molecule was based on optical rotation. Linearly polarized light of a particular wavelength undergoes right- or left-helical rotation as it propagates through a solution or a crystal of a chiral compound. A solution or solid that rotates linearly polarized light is said to be optically active. In fact, the discovery of optical activity by the physicist Biot in 1815 and Louis Pasteur's insight that it was related to molecular asymmetry gave us a general means to identify enantiomers. Consequently, the handedness of a molecule was identified by its optical activity. If the

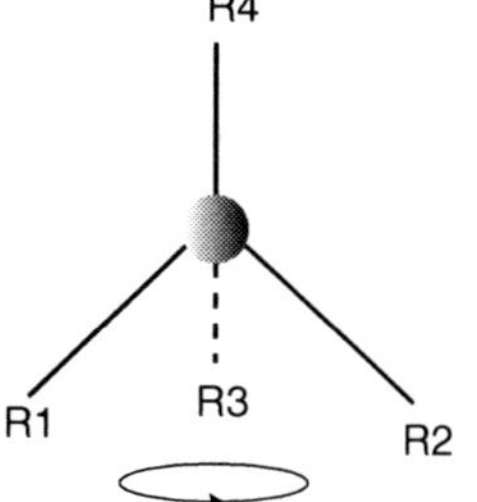

Fig. 2.2. An example of *R*-enantiomer in accordance with Cahn–Ingold–Prelog rules. The four groups, R1 through R4, attached to the central carbon are ranked, R1 > R2 > R3 > R4. Pointing the lowest ranked group away from the viewer, the clockwise rotation R1 → R2 → R3, defines the molecule as *R*-enantiomer. If the rotation is in the counter-clockwise direction, then it is a *S*-enantiomer.

References pp. 43–45

light was rotated clockwise as it approaches the viewer (Fig. 2.3), the solution was designated "dextrorotatory" or "+" and the enantiomer causing such a rotation was called d-enantiomer; if the light rotated anti-clockwise, the rotation was called "levorotatory" or "−" and the corresponding enantiomer was called l-enantiomer. This designation does not give us the absolute molecular structure, i.e. one cannot say which one of the two mirror-image structures of the molecule gives rise to the observed optical rotation. In fact, the optical rotation for a given compound depends on the wavelength of the light and it can be dextro at one wavelength and levo at another. In addition, for a given wavelength, the direction of rotation can change with the solvent. Clearly the designation "*R* and *S*" specifies the molecular structure more precisely and it is preferred over the "l and d" designation. Since optical activity is readily measured and often used to analyze chiral products of a reaction, if the optical activity of enantiomer is known, it is specified by combining the *R*–*S* and l–d notation; thus *R*(−) indicates that the *R*-enantiomer is levorotatory.

In addition to the *R*–*S* designation, another designation that is widely used in the context of amino acids identifies the enantiomers as L and D. This designation refers to the absolute structure as shown in Fig. 2.4. This designation is independent of the side chain and hence all L-amino acids have a similar basic structure. However, depending on the side chain, an L-amino acid may be designated *R* or *S* (a drawback of the *R*–*S* nomenclature).

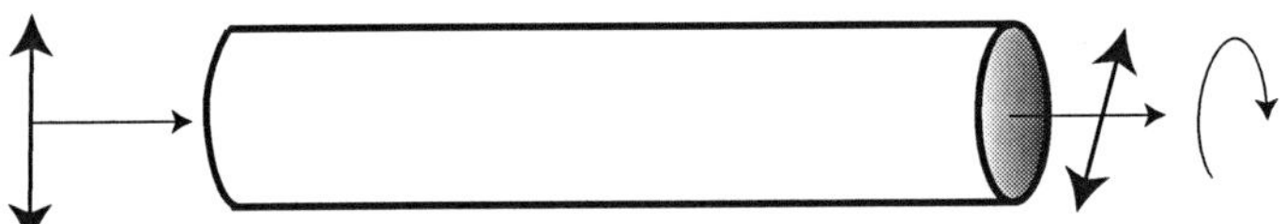

Fig. 2.3. Optical activity is the rotation of linearly polarized light as it passes through the sample. As shown in the figure, if the direction of polarization rotates clockwise as it approaches the viewer, the substance is dextrorotatory and is designated "d" or "+". If the rotation is in counter-clockwise direction, the substance is levorotatory or "l". or "−". Optical activity arises when the refractive indices of left- and right-circularly polarized light are unequal.

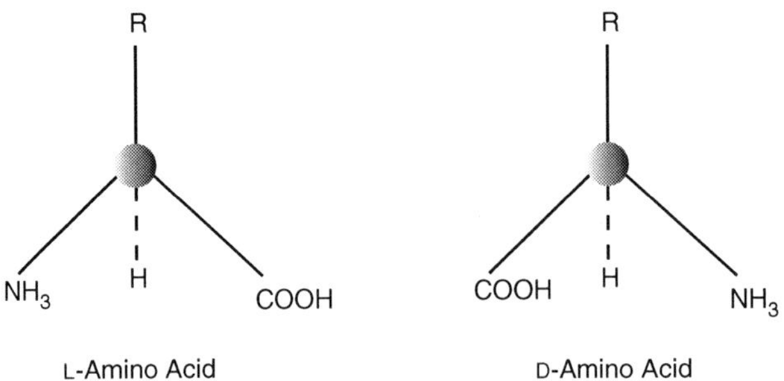

Fig. 2.4. The terminology "L- and D-amino acids" refers to the geometric structure as shown. Solutions of L-amino acids can be levo- or dextro-rotatory.

2.2.2 Optical rotation

The angle through which linearly polarized light is rotated, i.e. the optical rotation, is usually denoted by α. Optical rotation is due to the difference in the refractive indices of right- and left-circularly polarized light. Linearly polarized light may be considered as a combination of the right- and left-circularly polarized light, i.e. linearly polarized light may be resolved into right- and left-circularly polarized components. If the two components travel at different speeds, the resultant linear polarization undergoes a rotation. If n_R and n_L are the refractive indices of right- and left-circularly polarized light respectively, for light of wavelength λ, traversing a path in a chiral medium of length l, the optical rotation α is given by:

$$\alpha = \frac{\pi l}{\lambda}(n_L - n_R) \tag{1}$$

In most practical situations, to a good approximation, optical activity of a solution is directly proportional to its concentration and the path length (length of the polarimeter tube). In addition, α also depends on the wavelength, the solvent and the temperature. Consequently, specific rotation, $[\alpha]$, of light of wavelength λ, at a temperature T is defined as:

$$[\alpha]_\lambda^T = \frac{\alpha}{lc} \tag{2}$$

in which c is the concentration in g/mL and l is the path length in decimeters.

2.2.3 Enantiomeric excess

Optical activity of a solution is proportional to the difference in the amounts of the two enantiomers. This difference, which is a measure of the chiral asymmetry of the systems, is often quantified in terms of enantiomeric excess (EE), which is defined as:

$$\text{Enantiomeric Excess} = \frac{|[R]-[S]|}{[R]+[S]}, \tag{3}$$

in which $[R]$ and $[S]$ are the concentrations of the two enantiomers. It is a measure of asymmetry relative to the total amount of the substance, not an absolute measure of the difference in the amount of R and S. As we shall see in the following sections, in formulating a general theory of spontaneous chiral symmetry breaking, the difference ($[R]$–$[S]$) is mathematically more convenient to use than EE.

2.2.4 Crystal enantiomeric excess

In the nucleation of chiral crystals, asymmetries can arise due to chirally autocatalytic processes. In this case, the asymmetry is due to the difference in the number of levo- and dextro-rotatory crystals. For this reason, one might define a crystal enantiomeric excess (CEE) as:

$$\text{Crystal Enantiomeric Excess} = \frac{N_\text{l} - N_\text{d}}{N_\text{l} + N_\text{d}}, \tag{4}$$

in which N_l and N_d are the number of levo- and dextro-rotatory crystals. This quantity measures the asymmetry in nucleation, not in the total number of molecules.

2.2.5 Chirality measures

When we look at chiral objects, we often have a sense that one is "more chiral" than another. This notion could be intrinsic to the chiral objects themselves or in their interaction with other chiral objects. For example, when we compare two molecules that are in the shape of a twisted H, one could say the one with the larger twist angle is "more chiral", which is a measure intrinsic to the object. The specific rotation for a given wavelength and other well-specified conditions, a measure not intrinsic to the object, could be used to compare the chirality of molecules, the molecules with larger rotation being "more chiral".

It is desirable to have the measure be such that: (i) the value of the chirality measure changes continuously with the shape of the object and is equal in magnitude but opposite in sign for enantiomers and (ii) the chirality measure is zero if and only if the object is achiral. It turns out that it is impossible to define a universal measure with these two properties! This is because enantiomers of certain chiral objects can be transformed one to the other through a series of states which are all chiral—just as a left-hand glove can be transformed to a right-hand glove by turning it inside out one finger at a time. Examples of such transformations were noted by Mislow [16,17]. Molecules in which such a transformation can be realized are called "molecular gloves". During such a transformation, the corresponding chirality measure has to change continuously from positive to negative (or vice versa) and hence must cross zero. The transformation, however, consists entirely of chiral states. Hence, there must be a chiral state, of which the chirality measure is zero, contrary to the requirement (ii). If a *R*-enantiomer of a molecule can be transformed into a *S*-enantiomer through intermediate states which are all chiral, it means that there is no well-defined achiral state at which the change from *R* to *S* occurs; hence the point at which the change in designation is made is arbitrary—the notion of "left" and "right" cannot be made absolute.

Avoiding the ambiguity that arises with the sign of a chirality measure, the intrinsic chirality of an object can be viewed, for example, in terms of how different the two enantiomers are from each other. In this regard, several measures have been proposed [10–15]. Buda et al. [11] define a degree of chirality as: "the value of a real-valued, continuous function that is zero if, and only if, the object is achiral. The degree of chirality is the measure of an absolute quantity, and its value is therefore the same for both enantiomorphs of the measure object." An example of the degree of chirality is the maximum amount of overlap between a chiral object and its enantiomer. Clearly, in this measure, the degree of all achiral objects is zero and that of all chiral objects is nonzero.

Only chiral objects can distinguish between the enantiomers of chiral molecules. Could we not define a measure of chirality on the basis of the interaction of the reference chiral object with all chiral molecules? For example, could we take optical rotation (at specified wavelength, temperature and other conditions) as a measure of chirality? This approach too has difficulties because there are some chiral molecules the optical activity of which is, for all practical purposes immeasurable and there is no simple, reliable and theoretical way to predict what it would be [17]. Furthermore, a molecule that is classified as "left-handed" on the basis of optical rotation with one achiral solvent, may have to be classified as "right-handed" with another achiral solvent which indicates that this notion of "left" and "right", derived from the interaction of the molecule with right- and left-circularly polarized photons of a particular wavelength, is also dependent on the environment and is somewhat arbitrary.

Though there is no universal measure of chirality, it must be noted that there are situations in which quantifying chirality could be extremely useful in organizing and understanding chiral interactions [18–20].

2.3 CHIRAL ASYMMETRY IN NATURE

That nature does not treat the "left" and the "right" equally is not only remarkable but also puzzling because the fundamental laws have a high degree of symmetry in many respects. All basic laws of physics are invariant under translations in space and time and rotations in space. Electromagnetic phenomena are also invariant under mirror reflection, more precisely, parity (which is equivalent to mirror reflection and a rotation). But other fundamental interactions, at the nuclear level, are not invariant under parity. In this section we briefly survey the chiral asymmetries found at various levels of scale, from the nuclear level to the level of morphology of mammals.

2.3.1 Asymmetry in nuclear processes

Chiral asymmetry in nuclear processes responsible for β-radioactivity was discovered in 1957 by Chen-Shiung Wu et al. [21]. Some unstable nuclei "decay" to a lower state of energy by emitting electrons or positrons, also called β-particles. The isotope of cobalt, cobalt-60, for example, emits electrons. The nucleus of cobalt-60 has spin but a stationary particle with a spin is not chiral; it does not possess handedness unless there is a direction associated with the spin that could be used to define chirality. (In identifying the "north pole" using the right hand, we are imposing the chirality of the hand on the spinning sphere; a stationary spinning object is not chiral.) If there is an additional direction associated with a static spinning object, however, one could define left and right. In the case of radioactivity in cobalt-60, when the nuclei are placed in a magnetic field, their spin axes line up along the magnetic field. In cobalt-60 nuclei aligned in a magnetic field, more electrons were emitted from one of the two hemispheres thus indicating the handedness of this phenomenon. If the two hemispheres of a spinning sphere are not identical, then that object is chiral because the lack of identity of the two hemispheres could be used to define a direction, which along with the rotation defines a helix.

Further studies of β-radioactivity revealed that the emitted electrons themselves exhibit asymmetry. Since electrons have spin, one can associate left- or right-helicity to each electron by projecting its momentum vector along the spin axis. It was found that the electrons emitted by radioactive nuclei are predominantly left-helical. Chiral asymmetry in β-radioactivity appears in other ways as well. The emission of β-particles is accompanied by the emission of elusive particles called the neutrinos, which also have spin. All neutrinos generated in β-radioactivity are left helical, and antineutrinos are right helical. Neither right-helical neutrinos nor left-helical antineutrinos have been detected; they are presumed not to exist. If there were equal number of left-helical neutrinos and right-helical anti-neutrinos, then we might say that left–right symmetry is restored, but this would require a balance between matter and anti-matter. As far as we can tell, there is matter in the universe but not antimatter. The consequence is that the present universe is filled exclusively with left-helical neutrinos.

2.3.2 Asymmetry in atoms

We now move from the level of the nucleus to the level of the atom. Before the forces that held the particles in the atomic nucleus were discovered, physicists knew two fundamental forces in nature—gravity and electromagnetism. With the discovery of the nucleus, in which positively charged protons and uncharged neutrons were held together, new forces came to light. Soon after, studies in nuclear physics showed that there were basically two types of nuclear forces, which were termed, "strong" and "weak" forces. The weak forces were responsible for β-radioactivity, while the strong force was responsible for nuclear fission and fusion. Ever since James Clerk Maxwell realized that the laws of electricity and magnetism can be unified as "electromagnetic" laws, there has been a strong interest among physicists to unify different forces. Einstein, for example, tried very hard, but unsuccessfully, to unify gravity and electromagnetism. But the search for unification of fundamental forces continues and one of its greatest achievements is the unification of electromagnetic and weak forces, which necessitated the coining of the term "electro-weak" force. The formulators of the theory of electro-weak force, Steven Wienberg, Abdus Salam and Sheldon Glashow were awarded the Nobel Prize for physics in 1979. The exact nature of this unification is not important for us, but the chiral asymmetry that this unification revealed is. One of the predictions of the electro-weak theory is that the electron interacts with the nucleons (particles in the nucleus, protons and the neutrons) with a force that depends on its helicity. Left- and right-helical electrons are acted upon by different forces (Fig. 2.5). This fact brings the chiral asymmetry to the level of atoms, making them optically active. Atoms were thought to be spherically symmetric entities with no chiral attributes until the electro-weak force between the electron and the nucleons was discovered. A tube filled with a vapor of the element bismuth, for example, is found to be levo-rotatory [22]. Due to electro-weak interactions we only find levo-bismuth in nature, its mirror-image twin does not exist. The optical activity of atoms, though very small, has been measured and

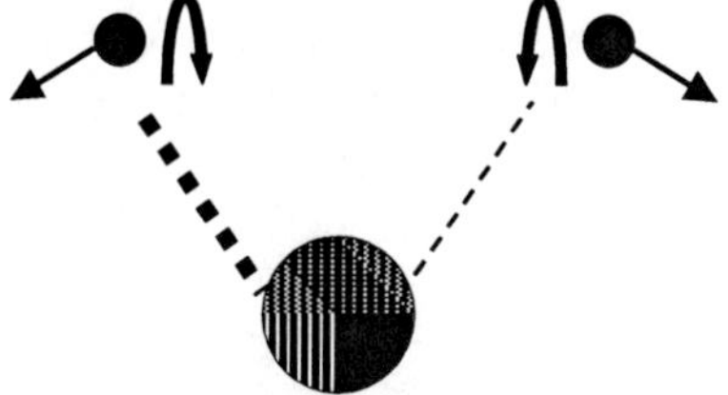

Fig. 2.5. Asymmetry in electro-weak interactions. The interaction between the electron and the nucleons depends on the electron helicity; the interaction energies of the left- and right-helical electrons are unequal. Due to electro-weak interaction, the ground-state energies of enantiomers are unequal.

found to be as predicted by the electro-weak theory, of the order of 10^{-7} radians, depending on the density of the vapor [22].

2.3.3 Asymmetry in molecules

The electro-weak interaction between the electron and the nucleons also brings chiral asymmetry to the level of molecules. In most chemistry text books it is stated that the ground-state energies of enantiomers are identical. This would be the case if the interaction between the electron and the nucleons were entirely governed by the laws of electromagnetism which are parity invariant. But since the electron–nucleon interaction has also a component of the parity violating weak-force, the ground-state energies of enantiomers must be different. This difference has been calculated by Mason and Tranter [23,24] using electro-weak theory and it is found to be of the order of 10^{-14} J/mol, too small to be detected by the current experimental methods. Calculations done on amino acids, indicate that the ground-state energies of L-amino acids are lower than those of D-amino acids. Since then, many more accurate calculations have been performed by Schewerdtfeger's group [25–30] and some recent theoretical calculations found that the effect in excited states could be larger by a factor of 10^3 [31,32]. This raises the hope that we might be able to measure electro-weak energy differences in enantomers in the laboratory.

2.3.4 Biomolecular asymmetry

Louis Pasteur's discovery in 1857 that microbes metabolized only the dextro-rotatory sodium-ammonium-tartrate is a momentous one. It not only presented us with one of the most fundamental and universal aspects of the biological chemistry, but also left us with a fundamental question regarding life—Is chiral symmetry essential for the evolution of life? With our present knowledge of biochemistry we know that all life is dominated by L-amino acids and D-riboses. In the words of Francis Crick, "The first great unifying principle biochemistry is that the key molecules have the same hand in all organisms" [33]. The food we eat, the medicines we take and the pesticides we use all have to conform to the chemistry dictated by this asymmetry. The evolutionary origin of this asymmetry remains an enigma. Though we have learned much about the processes that can spontaneously generate chiral asymmetry in the last fifteen years, we are still far from having plausible examples of processes that could have caused the dominance of L-amino acids and D-riboses.

2.3.5 Morphological asymmetry

The asymmetry in the morphology of living organisms is all too well-known—most of the spiral seashells are right helical; the organ placement in mammals is asymmetric—except for the rare one-in-ten-thousand inversion in humans (a condition called *situs inversus*). Chiral asymmetry of the brain is reflected in behavior—about 90% of humans in all societies and cultures are right-handed. A good compendium of asymmetries in the animal world was assembled by Neville in his Animal Asymmetries [34].

Morphological asymmetry is inherited. How does this happen? This question has a different significance compared to other inherited traits that are coded in a DNA sequence of A, T, G and C. Handedness cannot be coded in a linear sequence of letters, that is, the information about the placement of the liver on the right, for example, could not be coded in a sequence such as ATTGCGGTAC... while placement on the left be coded in a different sequence GTTACCTGA.... To understand why this is so, we only have to look at the process that place the liver on the right in a mirror to realize that the sequence ATTGCGGTAC... could also carry the code to place the liver on the left. The information about right or left has to come from some existing asymmetry. While asymmetry is easy to find at the molecular level of L-amino acids and D-riboses, we do not know how the molecular asymmetry gets translated to morphological asymmetry. The connection between structure of amino acids and organs is similar to the connection between bricks and the shape of a building. Clearly the shape of a building is quite independent from the shape of a brick; yet morphological asymmetry seems to have its origins at the molecular level. Elucidating this link is one of the greatest challenges of developmental biology [35].

2.3.6 Astrophysical asymmetries

On the astronomical scale, asymmetry was found in large regions of interstellar space in the form of circularly polarized light. Bailey et al. [4,36] reported the detection of infrared circularly polarized light in star-forming regions of Orion Nebula. Though this is not a global asymmetry, it nevertheless shows that in vast regions in the universe chiral asymmetry can arise and it may have important implications for the origin of molecular asymmetry in these regions.

An obvious place to look for chiral asymmetry on the astrophysical scale is in spiral galaxies. With respect to an observer on the earth, all galaxies are receding and hence have a well-defined direction of motion. For spiral galaxies, if we combine the direction of the spiral with the recession velocity vector, we can define the helicity or handedness of a spiral galaxy. The Carnegie Atlas of Galaxies [37] is a good resource to look at the statistics of spiral galaxies. A simple count of the left- and right-helical spiral galaxies shows an equal number, indicating no apparent asymmetry. However, if the statistics are extended to the different classes of spiral galaxies, an interesting trend seems to emerge [38]. Spiral galaxies are classified into ordinary spirals (S) and barred spirals (SB), a classification initiated by Edwin Hubble. The two classes of galaxies show opposite asymmetries that cancel each other.

If we assign +1 to "Z-shaped" and –1 to the "S-shaped" spirals in a sample of 540 galaxies that consists of both S and SB spirals, the sum was found to be –4 (–0.172 standard deviations from zero), indicating that there is no observable asymmetry in the total sample. But in a subsample of 378 galaxies of class S, the sum was found to be –26 (–1.34 standard deviations from zero), while in the subsample of 162 galaxies of class SB the sum was +22 (+ 1.73 standard deviations from zero). The probability of observing these sums as a consequence of random fluctuations in a system that has no real asymmetry is 0.09 for class S and 0.04 for class SB. With such low probabilities for the observed statistics, we may consider the above data as an indication that there is an asymmetry in the distribution of the subclasses of ordinary spiral galaxies and barred spiral galaxies but of opposite sign. More details and references to the previous work on this topic can be found in [38]. This observation should be considered only an indication and not a definite conclusion that there is chiral asymmetry in subclasses of spiral galaxies; to be certain, analysis based on a larger data set is essential.

2.4 THEORY OF SPONTANEOUS CHIRAL SYMMETRY BREAKING

The chiral asymmetry at the biomolecular level lies in stark contrast to the paucity of chemical reactions that spontaneously generate asymmetry. When chiral molecules are synthesized from achiral reactants, we find equal amounts of the two enantiomers. Since chemical reactions are governed by the electromagnetic laws that are mirror-symmetric, the rates of production of the two enantiomers are equal. The deviation from chiral symmetry due to the electro-weak component of the interaction is too small to influence chemical reactions on the laboratory volume and time scales. As we shall see later in this chapter, when very large scales of time and volume are involved, even very small asymmetries can have large effects. The mirror-symmetry of the electromagnetic laws implies that a process and its mirror image are equally realizable in nature; thus, if we ignore the electro-weak interactions, there can be no noticeable difference in chemical reaction rates of enantiomers. But, identical reaction rates need not always lead to equal amounts of the two enantiomers. In reactions that are autocatalytic, symmetric kinetic rate laws could produce asymmetric products. The products of a chemical process need not reflect the symmetries of the process that generated it. When this happens, the system is said to exhibit "spontaneous symmetry breaking". In mathematical terms, the solutions of a system of differential equations need not have the same symmetries as the system of equations; the solutions could "break the symmetries" of the system. This phenomenon could be illustrated using a simple model that is an extension of a model suggested by Sir Charles Frederick Frank in 1953 [39].

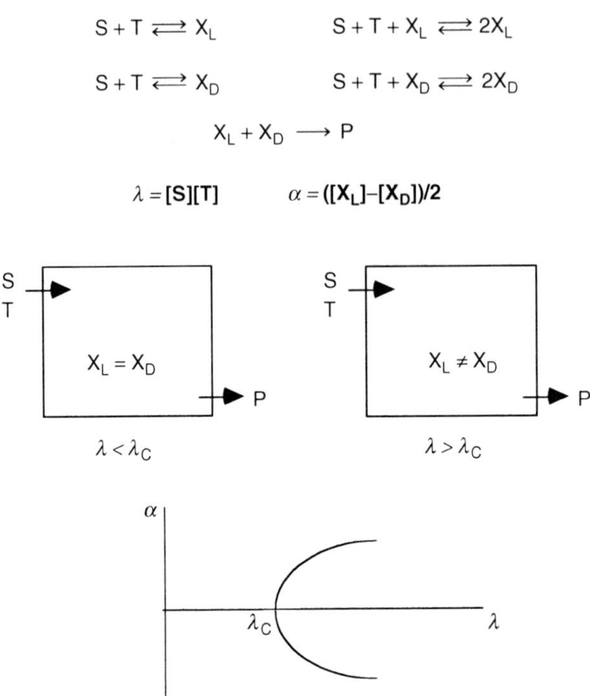

Fig. 2.6. An autocatalytic reaction scheme in which X_L and X_D have identical kinetic rate laws. However, in an open system, this reaction gives rise to a state of broken symmetry in which the concentrations of X_L and X_D are unequal. Lower part of the figure shows a "bifurcation diagram" used in describing the transitions to asymmetric states. The parameter $\alpha = (X_L - X_D)/2$ is a measure of the asymmetry. $\lambda = [S][T]$, the product of the concentrations of S and T, is the bifurcation parameter. When the value of λ exceeds λ_C the system becomes unstable and makes a transition to an asymmetric state in which α is not zero.

The modification to Frank's model was made to elucidate the thermodynamic nature of the model and to assess the impact of small external asymmetries on such models. This modified model is shown in Fig. 2.6. In this model, achiral molecules, S and T react to form a chiral molecule X_L or X_D. The molecule X is chirally autocatalytic: X_L catalyzes the production of X_L and X_D catalyzes the production of X_D. In addition, X_L and X_D react to form an inactive product P. The system consists of a "reaction chamber" into which there is an inflow of reactants S and T and an outflow of P. For a given set of flows, the system reaches a stationary state in which the concentrations of all the species remain constant. The flows maintain the concentrations of S and T at a constant value. If the flows are reduced to zero, the system will eventually reach thermodynamic equilibrium in which the concentrations of X_L and X_D are equal. If the flows are slowly increased from zero, the concentrations of S and T as well as X_L and X_D increase but still the concentrations of X_L and X_D remain equal until the concentrations of S and T reach a critical value. Below the critical value, any difference in the concentrations of X_L and X_D that might arise through a random fluctuation will decay to zero; here the symmetric state of equal X_L and X_D is stable. Above the critical value, the system becomes "unstable" in the sense that any small difference between the concentrations of X_L and X_D will grow; the system can no longer maintain the symmetry between X_L and X_D. If the concentrations of S and T surpass the critical value, the system will be forced to make a transition to an asymmetric state in which the concentrations of X_L and X_D are not equal. Which one will be greater is entirely random, but one of the two enantiomers will dominate. This symmetry-breaking transition is summarized in the second row of Fig. 2.6.

In the terminology of non-equilibrium chemistry, the system undergoes a non-equilibrium transition to a symmetry-breaking state. The thermodynamic state thus reached through a non-equilibrium transition is more ordered or has more structure. This structure is maintained by the "dissipative" or entropy-generating chemical processes; they were therefore called "disspative structures" by Ilya Prigogine, who was awarded the 1977 Nobel Prize in Chemistry for his contributions to non-equilibrium thermodynamics. There are numerous other examples of dissipative structures that are a result of symmetry-breaking transitions [40]. The above example also makes it clear that only a system far from the thermodynamic equilibrium can generate and sustain the asymmetry of form that we see in proteins and DNA. Though the racemization rates of amino acids are very small, a steady generation of L-amino acids is necessary to maintain the chiral purity of active proteins.

The above model can be used to illustrate general features that all chiral symmetry-breaking transitions will exhibit. As shown in Fig. 2.6, there is a quantity $\alpha = ([X_L] - [X_D])/2$ which is a measure of the asymmetry; $\alpha = 0$, represents a symmetric state in which two enantiomers are present in equal amounts; $\alpha \neq 0$, represents an asymmetric state in which one of the two enantiomers is in excess. α is more mathematically convenient than EE in analyzing symmetry-breaking transitions. The parameter $\lambda = [S][T]$ is a measure of the system's distance from the thermodynamic equilibrium because, through appropriate inflow of S and T, their concentrations are maintained at a non-equilibrium value. It is also called the bifurcation parameter. If λ is such that it corresponds to equilibrium values of [S] and [T], the system will be in a symmetric state corresponding to $\alpha = 0$. When λ reaches a critical value, indicated by λ_C, the systems begin to become unstable to small deviations of α from zero; if $\lambda > \lambda_C$, then random fluctuations in α will grow and drive it to either a positive or a negative value (a steady state) depending on the fluctuation. At $\lambda = \lambda_C$, two new asymmetric solutions "bifurcate" from the symmetric solution. In the vicinity of the critical point λ_C, it is possible to obtain an equation for the time evolution of α that will have the same general form for all systems that break chiral symmetry [41,42]. This equation is of the form:

$$\frac{d\alpha}{dt} = -A\alpha^3 + B(\lambda - \lambda_C)\alpha + \sqrt{\varepsilon} f(t). \tag{5}$$

In this equation, the coefficients A and B depend on the kinetic rate constants; the term $\sqrt{\varepsilon} f(t)$ represents random fluctuations in which $\sqrt{\varepsilon}$ is the root-mean-square value of the fluctuations and $f(t)$ a normalized random variable with zero mean. (As a first approximation, one may assume $f(t)$ to be a Gaussian white noise.) The fluctuations make the evolution of α stochastic or random. Normally, random fluctuations in concentrations are quite unimportant for predicting the time-variation of concentrations during the reaction and the final product distribution. The kinetic rate equations enable us to predict these within the accuracy of the experiments. This is not so in systems that spontaneously break chiral symmetry; the evolution of α is random, not deterministic. For example, if λ moves from a value below the critical point to a value above the critical point, α could evolve from zero to a positive or a negative value. Hence, we can only assign a probability, P_+ or P_-, that α will evolve to either value because the final state of α depends on a random fluctuation. In the absence of a chiral influence that favors one of the enantiomers, we must have $P_+ = P_- = 0.5$. Equation (5), which contains a term representing random fluctuation is an example of a Langevin equation. For such equations, with appropriate specification of the property of the random variable $f(t)$, it is possible to obtain an equation for the evolution of the probability distribution $P(\alpha,t)$, called the Fokker–Planck equation. We do not discuss these mathematical details in this introductory chapter but focus on the general theoretical aspects. Note that the mirror symmetry of the chemical reactions is manifest in the general Eq. (5) in that both α and $-\alpha$ are its solutions. When we consider steady states ($d\alpha/dt = 0$) if $\alpha = -\alpha = 0$, then the solution is symmetric; on the other hand, if $\alpha \neq 0$, the system has two possible solutions, α and $-\alpha$, each representing a state of broken symmetry. For any chemical system that is capable of spontaneous chiral symmetry breaking, the general equation of the form (5) can be derived using group theory, without reference to particular details of the kinetic mechanism; in addition, general expressions relating the coefficients A and B to the kinetic rate constants can be derived [42]. This theory is similar to the Ginzberg–Landau theory of second-order phase transitions.

2.5 SENSITIVITY OF CHIRAL SYMMETRY-BREAKING TRANSITIONS TO ASYMMETRIC INTERACTIONS

The above general formalism can be used to analyze the effects of small asymmetric interaction on symmetry-breaking transitions. For example, we can analyze how circularly polarized light might influence the symmetry-breaking transition if it enhances the reaction rate of one of the two enantiomers. Due to the asymmetric interaction, the probabilities of transition, P_+ and P_-, to asymmetric states will be different from 0.5. Let us say, that due to the asymmetric interaction, the forward rate constant of the reaction $S + T \rightarrow X_L$ is larger than that of $S + T \rightarrow X_D$ by a factor g, i.e. the rate constant $k_L = k_D\,(1 + g)$, in which k_L and k_D are the rate constants of the two reactions respectively. Given such a chiral influence, the task is to determine the probabilities P_+ and $P_- = (1 - P_+)$. Using the same general approach that leads to Eq. (5), the following modification of this equation can be obtained in the presence of the asymmetric interaction [41,42]:

$$\frac{d\alpha}{dt} = -A\alpha^3 + B(\lambda - \lambda_C)\alpha + Cg + \sqrt{\varepsilon}f(t), \qquad (6)$$

in which C is a constant that depends on the kinetic rate constants. As before, for any given set of reactions, one can obtain C using general expressions [42]. The effect of the

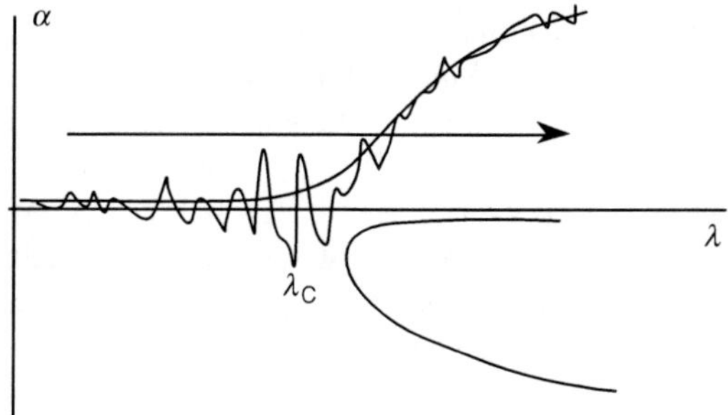

Fig. 2.7. In the presence of a chiral influence that favors one enantiomer (here corresponding to $\alpha > 0$), the bifurcation diagram in Fig. 2.6 is modified to the one shown above. As λ increase from a value below λ_C to a value above λ_C, the system makes a transition to the favored upper branch with a higher probability. In this figure, a fluctuating trajectory of α shows its evolution to the favored branch in which $\alpha > 0$. This process is very sensitive to small chiral influences. The probability of transition to the favored branch is given by the formula (7).

bias term Cg is countered by the random fluctuations whose strength, as measured by the root-mean-square value, is $\sqrt{\varepsilon}$. At first sight, it might seem that if $Cg \ll \sqrt{\varepsilon}$ then the effect of the chiral interaction P_+ would be negligible. But a careful analysis shows that the system could in fact be very sensitive to the chiral interaction even if $Cg \ll \sqrt{\varepsilon}$ [41]. Such sensitivity arises when the parameter λ sweeps through the critical value, λ_C, from a sub-critical to a super-critical value. Let us assume that this sweep of λ happens at an average rate of γ, that is, $\lambda = \lambda_0 + \gamma t$, in which $\lambda_0 < \lambda_C$ and t is the time (Fig. 2.7). It turns out that the sensitivity of the system to the chiral interaction, which is reflected in the deviation of P_+ from 0.5, depends on the sweep rate γ. Lower values of γ make the system more sensitive. The analytical expression for P_+ in the presence of the chiral interaction is given in Ref. [41]:

$$P_+ = \frac{1}{\sqrt{2\pi}} \int_{-\infty}^{N} e^{-x^2/2} dx \quad \text{in which} \quad N = \frac{Cg}{\sqrt{\varepsilon/2}} \left(\frac{\pi}{B\gamma} \right)^{1/4}. \tag{7}$$

This result, which has been verified in electronic systems [43], shows that even when $Cg \ll \sqrt{\varepsilon}$, the value of P_+ could be as high as 0.98 depending on the value of γ. The theory can also be generalized to symmetry-breaking transitions for symmetries other than the mirror symmetry [44].

The significance of this results is in what it implies for the relation between symmetry-breaking at different levels. If symmetry-breaking transitions occur at various levels, the occurrence at one level might depend very sensitively on a small asymmetry at a lower level. This type of theory has the potential of revealing how asymmetries at various levels could be interconnected. Also, this theory is general enough that it applies at all levels, wherever spontaneous chiral symmetry breaking occurs.

One of us has suggested the use of such a theory in analyzing the higher-level chiral asymmetries in macro molecules such as proteins [8]. A complex chiral structure, such as a protein molecule could be analyzed in the following way. We begin with identifying the basic chiral objects in the system—in this case, the amino acids. To see the chirality at the next level, we replace each basic chiral unit by an achiral unit such as a sphere. In this string of spheres, we now identify the basic chiral units. In the helical parts of the protein, for example, there is chirality at the next level, i.e. in the secondary structure. As is well known, all the helices in proteins are right-handed because the amino acids are all L-enantiomers. Now the folding of a string of spheres can itself be seen as a symmetry-breaking transition; the folding can give rise to right- or left-handed helix with equal probability if there is no chiral bias. In this transition, if we take into account that the spheres represent L-amino acids, then there is a chiral bias arising from a lower level and influencing the helices to go right-handed. From the above theory we may expect the transition to a helix be very sensitive to even small chiral biases (in the case of L-amino acids, the bias is probably not small). To identify the chirality at the next level, that is, at the level higher than the helices, we replace each of the helices by a cylinder and each of the beta-sheets with a ribbon. In this string of cylinders and ribbons, we can now identify basic chiral units depending on how this string folds. Thus, the chirality of the protein can be seen at various levels and it might well show a systematic asymmetry at higher levels. If this is so, using the above theory, or another appropriate theory, we can see how an asymmetry at one level influences the next higher level. This gives us a way to understand how asymmetry propagates from one structural level to another. It may well be that in some macromolecules the asymmetry slowly vanishes at higher levels.

An interesting consequence of the theory of spontaneous chiral symmetry-breaking is that it provides a relationship between the strength of a chiral influence and the time scale at which it can influence a symmetry-breaking transition. By applying this theory we find, on a time scale of 15,000 years, electro-weak parity violation could influence the outcome of a symmetry-breaking process in a chemical system whose reactor volume is 1 km × 1 km × 10 m. Thus, in the context of chemical evolution in the oceans, such processes are relevant. Though there have been many suggestions, at this time we do not have much understanding of when and at what stage of life's evolution chiral asymmetry arose. Hence we can only say that the possible link between the dominance of L-amino acids and electro-weak asymmetry cannot be ruled out on the basis that the later's chemical influence is too small. Our observation, however, is not a proof that such a link exists.

2.6 EXAMPLES OF SPONTANEOUS CHIRAL SYMMETRY BREAKING

The theory presented in the last two sections has been formulated nearly two decades ago but laboratory examples on spontaneous chiral symmetry emerged in the last 15 years. The simplest example of spontaneous chiral symmetry-breaking can be demonstrated in stirred crystallization of $NaClO_3$ [45]. Though the molecules are not chiral, the crystals of this compounds are chiral. The two enantiomeric forms of crystals of $NaClO_3$ can be easily detected through their optical rotation. An aqueous solution of $NaClO_3$ is not optically active, it has no chirality. But if we prepare a saturated solution of $NaClO_3$ and allow the solvent to evaporate so that crystals form, each crystal is either levo- or dextro-rotatory (l- and d-crystals). When crystallization is performed,

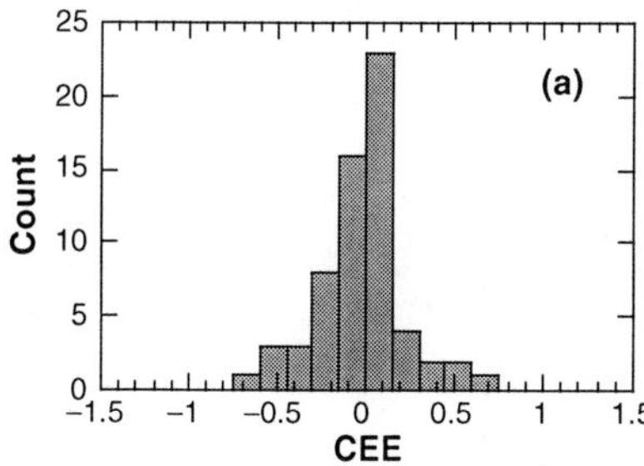

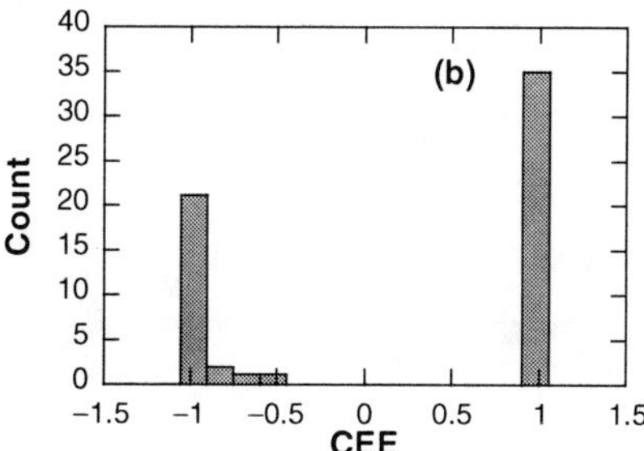

Fig. 2.8. Histogram of crystal enantiomeric excess (CEE) in unstirred (a) and stirred (b) crystallization of $NaClO_3$. CEE $= (N_l - N_d)/(N_l + N_d)$ in which N_l and N_d are levo- and dextro-rotatory crystals respectively [45]. In unstirred crystallization, the distribution is centered at zero but for stirred crystallization it is bimodal with peaks close to ± 1.

as it is usually done, in a crystallization dish, without in anyway disturbing the solution, statistically equal number of l- and d-crystals are found. This was as expected—no asymmetry. Spontaneous chiral symmetry breaking occurs if the crystallization is done while the solution is continuously stirred. In almost every stirred crystallization, more than 98% of the crystals were either l or d! Fig. 2.8 shows a typical experimental result when a large number of stirred and unstirred crystallizations are performed. This was the first clear experimental demonstration of the phenomenon of spontaneous chiral symmetry breaking in batch or a macroscopic scale and it is distinct from obtaining asymmetry in crystallization due to chiral impurities. (The nucleation of a chiral crystal could itself be thought of chiral symmetry breaking on a microscopic scale). The details of the mechanism, the required chiral autocatalysis and mutual competition, were elaborated in our later studies [46,47]. Though a fundamental theory of the chirally autocatalytic nucleation in stirred systems does not exist at present, there are good empirical theories for this step of the kinetics. Using one of these empirical rate laws, it is possible to simulate most of the experimental results through stochastic computer codes, including the stochastic behavior such as the random variation of the crystal enantiomeric excess as a function of the stirring rate [47].

The spontaneous symmetry-breaking that was found in $NaClO_3$ is not an isolated phenomenon. Later work showed that this phenomenon is general and it can occur in several compounds; it also occurs in the crystallization from a melt. Other similar chemical systems were also identified where the first example of chiral autocatalysis at the molecular level has been demonstrated in the synthesis of a chiral cobalt complex (Fig. 2.9) [48,49]. A detailed discussion of these examples and related references can be found in a recent review [50].

One of the key reaction steps in the generation of chiral asymmetry is chiral autocatalysis. Chirally autocatalytic reaction are hard to find in organic reactions. Soai et al. have reported a family of such reactions which are proven to be extremely sensitive to chiral influences [51–53]. Even very small amount of initial asymmetry is greatly amplified in these reactions [54–57]. In some cases, the sensitivity to even very small amounts of impurities is so great that it was found to be very difficult to eliminate it altogether in the laboratory [58].

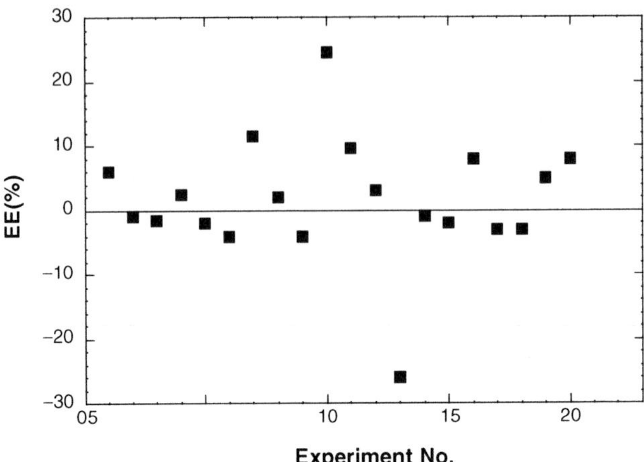

Fig. 2.9. Asymmetry generation in the synthesis of a Co-complex [48]. Though the distribution is not bimodal, but the large random EE shows the existence of chiral autocatalysis.

Spontaneous chiral symmetry-breaking can also be observed in Langmuir monolayers [59] and in molecular aggregation [60]. In the aggregation of molecules, Ribo et al. [60] reported chiral selection induced by vortex motion. They also reported that sensitivity to vorticity depended on the rate at which the system sweeps through the aggregation point, a feature predicted by the theory presented in Section 2.5. Purrello's group has reported the influence of chiral additives on the assemblies of porphyrins [61–63]. The assemblies of porphyrins can be chiral though the porphyrins themselves are not chiral molecules. Since this is a case of chiral symmetry-breaking, we could expect a very high degree of sensitivity to chiral influences during the process of assembly. Purello's group has demonstrated that once a particular chirality is induced in the assembly by a chiral additive, the assembly maintains its chiral structure if the chiral additive is removed. At this point it is fair to say that there is no dearth of examples of systems that show chiral symmetry-breaking and we may expect interesting applications of the sensitivity of such systems to chiral influences in the future.

2.7 CONCLUDING REMARKS

We inhabit a world that exhibits chiral asymmetry. In the words of Louis Pasteur, "L'univers est dissymmétrique". But we do not know much about the exact origins of these asymmetries though our knowledge of chemical systems that spontaneously generate asymmetry has grown in the last 15 years. The origin of biomolecular asymmetry, for example, still remains lost in a cloud of speculations.

The asymmetries in nature at all spatial scales make us wonder if they are somehow related. Since the symmetry-breaking transitions are highly sensitive to chiral influences, it is possible that the asymmetries at various levels are related, the asymmetry at one level being the small influence that acts on the symmetry-breaking process at the next higher level. Thus, the asymmetry at the molecular level of amino acids and DNA somehow might propagate to higher level and influence the asymmetric organ placement during morphogenesis. While it is easy to see the propagation of asymmetry from one level to another at the primary and secondary structures of proteins, the asymmetry and its propagation is not so evident at higher levels. It is a real challenge to elucidate the path through which the molecular asymmetry finds its way to morphology of an organism [35].

Propagation of asymmetry in macromolecular aggregates is of much interest. As noted by Thomas et al. [64–66], the asymmetry of the monomer units does not always propagate to the higher level of aggregation. To understand the propagation of asymmetry at various levels we need new theoretical approaches and experimental methods. Given the high level of interest in this topic, we expect rapid progress in this field.

2.8 ACKNOWLEDGMENTS

We gratefully acknowledge the contributions of the undergraduate and graduate students and postdoctoral fellows to the work presented in this chapter. This work was supported by NSF (grant CHM-9527095) and by the Japan Society for the Promotion of Science.

REFERENCES

1 R. Hegstrom and D.K. Kondepudi, The handedness of the universe. Sci. Am., 262 (1990) 108–115.
2 R. Janoschek (Ed.), Chirality: From the Weak Boson to the α-Helix, Springer Verlag, New York, 1991.
3 S.F. Mason, Chemical Evolution, Clarendon Press, Oxford, 1991.
4 J.A. Bailey, A. Chrysostomou, J.H. Hough, T.M. Gledhill, A. McCall, S. Clark, F. Menard and M. Tamura, Circular polarization in star-forming regions: implications for biomolecular homochirality. Science, 281 (1998) 672–674.
5 L. Kelvin, Baltimore Lectures (p. 436) (cf. also Baltimore Lectures, Appendix H, 1904, 439), 1884.
6 R.P. Feynman, The Character of Physical Law, MIT Press, Boston, 1990.
7 M. Gardner, The New Ambidextrous Universe, W.H. Freeman, New York, 1990.
8 D.K. Kondepudi, Theory of Hierarchical Homochirality, in: G. Palyi, C. Zucchi and L. Caglioto (Eds.), Progress in Biological Chirality, Elsevier, Amsterdam, 2004.
9 R.S. Cahn, C. Ingold and V. Prelog, Specification of Molecular Chirality. Angew. Chem. Int. Ed., 5 (1966), 385–415.
10 D. Avnir and A.Y. Meyer, Quantifying the degree of molecular shape distortion—a chirality measure. J. Mol. Struct. (Theochem), 72 (1991) 211–222.
11 A.B. Buda, T.A. Derheyde and K. Mislow, On quantifying chirality. Angew. Chem.-Int. Ed. Engl., 31 (1992) 989–1007.
12 G. Gilat, On quantifying chirality – obstacles and problems towards unification. J. Math. Chem., 15 (1994) 197–205.
13 A.B. Harris, R.D. Kamien and T.C. Lubensky, Molecular chirality and chiral parameters. Rev. Mod. Phys., 71 (1999) 1745–1757.
14 K.B. Lipkowitz, D.Q. Gao and O. Katzenelson, Computation of physical chirality: an assessment of orbital desymmetrization induced by common chiral auxiliaries. J. Am. Chem. Soc., 121 (1999) 5559–5564.
15 H. Zabrodsky, S. Peleg and D. Avnir, Continuous symmetry measures. J. Am. Chem. Soc., 114 (1992) 7843–7851.

16 K. Mislow, Limitations of the symmetry criteria for optical inactivity and resolvability. Science, 120 (1954) 232–233.
17 K. Mislow and R. Bolstad, Molecular dissymetry and optical activity. J. Am. Chem. Soc., 77 (1955) 6712–6713.
18 S. Alvarez, S. Schefzick, K. Lipkowitz and D. Avnir, Quantitative chirality analysis of molecular subunits of bis(oxazoline)copper(II) complexes in relation to their enantioselective catalytic activity. Chem. Eur. J., 9 (2003) 5832–5837.
19 S. Alvarez and D. Avnir, Continuous chirality measures of tetracoordinate bis(chelate) metal complexes. Dalton Transactions, (2003) 562–569.
20 S.A. Kane, Quantitative chirality measures applied to domain formation in Langmuir monolayers. Langmuir, 18 (2002) 9853–9858.
21 C.S. Wu, Experimental test of parity conservation in beta decay. Phys. Rev., 105 (1957) 1413.
22 M.-A. Bouchiat and C. Bouchiat, Parity violation in atoms. Rep. Prog. Phys., 60 (1997) 1351–1396.
23 S.F. Mason and G.E. Tranter, The parity-violating energy differences between enantiomeric molecules. Mol. Phys., 53 (1984) 1091–1111.
24 S.F. Mason and G.E. Tranter, The electroweak origin of biomolecular handedness. Proc. R. Soc. Lond., A 397 (1985) 45–65.
25 J.K. Laerdahl and P. Schwerdtfeger, Fully relativistic ab initio calculations of the energies of chiral molecules including parity-violating weak interactions. Phys. Rev. A, 60 (1999) 4439–4453.
26 J.K. Laerdahl, R. Wesendrup and P. Schwerdtfeger, D- or L-alanine: that is the question. Chemphyschem, 1 (2000) 60–62.
27 J.K. Laerdahl, P. Schwerdtfeger and H.M. Quiney, Theoretical analysis of parity-violating energy differences between the enantiomers of chiral molecules. Phys. Rev. Lett., 84 (2000) 3811–3814.
28 P. Schwerdtfeger, J.K. Laerdahl and C. Chardonnet, Calculation of parity-violation effects for the C-F stretching mode of chiral methyl fluorides. Phys. Rev. A, 65 (2002).
29 P. Schwerdtfeger, J. Gierlich and T. Bollwein, Large parity-violation effects in heavy-metal containing chiral compounds. Angew. Chem. Int. Ed., 42 (2003) 1293–1296.
30 P. Schwerdtfeger and R. Bast, Large parity violation effects in the vibrational spectrum of organometallic compounds. J. Am. Chem. Soc., 126 (2004) 1652–1653.
31 R. Berger, Molecular parity violation in electronically excited states. Phys. Chem. Chem. Phys., 5 (2003) 12–17.
32 M. Quack, How important is parity violation for molecular and biomolecular chirality? Angew. Chem. Int. Ed., 41 (2002) 4618–4630.
33 F. Crick, Life Itself, Simon and Schuster, New York, 1981.
34 A.C. Neville, Animal Asymmetry, Edward Arnold, London, 1976.
35 D.M. Supp, S.S. Potter and M. Brueckner, Molecular motors: the driving force behind mammalian left-right development. Trends Cell Biol., 10 (2000) 41–45.
36 J.H. Hough, J.A. Bailey, A. Chrysostomou, T.M. Gledhill, P.W. Lucas, M. Tamura, S. Clark, J. Yates and F. Menard, Circular polarisation in star-forming regions: possible implications for homochirality. Adv. in Space Res., 27 (2001) 313–322.
37 A. Sandage and J. Bedke, The Carnegie Atlas of Galaxies, Carnegie Institution of Washington, Washington, DC, 1996.
38 D.K. Kondepudi and D.J. Durand, Chiral asymmetry in spiral galaxies? Chirality, 13 (2001) 351–356.
39 F.C. Frank, On spontaneous asymmetric synthesis. Biochem. Biophys. Acta, 11 (1953) 459.
40 G. Nicolis and I. Prigogine, Self-Organization in Nonequilibrium Systems, Wiley-Interscience, New York, 1977.
41 D.K. Kondepudi and G.W. Nelson, Weak neutral currents and the origin of biomolecular chirality. Nature, 314 (1985) 438–441.
42 D.K. Kondepudi and G.W. Nelson, Chiral-symmetry-breaking states and their sensitivity in nonequilibrium chemical-systems. Physica A, 125 (1984) 465–496.
43 D.K. Kondepudi, F. Moss and P.V.E. McClintock, Observation of symmetry-breaking, state selection and sensitivity in a noisy electronic system. Physica D, 21 (1986) 296–306.

44 D.K. Kondepudi and M.J. Gao, Passages through the critical-point and the process of state selection in symmetry-breaking transitions. Phys. Rev. A, 35 (1987) 340–348.
45 D.K. Kondepudi, R.J. Kaufman and N. Singh, Chiral symmetry-breaking in sodium-chlorate crystallization. Science, 250 (1990) 975–976.
46 D.K. Kondepudi, K.L. Bullock, J.A. Digits, J.K. Hall and J.M. Miller, Kinetics of Chiral-Symmetry Breaking in Crystallization. J. Am. Chem. Soc., 115 (1993) 10211–10216.
47 D.K. Kondepudi, K.L. Bullock, J.A. Digits and P.D. Yarborough, Stirring rate as a critical parameter in chiral-symmetry-breaking crystallization. J. Am. Chem. Soc., 117 (1995) 401–404.
48 K. Asakura, K. Kobayashi and Y. Mizusawa, Generation of optically active octahedral cobalt complex by stereospecific autocatalytic system. Physica D, 84 (1995) 72–78.
49 K. Asakura, D.K. Kondepudi and R. Martin, Mechanism of chiral asymmetry generation by chiral autocatalysis in the preparation of chiral octahedral cobalt complex. Chirality, 10 (1998) 343–348.
50 D.K. Kondepudi and K. Asakura, Chiral autocatalysis, spontaneous symmetry breaking, and stochastic behavior. Acc. Chem. Res., 34 (2001) 946–954.
51 T. Shibata, H. Morioka, T. Hayase, K. Choji and K. Soai, Highly enantioselective catalytic asymmetric automultiplication of chiral pyrimidyl alcohol. J. Am. Chem. Soc., 118 (1996) 471–472.
52 T. Shibata, T. Hayase, J. Yamamoto and K. Soai, One-pot asymmetric autocatalytic reaction with remarkable amplification of enantiomeric excess. Tetrahedron Asym., 8 (1997) 1717–1719.
53 K. Soai, T. Shibata, H. Morioka and K. Choji, Asymmetric autocatalysis and amplification of enantiomeric excess of a chiral molecule. Nature, 378 (1995) 767–768.
54 I. Sato, H. Urabe, S. Ishiguro, T. Shibata and K. Soai, Amplification of chirality from extremely low to greater than 99.5% ee by asymmetric autocatalysis. Angew. Chem. Int. Ed., 42 (2003) 315–317.
55 I. Sato, K. Kadowaki, H. Urabe, J.H. Jung, Y. Ono, S. Shinkai and K. Soai, Highly enantioselective synthesis of organic compound using right- and left-handed helical silica. Tetrahedron Lett., 44 (2003) 721–724.
56 I. Sato, K. Kadowaki, Y. Ohgo and K. Soai, Highly enantio selective asymmetric autocatalysis induced by chiral ionic crystals of sodium chlorate and sodium bromate. J. Mol. Cat., 216 (2004) 209–214.
57 K. Soai, T. Shibata and I. Sato, Enantioselective automultiplication of chiral molecules by asymmetric autocatalysis. Acc. Chem. Res., 33 (2000) 382–390.
58 D.A. Singleton and L.K. Vo, Enantioselective synthesis without discrete optically active additives. J. Am. Chem. Soc., 124 (2002) 10010–10011.
59 R.Viswanathan, J.A. Zasadzinski and D.K. Schwartz, Spontaneous chiral symmetry breaking by achiral molecules in a Langmuir-Blodgett film. Nature, 368 (1994) 440–443.
60 J.M. Ribo, J. Crusats, F. Sagues, J.M. Claret and R. Rubires, Chiral sign induction by vortices during the formation of mesophases in stirred solutions. Science, 292 (2001) 2063–2066.
61 M. De Napoli, S. Nardis, R. Paolesse, M.G.H. Vicente, R. Lauceri and R. Purrello, Hierarchical porphyrin self-assembly in aqueous solution. J. Am. Chem. Soc., 126 (2004) 5934–5935.
62 R. Purrello, Supramolecular chemistry – lasting chiral memory. Nat. Mater., 2 (2003) 216–217.
63 R. Lauceri, A. Raudino, L.M. Scolaro, N. Micali and R. Purrello, From achiral porphyrins to template-imprinted chiral aggregates and further. Self-replication of chiral memory from scratch. J. Am. Chem. Soc., 124 (2002) 894–895.
64 B.N. Thomas, C.M. Lindemann, R.C. Corcoran, C.L. Cotant, J.E. Kirsch and P.J. Persichini, Phosphonate lipid tubules II. J. Am. Chem. Soc., 124 (2002) 1227–1233.
65 B.N. Thomas, C.M. Lindemann and N.A. Clark, Left- and right-handed helical tubule intermediates from a pure chiral phospholipid. Phys. Rev. E, 59 (1999) 3040–3047.
66 B.N. Thomas, R.C. Corcoran, C.L. Cotant, C.M. Lindemann, J.E. Kirsch and P.J. Persichini, Phosphonate lipid tubules. I. J. Am. Chem. Soc., 120 (1998) 12178–12186.

© 2006 Elsevier B.V. All rights reserved.
Chiral Analysis
K.W. Busch and M.A. Busch, Eds.

CHAPTER 3

Physical properties and enantiomeric composition

Chong-Hui Gu[1] and David J.W. Grant[2†]

[1]*Biopharmaceutics R&D, Pharmaceutical Research Institute, Bristol-Myers Squibb Company, 1 Squibb Drive, P.O. Box 191, New Brunswick, NJ 08903, USA*
[2]*Department of Pharmaceutics, College of Pharmacy, University of Minnesota, Weaver-Densford Hall, 308 Harvard Street SE, Minneapolis, MN 55455-0343, USA*

3.1 INTRODUCTION

Compounds that lack inverse symmetry elements, such as a center, a plane, and an improper axis of symmetry in their molecular structures, are chiral [1]. A chiral compound may exist in two enantiomorphic forms, called enantiomers, which are mirror images of each other. An equimolar mixture of opposite enantiomers is called a racemate. Diastereomers are chiral compounds that possess multiple chiral centers, not all of which are superimposable. Opposite enantiomers, in principle, exhibit identical properties in an achiral environment. The racemate may possess properties that differ from those of the enantiomer, depending on the nature of the racemate [2]. Diastereomers usually behave differently even in an achiral environment. Because opposite enantiomers are isostructural and exhibit identical properties, it is difficult to separate them completely [3]. Moreover, a racemate contains at least two chemical components, namely the two enantiomers, and may contain a third component, such as a solvent. The presence of chiral impurities significantly alters the physicochemical properties of a host chiral compound [4–7]. In this chapter, the physical properties that are unique to chiral crystals are discussed with emphasis on the influence of chiral impurities on the properties of host chiral crystals.

3.2 NATURE OF RACEMATES

Based on the molecular interactions in the crystal lattice, three distinct types of crystalline racemate are recognized: conglomerate, racemic compound, or pseudoracemate (Fig. 3.1) [8,9]. A racemic conglomerate consists of an equimolar physical mixture

References pp. 72–77

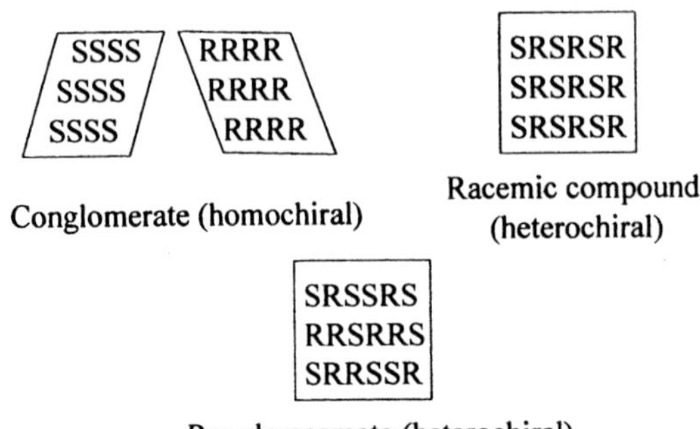

Fig. 3.1. Schematic representation of the molecular arrangements in three types of racemates: conglomerate contains homochiral molecules in each single crystal; racemic compound contains paired enantiomers in each crystal; pseudoracemate or solid solution contains randomly arranged enantiomers in each crystal. (Adapted from Ref. [9].)

of crystals of the two opposite enantiomers. Because each crystal in a conglomerate contains homochiral molecules, the opposite enantiomers can, in principle, be separated manually. Approximately 5–10% of the crystalline racemic species are conglomerates. A racemic compound contains equal number of molecules of the opposite enantiomers, usually paired up, in the unit cell of each crystal. A racemic compound is a co-crystal of opposite enantiomers, which makes it heterochiral. A racemic compound possesses physical properties different from those of the enantiomers. Racemic compounds are the most common racemic species and occur with 90–95% frequency. Pseudoracemates comprise a rare type of racemic species with $<1\%$ occurrence. A pseudoracemate is a solid solution containing equal number of molecules of the opposite enantiomers in a more or less random arrangement in each crystal. A pseudoracemate, like a racemic compound, is a heterochiral crystal. Many pseudoracemates exhibit the properties of plastic crystals and liquid crystals [1]. Although it is rare for opposite enantiomers to be completely miscible, thereby forming a pseudoracemate, limited miscibility among different crystalline species is inevitable. Even when a racemate is a conglomerate or a racemic compound, a solid solution may be found in the vicinity of either pure enantiomer or racemic compound, as illustrated in the phase diagram in Fig. 3.2(e). This type of solid solution is also called a terminal solid solution. The formation of terminal solid solutions poses a challenge in isolating pure enantiomers [1]. The solubility limit of one species in another may be determined by thermal analysis, which is discussed in Section 3.4.1.

A racemic species may coexist in any of the above three forms. However, only one form of racemate is thermodynamically stable under given conditions. This phenomenon is discussed in Section 3.6.

The different types of racemic species can be distinguished by their phase diagrams (Fig. 3.2) [1]. X-ray diffractometry (Fig. 3.4) [10], and spectrometric techniques (Fig. 3.5) may also distinguish these different types [1,10], which are discussed in the corresponding sections below. The molecular origin of the different types of racemic species resides in the different intermolecular interactions in their crystals, as shown in Fig. 3.1.

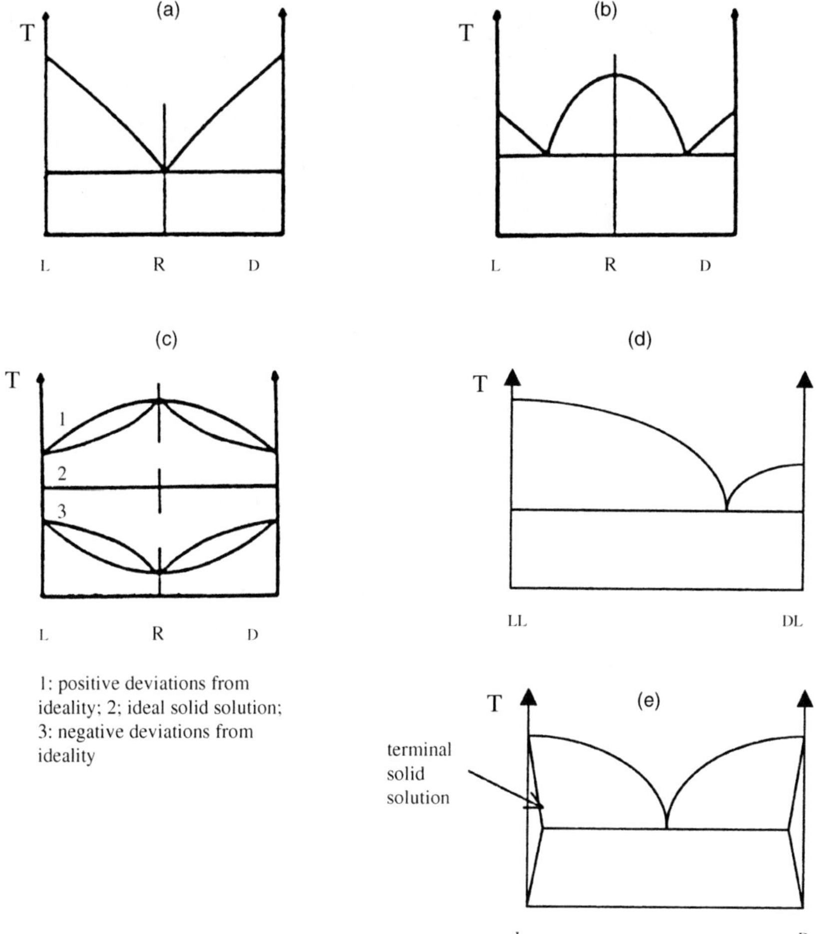

Fig. 3.2. Typical phase diagrams of melting point against composition for: (a–c) three types of racemic species, (a) racemic conglomerate, (b) racemic compound, (c) pseudoracemates; (d) a pair of diastereomers that form eutectic mixture; (e) terminal solid solutions involving opposite enantiomers [1]. (Reproduced by permission of John Wiley and Sons.)

3.3 OPTICAL ACTIVITY

Optical activity results from the refraction of right and left circularly polarized light to different extent by chiral molecules [2]. One type of observed chiral optical activity, termed optical rotatory dispersion, refers to the equal but opposite directions of rotation power of polarized light by the opposite enantiomers. Another type of observed chiral optical activity, termed circular dichroism, arises from the anisotropic absorption of polarized light by chiral samples containing an excess of one enantiomer. Racemates are optically inactive. However, a single crystal from a conglomerate may exhibit optical activity, while the conglomerate as a whole is devoid of optical activity [2]. Optical activity is primarily applied to assign the configuration and conformation of a molecule [11]. The enantiomeric composition may also be determined from the optical activity [12]. The ratio of the experimental optical rotation of a sample, $[\alpha]_{exp}$, to that of the pure enantiomer, $[\alpha]_{pure}$, is known as the optical purity, whose value is equal to the enantiomeric excess (%ee, Eq. (1)) [12].

$$\%ee = \frac{[\alpha]_{exp}}{[\alpha]_{pure}} \times 100\% = \frac{R - S}{R + S} \times 100 \quad (1)$$

3.4 SOLID-STATE ANALYSIS OF PROPERTIES AND COMPOSITIONS OF CHIRAL COMPOUNDS

3.4.1 Thermal properties

The melting property of binary mixtures of two enantiomers is a function of their composition and is characterized by the binary melting point phase diagram of the two enantiomers, which also reveals the nature of the racemate. Fig. 3.2(a) represents the phase diagram of enantiomers that form conglomerates with no mutual solubility in the solid state, and corresponds to the typical behavior of a eutectic mixture. The liquidus line in the diagram can be calculated by the modified Schröder–Van Laar equation [13], which states the relationship between the composition of mixtures and the terminal melting temperature,

$$\ln x = \frac{\Delta H_A^f}{R}\left(\frac{1}{T_A^f} - \frac{1}{T^f}\right) - \frac{C^l - C_A^S}{R}\left(\ln \frac{T_A^f}{T^f} + 1 - \frac{T_A^f}{T^f}\right), \quad (2)$$

where x is the mole fraction of the more abundant enantiomer, ΔH_A^f is the molar enthalpy of fusion of the pure enantiomer, R is the gas constant, T_A^f and T^f are the melting points of the pure enantiomer and the mixture, respectively, and C^l and C_A^s are the heat capacities of the liquid and solid enantiomer, respectively. Because the heat capacity difference between the liquid and the solid is relatively small, the term that contains the heat capacities in Eq. (2) is often much smaller than the other terms and is therefore usually neglected [10].

The phase diagram of enantiomers that form a racemic compound is represented by Fig. 3.2(b), which is similar to the typical phase diagram of two solids forming an intermolecular complex or compound. Fig. 3.2(b) indicates that the racemic compound is a new solid phase, which forms a eutectic mixture with either enantiomer. The melting point of the racemic compound can be either higher or lower than that of the corresponding enantiomer. The liquidus line between the pure enantiomer and the eutectic point can be calculated by the Schröder–Van Laar equation (Eq. (2)), whereas the liquidus line between the two eutectic points can be calculated by the Prigogine–Defay equation, which relates the solid composition to the melting temperature,

$$\ln 4x(1 - x) = \frac{(2\Delta H_R^f)}{\mathrm{R}}\left(\frac{1}{T_R^f} - \frac{1}{T^f}\right), \quad (3)$$

where x is the mole fraction of more abundant enantiomer and ΔH_R^f is the enthalpy of fusion of the racemic compound, and T_R^f and T^f are the melting points of the racemic compound and the solid mixture, respectively [14]. In this equation, the heat capacity terms are ignored for the same reason as that stated under Eq. (2). The chiral purity can be estimated by applying either the Schröder–Van Laar equation or the Prigogine–Defay equation or by measuring the area of the eutectic peak [1].

The third type of racemate, termed a pseudoracemate [1], for which the two enantiomers exhibit complete solid solubility, gives one of the melting curves shown in Fig. 3.2(c). The phase diagram in Fig. 3.2(c)(2) is that of an ideal solid solution, which means that the enthalpy of mixing is zero and the entropy of mixing is equal to R ln 2. The other types of pseudoracemate comprise those giving positive deviations (Fig. 3.2(c)(1)) and those giving negative deviations (Fig. 3.2(c)(3)) from the ideal solid solubility behavior [1]. The deviations suggest that the intermolecular energies between the two opposite enantiomers may be less than or greater than those between molecules of like chirality, which are associated with an enthalpy of mixing that is either positive or negative, respectively.

When two chiral compounds are diastereomers, their binary melting phase diagrams can be analyzed similar to those of enantiomers discussed above. Phase diagrams of diastereomers do not exhibit the symmetry shown by enantiomers. The most common phase diagram of diastereomers is that of an eutectic (Fig. 3.2(d)) [1].

Besides the melting phase diagram, the phase diagram based on the enthalpy of fusion, which is usually determined by differential scanning calorimetry (DSC), can also be used to infer the type of racemate [15,16]. The DSC curve may be analyzed by applying a multiple non-linear regression program to deconvolute the overlapping endothermic peaks [16]. The enthalpy of fusion so obtained may be plotted against the composition. It is easier to determine the eutectic composition from the phase diagram based on the enthalpy of fusion than from that based on the melting point [15].

Thermal analysis has been applied to determine the enantiomeric composition quickly [15,16], even though chromatography is a more accurate method. Often there is a linear relationship between the area of the eutectic peak or the peak asymmetry of the DSC endotherm and the enantiomeric composition. Then the solid solubility limit of one component in another may be estimated by extrapolating the relationship to the composition at which either the eutectic peak area or the peak asymmetry is zero [17]. It is important to be aware of the effect of a chiral impurity on the thermal properties of host crystals, which may decrease the melting point and enthalpy of fusion, as discussed in Section 3.8.

3.4.2 Thermodynamic stability relationship between racemates

It is of theoretical and practical importance to estimate the thermodynamic stability relationship between racemates. The thermodynamic stability of a conglomerate *versus* a racemic compound may be derived from thermal data using thermodynamic cycles, as developed by Jacques et al. [1]. Recently, Li et al. [10] modified this approach and derived the enthalpy, entropy, and Gibbs free energy of racemates with respect to the individual enantiomers from thermal data. The formation of a racemic compound is given by:

$$\mathrm{D_S} + \mathrm{L_S} = \mathrm{R_S}, \tag{4}$$

where the left-hand-side represents the racemic conglomerate, which is an equimolar mixture of opposite enantiomers, D and L, and the right-hand-side represents the racemic compound, R, both in the solid state (subscript S). The free energy of formation of the racemic compound at a given absolute temperature, T, is given by:

$$\begin{aligned}\Delta G^{\mathrm{o}}_{\mathrm{T}} = & -\Delta S^{\mathrm{f}}_{\mathrm{A}}(T^{\mathrm{f}}_{\mathrm{R}} - T^{\mathrm{f}}_{\mathrm{A}}) - T^{\mathrm{f}}_{\mathrm{A}} R \ln 2 + (\Delta S^{\mathrm{f}}_{\mathrm{A}} - \Delta S^{\mathrm{f}}_{\mathrm{R}} + R \ln 2)(T^{\mathrm{f}}_{\mathrm{A}} - T) \\ & + (C^{\mathrm{l}} - C^{\mathrm{s}}_{\mathrm{R}})\left[T^{\mathrm{f}}_{\mathrm{R}} - T^{\mathrm{f}}_{\mathrm{A}} - T \ln \frac{T^{\mathrm{f}}_{\mathrm{R}}}{T^{\mathrm{f}}_{\mathrm{A}}}\right] + (C^{\mathrm{s}}_{\mathrm{A}} - C^{\mathrm{s}}_{\mathrm{R}})\left[T^{\mathrm{f}}_{\mathrm{A}} - T - T \ln \frac{T^{\mathrm{f}}_{\mathrm{A}}}{T}\right], \\ & \text{when } T^{\mathrm{f}}_{\mathrm{A}} < T^{\mathrm{f}}_{\mathrm{R}}\end{aligned} \tag{5}$$

$$\begin{aligned}\Delta G^{\mathrm{o}}_{\mathrm{T}} = & -\Delta S^{\mathrm{f}}_{\mathrm{A}}(T^{\mathrm{f}}_{\mathrm{R}} - T^{\mathrm{f}}_{\mathrm{A}}) - T^{\mathrm{f}}_{\mathrm{A}} R \ln 2 + (\Delta S^{\mathrm{f}}_{\mathrm{A}} - \Delta S^{\mathrm{f}}_{\mathrm{R}} + R \ln 2)(T^{\mathrm{f}}_{\mathrm{R}} - T) \\ & + (C^{\mathrm{l}} - C^{\mathrm{s}}_{\mathrm{A}})\left[T^{\mathrm{f}}_{\mathrm{R}} - T^{\mathrm{f}}_{\mathrm{A}} - T \ln \frac{T^{\mathrm{f}}_{\mathrm{R}}}{T^{\mathrm{f}}_{\mathrm{A}}}\right] + (C^{\mathrm{s}}_{\mathrm{A}} - C^{\mathrm{s}}_{\mathrm{R}})\left[T^{\mathrm{f}}_{\mathrm{R}} - T - T \ln \frac{T^{\mathrm{f}}_{\mathrm{R}}}{T}\right], \\ & \text{when } T^{\mathrm{f}}_{\mathrm{A}} > T^{\mathrm{f}}_{\mathrm{R}},\end{aligned} \tag{6}$$

where ΔS^{f} is the entropy of fusion, T^{f} is the melting temperature, T is the temperature of interest, C is the heat capacity at constant pressure, the subscripts A and R refer to enantiomer and racemic compound, respectively, as mentioned above, while the superscripts l and s refer to liquid and solid states, respectively [1,10]. The last three terms in Eqs. (5) and (6) are relatively small and may be neglected to simplify the use of the equation.

When the free energy of formation is negative at a given temperature, the racemic compound is more stable than the conglomerate at that temperature. Equations (5) and (6) suggest that the negative free energy of formation at the melting temperature is proportional to the difference in melting point, ΔT^{f}, between the racemic compound and enantiomer [10]. This linearity suggests that greater the melting temperature of a racemic compound, as compared with that of its enantiomer, the more stable is the racemic compound, which is supported by a study of 23 chiral drugs [10]. Equations (5) and (6) also enable calculation of the transition temperature of a racemate compound and conglomerate, at which the free energy of formation is zero.

3.4.3 Crystal structure of chiral compounds

A crystal is composed of atoms or molecules arranged in a periodic pattern in three dimensions. There are 230 possible repetitive arrangements of objects in three dimensions, corresponding to 230 space groups. A racemic compound may crystallize in any one of the 230 space groups [1,18]. However, a survey of 792 cases of racemates by Bel'skii and Zorkii revealed that racemic compounds crystallize preferentially in certain space groups: 56% in the $P2_1/c$ space group, 15% in the $C2/c$ space group, and

13% in the $P\bar{1}$(bar) space group, while 16% crystallize in the other 161 achiral space groups [19]. Unlike a racemic compound, an enantiomer can crystallize only in 66 space groups that are devoid of inverse symmetry elements. Among these 66 space groups, a few are preferred. Among 430 cases of enantiomers, 67% crystallize in the $P2_12_12_1$ space group, 27% in the $P2_1$ space group, 1% in the $C2$ space group, and 5% in the other 63 space groups [19]. While an enantiomer inevitably crystallizes in a chiral space group, certain racemic compounds and certain achiral molecules crystallize in chiral space groups, i.e. as chiral crystals, although with a low frequency, i.e. 5% [20].

The crystal structure and the interactions between the molecules in the structure dictate the physical property of a crystal [21]. For example, differences between the powder X-ray diffraction (PXRD) patterns, infrared (IR) spectra, and solid-state nuclear magnetic resonance (NMR) spectra of the racemic compound and enantiomers are all the result of the different molecular packing and consequently the different molecular interactions in the crystals. In addition, the solubility and thermal properties reflect the lattice free energy of the crystal [22]. The crystal structure can be determined by single crystal X-ray diffraction. From the solved crystal structure, the intermolecular interactions in the crystal may be visualized and calculated to correlate and predict crystal properties [23], as illustrated by the following examples that compare the stability of the racemic compound with that of the enantiomer [24–26].

The question "Is the racemic compound or the racemic conglomerate (corresponding enantiomers) more stable?" may be answered by comparing the stability of the crystals of racemic compounds with those of the enantiomers, from considerations of compactness, symmetry, and intermolecular forces. By comparing the densities of eight pairs of chiral compounds at room temperature, Wallach [27] formulated a rule that crystals of a racemic compound tend to be denser than the corresponding homochiral crystals. This rule implies that molecules in racemic crystals tend to be more tightly packed than in the corresponding homochiral crystals. It is generally true that more tightly packed crystals are more stable at low temperatures. However, Brock et al. [28] found that Wallach's rule is statistically biased, because it excludes those racemic compounds that cannot be isolated, because they are markedly less stable than the corresponding enantiomers. Nevertheless, the racemic compounds are statistically more probable than the homochiral mixtures, i.e. conglomerates, because about 90% of chiral compounds form racemic compounds, while only 10% form conglomerates [29]. The greater occurrence of racemic compounds may be attributed to the greater compactness of their crystal structures than those of the corresponding conglomerates. The closer packing of the racemic compound may be explained by the fact that enantiomers can crystallize only in asymmetric space groups, which are devoid of inverse symmetry elements, while almost all racemates crystallize in those space groups that possess elements of inverse symmetry [1,18]. The differences in compactness and symmetry are associated with different molecular arrangement and intermolecular interactions, which fundamentally govern the stability of crystals.

To compare crystal stability from considerations of the strength of the molecular interactions, the crystal structure can be analyzed qualitatively, i.e. by comparing the molecular packing mode and the geometry of the hydrogen bond network, or quantitatively, i.e. by comparing the crystal lattice energy [30,31]. Both qualitative and quantitative analyses enable us to interpret the physical properties at the molecular level.

Qualitative structure analysis has been employed to study the structural basis of the formation of racemic compounds and conglomerates, using as model systems a series of salts formed by a chiral base and a series of structurally related acids. Kinbara et al. [24] compared the crystal structures of a series of salts of chiral primary amines with achiral carboxylic acids (Fig. 3.3). They found that both the formation and assembly of a characteristic columnar hydrogen-bond network, in which the ammonium cations and the carboxylate anions are aligned around a 2-fold screw axis, constituting a 2_1 column, are essential in the formation of conglomerates from these salts (Fig. 3.3). Similar analysis was applied to diastereomers to rationalize the resolution efficiency by visually comparing the crystal structures of a series of diastereomeric pairs. High resolution efficiency was achieved with those pairs, for which one crystal is stabilized by both hydrogen bonding and van der Waals interactions [25,32,33], or with stronger hydrogen bonding structures [34].

Li et al. quantitatively compared the lattice energy of seven pairs of racemic compounds with the corresponding enantiomeric crystals [26]. Each racemic compound showed a greater lattice energy than that of the corresponding enantiomer, to which the van der Waals interactions mainly contribute. The other two types of interactions, electrostatic interactions and hydrogen bonding, fail to show correlation with the overall lattice energy. However, in chiral crystalline salts, the contribution of electrostatic interactions to the lattice energy increases, while the contribution of van der Waals interaction decreases, leading to greater similarity of lattice energies between the racemic compound and the corresponding enantiomer. These energetic considerations may explain the fact that the probability of forming a conglomerate is increased by the formation of a salt [29].

3.4.4 Powder X-ray diffraction (PXRD) patterns

Crystal structures are usually determined by single crystal X-ray diffractometry, which provides fundamental information about crystal properties. However, PXRD is more routinely used to reveal differences in crystal structure and consequently to demonstrate the existence of polymorphs and solvates. PXRD can reveal the type of racemate. Opposite enantiomers exhibit identical PXRD patterns. If a racemate shows a PXRD pattern similar to that of the pure enantiomer, it is a conglomerate, whereas if the pattern is different, it is a racemic compound (Fig. 3.4(a)) [10]. In the case of pseudoracemates, the lattice parameters are often very close to those of the corresponding single enantiomer, leading to subtle difference between the PXRD patterns (Fig. 3.4(b)) [10]. PXRD can also quantify the enantiomeric composition in a racemic compound, which was applied to racemic ibuprofen [35,36]. A standard curve was plotted by analyzing physical mixtures of (*S*)-ibuprofen and racemic (*RS*)-ibuprofen in different proportions, using lithium fluoride as an internal standard. A linear relationship was obtained between the intensity ratio of a diffraction peak of racemic ibuprofen to the internal standard and the weight fraction of the racemate. The

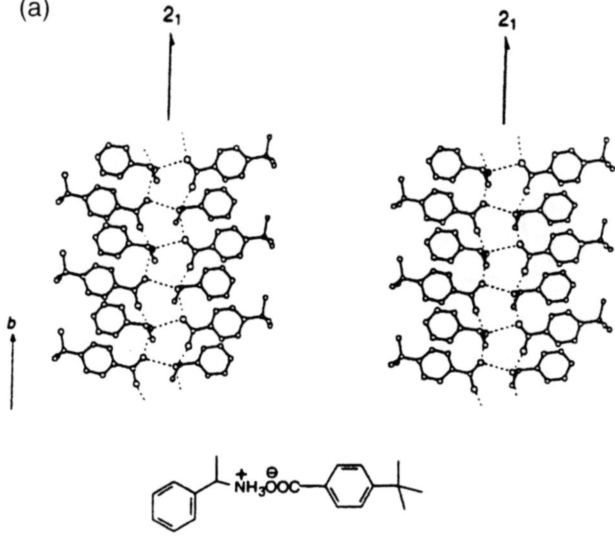

Salt of 1-phenylethylamine with *p*-butyl benzoic acid

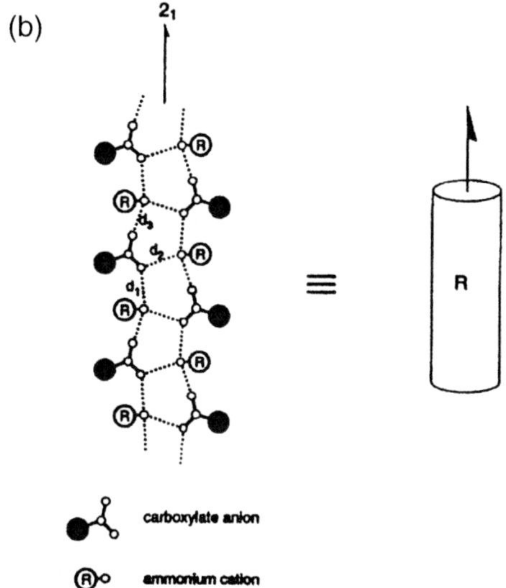

Fig. 3.3. (a) Hydrogen-bond network in the crystal of the salt, 1-phenylethylamine with *p*-butylbenzoic acid, represented by the dotted lines. Two ammonium cation and carboxylate anion pairs form a unit through the hydrogen bonds between the ammonium hydrogens and the carboxylate oxygens. This unit forms an infinite columnar structure around a 2-fold screw axis along the *b* axis (2_1-column). (b) Schematic representation of a 2_1-column formed in conglomerate salts of chiral primary monoamines with achiral monocarboxylic acids [24]. (Reproduced by permission of American Chemical Society.)

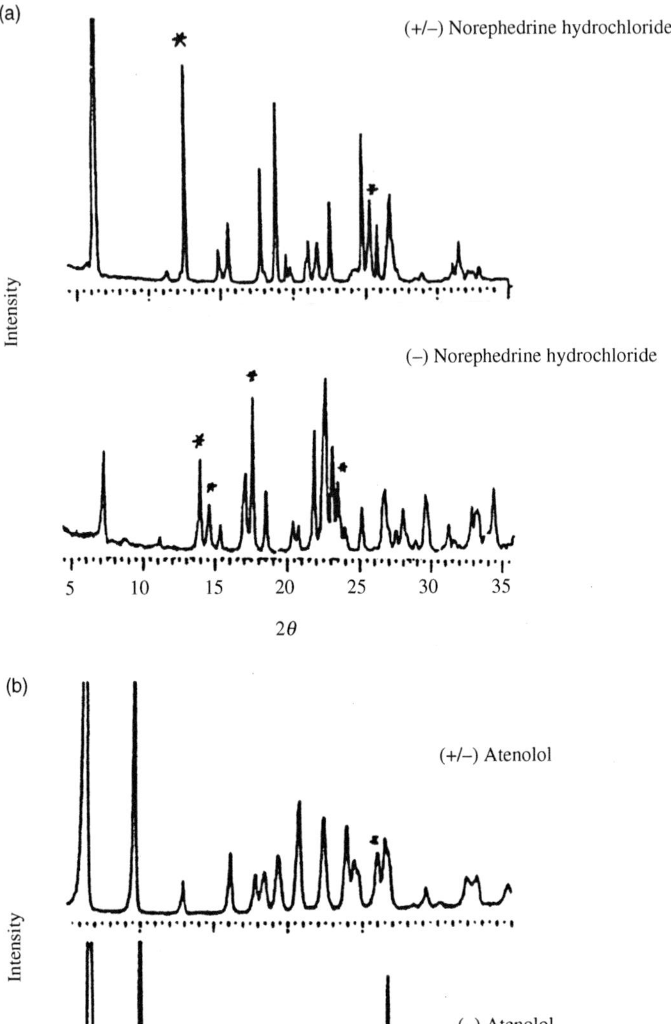

Fig. 3.4. Powder X-ray diffraction (PXRD) patterns of (a) the racemic compound (upper traces) and an enantiomer (lower traces) of norephedrine hydrochloride, and (b) the pseudoracemate (+/–) and an enantiomer (–) of atenolol. The differences between the pseudoracemate and enantiomer of atenolol are subtle and are characterized by a shift of peak position indicated by an asterisk [10]. (Reproduced by permission of American Pharmaceutical Association.)

experimentally determined concentration was found to range between 98 and 104% of the true value with a minimum detection limit of approximately 3% (w/w) for both the enantiomer and the racemate [35].

3.4.5 Infrared, Raman, and solid-state nuclear magnetic resonance spectra

Infrared (IR) and Raman spectroscopy are widely used for characterizing chiral crystals. Because opposite enantiomers exhibit identical IR and Raman spectra, the spectra of a conglomerate are identical to those of the corresponding single enantiomer. Because a racemic compound has a crystal structure that differs from that of the pure enantiomer, the IR and Raman spectra of the racemic compound also differ from those of the enantiomer [1]. Consequently, the type of racemate can be identified by comparing the spectra of the racemate with those of the enantiomer. Both IR and Raman spectra can be used to quantify the amount of enantiomer in a racemate, provided that the racemate is a racemic compound [37].

Solutions of opposite enantiomers show similar solution nuclear magnetic resonance (NMR) spectra. However, they are distinguishable by the use of chiral shift reagents, which enable the chiral composition to be determined. In the solid state, ^{13}C NMR spectroscopy is used to probe the short-range molecular interactions, and particularly for identifying molecular conformations in a crystal lattice [38]. Solid-state NMR spectroscopy can identify the type of racemate, because the spectrum of a conglomerate is identical to that of the pure enantiomer, while that of the racemic compound differs from that of the enantiomer (Fig. 3.5) [10]. Solid-state NMR spectroscopy has also been used to determine the content of a single enantiomer in a mixture containing a racemic compound and a pure enantiomer [39].

3.5 SOLUBILITY AND DISSOLUTION

The nature of a racemate can be determined from a ternary solubility phase diagram (Fig. 3.6), which is especially suitable for thermally labile compounds, whose melting phase diagram cannot be obtained. Furthermore, knowledge of the solubility behavior of racemates and enantiomers is essential in the resolution of racemates via crystallization [1]. The theoretical phase diagram of a conglomerate is represented by Fig. 3.6(a), which corresponds to eutectic behavior [2]. Most solubility diagrams of diastereomers also show eutectic behavior.

A representative ternary phase diagram showing the solubility of a racemic compound is shown in Fig. 3.6(b). The solubility of the racemic compound can either be greater or less than that of the enantiomer. If the solubility of the enantiomer is less than that of the corresponding racemic compound, rapid crystallization in the presence of seed crystals may provide pure enantiomer without allowing the system to reach its solubility equilibrium (the eutectic composition). However, most racemic compounds exhibit a lower solubility than the pure enantiomer [40]. Typically, the solubility of the enantiomer is 2–10 times that of the racemic compound [41]. The different solubilities

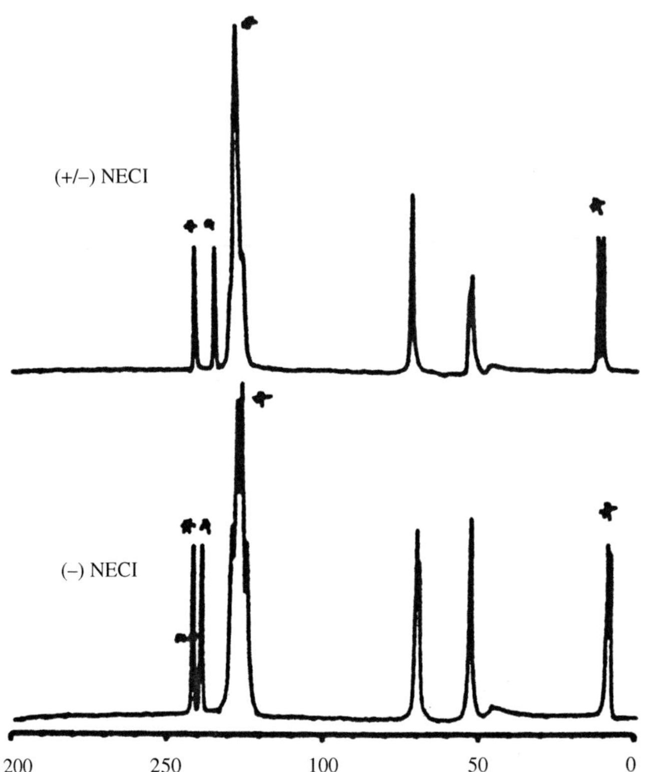

Fig. 3.5. ^{13}C solid-state NMR spectra of the racemic compound (±) NECl, and an enantiomer (–) NECl, of norephedrine hydrochloride [10]. (Reproduced by permission of American Pharmaceutical Association.)

of the racemate and enantiomer will lead to different dissolution rates. Regardless of the rate-controlling step during dissolution, the dissolution rate under constant hydrodynamic conditions increases with increasing solubility [42].

In the pharmaceutical industry, solubility and dissolution rate are the key properties of a drug substance [43]. For a drug substance whose absorption is controlled by the dissolution rate or solubility, the greater dissolution rate or solubility of the enantiomer may lead to significantly greater bioavailability over the racemic drug. This difference favors the administration of the enantiomeric drug. For example, the enantiomeric drug in a topical formulation, being more soluble, exhibits a significantly greater rate of skin permeation than the corresponding racemic drug [44]. However, the presence of a chiral impurity in the host crystals may significantly affect the dissolution rate, which is discussed in Section 3.8.

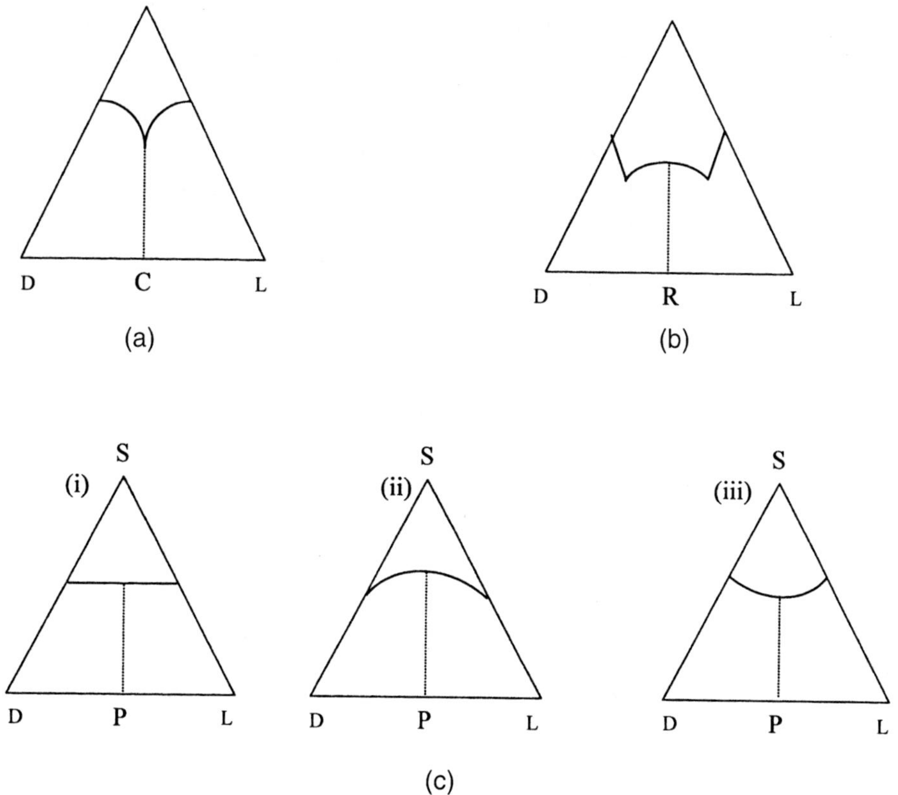

Fig. 3.6. Ternary phase diagram showing the solubility of the racemic species: (a) conglomerate (C), (b) racemic compound (R), (c) pseudoracemate (P), (i) ideal; (ii) positive deviations; (iii) negative deviations. D and L represent the enantiomers, S represents the solvent, and r represents the racemic species other than R [2]. (Reproduced by permission of John Wiley and Sons.)

3.6 POLYMORPHISM AND PSEUDOPOLYMORPHISM OF CHIRAL CRYSTALS

Polymorphism refers to the ability of a substance to crystallize in more than one crystal structure [45]. Crystals with the same chemical composition but with different arrangements and/or conformation of the molecules in the crystal lattice are termed polymorphs. A compound may also form solvates (sometimes termed pseudopolymorphs), which are crystals that contain the solvent of crystallization as part of the crystal lattice structure [46]. When the incorporated solvent is water, the solvate is called a hydrate. Polymorphs often exhibit significantly different physicochemical properties [45], such as those that arise from differences in crystal packing, thermodynamic properties, kinetic properties, surface properties, mechanical properties, and spectroscopic transitions. The significant differences in solubility, dissolution rate, processing properties, and stability have stimulated great interest in the screening and characterization of polymorphs in many industries, such as the pharmaceutical and food industries [19]. Among chiral compounds, the occurrence of polymorphism not only affects their physicochemical properties but also results in interconversion between different types of racemate. The polymorphism of an enantiomer, when no racemization occurs, may be treated in the same way as that of an achiral compound. However, the characterization of the polymorphs of a racemate may be rather complicated, if different types of racemates (Fig. 3.2(a)–(c)) are considered to be polymorphs. For example, racemic sodium ibuprofen has three polymorphs. The stable polymorph at room temperature is the conglomerate, while two metastable polymorphic racemic compounds are obtained by rapidly cooling the melt of the conglomerate [47]. Several theoretically possible polymorphs of chiral compounds are listed in Table 3.1 [1,47–55]. The most common types of polymorphs are polymorphic enantiomers, polymorphic racemic compounds, and the existence of both a racemic conglomerate and a racemic compound of a chiral drug.

Based on the thermodynamic stability relationships between the polymorphs, a polymorphic pair may belong to one of the two types, namely monotropy or enantiotropy [56,57]. In monotropy, one polymorph is always more stable than the other at all temperatures below their melting points. In enantiotropy, the polymorphs have a transition point below the melting points, such that, on crossing the transition point, the relative stability of the polymorphs is inverted. This thermodynamic distinction is based on the assumption that the pressure remains constant at atmospheric pressure. An alternative definition is that, if the pressure–temperature phase diagram does not allow a polymorph to be in equilibrium with its vapor phase below the critical point, it is an unstable monotrope, otherwise it is an enantiotrope [41]. This definition recognizes that some monotropes may be thermodynamically stable at elevated pressures and temperatures, e.g. diamond, which is the metastable polymorph of carbon under

Table 3.1 Several theoretically possible polymorphic systems of enantiomers and the corresponding racemates, with examples

Enantiomers	Racemic species	Examples
Individual enantiomers exihibit polymorphism	Exhibit polymorphism without changing nature of racemates	Carvoxime [1] Nicotine derivatives [50]
	Exhibit polymorphism with different nature of racemates	Nitrendipine [52] Nimodipine [51]
	Racemic compound with no polymorphism	Difficult to prove the absence of polymorphism
	Exhibit polymorphism as racemic compounds	Mandelic acid [53] Tazoferone [55]
Exhibit polymorphism with transition between racemic compound and solid solution	Exhibit polymorphism with transition between conglomerate and racemic compound	Sodium ibuprofen [47] Corycavidine [48] Camphoroxime [49]
	Exhibit polymorphism with transition between conglomerate and solid solution	*cis*-Π-Camphanic acid [54]
	Exhibit polymorphism as solid solutions	No example available

ambient conditions. The thermodynamic stability relationship between enantiotropic polymorphs at a given temperature is determined by the transition temperature, at which the free energy difference between the two polymorphs is zero. The transition temperature may be estimated either from solubilities or from intrinsic dissolution rates (dissolution rate per surface area) determined at several different temperatures [42], or from melting data [58], or from both enthalpy (heat) of solution and solubility data determined at any one temperature [59].

Polymorphs may be discovered by crystallization from supersaturated solutions in various solvents, by solvent-mediated polymorphic transformation, by crystallization of amorphous solids under different conditions, or by polymorphic transformations induced by heat, with or without mechanical stress [60]. The most reliable evidence for polymorphism is a clear difference in the crystal structure determined by single crystal X-ray diffractometry. However, techniques, such as PXRD, variable temperature X-ray diffractometry (VTXRD), DSC, hot-stage microscopy, Fourier transform IR (FTIR), Raman, and solid-state NMR spectroscopy, may be used to identify polymorphs. Solvates may also be characterized by the above techniques. In addition, the stoichiometric number of solvent molecules per molecule of drug substance in the crystal lattice of solvates may be determined by thermogravimetric analysis (TGA), gas chromatography, or in the case of a hydrate, Karl Fischer titrimetry. Recently, TG-MS or TG-FTIR, which couple TGA with mass spectroscopy (MS) or with FTIR, respectively, have been developed to identify and to quantify the chemical composition of solvates [61].

It is often necessary to apply several techniques to characterize both the enantiomer and the racemate and to characterize the type of polymorphism. For example, three monotropically related modifications were found for (*RS*)-nitrendipine [52]. The melting phase diagram showed that the thermodynamically stable form (I) is a racemic compound, while the other polymorphs, form II and III, are both conglomerates. These conclusions were confirmed by the IR spectra and PXRD patterns. Study of the pure enantiomer also revealed that three modifications exist, among which enantiomeric form I corresponds to the metastable conglomerate form II, enantiomeric form II corresponds to conglomerate form III, while enantiomeric form III is not related to any of the racemic modifications. Therefore, both the enantiomer and the racemate are polymorphic in this system. When the designations of the enantiomeric polymorphs and the conglomerate polymorphs do not correspond, careful attention to the nomenclature is necessary.

Understanding the polymorphism of a racemate is essential for the success of resolution by direct crystallization, which separate opposite enantiomers without the use of chiral reagent or enantioselective chromatography [2,62]. Two direct methods of resolution are available. The first method involves the formation of conglomerates, from which the opposite enantiomers can then be separated mechanically. The second method, called entrainment or preferential crystallization, utilizes the different rates of crystallization of opposite enantiomers to achieve separation. Entrainment may need to be performed several times to achieve acceptable resolution. For both the methods, the prerequisite for a successful resolution is that a racemic compound, if it exists, should not crystallize [1]. With the knowledge of the transition temperature between conglomerate and racemic compound, preferential crystallization can be performed in the

References pp. 72–77

temperature range at which the conglomerate is more stable and hence exhibits lower solubility. The formation of a conglomerate in a metastable state may also be employed for preferential crystallization by inhibiting the crystallization of the more stable racemic compound, e.g. resolution of 1,1′-binaphthyl at room temperature [63]. The existence of the metastable racemic compound, (±)-α-methylbenzylammonium (±)-2-phenylpropionate, is found to limit the enantiomeric excess to a maximum of 5.2% by preferential crystallization [64]. Impurities play an important role in the rate of crystallization and thereby influence the outcome of resolution, which is described in detail in Section 3.8.

Solvates and hydrates are commonly formed by chiral compounds. The incorporated solvent may be present either in stoichiometric amounts, e.g. histidine monohydrate [49], or in nonstoichiometric amounts, e.g. cromolyn sodium hydrates [65]. The relative stability of the anhydrate (unsolvated crystal) and the hydrate is determined by the thermodynamic activity of the solvent and the temperature [66,67]. The critical water activity, at which the hydrate and anhydrate are equally stable at a given temperature, is expressed by the following equation:

$$a_{\text{water}} = \frac{p_t}{p_s} = \exp\left(\frac{\Delta H_{\text{tr}}}{nRT_t} - \frac{\Delta H_{\text{tr}}}{nRT}\right), \tag{7}$$

where p_t is the critical water vapor pressure for equilibrium between the hydrate and anhydrate, and p_s is the saturated water vapor pressure, both at temperature T. ΔH_{tr} is the enthalpy (heat) of transition from the hydrate to the anhydrate and is equal to the difference in enthalpy of aqueous solution between the hydrate and anhydrate, and n is the number of moles of water associated with one mole of drug substance in the hydrate [68,69]. T_t is the transition temperature when the hydrate and anhydrate are equally stable at the saturated water vapor pressure, p_s. Equation (7) assumes that the dissolved solute exerts a negligible effect on the water vapor pressure. If, however, the dissolved concentration is large enough to lower the water vapor pressure, the necessary correction can be calculated by means of Raoult's law. At T, the hydrate is more stable than the anhydrate when the water activity of the medium exceeds a_{water} in Eq. (7). Equation (7) is also applicable to the stability of other solvates at the activity of the corresponding solvent. The activity of a solvent can be reduced to the desired value by dilution with a miscible cosolvent. The resulting solvent mixtures can then be used to prepare solvates with different degrees of solvation [70].

In chiral systems, the racemic compound and the enantiomer may exhibit different degrees of solvation under given conditions. There are many examples of a change in the degree of solvation as a result of a change in the nature of the racemate [49]. For example, enantiomeric histidine hydrochloride forms a monohydrate when crystallized from water, whereas racemic histidine hydrochloride forms a dihydrate. The racemic dihydrate transforms to the conglomerate monohydrate above 45°C (Fig. 3.7) [49]. It would be informative to perform similar studies on chiral drugs to determine whether the formation of a solvate can be used to transform a racemic compound into a conglomerate and thereby facilitate resolution. For solvates with different degrees of solvation, van't Hoff proposed an empirical rule that the least solvated compound is the most stable at a relatively high temperature [71]. The rule follows the principle

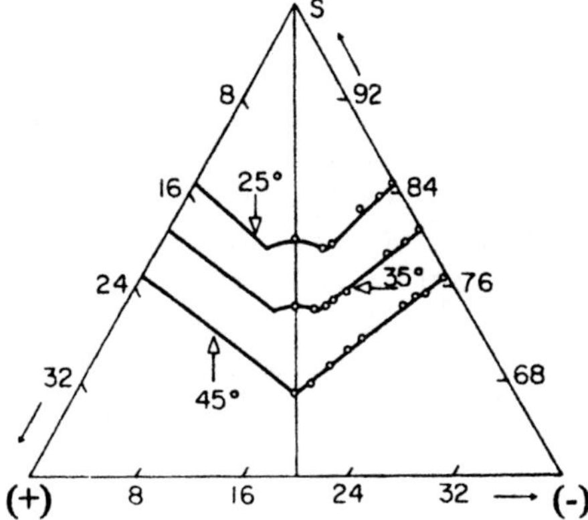

Fig. 3.7. Ternary phase diagram showing the solubility of histidine hydrochloride enantiomers, (+) and (−), in water (*S*) at various temperatures [49] (Reproduced by permission of the copyright owner, Société Francaise de Chimie.). At 25 and 35°C, the most stable racemate is a racemic compound, while the conglomerate becomes the most stable racemate at 45°C.

of Le Chatelier, but exceptions to the rule may be found. The transition temperature of solvated forms, with different degrees of solvation in a solution of the corresponding solvent may be estimated by the following equation, in which the heat capacity difference between the two solvates is ignored:

$$T_t = \frac{\Delta H_{tr}}{(\Delta H_{tr}/T) - R \ \ln \ (C_B/C_A)}, \tag{8}$$

where ΔH_{tr} is heat of transition from solvate A to B, which can be determined from heat of solution data, and c_A and c_B are the solubilities of A and B, respectively determined at temperature T [59]. To determine the transition temperature between an enantiomer and a racemic compound, Eq. (8) needs to be modified assuming that dissolved opposite enantiomers form an ideal solution:

$$T_t = \frac{\Delta H_{tr}}{(\Delta H_{tr}/T) - R \ln (c_R/c_E) + R \ln 2} \tag{9}$$

and c_R and c_E are the solubilities of racemates and enantiomers, respectively determined at temperature T.

Another method of determining the transition temperature is the use of DSC with controlled solvent activity to measure the desolvation temperature [68]. Although the DSC method may often be applicable to a solvate and nonsolvate, it may be difficult to apply it to two solvates with different degrees of solvation, because of the difficulty in controlling the desolvation process in DSC.

Diastereomeric pairs also form solvates with different degrees of solvation. This phenomenon is useful for achieving a high efficiency of resolution, because solvates with different degrees of solvation may have significantly different solubilities. The solubility of a solvate with a higher degree of solvation in a given solvent is usually lower than that with a lower degree of solvation in the same solvent. On the other hand, in a solvent that is miscible with the solvating solvent, the solubility of the more highly solvated solid is usually higher than that of the solid with a lower solvation [22]. Therefore, by choosing the appropriate solvent, the solubility ranking of a diastereomeric pair can be changed to obtain the desired enantiomer. For example, in the resolution of DL-leucine by forming salts with (*S*)-(−)-phenylethanesulfonic acid (PES), L-leucine-(−)-PES (LS) is less soluble in acetonitrile–methanol mixture than its diastereomer, D-leucine-(−)-PES (DS). However, in acetonitrile–water mixture, DS monohydrate is formed and is less soluble than LS, which does not form a hydrate [72].

It is important to be aware that polymorphic transitions may occur during processing and handling, such as grinding, milling, and compression [73]. Both mechanical forces, especially shear, and relative humidity (RH) are believed to exert influences. The external mechanical energy may transform to structural energy, facilitating the transition, while the water content of a drug may increase under higher RH and will significantly increase the molecular mobility [74]. Hydration and dehydration may occur during these processes. It was found that physical mixtures of the enantiomers of malic acid, tartaric acid, and serine transformed to racemic compounds after grinding or during storage at 53% RH and at 40°C [75].

3.7 RACEMIZATION

Although, historically, most synthetic chiral drugs have been formulated as racemates, a recent trend in the pharmaceutical industry is to develop pure enantiomers, thanks to the advances in asymmetric synthesis and technologies for chiral separation [76]. When utilizing a single enantiomer, possible racemization should be monitored. Racemization is the formation of a racemate from the corresponding pure enantiomer [2]. In most cases, racemization involves the breaking of covalent bonds through a chemical reaction. However, racemization may be achieved by an increase in temperature, which induces changes in molecular arrangement (bond angles and lengths). Pasteur discovered that the enantiomer of the cinchonine salt of (+)-tartaric acid was converted to the racemic salt of tartaric acid after heating to 170°C. It has since been reported that biaryls [77], chiral substituted cyclooctatetraenes [78], and allylic sulfoxides [79] (Fig. 3.8) also undergo thermally induced racemization [2]. It is therefore important to monitor the change in chirality during heating of these classes of compounds. The racemization of corycavidine (Fig. 3.8) was characterized by X-ray crystallography, DSC, and MS [48]. After melting and cooling, (−)-corycavidine is converted to the racemic compound, (±)-corycavidine. By determining the racemization rate in solution at different temperatures and by applying the Arrhenius equation, the activation energy for racemization was determined to be 4.7 kcal/mol.

3.8 INFLUENCE OF CHIRAL IMPURITY ON PHYSICAL PROPERTIES OF HOST CHIRAL CRYSTALS

The presence of impurities is ubiquitous. In particular, a chiral crystal often coexists with its opposite enantiomer, and a racemic compound often coexists with excess enantiomer.

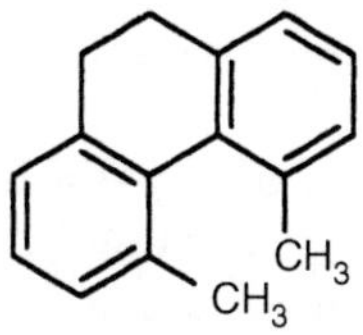

(a) 1,10 - Dimethylphenanthrene

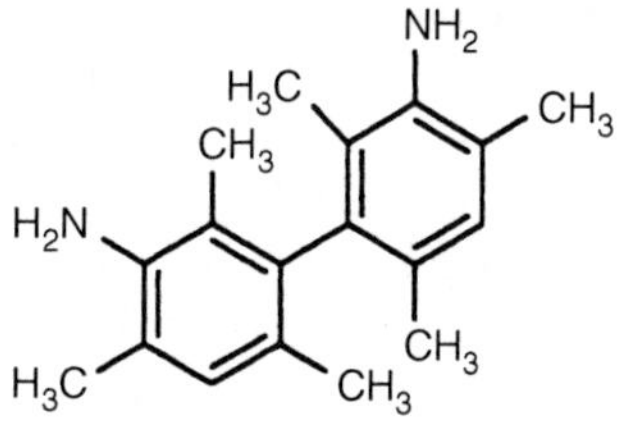

(b) 2,3,4,6,2,',3',4',6' - Octamethylbiphenyl

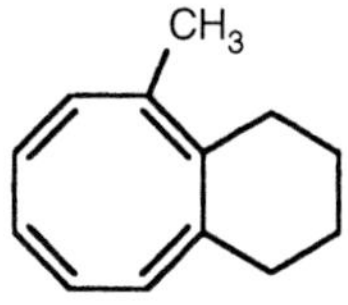

(c) Cyclooctatetraene

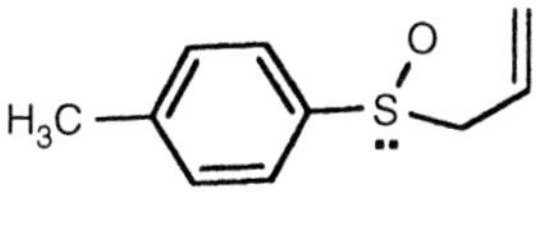

(d) Allyl-*p*-tolysulfoxide

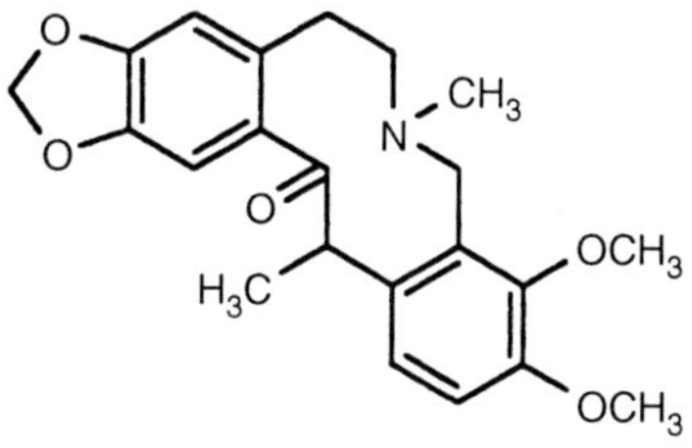

(e) Corycavidine

Fig. 3.8. Molecular structures of compounds that undergo thermally induced racemization.

Each may exist with diastereomers because of the difficulty of complete separation of chiral molecules [3].

The presence of an impurity can significantly affect the physical properties of the host crystals, either during the process of preparing the crystals by crystallization or by concomitant incorporation into the host crystals [80]. Crystallization is a method that is widely used to separate and to purify materials [81], as well as for chiral resolution [1,2]. It is well known that impurities, including crystallization solvents, can change the external properties of the resulting crystals, e.g. the morphology, or the internal structure, e.g. the polymorphic form [81–83]. Understanding the influence of impurities on the properties of the resulting crystals helps to avoid undesirable effects and even to engineer crystals with desirable properties.

During crystallization, embryonic nuclei with structures resembling each of the mature phases may be present, but only those embryos growing beyond a certain critical size develop into nuclei that are stable enough to grow into crystals [83–85]. Based on this assumption, impurities in the crystallization medium may recognize, and interact with, a specific crystalline phase and selectively inhibit the development of the nuclei and/or growth of crystals of that phase. Therefore, another phase may crystallize from the solution in a kinetically controlled process. By carefully comparing the crystal structure of a racemic compound and its enantiomer, based on the principle of molecular recognition at interfaces, "tailor-made" chiral impurities have been designed to selectively inhibit the crystallization of undesired chiral species to isolate a single enantiomer [85]. The enantiomeric *S*-amino acid was found to selectively interact with the face of (*S*)-glutamic acid HCl salt and to inhibit its growth but to exert no effect on (*R*)-glutamic acid HCl salt [86]. After attaching (*S*)-lysine to a polymer backbone, the resulting poly-(N^{ε} methylacryloyl-lysine) inhibits nucleation of (*S*)-glutamic acid HCl salt at 0.1% concentration [85]. In the case of histidine, the imidazole ring of (*S*)-histidine points to the {011} face of the racemic compound, which can interact with the aromatic side group of poly-(*p*-acrylamido-*S*-phenyalanine) and the latter can inhibit the crystallization of (*RS*)-histidine to yield (*R*)-histidine (Fig. 3.9) [87]. Chiral impurities have also been used to remove epitaxial twinning of conglomerates by selectively inhibiting the crystallization of a specific enantiomer. Crystallization of the HCl salt of methionine or cysteine in the presence of 1% (*S*)-poly(acrylic acid) or (*S*)-polymethylacrylic acid, which is grafted with enantiomerically pure lysine *via* the ε-amine group, resulted only in crystals composed of molecules with handedness opposite to that of the grafted lysine [88].

In another interesting case, although nucleation of the racemate was inhibited by an impurity, the enantiomeric nuclei served as seeds for crystallization of the racemate. In the presence of 1% *R*-threonine, the nucleation of both the (*R,S*) and (*S*) forms of alanine was inhibited, leading to the nucleation of (*R*)-alanine. The opposite {001} faces of (*R*)-alanine then served as templates for the epitaxial growth of (*R,S*)-alanine, yielding the twin crystals of the racemate [89].

Chiral impurities may also selectively inhibit or induce the crystallization of a particular polymorph. Glycine has three known polymorphs, α, β, and γ, for which the thermodynamic stability is in the order, $\gamma > \alpha > \beta$. Racemic hexafluorovaline inhibited the crystallization of the α form from aqueous solution, leading to preferred

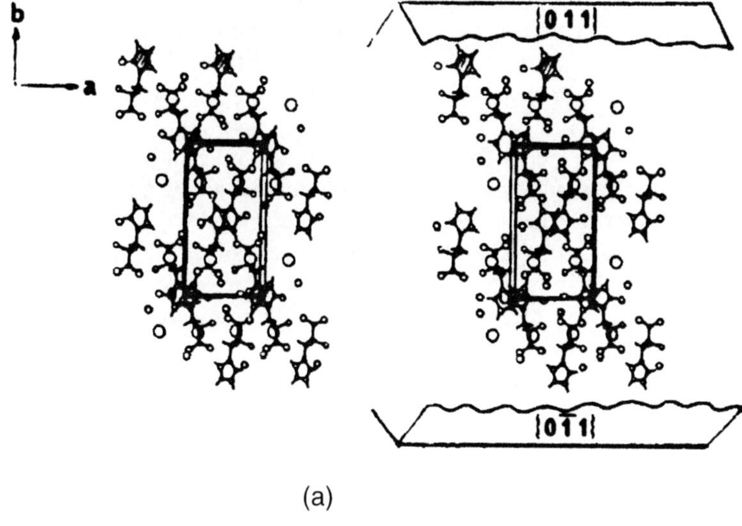

(a)

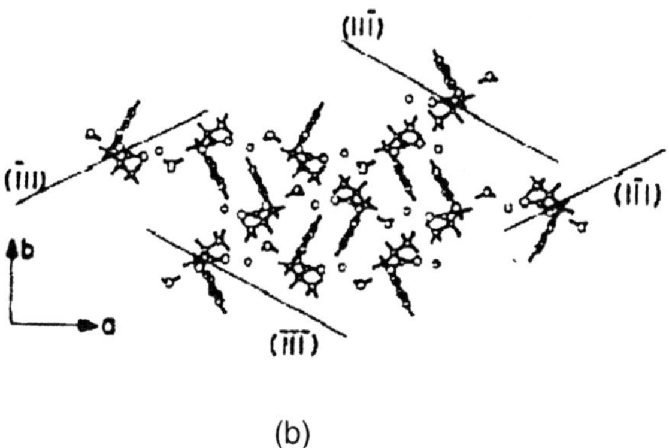

(b)

Fig. 3.9. Crystal structures of histidine hydrochloride, (a) racemic compound, (b) enantiomer. The shaded molecules of (*S*) configuration exposed in the {011} faces of racemic compound may interact with the (*S*)-amino acid moiety grafted onto a polymer backbone, which will inhibit further growth. In the enantiomer, the imidazole ring of histidine is not exposed to the surface [87]. (Reproduced by permission of American Chemical Society.)

growth of the γ form [90]. This result can be explained by the differences between the crystal structures of the α and γ forms. Hexafluorovaline effectively inhibits growth of the faces with exposed NH_3^+ groups but not the faces with exposed CO_2^- groups. Therefore, in the centrosymmetric α form, hexafluorovaline inhibits the growth of all the {011} faces and thus blocks crystal growth along both the *b*- and *c*-directions. In the polar γ form, only the {010} faces are blocked and the crystal can grow in the polar *c*-direction at the carboxylate end. In contrast, the α form remains the preferred form for crystallization in the presence of racemic mixtures of common α-amino acids, up to 6%, because they block only the growth of {010} faces [91]. As expected, the morphology of α glycine is modified in the presence of other amino acids [90,91].

The formation of diastereomeric salts followed by separation *via* crystallization is the most widely used method in the industry for chiral resolution. Recently, a new method, Dutch Resolution, which uses families of resolving agents, was reported to increase the success rate of isolating a salt with a high diastereomeric excess [92]. The principle of this method is the use of structurally related compounds as selective crystallization inhibitors. In practice, equimolar proportions of three structurally related resolving agents are added, leading to diastereomeric salts containing mixtures of resolving agents in nonstoichiometric ratios. Interestingly, a survey of 46 examples of resolution found ten cases, in which no detectable amount of one or more of the three resolving agents was incorporated in the salt [93]. These unincorporated agents served as nucleation inhibitors, as discussed above, and thereby increased the resolution efficiency. Based on this principle, a modified Dutch Resolution has been suggested, which uses small amounts of a certain structurally related resolving agent as an additive that inhibits the nucleation of undesirable diastereomers but is itself minimally incorporated [94,95].

The effects of impurities on crystallization are related to the strength of the intermolecular interactions between the impurity and the surface of the nuclei or crystals. The inhibitory effect of impurities on crystallization may be predicted from the surface binding energy of the impurity molecule on to the crystal surface of the host crystals, which can be calculated by molecular modeling. In the systems, adipic acid [96] and sulfamerazine [97], surface binding energies of impurities to the host crystal surface are correlated with the inhibitory effect on crystallization. The scale of inhibitory effect may be estimated from a model based on the assumption that the impurity is adsorbed one-dimensionally on the step lines and the growth rate of the face is proportional to the step-advancement velocity [97,98]. The ratio of the growth rate in the presence of impurity to that in its absence (f), is related to the percentage of the surface covered by the impurity molecules (θ), according to the following relationship,

$$f = 1 - \alpha\theta, \tag{10}$$

where α is the inhibitory effectiveness factor of the impurity. The parameter (θ), can be estimated from an empirical Langmuir equation [99]. At equilibrium, the fraction of the surface covered (θ), may be expressed by:

$$\theta = \frac{\theta_{\max} kc}{1 + kc}, \tag{11}$$

where $\theta_{\max}$ ($0 < \theta_{\max} \leq 1$) is the maximum proportion of the surface available for adsorption, c is the concentration of the impurity in solution, and k is the ratio of the adsorption rate coefficient to the desorption rate coefficient. If $\alpha > 1$, the crystal growth rate may decrease to zero, when $\alpha\theta \geq 1$.

The structurally related impurities can also be incorporated into the host crystals by random events, by physical entrapment, or by the formation of a terminal solid solution [80]. Although impurities exert effects on the nucleation step, the chance of impurities being occluded in the nuclei of the host is rather small, because of the relatively low concentration of the impurities in the solution. However, appreciable amounts of impurities in the crystallization medium can be taken up during crystal growth [100]. The process of crystal growth from solution includes the diffusion of a solute molecule to the surface and the incorporation of the solute into a growth site, preferentially a kink site. The concept of molecular recognition at interfaces, discussed above, also suggests that the surface of a growing crystal can be thought of as composed of active sites which interact stereospecifically with molecules in solution, in a manner similar to enzyme–substrate interactions [83,85]. Based on the principles of crystal growth and the concept of molecular recognition, the impurity molecules are first adsorbed at selective crystal growth surfaces but are rejected at other surfaces. The adsorbed impurities may change the growth rate of that specific surface. If the impurities can interfere with the addition of solute species to the growth steps, the growth rate of this specific surface will, in most cases, be lower than those of other surfaces. The overgrowth of other faces may bury the adsorbed impurities in the host crystals. Another possible mechanism of impurity incorporation into the host crystal lattice is that, when the crystal growth rate is too fast to desorb the impurities from the surface, the impurities are incorporated into the crystal lattice by misrecognition [101]. Because the rate of adsorption and desorption, the step velocity and height, and related properties, vary from face to face, impurity incorporation also varies from face to face. The selective incorporation of an impurity can sometimes give rise to the "hour glass" phenomenon, a term used by Buckley to describe the hour glass-like distribution of dye impurities in the host crystal [100]. It is also possible that the impurity and the host form a thermodynamically stable terminal solid solution, which has a lower solubility and therefore crystallizes preferentially [1,4].

The extent of impurity incorporation is controlled by both the solid solubility of the impurity in the host crystals and the crystallization kinetics, such as the crystal growth rate and degree of supersaturation [102]. The affinity of the impurity for the host crystals is characterized by the segregation coefficient, which is defined as the ratio of the concentration of the impurity in the solution to that in the host crystals, both with respect to the host molecules. In most systems, the segregation coefficient is less than unity, which means that the impurity tends to be rejected by the growing crystals [103]. For these systems, the extent of impurity incorporation is greater when the crystal growth is faster and the degree of supersaturation is greater. In a single batch, the larger crystals are found to incorporate more impurities than the smaller crystals for those systems for which the segregation coefficient is less than unity [104].

Once a molecule of impurity is incorporated into the host crystal lattice, it will interact with the neighboring host molecules in a way that is different from that between host molecules [105,106], because of its different shape and structure. Thus, the incorporation of impurity may distort the lattice, leading to changes in the thermodynamic properties of the host crystals, including the enthalpy, entropy, and free energy. Previous studies in pharmaceutical systems have demonstrated that the incorporation of a structurally related compound causes the enthalpy and entropy of fusion to decrease with increasing incorporation of impurity into the host crystals, corresponding to increases of lattice energy and lattice disorder brought about by the incorporation of impurity [107–109]. The relationship between the free energy change and the changes of enthalpy and entropy can be expressed by the following fundamental equation,

$$\Delta G = \Delta H - T \times \Delta S \qquad (12)$$

Equation (12) shows that the changes in enthalpy and entropy exert opposing effects on the free energy. When the changes of enthalpy and entropy are of the same sign, the direction of the free energy change depends on the relative magnitude of the terms, ΔH and $T \times \Delta S$. However, when enthalpy–entropy compensation occurs in the system [110], parallel changes in entropy and enthalpy will be associated with a parallel, but smaller, change in free energy. If stable solid solutions are formed, the free energy of the solid solution must be lower than that of the pure components. However, if the solid solution is formed under nonequilibrium conditions, the solid solutions formed may be metastable, as a result of their higher free energy than that of the pure components. Because the incorporation of an impurity often leads to increases in both the enthalpy and entropy, the free energy change of the host crystals will be determined by the relative magnitudes of the changes in these two terms and the temperature.

The effects of the incorporated impurity on the intrinsic dissolution rate (IDR) of the host are profound and vary from system to system. Regardless of the mechanism of dissolution, the IDR increases with an increase of solubility, which depends on the partial molar free energy (chemical potential) of the compound [22]. Depending on whether the mole fraction of the guest in the host crystal lattice is less than or greater than the solid solubility limit, either a stable or a metastable solid solution may be formed. When a stable solid solution is formed, which implies that the guest and the host in the impure crystals have lower chemical potentials than in the respective pure crystals, the impure crystals will have a lower solubility than the pure crystals and therefore a lower IDR under constant hydrodynamic conditions [111]. Conversely, if the concentration of the incorporated impurity exceeds the solid solubility, a metastable solid solution will be formed, in which case the impure crystals are expected to have a higher IDR than the pure crystals [7]. Furthermore, the impurity itself may exert an inhibitory effect on the dissolution rate of the host, either by segregating at dislocations [112], where the dissolution process often begins, or by entering the dissolution medium and acting as an inhibitor [113].

The mechanical properties of crystals are also related to the lattice imperfections of the crystals [114]. The presence of crystal lattice imperfections renders the crystal more plastic and improves the tableting properties [115]. However, when the density of the crystal imperfections is too high, the defects may interact with each other under the stress of compression, causing hindrance to plastic flow and a corresponding increase in hardness.

The solid-state reactivity of the host may also be changed by the incorporation of the impurity. Generally, crystals containing a high density of defects have higher reactivities than perfect crystals, because the defective crystals have higher molecular mobility which may facilitate the molecular loosening process. Sometimes, the reaction starts at the defect points produced either by the impurity or by structural imperfections [116]. It has been demonstrated in solid-state photochemical reactions that the absorbed quanta of light energy may be transferred from the more perfect parts of the crystal to the impurity molecules or defect sites so that reaction occurs at these sites [117]. Although few studies comparing the reactivity of doped and undoped crystals have been carried out, many experiments have shown that crystals containing impurities or defects brought about by mechanical processing (e.g. grinding, milling, and compression) have a greater reactivity in the solid state than more perfect crystals [118,119]. From a thermodynamic point of view, the crystals with defects may have a higher free energy and this structurally stored energy may be transformed into chemical energy, enhancing the reaction rate, as described by the Eyring equation [114]. Because the solid-state reactivity of a drug and excipient is directly related to the stability and the shelf life of a formulation, the incorporation of an impurity and the subsequent disruption of the host crystals may cause an interbatch variation of stability.

To quantify the disruptive effect of incorporated impurities on the host crystal lattice, Grant and York introduced the concept of *disruption index*, which is defined as the rate of change of the difference between the entropy of the solid and that of the liquid, with respect to the ideal entropy of mixing of the components, namely, the guest and the host [120]. The disruption index can be estimated from melting data based on the following equation:

$$\Delta S^{\mathrm{f}} = \Delta S_0^{\mathrm{f}} - (b - c) \times \Delta S_{\mathrm{ideal}}^{\mathrm{m}}, \tag{13}$$

where ΔS^{f} is the entropy of fusion of doped crystals, ΔS_0^{f} is the entropy of fusion of pure, undoped crystals, $\Delta S_{\mathrm{ideal}}^{\mathrm{m}}$ is the entropy of ideal mixing, and the value of $(b - c)$ is the disruption index. The disruption index can also be estimated from the heat of solution and solubility data [121].

Studies of the ephedrine and pseudoephedrine systems have demonstrated that appreciable amount of chiral impurities (guests), including the opposite enantiomer [6,7], excess enantiomer of a racemate [5], and a diastereomer [4], may be incorporated into the lattice of the host chiral crystals. In all systems, the presence of impurities reduces the enthalpy and entropy of fusion of the host crystals, indicating increases in lattice energy and disorder. The effect of incorporated impurities on the intrinsic dissolution rate varies from system to system and is influenced by the amount of impurities incorporated, which may be used to deduce the change in free energy of the host crystals after the incorporation of impurity [4].

3.9 CONCLUDING REMARKS

On account of chirality, chiral compounds exhibit special physical properties that require special characterization. Both the enantiomer and the corresponding racemate should be characterized to fully understand a chiral system. The nature of the racemate is a key property of a chiral compound and may be determined by thermal analysis, the solubility phase diagram, and spectroscopic methods. Classical characterization tools, such as thermodynamics and spectroscopy, are now routinely used to characterize chiral compounds. Meanwhile, crystal structure analysis is more extensively employed for physical characterization, because the internal structure governs the solid-state properties. With developments in X-ray crystallography and computational chemistry, molecular interaction and molecular packing will be increasingly emphasized in physical characterization. Currently, computational methods are mainly used retrospectively to explain experimentally observed properties, while computational prediction of physical properties is still in its infancy. In limited cases, deducing qualitative information on properties from calculation of molecular interaction and crystal structures appears to be achievable. The inevitable presence of chiral impurities significantly affects the properties of host chiral crystals, either during the crystallization process or through incorporation into the host crystal lattice. The effect of impurities may be elucidated by measuring thermodynamic quantities and by studying the interactions between the host and guest molecules and the intermolecular interactions at the crystal surfaces. Considering the significant influence of chiral impurities, the physical properties of chiral crystals should preferably be determined as a function of enantiomeric composition.

REFERENCES

1 J.C. Jacques, A. Collet and S.H. Wilen, Enantiomers, Racemates, and Resolutions, John Wiley & Sons, New York, 1981.
2 E.L. Eliel, S.H. Wilen and L.N. Mander, Stereochemistry of Organic Compounds, John Wiley & Sons, New York, 1994, pp. 153–295.
3 S.P. Duddu, R. Mehvar and D.J.W. Grant, Liquid chromatographic analysis of the enantiomeric impurities in various (+)-pseudoephedrine samples. Pharm. Res., 8 (1991) 1430–1433.
4 C. Gu and D.J.W. Grant, Effects of crystallization in the presence of the diastereomer on the crystal properties of (SS)-(+)-pseudoephedrine hydrochloride. Enantiomer, 5 (2000) 271–280.
5 Z.J. Li and D.J.W. Grant, Effects of excess enantiomer on the crystal properties of a racemic compound: ephedrinium 2-naphthalenesulfonate. Int. J. Pharm., 137 (1996) 21–31.
6 S.P. Duddu, F.K.Y. Fung and D.J.W. Grant, Effect of the opposite enantiomer on the physicochemical properties of (−)-ephedrinium 2-naphthalenesulfonate crystals. Int. J. Pharm., 94 (1993) 171–179.
7 S.P. Duddu, F.K.Y. Fung and D.J.W. Grant, Effects of crystallization in the presence of the opposite enantiomers on the crystal properties of (*SS*)-(+)-pseudoephedrinium salicylate. Int. J. Pharm., 127 (1996) 53–63.
8 H.W.B. Roozeboom, Löslichkeit und Schmelzpunkt als Kriterien für racemische Verbindungen, pseudoracemische Mischkrystalle und inaktive Konglomerate. Z. Phys. Chem., 28 (1899) 494–517.
9 Z.J. Li and D.J.W. Grant, Relationship between physical properties and crystal structures of chiral drugs. J. Pharm. Sci., 86 (1997) 1073–1078.
10 Z.J. Li, M.T. Zell, E.J. Munson and D.J.W. Grant, Characterization of racemic species of chiral drugs using thermal analysis, thermodynamic calculation, and structural studies. J. Pharm. Sci., 88 (1999) 337–346.
11 K. Nakanishi, N. Berova and R.W. Woody, Circular Dichroism: Principles and Applications, VCH, New York, 1994, pp. 1–31.
12 H. Kagan, Is there a preferred expression for the composition of a mixture of enantiomers? Recl. Trav. Chim. Pays-Bas, 114 (1995) 203–205.
13 I. Schröder, Über die Abhängigkeit der Löslichkeit eines festen Körpers von seiner Schmelztemperatur. Z. Phys. Chem., 11 (1893) 449–465.
14 I. Prigogine and R. Defay, Chemical Thermodynamics (English Translation), 4th Ed. Longmans, London, 1967.

15 M. Elsabee and R.J. Prankerd, Solid-state properties of drugs. III. Differential scanning calorimetry of chiral drug mixtures existing as racemic solid solutions, racemic mixtures and racemic compounds. Int. J. Pharm., 86 (1992) 221–230.
16 M. Elsabee and R.J. Prankerd, Solid-state properties of drugs. II. Peak shape analysis an deconvolution of overlapping endotherms in differential calorimetry of chiral mixtures. Int. J. Pharm., 86 (1992) 211–219.
17 G. Zhang, Influence of Solvents on Properties, Structures, and Crystallization of Pharmaceutical Solids. Ph.D. Thesis, University of Minnesota, Minneapolis, 1998, pp. 70–122.
18 H. Brittain, Crystallographic consequences of molecular dissymmetry. Pharm. Res., 7 (1990) 683–690.
19 V.K. Bel'skii and P.M. Zorkii, Distribution of molecular crystals by structural classes. Soviet Phys. Crystallogr. (English translation), 15 (1971) 607–610.
20 H. Koshima and M. Miyauchi, Polymorph of a cocrystal with achiral and chiral structures prepared by pseudoseeding: tryptamine/hydrocinnamic acid. Cryst. Growth Design, 1 (2001) 355–357.
21 S. Datta and D.J.W. Grant, Crystal structures of drugs: advances in determination, prediction and engineering. Nature Reviews – Drug Discovery, 3 (2004) 42–57.
22 D.J.W. Grant and T. Higuchi, Solubility Behavior of Organic compounds, John Wiley & Sons, New York, 1990, pp. 12–133.
23 F.J.J. Leusen, Rationalization of Racemate Resolution–A Molecular Modeling Study. Ph.D. Thesis, University of Nijmegen, The Netherlands, 1993, pp. 1–25.
24 K. Kinbara, Y. Hashimoto, M. Sukegawa, H. Nohira and K. Saigo, Crystal structures of the salts of chiral primary amines with achiral carboxylic acids: recognition of the common-occurring supramolecular assemblies of hydrogen-bond networks and their role in the formation of conglomerate. J. Am. Chem. Soc., 118 (1996) 3441–3449.
25 K. Kinbara, Y. Hashimoto, H. Nohira and K. Saigo, Chiral discrimination upon crystallization of the diastereomeric salts of 1-arylethylamines with mandelic acid of *p*-methoxymandelic acid: Interpretation of the resolution efficiencies on the basis of the crystal structures. J. Chem. Soc. Perkin Trans., 2 (1996) 2615.
26 Z.J. Li, W.H. Ojala and D.J.W. Grant, Molecular modeling study of chiral drug crystals: lattice energy calculation. J. Pharm. Sci., 90 (2001) 1523–1539.
27 O. Wallach, Zur Kenntniss der Terpene und der ätherischen Oele. Liebigs Ann. Chem., 286 (1895) 90–143.
28 C. Brock, W.B. Schweizer and J.D. Dunitz, On the validity of Wallach's rule: on the density and stability of racemic crystals compared with their chiral counterparts. J. Am. Chem. Soc., 113 (1991) 9811–9820.
29 A. Collet, L. Ziminski, C. Garcia and F. Vigne-Maeder, Chiral discrimination in crystalline enantiomer systems: facts, interpretations, and speculations, in: J. Siegel (Ed.), NATO ASI Series-Supramolecular Stereochemistry, Kluwer Academic, The Netherlands, 1995, pp. 91–110.
30 K. Lipkowitz and D. Boyd, Reviews in Computational Chemistry, VCH publishers, New York, 1991, pp. 355–372, 383–392.
31 J. Krieger, New software expands role of molecular modeling technology. Chem. Eng. News, 4 (1995) 30–40.
32 K. Kinbara, Y. Kobayashi and K. Saigo, Chiral discrimination of 2-arylalkanoic acids by (1*S*,2*R*)-1-aminoindan-2-ol through the formation of a consistent columnar supramolecular hydrogen-bond network. J. Chem. Soc. Perkin Tran., 2 (2000) 111–119.
33 K. Kinbara, Y. Kobyashi and K. Saigo, Systematic study of chiral discrimination upon crystallization. Part 2. Chiral discrimination of 2-arylalkanoic acids by (1*R*,2*S*)-2-amino-1,2,-diphenylethanol. J. Chem. Soc. Perkin Trans., 2 (1998) 1767–1775.
34 K. Kinbara, K. Yoshiyuki and K. Saigo, Chiral discrimination of 2-arylalkanoic acids by (1*S*,2*S*)-1-aminoindan-2-ol and (1*S*,2*S*)-2-aminoindan-1-ol: correlation of the relative configuration of the amino and hydroxy groups with the pattern of a supramolecular hydrogen-bond network in the less-soluble diastereomeric salt. Chirality, 15 (2003) 564–570.

35 N. Phadnis and R. Suryanarayanan, Simultaneous quantification of an enantiomer and the racemic compound of ibuprofen by X-ray powder diffractometry. Pharm. Res., 14 (1997) 1176–1180.

36 P.G. Stahly, A.T. McKenzie, M.C. Andres, C.A. Russell, S.R. Byrn and P. Johnson, Determination of the optical purity of ibuprofen using X-ray powder diffraction. J. Pharm. Sci., 86 (1997) 970–971.

37 J.P. Mathieu, Vibration spectra and polymorphism of chiral compounds. J. Raman Spectrosc., 1 (1973) 47–51.

38 D.E. Bugay, Magnetic resonance spectrometry, in: H.G. Brittain (Ed.), Physical Characterization of Pharmaceutical Solids, Marcel Dekker, New York, 1995, pp. 93–126.

39 M.J. Potrzebowski, What high-resolution solid-state NMR spectroscopy can offer to organic chemists. European J. Org. Chem., 8 (2003) 1367–1376.

40 A.J. Repta, M.J. Baltezor and P.C. Bansal, Utilization of an enantiomer as a solution to a pharmaceutical problem: application to solubilization of 1,2-di(4-piperazine-2,6-dione)propane. J. Pharm. Sci., 65 (1976) 238–242.

41 C. Gu and D.J.W. Grant, Physical properties and crystal structures of chiral drugs, in: M. Eichelbaum, B. Testa and A. Somogyi (Eds.), Stereochemical Aspects of Drug Action and Disposition, Springer-Verlag, Berlin, Germany, 2003, pp. 113–139.

42 D.J.W. Grant and T. Higuchi, Solubility Behavior of Organic Compounds, John Wiley and Sons, New York, 1990, pp. 22–36.

43 M.J.D. Nerurkar, S.P. Duddu, D.J.W. Grant and J.H. Rytting, Properties of solids that affect transport, in: G.L. Amidon, P.I. Lee and E.M. Topp (Eds.), Transport Processes in Pharmaceutical Systems, Drugs and the Pharmaceutical Sciences, 102, Marcel Dekker, New York, NY, 2000, pp. 575–610.

44 L. Wearley, B. Antonacci, A. Cacciapuoti, S. Assenza, I. Chaudry, C. Eckhart, N. Levine, D. Loebenberg, C. Norris, R. Parmegiani, J. Sequeira and T. Yarosh-Tomaine, Relationship among physicochemical properties, skin permeability, and topical activity of the racemic compound and pure enantiomers of a new antifungal. Pharm. Res., 10 (1993) 136–140.

45 D.J.W. Grant, Theory and origin of polymorphism, in: H.G. Brittain (Ed.), Polymorphism in Pharmaceutical Solids, Marcel Dekker, New York, 1998, pp. 1–33.

46 S.R. Byrn, R.R. Pfeiffer and J.G. Stowell, Solid-State Chemistry of Drugs, 2nd Ed. SSCI, West Lafayette, IN, 1999.

47 G.G.Z. Zhang, S.Y.L. Paspal, R. Suryanarayanan and D.J.W. Grant, Racemic species of sodium ibuprofen: characterization and polymorphic relationships. J. Pharm. Sci., 92 (2003) 1356–1366.

48 M. Kamigauchi, M. Yoshida, K. Saiki, M. Sugiura, J. Nishijo, Y. In and T. Ishida, Structural/physicochemical properties of corycavidine, a key intermetabolite in the biosynthesis of isoquinoline alkaloids, elucidated by X-ray crystallography, solution conformation and thermal behavior analyses, and energy calculations. Bull. Chem. Soc. Jpn., 73 (2000) 1233–1241.

49 J. Jacques and J. Gabard, Étude des mélanges d'antipodes optiques. III. Diagrammes de solubilité pour les divers types de racémiques. Bull. Soc. Chim. Fr., (1972) 342.

50 L. Langhammer, Binary systems of enantiomeric nicotine derivatives. Acta Pharm., 308 (1975) 933–939.

51 A. Grunenberg, B. Keil and J.-O. Henck, Polymorphism in binary mixtures, as exemplified by nimodipine. Int. J. Pharm., 118 (1995) 11–21.

52 A. Burger, J.M. Rollinger and P. Brüggeller, Binary system of (*R*)- and (*S*)-nitrendipine-polymorphism and structure. J. Pharm. Sci., 86 (1997) 674–679.

53 M. Kuhnert-Brandstätter and R. Ulmer, Beitrag zur thermischen analyse optischer antipoden: Mandelsäure. Mikrochim. Acta [Wien], (1974) 927–935.

54 M.J. Brienne and J. Jacques, Une γ-lactone d'un type rare: l'acide trans π-camphanique. Tetrahedron, 62 (1970) 5087–5100.

55 S.M. Reutzel-Edens, V.A. Russell and L. Yu, Molecular basis for the stability relationships between homochiral and racemic crystals of tazofelone: a spectroscopic, crystallographic, and thermodynamic investigation. J. Chem. Soc. Perkin Tran., 2 (2000) 913–924.

56 A. Burger and R. Ramberger, On the polymorphism of pharmaceuticals and other molecular crystals. I Theory of thermodynamic rules. Mikrochim. Acta [Wien], 2 (1979) 259–271; II Applicability of thermodynamic rules. Mikrochim. Acta [Wien], 2 (1979) 273–316.
57 D. Giron, Thermal analysis and calorimetric methods in the characterization of polymorphs and solvates. Thermochim. Acta, 248 (1995) 1–59.
58 L. Yu, Inferring thermodynamic stability relationship of polymorphs from melting data. J. Pharm. Sci., 84 (1995) 966–974.
59 C. Gu and D.J.W. Grant, Estimating thermodynamic stability relationships of polymorphs from heats of solution and either solubility or dissolution rate. J. Pharm. Sci., 90 (2001) 1277–1287.
60 J.K. Guillory, Generation of polymorphs, hydrates, solvates, and amorphous solids, in: H.G. Brittain (Ed.), Polymorphism in Pharmaceutics, Marcel Dekker, New York, 1998, pp. 183–226.
61 J.M. Rollinger, C. Novak, Z. Ehen and K. Marthi, Thermal characterisation of torasemide using coupled techniques. J. Therm. Anal. Calorimetry, 73 (2003) 519–526.
62 R.F. Tamura, L.Z. Daisuke, K. Misaki, H. Miura, H. Takahashi, T. Ushio, T. Nakai and K. Hirotsu, Mechanism of preferential enrichment, an unusual enantiomeric resolution phenomenon caused by polymorphic transition during crystallization of mixed crystals composed of two enantiomers. J. Am. Chem. Soc., 124 (2002) 13139–13153.
63 R.M. Kuroda and F. Stephen, The crystal and molecular structure of *R*-(−)-1,1′-binaphthyl: the conformational isomerism and a comparison of the chiral with the racemic packing mode. J. Chem. Soc. Perkin Trans., 2 (1981) 167–171.
64 F. Dufour, C. Gervais, M.-N. Petit, G. Perez and G. Coquerel, Investigations on the reciprocal ternary system (±)-2-phenylpropionic acid-(±)-α-methylbenzylamine. Impact of an unstable racemic compound on the simultaneous resolution of chiral acids and bases by preferential crystallization. J. Chem. Soc. Perkin Tran., 2 (2001) 2022–2036.
65 L.R. Chen, V.G. Young Jr., D. Lechuga-Ballesteros and D.J.W. Grant, Solid-state behavior of cromolyn sodium hydrates. J. Pharm. Sci., 88 (1999) 1191–1200.
66 E. Shefter and T. Higuchi, Dissolution behavior of crystalline solvated and nonsolvated forms of some pharmaceuticals. J Pharm Sci., 52 (1963) 781–791.
67 R. Khankari and D.J.W. Grant, Pharmaceutical hydrates. Thermochim. Acta, 248 (1995) 61–79.
68 J. Han and R. Suryanarayanan, A method for the rapid evaluation of the physical stability of pharmaceutical hydrates. Thermochim. Acta, 329 (1999) 163–170.
69 L. Chen and D.J.W. Grant, Extension of Clausius-Clapeyron equation to predict hydrate stability at different temperatures. Pharm. Dev. Technol., 3 (1998) 487–494.
70 H. Zhu, C.M. Yuen and D.J.W. Grant, Influence of water activity in organic solvent plus water mixtures on the nature of the crystallizing drug phase 1. Theophylline Int. J. Pharm., 135 (1996) 151–160.
71 J.H. van't Hoff and H.M. Dawson, The racemic transformation of ammonium bimalate. Berichte der Deutschen Chemischen Gesellschaft, 31 (1898) 528–535.
72 R. Yoshioka, K. Okamura, S. Yamada, K. Aoe and T. Date, The role of water-solvation in the optical resolution of DL-leucine with (*S*)-(−)-2-phenylethanesulfonic acid – characterization and X-ray crystal structures of their diastereomeric salts. Bull. Chem. Soc. Jpn., 71 (1998) 1109–1116.
73 L. Yu, S.M. Reutzel and G.A. Stephenson, Physical characterization of polymorphic drugs: an integrated characterization strategy. Pharm. Sci. Tech. Today, 1 (1998) 118–127.
74 G. Zografi, States of water associated with solids. Drug Dev. Ind. Pharm., 14 (1988) 1905–1926.
75 S. Piyarom, E. Yonemochi, T. Oguchi and K. Yamamota, Effects of grinding and humidification on the transformation of conglomerate to racemic compound in optical active drugs. J. Pharm. Pharmacol., 49 (1997) 384–389.
76 S. Stinson, Counting on chiral drugs. Chem. Eng. News, (1998) 83–103.
77 S.E. Biali, K. Bart, Y. Okamoto, R. Aburatani and K. Mislow, Structure, resolution, and racemization of decakis(dichloromethyl)biphenyl. J. Am. Chem. Soc., 110 (1988) 1917–1922.
78 L.A. Paquette and T.Z. Wang, Dynamic behavior of 1,2-annulated cyclooctatetraenes. Kinetic analysis of transition-state steric congestion involving neighboring alicyclic substituents. J. Am. Chem. Soc., 110 (1988) 8192–8197.

79 P.C. Bickart, W. Frederick, J. Jacobus, E.G. Miller and K. Mislow, Thermal racemization of allylic sulfoxides and interconversion of allylic sulfoxides and sulfenates. Mechanism and stereochemistry. J. Am. Chem. Soc., 90 (1968) 4869–4876.

80 C. Gu, Influence of Solvent and Impurity on the Crystallization and Properties of Crystallized Products. Ph.D. Thesis, University of Minnesota, Minneapolis, 2001.

81 J.W. Mullin, Crystallization, 3rd Ed. Butterworth-Heinemann, London, 1993, pp. 202–260.

82 A.S. Myerson, D.A. Green and P. Meenan, Crystal Growth of Organic Materials. American Chemical Society, Washington DC, 1996, pp. 66–109.

83 I. Weissbuch, R. Porovitz-Biro, M. Lahav and L. Leiserowitz, Understanding and control of nucleation, growth, habit, dissolution and structure of two- and three-dimensional crystals using 'tailor-made' auxiliaries, Acta Cryst. B., 51 (1995) 115–148.

84 I. Weissbuch, L. Addadi, M. Lahav and L. Leiserowitz, Molecular recognition at crystal interfaces. Science, 253 (1991) 637–645.

85 I. Weissbuch, M. Lahav and L. Leiserowitz, Towards stereochemical control, monitoring, and understanding of crystal nucleation. Cryst. Growth Design, 3 (2003) 125–150.

86 T. Bushe, D.K. Kondepudi and B. Hoskins, Kinetics of chiral resolution in stirred crystallization of D/L-glutamic acid. Chirality, 11 (1999) 343–348.

87 I. Weissbuch, D. Zbaida, L. Addadi, M. Lahav and L. Leiserowitz, Design of polymeric inhibitors for the control of crystal polymorphism. Induced enantiomeric resolution of racemic histidine by crystallization at 25°C. J. Am. Chem. Soc., 109 (1987) 1869–1871.

88 M. Berfeld, D. Zbaida, L. Leiserowitz and M. Lahav, Tailor-made polymers for the removal of lamellar twinning. Resolution of α-amino acids by entrainment. Adv. Mater., 11 (1999) 328–331.

89 I. Weissbuch, I. Kuzmenko, M. Vaida, S. Zait, L. Leiserowitza and M. Lahav, Twinned crystals of enantiomorphous morphology of racemic alanine induced by optically resolved α-amino acids; a stereochemical probe for the early stages of crystal nucleation. Chem. Mat., 6 (1994) 1258–1268.

90 I. Weissbuch, L. Leiserowitz and M. Lahav, 'Tailor-made' and charge-transfer auxiliaries for the control of the crystal polymorphism of glycine. Adv. Mat., 6 (1994) 952–956.

91 L. Li, D. Lechuga-Ballesteros, B.A. Szkudlarek and N. Rodriguez-Hornedo, The effect of additives on glycine crystal growth kinetics. J. Coll. Interface Sci., 168 (1994) 8–14.

92 Q.B. Broxterman, E. Van Echen, L.A. Hulshof, B. Kaptein, R.M. Kellogg, A.J. Minnaard, T.R. Vries and H. Wynberg, Dutch Resolution, a new technology in classical resolution. Chimica. Oggi. 16 (1998) 34–37.

93 T.R. Vries, H. Wynberg, E. Van Echten, J. Koek, W. Ten Hoeve, R.M. Kellogg, Q.B. Broxterman, A. Minnaard, B. Kaptein, Bernard, S. Van der Sluis, L. Hulshof and J. Kooistra, The family approach to the resolution of racemates. Angew. Chem. Int. Ed., 37 (1998) 2349–2354.

94 J.W. Nieuwenhuijzen, R.F.P. Grimbergen, C. Koopman, R.M. Kellogg, T.R. Vries, K. Pouwer, E. van Echten, B. Kaptein, L.A. Hulshof, Q.B. Broxterman, B.V. Syncom and N. Groningen, The role of nucleation inhibition in optical resolutions with families of resolving agents. Angew. Chem. Int. Ed., 41 (2002) 4281–4286.

95 R.M. Kellogg, J.W. Nieuwenhuijzen, K. Pouwer, T.R.Vries, Q.B. Broxterman, R.F.P. Grimbergen, B. Kaptein, R.M. La Crois, E. de Wever, K. Zwaagstra, A.C. van der Laan, B.V. Syncom and N. Groningen, Dutch Resolution: separation of enantiomers with families of resolving agents. A status report. Synthesis, 10 (2003) 1626–1638.

96 A.S. Myerson and S.M. Jang, A comparison of binding energy and metastable zone width for adipic acid with various additives. J. Cryst. Growth, 156 (1995) 459–466.

97 C. Gu, K. Chatterjee, V. Young and D.J.W. Grant, Stabilization of a metastable polymorph of sulfamerazine by structurally related additives. J. Cryst. Growth, 235 (2002) 471–481.

98 R.J. Davey, The effect of impurity adsorption on the kinetics of crystal growth from solution. J. Cryst. Growth, 34 (1976) 109–119.

99 Z. Adamczyk, B. Siwek, M. Zembala and P. Belouschek, Kinetics of localized adsorption of colloid particles. Adv. Colloid Interface Sci., 48 (1994) 151–280.

100 H.E. Buckley, Crystal Growth. Wiley, London, 1951.

101 W.R. Wilcox, Fundamentals of solution growth, in: H. Arend and J. Hullinger (Eds.), Crystal Growth in Science and Technology, Plenum Press, New York and London, 1989, pp 119–132.
102 D.L. Klug, The influence of impurities and solvents on crystallization, in: A.S. Myerson (Ed.), Handbook of Industrial Crystallization, Butterworth-Heinemann, Boston, 1993, pp. 65–87.
103 K. Sangwal and T. Palczýnska, On the supersaturation and impurity concentration dependence of segregation coefficient in crystals grown from solutions. J. Cryst. Growth, 212 (2000) 522–531.
104 C. Gu and D.J.W. Grant, Relationship between particle size and impurity incorporation during crystallization of (+)-pseudoephedrine hydrochloride, acetaminophen and adipic acid from aqueous solution. Pharm. Res., 19 (2002) 1068–1070.
105 J.D. Wright, Molecular Crystals, 2nd Ed. Cambridge University Press, Cambridge, UK, 1995, pp. 42–65.
106 S.P. Duddu and D.J.W. Grant, The use of thermal analysis in the assessment of crystal disruption. Thermochim. Acta, 248 (1995) 131–145.
107 K.Y. Chow, J. Go, M. Mehdizadeh and D.J.W. Grant, Modification of adipic acid crystals: influence of growth in the presence of fatty acid additives on crystal properties. Int. J. Pharm., 20 (1984) 3–24.
108 A.H.L. Chow, P.K.K. Chow, Z. Wang and D.J.W. Grant, Modification of acetaminophen crystals: Influence of growth in aqueous solutions containing *p*-acetoxyacetanilide on crystal properties. Int. J. Pharm., 24 (1985) 239–258.
109 A.H.L. Chow and C.K. Hsia, Modification of phenytoin crystals: influence of 3-acetoxymethyl-5, 5-diphenylhydantoin on solution-phase crystallization and related crystal properties. Int. J. Pharm., 75 (1991) 219–230.
110 E. Tomlinson, Enthalpy-entropy compensation analysis of pharmaceutical, biochemical and biological systems. Int. J. Pharm., 13 (1983) 115–144.
111 F. Giordano, R. Bettini, C. Donini, A. Gazzaniga, M.R. Caira, G.G.Z. Zhang and D.J.W. Grant, Physical properties of parabens and their mixtures: solubility in water, thermal behavior, and crystal structures. J. Pharm. Sci., 88 (1999) 1210–1216.
112 J.J. Gilman, W.G. Johnston and G.W. Sears, Dislocation etch pit formation in lithium fluoride. J. Appl. Phys., 29 (1958) 747–754.
113 H. Bundgaard, Influence of an acetylsalicylic anhydride impurity on the rate of dissolution of acetylsalicylic acid. J. Pharm. Pharmacol., 26 (1974) 535–540.
114 R. Hüttenrauch, S. Fricke and P. Zielke, Mechanical activation of pharmaceutical systems. Pharm. Res., 2 (1985) 302–306.
115 D. Law, Influence of Composition of the Crystallization Medium on the Physical Properties and Mechanical Behavior of Adipic Acid Crystals. University of Minnesota, Minneapolis, 1994, pp. 153–177.
116 J.T. Carstensen, Drug Stability, Principles and Practices, Marcel Dekker, New York, 1995, pp. 230–280.
117 M.D. Cohen, The photochemistry of organic solids. Angew. Chem. Int. Ed., 14 (1975) 386–393.
118 E.Y. Shalaev, M. Shalaeva, S.R. Byrn and G. Zografi, Effects of processing on the solid-state methyl transfer of tetraglycine methyl ester. Int. J. Pharm., 152 (1997) 75–88.
119 M. Otsuka and N. Kaneniwa, Effect of grinding on the crystallinity and chemical stability in the solid state of cephalothin sodium. Int. J. Pharm., 62 (1990) 65–73.
120 P. York and D.J.W. Grant, A disruption index for quantifying the solid state disorder induced by additives or impurities. I. Definition and evaluation from heat of fusion. Int. J. Pharm., 25 (1985) 57–72.
121 D.J.W. Grant and P. York, A disruption index for quantifying the solid state disorder induced by additives or impurities. II. Evaluation from heat of solution. Int. J. Pharm., 28 (1986) 103–112.

© 2006 Elsevier B.V. All rights reserved.
Chiral Analysis
K.W. Busch and M.A. Busch, Eds.

CHAPTER 4

Emerging high-throughput screening methods for asymmetric induction

M.G. Finn

Department of Chemistry and The Skaggs Institute for Chemical Biology, The Scripps Research Institute, La Jolla, CA, USA

4.1 INTRODUCTION

The importance and value of chiral molecules to the drug discovery industry [1–3] has provided the impetus for much research in asymmetric organic synthesis. Drug discovery as practised in recent years is sometimes a poorly compatible combination of searches for general function—for classes of compounds having a desired type of reactivity or binding ability—and for particular structures having the magic combination of potency, biodistribution, and metabolism. The sense that one is searching for a needle in a haystack has led to ever faster and more sophisticated ways of sorting hay.

Chirality usually enters such efforts relatively late in the discovery process. While many drug molecules are chiral and most of these have the desired function in only one enantiomer, the synthesis of large and diverse libraries of enantiopure compounds is not practical. Accordingly, initial screens for drug-like properties are almost always done on racemates. Enantiopure structures of interest are then identified after rounds of winnowing, and the challenge of asymmetric synthesis is directed at particular targets of high value.

The physical separation of enantiomers, either by classical resolution techniques or by chromatography over chiral stationary phases, is often a preferred method of preparation of enantiopure molecules. Indeed, there are many examples of resolution by crystallization in industrial process or production laboratories, often using inexpensive chiral carboxylate esters such as tartrate as precipitating additives. But the direct and efficient synthesis of the desired enantiomer of a target compound is the optimal goal in any chiral enterprise, whether on a large or small scale.

Only a few reactions have been developed that provide high levels of asymmetric induction for a wide range of substrates. The chances are therefore relatively slim that such methods will suffice without change for an arbitrary target structure. Since general solutions to catalytic asymmetric problems are few and far between, the development of

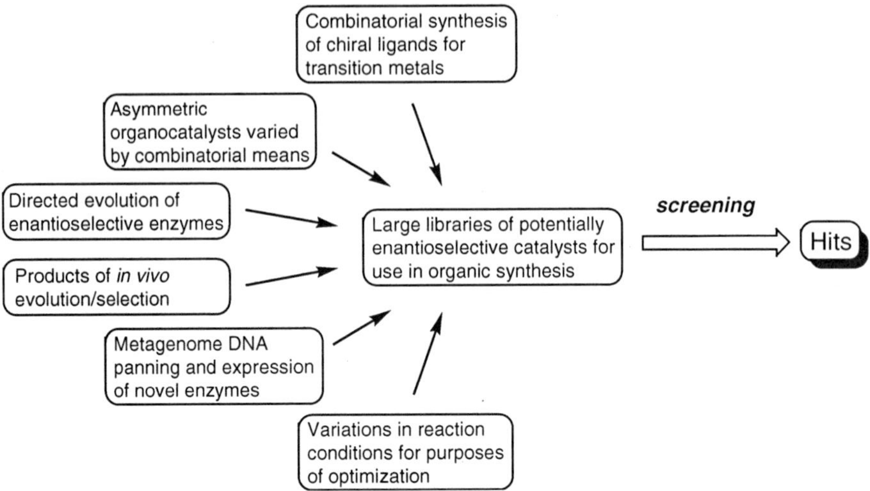

Fig. 4.1. The combinatorial catalysis discovery bottleneck.

effective catalysts for particular applications, either by the invention of new systems or the modification of known ones, is a matter of design and testing. The designed control over catalytic transition states is still beyond our routine grasp, and so attention has turned to making trial-and-error as easy to perform as possible—sorting evanescent hay, as it were.

Fig. 4.1, adapted from a representation of Reetz [4,5], illustrates the situation. Transition metal complexes and organic moieties can provide one or more functions (nucleophilicity, electrophilicity, and bond rearranging ability) having a catalytic effect, and these engines can be encased in chiral environments. In addition, the tools of molecular biology and enzymology can be adapted to provide large numbers of candidate enzymes, catalytic antibodies, and polynucleotides with the potential for asymmetric transformations. Therefore, the problem of chiral catalyst discovery or optimization has been viewed as being constrained most severely by the testing of candidates to pick out the best ones. In other words, screening has been a limiting step, complicated by the fact that it is the relative amounts of enantiomers, identical in most chemical respects, that is the desired information. Such information is not easy to come by, especially when speed is of the essence.

This chapter surveys techniques that are being developed for the rapid measurement of enantiomeric excess of chiral nonracemic organic compounds, and thus serves as an introduction to later chapters of this volume. While usually intended for application to the screening of candidate chiral catalysts, some methods are so new as to have not yet been applied to such practical targets, but represent in the opinion of the author promising technologies that warrant the attention of chemists in the field. With a few exceptions, I focus here on the developments reported since 2002, in order to avoid repeating the content of an earlier review [6]. It must be said that the field has advanced incrementally since then, and much of the earlier discussion is still quite relevant to current efforts in combinatorial asymmetric catalyst development. Other recent treatments of methods [7] and catalyst screens [8,9] have appeared.

Nomenclature note: The term enantiomeric excess [$100(R-S)/(R+S)$, where R and S are the amounts of the major and minor enantiomers, respectively], often abbreviated as "ee" or "e.e.," has gained wide currency and has the advantage of being unambiguous [10]. However, I find it awkward and confusing, particularly in an age in which optical rotation is seldom used to determine the relative enantiomeric content (enantiomeric excess corresponds precisely to the ratio of the optical rotation of a sample of interest to the optical rotation of the authentic pure enantiomer, expressed as a percentage). I therefore use here the simpler *enantiomeric ratio* or "e.r.," expressed as R/S:1, if R is in excess. Thus, a 5:2 molar ratio of *S*- to *R*-enantiomers of a compound can be described as representing an e.r. of 2.5:1, rather than 43% e.e. Racemates have an e.r. of 1:1.

4.2 STRATEGIES

Two general strategies are possible for the determination of the outcome of a candidate asymmetric reaction (Fig. 4.2): (A) coding of enantiomeric information into the starting substrate, or (B) readout of the enantiomeric information contained in the product.

The first strategy is characterized by operations on separated enantiomers. For example, parallel measurements can be made on the enantiomers and compared (such as the measurement of rates of reaction to determine the magnitudes of kinetic resolution). Enantiomers can also be distinguished ahead of time by tags so that they can be followed in the same pot. This approach can be applied only to reactions in which the chiral centers of interest are pre-existing in the substrate, as for kinetic resolution and desymmetrization reactions, and not to reactions in which a chiral center is created from a prochiral compound. Thus, kinetic resolutions of enantiomers become operations on "pseudo-enantiomers" which contain chiral centers of opposite configuration and some other tag or label to mark that difference. The loss of generality and the need to make enantiopure molecules ahead of time are compensated by analytical convenience, a frequent insensitivity to impurities, and higher throughput. Manfred T. Reetz and coworkers have been the most productive practitioners of this strategy.

In the second class may be found measurements such as the direct detection of optical activity (polarimetry, circular dichroism), the physical separation and quantitation of product enantiomers by chromatography over chiral stationary phases, and the use of chiral modifiers to turn enantiomeric information into diastereomeric information. As described below, the most elegant and useful developments in this field are found in the engineering of such diastereomeric interactions and of ways to distinguish the adducts from impurities.

The choice of proper technique for any particular problem depends, as always, on the details [11]. Among the most important questions are: how many reactions do you really have to screen per day? Do you need to profile reactions with many different substrates or must you optimize the transformation of a single substrate? What is the level of accuracy required in the determination of the enantiomeric ratio? (Do you need to distinguish bad from good catalysts, or good from better ones?) What instrumentation do you have? On what scale will you perform your tests? Can you tolerate a purification step before analysis, or must you examine the "crude" reaction mixtures?

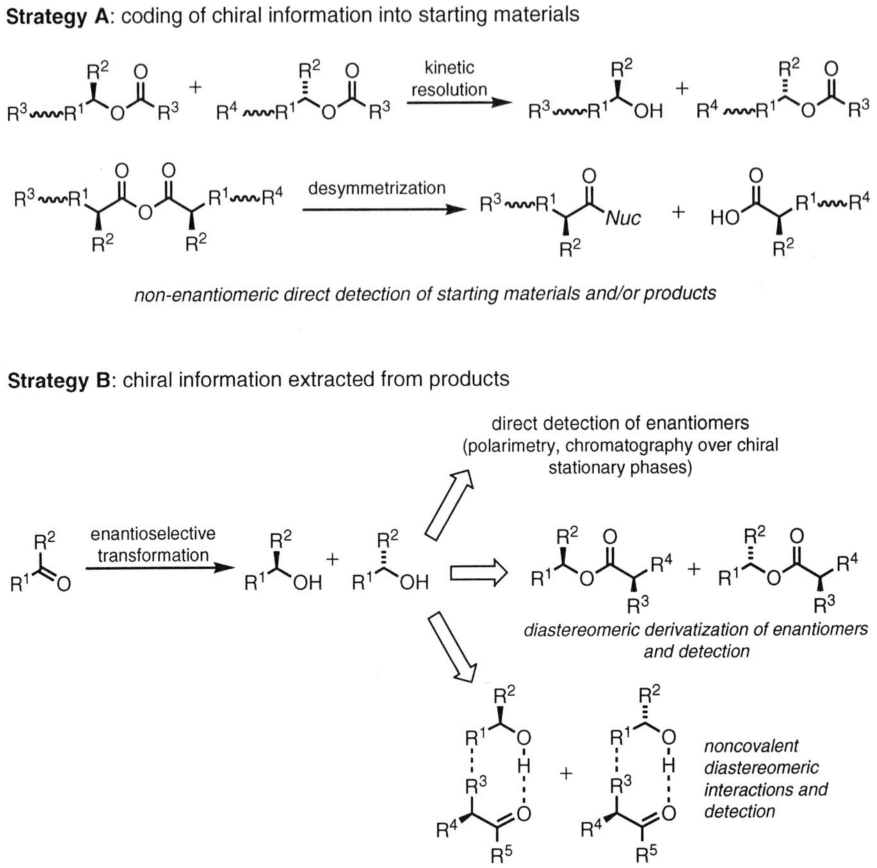

Strategy A: coding of chiral information into starting materials

kinetic resolution

desymmetrization

non-enantiomeric direct detection of starting materials and/or products

Strategy B: chiral information extracted from products

enantioselective transformation

direct detection of enantiomers (polarimetry, chromatography over chiral stationary phases)

diastereomeric derivatization of enantiomers and detection

noncovalent diastereomeric interactions and detection

Fig. 4.2. Illustrations of strategies for the analysis of enantioselective reactions. Like colors mark substituents and stereogenic centers that are correlated with each other.

The answers to these and other questions will steer the investigator to the most advantageous method, or lead him to invent one of his own. There is certainly satisfaction to be gained in devising new ways to trick a reaction or a substance into revealing its most subtle secret—its enantioselectivity or chirality. The remainder of this chapter is organized by the analytical instrumentation used in this joyful act of molecular voyeurism.

4.3 METHODS

4.3.1 Separation or direct detection of enantiomers

While this chapter is mostly concerned with other techniques, the use of chiral stationary phases in chromatography remains the gold standard of accuracy against which all new methods are compared. While usually not a "high-throughput" procedure, advances in columns and instrumentation have made it sufficiently rapid for many, if not most, situations [12].

The large majority of chiral compounds can be resolved by some form of chromatography, usually HPLC or GC over chiral stationary phases, but optimization of the analytical procedure to achieve this goal for a new molecule often requires expensive and time-consuming testing of different columns and elution conditions. The groups of Kagan and Gennari have recently demonstrated that the outcome can be well worth the effort. Thus, the action of candidate catalysts on a panel of substrates rather than a single substrate—of importance for both reaction discovery and optimization — was examined efficiently by arranging it so that the various products could be cleanly separated in the chiral analytical step [13–15].

Capillary electrophoresis using chiral additives is also worthy of note, and is especially useful for polar compounds not well suited to HPLC or GC [16]. New materials for enantiomer resolution continue to be developed, such as an interesting reversible hydrogel containing guanosine as the chiral selector [17]. Of particular significance is the development by the Reetz laboratory of a parallel capillary array electrophoresis method for the determination of the enantiomeric content of chiral amines using cyclodextrin-based electrolytes, with capabilities for tens of thousands of measurements in a reasonable time [18].

Lastly, the intrinsic chiroptical properties of enantiomers in the vibrational spectroscopy regime has been newly exploited by Nafie and coworkers to measure enantiomeric ratios by Fourier transform near IR vibrational circular dichroism [19,20].

4.3.2 Absorbance or fluorescence spectra with nonchiral reporter molecules derived from enantioenriched substrates

In terms of ease, speed, and sensitivity, electronic spectroscopy is the most convenient general analytical technique available. Indeed, absorption or emission measurements have long been used in general strategy A in Fig. 4.2: if the reactions of separate enantiomers can be followed spectrophotometrically, the effectiveness of chiral catalysts in kinetic resolution can be very easily and rapidly determined. Thus, the groundbreaking program of Reetz and coworkers on the evolution of enantioselective lipase enzymes was started by the screening of thousands of hydrolysis reactions of enantiopure *p*-nitrophenolate esters using UV-visible spectroscopy, relying on the generation of the strongly absorbing *p*-nitrophenol product [21]. Conceptually, similar approaches include the generation of a visible signal by a pH reporter molecule, converting a released acid or base from separate enantiopure substrates into quantitative signals reporting on the rates of the reactions [22–24]. In one case, supercritical fluid chromatography has been used to purify product analytes in a rapid fashion [25]. The testing of multiple reactions in the same vessel can be accomplished by the segregation of the substrates on beads along with indicator molecules. In this way, the color of each bead reports on the conversion of its substrate to product, and can be used to identify active and enantioselective catalysts [26–28].

4.3.3 Absorbance or fluorescence spectra with chiral reporter molecules

In order to make use of electronic spectroscopy in "strategy B" situations, chiral information must be transformed into the absorption or emission of visible light. In general, this is accomplished by the design of chiral molecules which bind to chiral analytes of interest giving rise to a signal dependent on the relative diastereomeric nature of the interaction. The signal is usually (but not always [29,30]) the inducement or quenching of fluorescence [31]. The use of covalent bond formation to make dye-tagged diastereomers of amino acids, with concomitant transmission of chiral information by way of kinetic resolution, was implemented in microarrays by Shair and coworkers in 2001 [32], but has received little use since that time. The many reported attempts to achieve similar results by noncovalent interactions can be said to define a subdiscipline of fluorescent chiral sensing [33].

Fluorescent sensors have recently been reported for chiral acids and amino acid derivatives [34,35]. Structures **1**[31] and **2**[36] are good examples, developed for the enantiodiscrimination of mandelic acids and *trans*-1,2-diaminocyclohexane, respectively. The fluorescent emission of each sensor molecule becomes more intense when binding its chiral guest. A related approach is provided by the displacement of an indicator **4** from a receptor complex **3** by the chiral analyte [37]. In this case, the chiral information is revealed by a differential absorbance depending on the enantiomeric content of the analyte. The generation of induced chiroptical signals from the binding of chiral analytes to helical polymers [38] and liquid crystalline materials [39,40] are also exciting strategies for enantiomer detection.

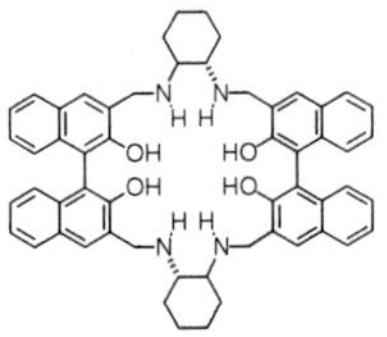

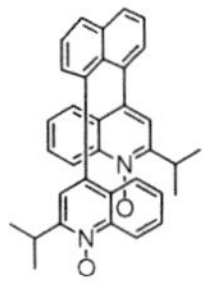

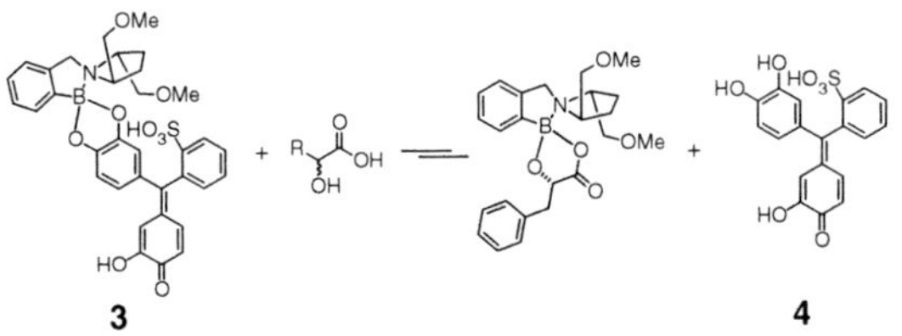

Many of these sensing schemes have so far achieved only proof-of-concept status with added chiral molecules. However, **1** has been employed to screen catalysts for the chiral hydrocyanation of an aromatic aldehyde bearing a long-chain aliphatic group in the *para* position [41], and **2** has been used to analyze diamine kinetic resolution reactions [36]. The selectivity of enantiomeric recognition and the difference in optical properties upon binding do not have to be dramatic in order for the sensor to be useful—what matters most is the signal-to-noise, where "noise" refers to any complicating factor such as a change in fluorescence because of the presence of impurities or variations in analytical conditions. This is often the limiting feature: in the case of **1**, the problem was solved by the hydrophobic tail, which made it easy to purify the products before analysis. Sensor **2** was found to be impervious to the products and other contaminants in the reaction mixtures. Another common drawback of the molecular sensor approach is a lack of generality: different sensors are usually required for even small changes in analyte structure. Even when such problems have been overcome, the number of reactions examined has so far been quite limited.

4.3.4 Mass spectrometry with nonchiral reporter molecules derived from enantioenriched substrates

Because of its high sensitivity, the routine availability of instrumentation, and the fact that labeling with chromophores is not required, mass spectrometry has become a popular analytical tool for enantioselective catalytic screening. The problems and solutions parallel to those of optical spectroscopy: the translation of chiral information into mass information has been performed in the modes of both strategies A and B (Fig. 4.2). The encoding of chiral information into mass-tagged pseudoenantiomers (a "strategy A" approach) was initiated by the Reetz laboratory [42,43]. The idea is so simple and cleanly implemented that only adaptations particular to the molecules of interest are required.

Markert and Pfaltz recently achieved an especially clever fusion of this technique with the method of Chen and coworkers [44,45] in order to determine the enantiodiscriminating power of candidate catalysts in the gas phase [46]. Pseudoenantiomeric substrates such as **5a** and **5b** (Fig. 4.3) were fed to a chiral catalyst of interest and the ratio of diastereomeric intermediates (**6a**, **6b**) was assayed by mass spectrometry at a rate of approximately 10 samples per hour. When the rate of formation of the intermediates determine the enantioselectivity of the overall catalytic reaction (which is not always the case [47]), the method is likely to prove to be quite powerful if charged intermediates are formed.

4.3.5 Mass spectrometry with chiral reporter molecules

It is the "strategy B" scenarios—in which chiral information must be extracted from true enantiomers—that have sparked the most intensive attention in the mass spectrometry-based detection of enantiomers. Advances in this area have been recently reviewed [48]. Enantiopure chiral additives showing different affinities for the formation of noncovalent or covalent complexes with the antipodes of the chiral analyte are used. When the additives ("reporters") are tagged with groups of different mass to distinguish one absolute configuration from the other, the e.r. of the analyte can be revealed by the

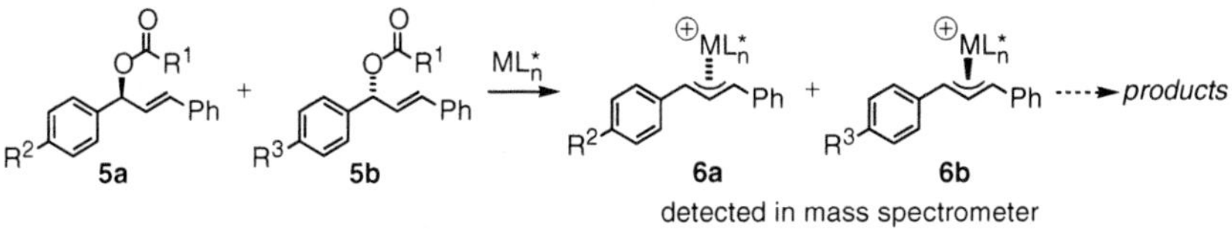

Fig. 4.3. Mass spectrometry screening catalytic activity and enantioselectivity by the detection of key intermediates; ML^*_n denotes a catalytic metal complex bearing enantiopure ligand(s).

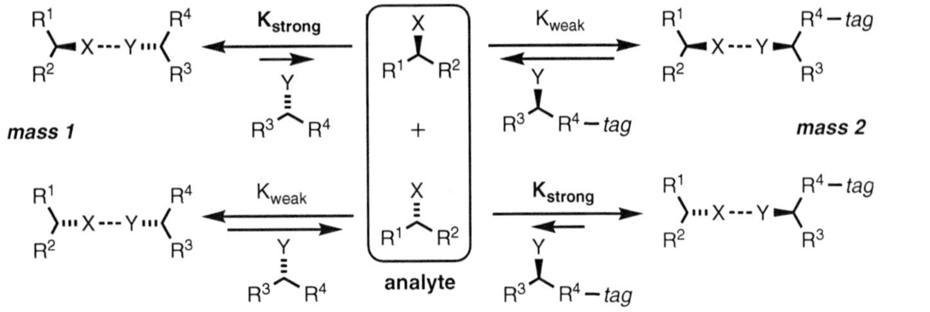

Fig. 4.4. Measurement of e.r. with the use of mass tagged reporter molecules based on differential energies of noncovalent association. X and Y represent interacting moieties on the analyte and reporter, respectively.

analysis of the interactions with the pseudoenantiomers of the reporter. Calibration experiments to determine the differential diastereomeric affinity (or rate of bond formation), and any differences in ionization or detection efficiency that may exist between the diastereomeric adducts, are required [49]. Fig. 4.4 outlines this approach for the formation of noncovalent adducts (in terms of differential equilibrium constants K_{strong} and K_{weak} for matched and mismatched cases), but exactly the same analysis applies to the formation of covalent bonds (in terms of differential rate constants k_{fast} and k_{slow} [50]. Mass tags can be isotopic labels or unobtrusive functional groups that do not change the chemistry of the tagged compound.

Illustrative examples of noncovalent interactions in this approach are provided by the use of Pirkle-style compounds as chiral additives [51,52], but other structures have also been employed [53–58]. The proper choice of chiral reporter includes an apprecia-tion of ionization efficiency: good reporters confer high sensitivity to the detection method [59]. A mechanistically complementary approach has been taken by Cooks and coworkers, who have employed enantiopure transition metal centers that interact with chiral analytes *in situ*. The enantiomeric content of the analyte is revealed by differential rates of *dissociation* of the diastereomeric complexes in the gas phase [60–63].

4.3.6 Detection methods based on antibodies and enzymes

The ability of antibodies and enzymes to bind or process one enantiomer of a molecule with especially high selectivity allows them to be used as powerful sensors to detect and quantify the presence of enantioenriched compounds. Such methods are gaining increasing application in recent years since it has become routine to generate new proteins with the desired function by immunization or library selection techniques. Interestingly, a much higher proportion of reports of protein-based, as opposed to small-molecule based, e.r. detection methods include their actual use in screening libraries of reactions. Perhaps this is due to the fact that the enantiorecognition abilities of enymes and antibodies are well accepted, and therefore the challenge to be met and described is the engineering required to translate such recognition into a practical readout.

Proteins, like optical spectroscopy and mass spectrometry, have been used as tools for the measurement of enantiomeric ratios in both "strategy A" and "strategy B" ways. As an example of the former, Berkowitz, Li, and others employed efficient enzymes (such as dehydrogenases) to convert a reaction product (ethanol) into an easily detected signal (NADH). The enzyme resides in a different compartment than the reaction of interest; in this case in the aqueous layer of a two-phase mixture, and the key by-product diffuses from the "reaction" to the "detection" compartment. The method provides a quantitative readout of the rate of the ethanol-producing reaction in the organic phase, eliminating the need to take aliquots and to label the substrate with a chromophore or mass tag [64]. When performed on separate enantiomers by enantiopure catalysts, it reported efficiently on both the rate and enantioselectivity of the reaction of interest [65,66].

Lipases feature prominently in the field of combinatorial enantioselective catalyst development, as both the subjects of such investigations and the tools to enable them. Under the heading of "strategy B," Onaran and Seto used enantiospecific lipases for e.r. determination of allylic acetates [67]. The enzyme processes only one enantiomer of the analyte and the rate of the enzymatic reaction was read out with a pH indicator. The enantiomeric content of the analyte was then determined by relating the observed rate to the known Michaelis–Menten properties of the enzyme [67].

The enantiorecognition abilities of antibodies have been exploited most commonly for e.r. determination. A competition form of the enzyme-linked immunosorbent assay (ELISA) method, ubiquitous in molecular biology and biochemistry, has been adapted by Taran and coworkers, taking advantage of the routine generation of antibodies to any molecular target [68]. Thus, an antibody raised against the analyte of interest can be immobilized to the surface of a microtiter plate and a form of the analyte tagged with a reporter enzyme that generates an amplified signal is bound to the immobilized enzyme. Analyte in solution can then displace the analog and wash away the reporter, thus converting analyte concentration into a quantitative change in signal. In this way, a polyclonal antibody binding the racemate was used to determine the overall yield of a family of catalytic reactions, and enantiospecific antibodies were used to determine enantioselectivities [68].

A related recent publication from the Taran laboratory describes the use of antibodies raised to each reactant of a bimolecular fusion reaction of interest. One antibody can then be anchored to the microtiter plate and will bind its hapten motif. When that motif has been incorporated in the desired product, the second antibody (conjugated to a reporter enzyme) is specifically adsorbed in a sandwich assay and generates a signal [69]. While not yet employed for catalysis screening, this method has great promise since the specificity and strength of the antibody–hapten interaction often makes sandwich ELISA assays insensitive to impurities.

Other variations on the antibody capture theme include the immobilization on a porous membrane of an antibody that binds a particular analyte enantiomer. The analyte of interest is competed with an analog bearing a biotin tag, and the amount of binding is determined with a secondary avidin-peroxidase conjugate [70]. An interesting twist was provided by Matsushita and coworkers, who used a monoclonal antibody that binds both enantiomers of an amino acid derivative, but gives a fluorescent signal

References pp. 90–94

with only one of them. After calibration, the antibody was used to screen a small organometallic catalyst library to identify the most enantioselective variants [71]. Lastly, similar in principle to these ELISA-based methods is the use of paramagnetic nanoparticles which have strong effects on NMR spectra. If such particles are coated with the analyte, they form aggregates with divalent IgG antibodies raised against the analyte, changing the NMR properties of the particles. The presence of the analyte in solution can thereby be detected by competitive binding with the antibody, sensitively affecting the aggregation state of the mixture [72].

4.3.7 Detection methods based on other instrumentation

Analytical challenges such as enantiomer detection bring out the best in gadget design. Several classes of instrumentation mentioned in a previous review [6] are not considered here because they have not been substantially improved in the past few years, but others have received continued attention. For example, the screening of catalysts by thermography has been reviewed [73], and recently translated to organogel media [74], although no new applications to enantioselective catalyst identification or e.r. determination have appeared. More promising has been the application of polarimetry by Willson and coworkers in the style of high-throughput thermography—i.e., imaging parallel reactions in a microtiter plate—to obtain information on rate and enantioselectivity [75]. While this method requires highly specialized instrumentation, its initial application to 1536-well plates and a large collection of chiral reactions bodes well for its eventual adoption by other laboratories.

Since enantiomer separation by electrophoresis is now well established, the miniaturization of the technique would make it more amenable to true high-throughput applications. Accordingly, microchip-based electrophoresis methods have been pursued vigorously by Belder and coworkers [76–78]. While not yet reportedly applied to reaction screening, this is likely to be imminent.

The high information content and familiarity of NMR to chemists makes it ideal for reaction analysis, but NMR cannot be performed in parallel fashion and so is not regarded as a high-throughput technique. Nevertheless, the Reetz laboratory has reported the use of a flow cell and a "strategy A" pseudoenantiomer label approach to acquire data at a rate corresponding to 1400 e.r. measurements per day [5]. If such methods become widely distributed, the power of NMR chiral resolution agents [79] should make this an attractive choice for medium-throughput requirements at least.

Lastly, infrared spectroscopy has been brought squarely into the set of tools available to the combinatorial chemist seeking to determine enantiomer ratios. The Reetz group has adapted the "strategy A" approach by coding chiral information into substrates bearing a carbonyl group close to the stereogenic center. If the carbonyl group of one pseudoenantiomer is labeled with ^{13}C, the enantioselectivity of a kinetic resolution reaction involving a transformation of the carbonyl moiety can be ascertained by convenient measurement of the intensities of shifted (labeled) *versus* unshifted (unlabeled) carbonyl stretching bands. The use of a microtiter-plate IR spectrometer allowed these workers to achieve up to 10,000 e.r. measurements per day. A particularly intriguing "strategy B" method was reported by Oliveira and coworkers in which the interactions of enantiopure carbohydrates with amino acids were examined by near-infrared spectroscopy [80]. With appropriate calibration, the e.r. of the amino acids could be determined without the need to design a particular molecular interaction, relying on the high information content of near-IR vibrational modes.

4.3.8 The wave of the future?

Any dispassionate review of the field of combinatorial catalyst discovery must conclude that the anticipated revolution has not yet occurred, and that the availability of screening methods may no longer be the primary roadblock. Published reports of analytical techniques (including most of those cited above, and even those that advertise "high-throughput" efficiencies) almost invariably describe the testing of at most a few hundred candidates [8,81], rather than the thousands or millions of trials that characterize screening experiments in drug discovery.

The lone exception is instructive. In 1997, Reetz and coworkers described the evolution of an enantioselective lipase catalyst in four rounds of screening, involving the testing of a total of 7600 candidates by a microtiter-plate absorbance assay involving separate enantiopure substrates [21,82]. This pioneering work brought the idea of *in vitro* evolution of enzyme function to chemists, and subsequent studies encompassed sets of candidates numbering in the 10,000–20,000 range [83,84]. Biocatalysts (enzymes) have so far provided the only real need for many thousands of enantioselective screening experiments for two reasons: (a) it is easier to generate very large numbers of candidate catalysts by molecular biology techniques than by chemical synthesis, and (b) the structural space of accessible proteins is immeasurably more vast than that of small molecule catalysts when the latter are built on a limited number of structural motifs.

True evolutionary selection for asymmetric synthesis, in which the production of a novel chiral molecule is linked to the survival of the mutatable catalyst that produces it, has yet to be achieved in the laboratory. It is likely that "test tube evolution" methods such as phage display [85] will be the first to be adapted toward this end, since the production of functional enzymes on phage particles is now routine. As described above, the Reetz laboratory has taken the first steps in this direction, testing candidate catalysts one at a time in various high-throughput ways, and many useful enantioselective enzymes will doubtless be developed in this manner. However, to bring the full power of selection to bear, one must be able to test many catalysts in the same solution and physically separate the effective (enantioselective) ones from the ineffective majority. Urlacher and colleagues have described the use of a pH-sensitive fluorescent protein expressed in cells that simultaneously express candidate enzymatic catalysts. The catalyzed reaction produces a carboxylic acid, changing the cellular pH and allowing for the rapid isolation of cells encoding active catalysts [86]. While not yet employed to detect enantioselective transformations, the use of a product-responsive dye coupled with fluorescence-activated cell sorting (FACS) is an extraordinarily promising way to select catalysts that can be encoded by genetic means. Other exciting work has also appeared for non-enantioselective transformations in recent years [87,88].

References pp. 90–94

Yet, while biological molecules, techniques, and conventions are bound to make an increasing mark on the discovery of practical enantioselective catalysts, most reactions and reaction conditions of organic chemistry will be outside the scope of proteins for the foreseeable future. It is therefore up to the chemists to take better advantage of the analytical tools that are now available by broadening the scope and increasing the numbers of enantioselective catalysts tested for targets of importance.

REFERENCES

1 A.M. Rouhi, Taking a measure of chiral riches. Chem. Eng. News, June 10 (2002) 51–53.
2 A.N. Collins, G.N. Sheldrake and J. Crosby (Eds.), Chirality in Industry: The Commercial Manufacture and Applications of Optically Active Compounds, Wiley, Chichester, 1992.
3 A.N. Collins, G.N. Sheldrake and J. Crosby (Eds.), Chirality in Industry II: Developments in the Commercial Manufacture and Applications of Optically Active Compounds, Wiley, Chichester, 1997.
4 P. Tielmann, M. Boese, M. Luft and M.T. Reetz, A practical high-throughput screening system for enantioselectivity by using FTIR spectroscopy. Chem. Eur. J., 9 (2003) 3882–3887.
5 M.T. Reetz, A. Eipper, P. Tielmann and R. Mynott, A practical NMR-based high-throughput assay for screening enantioselective catalysts and biocatalysts. Adv. Synth. Catal., 344 (2002) 1008–1016.
6 M.G. Finn, Emerging methods for the rapid determination of enantiomeric excess. Chirality, 14 (2002) 534–540.
7 V. Charbonneau and W.W. Ogilvie, High-throughput screening methods for asymmetric synthesis. Mini-Rev. Org. Chem., 2 (2005) 313–332.
8 J.P. Stambuli and J.F. Hartwig, Recent advances in the discovery of organometallic catalysts using high-throughput screening assays. Curr. Opin. Chem. Biol., 7 (2003) 420–426.
9 J.F. Traverse and M.L. Snapper, High-throughput methods for the development of new catalytic asymmetric reactions. Drug Disc. Today, 7 (2002) 1002–1012.
10 E.L. Eliel, Infelicitous stereochemical nomenclature. Chirality, 9 (1997) 428–430.
11 A brief discussion of the various options for e.r. screening in the context of a combinatorial (bio)catalyst development problem is presented by Reetz in Ref. [84].
12 The following examples are representative. (a) T. Ireland, F. Fontanet and G.-G. Tchao, Identification of new catalysts for the asymmetric reduction of imines into chiral amines with polymethylhydrosiloxane using high-throughput screening. Tetrahedron Lett., 45 (2004) 4383–4387. (b) C. de Bellefon, N. Pestre, T. Lamouille, P. Grenouillet and V. Hessel, High-throughput kinetic investigations of asymmetric hydrogenations with microdevices. Adv. Synth. Catal., 345 (2003) 190–193.
13 T. Satyanarayana and H.B. Kagan, The multi-substrate screening of asymmetric catalysts. Adv. Synth. Catal., 347 (2005) 737–748.
14 I. Chataigner, C. Gennari, U. Piarulli and S. Cecarelli, Discovery of a new efficient chiral ligand for copper-catalyzed enantioselective Michael additions by high-throughput screening of a parallel library. Angew. Chem. Int. Ed., 39 (2000) 916–918.
15 C. Gennari, S. Ceccarelli, U. Piarulli, C.A.G.N. Montalbetti and R.F.W. Jackson, Investigation of a new family of chiral ligands for enantioselective catalysis via parallel synthesis and high-throughput screening. J. Org. Chem., 63 (1998) 5312–5313.
16 L.G. Blomberg and H. Wan, Determination of enantiomeric excess by capillary electrophoresis. Electrophoresis, 21 (2000) 1940–1952.
17 V.A. Dowling, J.A.M. Charles, E. Nwakpuda and L.B. Mcgown, A reversible gel for chiral separations. Anal. Chem., 76 (2004) 4558–4563.
18 M.T. Reetz, K.M. Kuhling, A. Deege, H. Hinrichs and D. Belder, Super-high-throughput screening of enantioselective catalysts by using capillary array electrophoresis. Angew. Chem. Int. Ed., 39 (2000) 3891–3893.

19 C.N. Guo, R.D. Shah, R.K. Dukor, X.L. Cao, T.B. Freedman and L.A. Nafie, Enantiomeric excess determination by Fourier transform near-infrared vibrational circular dichroism spectroscopy: simulation of real-time process monitoring. Appl. Spectrosc., 59 (2005) 1114–1124.

20 C. Guo, R.D. Shah, R.K. Dukor, X. Cao, T.B. Freedman and L.A. Nafie, Determination of enantiomeric excess in samples of chiral molecules using Fourier transform vibrational circular dichroism spectroscopy: simulation of real-time reaction monitoring. Anal. Chem., 76 (2004) 6956–6966.

21 M.T. Reetz, A. Zonta, K. Schimossek, K. Liebeton and K.-E. Jaeger, Creation of enantioselective biocatalysts for organic chemistry by in vitro evolution. Angew. Chem. Int. Ed., 36 (1997) 2830–2832.

22 L.E. Janes, A.C. Löwendahl and R.J. Kazlauskas, Quantitative screening of hydrolase libraries using pH indicators: identifying active and enantioselective hydrolases. Chem. Eur. J., 4 (1998) 2324–2331.

23 R.F. Harris, A.J. Nation, G.T. Copeland and S.J. Miller, A polymeric and fluorescent gel for combinatorial screening of catalysts. J. Am. Chem. Soc., 122 (2000) 11270–11271.

24 F. Moris-Varas, A. Shah, J. Aikens, N.P. Nadkarni, J.D. Rozzell and D.C. Demirjian, Visualization of enzyme-catalyzed reactions using pH indicators: rapid screening of hydrolase libraries and estimation of the enantioselectivity. Bioorg. Med. Chem., 7 (1999) 2183–2188.

25 L. Di, O.J. Mcconnell, E.H. Kerns and A.G. Sutherland, Rapid, automated screening method for enzymatic transformations using a robotic system and supercritical fluid chromatography. J. Chromatog. B, 809 (2004) 231–235.

26 P. Krattiger, C. Mccarthy, A. Pfaltz and H. Wennemers, Catalyst-substrate coimmobilization: a strategy for catalysts discovery in split-and-mix libraries. Angew. Chem. Int. Ed., 42 (2003) 1722–1724.

27 L. Lingard, G. Bhalay and M. Bradley, Dyad beads and the combinatorial discovery of catalysts. Chem. Commun., (2003) 2310–2311.

28 M. Meldal, One bead two compound libraries for detecting chemical and biochemical conversions. Curr. Opin. Chem. Biol., 8 (2004) 238–244.

29 K.W. Busch, I.M. Swamidoss, S.O. Fakayode and M.A. Busch, Determination of the enantiomeric composition of guest molecules by chemometric analysis of the UV-visible spectra of cyclodextrin guest-host complexes. J. Am. Chem. Soc., 125 (2003) 1690–1691.

30 Y.F. Xu and M.E. Mccarroll, Determination of enantiomeric composition by fluorescence anisotropy. J. Phys. Chem. A, 108 (2004) 6929–6932.

31 Z.-B. Li, J. Lin and L. Pu, A cyclohexyl-1,2-diamine-derived bis(binaphthyl) macrocycle: enhanced sensitivity and enantioselectivity in the fluorescent recognition of mandelic acid. Angew. Chem. Int. Ed., 44 (2005) 1690–1693.

32 G.A. Korbel, G. Lalic and M.D. Shair, Reaction microarrays: a method for rapidly determining the enantiomeric excess of thousands of samples. J. Am. Chem. Soc., 123 (2001) 361–362.

33 L. Pu, Fluorescence of organic molecules in chiral recognition. Chem. Rev., 104 (2004) 1687–1716.

34 M.-H. Xu, J. Lin, Q.-S. Hu and L. Pu, Fluorescent sensors for the enantioselective recognition of mandelic acid: signal amplification by dendritic branching. J. Am. Chem. Soc., 124 (2002) 14239–14246.

35 S.Y. Liu, Y.B. He, G.Y. Qing, K.X. Xu and H.J. Qin, Fluorescent sensors for amino acid anions based on calix[4]arenes bearing two dansyl groups. Tetrahedron Asymm., 16 (2005) 1527–1534.

36 G.E. Tumambac and C. Wolf, Enantioselective analysis of an asymmetric reaction using a chiral fluorosensor. Org. Lett., 7 (2005) 4045–4048.

37 L. Zhu and E.V. Anslyn, Facile quantification of enantiomeric excess *and* concentration with indicator-displacement assays: an example in the analyses of α-hydroxyacids. J. Am. Chem. Soc., 126 (2004) 3676–3677.

38 E. Yashima, K. Maeda and T. Nishimura, Detection and amplification of chirality by helical polymers. Chem. Eur. J., 10 (2004) 42–51.

39 R.A. Van Delden and B.L. Feringa, Colour indicator for enantiomeric excess and assignment of the configuration of the major enantiomer of an amino acid ester. Chem. Commun., (2002) 174–175.

40 R. Eelkema, R.A. Van Delden and B.L. Feringa, Direct visual detection of the stereoselectivity of a catalytic reaction. Angew. Chem. Int. Ed., 43 (2004) 5013–5016.

41 Z.-B. Li, J. Lin, Y.-C. Qin and L. Pu, Enantioselective fluorescent recognition of a soluble "supported" chiral acid: toward a new method for chiral catalyst screening. Org. Lett., 7 (2005) 3441–3444.

42 M.T. Reetz, M.H. Becker, H.-W. Klein and D. Stöckigt, A method for high-throughput screening of enantioselective catalysts. Angew. Chem. Int. Ed., 38 (1999) 1758–1761.

43 W. Schrader, A. Eipper, D.J. Pugh and M.T. Reetz, Second-generation MS-based high-throughput screening system for enantioselective catalysts and biocatalysts. Can. J. Chem., 80 (2002) 626–632.

44 C. Hinderling, C. Adlhart and P. Chen, Mechanism-based high-throughput screening of catalysts. Chimia, 54 (2000) 232–235.

45 C. Hinderling and P. Chen, Rapid screening of olefin polymerization catalyst libraries by electrospray ionization tandem mass spectrometry. Angew. Chem. Int. Ed., 38 (1999) 2253–2256.

46 C. Markert and A. Pfaltz, Screening of chiral catalysts and catalyst mixtures by mass spectrometric monitoring of catalytic intermediates. Angew. Chem. Int. Ed., 43 (2004) 2498–2500.

47 J. Halpern, Mechanism and stereoselectivity of asymmetric hydrogenation. Science, 217 (1982) 401–407.

48 K.A. Schug and W. Lindner, Chiral molecular recognition for the detection and analysis of enantiomers by mass spectrometric methods. J. Separ. Sci., 28 (2005) 1932–1955.

49 An additional chromatography step to separate diastereomers may be substituted for these calibration measurements: H. Taji, M. Watanabe, N. Harada, H. Naoki and Y. Ueda, Diastereomer method for determining %ee by ^{1}H NMR and/or MS spectrometry with complete removal of the kinetic resolution effect. Org. Lett., 4 (2002) 2699–2702.

50 J. Guo, J. Wu, G. Siuzdak and M.G. Finn, Measurement of enantiomeric excess by kinetic resolution and mass spectrometry. Angew. Chem. Int. Ed., 38 (1999) 1755–1758.

51 B.N. Brewer, C. Zu and M.E. Koscho, Determination of enantiomeric composition by negative ion electrospray ionization-mass spectrometry using deprotonated *N*-(3,5-dinitrobenzoyl)amino acids as chiral selectors. Chirality, 17 (2005) 456–463.

52 M.E. Koscho, C. Zu and B.N. Brewer, Extension of chromatographically derived chiral recognition systems to chiral recognition and enantiomer analysis by electrospray ionization mass spectrometry. Tetrahedron Asymm., 16 (2005) 801–807.

53 H.J. Lu and Y.L. Guo, Evaluation of chiral recognition characteristics of metal and proton complexes of di-*O*-benzoyl-tartaric acid dibutyl ester and L-tryptophan in the gas phase. J. Am. Soc. Mass Spectrom., 14 (2003) 571–580.

54 H.J. Lu and Y.L. Guo, Chiral recognition of borneol by association with zinc(II) and L-tryptophan in the gas phase. Anal. Chim. Acta, 482 (2003) 1–7.

55 G. Grigorean, J. Ramirez, S.H. Ahn and C.B. Lebrilla, A mass spectrometry method for the determination of enantiomeric excess in mixtures of D,L-amino acids. Anal. Chem., 72 (2000) 4275–4281.

56 Z.P. Yao, T.S.M. Wan, K.P. Kwong and C.T. Che, Chiral analysis by electrospray ionization mass spectrometry/mass spectrometry. 2. Determination of enantiomeric excess of amino acids. Anal. Chem., 72 (2000) 5394–5401.

57 Z.P. Yao, T.S.M. Wan, K.P. Kwong and C.T. Che, Chiral analysis by electrospray ionization mass spectrometry/mass spectrometry. 1. Chiral recognition of 19 common amino acids. Anal. Chem., 72 (2000) 5394–5401.

58 M. Sawada, Y. Takai, H. Yamaoka, H. Imamura and M. Shizuma, The determination of enantiomeric excess of organic amine compounds by chiral recognition FAB mass spectrometry: the enantiomer-labeled host method. Adv. Mass Spectrom., 15 (2001) 359–360.

59 C. Zu, B.N. Brewer, B. Wang and M.E. Koscho, Tertiary amine appended derivatives of *N*-(3,5-dinitrobenzoyl)leucine as chiral selectors for enantiomer assays by electrospray ionization mass spectrometry. Anal. Chem., 77 (2005) 5019–5027.

60 W.A. Tao, L. Wu and R.G. Cooks, Rapid enantiomeric determination of α-hydroxy acids by electrospray ionization tandem mass spectrometry. Chem. Commun., (2000) 2023–2024.

61 L.M. Wu, R.L. Clark and R.G. Cooks, Chiral quantification of D-, L-, and meso-tartaric acid mixtures using a mass spectrometric kinetic method. Chem. Commun., (2003) 136–137.

62 L.M. Wu and G. Cooks, Chiral and isomeric analysis by electrospray ionization and sonic spray ionization using the fixed-ligand kinetic method. Eur. J. Mass Spectrom., 11 (2005) 231–242.

63 L.M. Wu, E.C. Meurer and R.G. Cooks, Chiral morphing and enantiomeric quantification in mixtures by mass spectrometry. Anal. Chem., 76 (2004) 663–671.

64 D.B. Berkowitz, W.J. Shen and G. Maiti, In situ enzymatic screening (ISES) of P,N-ligands for Ni(0)-mediated asymmetric intramolecular allylic amination. Tetrahedron Asymm., 15 (2004) 2845–2851.

65 S. Dey, K.R. Karukurichi, W.J. Shen and D.B. Berkowitz, Double-cuvette ISES: *in situ* estimation of enantioselectivity and relative rate for catalyst screening. J. Am. Chem. Soc., 127 (2005) 8610–8611.

66 Z. Li, L. Butikofer and B. Witholt, High-throughput measurement of the enantiomeric excess of chiral alcohols by using two enzymes. Angew. Chem. Int. Ed., 43 (2004) 1698–1702.

67 M.B. Onaran and C.T. Seto, Using a lipase as a high-throughput screening method for measuring the enantiomeric excess of allylic acetates. J. Org. Chem., 68 (2003) 8136–8141.

68 F. Taran, C. Gauchet, B. Mohar, S. Meunier, A. Valleix, P.Y. Renard, C. Creminon, J. Grassi, A. Wagner and C. Mioskowski, Communications: high-throughput screening of enantioselective catalysts by immunoassay. Angew. Chem. Int. Ed., 41 (2002) 124–127.

69 P. Vicennati, N. Bensel, A. Wagner, C. Créminon and F. Taran, Sandwich immunoassay as a high-throughput screening method for cross-coupling reactions. Angew. Chem. Int. Ed., 44 (2005) 6863–6866.

70 O. Hofstetter, J.K. Hertweck and H. Hofstetter, Detection of enantiomeric impurities in a simple membrane-based optical immunosensor. J. Biochem. Biophys. Methods, 63 (2005) 91–99.

71 M. Matsushita, K. Yoshida, N. Yamamoto, P. Wirsching, R.A. Lerner and K.D. Janda, High-throughput screening by using a blue-fluorescent antibody sensor. Angew. Chem. Int. Ed., 42 (2003) 5984–5987.

72 A. Tsourkas, O. Hofstetter, H. Hofstetter, R. Weissleder and L. Josephson, Magnetic relaxation switch immunosensors detect enantiomeric impurities. Angew. Chem. Int. Ed., 43 (2004) 2395–2399.

73 N. Millot, P. Borman, M.S. Anson, I.B. Campbell, S.J.F. Macdonald and M. Mahmoudian, Rapid determination of enantiomeric excess using infrared thermography. Org. Proc. Res. Devel., 6 (2002) 463–470.

74 K.-J. Johansson, M.R.M. Andreae, A. Berkessel and A.P. Davis, Organogel media for on-bead screening in combinatorial catalysis. Tetrahedron Lett., 46 (2005) 3923–3926.

75 P.R. Gibbs, C.S. Uehara, P.T. Nguyen and R.C. Willson, Imaging polarimetry for high throughput chiral screening. Biotechnol. Prog., 19 (2003) 1329–1334.

76 D. Belder and M. Ludwig, Microchip electrophoresis for chiral separations. Electrophoresis, 24 (2003) 2422–2430.

77 M. Ludwig, F. Kohler and D. Belder, High-speed chiral separations on a microchip with UV detection. Electrophoresis, 24 (2003) 3233–3238.

78 N. Piehl, M. Ludwig and D. Belder, Subsecond chiral separations on a microchip. Electrophoresis, 25 (2004) 3848–3852.

79 For reviews, see: (a) T.J. Wenzel and J.D. Wilcox, Chiral reagents for the determination of enantiomeric excess and absolute configuration using NMR spectroscopy. Chirality, 15 (2003) 256–270. (b) R. Rothchild, NMR methods for determination of enantiomeric excess. Enantiomer, 5 (2000) 457–471. For recent examples, see: (c) D.Z. Wang and T.J. Katz, A [5] HELOL-analogue that senses remote chirality in alcohols, phenols, amines, and carboxylic acids. J. Org. Chem., 70 (2005), 8497–8502. (d) J. Chin, D.C. Kim, H.J. Kim, F.B. Panosyan and K.M. Kim, Chiral shift reagent for amino acids based on resonance-assisted hydrogen bonding. Org. Lett., 6 (2004) 2591–2593.

80 C.D. Tran, V.I. Grishko and D. Oliveira, Determination of enantiomeric compositions of amino acids by near-infrared spectrometry through complexation with carbohydrate. Anal. Chem., 75 (2003) 6455–6462.

81 E. Vedejs and M. Jure, Efficiency in nonenzymatic kinetic resolution. Angew. Chem. Int. Ed., 44 (2005) 3974–4001.

82 M.T. Reetz and K.-E. Jaeger, Superior biocatalysts by directed evolution. Top. Curr. Chem., 200 (1999)

83 M.T. Reetz, Directed evolution of selective enzymes and hybrid catalysts. Tetrahedron, 58 (2002) 6596–6602.

84 M.T. Reetz, Changing the enantioselectivity of enzymes by directed evolution. Meth. Enzym., 388 (2004) 238–256.

85 G.P. Smith and V.A. Petrenko, Phage display. Chem. Rev., 97 (1997) 391–410.

86 S. Schuster, M. Enzelberger, H. Trauthwein, R.D. Schmid and V.B. Urlacher, pH luorin-based *in vivo* assay for hydrolase screening. Anal. Chem., 77 (2005) 2727–2732.

87 J.-L. Jestin, P. Kristensen and G. Winter, A method for the selection of catalytic activity using phage display and proximity coupling. Angew. Chem. Int. Ed., 38 (1999) 1124–1127.

88 A.A. Henry and F.E. Romesberg, The evolution of DNA polymerases with novel activities. Curr. Opin. Biotech., 16 (2005) 370–377.

PART II

© 2006 Elsevier B.V. All rights reserved.

CHAPTER 5

Analysis of chiral chromatographic separations by molecular modeling

Kenny B. Lipkowitz

Department of Chemistry, Howard University,
525 College Street, NW, Washington DC 20059, USA

5.1 INTRODUCTION

In an article entitled "Infelicitous Stereochemical Nomenclature," Eliel bemoaned the misuse of stereochemical terms by scientists [1]. One of those terms, chiral chromatography, is the focus of this chapter. Although being infelicitous in a strict sense, it has become the lingua franca of most bench chemists for a good reason: it works and most users of the method do not care what it is called. The idea that one could perform a direct separation of enantiomers on a chiral stationary phase (CSP) was understood many years ago, but only within the last 15 years, there have been major practical advances in the concept, most of which are described in a set of books chronicling the progress in that field [2]. Nowadays, one can purchase CSPs for use in planar-, liquid-, gas-, super- and sub-critical fluid phase chromatographies. The method is now so well accepted and fully integrated into the daily routine of bench chemists working in the area of asymmetric induction that they would be at a loss of what to do without this technology.

Many of the advances in CSP design were based on chemical intuition and insight gained from chiral recognition studies of soluble CSP analogs with nuclear magnetic resonance (NMR) spectrometry and from X-ray crystallography when possible. Complementing these experimental studies were computational studies directed towards rationalizing how chiral discrimination takes place and with the hopes of designing improved CSPs. In this chapter, we describe several of the techniques used for modeling chiral recognition in chromatography and provide several demonstrative examples of each method mentioned.

5.2 ATOMISTIC MODELING

A model is a likeness, a semblance or a representation of reality. Ideally one uses a model to simplify an otherwise too complex observation and to provide a rationalization for that observation. In the area of chiral chromatography, hand-held mechanical models were used to visualize and rationalize why one CSP could resolve some racemates but not others and to assist in designing new materials that would serve as enhanced CSPs. To some developers and users of CSPs, there was no need for anything other than such mechanical models but for other researchers the need for using more refined models, i.e. those that better expressed the basic physics of molecular interactions, was apparent. Even though soluble CSP–analyte complexes could be studied by NMR spectrometry, only the diastereomeric complex corresponding to the more stable, and longer retained analyte on the column would typically give rise to the intermolecular nuclear Overhauser enhancements (NOE) needed to describe the shape of the complex. Similarly, only the more stable complexes could typically be crystallized for X-ray analysis. Relatively little was known about the energies and structural features of the less stable diastereomeric complexes (corresponding to the first eluted analyte). Accordingly, several research groups relied on computational models to discern differences (and similarities) of the competing diastereomeric complexes that form as the racemic analytes migrate through a chiral chromatographic column.

Two broad categories of models were used for this purpose. One model relies on regression methods to fit a set of independent variables with a dependent variable. The independent variables in this case are "molecular descriptors" that will be described below. The independent variables are the relative retention times of the analytes on the column given as capacity factors, k, or their ratio given as α which is the separation factor (related to the differential free energy of analyte binding as $\Delta\Delta G = -RT \ln \alpha$), a value assumed to be a measure of chiral discrimination. The second type of modeling uses empirical force fields (FFs) (or quantum-based computational methods) to compute the relative energies of the diastereomeric complexes directly. In both types of models atomic level detail is taken into account, so, both methods are atomistic modeling methods. These atomistic computer models were used to provide information that is not amenable to experimentation (see below). Moreover, in contrast to mechanical models, results from NMR spectroscopy, or from a single crystal structure, these models were meant to quantify the amount of a particular kind of interaction that takes place during chiral discrimination. Examples of these methodologies and their applications to chiral analysis in chromatography are given below.

5.3 STRUCTURE–PROPERTY RELATIONSHIPS

It has long been understood that relationships exist between molecular structure and molecular properties (or function). This relationship serves as the underpinning for the predictive ability of modern chemistry. For example, a plot of experimental boiling points for a homologous series of normal alkanes *versus* the number of carbon atoms in the chain is presented in Fig. 5.1. A clear relationship exists. While one could fit the data to a higher order polynomial, a simple straight line fit provides a remarkable correlation. If, instead of plotting the number of carbons on the abscissa one were to plot the molecular weight of each alkane, a similar plot would be obtained. But why use the number of carbons or the molecular weight? Could not one use for the

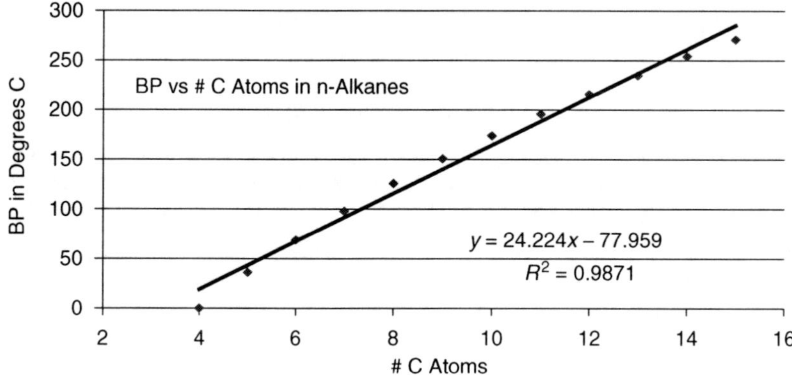

Fig. 5.1. Example of structure–property relationships. Experimental boiling point as a function of the number of carbon atoms in a homologous series of normal alkanes.

abscissa, say, the average end-to-end distance of the chain, or the surface area of the alkane, or the molar volume of the chain? One could, and similar relationships would be derived, all having the ability to predict a missing homolog in the series by interpolation or the next few homologs in the series by extrapolation.

What we are doing in these examples is to describe some attribute of the molecule as a single numerical value that can be used for making such structure–property relationships. These attributes are called molecular descriptors and there are no rules about what kind of descriptors should be used, nor are there rules about how many descriptors can be used for deriving structure–function relationships. Scientists simply want to rationalize the connections between molecular structure and physicochemical properties. The question is: what do we mean by a molecule's structure?

5.4 DESCRIBING MOLECULAR STRUCTURE

There exist different interpretations for the term "molecular structure;" for certain, the term molecular structure does not necessarily mean the same thing as molecular geometry. One view of molecular structure is topological. That is, the atoms and their connectivity define a molecular structure. The second view is topographical. Here the three-dimensional (3D) shape of the molecule is considered. For structure–property relationships both types of "molecular structure" descriptor can be used.

5.4.1 Graph theory

A molecular structure can be represented as a hydrogen-suppressed molecular graph, an example of which is $(CH_3)_2CHCH_2C(CH_2CH_3)_2CH_3$ depicted below.

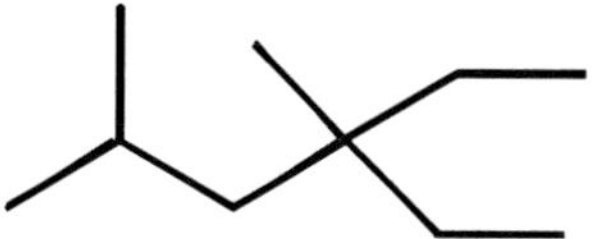

The vertexes of this graph represent skeletal atoms and the edges of the graph represent a pair of adjacent atoms, usually meaning a chemical bond. The idea of chemical graph theory is to distill from the chemical structure those elements that provide structure variables in the form of numerical indexes. The zeroth level structure descriptor is to simply count the number of vertexes (atoms). This worked well for the boiling point example in Fig. 5.1 but it works well only for linear alkanes where properties are anticipated to be proportional to the number of atoms in the chain. For branched hydrocarbons or more complex systems, more information-rich descriptors are needed and many such descriptors have been developed an example of which is the path-1 molecular connectivity index, ${}^{1}\chi_{\mathrm{p}}$ developed by Randic [3].

$$ {}^{1}\chi_{\mathrm{p}} = \sum_{\text{edges}} \frac{1}{\sqrt{mn}} \tag{1} $$

Here m and n are the degrees of the adjacent vertexes joined by each edge. It is instructive to see how this is done with a simple example of 2-methylbutane.

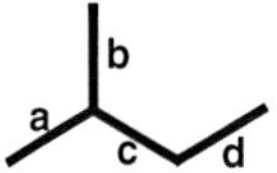

This graph has 4 edges, thus 4 terms—one term for each edge. For each term we just count the number of bonds attached to each atom participating in the bond. The four-term equation for this is:

$$ {}^{1}\chi_{\mathrm{p}} = \frac{1}{\sqrt{1 \cdot 3}} + \frac{1}{\sqrt{1 \cdot 3}} + \frac{1}{\sqrt{3 \cdot 2}} + \frac{1}{\sqrt{2 \cdot 1}} = 2.27 \tag{2} $$

The first term corresponds to the edge labeled **a** above. The first atom (m) is connected to one other atom so it has a value of 1. The second atom (n) defining bond **a** is connected to three atoms so it has a value of 3. Hence the square root of 1·3 exists in the first denominator. The second term in the above equation corresponds to the edge labeled as **b**. The third term is for the edge labeled **c**. Here an atom is connected to 3 other atoms and the other atom comprising edge **c** is connected to 2 atoms. This is done for all edges in the graph and eventually a single numerical value is derived that encodes information about the molecule. This index reflects both size and branching of the molecule. As the molecular size increases, the number of terms increases and ${}^{1}\chi_{\mathrm{p}}$ increases. As branching increases, the denominator increases thus reducing the value of ${}^{1}\chi_{\mathrm{p}}$. Many other topological descriptors exist, some of which account for the types of atoms involved and some of which can provide information about 3D shape and charge distributions [4].

5.4.2 3D descriptors

Other types of molecular descriptors that convey information about molecular geometry exist. Many of these descriptors are derived from experimentation while others are derived from theory. Simple descriptors including molecular length to width ratios, ovality ratios of principle moments of inertia and the like portray molecular shapes, while accessible surface areas and electrostatic potential surfaces provide information about lipophilicity and electronic influences of substituents. Polarizabilities, dipole moments, ionization potentials, electron affinities and log P, the partitioning of a molecule between water and a hydrophilic medium like 1-octanol, can all be derived experimentally or as more commonly done nowadays, computationally. There are thousands of such descriptors that have been and can be used for relating molecular structure with function or property. Below we describe how these descriptors have been used for assessing retention times of chiral analytes on CSPs.

5.5 QUANTITATIVE STRUCTURE ENANTIOSELECTIVITY RELATIONSHIPS (QSERR)

Quantifying the relationship between structure and activity is referred to as quantitative structure activity relationships (QSAR) [5]. In traditional QSAR, one attempts to correlate the activities of a set of molecules (that are presumed to carry out their tasks in a similar manner) with one or more attributes of those molecules. These attributes are embedded in molecular descriptors described above that typically describe hydrophobic, steric and electronic features of each molecule but a descriptor can be anything that describes some feature of a molecule. These descriptors are regressed onto the activity data to generate a mathematical model (see Eq. (3)) that can be used to predict the activity of analogs that are yet to be synthesized.

$$\log(1/C) = b_0 + \sum_i b_i D_i \qquad (3)$$

In Eq. (3), C is the concentration of a compound that elicits a response to an assay of some sort (e.g. an LD_{50}), b_0 is a constant, and b_i are the coefficients weighing the importance of the individual molecular descriptors, D_i used in the study. A least squares multiple linear regression is commonly used to identify the coefficients which provide the best fit from a pool of purportedly relevant molecular descriptors to a set of experimental data.

Studies directed towards providing a quantitative relationship between structure and enantioselectivity are referred to as quantitative structure enantioselectivity relationship (QSERR) studies. Applications in the area of chiral chromatography have been published. In these publications, the authors typically consider the capacity factors, k_S and k_R, for each enantiomeric analyte separately. The idea here is to seek differences between regression coefficients of the best mathematical models derived by the modeler to rationalize the different retention times of the antipodal analytes. In some instances, different descriptors can appear in the two equations indicating, perhaps, different retention mechanisms that in turn are responsible for chiral separation.

An example of this comes from two independent research groups modeling the enantioselective binding of sulfoxides to different types of CSPs. In the work of Altomare et al. [6], a π-acidic high performance liquid chromatography (HPLC) stationary phase composed of *N*,*N*′-(3,5-dinitrobenzoyl)-*trans*-1,2-diaminocyclohexane

References pp. 126–129

(DACH-DNB) was used and in the work of Montanari et al. [7], cellulose tris-phenylcarbamates and amylose tris-phenylcarbamates were examined. In both cases most of the descriptors used as independent variables were computed.

In the study by Altomare, the π-donating and the π-accepting ability of the aryl rings of aromatic sulfoxides was described by electrophilic and nucleophilic superdelocalizabilities. Verloop (L, B_1, B_5) and Charton (υ) descriptors were used to evaluate steric influences and the S=O vibration frequency (ν) was measured experimentally. The hydrophilicity of each analyte (log P) was computed. Other descriptors like net charge on sulfoxide oxygen (q_O), net charge on sulfur (q_S), Hammett electronic substituent constant (σ), Taft hydrophobic constant (π) and molar refractivity (MR) were used. Even though this set of descriptors is small in comparison to most QSAR studies published, the authors relied on partial least squares (PLS) regression techniques to derive their models. PLS has the advantage over stepwise multiple regression methods of avoiding collinearity issues and use of more variables than compounds. Because one can compute so many different types of descriptors so easily it is not uncommon to find hundreds of descriptors for only a few experiments. The problem of "variable selection" is currently an intensely debated, contentious issue in the QSAR community and different strategies are used to select the most significant descriptors from among the many possibilities. Though contentious and problematic, all researchers in the field are keenly aware of the need for statistically meaningful variable selection methods. A tutorial on the topic is worth reading before conducting such studies [8].

The best models derived by Altomare et al. for each antipode are given in Eqs. (4) and (5).

$$\log k_S = 0.771\, S_{Ph}^{HOMO} - 0.048 \log P - 10.2\, q_O - 6.32 \tag{4}$$

$$\log k_R = 0.818\, S_{Ph}^{HOMO} - 0.037 \log P - 12.8\, q_O - 8.13 \tag{5}$$

The retention is governed in part by the pi-basic nature of the analytes (described by the superdelocalizability of the phenyl ring), by the net charge on the S=O oxygen and by the system's lipophilicity. Because the same terms exist in both these equations and because the regression coefficients associated with each independent variable are so similar, the retention times of both enantiomers is influenced almost identically. In other words, these equations are not able to discern the differences in retention times for the mirror image isomers. Moreover, the correlation coefficients (r^2) can account for only ~75% of the variance in the observed capacity factors, and, the predictive abilities of these models (r^2_{cv}) are poor, being on the order of 0.65. Based on this, an alternative modeling method called comparative molecular field analysis (CoMFA, described below) was implemented.

Montanari et al. evaluated the separations of 23 sulfoxides on two amylose HPLC CSPs containing C_6H_5- and 3,5-dimethyl-C_6H_3-tris-carbamate groups and on similarly substituted cellulose CSPs [7]. A total of 299 descriptors were calculated for that study and like Altomare et al. these authors relied on Principle Components Analysis (a technique akin to PLS that reduces the total number of variables for use in model construction). Following variable selection with forward selection and backward elimination of variables, the authors used only the most important descriptors for a multiple regression analysis. The most robust CSPs were found to be the tris(3,5-dimethylphenylcarbamates) of cellulose and amylose with the former being able to separate the greater number of enantiomers. The best mathematical models for the tris-phenylcarbamate of cellulose CSP are given in Eqs. (6) and (7).

$$\log k_1 = 0.998J - 0.144 \tag{6}$$

$$\log k_2 = -0.674J + 0.236\mu - 0.007\,\mathrm{MV} + 2.442 \tag{7}$$

In these equations, J is a topological descriptor that encodes branching information. It is to be noted that a simple topological index gives rise to an equation that that can account for over 90% of the variance in $\log k_1$. This seems to be a good predictive model as well, with $r^2_{\mathrm{cv}} = 0.876$. In Eq. (7), μ is the dipole moment and MV the molar volume. Aside from having additional terms contributing to Eq. (7), we note that the sign of the J coefficient is reversed from that in Eq. (6) suggesting that the retention of this enantiomer is affected in an inverse manner compared to its antipode on this particular column (and under those experimental conditions).

The best QSERR models for the tris-3,5-dimethylphenylcarbamate of cellulose CSP are given by Eqs. (8) and (9) for k_1 and k_2, respectively.

$$\log k_1 = 0.006\,\mathrm{MM} - 0.2099\,\log P - 5.050\,{}^0\chi_{\mathrm{R}} + 0.913\,\mathrm{P2p} + 3.397 \tag{8}$$

$$\log k_2 = 0.006\,\mathrm{MM} - 0.386\,\log P - 5.11\,{}^0\chi_{\mathrm{R}} + 0.232\,\mathrm{L2u} + 3.408 \tag{9}$$

For Eq. (8), $r^2 = 0.916$ and $r^2_{\mathrm{cv}} = 0.878$. For Eq. (9) $r^2 = 0.908$ and $r^2_{\mathrm{cv}} = 0.840$ suggesting high-quality models. MM is the molar mass, ${}^0\chi_{\mathrm{R}}$ is a zeroth order Randic connectivity index and the P2p and L2u are polarizability shape descriptor and a directional dimension respectively. QSERR equations using other descriptors for the other CSPs studied by Montanari were derived in a similar way. While such models are useful for predicting whether or not sulfoxides related to those in the authors' study could be separated, it is evident that only "fuzzy" and somewhat hand-waving rationalizations about what is happening between the CSP and these analytes can be derived from such mathematical models. This is not surprising because some of the descriptors themselves are "fuzzy." Follow-up studies concerning the binding of acetic acid methyl esters by Carotti et al. [9] and of oxadiazolines by Altomare et al. [10] on the DACH-DNB CSP have appeared. The same types of descriptors were used in both studies. The key finding from these studies is that the balance of intermolecular forces determining retention factors are not the same as those forces affecting chiral separation.

QSERR studies of benzodiazepine retention times on a human serum albumin-based CSP were carried out collaboratively by Kaliszan et al. [11,12]. In their studies they intentionally used a minimum number of descriptors, thus obviating the need for using the PLS method that obfuscates the physical meaning of the initially selected molecular descriptors. Despite using a small number of diazepines in their analyses, Kaliszan and Wainer were able to derive statistically significant models that allowed them to determine structural and electronic features responsible for retention of analytes on the protein CSP. In their second article [12], Kaliszan et al. demonstrated differences

References pp. 126–129

in retention mechanisms of drug-like molecules in affinity, reversed-phase and adsorption HPLC modes.

Booth and Wainer developed extremely simple models for rationalizing the enantioselective binding of α-alkylarylcarboxylic acids to a tris(3,5-dimethylphenylcarbamate) amylose CSP [13]. Their descriptors included the numbers of hydrogen bond donors, the numbers of hydrogen bond acceptors and the degree of aromaticity of the analytes. As with most such QSERR studies, regressions were made to the capacity factors (retention times) rather than separation factors. What is interesting in this article is their finding that both enantiomers of analytes can fit into a chiral groove on the helical CSP and both can achieve the same kind of intermolecular interactions between the analyte and CSP. However, the less retained enantiomer can achieve this fit only by adopting a higher energy conformation than that of the longer retained (more tightly bound) enantiomer. For both of the analytes, the steric bulk of the alpha substituent is projected away from the CSP and, accordingly, steric factors are not important (other than to influence conformations). This is an example of a "conformationally driven" rather than a more typical sterically driven chiral separation in chromatography. Several other QSERR publications have appeared including that of Calleri et al. who developed models for the retention mechanism on a riboflavin-based CSP [14] and the work by Suzuki et al. [15] who carried our QSERR and CoMFA (see below) models for the separation of secondary alcohols on Pirkle-type CSPs. In this latter study, four different (*R*,*R*)-*N*-3,5-dinitrobenzoyl-1,2-diphenyl-1,2-diamino-linked CSPs were evaluated for their ability to separate 42 chiral alkylarylcarbinols. Twenty-seven descriptors were used in the authors' assessment. Standard multiple linear regression (MLR) methods were implemented to derive models correlating log k_1, log k_2 and log α with the most important descriptors for each CSP. Concerning the retention behavior, the two most important descriptors are the molecular polarizability and the energy of the first vacant molecular orbital (LUMO) that expresses the analyte's penchant to form hydrogen bonds and/or pi-stacking to the CSP. For the chiral recognition models, in contrast, the authors found that log P and the sum of charges on the carbon atoms of the phenyl group of the analyte attached to the stereogenic center are most important. What is significant about this latter study in terms of computer modeling was the use of artificial neural networks (ANN) in deriving the models. The ANNs account for non-linear behavior that the MLR models cannot pick up. These authors also relied on CoMFA to develop models for chiral ediscrimination.

5.6 COMPARATIVE MOLECULAR FIELD ANALYSIS (CoMFA)

Comparative molecular field analysis is a complement to traditional QSAR methods [16]. The genius of this method is that it addresses the interactions each molecule in the data set "feels" external to itself. This is done by placing each molecule, oriented the same way, at the center of a 3D grid and evaluating at each grid point the interaction energy between a probe molecule (or atom) with each aligned molecule. The interaction energy at each grid point is then a descriptor that is used in the regression. The resulting regression equation is given in Eq. (10), where the *c*s are the coefficients, S_1 is the steric energy at grid point 1, S_2 is the steric energy at grid point 2 (continued for all n grid points) and the E and H refer to the electronic and hydrophobic interaction energies at the same grid points derived for the steric energies.

$$\begin{aligned}\text{Activity} = y + c_1S_1 + c_2S_2 + \cdots + c_nS_n + c_1E_1 + c_2E_2 \\ + \cdots + c_nE_n + c_1H_1 + c_2H_2 + \cdots + c_nH_n + \text{other descriptors}\end{aligned} \tag{10}$$

Because there are thousands of grid points being evaluated, there are thousands of descriptors being generated requiring the use of PLS or PCA for regressions. An example of the alignment of two closely related molecules from a large dataset is depicted in Fig. 5.2.

CoMFA has been used for chiral analysis in chromatography. One advantage of using this type of QSAR is that one can visualize the large number of computed descriptor coefficients by making iso-value contour maps of those coefficients at grid points surrounding the aligned data set. In Fig. 5.6, the regions of space where steric bulk should enhance or destroy stereoinduction are plotted, as an example.

In the CoMFA models derived by Altomare et al. only steric and electrostatic fields were used. The relative contribution of the steric and electrostatic components is 50:50 for the separation of sulfoxides on the DACH-DNB CSP suggesting an equal dependence on steric and Coulombic contributions to chiral separation [6]. For the separation of oxadiazolines on that CSP, steric factors are more dominant [10]. For the

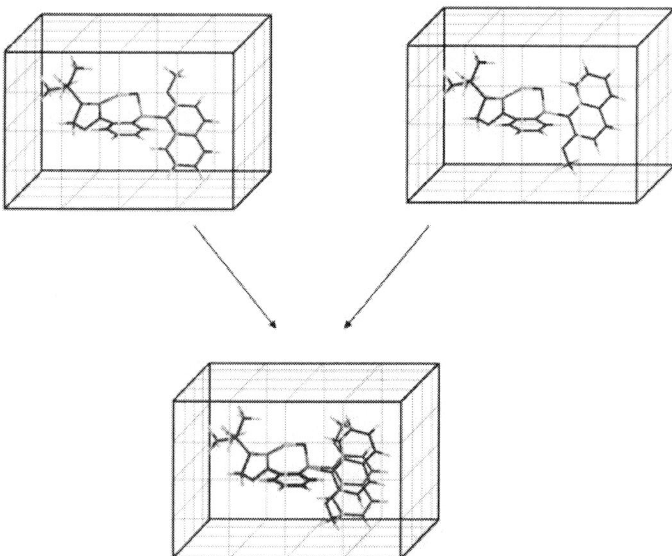

Fig. 5.2. CoMFA interaction points. Depicted here are two closely related molecules (of a large set of such structures) embedded in a 3D grid. The molecules are then aligned in a common orientation. At each intersecting grid point various test probes evaluate steric interactions, Coulombic interactions, hydrogen bonding interactions, hydrophobic interactions, etc. around the aligned set of molecules.

enantioselection of carbinols on the four Pirkle-like CSPs studied by Suzuki et al. [15], the contributions were ~45% steric and ~55% Coulombic in nature. In all of these and other such CoMFA studies, the research groups were interested primarily in understanding how a single CSP discriminates amongst a set of related analytes. We describe now, a unique CoMFA study where the CoMFA strategy is turned upside down in a way where a set of related CSPs is allowed to interact with a single analyte molecule.

In a collaborative effort between Lindner and Lipkowitz, a computer-aided molecular design (CAMD) strategy using CoMFA was undertaken to derive a predictive model for the separation of the amino acid leucine (in a derivatized form) [17]. The idea was to design a CSP that has enhanced enantioseparability using computational methods. To make this chapter a pedagogically driven review, we illustrate how a CoMFA analysis is done, step-by-step.

Step 1: Generate a database of information

Lindner had been working with alkaloid natural products for a variety of separations and found good performance with quinine-based CSPs for LC in which the carbinol is transformed to a carbamate functionality. Fig. 5.3 shows the basic structure of the CSPs (1), where *R* is a variable group. The figure also demonstrates how the alkaloids are tethered to the stationary support. The part of the CSP enclosed in the box is the portion that was modeled. The analyte, as a 3,5-dinitrobenzoyl derivative (2), is also depicted. The different CSPs synthesized and tested are illustrated in Fig. 5.4. The separation factor, α, for each CSP is given. This dataset was used for the derivation of the CoMFA model.

Step 2: Generate 3D molecular structures computationally

A 3D representation of each CSP analog was determined by carrying out a complete conformational analysis with a FF. It was assumed that the lowest energy structure of each molecule best represents the shape of the CSP.

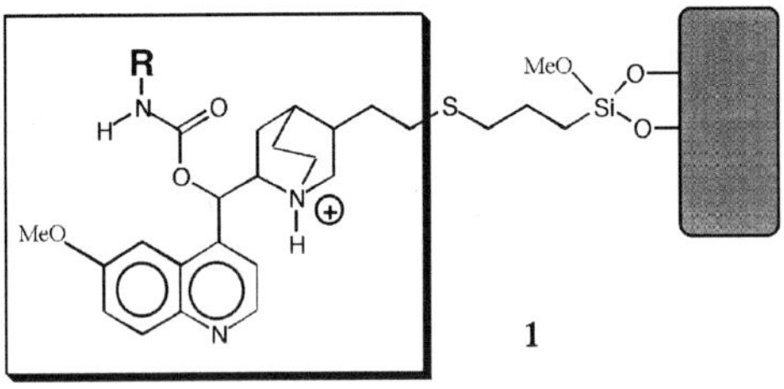

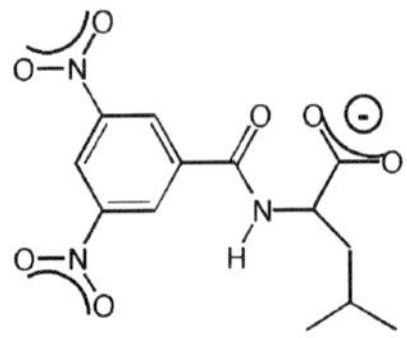

Fig. 5.3. The quinine-based CSP, 1 and the analyte, 2. The intermolecular interaction of *R*/*S* analytes with the different CSPs leads to a zwitterion as shown.

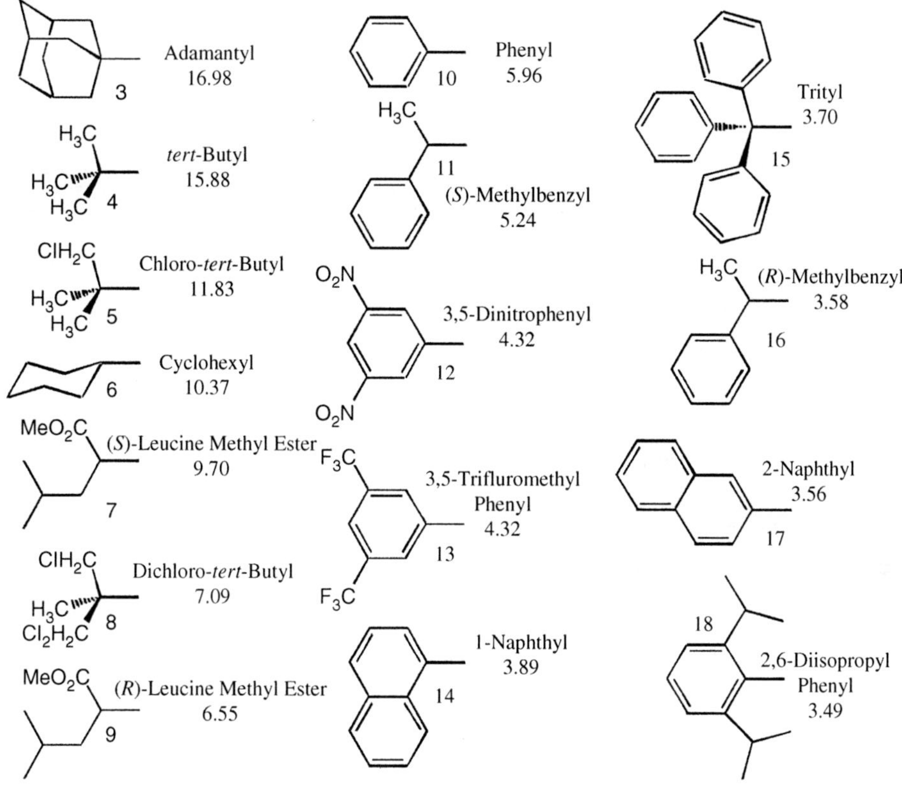

Fig. 5.4. The set of carbamate derivatives used for the CoMFA analysis. Below each substituent (R in Fig. 5.3) is the experimental separation factor, α.

Step 3: Overlay the CSPs in a common alignment

A least squares fitting of all structures to a common template (see original article for details) was performed. That ensemble of aligned CSP analogs was then embedded into a uniform 3D grid (Fig. 5.5).

Step 4: Select variables

At each grid point test probes are placed and the intermolecular electrostatic energies and the intermolecular steric energies are computed. The energies at each grid point (each descriptor) depend on the type of probe used. The number of grid points dictates the number of descriptors generated. The novice modeler would anticipate that a fine mesh of grid points is more desirable than a coarse grid but this assertion is not true. Only a small number of informative descriptors that correlate with α will be selected. Selection of the most informative descriptors and elimination of the background "noise" becomes problematic for such large datasets and it has been shown that retaining irrelevant variables in a model can have a detrimental effect on its predictive ability. A coarse grain gridding of ~2 Å between points was used by the authors. Approximately

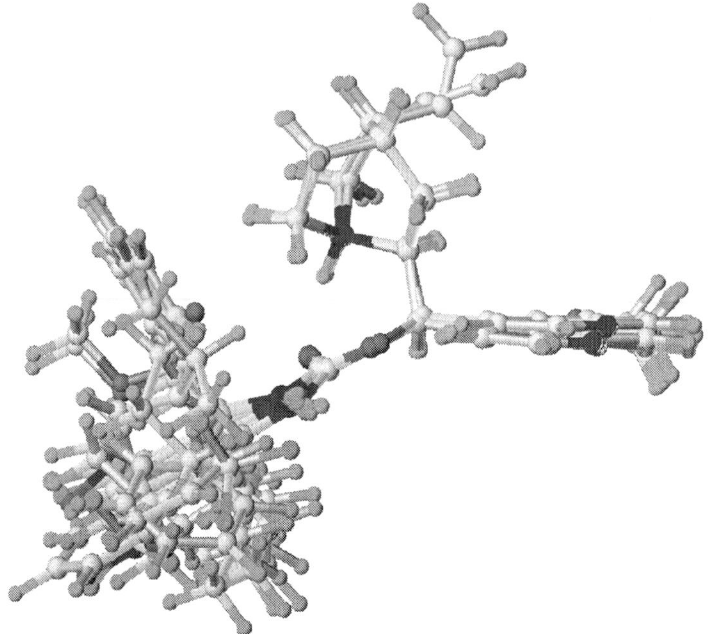

Fig. 5.5. Aligned CSP analogs embedded in a uniform 3D grid (grid not shown for clarity).

1200 grid points were used giving rise to ~2400 descriptors (1200 each for the steric and electrostatic probes).

Step 5: Derive the QSAR

PLS was used to derive the model. The leave-one-out (LOO) method was used to cross-validate the results. The correlation coefficient squared between calculated and observed separation factors (α) was very high ($r^2 = 0.995$), but more significantly the $r^2_{cv} = 0.671$. Thus the derived model fits well to the experimental data but more significantly it has high predictive capabilities as well (this is what the authors were seeking in their study).

Step 6: Interpret the results/suggest new candidates with improved performance

Lindner and Lipkowitz found from their model that the steric field descriptors accounted for more than 90% of the variance in their experimental data while the electrostatic field descriptors explain less than 10% of the variance. Given that the analytes associate with the CSP by forming salt bridges, this may appear counter-intuitive. It is noted that the dominant force holding the transient diastereomeric complexes together *are* electrostatic. However, the forces most responsible for chiral discrimination are not.

To visualize where in space the steric terms contributing to the separation exist, isocontour maps were generated. Fig. 5.6 shows where differences in fields are most highly influencing chiral recognition.

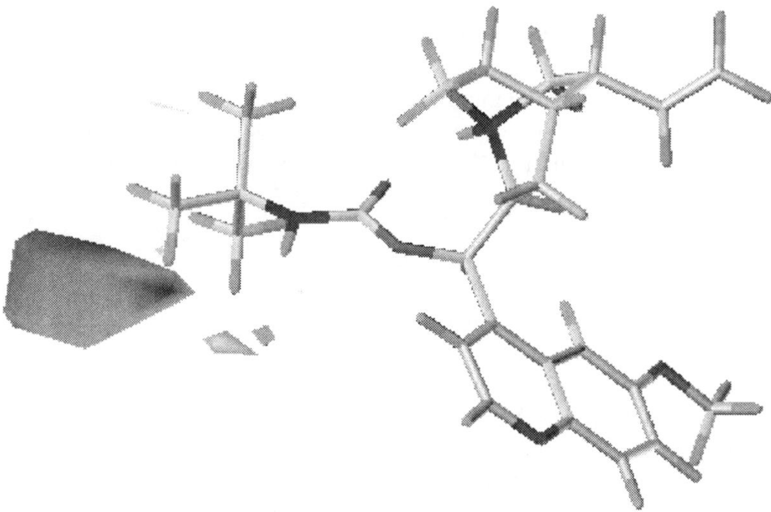

Fig. 5.6. Region around the aligned CSPs where steric bulk is predicted to enhance chiral discrimination. For clarity, only one of the 16 aligned CSPs is shown.

In this figure only steric fields have been depicted, as this is the dominant field in the authors' model. In the figure, the regions around the aligned CSPs are highlighted showing where addition of steric bulk is predicted to enhance the enantioselectivity of this class of CSP. That region is at the carbamoyl group, behind the carbamate's N–H bond. A new CSP based on this CAMD approach was synthesized after Lindner and Lipkowitz's article was published, affixed to a solid support and found to have enhanced discriminatory ability.

The QSERR studies described above provide mathematical models that can reproduce stereoselective data in chromatography, and when developed and validated properly, can be predictive as well. Because of this they have great value for molecular design, akin to the QSAR models that are developed by drug design groups in all major pharmaceutical companies. These models also provide insights about what types of forces, e.g. steric, geometric, hydrophobic and so on, give rise to chiral selection. Unfortunately such models do so in a somewhat obfuscated manner. What most bench chemists want is an explanation of chiral selectivity based on robust theories that evaluate stereodifferentiating energies, and which depict (with a graphical model) what the binary complex looks like between selector and selectand, when interactions occur between CSP and analyte. These energy-based modeling techniques are described in the next section of this review.

5.7 COMPUTATION OF SELECTOR–SELECTAND INTERACTIONS

Two methods exist for computing energies: those based on quantum mechanical (QM) theories and those that rely on FFs. Before reviewing articles that use these methods, we point out that there exist many assumptions and approximations commonly made while computing selector–selectand interaction energies. Most scientists often truncate the size of the CSP or analyte, omit buffers, counter-ions and even solvent environments to make the system computationally tractable. Moreover, they do not attempt to compute absolute free energies, but rather determine differential free energies. In all systems where enantioselective binding takes place there exist two competing equilibria:

$$\mathrm{CSP}^R + \mathrm{A}^R \rightleftharpoons \mathrm{CSP}^R \cdot \mathrm{A}^R \tag{11}$$

$$\mathrm{CSP}^R + \mathrm{A}^S \rightleftharpoons \mathrm{CSP}^R \cdot \mathrm{A}^S \tag{12}$$

In these equations, CSP is the chiral stationary phase molecule serving as a selector, A is the analyte molecule that is the selectand, and the superscripted R and S are stereochemical descriptors. In both equilibria, the binary complexes are weakly bound diastereomeric complexes held together by van der Waals forces, hydrogen bonding, charge transfer forces, etc. The enantiodiscriminating forces are very small compared to the binding forces, typically by 2–3 orders of magnitude. Thus, one is faced with the challenging task of computing very small energy differences, often less than 500 cal/mol.

Because the left-hand side of both equilibria are equivalent, i.e. one has the same CSP, and, by virtue of an enantiomeric relationship, (R)-analyte = (S)-analyte (they have the same shapes, same energies and the same extent of solvation in the unbound state), all one needs to do is to compute the free energies of the bound state complexes on the right-hand side of Eqs. (11) and (12). The differences between the free energies of those complexes, $\Delta\Delta G$, can be used to predict enantiodiscriminating binding, and to provide a well-founded rationale for the observed selectivity. This double difference method assumes that, if the competing binding mechanisms are similar enough, influences of polar effects, solvent effects and entropy differences cancel, thus making differences in computed energies, either from QM or FF calculations, comparable to differential free energies ($\Delta\Delta G$), derived from chromatographic experiment.

Where this panacea of "cancellation of errors" fails is when one inadequately samples the conformations, the orientations and the positions of analyte with respect to the CSP. Merely taking an analyte molecule (in an arbitrarily defined conformation) and docking it somehow to a CSP molecule followed by energy minimization (for R and then for S isomers) is a poor way to compute energies for comparison with experiment. In this case, multiple calculations are required. Most published articles concerning the calculation of enantioselection take advantage of the double difference approach but unfortunately they do not adequately sample enough structures to make their results meaningful for comparison with experiment. In this chapter, we focus on studies where this issue has been addressed; i.e. we consider publications where methods have been developed or used to sample multiple binding configurations in an attempt to find the most stable structures or for statistical averaging.

5.8 QUANTUM MECHANICS-BASED STUDIES

QM allows one to compute structures, energies and properties of molecules that are of relevance to researchers in separation science. For example, heats of formation can be computed from quantum theory, as can electronic spectra for comparison to experiment. Likewise electronic properties such as dipole and higher order moments, molecular electrostatic potentials, and so on can be computed. In the real world of experimentalists, however, all these attributes of QM become less useful than desired because there are severe limitations on such calculations, the most significant being that QM calculations are very time consuming for the central processing unit (CPU) of a computer. Many approximations with concomitant re-parameterizations of the Hamiltonian have consequently been made to speed up these calculations, but, a loss in rigor and accuracy is introduced by doing this.

Because of this CPU bottleneck, relatively few chiral recognition studies in chromatography have been published. Most of those publications did not involve adequate sampling of selector–selectand configurations to allow the author to predict separation factors. Instead, with one notable exception described below, diastereomeric complexes determined from NMR or crystallography experiments, or from computations using FFs, were used as input to the QM calculations. For example, in a series of articles Topiol and his collaborators used both *ab initio* and semiempirical levels of QM to evaluate the significance of three-point binding interactions in Pirkle-type CSPs [18–21]. Those QM results were compared with those from FF methods and the results were used to refute some ideas presented by Pirkle. Structures and energies of the competing diastereomeric complexes were computed. Molecular electrostatic potentials (MEPs) were also computed and comparisons between such maps were made in an attempt to seek the long-range recognition elements responsible for enantiodifferentiation. Comparisons between the dinitrobenzoyl fragments of a model Pirkle-type CSP interacting with *R* and *S* methyl *N*-(2-naphthyl)alininates revealed no difference in MEPs suggesting that pi-interactions alone are not responsible for enantiomer separation. MEP maps of the CSP amide moiety likewise revealed no recognition element. What the authors did find from their QM studies is that all interactions found in the more stable diastereomeric complex exist in the less stable complex as well, thus negating simplistic models of chiral recognition based on 3-point *versus* 2-point interaction schemes.

To model the interaction between a CSP and an analyte, one must account for: (1) the shapes of the two molecules in the binary complex, (2) the position of the two molecules relative to one another (i.e. the analyte should be at its proper binding site on or around the CSP) and (3) the orientation of the two molecules with respect to each other. This is just a simple way of saying that some sort of ensemble average is needed. The two most frequently used strategies for exploration of the diastereomeric potential energy surfaces are (1) motif-based methods and (2) full searches using grid-based searching methodologies where all possibilities are considered. In some cases, NOE information from NMR studies have been used. The CPU bottleneck described earlier has kept most researchers from using QM for such computing. One group, however, has demonstrated the use of semiempirical QM with extensive searching to reveal how chiral discrimination takes place and to design a new CSP based on that knowledge.

Däppen et al. [22] developed a rigorous computational protocol involving the idea of "motif-based docking" which they invented. Here, one takes advantage of the known interactions that can take place between two molecules. For example, if one molecule has a pi-acidic ring and the other a pi-basic ring, the modeler would lay one ring over the other to account for the charge transfer complex that will probably occur. Simultaneously, one would associate a hydrogen bond donor on one molecule with a hydrogen bond acceptor on the other and likewise attempt to match electrostatics between molecules. The authors had determined experimentally that the *R* enantiomer of analyte **1** is bound more tightly to a CSP, modeled as **2,** than its antipode.

Because these authors had determined experimental $\Delta\Delta H$ binding energies, they focused their efforts on computing enthalpies rather than free energies. The following steps were carried out to accomplish this. First, they performed a complete conformational analysis of compounds **1** and **2** using MM and/or QM techniques. Second,

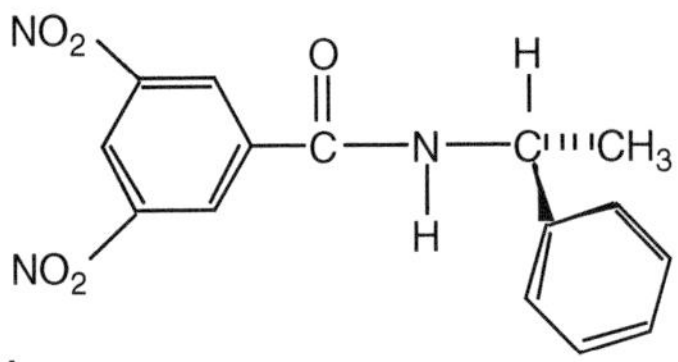

NO_2 O H C N C CH_3 H NO_2 **1**

H_2N H O C NH C $C(CH_3)_3$ CH_3 **2**

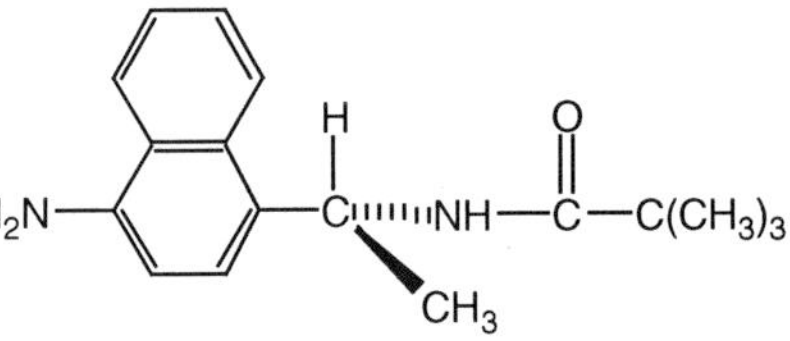

those conformers were used to construct the initial diastereomeric complexes. Last, they developed an *a priori* classification concerning the possible interactions between parts of the two molecules so as to reduce the complexity of the problem. Every low energy conformer of **1** was then combined with every low energy conformer of **2,** in an orientation such that one or more of the assumed binding motifs was realized.

In their problem, three basic binding motifs were considered relevant. These were hydrogen bonds between the selector and the selectand, dipole stacking of the amide moieties, and pi–pi stacking (charge transfer) between the dinitro-pi-acid moiety of **1** with the pi-basic aminonaphthyl ring in **2**. The authors found only three stable conformations of CSP and two stable conformers of analyte, thus giving rise to a relatively small number of initial binary complexes to consider. From this, a small number of possible diastereomeric complexes were generated. Eventually a Boltzmann weighted average energy for the *RR* complex was computed to be −7.35 kcal/mol at 300 K while that of the corresponding *RS* complex was −5.35 kcal/mol, the 2 kcal/mol energy difference corresponding well with the observed value of 1.22 kcal/mol. Based upon this success, the authors then examined various structures to rationalize why one diastereomer is more stable than the other, as well as to gain insights about what is needed to improve the stereoselectivity of binding. Finally, they delineated an extensive design protocol for the creation of improved CSPs.

Other researchers have used QM methods for the analysis of chiral chromatographic separations but those scientists did not use as complete a sampling method as prescribed above. For example, in the computational studies of Zborowski and Zuchowski, semiempirical QM calculations relied on limited sampling of thiobarbituric acids with the interior of a sevenfold symmetric cyclodextrin ring [23]. Although the computed results agreed with observed retention orders, their approach will invariably fail for other analytes because so many restrictions are being placed on the sampling of structures that occur under actual chromatographic conditions. More details about computing stereoselective binding can be found in a review that focuses on cyclodextrins [24]. In a more recent article, Del Rio et al. assessed the validity and applicability of the simple three-point paradigm for the Whelk-O1 CSP, modeled as CSP analog **3**, discriminating between a series of enantiomeric *p*-substituted ethylamine derivatives **4** [25].

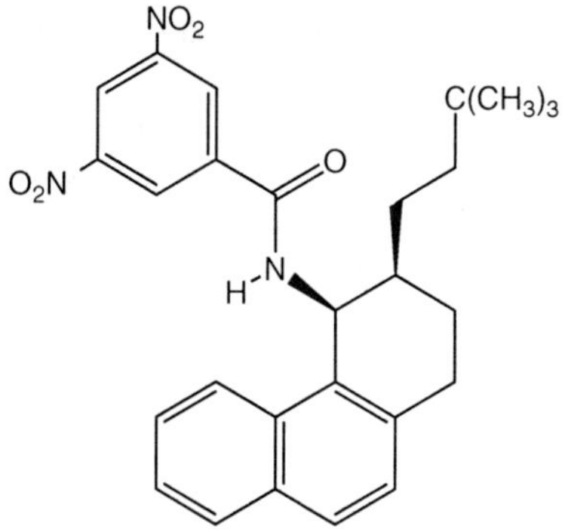

X = H, F, Cl, Br, I

3 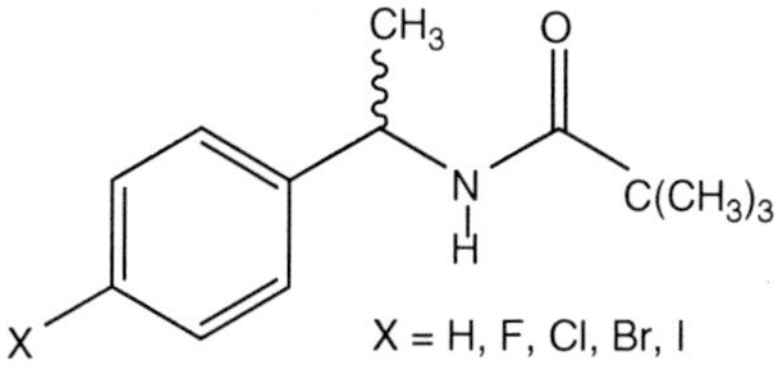 **4**

Using a FF-based method, the authors computed a statistically averaged $\Delta\Delta H$ that agreed with elution orders. Application of density functional theory (DFT), a QM method, with no geometry optimization of the most stable structures gave a qualitative agreement of those elution orders. A natural bond order (NBO) study was carried out by the authors in an attempt to define the factors contributing to the stability of the transient diastereomeric complexes. They were unable to derive a consistent explanation about CH-pi and aromatic stacking effects on stereoselection.

The reader should note that even though quantum mechanics is touted by many researchers as a "first principles" method, the many assumptions and approximations made when implementing these theories is less rigorous than one would hope. Almost all of the methods cannot compute attractions from dispersion (i.e. London forces), the faster semiempirical QM methods reproduce poorly the hydrogen bonds, and so on. While very high levels of quantum theory can avoid these shortcomings, the CPU times needed to attain this are so large as to make QM an intractable solution to problems in chromatography. A non-quantum mechanical method that can compute such interactions accurately, and at a fraction of the CPU time required by QM methods, involves using FFs. These FFs also have their limitations, the most notable being that they need to be parameterized for each unique functional group existing in the molecules being investigated. Nonetheless, FFs have been used extensively to calculate selector–selectand binding for studies of chiral discrimination in chromatography.

5.9 FORCE FIELD (FF)-BASED STUDIES

FFs are used in computational methods like molecular mechanics (simple energy minimization) and for molecular simulations where many different structures (configurations) are collected and averaged using Monte Carlo (MC) or molecular dynamics (MD) methods as examples. Lipkowitz et al. fully automated the search of structures participating in analyte binding to a given CSP [26]. The approach developed for sampling the microstates needed for statistical thermodynamic averaging is illustrated in Fig. 5.7. Here the CSP molecule's center of mass (or any atom) is placed at the origin of a coordinate system. An origin is likewise selected on the analyte and the position of the analyte relative to the CSP is given in polar coordinates.

At each longitude, Θ, and latitude, Φ, a large number of orientations of the analyte with respect to the CSP are generated. All values of Θ and Φ are then evaluated this way by moving the analyte around the CSP molecule in very small increments so that all minima on the intermolecular potential energy surface can be located. In essence, what was done was to roll the analyte molecule over the van der Waals surface of the CSP, sampling configurations for the statistical averaging to be carried out.

The rolling motion involved rigid body structures. It was clear that using only the lowest energy conformations (obtained from a conformational analysis) of each molecule would be inadequate for the problem because high-energy conformers might be involved in both the binding and recognition processes. Hence, it was necessary to account for all the reasonable shapes of CSP (meaning all low energy conformations) as well as of the analyte. This was done by first carrying out a conformational search

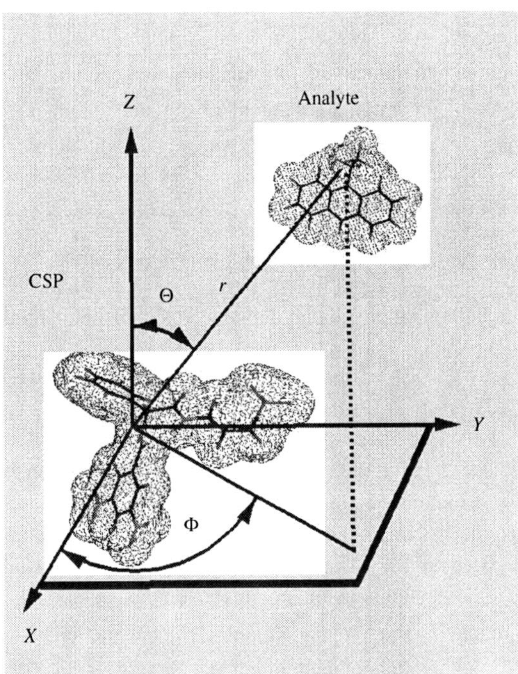

Fig. 5.7. Position of analyte with respect to CSP is given in spherical coordinates, (r, Θ, Φ). Here, r is the distance between arbitrarily selected origins and Θ and Φ describe the latitude and longitude of analyte with respect to CSP. The relative orientations of the two molecules at each latitude and longitude are defined by their Euler angles. Sampling is done as the van der Waals surfaces just touch one another.

followed by a Boltzmann weighting of each conformer of CSP and of analyte. If there existed m conformational states of CSP and n conformational states of analyte, $m \times n$ initial binary complexes were constructed, each of which were rolled around one another as described above. For small Pirkle-like CSPs and correspondingly small analytes, the authors typically created several million configurations for the *RR* complex and an equivalent number for the *SR* complexes, the energies of which were evaluated with a FF. The averaged energies of these complexes were computed using Eq. (13):

$$\overline{E} = \sum_{h=1}^{l} \sum_{i=1}^{m} \left(\frac{e^{-E_{CSP,h}/kT}}{\sum_{h'=1}^{l} e^{-E_{CSP,h'}/kT}} \right) \left(\frac{e^{-E_{A,i}/kT}}{\sum_{i'=1}^{m} e^{-E_{A,i'}/kT}} \right) \sum_{j=1}^{n} g_{hij} \left(\frac{e^{-\varepsilon_{hij}/kT}}{\sum_{j'=1}^{n} e^{-\varepsilon_{hij'}/kT}} \right) \tag{13}$$

The terms in brackets are probabilities. The first term accounts for the probability of finding the CSP in a given conformation, the second term accounts for the probability of finding the analyte in a given conformational state, and the last term is the likelihood of finding the analyte at some position and particular orientation with respect to the CSP. Because all microstates are being sampled this way, their energies and probabilities are known, so that one can also determine entropies as well as enthalpies of binding. Hence the final energies computed are free energy differences that can be compared directly with the experimental values. The methodology developed by these authors was applied to various Type I CSPs for analysis with good results [27–29]. Based upon those successful simulations the authors answered the following scientific questions of relevance to separation science: (1) Where does the analyte tend to bind around the CSP, and, do both analytes bind to the same or to different binding sites? (2) What are the intermolecular forces responsible for complexation? (3) What are the enantiodiscriminating forces, and, are they the same as the forces responsible for complexation? (4) What fragment or fragments of a given CSP are doing most of the work holding the complexes together, and, what fragments are most enantiodiscriminating? Are they the same or different fragments? (5) What conformations of the CSP are most enantiodiscriminating and can this shape be selected and modified to amplify stereoselection? (6) What role does a solvent play in selectivity and how does entropy influence selectivity? These and other questions were posited and answered using molecular simulations.

Other groups at that time were also pursuing systematic searches. Most notable among these is Gasparrini, who, with Misiti and others created and used various CSPs for chiral separations [30]. In particular they developed phases based on chiral, *trans*-1,2-diaminocyclohexane (DACH), **5**.

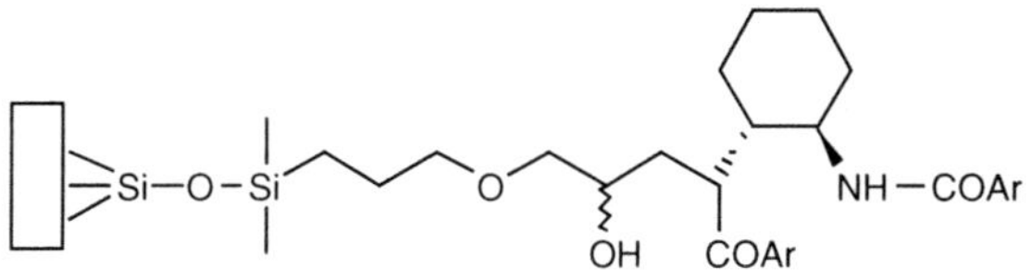

5a Ar = 3,5(NO_2)$_2$$C_6H_3$ **5b** Ar = 3,5(Cl)$_2$$C_6H_3$ **5c** Ar = C_6F_5

Gasparrini's computational approach is called the Global Molecular Interaction Evaluation (Glob-Moline) whose flow chart is outlined in Fig. 5.8.

Like Lipkowitz, Gasparrini considers all the conformations of both the molecules and computes true free energies from this search strategy. It is to be noted that Gasparrini's simplex optimization of rigid guest with rigid host positions is similar to Lipkowitz's method using a very fine grid. Eventually, though, the structures located by Gasparrini

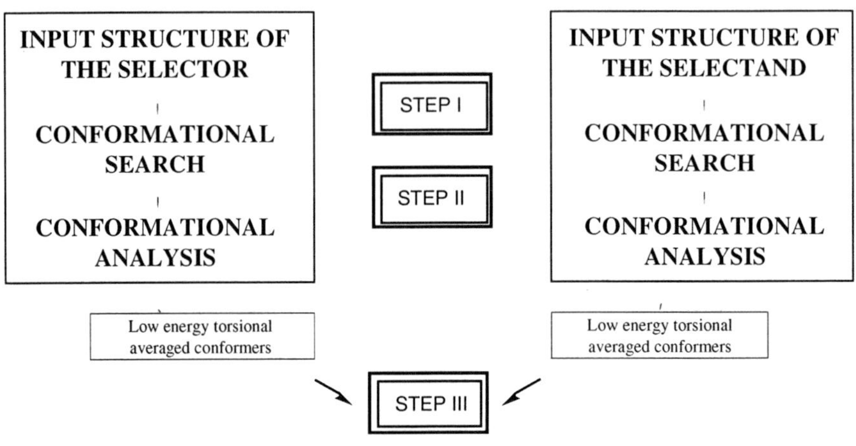

GRID ROUTINE

- Intermolecular energy evaluation (H-bonding, electrostatic and van der Waals interactions) for a regular distribution of points on the selector and selectand surface (According to the Lipkowitz procedure), INTERMOLECULAR POTENTIAL ENERGY SURFACE
- Evaluation of a statistical mechanics interaction energy (ΔH, ΔS, ΔG)
- FORCE FIELD: MMX, MM2-91, AMBER etc.

SELECTING ROUTINE

- Search of energy minima on the Intermolecular Potential Energy Surface.

SIMPLEX OPTIMIZATION PROCEDURE

- A sequential optimization procedure, which considers only nonbonding interactions and which treats the molecules as rigid bodies, was used to locate stable orientations of the selector and selectand: the procedure is applied only to the minima obtained by the selecting routine
- Evaluation of a statistical mechanics averaged steric energy on the stable orientations located by the SIMPLEX OPTIMIZATION PROCEDURE.

FLEXIBLE DOCKING

- The last procedure for docking uses a SIMPLEX routine (for the optimization of the relative orientations) in conjunction with intramolecular minimization in an iterative fashion (Rogers LB. Procedure)
- Two docked pairs having the same or very similar energies for the optimized files were considered to be the same if the average difference in the positions of each atom was within 1.0 Å
- Evaluation of a statistical mechanics averaged interaction energy (ΔH, ΔS, ΔG) on the resulting stable orientations

Fig. 5.8. The Glob-Moline flowchart of Gasparrini.

are fully geometry optimized accounting for induced fit changes of structure. This methodology provides meaningful results when compared to experiment.

Other research groups at that time were also actively developing FF-based sampling techniques. For example, Still and Rogers abandoned their motif-based search strategy and began developing improved grid searching methods [31,32]. The system Still and Rogers focused on was the *R*-phenylglycine phase modeled as **6**.

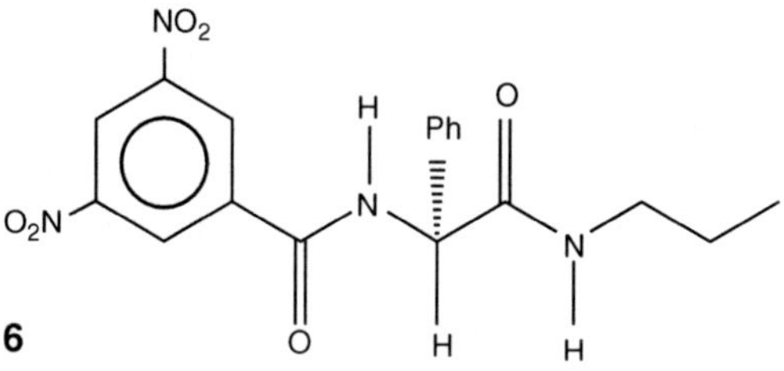

Three aminoethanes containing pi-basic aryl rings whose retention orders and separation factors had been determined experimentally, were examined. In all cases the *S* enantiomer is retained longer on the *R* CSP. First, a conformational analysis using the MM2 FF was carried out. Then, using the most stable structures, the CSP and the analyte were docked using an automated search algorithm. A simplex optimization procedure was used to minimize the non-bonded contact energies of the rigid bodies in the complex. A large number of starting orientations were obtained and a screening process was devised to remove equivalent or near-equivalent structures. Finally, full geometry optimizations were carried out on the docked starting structures that were within 4 kcal/mol of the lowest one. Their method not only predicted the correct elution order for each enantiomeric pair, but also gave reasonable α values.

Another important development came from the work of Aerts who pointed out that using rigid bodies in such searches is valid only in the realm of weak binding; for systems like tartrate ions binding with positively charged diamines, there would be significant induced-fit changes that had to be accounted for by full geometry relaxation [33]. Aerts developed an alternative modeling method for the prediction of enantioselectivity. His approach for sampling structures was to use high-temperature MD trajectories to generate a large number of conformations and orientations of the two molecules in the complex followed by energy minimization of those structures. At 1500 K, explorable only *via* computation not experiment, it is presumed that one can overcome high potential barriers separating conformers, so that new conformations can be generated while simultaneously creating random orientations of the two molecules with respect to one another. A hard wall constraint was imposed by Aerts to prevent the two molecules from flying apart during this process. Both molecules take all possible conformations this way and the results become independent of the starting conformation, thus providing an advantage over other search strategies.

Bartle and his collaborators proposed a computational method that accounts for the matrix to which these brush-like stationary phases are attached [34]. All prior studies neglected the stationary phase matrix and treated the CSP as if it were freely available to the analyte from all directions. Bartle contends that regions of the CSP near the matrix, to which the CSP is grafted are less accessible to the analyte than the other regions. His group did not explicitly model the matrix, but instead included a penalty function for analyte approach. The analyte molecule is moved stepwise around the CSP in a grid-like fashion in an automated and systematic way, responding to that penalty function. At each point around the CSP several analyte orientations are considered. Those initial geometries of the binary complex are allowed to relax by minimizing the complex's energy. The authors carried out several thousand minimizations using this sampling strategy and then computed a Boltzmann-weighted average energy. Comparison of the weighted averages allowed for predicting elution orders and times. The authors also addressed how the separation factor α, is affected by the steric hindrance of the support and found that resolution becomes increasingly difficult as the steric effect increases. Their interpretation is that fewer binding sites become available to discriminate between the enantiomers but other interpretations can be envisaged. While the authors implemented only a rudimentary penalty function, their results are good and tend to support their methodology.

Other articles where exhaustive search strategies were used to assess enantiomer binding to non-polymeric CSPs include the work of Schefzick et al., who used stochastic MD in their searches [35], Gasparrini et al., who used MC searching methods [36] and Lee et al. who also implemented MC methods [37]. A significant number of additional publications appeared in the literature using either FF or QM energy calculations but they are not mentioned here because they rely on manual docking or they use a very limited number of structures to compute energies of interaction.

The chromatographic systems studied by FF methods above are relatively small and were usually synthesized *de novo* by bench chemists. We now consider polymeric systems, often taken from the chiral pool and then modified somewhat to optimize separations, beginning with linear polymers and then considering cyclic polymers. Again we focus on publications involving substantive sampling methodologies and omit here those reports based upon simple docking followed by energy minimization.

The most detailed and most comprehensive computational study of chiral analytes binding to CSPs that are linear polymers comes from Yashima et al. Their molecular modeling on such polymers involved evaluating binding energies of ±*trans*-stilbene oxide, a diphenyl epoxide, and ±*trans*-1,2-diphenylcyclopropane (the same system lacking a hydrogen bond binding site) binding to a cellulose triscarbamate polymer (CTPC) [38]. This particular polymer was selected for study computationally because it resolves the aforementioned enantiomers with amazing efficiency, but more importantly because this polymer is amenable to NMR spectral analysis using polar solvents so that structural comparisons can be made with the computational results.

The polymer was constructed by fully optimizing a triscarbamate monomer containing methoxy groups at the 1- and 4-positions. To do this they used a MM method and the structure that was obtained was considered to be a suitable "monomer" for further elaboration. That structure was polymerized, computationally, by concatenating those monomers to form an octamer with a left-handed, threefold (3/2) helix similar to a CTPC structure previously proposed in the literature based on X-ray fiber diffraction studies. That octamer was then fully geometry optimized and MD was carried out. From the MD trajectory file, a set of new structures that were energy

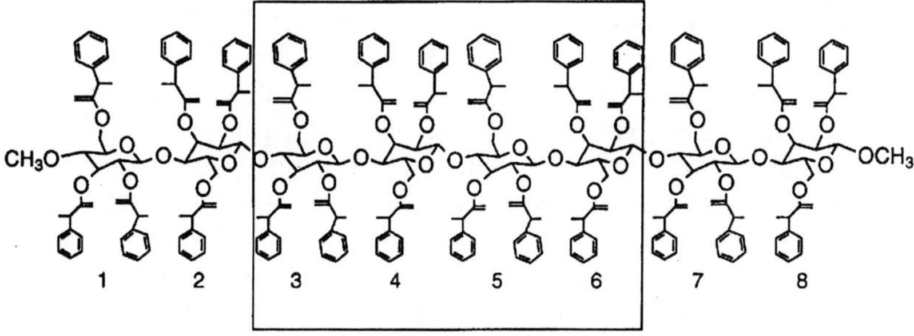

Fig. 5.9. Octameric polymer of cellulose tris-phenylcarbamate used for modeling by Yashima et al. [38]. All hydrogen atoms and the N and O atoms on the carbamate moieties are omitted for clarity.

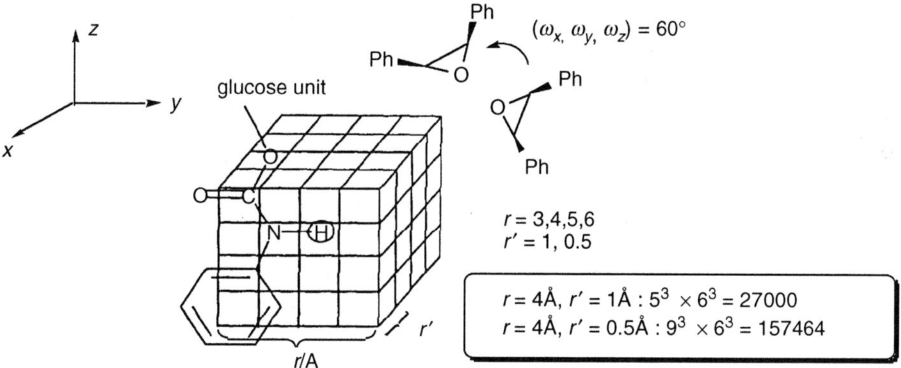

Fig. 5.10. Cubic sampling box of dimension r and mesh size r' in Å. Analyte is rotated in 60° increments about x, y and z axes at each grid point. Energies are computed with the CHARMM force field (FF).

minimized was obtained, but no lower energy structures could be found. A schematic diagram depicting the polymer is shown in Fig. 5.9. The simulations were limited to the four central monomers within the box to avoid "end effects."

Based upon previous experimental work, it was hypothesized that the most important adsorbing site for the stilbene oxide involves the N–H hydrogens of the carbamate groups. Accordingly, the authors set up sampling boxes centered on the carbamate's N–H hydrogen. A schematic of the sampling grid is depicted in Fig. 5.10. Note that a sampling box was set up for all three amides on each monomer, and that only monomers 3–6 were sampled to avoid the influence of the end groups (experimentally these polymers have a degree of polymerization ~100). Also note that different grid sizes were used (r = 3–6 Å) but the grid mesh was initially fixed at 1 Å.

At each grid point the (R,R)-(+)-epoxide or the (S,S)-(−)-epoxide was rotated in 60°-intervals around the x, y and z axes, individually. The calculations involved a rigid analyte interacting with a rigid CSP. Using this sampling strategy, the authors were able to tabulate the minimum interaction energy between analyte and each carbamate moiety on each monomer. Those energies varied substantially, depending on the glucose as well as the position of the carbamate within a particular monomer. Next, a finer grid mesh of 0.5 Å was used and the lowest interaction energies were thus determined. Both the grid meshes provided results in agreement with retention orders (the *R,R* enantiomer elutes before the *S,S* enantiomer) but the computed differential interaction energies were substantially overestimated. These energies are not averaged in any way and do not correspond to free energies. Rather, they are presumed to be the global minima of the diastereomeric complexes. Nonetheless, the authors were able to find the most probable analyte-binding site, that, for the epoxide, is in a chiral groove existing along the main chain of the polymer backbone. Moreover, the authors were able to discern the most important types of interactions including hydrogen bonding to the oxirane ring and pi-stacking of carbamates with the analyte phenyl groups. For the diphenylcyclopropane where the hydrogen bonding is absent, these researchers found no enantioselective preference from their calculations, and none was observed experimentally.

A more detailed computational investigation concerning the chiral discrimination mechanism of phenyl carbamate derivatives of cellulose was performed by Yamamoto et al. [39]. Two commercially available CSPs were studied: cellulose tris(phenylcarbamate) (CTPC) and cellulose tris(3,5,-dimethylphenylcarbamate) (CDMPC) each of which were interacting with the enantiomers of *trans*-stilbene oxide and with benzoin. The oxirane is separable on the CTPC column having the *S,S* antipode longer retained but on the CDMPC column the *R,R* isomer is more strongly affixed to the column (reversal in enantioselection). The benzoin is separable only on the CDMPC column with the *R* isomer longer retained.

The authors constructed computer models of both CSPs using a somewhat different strategy (albeit very similar) to that described above. A 9 mer was generated and molecular associations with the analyte molecules were directed toward the interior units of the oligomer chain to avoid end effects as before. Both polymers are similar in that they are left-handed helices with polar carbamate groups inside the polymer chain and having hydrophobic groups outside of the chain. Importantly, though, the aryl rings of CDMPC are arranged parallel to the helical axis but differently than in CTPC due to the steric hindrance of the methyl groups. It is speculated that this may be responsible for the reversal in enantioselectivity when the two CSPs are compared.

Three computational strategies were used to dock analytes around the CSPs. The first is a deterministic grid search where analyte molecules were placed around the NH and C=O functionality as described above, while the other three methods involved analyte placements based on stochastic moves (MC) in a polar coordinate system. Most of the structures generated in this latter way resulted in the analyte molecules being on the exterior of the polymer surface and not allowing for close contact with the purported NH and C=O binding sites on the interior of the polymer (structures too close to the rigid polymer had very high molecular mechanics energies and were automatically excluded from further consideration). To overcome this problem and to generate diastereomeric complexes where the enantiomers could be generated inside the polymer surface, the authors carried out a series of van der Waals slow-growth calculations where the van der Waals radii of analyte molecules were first reduced in size so that they could approach the interior of the polymer surface and then the radii were alchemically grown back to their standard sizes, incrementally, while the whole system was geometry optimized, beginning from those internally docked positions.

The results from these calculations were in agreement with the experiment. Tabulations of intermolecular energy components, energies of interactions with various sites along the polymer chain, and assessments of lowest energy structures as well as averaged structures were presented and discussed. The results indicate that the polar carbamate residues of cellulose derivatives are important adsorbing sites for polar racemates and may play a role in chiral discrimination. It was also speculated that under reverse phase conditions using water eluents, hydrophobic cavities between polymer chains might be important for chiral discrimination. Other examples of FF-based calculations of analytes binding to linear polymers used as CSPs in chromatography exist. A mechanistic study of an amylose tris(3,5-dimethylphenyl)carbamate CSP involved MD simulations and is cited here for comparison with the cellulose systems described above [40].

In comparison to other kinds of CSPs that are used heavily by bench chemists, there is a paucity of computational work involving good sampling methods on linear polymers. Cyclic polymers, in contrast, especially cyclodextrins (CDs) in their native and derivatized states, have been studied extensively with FFs to rationalize both where and how enantioselection takes place. The literature associated with host–guest binding and chiral recognition in CDs is enormous and well beyond the scope of this chapter. Accordingly, we provide here only FF-based studies using advanced sampling methods on host–guest complexes that have been resolved with CD stationary phases.

The first such example was done by Lipkowitz et al. [41] who provided NMR data and modeling results to understand the separation of tryptophan by α-CD. A rigid body approach described earlier was not used for this CSP because cyclodextrins are flexible and are prone to induced-fit structural changes. Accordingly, MD calculations, which account well for this flexibility, were carried out using the same explicit solvents as in the chromatographic and NMR studies. Using the CHARMM FF for MD, their simulations reproduced the correct retention order and separation factor from the chromatography experiments, and their simulation results also agreed with both intermolecular and intramolecular NOE observations from their NMR experiments.

Because intermolecular hydrogen bonding was considered to be important, the number and the kinds of intermolecular H-bonds between CSP and analyte were evaluated. Not only does the more retained *R* enantiomer form a greater number of H-bonds than does the *S* isomer, but these H-bonds were found to be simultaneous, multiple-contact H-bonds between CSP and analyte. A schematic representation summarizing these findings is depicted in Fig. 5.11.

Three key features emerge from this representation. First, both complexes are highly localized on the interior of the CD with the *R* analyte binding to one side of the macrocycle and the *S* analyte to the other. Second, the *R* enantiomer forms nearly twice as many intermolecular H-bonds as does *S* and, as pointed out above, they are of the multiple-contact type. Third, the hydrogen bonding occurs primarily from tryptophan's carboxylate and indole N–H but not the ammonium group. Based on this, the authors confirmed an earlier recognition model but pointed out that a high degree of localization is tantamount to a tight fit in the CD cavity. These studies were followed by an assessment of mandelic acid binding to β-cyclodextrin using chromatography, NMR spectroscopy and MD simulations to rationalize the observed enantioselection [42].

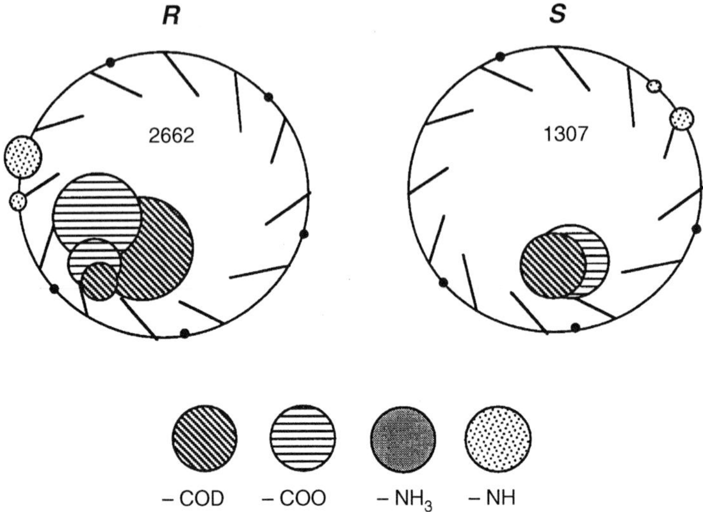

Fig. 5.11 Schematic depiction of intermolecular hydrogen bonds of (*R*)- and (*S*)-tryptophan with α-cyclodextrin. The large circle represents the CSP. The small black dots on that circle are the acetal linker oxygens, and the lines attached to the circle represent the unidirectional C2 and C3 hydroxyl groups. The cross-hatching indicates the atoms or groups of atoms on the tryptophan that are forming the intermolecular hydrogen bonds. The size of the cross-hatched circle corresponds to the number of hydrogen bonds formed during the simulation. The centers of the cross-hatched circles are placed at the mean positions of the hydrogen bond contacts. There exist 2662 and 1307 hydrogen bonding interactions for *R*- and *S*-analyte, respectively.

A key point associated with these studies is that the authors were careful to use the same solvent conditions and temperatures in the modeling studies as was used in the NMR and chromatography experiments. We point out here that not all groups have been successful in their MD simulations, especially Dodziuk and her collaborators who have questioned repeatedly the validity of such calculations [43].

Molecular modeling of gas chromatographic separations with CDs was first published by Köhler et al. [44]. Experimentally, his group found the *S* enantiomer of methyl-2-chloropropionate to be more retained on Lipodex D ([heptakis(3-*O*-acetyl-2,6-di-*O*-pentyl)-β-CD] coated on a capillary column) at 333 K. To discern the structural features of the transient complexes with MD simulations, analytes were placed into the interior of the derivatized β-CD cavity in two orientations, "up" or "down." The authors found the "down" orientation led to immediate expulsion of guest from the host cavity upon warming and equilibration of their selector–selectand complex. The *R*-"up" orientation migrated outside of the cavity but remained near the hydrophobic side-chains and the *S*-"up" complex was found to be most stable. These results are consistent with the GC results and also with NOE intensities from NMR studies that further showed the methyl ester to be near the C3 groups of the CSP. In a follow-up article the authors carried out longer simulations [45]. They were able to deduce the time-averaged orientation of the analyte in the CSP host cavity, evaluate detailed structural features of the guest–host complex, and to describe important intermolecular distances between the enantiomeric analytes and chiral cavity.

Koen de Vries et al. also used MD for rationalizing qualitative gas chromatographic trends [46]. They evaluated experimentally the thermodynamic parameters (ΔG, $\Delta\Delta G$, ΔH, $\Delta\Delta H$, etc.) for complexes of six analytes on a variety of derivatized CD columns. Their interpretation of the computational results is that one enantiomer fits the CD cavity better than the other, thereby resulting in a larger interaction energy and greater loss of mobility. Kobor et al. [47] also used molecular modeling to examine how polar and nonpolar analytes bind to derivatized CDs used as selectors in gas chromatography. Their goals were to systematically explore potential GC-compatible chiral selectors that might be more universal with regard to their application as stationary phases. A MC docking algorithm was used to generate guest–host complexes that were then geometry optimized by molecular mechanics. Based on their modeling studies and experimental observations the authors made several conclusions about how the analytes fit into the macrocyclic cavity as well as the influence of CSP rigidity on chiral discrimination.

Black et al. [48] have also used molecular modeling tools to predict the retention order of polar and nonpolar analytes that are resolvable on cyclodextrin stationary phases in gas chromatography. The cyclodextrins considered include permethylated α-cyclodextrin and native β-cyclodextrin. The analytes studied include polar and nonpolar analytes. The authors extracted the low-energy structures from grid searches and fully optimized the entire complex of each. A large number of degenerate structures were found that were subsequently deleted, based on a criterion of similarity. Boltzmann-weighted energies from this sampling method gave good results; four of the five examples were correctly predicted and in some instances the separation factors had the same trends as the experiments. The structural changes of the cyclodextrin cavities were found to be small, yet large enough to better accommodate the guest molecules and to change the statistically weighted energies to agree with the experiment. The authors conclude that the induced-fit behavior of cyclodextrin is important. MD simulations were used by Nie et al. [49] to evaluate the binding site of phenethylamine binding to a derivatized CD under gas chromatographic conditions and both MC and MD strategies were used by Kim et al. [50] to understand the chiral recognition of propranolol by a β-cyclodextrin in condensed phase conditions.

The issue of chiral discrimination by CDs in the gas phase *versus* condensed, aqueous phase was controversial; most of the NMR studies done in aqueous conditions showed that analytes bind to the interior of the CD cavity due to hydrophobic forces. In gas chromatography, hydrophobic forces do not exist and questions concerning binding and discrimination on the exterior of the CD were raised by a number of separation scientists, an issue that was subsequently addressed theoretically by Lipkowitz. The computational protocol they developed for this project [51] takes advantage of the good local sampling ability of MD together with global MC-like structural moves where the analyte is forced to visit all possible binding sites of the CSP. Thus, multiple trajectories were started from different beginning points and the data from all trajectories was averaged statistically. Each trajectory was itself quite lengthy (10 ns) and typically five trajectories were used (total of 50 ns) for the R and also for the S analyte. A spherical reflective wall around the cyclodextrin was used to ensure multiple collisions would occur between the selector and the selectand. That way, when the analyte detaches itself from the cyclodextrin, it would hit the wall and be shoved back gently to re-encounter the cyclodextrin that would be in a different conformation (as would be the analyte) for each new encounter. The correct retention orders and energetics of chiral discrimination in all examples were reproduced, thus validating the method. To visualize the preferred gas-phase binding sites, the authors developed plots similar to that shown in Fig. 5.12.

This figure shows the distribution of the center of mass of 2-methylbutanoic acid, one of several polar and nonpolar analytes studied, relative to permethyl-β-cyclodextrin over a 50 ns simulation time period. It is clear that all regions inside and around the macrocycle have been sampled and that the interior of the cavity has a higher probability of binding than the exterior. For all analytes examined, the authors found that interior binding is preferred because dispersion interactions (which are treated well with FFs) are maximized there. Those authors then generated a method for determining which region

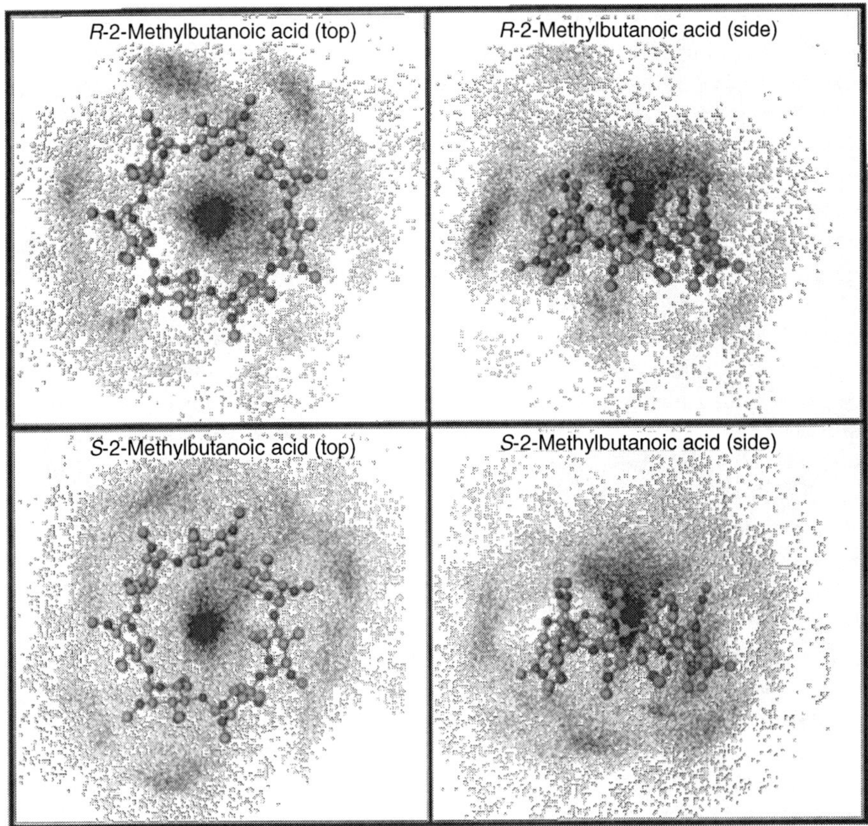

Fig. 5.12. A history of the location of analytes binding to a permethylated cyclodextrin CSP from a simulation. Each dot corresponds to the analyte's center of mass position sampled along the MD trajectory. Preferential binding to the interior of the CSP is evident for both enantiomeric analytes.

inside or around the host is most enantiodiscriminating for a given analyte [52] and eventually posited a hypothesis about maximizing chiral discrimination in chromatography [53]. There exist a large number of computational studies employing FFs to model chiral discrimination in cyclodextrins [24].

Other macrocyclic cavity-containing, polymeric CSPs have been used for chiral separations in addition to cyclodextrins and they too have been modeled with FFs. Several publications where MD and/or MC simulations were used include studies of crown ethers [54–57], and cage-like receptors [58,59].

5.10 SUMMARY AND PROSPECTUS

In this chapter, we described different types of modeling techniques that have been applied to the analysis of chiral discrimination in chromatography. All of the methods described involve an atomistic view of the molecules involved but the modeling techniques themselves are very different from one another. One type of modeling involves the correlation of molecular descriptors with chromatographic retention times. These types of models define quantitative structure–property relationships (QSPR) that, in the area of chromatography, are also referred to as quantitative structure–enantioselectivity relationships (QSERR). Molecular descriptors and graph theoretical descriptors can be used effectively to derive such relationships and when done properly, one can derive mathematical models that are predictive and allow the modeler to be involved in computer-aided molecular design (CAMD). A related type of modeling using comparative molecular field analysis (CoMFA) was introduced and in a tutorial manner shown how it can be used for understanding chiral recognition and for CAMD.

The other type of atomistic modeling involves the direct computation of energies between selector and selectand. Quantum-based and FF-based methods were described. The majority of studies in this area involve application of FFs because they represent accurately the intra- and intermolecular forces responsible for binding and discrimination and because they are typically three orders of magnitude faster than QM-based methods. Accordingly, one can use the FFs to sample many configurations for statistical averaging. Only FF-based studies that use grid searches, or MC or MD techniques were described in this chapter. Omitted are the many examples of manual docking followed by simple energy minimizations because those protocols tend to give capricious results at best and misleading results at worst.

The future of computational chemistry in the area of separation science is very promising, if, and only if, rigorous application of existing theories are implemented by the modeler. In that regard, the QSERR models that are derived must be validated on an external test set, separate from the training set of compounds used to derive those models. In the area of energy calculations it is important to include as much of the system being modeled in the computer experiments. Solvents, immobilized surfaces, temperatures and the like should be included in these simulations, and, MC or MD sampling protocols should be used to derive statistically meaningful averages. Computing hardware and storage devices are at a point of development where all of this is now possible. The equipment exists but the cognition of how to use it may not; this chapter was written to inform the reader about different computational methods, pitfalls to avoid and to review the literature to rectify that shortcoming. The author of this chapter looks forward to reading articles describing such work in the future.

5.11 ACKNOWLEDGMENTS

Results from the author of this chapter described here were funded primarily by grants from the Petroleum Research Fund administered by the American Chemical Society and from the National Science Foundation. The many years of support by these organizations is greatly appreciated as are the interactions with many students, postdoctoral associates and visiting collaborators with whom he had the pleasure to meet.

REFERENCES

1 E. Eliel, Infelicitous stereochemical nomenclature. Chirality, 9 (1997) 428–430.

2 G. Subramanian (Ed.), Chiral Separation Techniques: A Practical Approach, 2nd. Ed. Wiley, New York, 2000 and other books on the topic cited therein.

3 M. Randic, Topological indices, in: P.v.R. Schleyer, Editor-in-chief, Encyclopedia of Computational Chemistry, Wiley, Chichester, 1998, pp. 3018–3032.

4 L.H. Hall and L.B. Kier, The molecular connectivity chi indexes and kappa shape indexes in structure-property modeling, in: K.B. Lipkowitz and D.B. Boyd (Eds.), Reviews in Computational Chemistry, Vol. 2, VCH, Weinheim, 1991, pp. 367–422.

5 H. Kubinyi, Quantitative structure-activity relationships in drug design, in: P.v.R. Schleyer, Editor-in-chief, Encyclopedia of Computational Chemistry, Wiley, Chichester, 1998, pp. 2309–2320.

6 C. Altomare, A. Carotti, S. Cellamare, F. Fanelli, F. Gasparrini, C. Villani, P.-A. Carrupt and B. Testa, Enantiomeric resolution of sulfoxides on a DACH-DNB chiral stationary phase: A quantitative structure-enantioselective retention relationship (QSERR) study. Chirality, 5 (1993) 527–537.

7 C.A. Montanari, Q.B. Cass, M.E. Tiritan and A.L. Soares de Souza, A QSERR study on the enantioselective separation of enantiomeric sulphoxides. Anal. Chim. Acta, 419 (2000) 93–100.

8 D.J. Livingstone and D.W. Salt, Variable selection – spoilt for choice? in: K.B. Lipkowitz, R.M. Larter and T.R. Cundari (Eds.), Reviews in Computational Chemistry, Vol. 21, Wiley-VCH, New York, 2005, pp. 287–348.

9 A. Carotti, C. Altomare, S. Cellamare, A.M. Monforte, G. Bettoni, F. Loiodice, N. Tangari and V. Tortorella, LFER and CoMFA studies on optical resolution of α-alkyl α-aryloxy acetic acid methyl esters on R,R-DACH-DNB chiral stationary phases. J. Comput.-Aided Mol. Des., 9 (1995) 131–138.

10 C. Altomare, S. Cellamare, A. Carotti, M.L. Barreca, A. Chimirri, A.-M. Monforte, F. Gasparrini, C. Villani, M. Cirilli and F. Mazza, Substituent effects on the enantioselective retention of anti-HIV 5-aryl-Δ^2-1,2,4-oxadiazolines on R,R-DACH-DNB chiral stationary phase. Chirality, 8 (1996) 556–566.

11 R. Kaliszan, T.A.G. Noctor and I.W. Wainer, Quantitative structure-enantioselective retention relationships for the chromatography of 1,4-benzodazepines on a Human Serum Albumin based HPLC chiral stationary phase: an approach to the computational prediction of retention and enantioselectivity. Chromatographia, 33 (1992) 546–550.

12 R. Kaliszan, A. Kaliszan, T.A.G. Noctor, W.P. Purcell and I.W. Wainer, Mechanism of the retention of benzodiazepines in affinity, reversed-phase and adsorption high-performance liquid chromatography in view of quantitative structure-retention relationships. J. Chromatogr., 609 (1992) 69–81.

13 T.D. Booth and I.W. Wainer, Investigation of the enantioselective separations of α-alkylarylcarboxylic acids on an amylose tris(3,5-dimethylphenylcarbamate) chiral stationary phase using quantitative structure-enantioselective retention relationships. Identification of a conformationally driven chiral recognition mechanism. J. Chromatogr. A, 737 (1996) 157–169.

14 E. Calleri, E. DeLorenzi, D. Siluk, M. Markuszewski, R. Kaliszan and G. Massolini, Riboflavin binding protein – chiral stationary phase: Investigation of retention mechanism. Chromatographia, 55 (2002) 651–658.

15 T. Suzuki, S. Timofei, B.E. Iuoras, G. Uray, P. Verdino and W.M.F. Fabian, Quantitative-structure-enantioselective retention relationships for chromatographic separation of alkylarylcarbinols on Pirkle type chiral stationary phases. J. Chromatogr. A, 922 (2001) 13–23.

16 H. Kubinyi, Comparative molecular field analysis (CoMFA), in: P.v.R. Schleyer, Editor in chief, Encyclopedia of Computational Chemistry, Wiley, Chichester, 1998, pp. 448–460.

17 S. Schefzick, M. Lämmerhofer, W. Lindner, K.B. Lipkowitz and M. Jalaie, Comparative molecular field analysis of quinine derivatives used as chiral selectors in liquid chromatography: 3D QSAR for the purposes of molecular design of chiral stationary phases. Chirality, 12 (2000) 742–750.

18 S. Topiol, M. Sabio, J. Moroz and W.B. Walton, Computational studies of the interactions of chiral molecules: complexes of methyl N-(2-naphthyl)alaninate with N-(3,5-dinitrobenzoyl)leucine n-propylamide as a model for chiral stationary-phase interactions. J. Am. Chem. Soc., 110 (1988) 8367–8376.

19 S. Topiol and M. Sabio, Computational chemical studies of chiral stationary phase models. Complexes of methyl N-(2-naphthyl)alaninate with N-(3,5-dinitrobenzoyl)leucine n-propylamide. J. Chromatogr., 461 (1989) 129–137.

20 M. Sabio and S. Topiol, Computational chemical studies of chiral stationary-phase models: the nature of the pi interaction in complexes of methyl N-(2-naphthyl)alaninate with N-(3,5-dinitrobenzoyl)leucine n-propylamide. International J. Quantum Chem., XXXVI (1989) 313–317.

21 M. Sabio and S. Topiol, A molecular dynamics investigation of chiral discrimination complexes as chiral stationary-phase models: methyl N-(2-naphthyl)alaninate with N-(3,5-dinitrobenzoyl)leucine n-propylamide. Chirality, 3 (1991) 56–66.

22 R. Däppen, H.R. Karfunkel and F.J.J. Leusen, Computational chemistry applied to the design of chiral stationary phases for enantiomeric separation. J. Comput. Chem., 11 (1990) 181–193.

23 K. Zborowski and G. Zuchowski, Enantioselective chromatography of alkyl derivatives of 5-ethyl-5-phenyl-2-thiobarbituric acid studied by semiempirical AM1 method. Chirality, 14 (2002) 632–637.

24 K.B. Lipkowitz, Applications of computational chemistry to the study of cyclodextrins. Chem. Rev., 98 (1998) 1829–1873.

25 A. Del Rio, J.M. Hayes, M. Stein, P. Piras and C. Roussel, Theoretical reassessment of Whelk-O1 as an enantioselective receptor for 1-(4-halogeno-phenyl)-1-ethylamine derivatives. Chirality, 16 (2004) S1–S11.

26 K.B. Lipkowitz, D.A. Demeter, R. Zegarra, R. Larter and T. Darden, A protocol for determining enantioselective binding of chiral analytes on chiral chromatographic surfaces. J. Am. Chem. Soc., 110 (1988) 3446–3452.

27 K.B. Lipkowitz, S. Antell and B. Baker, Enantiodifferentiation in Rogers' BOC-D-Val chiral stationary phase. J. Org. Chem., 54 (1989) 5449–5453.

28 K.B. Lipkowitz, B. Baker and R. Zegarra, Theoretical studies in molecular recognition: Enantioselectivity in chiral chromatography. J. Comput. Chem., 10 (1989) 718–732.

29 K.B. Lipkowitz and B. Baker, A computational analysis of chiral recognition in Pirkle phases. Anal. Chem., 62 (1990) 770–774.

30 S. Alcaro, F. Gasparrini, O. Incani, S. Mecucci, D. Misiti, M. Pierini and C. Villani, A "quasi-flexible" automatic docking processing for studying stereoselective recognition mechanisms. Part I. Protocol validation. J. Comput. Chem., 21 (2000) 515–530.

31 M.G. Still and L.B. Rogers, Effects of computational variations for determining binding energies of diastereomeric complexes when using MM2. J. Comput. Chem., 11 (1990) 242–248.

32 M.G. Still and L.B. Rogers, Computational studies of the chiral separations of three N-acyl-1-aryl-1-aminoethanes on an (R)-N-dinitrobenzoylphenylglycine stationary phase. Talanta, 37 (1990) 599–612.

33 J. Aerts, An improved molecular modeling method for the prediction of enantioselectivity. J. Comput. Chem., 16 (1995) 914–922.

34 A.M. Edge, D.M. Heaton, K.D. Bartle, A.A. Clifford and P. Myers, A computational study of the chromatographic separation of racemic mixtures. Chromatographia, 41 (1995) 161–166.

35 S. Schefzick, W. Lindner, K.B. Lipkowitz and M. Jalaie, Enantiodiscrimination by a quinine-based chiral stationary phase: A computational study. Chirality, 12 (2000) 7–15.

36 F. Gasparrini, D. Misiti, M. Pierini and C. Villani, A chiral A_2B_2 macrocyclic minireceptor with extreme enantioselectivity. Org. Lett., 4 (2002) 3993–3996.

37 O.-S. Lee, Y.H. Jang, Y.G. Cho, M.H. Hyun, H.-J. Kim and D.S. Chung, Theoretical and experimental studies of chiral recognition in charged Pirkle phases. Chem. Lett., (2001) 232–233.

38 E. Yashima, M. Yamada, Y. Kaida and Y. Okamoto, Computational studies on chiral discrimination mechanism of cellulose triphenylcarbamate. J. Chromatogr. A, 694 (1995) 347–354.

39 C. Yamamoto, E. Yashima and Y. Okamoto, Computational studies on chiral discrimination mechanism of phenylcarbamate derivatives of cellulose. Bull. Chem. Soc. Jpn., 72 (1999) 1815–1825.

40 Y. Bereznitski, R. LoBrutto, N. Variankaval, R. Thompson, K. Thompson, P. Sajonz, L.S. Crocker, J. Kowal, D. Cai, M. Journet, T. Wang, J. Wyvratt and N. Grinberg, Mechanistic aspects of chiral discrimination on an amylose tris(3,5-dimethylphenyl)carbamate. Enantiomer, 7 (2002) 305–315.

41 K.B. Lipkowitz, S. Raghothama and J.-A. Yang, Enantioselective binding of tryptophan by α-cyclodextrin. J. Am. Chem. Soc., 114 (1992) 1554–1562.

42 K.B. Lipkowitz and C.M. Stoehr, Detailed experimental and theoretical analysis of chiral discrimination: Enantioselective binding of R/S methyl mandelate by β-cyclodextrin. Chirality, 8 (1996) 341–350.

43 H. Dodziuk, O. Lukin and K.S. Nowinski, Modelling of molecular and chiral recognition by cyclodextrins. Is it reliable? Part 2. Molecular dynamics calculations in vacuum pertaining to the selective complexation of decalins by β-cyclodextrin. Pol. J. Chem., 74 (2000) 997–1001.

44 J.E.H. Köhler, M. Hohla, M. Richters and W.A. König, Cyclodextrin derivatives as chiral selectors- Investigation of the interaction of (R,S)-methyl-2-chloropropionate by enantioselective gas chromatography, NMR spectroscopy, and molecular dynamics simulation, Angew. Chem. Int. Ed. Engl., 31 (1992) 319–320.

45 J.E.H. Köhler, M. Hohla, M. Richters and W.A. König, A molecular dynamics simulation of the complex formation between methyl (R)/(S)-2-chlorpropionate and heptakis(3-O-acetyl-2,6-di-O-pentyl)-β-cyclodextrin. Chem. Ber., 127 (1994) 119–126.

46 N. Koen de Vries, B. Coussens and R.J. Meier, The separation of enantiomers on modified cyclodextrin columns: Measurement and molecular modeling. J. High Resolut. Chromatogr., 15 (1992) 499–504.

47 F. Kobor, K. Angermund and G. Schomburg, Molecular modeling experiments on chiral recognition in GC with specially derivatized cyclodextrins as selectors. J. High Resolut. Chromatogr., 16 (1993) 299–311.

48 D.R. Black, C.G. Parker, S.S. Zimmerman and M.L. Lee, Enantioselective binding of α-pinene and some cyclohexanetriol derivatives by cyclodextrin hosts: A molecular modeling study. J. Comput. Chem., 17 (1996) 931–939.

49 M.Y. Nie, L.M. Zhou, Q.H. Wang and D.Q. Zhu, Molecular dynamics study for chiral discrimination of α-phenylethylamine by modified cyclodextrin in gas chromatography. Chinese Chem. Lett., 11 (2000) 347–350.

50 H. Kim, K. Jeong, S. Lee and S. Jung, Molecular modeling of chiral recognition of propranolol enantiomers by a β-cyclodextrin. Bull. Korean Chem. Soc., 24 (2003) 95–98.

51 K.B. Lipkowitz, G. Pearl, B. Coner and M.A. Peterson, Explanation of where and how enantioselective binding takes place on permethylated β-cyclodextrin, a chiral stationary phase used in gas chromatography. J. Am. Chem. Soc., 119 (1997) 600–610.

52 K.B. Lipkowitz, B. Coner and M.A. Peterson, Locating regions of maximum chiral discrimination: A computational study of enantioselection on a popular chiral stationary phase used in chromatography. J. Am. Chem. Soc., 119 (1997) 11269–11276.

53 K.B. Lipkowitz, B. Coner, M.A. Peterson, A. Morreale and J. Shackelford, The principle of maximum chiral discrimination: Chiral recognition in permethyl-β-cyclodextrin. J. Org. Chem., 63 (1998) 732–745.

54 J.S. Bradshaw, P. Huszthy, C.W. McDaniel, C.Y. Zhu, N.K. Dally, R.M. Izatt and S. Lifson, Enantiomeric recognition of organic ammonium salts by chiral dialkyl-, dialkenyl-, and tetramethyl-substituted pyridino-18-crown-6 and tetramethyl-substituted bis-pyridino-18-crown-6 ligands: Comparison of temperature-dependent ^1H NMR and empirical force field techniques. J. Org. Chem., 55 (1990) 3129–3137.

55 S. Hwang, O.-S. Lee and D.S. Chung, Free energy perturbation studies of pyridino-18-crown-6 ethers. Chem. Lett., 5 (2000) 1002–1003.

56 O.-S. Lee, S. Hwang and D.S. Chung, Free energy perturbation and molecular dynamics simulation studies on the enantiomeric discrimination of amines by dimethyldiketopyridino-18-crown-6. Supramolec. Chem., 12 (2000) 255–272.

57 E. Bang, J.-W. Jung, W. Lee, D.W. Lee and W. Lee, Chiral recognition of (18-crown-6)-tetracarboxylic acid as a chiral selector determined by NMR spectroscopy. J. Royal Chem. Soc., Perkin Trans., 2 (2001) 1685–1692.

58 R.J. Pieters, J. Cuntzee, M. Bonnet and F. Diederich, Enantioselective recognition with C_3-symmetric cage-like receptors in solution and on a stationary phase. Journal of the Royal Chemical Society, Perkin Transactions, 2 (1997) 1891–1900.

59 F. Gasparrini, C. Villani, M. Pierini, S. Alcaro, S. Mecucci and D. Misiti, Enantioselective Recognition Mechanism of a New C_3-macrotricyclic Receptor by Molecular Modeling. 8th International Symposium on Chiral Discrimination, Edinburgh, Scotland, July 2, 1996.

Note: This chapter was completed and submitted in July, 2004; no publications since then have been included here.

© 2006 Elsevier B.V. All rights reserved.
Chiral Analysis
K.W. Busch and M.A. Busch, Eds.

CHAPTER 6

Gas chromatographic enantioseparation of chiral pollutants—techniques and results

Walter Vetter[1] and Kai Bester[2]

[1]*Institute of Food Chemistry, University of Hohenheim, Garbenstr. 28, D-70599 Stuttgart, Germany*
[2]*Institute for Environmental Analytical Chemistry, Department of Chemistry, University Duisburg-Essen Universitätsstr. 15, D-45141 Essen, Germany*

6.1 DIRECT GAS CHROMATOGRAPHIC ENANTIOSEPARATION OF CHIRAL POLLUTANTS

Immediately after the introduction of chiral stationary phases (CSPs) based on *O*-terminated cyclodextrins in the 1980s, direct gas chromatographic (GC) enantio-separations took a breathtaking development from the first report to the marketing of suitable columns by diverse suppliers (Section 6.2). Within very few years this technique made its way into most analytical laboratories. Initial studies focused on pharmaceuticals, flavor and other naturally occuring compounds which can be analyzed enantioselectively by gas chromatography. In 1989, König et al. extended the range of compounds by the GC enantioseparation of a racemic α-hexachlorocyclohexane (α-HCH) standard [1]. This major isomer in the technical mixture of the chloro-pesticide hexachlorocyclohexane has been applied to soils on a million-ton-scale since the 1940s. Volatility and persistency led to global distribution and contamination of various environmental systems including those of the Arctic and Antarctic. König et al. predicted in their revolutionary article that their GC technique may be suitable for studying the enantioselective transformation of this and further chiral pollutants in the environment [1].

α-HCH and other chiral pollutants were applied as racemates and in this composition they were released into the environment. Once spread and in contact with biological systems, the two enantiomeric forms of a chiral pollutant may show slightly different behaviour, and these minute differences may lead to a change of the enantiomeric composition. Finally, the initial racemic mixture may show an excess of one enantiomer over the other which could be very small (~1%) or extremely large (100%); it could once affect the (+)- and in another situation the (−)-enantiomer. In 1991, Kallenborn et al. and Faller et al. proved the enantioselective transformation of α-HCH in the environment, thus distinguishing enzymatic from nonenzymatic processes [2,3]. However, the biological properties (e.g. stability in a chiral surrounding as the environment is) *can* (but must not) be different. The phenomenon of enantioselective processes for a certain chiral pollutant in the environment can be compared with its interaction with different CSPs. On one CSP the enantiomers are resolved, on the next one not. On one CSP the (+)-enantiomer elutes second (interacts more with the CSP), on another CSP the (+)-enantiomer elutes first. Thus, it appears plausible that no single GC column coated with a CSP will solve all problems, though some GC columns are more flexible and can separate more enantiomers than other ones (Section 6.2). Enantioselective GC columns can be combined with all standard injection and detection techniques, but CSP-coated columns are less stable than conventional columns are (Section 6.2.6).

Prerequisite for enantioseparation is the interaction of the enantiomers with another chiral system followed by the formation of associations which behave like diastereomers. Diastereomers have different physical and chemical properties. Already in the 1930s, first simple (planar) three-point association models have been suggested for chiral molecules with the active site of (chiral) enzymes [4]. Interaction of enantiomers with a chiral selector (e.g. an enzyme system) can cause different biological properties such as different toxic action and metabolism of the enantiomers. Given the fact that enantiomers of pesticides may differ in their pesticidal activity, voices arose claiming that in such a situation only the use of the active enantiomer should be allowed (for instance, the herbicide *S*-phosphinotricin is biologically active but not the *R*-form [5]). In the worst case scenario, 50% of non-active pesticides (i.e. tons of unnecessary chemicals) would end up as waste in the environment which does not appear to be acceptable today. Note that similar considerations on isomers led to the substitution of technical hexachlorocyclohexane in many parts of the world with the pure active ingredient (nonchiral) lindane which only amounted for ~15% of technical HCH. Consequently, the research on enantiopure pesticides has increased and would further increase in future. Readers with more interests in modern plant protection and stereoselectivity of pesticides will find valuable information in the the article of Ramos Rombo and Bellus [5] and the book by Ariens and van Rensen [6].

Although initially not in the focus of crop protection chemists, chirality of pollutants has been discovered long before the introduction of suitable gas chromatographic tools. For instance, chirality of α-HCH has been realized as early as in 1949 when Cristol predicted that this most abundant isomer in technical hexachlorocyclohexane has aaeeee-conformation of the chlorine substituents. Only this HCH isomer is chiral and Cristol based his structure assignment of α-HCH by proving this feature [7]. However, this pioneering work was cumbersome—gram amounts had to be isolated, treated for several days with the chiral base brucine for enantioselective degradation; the remainder had to be washed before the specific rotation of polarized light could be determined [7]. A similar structural proof was recently obtained by the application of enantioselective GC, since only one out of three potential isomers was chiral [8]. Thus enantioselective gas chromatography is a big step ahead considering the work of Cristol.

Atropisomerism of 19 PCB congeners has been postulated in 1974 [9], and even years after the experimental proof, important reviews on PCBs did not focus on chirality and its possible consequences [10,11]. Only recently, it was shown that atropisomeric PCBs belong to the class of multiple *ortho*-substituted PCBs suspected to be neurotoxic [12]. Enantiomers of PCB 84 were less potent and efficacious than both pure enantiomers in two neurochemical measures [13].

Another good example is the enantiomers of heptachlor which were shown to have different insecticidal activity but racemic heptachlor was a more active insecticide than the individual heptachlor enantiomers [14,15]. McBlain reported that (−)-*o,p'*-DDT is much stronger estrogenic than (+)-*o,p'*-DDT [16]. It may be worth reconsidering synergistic effects—if this is a theme for chiral pollutants, one may conclude that a change from racemic to non-racemic composition may be accompanied with synergistic effects.

The 1990s were mainly dedicated to the refinement of the analytical methods suitable for the enantioselective GC determination of chiral pollutants at ultra-trace levels in environmental samples. This included the selection of well-suited CSPs and the collection of enantioselective data in environmental samples for mostly all known chiral pollutants. These data indicate that enantioselective processes are rather the common case than an exception for chiral pollutants. Previous summarizing articles in the field were those of Vetter and Schurig (1997) [17], Kallenborn and Hühnerfuss (2001) [18], Vetter (2001) [19], Hegeman and Laane (2002) [20], Müller and Kohler (2004) [21], and Ali and Aboul-Enein (2004) [22].

The first years of the new millennium added valuable applications including enantioselective analyses for the elucidation of processes that would not have been able to be unraveled without this technique. For instance, enantioselective studies prove the evaporation of chloropesticides from soil [23]; they were also used to distinguish individual contributions of two point sources [24,25], as well as for the amount of a chiral-pollutant-fish processing in a specific environment [26], just to mention some. The results obtained in this interesting field of research are presented in detail in Section 6.4. Before that, we focus on the gas chromatographic techniques employed including the applied CSPs.

6.2 GAS CHROMATOGRAPHIC ENANTIOSEPARATIONS WITH MODIFIED CYCLODEXTRINS

Direct enantioseparations by gas chromatography date back to the mid-1960s when Emanuel Gil-Av, Binyamin Feibush, and Rosita Charles-Sigler succeeded in enantioseparations by using derivatized amino acids and dipeptides such as *N*-trifluoroacetyl-derivatized L-isoleucine as CSPs [27,28]. An important step forward was the introduction of capillary columns coated with polysiloxane-based *N*-propionyl-L-valine-*tert*-butylamide (Chirasil-L-Val) by Frank et al. in 1977 [29]. König et al. widened the scope with the derivatization of analytes with isocyanates and other reagents [30] while Schurig introduced metal-complexation gas chromatography [31].

These attempts surely improved the possibilities of gas chromatographic enantioseparations but the usability was restricted to a rather narrow range of chiral target compounds. The situation dramatically changed within few years by the introduction of cyclodextrin based CSPs. In 1983, Koscielski et al. published the first successful gas chromatographic enantioseparation using cyclodextrins which was achieved on packed columns coated with native α-cyclodextrin dissolved in formamide [32]. Despite a good separation factor α (Section 6.3.2) for the racemates of α- and β-pinene, the columns had a limited life time and efficiency was poor. Shortly after, Juvancz et al. demonstrated that permethylated β-cyclodextrin (β-PMCD) can be employed in capillary columns for high-resolution enantioseparations [33]. Immediately, this initial work stimulated the syntheses of a steadily growing number of cyclodextrin derivatives.

6.2.1 Native cyclodextrins

In 1891, cyclodextrins were isolated as enzymatic degradation products of starch by Villiers [34]. Some 100 years ago, Schardinger identified the role of *Bacillus macerans* in their formation [35]. The reaction is enzyme-catalyzed by cyclodextrin-glucanosyltransferase (CGTase) from *B. macerans* and *B. megaterium* among other bacterial strains [36]. The helical structure of the linear amylose component of starch favors formation of rings with six to eight α-(1→4)-linked D-glucopyranose units (as found in starch). Rings with less α-D-glucose units are not formed due to steric hindrance, and rings with more than eight units are scarce. Depending on the number of α-D-glucose units, we distinguish α-cyclodextrin (6 glucose units), β-cyclodextrin (7 glucose units), and γ-cyclodextrin (8 glucose units) (Fig. 6.1a). Nowadays, commercial CGTases more or less exclusively form α-, β-, or γ-cyclodextrin. This rather simple biotechnological method should be seen in the context that total chemical synthesis of cyclodextrins starting from maltose required 21 reaction steps [37].

The native cyclodextrins have hydroxy groups on C2, C3, and C6 (the oxygen on C5 is on the glucose ring). The geometrical shape of cyclodextrins is a torus with a wide and a narrow rim (Fig. 6.1b). C2 and C3 substituents are at the wider rim and the primary hydroxy groups on C6 are on the narrower opening (Fig. 6.1c). Already some 80 years ago, Pringsheim showed that cyclodextrins form stable complexes with suitable substrates [38]. Cyclodextrins (and derivatives) were first used in liquid chromatography for enantioseparation (see Chapter 7 in this book).

Derived from the natural product starch, cyclodextrins exclusively contain α-D-glucoses. Cyclodextrins with reversed selectors (here, α-L-glucoses), which would reverse the elution orders of enantiomers at otherwise identical conditions, are not available [17]. Such peak reversals have been shown in enantioselective HPLC on Pirkle phases with reversed chiral selectors. The native cyclodextrins are crystalline and very polar and thus not suitable for the use as GC stationary phases. The suitability can be obtained by (partial) derivatization of the 18 (α-cyclodextrin), 21 (β-cyclodextrin) or 24 (γ-cyclodextrin) hydroxy groups on C2, C3, and C6, respectively (see Section 6.2.2).

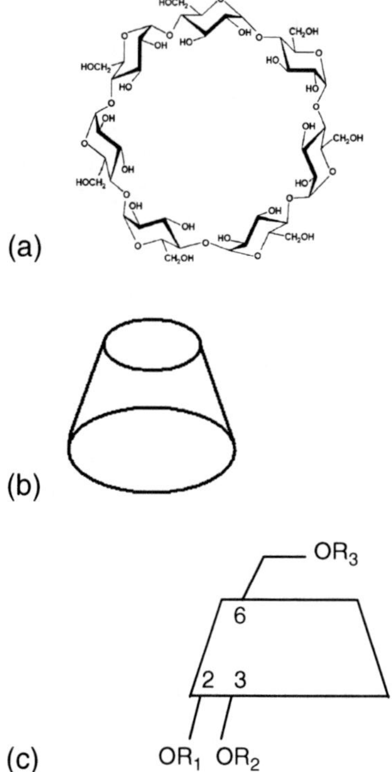

Fig. 6.1. (a) Structure of β-cyclodextrin, (b) Simplified geometric form of the torus, and (c) 2D presentation of the torus indicating one of the secondary carbons C2 and C3 on the wide rim and one of the primary carbons C6 on the narrow ring all of which can be modified by alkylation, siliylation, or other substituents.

6.2.2 Modified cyclodextrins in gas chromatography

Derivatization of hydroxy groups deforms the shape and the openings of torus body. The elliptically distorted torus of a modified cyclodextrin will in certain cases facilitate association complexes according to the key–lock-principle [39]. A drawback of the permethylated cyclodextrins (see above) is that this CSP is only liquid at higher temperatures due to its high melting point. Thus, it could not be de-installed and stored without losing the previous separation characterstics. From 1988 onwards, König et al. demonstrated that perpentylated and partially pentylated cyclodextrins are fluid at room temperature and could be used for the enantioseparation of many classes of compounds using capillary columns [36]. In parallel, Schurig and Novotny dissolved β-PMCD in moderately polar polysiloxanes [40], which has become the common state-of-the-art. Due to the polarity of *O*-terminated cyclodextrins, medium polar polysiloxanes were necessary such as OV-1701 (86% methyl-, 7% cyanopropyl-, 7% phenylpolysiloxane), PS086 (85% dimethyl-, 15% diphenylpolysiloxane), or 65%-methyl, 35%-phenylpolysiloxane [39,41]. Diluting the chiral selector combined the enantioselectivity of modified cyclodextrins with the excellent gas chromatographic properties of polysiloxanes [17]. Moreover, CSPs could be applied irrespective of their melting points and phase transitions [39].

Importantly, the hydroxyl groups on C2, C3, and C6 have different chemical properties—the hydroxy groups on C2 are the most acidic, the hydroxy groups on C6 are sterically favored, and those on C3 are least reactive (apparently due to internal hydrogen bonds) [36]. Thus, the hydroxyl-groups on C2, C3, and C6 of α-, β-, and γ-cyclodextrins could be regioselectively derivatized [36]. The regioselective modification along with the approach of diluting the CSP in a polysiloxane led to more polar cyclodextrins and partly substituted, well-defined derivatives such as 2,6-di-*O*-pentyl-β-cyclodextrin and 2,6-di-*O*-pentyl-3-*O*-trifluoroacetyl-β-cyclodextrin [42,43].

Important for the separation of chiral pollutants was the introduction of randomly *tert*-butyldimethylsilylated β-cyclodextrin (β-BSCD) [44]. Since the full derivatization with the bulky *tert*-butyldimethylsilyl-group is not possible, regioselectively silylated CSPs such as 2,3-di-*O*-acetyl-6-*O*-*tert*-butyldimethylsilyl-β-cyclodextrin and 6-*O*-*tert*-butyldimethylsilyl-2,3-di-*O*-methyl)-β-cyclodextrin (β-TBDM) [45,46] and 2,6-di-*O*-*tert*-butyldimethylsilylated products have been synthesized as well [47] (see below).

Since 1990, several techniques to achieve the covalent bonding of CSPs to the polysiloxane backbone were developed [48–51]. At present, polysiloxane-linked β-PMCD is commercially available ("Chirasil-Dex," Varian). However, it should be noted that the enantioseparation characteristics of polysiloxane diluted and polysiloxane linked cyclodextrins is significantly different (see below). Thus, chemically bonded has been classified by the suffix "CB" throughout this chapter.

To date, more than 1500 cyclodextrin derivatives have been synthesized [38], and 50 different CSPs with modified cyclodextrins are known for GC analysis [36,39,52–54] which allowed for the (partial) enantioseparation of most of the chiral compounds suitable for GC analysis.

6.2.3 Nomenclature of modified cyclodextrins

In accordance with the following rules of the International Union of Pure and Applied Chemistry (IUPAC), substituents on three hydroxy groups are ordered with respect to the first letter in the name following "*O*" with the exception of prefixes like "*n*" in *n*-pentyl and "*tert*-" in "*tert*-butyl". Numbering of the positions is not taken into account (example: "6-*O*-*tert*-**b**utyl-2-*O*-**e**thyl-3-*O*-**m**ethyl"). In the case of identical first letters (e.g. buty**l**, buty**r**yl), the first different letter determines the order. Prefixes "di" or "tri" are not taken into account if they are in front of "*O*".

Free hydroxy groups are not mentioned, and the replacement of the hydroxy group with hydrogen is noted as "deoxy". The order of typical substituents in GC of chiral pollutants is therefore *tert*-**b**utyl, **b**utyryl, **d**eoxy, **e**thyl, **m**ethyl, *n*-**p**entyl, **t**hexyl (*tert*-hexyl), **tr**ifluoroacetyl.

A frequent question is on the meaning of *hexakis*, *heptakis*, and *octakis*. While α-/β-/γ-refers to the number of glucose units in the cyclodextrin (see above), *hexakis*, *heptakis*, and *octakis* refer to the number of modification at carbons C2, C3, and C6 of the respective cyclodextrin. For instance, a β-cyclodextrin with pentyl in 6-position and methyl in 2- and 3-position would be *heptakis*(2,3-di-*O*-methyl,6-*O*-pentyl)-β-cyclodextrin. If one of the pentyl substituents is replaced by methyl, the correct chemical name would be *hexakis*(2,3-di-*O*-methyl,6-*O*-pentyl)*monakis*(2,3,6-tri-*O*-methyl)-β-cyclodextrin and so on. However, the use of hexakis/heptakis/octakis is only justified for well-defined, fully modified cyclodextrins (purity of the CSPs proven by LC and NMR) which is not justified in each case and often unknown. Therefore, and for simplification, the use of these terms has been avoided in this chapter.

6.2.4 Quality of modified cyclodextrins

With 3-*O*-butyryl-2,6-di-*O*-pentyl-γ-cyclodextrin directly coated on the inner walls of open tubular glass columns, König et al. succeeded in the first GC enantioseparation of a chiral organochlorine [1]. Although many organochlorine standards could be enantioseparated on these types of columns, problems arose in the field of trace analysis with environmental samples (see below). Modified cyclodextrin phases directly coated on the inner walls of glass or fused silica columns were only used in a few earlier studies [1,2,55–61]. By contrast, polysiloxane-diluted CSPs prove to be well suited for residue analysis (e.g. they have better peak shapes due to shielding/blocking of more polar sites [62]). Today, polysiloxane-diluted modified cyclodextrins according to Schurig and Nowotny is state-of-the-art, and if not specified, all remarks to columns refer to this kind of columns with a CSP. Today, capillary columns coated with modified cyclodextrins are readily available from many suppliers. However, as for most chemical syntheses, derivatization of CSPs can hardly be performed quantitatively. Purification steps either by TLC [46], adsorption chromatography on silica [36], or high performance liquid chromatography (HPLC), and ^{1}H-NMR investigation of the products are necessary to obtain well-defined CSPs. Currently, information on the purity of CSPs is often not provided by column suppliers, which is deplorable. For instance, five commercial products of 2,6-di-*O*-methyl-β-cyclodextrin were of different purity and also varied significantly in their enantioselectivity in capillary electrophoresis [63]. Purity-dependent variations in enantioseparations have also been reported for chiral pollutants [64–69]. This should always be kept in mind when CSPs with formally identical description are compared. The more simple CSPs such as permethylated CDs appear to be mostly of good quality. In the case of *tert*-butyldimethylsilylated β-cyclodextrin, the bulky *tert*-butyldimethylsilyl substituents (a protecting group for hydroxy groups in organic syntheses), particularly on the secondary OH groups of the glucose units prevent persilylation [44]. Lab-to-lab and even batch-to-batch variations (of the mixed products) became a serious problem resulting in the worst case in peak reversals of the enantiomers of α-HCH [64,70]. Oehme et al. studied the composition of a β-BSCD phase by electrospray LC/MS and found only a degree of substitution of 33–48% [65]. The product currently used is the only commercially available β-BSCD (BGB 172) which consists of >20 products [71]. Fractions of β-BSCD eluted the enantiomers of α-HCH, *cis*- and *trans*-chlordane in different orders [65].

Despite all problems related to the β-BSCD column, there is only one commercial supplier of this column (BGB Analytik, Switzerland), and a constant quality of this column is warranted. In general, the regioselective introduction of the bulky *tert*-butyldimethylsilyl group at the primary C6-hydroxy groups particularly improves the enantioseparation of nonpolar selectands in comparison with CSPs with small groups on the primary C6 carbons [72]. Consequently, 6-*O*-*tert*-butyldimethylsilyl-2,3-di-*O*-methyl)-β-cyclodextrin (β-TBDM) has become a frequently employed CSP [73,74]. However, differences in the quality of these compounds were also reported [68], and the purified product almost quantitatively lost enantioselectivity (see below). For most chiral organohalogen compounds, the randomly silylated β-TBDM column is superior to the pure product [68]. This column is marketed by BGB Analytik as BGB 176. These examples demonstrated that purified CSPs often were less suitable for the enantioseparation than the crude synthesis products [64].

Significant differences were also found for polysiloxane-diluted β-PMCD (for instance Hydrodex β-PM, Macherey-Nagel, and Betadex, Supelco) and the chemically bonded analog (CB-β-PMCD). In fact both columns showed almost complementary enantioseparation feature for chiral pollutants [69]. The commercially available CB-β-PMCD phase (CP-Chirasil-DEX CB, Varian) is anchored to approximately every 90th silicone unit of the polysiloxane backbone [75]. The most probable linkage *via* an octyl bridge is at C6 on one of the glucose units of the cyclodextrin [75], but to some amount, there might be a linkage to C2 as well [76]. It should be kept in mind that exactly one of the seven C6–OH-groups has to be linked to the polysiloxane while the remaining six C6–OH groups have to contain methyl groups [76]. On the other hand, immobilization of the CSP enabled using a non-polar polysiloxane (e.g. CP-Sil 5, 100% dimethylpolysiloxane) which benefits from the decreased overall polarity of the resulting stationary phase. According to this, CB-β-PMCD phases marketed by Varian meet all demands of both ISO standards and customers. Long-term reproducibility of CB-β-PMCD can be obtained by initial synthesis of high amounts since the resulting CB-β-PMCD is stable during storage. This batch will be suitable for production of reproducible columns for a decade or longer [76].

Despite the many types available, there is no single CSP that enantioseparates all chiral compounds but the greatest success in the field of chiral pollutants (the situation is different in other compounds with lower boiling point such as flavor compounds) to date was achieved on randomly modified cyclodextrins. It is still not clear which mode of action applies for randomly modified cyclodextrins. Two borderline cases are mentioned: (i) one product is responsible for the enantioseparations obtained with the CSP (the by-products only are mostly present by accident). In that case, however, low amounts of the CSP in the polysiloxane (probably much less than 10%) would be sufficient for well-defined, mixed chiral stationary phases (interested readers may refer to [77] and citations therein). (ii) Several products are responsible for the unique enantioseparations so that the CSP behaves like a mixed CSP. Combination of two or three suitable well-defined CSPs with additive enantioselectivity (a compound that is not enantioseparated on the one CSP is enantioseparated on the other CSP) may then serve to obtain one column suitable for the separation of the most important chiral organochlorines, which is the ultimate dream of trace analysts.

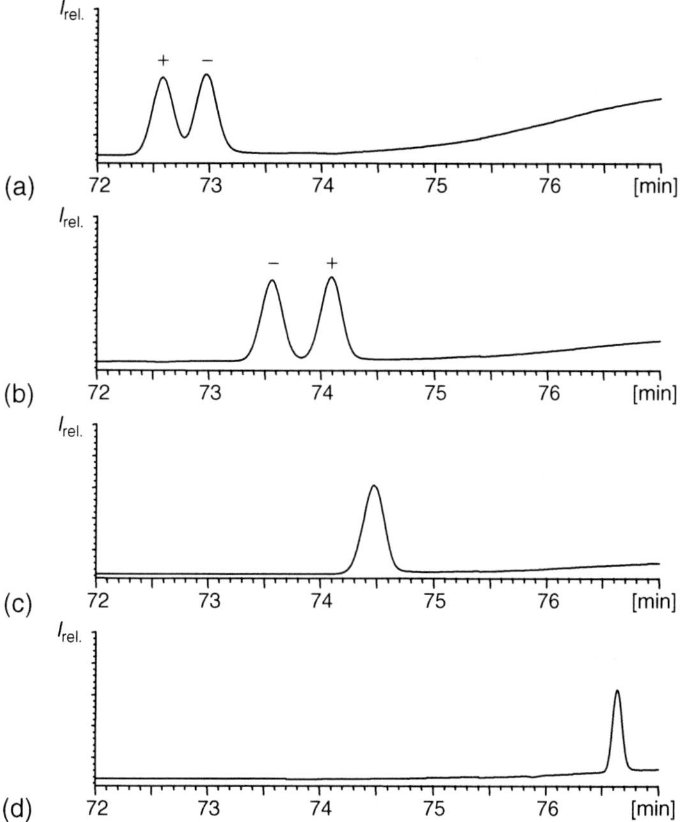

Fig. 6.2. Enantioselective analysis of *cis*-chlordane on (a) 10% β-PMCD, (b) 10% γ-PMCD, as well as both (c) 5% or (d) 10% of β- and γ-PMCD Ref. [79].

Only little activities can be observed in this field. Buser and Müller diluted β-BSCD and α-PECD in PS086 and achieved no enantioseparation of heptachlor and heptachlor epoxide although the compounds were separated on each of the CSPs [78]. Detailed information, however, on the relative amounts etc, were not given. Generally, the enantioselectivity should be additive. Fig. 6.2 shows the enantioseparation of *cis*-chlordane on β- and γ-PMCD as well as 1:1 mixtures on the two individual CSPs. As can be seen, the mixing yielded almost mathematically evened the events on the isolated columns. Since the elution order was reversed at comparable α- values on the individual column (Fig. 6.2a and Fig. 6.2b), the enantiomers co-eluted when both CSPs were mixed 1:1 (Fig. 6.2c) [79]. Note as well the obvious influence of the increased polarity of the GC column when double amounts of the CSPs were diluted in the (less polar) polysiloxane.

One of the basic questions is whether the impure products contain a (small) amount of product that resolves the racemate or whether additional effects/interactions support the enantioselectivity of the chiral selector. It was speculated that elevated GC oven temperatures may lead to the vibration of the cyclodextrin torus. Such vibrations may be moderated by intermolecular hydrogen bonds [19]. If so, less vibrating torus bodies will make associations with the chiral selector easier. These interactions would be particularly significant at high temperatures which would partly explain why these columns are particularly important for the enantioseparation of chiral pollutants which require comparably higher elution temperatures than flavor compounds which are mostly enantioseparated on columns coated with well-defined, fully derivatized CSPs.

6.2.5 CSPs suggested for the enantioseparation of chiral pollutants

Problems related to the quality of the CSPs have already been mentioned. Therefore, enantioenriched standard compounds should be used to verify elution orders. For testing of the column performance, chiral test mixtures may be used [65,67,80]. In recent years, commercially available CSP-coated capillary columns substituted the hand-made research columns often applied in the initial phase. This has led to more comparable results in recent years. However, only a few CSPs are currently in frequent use and they are mostly either permethylated or partially silylated. These types of columns are described in the following two sections and further columns are summarized below. An overview on the utilization of the CSPs which are used most often is given in Table 6.1.

6.2.5.1 Permethylated CSPs

Permethyl-β- or γ-cyclodextrin are easily available from many suppliers under trademarks such as Betadex and Gammadex (Supleco), Cyclodex-B (J&W) Chiraldex B-PA (Astec), Rt-βDEXest (Restek), CP-Cyclodextrin-β-2,3,6-M-19 (Varian). Several reversals of the enantiomer elution order have been observed between β- and γ-PMCD (see also Fig. 6.2) [73,81–83]. α-PMCD is one of the few examples for the application of modified cyclodextrins with 6 glucose units [84]. Furthermore, CB-β-PMCD (CP-Chirasil-DEX, Varian) has also been widely used for the enantioseparation of chiral pollutants. Note that (conventional polysiloxane diluted) β-PMCD and CB-β-PMCD enantioseparated different PCB atropisomers (see Table 6.1) and only the latter CTTs [85]. Both types of permethylated β-cyclodextrin have some drawbacks for chlordane-related compounds since *cis*- and *trans*-chlordane resulted in three peaks and oxychlordane enantiomers coeluted [73]. The *cis-/trans*-chlordane problem can be solved by tandem column technique [86] which is, in this case, somewhat cumbersome in view of the fact that several alternative CSPs manage this separation problem. CB-γ-PMCD has been used some years ago but is no longer commercially available. Both β-PMCD and CB-β-PMCD are also suggested for α-HCH and PCCHs.

6.2.5.2 Silylated CSPs

Silyl-containing substituents are often used as protection groups for reversible reaction intermediates on the way to regioselectively modified cyclodextrins. The typical substituent is the *tert*-butyldimethylsilyl group. 6-*O*-thexyldimethylsilylated CSPs (thexyl = *tert*-hexyl or 2-(2,3-dimethyl)-butyl) have been introduced and used by König et al. for both analytical and preparative GC [87,88]. The problems related to *tert*-butyldimethylsilylated β-cyclodextrin (β-BSCD) have been mentioned above. However, the batch-to-batch variations observed during the initial phase seem to be solved to some extend and the so-called BGB-172 (BGB Analytik) column and slight modification of it (PS086 instead of OV 1701) is the most frequently used column next to hand-made column of this type by different research groups. However, elution orders of enantiomers should be controlled by the injection of enantioenriched standards. β-BSCD shows unique separation characteristics for enantiomers of toxaphene. Additionally, enantioresolution of many chlordane-related compounds, atropisomeric PCBs, and methylsulfonyl-PCB-metabolites make the BGB-172 column to be the most applied in the present field (see Table 6.1).

As an alternative, 6-*O*-*tert*-butyldimethylsilyl-2,3-di-*O*-methyl)-β-cyclodextrin (β-TBDM) has been used in several works. Probably the most pronounced differences in the quality are available ranging from very pure CSP to something classified as randomly siliylated. The pure product (commercially available for instance from Macherey-Nagel, Hewlett-Packard) separated the enantiomers of some PCB atropisomers but the randomly silylated β-TBDM (BGB 176, BGB Analytik) separated different PCB atropsiomers and several CTTs. The major point is that the pure β-TBDM enantioresolved only a few chiral organochlorines while the raw product separated the enantiomers of the most of compounds [68,89]. In fact, this column seems to resolve a multitude of analytes such as pesticides (mecoprop, pyrethroides, organophosphate insecticides) as well as fragrances (HHCB, AHTN and their metabolites (Table 6.1). Additionally, they can be used at considerable temperatures such as 230°C. Also a multitude of natural compounds lactones and alcohols can be resolved by using this column. Attempts have been undertaken to gain fractions of the raw product with high enantioselectivity [69].

2,3-di-*O*-propionyl-6-*O*-*tert*-butyldimethylsilyl-β-cyclodextrin is also commercially available but no data exists for chiral pollutants discussed in this chapter. Well-defined 6-*O*-thexyl-derivatized CSPs offered good enantioseparation of atropisomeric PCBs and their methyl sulfonyl metabolites (see Table 6.1). CTTs and many chlordane-compounds have not yet been tested.

6.2.5.3 Other types of modified CSPs

Well-defined pentyl-containing CSPs were introduced by König et al. (for details see [36]), and some of them are commercially available from Macherey-Nagel (the polysiloxane-diluted are labelled Hydrodex®) and from Astec. The most important chlordane-related components were enantioseparated on regioselectively pentylated CSPs. Several atropisomeric PCBs have been separated on this columns as well (see below). Perethylated-α-cyclodextrin has been used by Buser and Müller but information on the purity of the CSP was not provided [78]. Jaus and Oehme studied perethylated CDs and found α-cyclodextrin easiest for perethylation [67]. They also fractionated raw products of perethyl-γ-cyclodextrin and some fractions showed good to excellent separation characteristics [67]. HPLC-MS characterization of the CSPs was carried out [67]. Well-defined, partly substituted CSPs (OH-groups on C6 etc.) have been marketed

Table 6.1 CSPs that have been used for the enantioseparation of chiral pollutants*

Chiral selector	Abbreviation	Manufacturer trademark	Components (remarks)
3-*O*-butyryl-2,6-di-*O*-pentyl-γ-CD	3by26pe-γ-CD	Lipodex E (MN)	β-PCCH, γ-PCCH, α-HCH, *trans*-chlordane, PCB95, PCB 136 [253]
2-*O*-methyl-3,6-di-*O*-pentyl-β-CD	2me36pe-β-CD	W. A. König	α-HCH, oxychlordane, heptachlor epoxide, heptachlor *cis*-chlordane/*trans*-chlordane: four peaks
2,6-di-*O*-methyl-3-*O*-pentyl-β-CD	26me3pe-β-CD	W. A. König	Heptachlor (+ < −) [88] *trans*-heptachlor epoxide *cis*-chlordane (+/−) [88], *trans*-chlordane (±) [88]
2,6-di-*O*-methyl-3-*O*-pentyl-γ-CD	26me3pe-γ-CD	W. A. König	*cis*-Heptachlor epoxide (- < +) [88], *trans*-heptachlor epoxide (+/−) [334] α-HCH (−/+) [88], *o,p′*-DDT atropisomeric PCBs (45, 88, 95, 132, 135, 139) [88]
3-*O*-methyl-2,6-di-*O*-pentyl-β-CD	3me26pe-β-CD		*cis*-chlordane, *trans*-chlordane
6-*O*-methyl-2,3-di-*O*-pentyl-γ-CD	6me23pe-γ-CD	W. A. König	Atropisomeric PCBs
3-*O*-acetyl-2,6-di-*O*-pentyl-β-CD	3ac26pe-β-CD	Lipodex D (MN)	α-HCH [60]
6-*O*-thexyl-2,3-di-*O*-methyl-β-CD	23me6tx-β-CD	W. A. König	Atropisomeric PCBs (45, 84, 95, 91, 136, 131, 149, 174, 176, 175, 183) [87] $MeSO_2$-PCBs (3-91, 4-91, 3-149, 3-132, 4-132, 4-149, 3-174, 4-174) [184]
2,3,6-tri-*O*-methyl-β-cyclodextrin	β-PMCD	Several sources	(+/−) α-HCH; β-PCCH, γ-PCCH, δ-PCCH, further PCCHs *cis*-chlordane (+/−), *trans*-chlordane (+/−) [123], MC5, *trans*-heptachlor epoxide (+/−) [74]
2,3,6-tri-*O*-methyl-γ-cyclodextrin	γ-PMCD	Several sources	α-HCH (+/−) [83, 222], *cis*-chlordane (+/−), *trans*-chlordane (+/−) [84] *cis*-heptachlor epoxide (+/−), *trans*-heptachlor epoxide (+/−) [84]
CB-2,3,6-tri-*O*-methyl-γ-cyclodextrin	CB-γ-PMCD	γ-Chirasil-Dex	α-HCH (+/−) [73], β-PCCH (+/−) [61]
CB-2,3,6-tri-*O*-methyl-β-cyclodextrin	CB-β-PMCD	β-Chirasil-Dex (Chrompack) and V. Schurig	α-HCH, *cis*-chlordane, *trans*-chlordane, oxychlordane ($R = 0.5$), heptachlor ($R = 0.6$) *cis*-heptachlor epoxide not separated [147], *trans*-heptachlor epoxide ($R = 0.9$) [147] photo-heptachlor, photo-heptachlor epoxide, photochlordene [147] photo-dieldrin [147] op-DDT ($R = 0.5$), op-DDD ($R = 0.6$) PCB 95, PCB 91, PCB 139, PCB 144, PCB 84 (+/−), PCB 132 (+/−), PCB 135 (−/+), PCB 136 (+/−) PCB 149 (+/−,) PCB 174 (−/+), PCB 176 (+/−), PCB 131 (+/−), PCB 144 (+/−), PCB 175 (+/−), PCB 183 (+/−), PCB 196 (+/−) [158]

2,3,6-tri-*O*-methyl-α-cyclodextrin	α-PMCD	AlphaDex 120 (Supelco)	*cis*-Chlordane (+/−), *trans*-chlordane (−/+) [84] *cis*-Heptachlor epoxide (±), *trans*-heptachlor epoxide (−/+) [84]
β-permethyl-acetyl	β-PMA	Astec	PCB 84, PCB 131, PCB 132, PCB 135, PCB 136, PCB 144, PCB 174, PCB 175, PCB 176
Random 2,3,6-tri-*O*-*tert*-butyldimethylsilyl-β-CD	β-BSCD	BGB 172 M. D. Müller M. Oehme	α-HCH (elution order subject to changes), α-MHCH [335] heptachlor epoxide (+/−) [251,202] trans-heptachlor epoxide heptachlor [23] (+/−), oxychlordane (+/−) [120] *o,p*′-DDT (−/+), *o,p*′-DDD, 4 further chiral minor comp. in technical DDT [200] B8-1413, B8-2229, B9-1679, B9-1025 *cis*-chlordane, *trans*-chlordane: elution order may change [91] photodieldrin [91] U82, MC5, MC6, MC7, chlordane (−/+) [120] 3-91, 4-91, 3-132, 4-132, 3-149, 4-149 not sep., 3-174, 4-174, 4-95 not sep. [185] dimethenamid, metalaxyl, metolachlor [314], methylated mecoprop [308]
Random 6-*O*-*tert*-butyldimethylsilyl-2,3-di-*O*-methyl-β-CD	Random-PD-β-TBDM	G. Hottinger M. D. Müller	α-HCH, careful check of elution order required [61,68] *trans*-chlordane (−/+), *cis*-chlordane (+/−) [74] heptachlor epoxide (+/−), *trans*-heptachlor epoxide (−/+) [74] oxychlordane (−/+) [73,74], heptachlor (+/−) [74], MC5 *o,p*′-DDT; *o,p*′-DDD not separated [200]
6-*O*-*tert*-butyldimethylsilyl-2,3-di-*O*-methyl-β-CD	Well-defined PD-β-TBDM	König, MN Hydrodex	α-HCH, *cis*-chlordane, oxychlordane ($R=0.8$) [147], *trans*-heptachlor epoxide ($R=0.7$) [147] photo-heptachlor, photoheptachlor epoxide, photo-dieldrin MC5, U82 (partly) [78], MC6 atropisomeric PCBs, bromocyclene [306], allethrin, bioallethrin, methamidophos, acephate, triclofon, bromacil [87], HHCB, AHTN, ATII, AHDI [331], HHCB-lactone [326]
6-*O*-*tert*-butyldimethylsilyl-2,3-di-*O*-methyl-β-CD	β-TBDM	β-DEX 325 Supelco	Methyl sulphonyl PCB 3-174, 4-174 (partly separated) [185]
2,3,6-tri-*O*-ethyl-β-CD	α-PECD	M. D. Müller	*cis*-Chlordane (+/−) [120], heptachlor (−/+) [120] heptachlor epoxide (+/−) [120], photoheptachlor [78] oxychlordane not separated [120]

(*continued*)

Table 6.1 Continued

Chiral selector	Abbreviation	Manufacturer trademark	Components (remarks)
2,3-di-*O*-methyl-β-CD	23me-β-CD B-DM	Astec	PCBs [92]
2,6-di-*O*-penthyl-3-*O*-trifluoracetyl-γ-CD	26pe3tfa-γ-CD (G-TA)	Astec	PCBs [92]
(S)-hydroxypropyl methyl ester-β-CD	B-PH	Astec	PCBs [92]
FS-Hydrodex-β-3P		Macherey-Nagel	α-HCH, *cis*-chlordane, *trans*-chlordane ($R=0.8$) [147] heptachlor ($R=0.6$), *trans*-heptachlor epoxide ($R=0.8$) [147] photo-heptachlor, photo-dieldrin [147]
Permethyl-trifluoroacetoxypropyl-γ-CD	G-PT	Astec	PCBs [92]
	Chirasil-Dex-TFA		α-HCH, *o,p'*-DDT [39]
Permethyl-trifluoroacetoxypropyl-γ-CD, tandem with XTI-5	G-PT	Astec/Restec	*o,p'*-DDT, *o,p'*-DDD [93]
2,3,6-tri-*O*-trifluoroacetyl)-β-CD	G-TA	Astec	

*Modified from [19]

by Astec. Trifluoroacetated γ-cyclodextrin (G-TA, Astec) has been used for several PCB atropsomers and for *cis*- and *trans*-chlordane [90]. Finally, it should be mentioned that CSPs which are not based on cyclodextrins have only been used in selected cases [19]. More polar CSPs (Astec, USA) have also been applied for enantioseparations of chiral pollutants [91–94].

6.2.6 Practical recommendations

The dimensions of commercial columns are typically 10–30 m × 0.25 mm (i.d.) while the film thickness (d_f) is usually ~0.25 μm. However, these conditions may be accompanied by coelutions of the enantiomers of the analyte with other compounds in the sample. Reduction of the internal diameter of the columns to 0.1 and 0.05 mm (i.d.) requires smaller films of the stationary phase in order to keep the phase ratio β constant [17].

Polysiloxane-diluted CSPs can be partially dissolved when large amounts of solvents are injected in the splitless or on-column modes mandatory for residue analysis of chiral pollutants. Formation of droplets at the column inlet will then result in a loss of efficiency. For this reason, large volume injection techniques are currently not used. For improving the lifetime of CSPs, low injection volumes and low-boiling solvents should be used. Cyclodextrin phases have a lower capacity (compared to conventional or achiral stationary phases), and low amounts should be injected to avoid overload. When derivatizing reagents are used, they should be removed prior to the injection since they may react and destroy the CSP. Extensive use at high temperature may shorten the lifetime of the CSP as well. The maximum operation temperatures are relatively low (usually in the order 220–250°C), which is more than sufficient for the enantioseparation of chiral pollutants. However, long run times are required when real samples are injected as they may contain high-boiling compounds which are easily eluted from conventional columns which can be used up to temperatures above 300°C.

6.2.7 Quality assurance (QA) in enantiofraction (EF) determinations

Due to the partly huge number of halogenated compounds in environmental samples (or multicomponent mixtures), the enantiomer separation of target compounds in real samples is a much greater challenge compared to standards. Three major points complicate the determination of EFs in the sample.

(i) The analytical GC performance cannot be optimized in the same way it is custom for the respective compounds on achiral columns.
(ii) Minor coelutions already falsify the EF.
(iii) There is often only one CSP available/in use for the determination of the EF. Under some circumstances, different EFs have been obtained for the same compound in the same sample but on different CSPs.

Consequently, QA is an important issue in environmental trace analysis. Where quantitative nonchiral work in environmental chemistry usually requires 50-m columns, enantioselective columns are 30 m at the most, and each resolved pair of enantiomers increases the number of peaks in a gas chromatogram. Due to the weak van-der-Waals interaction between the chiral selector and the chiral analyte, increase of temperature has an adverse impact on the enantioresolution of the target compound. A sufficient resolution of enantiomers usually requires specific/distinct optimized GC conditions so that there is little flexibility for improving the system toward the separation of potential interferences. Thus, the problem of co-elution can not often be solved by the complex GC oven programs that are used in achiral analyses. When only one CSP is available, a potential interference has to be separated by either the selective fractioning of the sample (i.e. enrichment of the target compound) before the GC analysis (e.g. by liquid chromatographic fractioning), during the GC analysis (e.g. by the application of tandem columns, multidimensional GC, or even GC × GC), or after the GC analysis by using a high-selective detection system (MS or MS/MS). Generally, the detection system should have a high selectivity and except for *very* limited cases, the electron capture detector can not be recommended for enantioselective determinations in environmental samples. Very often (due to the different vapour pressures of interesting chiral contaminants), multiple enantioselective determination in a single run is scarce. Thus, often only one result can be obtained in one chromatographic run, which is usually very long due to the low maximum operation temperatures of CSPs. Additionally, enantioselective measurements block the instrument for other purposes. When QS requires use of GC/MS or even more sophisticated techniques, instrument time may be limited. This means that most samples are only analyzed and measured once. To make sure that the accuracy is high, initial studies on both precision and accuracy have to be carried out.

6.2.7.1 Precision and accuracy of enantioselective determinations

Precision (low standard deviation determined during multiple injections) can be easily tested with racemic standards. About twenty publications note results on repetitive analysis and reached values ranging from EF = 0.493 to 0.505. The accuracy of the mean given as the standard deviation was from ±0.002 to ±0.004. Consequently, single measurements of individual compounds around the racemate usually have an accuracy of ~±0.004 EF units. However, these minute variations should be seen in the right contence. Ulrich and Hites showed that their enantioselective GC/MS data matched the ee of chiroptic measurements [95]. Suitability of internal standards for EF determinations (perdeuterated α-HCH in the determination of α-HCH) was shown [73] but not pursued further.

It appears that the quality was improved in comparison with the measurements in the pioneering years. When the GC determination of a standard deviates from the racemate, the calculated values in samples has to be corrected [96]. In other words, the analytical EF of the racemate has then to be defined being different from 0.500 [97]. Hoekstra et al. performed repetitive injections of a pure PCB 95 standard and found an EF of 0.493 ± 0.010 [98]. The mean value of PCB 95 in Arctic fox was 0.496 ± 0.014. Thus, there was a slight excess of the first eluting enantiomer in the samples although the value suggests depletion of the first eluting enantiomer [98]. However, such small deviations from the racemate in samples should not be considered as nonracemic unless comprehensive statistical data has been provided.

The above-documented data was mainly achieved under ideal conditions (no coelution, concentrated solutions with good S/N ratio of a racemic standard). Precision decreases if the target compound is found at low concentrations and taking into account only the accuracy of the instrumental performance. This situation does never exist for environmental samples. In reality, EFs are often determined at much lower concentration than found in the standard used for determining the precision of EF values. Therefore, the precision should be determined by analyzing multiple extracts from several subsamples of one real world sample. Multiple injections are normally not sufficient to prove the precision in enantioselective determination. As for nonchiral analyses, the limit of detection and limit of determination of EFs should be established as well.

In addition, worse S/N may occur when one enantiomer is significantly depleted. In the case of a racemic compound found at 100 pg in a chromatogram, both enantiomers account for 50 pg. An interference with as low as 5 pg of one of the enantiomers (caused by interference and/or matrix effects) leads to an error of 10% or $\sim 0.500 \pm 0.024$. Both scenarios (instrumental variations and interferences) together, may in the worst case, account for an insecurity of $\sim 0.500 \pm 0.027$. Typically, results of measurements in samples in this range should be accepted to be racemic unless there is no data available that supports the nonracemity of the compound in the sample which is not easy to provide. Generally, samples should be screened previously by achiral analyses in order to identify potential interferings with the detection method in the sample.

In the case of 50 pg of a sample with an EF of 0.300, an interference of enantiomer-1 with 1 pg already yields EF = 0.314. It is evident that the insecurity rises when one enantiomer is (significantly) depleted. However, in such a case the basic result, i.e. the significant depletion of one enantiomer relative to the other, is much more reliable. GLP and intercalibration procedures in the analysis of organochlorine compounds in the high-ppt to the low-ppm range accept deviations of 30% from the real value. This can also be seen in the concentration ranges given for compounds certified in commercially available samples. Another option is spiking of the compound with the racemate followed by subtraction of the spiked amount. A suitable addition would be quantitative enantioselective GC.

A major problem in evaluation of data is that the samples of researchers are usually unique. Furthermore, no samples with certified EFs are available in the market. To overcome such situations, researchers have started to report EFs in standard reference material (include certified concentrations of POPs) which are commercially available to the whole scientific community. For method validation, several authors have published the EFs in commercially available materials. These include cod liver oil (SRM 1588a), whale blubber (SRM 1945), sediment (CRM EC-5), lake trout (CRM DF-2525), and pilot whale (NIST IV) (Table 6.2). Note that these EFs have not been approved in intercalibration procedures but the use of these materials is currently the best choice for method optimization. The data set available confirms a good agreement of the EFs. It should be mentioned that even when two results were presented, there is a chance that both research groups used the same column. Under these circumstances,

Table 6.2 Enantiofractions (EFs) of chiral organochlorines in standard reference materials

Chiral organochlorines	SRM 1588a	SRM 1945	CRM EC-5	CRM DF-2525	NIST IV
α-HCH	0.500 ± 0.003 [113]; 0.500 [60]	0.574 ± 0.007 [113]	0.446 ± 0.005 [113]	0.500 ± 0.005 [113]	0.531 ± 0.006 [113]
o,p'-DDT	0.371 ± 0.013 [113]; 0.412 [86]	0.473 ± 0.012 [113]	ND [113]	0.415 ± 0.081 [113]	0.599 ± 0.015 [113]
cis-CD	0.468 ± 0.002 [113] 0.487 [74]; 0.476 [86]	0.172 ± 0.002 [113]	0.489 ± 0.003 [113]	0.335 ± 0.003 [113]	0.072 ± 0.020 [113]
trans-CD	0.532 ± 0.001 [113] 0.531 [74]; 0.526 [86]	0.828 ± 0.005 [113]	0.497 ± 0.002 [113]	0.626 ± 0.003 [113]	0.887 ± 0.003 [113]
Oxy	0.571 ± 0.001 [113] 0.571 [74]; ~0.583 [86]	0.625 ± 0.007 [113]	ND [113]	0.647 ± 0.001 [113]	0.672 ± 0.002 [113]
Hepox	0.608 ± 0.006 [113]; 0.615 [74]	0.622 ± 0.004 [113]	ND [113]	0.631 ± 0.009 [113]	0.642 ± 0.005 [113]
MC5	0.540 ± 0.001 [113] 0.465 (½) [74]; 0.524 (½) [86]	0.852 ± 0.001 [113]	0.481 ± 0.003 [113]	0.629 ± 0.003 [113]	0.932 ± 0.010 [113]
MC6	0.408 ± 0.003 [113]; 0.474 [86]	0.489 ± 0.071 [113]	ND [113]	ND [113]	0.496 ± 0.021 [113]
MC7	0.478 ± 0.003 [113]	0.517 ± 0.016 [113]	0.480 ± 0.013 [113]	0.393 ± 0.003 [113]	0.573 ± 0.030 [113]
PCB 91	0.518 ± 0.005 [113]	0.495 ± 0.013 [113]	0.503 ± 0.002 [113]	0.312 ± 0.006 [113]	0.377 ± 0.001 [113]
PCB 95	0.451 ± 0.002 [113] 0.448 ± 0.001 [25]	0.422 ± 0.001 [113]	0.488 ± 0.001 [113]; 0.483 ± 0.001 [113]; 0.489 ± 0.001 [171]; 0.487 ± 0.001 [25]	0.320 ± 0.010 [113]	0.454 ± 0.001 [113]
PCB 136	0.121 ± 0.013 [113]	0.507 ± 0.009 [113]	0.492 ± 0.002 [113]; 0.501 ± 0.002 [25]; 0.498 ± 0.001 [171]	0.380 ± 0.006 [113]	0.500 ± 0.003 [113]
PCB 149	0.622 ± 0.008 [113] 0.623 ± 0.004 [25]	0.528 ± 0.009 [113]	0.511 ± 0.003 [113]; 0.520 ± 0.004 [113]; 0.505 [171]; 0.511 ± 0.001 [25]	0.607 ± 0.004 [113]	0.573 ± 0.001 [113]
PCB 174	0.824 ± 0.009 [113]	0.573 ± 0.017 [113]	0.508 ± 0.005 [113]	0.532 ± 0.007 [113]	NA [113]
PCB 176	0.481 ± 0.004 [113]	0.431 ± 0.002 [113]	0.494 ± 0.004 [113]	0.504 ± 0.004 [113]	0.489 ± 0.020 [113]
PCB 183	0.484 ± 0.009 [113]	0.505 ± 0.006 [113]	0.502 ± 0.012 [113]	0.458 ± 0.014 [113]	0.486 ± 0.005 [113]
U82	0.475 ± 0.002 [113]; 0.474 [86]	0.490 ± 0.004 [113]	0.498 ± 0.032 [113]	0.528 ± 0.005 [113]	0.513 ± 0.005 [113]

a potential coelution would be misinterpreted on both the occasions. However, the advantage of these studies is evident.

6.2.7.2 Techniques suitable for improving the security in EF determinations

Organochlorine compounds are present in samples at particularly low levels and in a complex mixture. Consequently, the separation of the components requires the technique with the highest efficiency, i.e. high-resolution gas chromatography. Furthermore, highly sensitive (electron capture detection, GC/ECD) and highly selective detection techniques (GC/mass spectrometry with selected ion monitoring detection, GC/MS-SIM) are indispensible. Efficiency values in GC enantiomeric resolutions was noted to be 2000 to 13,000 plates per meter [51,54].

6.2.7.2.1 Possible measures for quality control. It should be stressed that even though some errors that are relevant for conventional analysis are not inherently relevant for chiral analysis, quality control is an issue that might be considered as underdeveloped for this branch of analytical chemistry. Recovery rates are only generally reported for the racemate and not the single compounds which should be improved. It is even more important to discuss standard deviation and thus the precision and the significance of difference from racemate for the respective method.

Precision. Ideally the precision should be determined by analyzing multiple extracts from several subsamples of one real world sample. Multiple injection is normally not sufficient to prove the precision in enantioselective determination.

Accuracy. Accuracy of enantioselective determination should be determined by spiking real world samples with known amounts of standards of known enantiomeric composition, to make sure no matrix effects interfere with the respective signals of the analytes. The apparent enantioratio (ER) or EF of an assumed racemic standard should be given. If the EF of the racemate deviated from the racemate ($EF \neq 0.5$), correction factors might be used, but should be discussed in the documentation of the respective work.

Linear range. Some detection systems, even GC-MS, do not exhibit a linear range for all compounds. Thus this needs to be reaffirmed for each system. Sometimes one enantiomer does not chromatograph as well as the other thus a racemate may not show as peak ratio 1:1 at low concentrations, while it does at higher ones.

6.2.7.2.2 Treatment before GC injection. Some efforts have been undertaken to fractionate samples prior to the enantioselective GC analysis. Loss of a chiral compound during the sample cleanup where only nonchiral chemicals and equipment were used will have no impact on the enantiomeric composition of the target compound. However, this is only warranted for racemic compounds. Partial enantiofractioning within a peak was observed for nonracemic compositions [99]. In this case, the racemate and the enantiomeric excess of one compound can be treated as individual compounds [100]. Note that the racemate and a pure enantiomer have different physicochemical properties (while those of the enantiomers are of course identical). In this case, the EF of the peak at the beginning of the cleanup chromatography may be slightly different from the EF at the end of the peak. Therefore, Liquid Chromatography (LC) separations should be carried out with quantitative collection of the target compound.

Kallenborn et al. used HPLC to cut α-HCH from the bulk of POPs prior to injection onto a GC/ECD system [2]. Only in such selected cases the use of ECD appears to be justified for the EF determination. However, Pfaffenberger et al. mentioned interferences from the sample matrix (analysis with GC/ECD) which were not observed on an achiral column although an additional cleanup step had been introduced [101]. QA in residue analysis recommends the use of ECD only if the result is confirmed on a second stationary phase with different polarity. This QS parameter is not easy to provide in enantioselective GC. A further problem is the poor linear range of the ECD (approximately two orders of magnitude). For PCBs, Chu et al. isolated atropisomeric PCBs from the excess of non-atropisomeric by using an HPLC system with a 5-PYE column [25]. Others have fractioned the samples by column chromatography using activated silica or mixes with charcoal for the separation of coelutions that prevented the EF determination of chiral pollutants [102–104].

6.2.7.2.3 Improvement during the GC analysis. Errors by coelutions in an optimized GC system (temperature, carrier gas flow) can, as soon as they are recognized, be eliminated by using of tandem techniques (i.e. the coupling of the CSP-column with a nonchiral column), or the introduction of a second dimension.

Introducing the tandem technique, Oehme et al. used the nonchiral phase in the first, and the chiral in the second dimension [86]. The tandem technique has been applied later by others as well, often with a switch of configuration (CSP used first) [70,93,94,105–107]. The latter may be beneficial since the enantioseparation is then performed at lower temperature although sharpness of the cut will then be lower.

A similar, more sophisticated approach is the use of multidimensional gas chromatography (MDGC) [108–111]. It has the advantage that only the interesting and not all analytes are transferred to the second column with the help of the heart-cut technique. Using a two-oven MDGC system, the achiral column can be used first (which allows sharper heart cuts) and enantioseparation of the eluate transferred to the CSP can be separated at lower temperature. MDGC has been used for the enantioseparation of complex organochlorines such as PCBs [108,109] and CTTs [111–113]. Initial work was performed in combination with an ECD, but more recently the high separation power of MDGC was combined with the high selectivity of MS detection [112,113].

Recently, comprehensive two dimensional GC (2D-GC) also nominated GC × GC to be applied for the enantioselective determination of atropisomeric PCBs in food [114]. GC × GC has been established as a routine method for the analysis of complex mixtures such as alkanes, PCBs, etc. [115–117]. A good introduction is given in [115].

GC × GC was made possible by the development of fast mass-analyzing systems and more importantly, modulation systems and devices for the comprehensive data treatment. In GC × GC, the secondary column is coupled directly to a primary column, with two alternatively "firing" cold spots at the connection (Fig. 6.3). Of course, this technique demands a very fast elution on the secondary column which is typically in

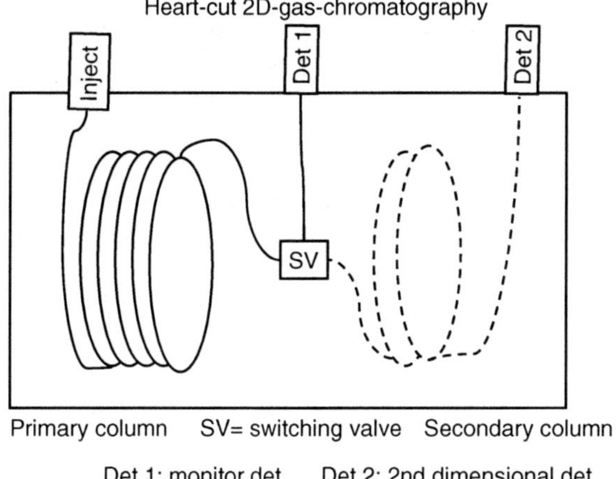

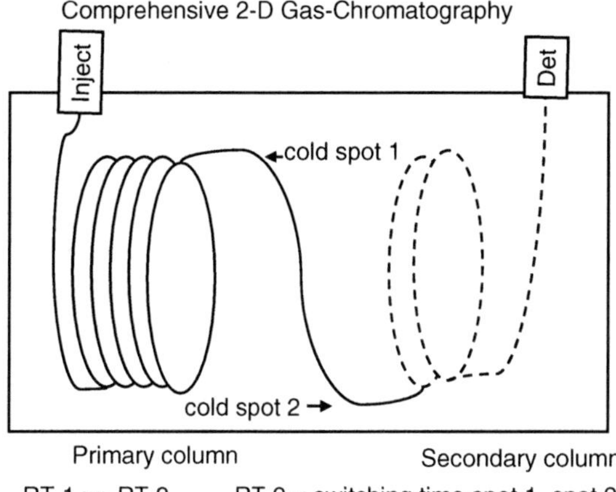

Fig. 6.3. Schematic presentation of (a) 2D-GC with heart-cut and (b) Comprehensive 2D-GC (GC × GC).

the order of about 5–10 s. Typically a conventional column ($L = 30$ m, i.d. = 0.25 mm, film = 0.25 μm) is coupled to a secondary column with typically $L = 0.5$–1 m; i.d.~0.1 mm, film = 0.1 μm. These aspects lead to very narrow peaks on the secondary column, thus data points with few milliseconds distance are necessary. This demands to fast detection systems such as time of flight mass spectrometers, very fast ECDs, or very fast quadrupole mass spectrometers. A good overview on GC × GC combined with fast time of flight mass spectrometers is given in [117]. The same time as the elution time on the secondary column (5 s) should be used as interval for the focussing cold spot. This setup relies on a fast secondary and a slow primary column, thus a rather slow elution on the first column is of an advantage. Though this is sometimes somewhat uncomfortable with "normal columns" this setup gives an ideal combination for enantioselective separations, as enantioselective separations normally tend to be pretty slow in comparison. Using this technique the separation power of a given GC system can be multiplied. Basically the GC × GC approach is a tandem chromatography with some kind of interruptor, thus the primary data treatment is very simple, but fast [115]. Sampling rates of at least 50 Hz or better 200 Hz are recommended. Only secondary data treatment results in a classical 2D plot which results from the primary chromatograms. However the same peak may appear in two to three secondary chromatograms, thus conventional integration takes place and the two or three integrals of one peak are added up to give the final result. An overview on this process is given by Ryan and Marriott [116].

In environmental samples there is always the risk that coeluting isomers or even just matrix compounds interfere with the original analyte. This is even worse in enantioselective analysis as the number of theoretical plates is more limited with chiral columns. Thus higher order of verification is highly welcome in enantioselective analysis, especially as the number of analyte as well as interfering peaks potentially increases due to chiral separations.

In Fig. 6.4a the enantioselective separation of HHCB and its isomer AHTN is demonstrated. This separation is a lengthy, but good one. However, comprehensive 2D-GC reveals in Fig. 6.4b several production impurities in the respective standard solutions as well as in the samples. These interferences could lead to false EFs as the mass fragments are the same. In this case only comprehensive 2D-GC was able to resolve the respective compounds in a way, that EFs can be calculated in a reliable way. A very good discussion on heart-cut techniques in comparison to comprehensive 2D-GC techniques for chiral analysis is given by Borajandi et al. (2005) [114]. Comprehensive 2D-GC has the advantage that all peaks are separated on a secondary column, while heart-cut techniques allow separations in different GC ovens and thus totally independent temperature programs. In [114] several coelution problems in chiral GC analysis of PCBs on single column systems are addressed.

6.2.7.2.4 Treatment after the GC separation. This addresses the selectivity of the detection system. When even MS is not selective enough, MS/MS techniques. GC/EI MS/MS with a triple quadrupole system was used for determination of GC separated toxaphene enantiomers and PBB 149 atropisomers [104,118], and GC/ECNI-MS/MS with a quadrupole for PBB 149 [119].

6.2.7.2.5 Sources of error. Elution orders of enantiomers on CSPs cannot be predicted [42] and should be checked with enantiopure or enantioenriched standards. When these are not available or have not been checked on a particular column, opposed EFs (this means that the deviation from 0.5 in one direction is now in the other direction, for instance 0.622 instead, of 0.378) will be obtained. Such peak reversals of enantiomer elution orders have been reported [65,70,73,83,120]. Even more, some authors noted EFs according to the order of the retention time regardless of the optical rotation was known [121]. For future work it would be important that it is clearly stated which reference system (EF_+ or EF_1, Section 6.3.1.2) was used.

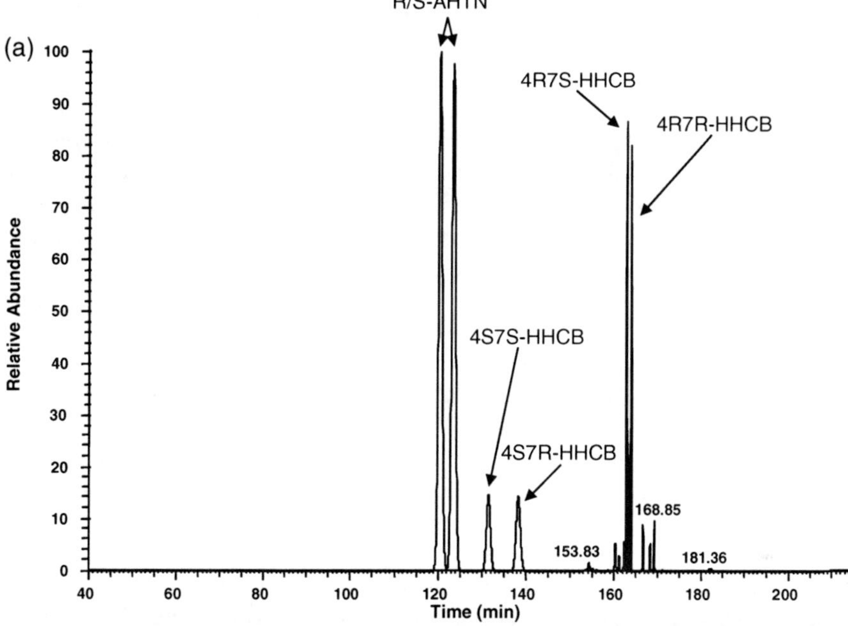

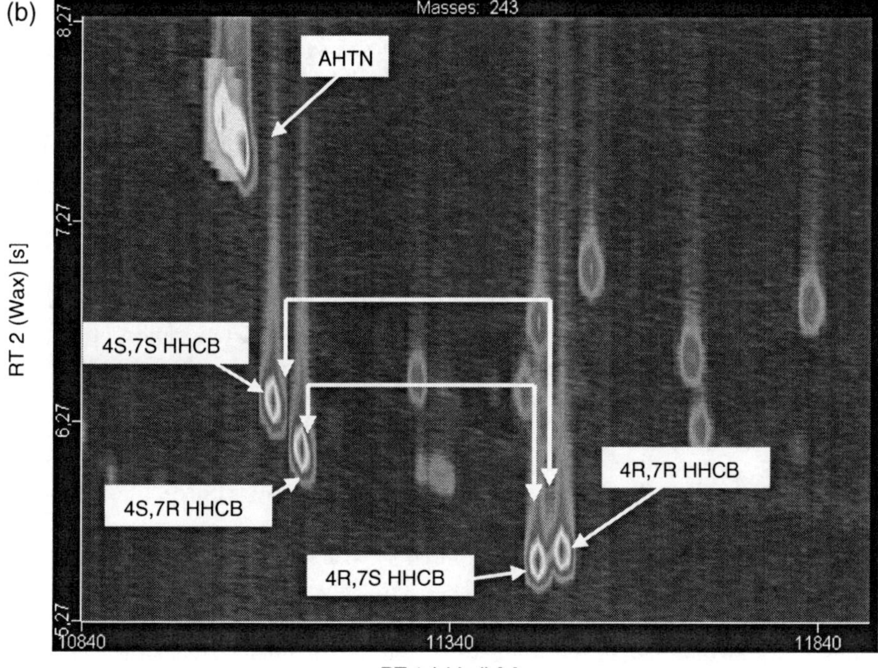

Fig. 6.4. Enantioselective determination of AHTN and HHCB. (a) one-dimensional capillary GC column (110°C, 1 min; at 5°C/min to 132°C, 140 min; at 2°C/min to 195°C, 25 min, and at 5°C/min to 230°C, 20 min) and (b) with a comprehensive 2DGC system.

Otherwise complications in interpreting EFs from different studies may occur. We paid attention to this phenomenon and have used the EF_+ which sometimes required that the values given in the original literature.

Enantiomerization of atropisomers at high temperatures during injection or chromatography (PCBs) is a problem entirely discussed in the PCB section. However, this can be avoided by using moderate GC injector and oven temperatures. There are no known cases where this problem occurred during the analysis of real samples.

Particularly for low abundant compounds the EF cannot directly be obtained by peak integration. For samples, Hites et al. used peak fitting for improving the ER [84]. Eitzer et al. used peak smoothing for improving the peak shape at low concentrations [122].

With regard to sample clean up, standard methods for the determination of chiral organochlorines have been applied. In some studies coelutions with other compounds were noted [73,123,124]. Deterioration of the resolution of oxychlordane enantiomers was observed which was caused by *p,p'*-DDE which eluted between the oxychlordane enantiomers [123]. Although the highly sensitive and selective SIM technique was applied, the enantioresolution was decreased and even the ER was reversed in dependence on the concentration of the interferent [123]. Complexation and saturation of the chiral selector caused by achiral interferents in samples was suggested as explanation [123]. Interference of one enantiomer with a compound that gave no response at the GC/ECNI-MS-SIM masses caused a concentration-dependent falsification of the results [125]. Thus, low amounts should be injected for a proper determination of ERs [126]. These reports—and less pronounced work of Ulrich and Hites [127]—are the only hints in the literature that the highly selective GC/MS-SIM technique might be subject to errors.

Different strategies can be suggested to minimize such problems including (i) measuring EFs at different sample concentration, (ii) spiking the sample with racemate followed by subtraction of the spiked amount, (iii) increasing the column capacity by increasing the separation temperature due to the increase in the average amount of analyte in the gas phase and its corresponding decrease in the stationary phase [126], (iv) confirmation of results on another CSP and (v) isolation or enrichment of the compound prior to or during the GC analysis. In the case of deterioration of enantiomeric resolution [123,126], R_s values (see Eq. (11)), or more simply, α values in standard and sample may be compared to exclude this phenomenon.

In some cases the determination of EFs in multicomponent mixtures was possible by application of multidimensional gas chromatography (determination of atropisomeric PCBs [108]) or single reaction monitoring experiments in MS/MS (determination of chiral CTTs [118]).

Due to the high elution temperatures required for the elution of chiral pollutants, the separation factors are low and in many cases no baseline separation of the enantiomers can be obtained. Under these circumstances, chromatographic quantitation of low amounts of an enantiomer in the presence of the mirror image can be difficult [128]. In this context deconvolution or peak fitting of the peaks have been suggested to improve the precision of EFs even when no baseline separation was obtained or in the presence of minor interferents [84,129].

6.3 DEFINITIONS OF TERMS TO DESCRIBE DEVIATIONS FROM THE RACEMATE

A *molecule* that has no plane of symmetry, center of symmetry, or an alternating axis at most an axis of rotation, is chiral and either "right-handed" or "left-handed". However, a chiral *substance* is racemic if it contains the same amounts of both enantiomer molecules [130]. This case applies for the technical application of chiral pollutants. If the enantiomers are present with different amounts the molecule is optically active (chirality alone is not prerequisite enough for optical activity). The use of the adjective "chiral" is sometimes ambiguous, and has to be specified with remarks on either being racemic or non-racemic [130]. As this term is usually linked to chiroptical measurements [131], the term non-racemic is more suitable for the present topic. Non-racemic may be further distinguished in pure enantiomers (enantiopure or homochiral compound) and an unequal mixture of both enantiomers/excess of one enantiomer (enantioenriched compound). According to Eliel and Wilen, chiral in combination with synthesis, recognition and chromatography is also not precise and should be avoided [130].

To avoid inconvenient long terms we abbreviate terms consisting of enantiomer/enantiomeric/enantiomerically in combination with a second term such as selective, ratio a. s. o. will be used by condensing both words and omitting the "mer..." part. The terms of choice will thus be enantiofraction instead of enantiomer fraction (EF), enantioratio instead of enantiomer(ic) ratio, enantioselective instead of enantiomer selective, enantiopure instead of enantiomerically pure, enantioenriched instead of enantiomerically enriched, enantioexcess instead of enantiomer excess a. s. o.

6.3.1 Suitable terms to describe deviations from the racemate for chiral pollutants

Chiral natural products are often enantiopure and challenges in pharmaceutical or stereoselective organic synthesis are directed towards the exclusive production of one enantiomer. By contrast, the situation is different for chiral pollutants. Most chiral pollutants were applied as racemates and the major task in this kind of research is showing often minute deviations of the target component from the racemate, and assigning these minute changes in the enantiomeric composition to responses with biological systems. Several terms have been used for defining the extents of deviations from the racemic composition.

6.3.1.1 The enantioratio (ER)

In the first decade (1991–2001), ER was the most often applied term to describe deviations from the racemic composition of an organochlorine in environmental analyses [17] (Eq. (1)):

$$ER = \frac{E_+}{E_-} \quad \text{or} \quad \frac{E_1}{E_2} \tag{1}$$

with E being the concentration of the levo (−), dextro (+) or the first (1) and the second (2) enantiomer eluting from the GC column.

Within the linear range of the detection system, the ER is directly obtained by integration of the GC signals. If the enantiomers' directions of optical rotation are known, the ER is formed by the quotient of the dextro to levorotary enantiomer ($ER_{(+/-)}$) [17]. In the case that enantiopure or enantioenriched standards are not available or the direction of light rotation has not been established, the ER is defined as the quotient of the first to the second eluted enantiomer ($ER_{(1/2)}$) [17]. Note that a value for the lack of the second enantiomer (E_2, $E_- = 0$) is mathematically not defined and has to be presented as the limit towards infinity. ERs extend from infinite (only E_+ or E_1) to lim→0 (only E_- or E_2) with ER = 1 for the racemate. Furthermore, ERs are not based on a numeric but on a log scale which may cause some problems [132,133]. e.g., the ER of a tenfold amount of the first eluted enantiomer is 10, the ER of a tenfold amount of the first second enantiomer is 0.1. The mean value would be $10^{(1+/-1)/2} = 10^0 = 1.0$ (and not the arithmetic mean 5.05). Note also that the reciprocal values of ER = 0.5 and 0.4 ($\Delta_{ER} = 0.1$) are 2.0 and 2.5 ($\Delta_{ER} = 0.5$) and ERs of 2.0 and 2.1 ($\Delta_{ER} = 0.1$) correspond with the reciprocal values of 0.5 and 0.48 ($\Delta_{ER} = 0.02$). Wiberg et al. displayed ERs for values >1 and 1/ER for values <1 in graphs [134]. In practical work, values of ERs should be presented within realistic limits. Usually it is reasonable to define ERs >1 with an accuracy of ± 0.1. For ERs < 1, it is often necessary to present two values after the period. The reciprocal value of 1.5 is 0.66; if 0.66 would be rounded to 0.7, the reciprocal value becomes 1.4. Therefore, an accuracy of at least 0.05 is recommended for the upper range and at least 0.01 for the lower range of an ER <1.

Schneider and Ballschmiter demonstrated that the comparision of ER values is only meaningful when the values are re-calculated [96]. This has lead to the invention of enantiofractions (see next section).

6.3.1.2 The enantiofraction (EF)

Currently, EF is the most often applied term used for deviations from the racemate (Eq. (2)):

$$EF_1 = \frac{E_1}{E_1 + E_2} \qquad EF_+ = \frac{E_+}{E_+ + E_-} \tag{2}$$

In pharmaceutical chemistry, EF is known as the chromatographic purity. EF presents the contribution of one enantiomer compared with the sum of the enantiomers. Again it is important to distinguish between the elution order or direction of rotation but since only one enantiomer is ruled, the suitable terms are EF_+ or EF_1, indicating whether the first eluting or the dextro enantiomer (by convention the more important enantiomer is dealt in a given EF. The meaning of the EF is the fraction of one enantiomer to the whole. An EF of 0.4 means that 0.4 parts (or 40%) are enantiomer-1 and 0.6 parts (60%) are enantiomer-2. In the EF scale, racemates are found at EF = 0.5 (50% of enantiomer 1), pure enantiomer-1 as 1.0 (100%) and pure enantiomer-2 as 0.0 (0%). With few exceptions, this system has not been used in environmental sciences [135]. In 2000, two research groups independently suggested that the EF is superior to ER [132,133]. Some of the drawbacks of ER have already been mentioned above [96]. More detailed

Table 6.3 Conversion of enantioratios (ERs) into enantiofractions (EFs) and *vice versa*

ER	EF	ER	EF
100	0.9901	0.95	0.4872
10	0.9091	0.9	0.4737
9	0.9000	0.85	0.4595
8	0.8888	0.8	0.4444
7	0.8750	0.75	0.4286
6	0.8571	0.7	0.4118
5	0.8333	0.65	0.3939
4	0.8000	0.6	0.3750
3	0.7500	0.55	0.3548
2.5	0.7143	0.5	0.3333
2.0	0.6666	0.45	0.3103
1.8	0.6429	0.4	0.2857
1.6	0.6154	0.35	0.2593
1.5	0.6000	0.3	0.2308
1.4	0.5833	0.25	0.2000
1.2	0.5455	0.2	0.1666
1.1	0.5238	0.1	0.0909

explanations can be found in the references mentioned above and in the work by Ulrich et al. [136].

EFs are suitable for the description of most enantioselective processes important for environmental issues and will be used in this chapter. Table 6.3 lists relevant ERs and the EFs calculated from them. There are two aspects that need to be mentioned. First, the EFs need to be presented with two, preferably with three significant numbers. These values such as 0.404 or 0.765 should be used and not be rounded to 0.4 or 0.40 and 0.8 or 0.77, respectively. Second, the recalculation of ER data into EF data as was carried out in this chapter might be accompanied with some minor errors due to the rounding of both values.

6.3.1.3 The enantioexcess (ee)

In pharmaceutical science, the enantiopurity of a component is usually expressed by the enantioexcess (ee). The ee expresses the excess of one higher concentrated enantiomer E_1 over the lower concentrated E_2 (Eq. (3)).

$$\mathrm{ee} = \frac{E_1 - E_2}{E_1 + E_2} \quad (3)$$

The magnitude of the ee ranges from ee = 0 for the racemic mixture to ee = 1 for pure E_1. In practice, ee is often quoted as a percentage. Since values of chiral pollutants are typically rather close to the racemate, the ee is only of minor relevance in environmental chemistry [96,137].

6.3.2 Thermodynamical considerations and classification of enantioseparations

General terms have been established to facilitate intercomparisons of enantioseparations. The separation factor α—originally developed for the separation of two components—has also been applied to describe the separation power of a chromatographic system (Eq. (4)):

$$\alpha = \frac{t_{R'2}}{t_{R'1}}, \tag{4}$$

where $t_{R'1,2}$ are the reduced retention times (total retention time minus the retention time in the mobile phase) on the stationary phase. Typical α values in capillary chromatography range from 1.00 to 1.2; α values above 1.2 are rather scarce. Note that α values are only valid for pure (undiluted) chiral stationary phases [39,138]. Since most of the currently used CSPs are diluted in an achiral medium-polar polysiloxane, retention occurs at both the achiral and the chiral part of diluted columns. The achiral polysiloxane does not influence Δt_R of enantiomers ($\alpha = 1$) because it increases the retention times of both enantiomers in the same extend. However, since α values decrease with higher retention times at a constant Δt_R, the α value for the chiral selector (considering that only enantioselective interactions occur in CSPs) would be too low. Consequently, α values in diluted CSPs are not to be referred to the CSP but to the entire column. Diluted CSPs are characterized by concentration-dependent and non-linear α values [39,138]. Experimental data imply that the optimum of interaction is often reached at low concentrations and no further improvement of selectivity is gained above 30% permethylated β-cyclodextrin or 50% for derivatives with high molecular weights (e.g. γ-cyclodextrins containing pentyl groups) in polysiloxanes [17].

In gas chromatography, enantioselectivity is governed by thermodynamics and is defined by the difference in the free enthalpy (Gibbs energy), $-\Delta_{2,1}(\Delta G)$, of the diastereomeric association complex of the cyclodextrin selector and the enantiomers of a racemic selectand. For temperature-dependent investigations the Gibbs–Helmholtz equation applies [139]:

$$-\Delta_{2,1}(\Delta G) = -\Delta_{2,1}(\Delta H) + T\Delta_{2,1}(\Delta S) = RT \ln K_2/K_1 \tag{5}$$

The enantioselectivity $-\Delta_{2,1}(\Delta G)$ is formed from the enthalpy ($-\Delta_{R,S}(\Delta H)$) and the product of the entropy term and the temperature ($T\,\Delta_{2,1}(\Delta S)$). K_2 and K_1 refer to the formation constants of the diastereomeric association between the chiral selector and the enantiomers 2 and 1. To a first approximation, K_2/K_1 can be substituted by the separation factor α which leads to Eq. (6):

$$-\Delta_{2,1}(\Delta G) = -\,\Delta_{2,1}(\Delta H) + T\Delta_{2,1}(\Delta S) = RT \ln \alpha \tag{6}$$

The separation factor α is thus proportional to the free enthalpy $\Delta_{2,1}(\Delta G)$. Furthermore, $-\Delta_{2,1}(\Delta H)$ and $\Delta_{2,1}(\Delta S)$ compensate each other in determining $-\Delta_{2,1}(\Delta G)$. Equation (6) may be rewritten as the van't Hoff plot:

$$\frac{-\Delta_{2,1}(\Delta G)}{T} = \frac{-\Delta_{2,1}(\Delta H)}{T} + \Delta_{2,1}(\Delta S) = R \ln \alpha \tag{7}$$

Equations (6) and (7) show that α is related to enantioselectivity and thus is a temperature-dependent quantity. α can therefore not be defined in temperature-programmed runs [39]. For this reason, this value is more of a theoretical relevance (see also section 6.3.3).

The enthalpy $\Delta_{2,1}(\Delta H)$ and entropy $\Delta_{2,1}(\Delta S)$ can be obtained by measuring the α values at different temperatures [39,47,126] and plotting $R \ln \alpha$ against $1/T$. These graphs give straight lines if $\Delta_{2,1}(\Delta H)$ is constant within the analyzed temperature range. If so, the slope is $\Delta_{2,1}(\Delta H)$ and the intercept is $\Delta_{2,1}(\Delta S)$ [126]. Koppenhöfer et al. predicted the existence of an isoenantioselective temperature T_{isoenan} at which enantiomers co-elute [140] (Eq. (8)):

$$T_{\text{isoenan}}=\Delta_{2,1}(\Delta H)/\Delta_{2,1}(\Delta S) \quad \text{for} \; -\Delta_{2,1}(\Delta G)=O \tag{8}$$

Below T_{isoenan}, enantioseparation is enthalpy controlled and the enantiomer 2 is eluted after enantiomer 1 while above T_{isoenan} enantioseparation is entropy controlled and enantiomer 2 is eluted after enantiomer 1. As the result of entropy/enthalpy compensation, reversal of the elution order of enantiomers may occur at different temperatures. The first peak reversal on compounds discussed in this chapter were found for the enantiomers of mecoprop and dichlorprop [141]. Further examples were described by the Maas et al. [47]. However, only van't Hoff plots that are linear over a wide temperature range around T_{isoenan} [142,143] fulfill the thermodynamic principle of entropy/enthalpy compensation [39].

Most enantioseparations are governed by enthalpy control and the elution order of enantiomers cannot be reversed. Increasing of the separation temperature decreases the α value until the enantiomers coelute ($\alpha = 1.000$). A further increase of the temperature will no more improve the α value. As a consequence, the temperature of the measurements should be lowered to an acceptable level in order to increase the separation factor α [54,126].

6.3.3 Chiral resolution

The thermodynamic relationship between the separation factor α and the van't Hoff Eq. (7) does not take into account the peak width (in thermodynamics, signals are not regarded as peaks with gauss-shape-like broadening at the bottom but as ideal vertical lines which are not defined in the horizontal dimension) [19]. Thus, an increase in the α value is not necessarily accompanied with an improved resolution of the enantiomers. For this reason, the chiral resolution R_S is an important factor to assess the efficiency of an enantioseparation (Eq. (9)):

$$R_S=\frac{2\Delta t'}{w_{b1}+w_{b2}}, \tag{9}$$

where w_{b1} and w_{b2} are the peak widths of the tangents at the baseline [144]. Note that R_S values on the basis of peak width at half height sometimes presented in the literature are not in agreement with the rules of the IUPAC [144]. Resolution values using the peak width at half height are usually a bit smaller (typical deviations are 0.1) than the IUPAC-conform R_S values derived from Eq. (9).

The maximum α value in enantioseparations is obtained with isothermal elution (which is in contrast to isomer separations) while the elution of the enantiomers within temperature ramps decreases α. In agreement with achiral gas chromatography, the peak width increases with increasing retention time. The lower the GC temperature the higher both the retention time and the chiral resolution values usually are. From a certain point on, the α value may still increase at lowering the elution temperature but the resolution R_S decreases due to the peak broadening [19].

In practice, most enantioseparations are performed using slow heating rates. GC oven programs for chiral pollutants may be developed in the following way. The GC oven is programmed from the starting temperature to ~120°C, and then the temperature is increased by 1–5°C/min until the enantiomers are eluted. Consecutively, the program should be refined until a sufficient chiral resolution is obtained (which may require heating rates between 0.5°C/min and 2°C/min). For testing any new chiral compound on a CSP, the temperature and program should be lowered until the target compounds elutes after ~100 min. In many cases, slow heating rates offer the possibility to separate more than one compound in one run.

6.4 CHIRAL POLLUTANTS

Racemic compounds consist of equal amounts of two enantiomers. Enantiomers are two chemically identical molecular species which differ from each other as non-superposable mirror images [145]. These non-superposable units have identical physical properties including identical mass spectra, retention times on achiral stationary phases, and melting points (yet, the melting point of single enantiomers is different to that of the racemate). For instance, the racemic heptachlor epoxide has a melting point of 157–160°C and pure (+)- and (−)-heptachlor epoxide enantiomers of 166°C [14].

However, chirality is not a general requirement for biological activity [5], and only a few chloropesticides are actually chiral, and if, chirality was unintended/unmotivated during the introduction of this compound some 50 years ago. In most cases, organochlorine compounds were synthesized by the chlorination of a suitable hydrocarbon backbone which often resulted in more or less complex mixtures. If the active compounds were achiral, the technical products at least contained major or minor chiral components (e.g. *o,p′*-DDT or α-HCH) so that more or less each substance class could be investigated enantioselectively. This holds also for PCBs and other chiral pollutants described in this book. The environmental fate of chiral pollutants could be studied enantioselectively with the GC equipment including CSPs on the basis of *O*-terminated cyclodextrins described in Section 6.2.

Sunlight, air and (pure) water are nonchiral media and will not change the ratio of enantiomers when involved in transformations of pollutants. This is important when the enantiomeric distribution of a chiral pollutant in air changes in dependence of the height above land or water (see below). Observed changes in the enantiomeric composition are then due to the mixing of two sources with different enantiomeric distribution. The reversed case is worth mentioning as well. In the case that a chiral pollutant is found in a non-racemic composition, and it is transformed by a nonchiral medium (for instance photochemically in air), the statistic likeability of transformation will be in the ratio found in the sample so that the EF will not be changed. The following example may serve for clarification. Assume that α-HCH in air is found at 150 pg/m^3 with an EF_+ of 0.666. This would translate into 100 pg/m^3 (+)-α-HCH and 50 pg/m^3 (−)-α-HCH. Assume further that photochemical transformation accounts for 10% of the original sample, i.e. 15 pg/m^3 (10 pg/m^3 (+)-α-HCH and 5 pg/m^3 (−)-α-HCH). Then, the resulting (remaining) concentration and EF_+ will be 135 pg/m^3 and 0.666, respectively. This is of course trivial. However, when we assume that one transformation product is resulting and that phototransformation enantioselectively leads from (+)-α-HCH (arbitrarily chosen) to the (+)-metabolite and (−)-α-HCH to the (−)-metabolite, then the photometabolite will be formed in non-racemic composition in air [146]. Non-enantioselective processes can be involved in the formation of chiral compounds. For instance, reductive dechlorination (as it may occur in air) may transform the nonchiral PCB 187 into the atropisomeric PCB 149 [146]. Likewise, the nonchiral chloropesticide dieldrin can be transformed into the chiral photodieldrin by sunlight [147]. The formed "chiralized" metabolites may then be enantioselectively transformed in contact with a chiral entity.

The diverse possibilities and their investigation is the major focus of the following subsections on chiral pollutants. Due to our own scientific background, we mainly focused on halogenated pollutants, including chloropesticide formulations (HCH, DDT, chlordane, and toxaphene), their chiral metabolites (e.g. oxychlordane, heptachlor epoxide), and minor side products in technical grade products, as well as the same range for atropisomeric PCBs (methylsulfonyl, hydroxy metabolites), their brominated analogs and further organochlorine industrial chemicals. The scope was expanded by the inclusion of modern pesticides and musk fragrances, which are amendable for GC enantioseparations.

6.4.1 Atropisomeric polychlorinated biphenyls (PCBs), their metabolites ($MeSO_2$-PCBs), and polybrominated analogs (PBBs)

PCBs have been used for different industrial applications at an estimated amount of 1.3 million tons, and 97% thereof in the northern hemisphere [148]. PCBs are produced by the controlled chlorination of biphenyl in the presence of a suitable catalyst. In dependence of the amount of chlorine applied to biphenyl, different technical PCB mixtures with varying average numbers of chlorine are produced. Each technical product consists of 50–70 PCB congeners [11]. Depending on the desired physico-chemical properties, the technical PCB products have been applied for many industrial purposes since the 1940s. Highest production rates were reported for tri- to pentachloro congeners [148], whereas the bulk of residues in the environment including those to be dealt in this section are hexa- and heptachlorobiphenyls. The use of PCBs was restricted soon after they have been detected by Jensen in fish, an eagle and hair from himself and his family [149]. Altogether, ~130 of the 209 possible PCBs have been identified in different technical formulations [150], and these may also be found in food and the environment. To avoid long and cumbersome chemical names, Ballschmiter and Zell suggested a coding system which has been accepted by the IUPAC [151].

Biological properties of PCBs depend on the number of *ortho*-substituents. An increasing number of bulky (chlorine) substituents in *ortho*-position leads to a twist of the bond between the two phenyl rings, thus they are accordingly no longer in one plane. The tendency for non-planarity due to bulky *ortho*-chlorine substituents is opposed by π-electron overlap, which produces maximum stabilization when the rings are coplanar [130]. The shift of the strong conjugation band in UV spectra is a good measure of the disturbance of the π-electron overlap in *ortho*-substituted biphenyls. Distortion from planarity gradually decreases the intramolecular π-conjugation of the two phenyl rings and leads to a blue shift of the UV absorption maximum [152].

6.4.1.1 Chirality, axial chirality, and atropisomerism

PCBs do not have sp^3-hybridized carbons and chirality results from the orientation of the two phenyl rings in space (coplanar PCB molecules cannot be chiral). The first criterion for chirality of PCBs is the lack of symmetrical element that can superimpose the molecule. Any substitution patterns which are at least different either at positions 2/6 or 3/5 are chiral when the rings are out of coplanarity (Fig. 6.5). Among di-*ortho* substituted phenyl rings only two (substitution in 2,3,6- and 2,3,4,6-positions) and among the mono-*ortho* substituted, eight (2; 2,3; 2,4; 2,5; 2,3,4; 2,3,5; 2,3,6; 2,4,5), and two non-*ortho*-ring-patterns (3; 3,4) nonsymmetric substitution patterns fulfill the requirement (Table 6.4). Combination of these 12 substitution patterns leads to $\sum_{n-1}^{12} n = 78$ chiral out of the 209 possible PCBs.

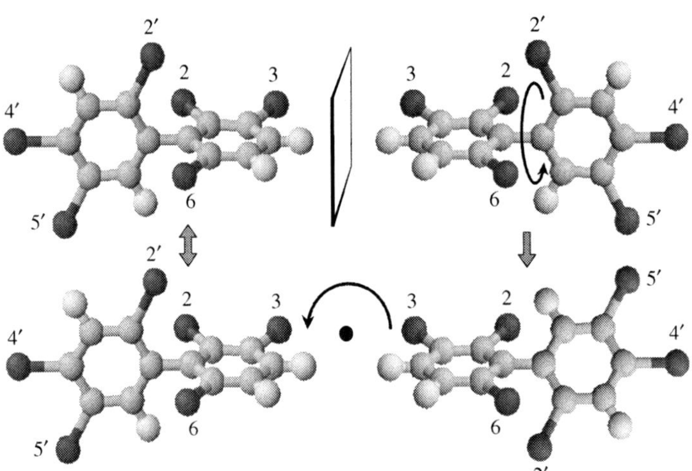

Fig. 6.5. Atropisomerism of PBB 149 [152]. Upper row: mirroring results at a plane results in two non-superposable molecules (enantiomers); right panel: intra-molecular rotation with 180° about the interannular phenyl–phenyl bond. This is impossible for the present molecule. This rotation may occur at elevated temperature; lower row: mirroring at a (hypotetic) point axis perpendicular to the paper sheet; left panel: identical molecules (chirality annulated by rotation about the intramolecular phenyl–phenyl bond).

Table 6.4 Chirality and atopisomerism of PCBs

Substitution	Chiral	Stable atropisomers if (+) appears in row	
		First ring 2,3,6*	First ring 2,3,4,6*
–	Nonchiral		
2	Chiral	+	+
3	(Chiral)**	–	–
4	Nonchiral	–	–
2,3	Chiral	+	+
2,4	Chiral	+	+
2,5	Chiral	+	+
2,6	Nonchiral	–	–
3,4	(Chiral)**	–	–
3,5	Nonchiral	–	–
2,3,4	Chiral	+	+
2,3,5	Chiral	+	+
2,3,6*	Chiral	+	+***
2,4,5	Chiral	+	+
2,4,6	Nonchiral	–	–
3,4,5	Nonchiral	–	–
2,3,4,5	Chiral	+	+
2,3,4,6*	Chiral	+***	+
2,3,5,6	Nonchiral	–	–
2,3,4,5,6	Nonchiral	–	–

*One of these two substitution patterns on one phenyl unit is madatory for PCB atropisomers.

**Chiral but no *ortho*-Cl.

***Identical.

Chirality alone is, however, not enough for resolving these 78 chiral PCBs into enantiomers. For this reason, the rings must not rotate aboute the interannular phenyl–phenyl–bond due to steric congression (if the molecule rotates about the phenyl–phenyl–bond, the one enantiomer is transferred into the other enantiomer, see Fig. 6.5). This is a matter of temperature (energy supplied) and the stability increases with number and size of bulky substituents in *ortho*-positions. In the case of PCBs and gas chromatographic investigation, hindered rotation is met by PCBs with three *ortho*-substituents (*ortho*-chlorine atoms on either 2,2′,6 or 2,6,6′ or 2,2′,6,6′) which is fulfilled by combining the two chiral di-*ortho* substitution patterns with di- or mono-*ortho*-substiution pattern. In other terms, hindered rotation at elevated temperatures (here: GC conditions) as well as chirality require that the respective PCBs have either a 2,3,6- or 2,3,4,6-substitution pattern on one ring. This is met by 19 of the 78 chiral PCB congeners most of which have been detected in technical mixtures (Table 6.4) [9,150]. These 19 PCBs are called PCB atropisomers (chirality plus hindered rotation). Atropisomerism is a type of stereoisomerism that may arise in systems where free rotation about a single covalent bond is impeded sufficiently so as to allow different stereoisomers to be isolated [153]. Consequently, chiral PCBs and atropisomeric PCBs are no synonyms. To give an example, the most prominent PCB 153 (2,2′,4,4′,5,5′-hexachlorobiphenyl) is chiral but forms no atropisomers even at

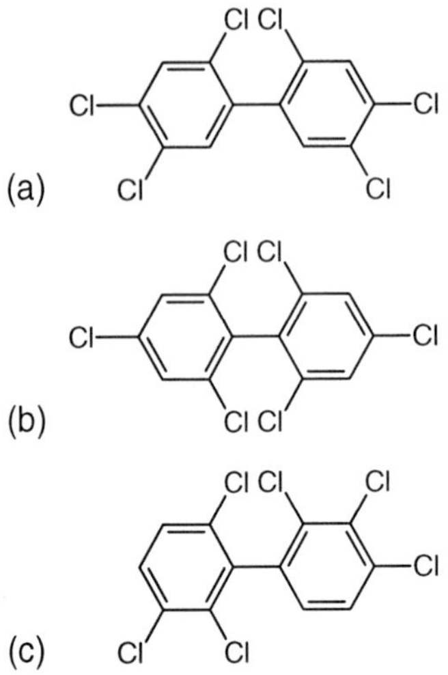

Fig. 6.6. Structure of (a) PCB 153 (chiral but no hindered rotation about the interannular phenyl–phenyl bond at physiological temperatures), (b) PCB 155 (hindered rotation about the phenyl–phenyl bond but nonchiral), and (c) PCB 132 (both hindered rotation and chiral; exists as a pair of atropisomers).

room temperature (Fig. 6.6). However, the brominated analog PBB 153 was recently shown to be partly resolved into enantiomers by enantioselective HPLC at 0°C [104]. Note as well that several di-*ortho*-PCBs were enantioresolved by HPLC [154]. However, their energy barrier is easily surmounted at the elevated temperatures mandatory for the GC elution. Atropisomeric PCBs do not have any sp^3-hybrided carbon, and enantiomerization (surmounting of the planarity) may occur without cleavage of any bond. As chirality of PCBs is temperature dependent, studies were necessary to check if enantiopure PCB atropisomers are prone to (partial) racemization under physiological ($T < 40$°C) and even GC ($T > 200$°C) conditions. Different studies confirmed that the 19 atropisomers discussed above are stable under the condition found in the environment [154,155]. However, some racemization (interconversion) was observed for chiral di-*ortho* substituted PCBs at 0°C [154].

Injection of pure enantiomers and analysis with a stopped-flow method (the GC carrier gas flow was stopped for a moment and the GC oven temperature was raised to 300 or 320°C) racemization of PCB 95, PCB 132, and PCB 149, i.e. three major atropisomeric PCBs in environmental samples, was proven [156,157]. Harju and Haglund also found significant racemization between 280 and 300°C; under these GC conditions, half lives of enantiomers were in the range of 6–12 min [158]. In this case, characteristic plateaus between the (resolved) enantiomers can be observed in the GC chromatograms [159–162]. However, these temperatures are not a problem for enantioselective chromatography which have to be carried out at significantly lower temperatures, but both the temperature of the GC injector port and the transfer line in a GC/MS system should be lowered accordingly (maximum temperature 260°C) in order to prevent any racemization of the 19 PCB atropisomers, and the study of enantioenrichment of these atropisomeric PCB in biological samples is feasible.

6.4.1.2 Enantioselective determination of PCB atropisomers in environmental samples
After some pioneering work with enantioselective liquid chromatography [155,163,164], GC enantioseparation of PCB atropisomers was achieved in 1993 [109,159]. In the meantime, the 19 atropisomeric PCBs have been successfully resolved but this required several GC stationary phases [19]. Haglund et al. determined the optical rotation of 12 of the 19 atropisomeric PCB enantiomers after HPLC fractionating and determined the elution order on the frequently applied CB-β-PMCD column [150,154,158]. In the case of PCB 84, PCB 132, PCB 136, PCB 131, PCB 144, PCB 149, PCB 175, PCB 176, PCB 183, and PCB 196 the (−)-enantiomers and in the case PCB 135 and PCB 174 the (+)-enantiomers eluted first [165]. This information was used and if appropriate EFs reported in the literature were converted into EF_{+}.

6.4.1.3 EFs of atropisomeric PCBs in samples
Application of MDGC using the heart-cut technique (Section 6.2.7) demonstrated that PCB atropisomers were racemic in a technical PCB mixture [108]. EFs of atropisomeric PCBs in Aroclor were later studied for getting insights on potential interferences [25]. PCB 136 showed the expected racemic composition in all Aroclors (0.501, respectively) but a 1:1:1 mix of the standards gave a slightly higher deviation (0.507) than typically found for individual standards (<0.505). Some deviations from the racemate were found for PCB 95 (EF = 0.490) in Aroclor 1242 and for PCB 149 (0.488) in Aroclor 1254. These examples indicate a small interference relative to neat standards, but these will not change the correct result unless the (potential) interferent is not significantly more stable than the atropisomeric PCB to be studied [25].

PCB 95, PCB 132, and PCB 149 were also racemic in the sediment of small river samples from Southern Germany [129,166]. Benická et al. measured EFs of 0.375 for PCB 95 while PCB 91 was racemic in other sediment samples [129]. Using one-column technique, Wong and Garrison tested seven CSPs and found two to five atropisomers quantificable on five CSPs using GC/MS [92]. These CSPs were used to determine the EF of PCB 91 in sediment from Lake Hartwell/GA, USA. EF of PCB 91 was 0.375 on CB-β-PMCD and the reciprocal value on β-TBDM (reversed elution order of PCB 91 atropisomers) [92]. Ren and Harrad investigated the influence of soil evaporation on the EF of PCB 149 in air. EFs of PCB 149 in soil ranged from 0.459 to 0.500 (arithmetic mean: 0.479) while PCB 149 was racemic in outdoor air [167]. No significant contribution of soil evaporation to outdoor air was found [167].

Soil samples from the UK taken at the same site within only a few meters showed surprisingly high variations in the EF when taken over several months. For instance, variations in time for PCB 95 ranged from EF = 0.350 to 0.430 [25]. Either seasonal variations or, more-likely, small-scale variations in the microbial community were made responsible for these variations [25]. Given this fact, the greatest error connected to these measurements occurs most likely due to sampling and not because of analytical errors. At 1.5 m above soil, PCBs in air exihibited higher difference to racemic standards than soil samples (Fig. 6.7). Thus, the authors concluded that the PCBs in air must

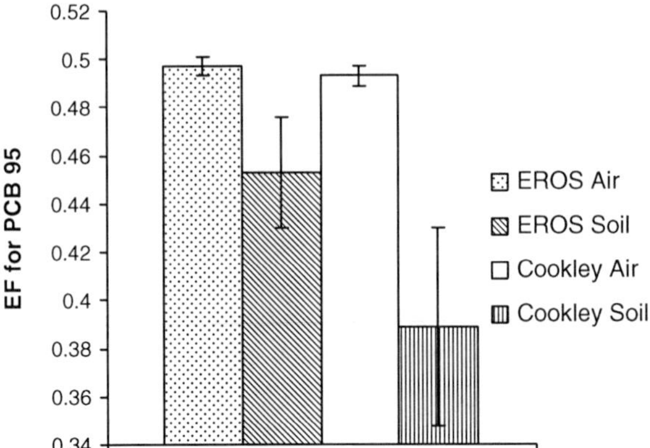

Fig. 6.7. EFs of PCB 95 in soil and air above soil from two sites in Great Britain [25]. The significantly less nonracemic composition in air indicates PCB 132 immission from other sources than the soil, suspected to arise from recent applications.

arise from other sources than evaporation from soil [25]. A clear mass balance of such measurements is currently not possible since other factors may affect such measurements as well. For example, photolytic processes may dehalogenate higher chlorinated PCBs in non-enantioselective processes to the PCB atropisomers. As an example, the nonchiral PCB 187 (2,2′,3,4′,5,5′,6-heptachlorobiphenyl) is converted into the chiral PCB 149 (2,2′,3,4′,5′,6-hexachlorobiphenyl) upon reductive dechlorination of the Cl on C5.

EF of atropisomeric PCBs in lake sediments (Lake Lartwell, SC) were reported by Wong et al. [168]. Active sediment bacteria when incubated with individual PCB 149 and PCB 132 were reductively dechlorinated but without enantioselectivity [169]. As mentioned previously, transformation of chiral compounds is not necessarily connected to enantioselective processes. Thus, changes in the EF clearly demonstrate biotransformation but *vice versa*, lack of changes in the EF does not rule out biotransformation. However, enantioselectivity was found for PCB 91 and 95 in both lake sediments and incubated sediment bacteria. In addition the EF was reversed in sediment of a near-by river so that both microbial sediment communities showed different enantioselectivity in biotransformation [168]. Interestingly, the EF changed with the depth of the sediment core section, including enantiomeric switch from EF > 0.5 in slices closer to the water surface and EF < 0.5 in deeper sediment [169]. Singer et al. [170] investigated transformation of four atropisomeric PCBs (PCB 45, PCB 84, PCB 91, and PCB 95) with five bacterial strains and found that enantioselectivity was varying with respect to strain, PCB congener, and cosubstrates [170]. Particularly, the unexpectedly great influence of the cosubstrates illustrates that the processes leading to changes in the EF are very complex. For instance, Harrad et al. found significantly month-to-month variations in the EFs of PCB 95 in soil which were different at different sites [171]. In light of these results, EF differences of atropisomeric PCBs in soil and air need to be carried out with statistic evaluation of the sample number.

Wong et al. studied the seven PCB atropisomers in an aquatic food web represented by seven species in Lake Superior. The study included four fish species whose tissues contained non-racemic proportions of the most PCB atropisomers [172]. EFs of atropisomeric PCBs in phytoplankton and zooplankton were racemic. The authors indicated that this is likely due to their low capacity to (enantioelectively) biotransform PCB atropisomers. Surprisingly, the EFs of mysids and *Diporeira* were non-racemic which points either towards non-racemic uptake or enantioselective biotransformation [172]. Since both water and sediments samples could not be analyzed, this question could not be answered. However, in some fish species analyzed, the prey was known and the significant change in the EF indicated enantioselective biotransformation processes [172]. Aquatic riparian biota form the same site confirmed changes in the EFs of PCB 91, PCB 95, PCB 136, and PCB 149 relative to sediments [173]. Under certain circumstances, determination of EFs may be used as an indicator for the preferred prey of predator fish. Trouts are both feeding on sculpins and lake herring. (−)-PCB 149 was enriched in lake herring and (+)-PCB 149 in sculpins, and both scenarios were found for the EF in lake trout [172]. Thus, the EF of PCB 149 in trout may indicate which diet was preferably consumed by lake trout [172].

$EF_{+} > 0.500$ (0.523–0.714) were determined for PCB 132 in human milk [108,174]. EFs of atropisomeric PCBs in human tissue varied in dependence of PCB congener and person from 0.500 to 0.747 (maximum value for enrichment of (+)-PCB 132 and E_1-PCB 95) [121]. No statistic relevant differences between brain, muscle, kidney and liver were observed. The EFs of PCB 132 in human liver were in the same range as previously found in (different samples of) human milk [174]. Deviations from the racemate correlated within the three PCBs (high non-racemity of PCB 95 meant also the same for PCB 132 and PCB 149). Daily intake appeared to be more relevant than gender or age [121]. In a feeding/elimination study with trout (for more details see Section 6.4.4), PCB 95 was racemic throughout the uptake and depuration phase but the EF of PCB 136 rapidly became non-racemic and the EF steadily decreased [175].

In liver of shark (*C. coelolepis*), PCB 95 and PCB 149 were racemic while EF_{+} of PCB 132 was 0.523–0.555 [110]. EFs of PCB 149 in blue mussels ranged from 0.500 to 0.545 indicating a weak enantioenrichment of the first eluting enantiomer (determined with pure β-TBDM) [176]. A slight enantioenrichment of (−)-PCB 149 was determined in blubber of Icelandic harbor seals (*Phoca vitulina*) and grey seals (*Halichoerus grypus*) and racemic composition in harbor seals from the North Sea and a grey seal from the Baltic [73,102,124]. In cetaceans stranded at the Mediterranean Sea, EFs of PCB 95 ranged from 0.459 to 0.500, EF_{+} of PCB 132 from 0.500 to 0.667, EF_{+} of PCB 135 from 0.387 to 0.500, EF_{+} of PCB 136 from 0.474 to 0.524, EF_{+} of PCB 149 from 0.488–0.588, EF_{+} of PCB 174 from 0.444 to 0.545, and EF_{+} of PCB 176 from 0.500 to 0.588 [177,178]. Blubber of cetaceans (*Balaena mysticetus*) feeding on invertebrates showed slight deviation from the racemate for PCB 95, PCB 136, PCB 149, and PCB 174 but significant changes for PCB 91 (EF 0.456), PCB 135 (EF 0.571), PCB 176 (EF 0.551), and PCB 183 (EF 0.475) [179]. The deviation from the racemate was usually more pronounced in the storage tissue blubber than in the metabolizing tissue liver. The EF of three PCBs (91, 95, and 149) deviated more from the racemate in longer (older) cetaceans than in shorter. Gender-specific variations were only found for PCB 91 [179]. Livers of Arctic foxes and wolverine from Canada showed slightly non-racemic proportions for PCB 91 (0.546 ± 0.060 and 0.497 ± 0.022), PCB 95 (0.496 ± 0.014 and 0.506 ± 0.006), PCB 136 (0.477 ± 0.011 and 0.411 ± 0.025), and PCB 149 (0.535 ± 0.007 and 0.461 ± 0.030) [98]. In most studies of higher organisms, only a weak enantioselective breakdown was found but some samples seem to present exceptions. These examples show that a clear interpretation of EFs of atropisomeric PCBs in biota is currently not possible.

6.4.1.4 Methylsulfonyl PCBs

Methylsulfonyl-PCBs are metabolites of PCBs. Their distribution in the environment is not well documented although their levels may be as high as 15% of the parent compounds. However, the analytical procedure is more demanding and the standard compounds are comparably expensive. A comprehensive summary of the state of the art in $MeSO_2$-PCB research can be found in a recent volume of the Handbook of Environmental Chemistry [180]. The theoretical number of $MeSO_2$-PCBs is 837 [181] and 170 of them may form stable atropisomers [182]. Methylsulfonyl groups are usually found in 3- or 4-positions of PCBs. Consequently, methylsulfonyl-derivatives of the 19 parent PCBs described above are candidates for enantioselective studies. A simple nomenclature of $MeSO_2$-PCBs is obtained in the following way: the carbon number of the methylsulfonyl group is listed (by the respective number) followed—separated by a hyphen—by the IUPAC number of the respective PCB. For instance, the 3-$MeSO_2$ metabolite of PCB 149 (3-$MeSO_2$-2,2′,3,4′,5′,6-hexachlorobiphenyl) is abbreviated 3-149 [180]. Recently, pure enantiomers of four key-$MeSO_2$-PCBs were separated into enantiomers and the absolute structures were determined. It was shown that both rings are almost perpendicular (twisted by 88–90°) and that the methylsulfone group is also perpendicular to the ring on which it is located [183]. Given the fact that formation of $MeSO_2$-PCBs mainly occurs in liver of higher organisms, different mammals species have been analyzed by enantioselective GC.

First enantioselective studies on the $MeSO_2$-PCBs indicated very strong enantioenrichment of 3-132 and 3-149 in human livers, ringed seals and polar bear adipose tissue [184,185]. For instance, the EFs of 3-149 in seal and polar bear were 0.242 and <0.090, respectively [185]. Enantioenrichment of further $MeSO_2$-PCBs was mentioned as well. Two further studies confirmed that liver samples of (marine) mammals almost exclusively contained one enantiomer [186,187]. Results obtained in rat liver after dosing rats with Clophen A50, as well as data in human liver illustrated that enantioselectivity of $MeSO_2$-PCBs is more pronounced than that of the parent PCBs [188]. In a rat study, one dosis of Aroclor was administered and the EF of atropisomeric $MeSO_2$-PCBs was determined after 1, 2, 4, and 8 weeks [189]. In all cases, the EF was significantly different from the racemate with often a tenfold excess of one of the atropisomers [189]. Interestingly, 4-$MeSO_2$-PCB 4-149 showed $EF = 0.820$ in liver and 0.240 in the lung and therefore reversed relative to the racemate. Since $MeSO_2$-PCBs were neither found in the controls nor in the administered PCBs, it is evident that the formation process was highly enantioselective for $MeSO_2$-PCBs [189]. Chu et al. studied the enantiocomposition of PCB 132 and PCB 149 and $MeSO_2$-metabolites in the same samples and found for the latter EFs of ~0.600 and 0.300, respectively; indicating that the proportion of the parent compound left was also transformed enantioselectively [190].

6.4.1.5 Polybrominated biphenyls (PBBs)

Polybrominated biphenyls (PBBs) have been applied as flame retardants in textiles, electronic equipment and plastics [191]. Technical PBB products have been marketed under trade names such as Firemaster BP-6®, Firemaster FF-1®, Bromkal 80® and Flammex-B®. Due to their structural analogy to polychlorinated biphenyls (PCBs), PBBs came into the focus of environmental research as early as in the 1970s [192,193]. Furthermore, in 1973 an accidental contamination of human food with PBBs in Michigan revealed the toxicological threat of this group of chemicals [194,195]. Despite a continuous reduction of the worldwide annual production since the late 1970s and slightly declining levels in the environment in the last decade, the ubiquitous presence of PBBs has been documented in a wide range of samples [191]. In environmental samples, ΣPBB concentrations are usually ~1% of the sum of the polybrominated diphenyl ethers (ΣPBDEs) [191]. In contrast to PCBs only a few individual PBB standards are commercially available but none of the 19 congeners known from PCBs to form stable atropisomers (see above). Ten PBBs were recently isolated from a technical product including several atropisomeric PBBs; six of these PBBs, including PBB 132 and PBB 149 (for substitution pattern see PCB chapter) were identified and enantioseparated by means of high performance liquid chromatography (HPLC) [152]. In addition, the enantiomers of PBB 149 were partially resolved by enantioselective gas chromatography (GC) [152]. In a follow up study the EFs of PBB 132 and PBB 149 were determined in one seabird sample [104,119]. It required a lot of work to obtain EF values for two compounds in one sample which gives only indicative information. However, the techniques used deserve to be discussed. For PBB 149 (the only atropisomeric PBB GC resolved to date), an interference with another brominated compound was observed during GC/ECNI-MS analysis. This prevented determination of the EF. Thus, more selective GC/EI-MS-MS and—for the first time for halogenated compounds—GC/ECNI-MS-MS methods were successfully applied (EF = 0.420–0.430) [104,119]. PBB 132 atropisomers which were not resolved by GC and thus quantitatively separated the two enantiomers by enantioselective HPLC. The isolated enantiomers were then quantified by nonchiral GC (Fig. 6.8). The quantities of the PBB 132 isomers in both fractions were almost identical so that PBB 132 was almost racemic. As was found for atropisomeric PCBs the EFs of atropisomeric PBBs varied depending on the structure.

6.4.2 DDT and DDT-related compounds

DDT (1,1,1-trichloro-2,2-*bis*(*p*-chlorophenyl)ethane or dichlorodiphenyltrichloroethane) was synthesized for the first time in the nineteenth century. However, its insecticidal properties were first discovered in 1939 by Paul Hermann Müller (Nobel Prize in Medicine, 1948, "for his development of DDT as an insecticide" [196]). Soon after the commercial release, DDT became a phenomenal success worldwide. To date,

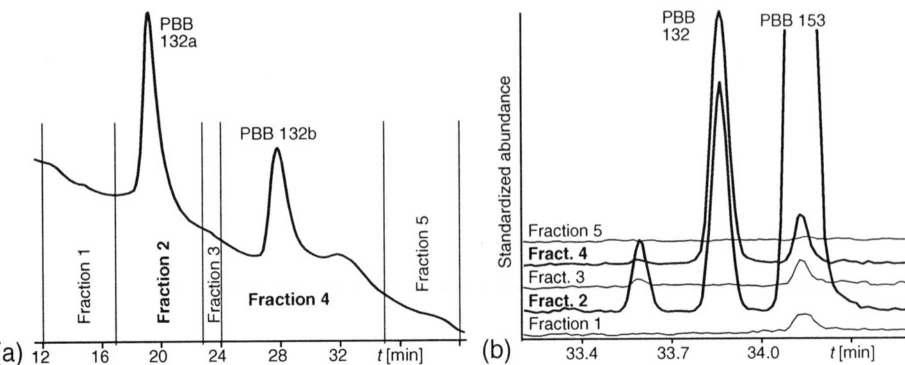

Fig. 6.8. Indirect determination of the EF of PBB 132 in a bird egg extract [104]. Since PBB 132 could not be GC-enantioseparated on any available CSP. Using enantioselective HPLC, the enantiomers were separately collected in fractions 2 and 4, and then quantified by conventional GC/MS. Controls (fractions 1,3,5) verified that the HPLC-enantioseparation was quantitative.

DDT remained the most effective agricultural pesticide ever developed [196]. DDT helped to prevent malaria by eradicating mosquitoes, and may have saved the lives of millions of people [197]. DDT skin powder was used to delouse nearly 1.5 million people during a typhus epidemic in Naples in 1943–45, and also in Japan [196]. On the other hand, the accumulation of DDT and its metabolites in the environment was the starting point for the recognition of organochlorine compounds as persistent and hazardous environmental contaminants.

The global usage of DDT was estimated at 2.6 million metric tons [198]. In contrast to other organochlorines classified as persistent organic pollutant (POP), use of DDT is still permitted in malaria endemic parts of the world—particularly tropical regions. Typical technical mixtures of DDT consist of 77.1% achiral *p,p′*-DDT, and 14.9% chiral *o,p′*-DDT (Fig. 6.9). Further compounds detected in technical DDT are 4% of the achiral *p,p′*-dichlorodiphenyldichloroethene (*p,p′*-DDE), 0.1% of *o,p′*-DDE, 0.3% of *p,p′*-dichlorodiphenyldichloroethane (*p,p′*-DDD), and 0.1% of the chiral *o,p′*-DDD (Fig. 6.9) along with ~3.5% unidentified compounds [199]. From the production rates mentioned above, it can be estimated that 390,000 tons of chiral *o,p′*-DDT and 2,600 tons *o,p′*-DDD have been directly released into the environment through the application of (technical) DDT. The latter is also a metabolite that is formed by the reductive dechlorination of *o,p′*-DDT. The chiral DDT compounds were present as racemates in the technical products [200].

6.4.2.1 Enantioselective determination of DDT-related compounds

EFs of *o,p′*-DDT in the North American atmosphere ranged from 0.410 to 0.500 [201–204]. The racemic composition in air samples from Mexico and Belize was attributed to recent use of the pesticide but due to the different EFs determined in soils from different geographic regions and the low number of samples analyzed this has not been proved statistically [203]. However, archived soil samples and samples

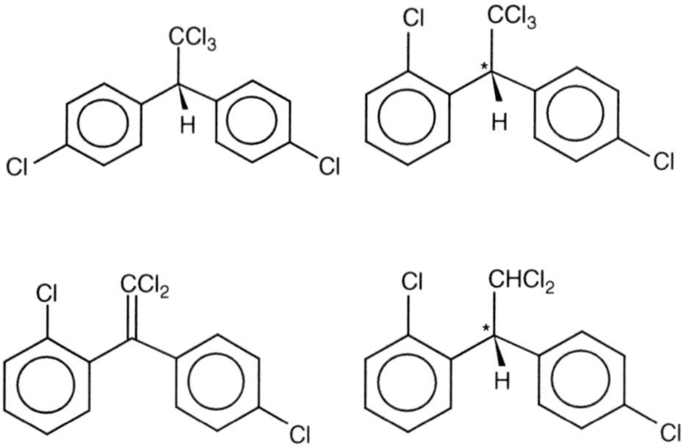

Fig. 6.9. Structures of non-chiral (left column) and chiral (right column) DDT-related compound. Clockwise from top left: *p,p′*-DDT, *o,p′*-DDT, *p,p′*-DDE, and *p,p′*-DDD.

from test plots treated with sludge from the UK showed non-constant EFs during the period 1972–1990. In either case, the older samples deviated more from the racemate than the newer, although ANOVA ($p < 0.05$) statistics did not show this to be significant [97]. A problem with the *o,p′*-DDT data was most likely the non-uniformal development of the EF through the years. *o,p′*-DDT was racemic in five silt loam soil samples but one showed an EF of 0.454 [202]. Enantioenriched (+)-*o,p′*-DDT was determined in Illinois soil while (−)-*o,p′*-DDT was more abundant in Ohio soil [201].

Garrison et al. studied the DDT degradation by an aquatic plant (*Elodea canadensis*) in a water-plant mix [93]. Half lives of *p,p′*-DDT and its chiral isomer *o,p′*-DDT were one to three days [94] and quantitative elimination was obtained within 5 or 24 days [93]. The only transformation products identified were *p,p′*-DDD and *o,p′*-DDD, which are formed by reductive dechlorination. However, it was found that 22% of the DDT was covalently bound within the plant. After gamma-irradiation, DDT elimination was slower by 40% which showed that live microbes were responsible for the transformation. Nevertheless, the observed processes were not enantioselective in the formation of *o,p′*-DDD from *o,p′*-DDT [94]. The work of Garrison et al. has been conducted enantioselectively using both capillary electrophoreses in the micellar electrokinetic chromatographic mode and GC [93].

Analysis of fish (channel catfish, buffalo, large mouth bass) captured from a river where the sediments have a history of severe DDT contamination indicated the existence of enantioselective processes—seventeen samples showed enrichment of the first eluting *o,p′*-DDT enantiomer, two of the second eluting enantiomer (elution order unknown although tentatively assigned to the signals) [93]. *o,p′*-DDT was racemic with slight variations in both salmon with and without M74 syndrome [205]. Preliminary results of Buser and Müller indicated for *o,p′*-DDT an EF of ~0.444 in human adipose tissue [200,206]. For EFs of *o,p′*-DDT in standard reference material see Table 6.2.

Despite the high production rates of DDT along with the high amounts of chiral *o,p′*-DDT released to the environment, the research on the enantioselective fate of DDT-related compounds is surprisingly low, although the enantiomers of *o,p′*-DDT and *o,p′*-DDD can be resolved on many CSPs. Partial explanations may be that *o,p′*-DDT is approximately five times faster transformed than *p,p′*-DDT, and that a major transformation pathway in biota (i.e. dehydrochlorination) is resulting in the nonchiral *o,p′*-DDE.

6.4.3 aaeeee-1,2,3,4,5,6-Hexachlorocyclohexane (α-HCH) and related compounds

Hexachlorocyclohexanes (HCHs) have been used as insecticides from the 1960s until recently. Basically the technical product is synthesized from benzene via photochlorination. By this reaction, a mixture of hexachlorocyclohexane isomers is gained. The respective isomers were labelled α-, β-, γ-, δ-, and ε-HCH in the order of their historic discovery. In the technical product, the main constituents are the α- and γ-isomers with relative abundances of about 60 and 15%, respectively [207]. It was found that only the γ-isomer possesses insecticidal properties. Consequently, only pure γ-HCH has been used in Europe and the United States in the past decades. However, in Asia and possibly the states of the former Soviet Union usage of the technical mixture continues [208].

Of all possible HCH isomers, only α-HCH is chiral. HCHs can be metabolized via dehydrochlorination processes in vertebrates but also by microbial communities [209]. Thus, the respective pentachlorocyclohexenes (PCCHs) are formed. In this process, α-HCH leads to the chiral β-PCCH, while nonchiral γ-HCH produces the chiral γ-PCCH isomer (Fig. 6.10). Similar reactions are possible via photo-reactions [56,57,209].

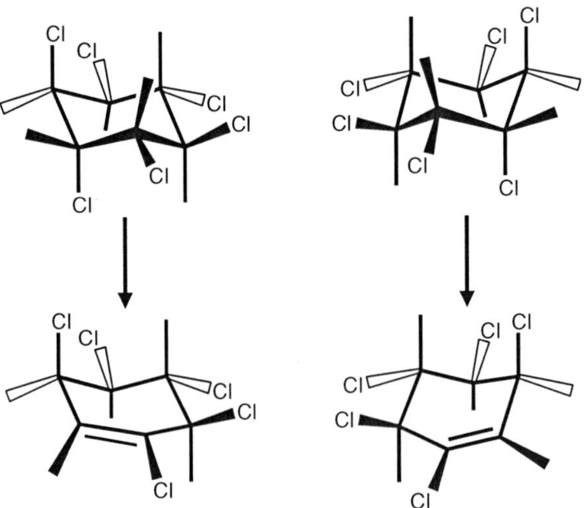

Fig. 6.10. Transformation pathway of α-HCH by dehydrochlorination. Formation of the respective β-pentachlorocyclohexene (PCCH)-enantiomers from the enantiomers of α-HCH.

6.4.3.1 Air–water exchange of HCHs

The first data on organochlorine compounds in atmospheric samples far away form potential sources were published as early as in 1973 [210]. These findings were verified by several other experiments also by other groups [207,208,211–215]. Some huge issues have been discussed since these early years:

(1) whether the respective compounds are atmospheric inventory and slowly settle in both terrestrial ecosystems and huge water bodies like the Great Lakes or the oceans,
(2) whether the oceans (and soil systems) are one huge storage from which the compounds are slowly released into the atmosphere,
(3) whether equilibrium between atmosphere and water bodies/soil has been reached and transport from one compartment to the other is rather due to temperature variations.

This issue can be treated by means of thermodynamic calculations if the respective system is in equilibrium (no changes of concentrations in time), and if the respective constants are known. However this problem could basically only be solved tentatively, as air–water equilibria constants (e.g. Henry's law constants) were not available with the needed accuracy [207,214,216]. Additionally Henry's law constant is dependent on many experimental details [217]. The same limitations held for complex models [218] as well as air–soil, or air–plant equilibria [219,220].

One big step forward came with enantioselective determinations. Under the assumption that the exchange from compounds from water to air is a process with no enantiomeric preference and in systems in which one enantiomer is enriched in one phase, it should be possible to observe exchange processes directly by determining gradients in enantiomeric distribution of chiral compounds in both phases. As neither water nor air is a chiral medium these assumptions should hold true.

In air, photochemical transformation is the major degradation pathway. This is considered to be a "non-enantioselective process" [57]. Consequently, Müller et al. and Falconer et al. found ERs in air which did not deviate significantly from the racemate as determined in the technical product [221,222]. However, α-HCH ERs in air ranging from 0.86 to 1.03 ($EF = 0.46$–0.51) were reported by Hühnerfuss et al. [59], and Jantunen et al. also found some deviation from the racemate in air ($EF = 0.47$–0.55) [223]. These, on the first view, contradictory results have been investigated by Bidleman et al. in detail. Some ERs in air could be compared to water samples from the same area. If one considers that transformation in air is non-enantioselective, the deviations from an EF of 0.5 must derive from evaporation of non-racemic α-HCH from water (or soil).

This approach on tracing chiral signatures for the determination of origin of HCH has been used to describe air–water gas exchange processes in the Canadian Arctic [83,222] and the Great Lakes [224]. In the latter study [224], samples from Lake Ontario exhibited for α-HCH, $EF = 0.46$. The same compound was determined in precipitation samples which were taken as samples for higher atmospheric regions [83,221]. α-HCH in these samples was racemic. These results were compared to air samples which were taken in 10 m heights during ship cruises on the lake. These air samples exhibited EFs of 0.49 in summertime and 0.5 in October. These findings were in accordance with the assumptions of air–water gas exchange of α-HCH: In wintertime, the Lakes serve as a sink due to lower temperature, while in summertime α-HCH evaporates from the lake which results in an increase of the atmospheric α-HCH concentrations [224].

Another experiment using enantioselective determination of α-HCH to determine volatilization or deposition in the North Sea was presented by Bethan et al. [225]. Total deposition and air samples indicated that direction of mass flow of α-HCH in the air–water system of this European shelf sea changes with the seasons. A similar situation was found for the Baltic Sea, when the same techniques were applied: in wintertime α-HCH was racemic in air samples, while in summertime the enantiomeric composition in air samples is getting near (EF = 0.47) to those found in water (EF = 0.45). Additionally the concentrations of HCHs in the air are rising during summertime [226].

In another experiment in the Southern Atlantic Jantunen et al. [217] determined HCH-concentration in water and air as well as EFs of α-HCH in both phases. The EFs were decreasing with higher geographical latitude and lower temperatures in both phases. Astonishingly the EFs were similar for air and water samples from the same region. These findings might indicate rather volatilization of HCHs from the water than deposition from the air as the enantiomeric enrichment probably originates from microbial processes in the sea. Both are examples in which enantioselective determination served to gain deeper insights into air–water–gas exchange for organic pollutants such as HCHs.

In initial reports on air–water–gas exchange, levels in water were approximately 1000 times higher than in air. Air–water–gas exchange is in equilibrium with volatilization compensating transport into the aqueous phase [227]. More emphasis on the use of chiral compounds is given in another article by Bidleman et al. [228]. These data were used to trace and distinguish two sources of α-HCH levels in air [223,224,229]. Bidleman and Falconer developed and optimized a formula to apportion two sources of chiral compounds on the basis of earlier efforts in this field [229] (see Eq. (10)):

$$F_a = \frac{(ER_m - ER_b)(ER_a + 1)}{(ER_a - ER_b)(ER_m + 1)} \tag{10}$$

with F_a being the fraction of the whole pesticide derived from source a, ER_m, ER_a, and ER_b are the enantioratio of the mixture m, source a, and source b.

6.4.3.2 Transformation of α-HCH in water

Nonracemic α-HCH has been determined in several marine water bodies [3,56,230]. However it seems that enantioselective depletion depends very much on the respective water body, as in some areas the (+)-enantiomer and in some areas the (−)-enantiomer dominates in the samples. However, Harner et al. [230] concluded that the enantioselective microbial degradation is the key parameter with half lives of 5.9 (+) and 22.8 (−) years for the elimination of HCHs from deep ocean waters and sedimentation with particulate matter and hydrolysis (half life 64 years) are not that relevant in the Arctic Ocean.

6.4.3.3 α-HCH in soil

This work and the perspective for application to similar problems to distinguish emission and immission is among the most promising results in enantioselective analysis. The authors checked Eq. (10) and were able to confirm their calculations within +2.1% to −9.8% which seems to be an excellent agreement of theory and practice (see also heptachlor epoxide, below). By this it was shown that enantioselective determinations are not only suitable for collecting data but also of outstanding importance for tracing the fate of organochlorines. However, this technique works only if the component is chiral and the ERs of the two sources vary. Furthermore, the significance of the ERs for the respective source must be proven.

Considering the air–soil exchange the data is diverse: while Leone et al. report mostly racemic mixtures of α-HCH in air above soils they also report some indications for either depletion of (+)- or (−)-α-HCH, depending on the sample region [231]. However the mean overall samples indicate racemic α-HCH. In a huge sampling campaign across North America, deviations towards lower EFs were determined on the sampling sites on the North-West coast of Canada, while racemic mixtures were determined in the prairie states. On the other hand EFs significantly different from the racemic mixture were determined in the samples from the Pacific Coast as inputs from marine areas [232].

Enantioenrichment of α-HCH has also been determined for groundwater at a disposal facility [233]. While well water exhibited EFs larger than 0.5, surface water was nearly racemic. However two ground water plumes were distinctly different. An acidic one originating from the core of the disposal site with pH ~4 and a neutral one (pH ~6–7) were analyzed. The acidic samples were the ones with the highest HCH concentrations (up to 100 μg/L), but exhibited racemic composition. The samples from the neutral plume held 0.02–1 μg/L with EFs up to 0.89. Obviously, enantioselective transformation preferably occurred in this experiment in soils with no pH alteration. The highest deviations from racemic EFs were determined in wells which had the largest distance to the source. The surface water samples held concentrations from 0.02 to 0.68 μg/L which is highly elevated against background concentrations. The surface water samples with high concentrations also had elevated EFs.

In archived soils from the UK background samples were compared to sludge amended soils [97] in respect of α-HCH, as well as chlordanes. Though several samples exhibited a clear nonracemic signature there was no clear trend in time. Neither did the sludge amended soils exhibit higher aberrations in EFs.

Enantioselective microbial α-HCH transformation was found in sewage sludge by anaerobic, methanogenic, mesophilic microorganisms that were not yet further characterized [82]. Half life for α-HCH was 35 h and the transformation of (+)-α-HCH was faster than that of (−)-α-HCH [82]. Buser and Müller estimated that ~85% of the transformation was biotic and also suggested that some strains of microorganisms may be even more enantioselective with respect to the transformation of α-HCH. Enantioselective transformation of γ-HCH in α-HCH is disputed [56,82,206].

6.4.3.4 α-HCH in biota

Enantiomeric enrichment of α-HCH was reported for biota samples extensively. However, it should be noted, that the enantiomeric distribution of α-HCH in the respective tissues of one organism is not homogeneous [2,58,234]. In fat and brain of rats, the (+)-enantiomer dominated over (−)-α-HCH whereas liver and blood samples were rather racemic. However, in eider duck samples the brain samples exhibited enantiomeric enrichment similar to the liver samples of these bird, but no other organs did exhibit similar enrichments. Possibly, the blood–brain barrier is indeed selective for some enantiomers [234].

Slightly elevated EFs (0.505–0.52) were determined in human liver samples stemming from the Belgian population [121].

Enantiomeric enrichment of α-HCH in fish tissue has been determined by several authors. However it was not clear whether this was due to enantiomeric enrichment in the food, or due to the fish metabolism itself. Wong et al. [175] fed fish with food contaminated with racemic α-HCH, which was not changed during the experiment. Thus, it was concluded that trout is not able to eliminate α-HCH stereoselectively, though elevated EFs were determined in wild catches. This finding is obviously due to stereoselective metabolization either in the water body or in the food chain.

Stereoselective enrichment (biomagnification) of α-HCH was determined in the simple food chain cod → seal → polar bear [235]. The biomagnification factors (BF) from cod to seal are 2.0 and 1.7 for the (+) and (−) enantiomer. From seal blubber to bear fat it was 1.4 and 0.71 respectively. Thus the (+) enantiomer was enriched in the food chain. The authors concluded, that the (−) enantiomer is metabolized more quickly than the (+) one in seals and bear.

6.4.4 Chlordane-related compounds

Chlordane is a versatile, broad spectrum, contact insecticide which was introduced in the USA in 1947 [236]. It belongs to the group of cyclodiene pesticides and has mainly been applied in the North America, with some limited use in Japan, some European countries (Sweden, Poland), and the former Soviet Union. In the USA, it was frequently used until 1983 mainly as a soil insecticide and from 1983 to 1988 for household uses in termite control [237]. The reported production rates which are one or two orders of magnitude below those of HCH, DDT, and toxaphene are not reflected in these proportions in the environment, where the role of chlordane is more pronounced. The synthesis starts with a Diels–Alder addition of hexachlorocyclopentadiene and cyclopentadiene which yields the non-chiral chlordene (Fig. 6.11(a) and (b)). Further chlorination of chlordene results in technical chlordane. The rather complicated chemical names of chlordane-related compounds (Table 6.5) are formally derived from indane or indene (Fig. 6.11(b)) with the following modifications: (i) there is an additional CH_2-bridge between carbons 4 and 7 (“4,7-methano”) and (ii) the lack of 2 or 3 double bonds (the latter only in the case of indene) in the six-membered ring (“3a,4,7,7,a-tetrahydro” or 2,3,3a,4,7,7a-hexahydro”).

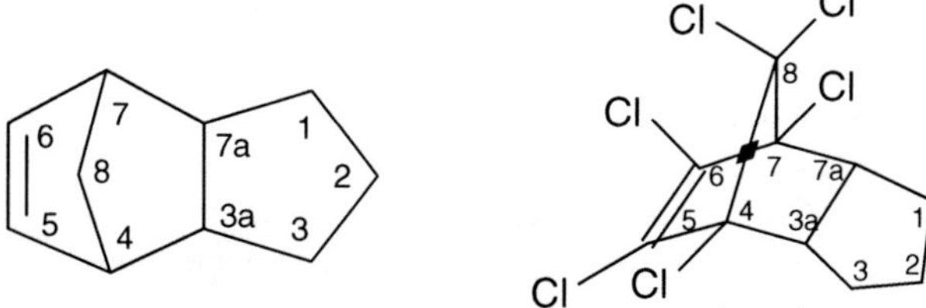

Fig. 6.11. Backbone of chlordane-related compounds. Systematic IUPAC nomenclature derives from indane with saturated carbons on C1, C2, C4, C7, C3a, and C7a (left). Most compounds have chlorine substituents in 4,5,6,7,8,8-positions (right).

Table 6.5 Structure of chlordane-related compounds [336]

Name	Cl substituents	Additional	3a,4,7,7a-Tetrahydro	Backbone 4,7-methanoindane
Chlordene	–	2,3-Dehydro*	+	+
cis-Chlordane	1-*exo*, 2-*exo*,4,5,6,7,8,8	–	+	+
trans-Chlordane	1-*exo*,2-*endo*,4,5,6,7,8,8	–	+	+
trans-Nonachlor	1-*exo*,2-*endo*,3-*exo*,4,5,6,7,8,8	–	+	+
cis-Nonachlor	1-*exo*,2-*exo*,3-*exo*,4,5,6,7,8,8	–	+	+
MC6, nonachlor III	1-*exo*,2,2,4,5,6,7,8,8			+
Oxychlordane	1-*exo*,2-*exo*,4,5,6,7,8,8	2,3-*exo*-Epoxy		+
Heptachlor	1-*exo*,4,5,6,7,8,8	2,3-Dehydro*		+*
cis-Heptachlor epoxide	1-*exo*,4,5,6,7,8,8	2,3-*exo*-Epoxy		+*
U82 [95]	1-*exo*,2-*endo*,3-*exo*,4,5,6,8,8			+
MC5	1-*exo*,2-*endo*,3-*exo*,4,5,7,8,8			+
MC7	1-*exo*,2-*exo*,3-*endo*,4,5,*X***,8,8			+

*"2,3-dehydro-methanoindane" is usually named "methanoindene."
**$X=6$ or 7, exact position unknown.

The synthetic mixture consists of >140 compounds [238]. The major compounds of the formulation, *cis*- and *trans*-chlordane, contribute with 15–20%, respectively, to the technical product [239]. A minor product in chlordane, heptachlor, was found to be more insecticidal active than the chlordane isomers, and it was later used separately as an insecticide. Miyazaki et al. synthesized enantiomers of heptachlor, *cis*- and *trans*-chlordane and found different insecticidal activity for selected enantiomers [14,15]. Chlordane is thought after as one of the most toxic organochlorines and as a potential human carcinogen [240]. In addition to *cis*- and *trans*-chlordane several other compounds of technical chlordane or their metabolites are found in environmental samples. Nonchiral *trans*- and *cis*-nonachlor are among the most prominent ones. Both *cis*- and *trans*-chlordane as well as *trans*-nonachlor are metabolized to the chiral oxychlordane [241]. It was mentioned that feeding studies with pigs carried out some 35 years ago resulted in a high excess of (+)-oxychlordane from racemic *trans*-chlordane but racemic oxychlordane was obtained from *cis*-chlordane [127]. However, chemical oxidation of pure *trans*-chlordane enantiomers led to enantiopure oxychlordane with the same optical rotation, respectively, while racemic *trans*-chlordane produced racemic oxychlordane [120]. Furthermore, *trans*-chlordane was faster converted into oxychlordane by rats [242]. These different sources, mechanisms, and interactions complicate a clear evaluation of oxychlordane EFs (see below). Another oxidative metabolite is (*exo*- or) *cis*-heptachlor epoxide (*endo* or *trans*-heptachlor epoxide is not formed in nature in significant amounts), which is the metabolite of heptachlor in biota. Some minor chlordane-related components are strongly bioaccumulative and play a more relevant role in biota than in technical grade chlordane. Chlordane-related compounds that have also been studied to some extend are the Miyazaki components MC5, MC6, MC7, MC8 [243], and U82 [244] (Table 6.5). In 1994, enantioenriched chlordane standards were obtained by preparative GC [245] and LC [120]. They are now commercially available and the elution orders of enantiomers from different CSPs are known. An enantioenriched solution of (+)–U82 was produced by the group of Oehme [66]. Initial studies confirmed that the technical synthesis of chlordane leads to a racemic mixture [123].

6.4.4.1 EFs of chlordane-related compounds in air, soil, and water

Initial measurements in air indicated both racemic and non-racemic composition of heptachlor and *cis*- and *trans*-chlordane in air [78,84,246]. Arctic air samples (1993–1998) from Canada, Russia, Sweden, and Finland virtually showed the same EFs for *trans*-chlordane (mean $EF_+ = 0.471$, 0.471, 0.477, and 0.463) and also for *cis*-chlordane (mean $EF_+ = 0.504$, 0.503, 0.507, and 0.511) (Fig. 6.12) [247]. Bidleman et al. also listed results from soil samples which showed EF_+ ranging from 0.423 to 0.496 (*trans*-chlordane) and 0.500–0.541 (*cis*-chlordane) indicating that the EFs in Arctic air represented a mixture of the soil variability [247]. Samples from 1999–2001 mostly confirmed the older data [248]. By contrast, air samples from the Great Lakes and other sites in North America showed a much larger variability [247]. Stored historic air samples from 1971 to 1973 (Sweden, Iceland, Slovakia) were

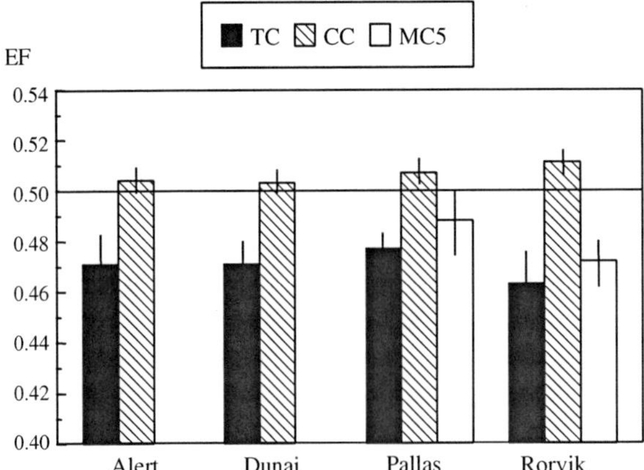

Fig. 6.12. EFs of *trans*-chlordane (TC), *cis*-chlordane (CC), and MC5 in air samples from different regions [247].

virtually racemic [248]. Although there are a few air samples known with EF ~ 0.5, these historic samples indicated that the EF of chlordane isomers has changed over the past 30 years, and it is plausible that microbial transformation in soil was the major source for this change. Further confirmation was found in age-dated lake sediment cores which were racemic in the past and nonracemic in recent times [248].

With few exceptions, there is a preference for an excess of (−)-*trans*-chlordane and (+)-*cis*-chlordane in soil and air [203,231,248]. By contrast, the EF in indoor samples from remote sites or following the use of chlordane as termiticide was racemic for both chlordane isomers [246,249]. The somewhat ambiguous data of chlordane in air led to the combined analysis of soil and air (above soil). Falconer et al. presented EF_+ of *cis*- and *trans*-chlordane in air and soil [250]. *trans*-chlordane was mostly racemic except directly above soil (corn belt region, USA) where an EF of 0.425 was measured which was comparable with the EF of 0.412 in soil; the same effect was observed for the (+)-*cis*-chlordane ($EF_+ > 0.500$) [250]. EFs of air samples at 0.5 m above the soil (experimental site) matched the EFs in soil. However, at 1.5 m and the EFs became more racemic and at 2.5 m they were virtually identical with background levels [24]. This indicates mixing with chlordane from another source (Section 6.4.3). Leone et al. studied the volatilization of chlordane from soils. The samples were taken directly (0.15 m) above soil and showed EFs for *cis*- and *trans*-chlordane similar to those mentioned above [231]. For *cis*- and *trans*-chlordane, the EF remained relatively stable when samples were collected between 0.15 and 1.75 m above soil (difference was ~0.5–1%), although the concentrations significantly decreased with increasing height above soil [231]. This indicated that at this site, chlordane in air was mainly from evaporation from soil and not from aerial transport from another site to the soil [231]. Alterations of the EF in dependence of the sampling height above ground were previously reported by Finizio et al. [23]. Heptachlor epoxide showed EFs in soil that were racemic (0.500) or—more frequently—significantly higher (up to 0.88) [201,202,204,246,247]. Experiments by Bidleman et al. on soil-air exchange revealed that soil is a source for atmospheric heptachlor epoxide [251]. The authors found that airborne heptachlor (source unknown) being racemic while soil-released heptachlor epoxide showed EFs ranging from 0.600 to 0.666. Oxychlordane in North American soils ranged from 0.370 to 0.650, thus having excess of both (+)- and (−)-oxychlordane, in dependance of the site and type of soil [201,202,204]. MC6 enantiomer distribution in soil and air (diverse sites) ranged from EF = 0.416 to 0.492 [247].

Archived soil samples and samples from soil test plots treated with sludge from the UK confirmed the direction of non-racemity of *cis*- ($EF_+ > 0.5$) and *trans*-chlordane ($EF_+ < 0.5$) mentioned above [97]. Results from an experimental field were presented by Eitzer et al. [122]. A solution containing 72 weight-percentage of chlordane was sprayed at a rate of 7.6 ml/m^2 in 1960 and covered with torf; in 1998, soil cores were collected, analyzed, and compared with data from "normal" soils. Three soils showed excess of (−)-*trans*- and (+)-*cis*-chlordane (EF_+ 0.408–0.465 and 0.541–0.548), respectively, while both compounds were racemic in other soils [24,122]. Mattina et al. investigated the transfer of chlordane from soil to plant [252]. Whole root, sap, and whole aerial showed similar EFs as was determined in soil (one result), varying by 6% relative to the soil EF_+ of 0.45 (*trans*-chlordane) and 10% for *cis*-chlordane ($EF_+ = 0.55$) [252].

6.4.4.2 *EFs of chlordane-related compounds in biological samples*

In higher organisms, residues of chlordane-related compounds are usually dominated by (nonchiral) *trans*-nonachlor and the chiral metabolites oxychlordane and heptachlor epoxide while *cis*- and *trans*-chlordane are often metabolized. In a feeding and elimination study with fish, trouts were fed for 40 days with organochlorine-spiked food. Following that, a depuration phase of 238 days with organochlorine-free food followed [175]. The EF of *trans*-chlordane was subject to changes throughout the experiment. However, the change occurred faster in liver (EF, 13 d = 0.558) than in carcass (EF, 13 d = 0.505), but over time the EFs became more and more equal. Biomagnification factors (BMF) were 1.0 for (−)-*trans*-chlordane, 2.7 for (+)-*trans*-chlordane, and 1.7 for *trans*-chlordane [175]. Furthermore, small amounts of the metabolite oxychlordane were detected after 20 days, which alone cannot explain the alteration of the *trans*-chlordane-EF [175]. Fish from Meramec River (Missouri) showed EFs of 0.300 for *trans*-chlordane and 0.472 for *cis*-chlordane (elution order not unequivocally proven) [90]. Furthermore, a 50-d feeding study indicated racemic uptake of chlordane isomers from spiked water followed by the decrease of the EF of *trans*-chlordane, mainly within the first week [90]. In the depuration phase, half lives of *trans*-chlordane enantiomers were 15 and 20 days, respectively [90]. Gender-dependent EFs of *cis*- and *trans*-chlordane were found in fish [253]. EFs of *cis*- and *trans*-chlordane in female cod muscle, gonad, and liver were <0.500, while male showed EFs > 0.500 [253]. Koske et al. found EFs of *cis*- and *trans*-chlordane in cod ranging from 0.333 to 0.500 and 0.444 to 0.778 [254]. Gender-dependency of chiral chlordanes (U82, *cis*- and *trans*-chlordane, MC5, MC7) were not found in Baltic herring [255]. However, significant deviations from the racemate were typically of these compounds. EF_+ of oxychlordane ranged from 0.565 and 0.615 in herring, salmon, and a grey seal from the Baltic, as well as human adipose tissue [123]. In the three Baltic species (salmon, herring, seal) heptachlor epoxide-EFs were < 0.5 (most pronounced in seal) but the human adipose tissue showed higher abundance of (+)-heptachlor epoxide whereas MC6 was racemic [123]. In the brain of a leopard seal, U82 and MC6 were racemic while (+)-U82 was slightly enriched in the brain of an elephant seal [85]. *trans*- and *cis*-chlordane as well as MC4 and MC7 were present in non-racemic composition. The EFs of MC5 deviated significantly on β-PMCD and β-BSCD, obviously due to interferences from isomers [123,256]. EFs of *cis*- and *trans*-chlordane in seal and human adipose differed significantly from the racemate [95]. Non-racemic proportions were also determined for the enantiomers of U82, MC5 and MC7 in seal. In top predators, these three compounds were more abundant than *cis*- and *trans*-chlordane [95].

U82 was racemic in fish but the first eluted enantiomer was strongly enriched in grey seal and depleted in an Antarctic penguin [123]. Very strong depletion of U82 was found in seal blubber and liver (EFs up to 0.956 [255]) but U82 was racemic in brain of Antarctic seals [85]. When enantioenriched heptachlor was applied to MFO enzyme systems of rats (+)-heptachlor was faster degraded to (+)-heptachlor epoxide than (−)-heptachlor to (−)-heptachlor epoxide [78]. The biomagnification study of Wiberg et al. in the short food chain cod—ringed seal—polar bear revealed food chain accumulation of both enantiomers of oxychlordane and heptachlor epoxide. However, BMF of (+)-oxychlordane was higher than (−)-oxychlordane both from cod to seal and seal to bear. In contrast the BMF of (+)-heptachlor epoxide was lower from cod to seal and higher from seal to bear [134]. For MC4, the BMF of the first eluted enantiomer was 5.2 times higher from cod to seal in comparison with the second eluted enantiomer [134]. Buser and Müller also studied the photo-metabolites of chlordane. These compounds are derived from chlordanes in UV or sunlight by formation of caged (similar to that of mirex) or half-caged backbones [78].

In terrestrial animals only EFs of the metabolites oxychlordane and heptachlor epoxide have been investigated so far. In roe-deer (*Capreolus capreolus*) livers from Northern Germany and Southern Germany high levels of the epoxides were measured by GC/ECD. In Northern Germany, oxychlordane EFs ranged from 0.900 to 0.944 (levels from 10 to 60 μg/kg fat) and heptachlor epoxide EFs ranged from 0.500 to 0.900 (levels from 10 to 100 μg/kg fat). In Southern Germany, the levels were in the same order and EFs in four samples confirmed an excess of the (+)-enantiomers with higher EFs for oxychlordane (0.666–0.750) [101]. An excess of the (+)-epoxides was also found in hare liver but the EFs (oxychlordane: 0.5–0.6; heptachlor epoxide: 0.714–0.787) were lower compared to roe deer though the levels were in the same order in both species [257]. Levels and EFs of oxychlordane were published in liver of polar foxes (*Alopex lagopus*) [258]. All samples exhibited EFs < 0.5 with the exception of the sample with the highest concentration of oxychlordane which had an EF > 0.5. The EFs were confirmed on two different CSPs on GC/ECD and also by GC/EI-MS. Apparently, the inversion of the EF was depending on the concentration of the analyte. Hoekstra et al. determined EFs in wolverine liver ($n = 12$) and arctic fox ($n = 20$) from the Canadian Arctic. *trans*-Chlordane and heptachlor epoxide were more non-racemic in arctic fox (0.890 *versus* 0.652 and 0.732 *versus* 0.554) while oxychlordane was more non-racemic in wolferine (0.712 *versus* 0.676) [98]. The EFs in arctic fox from Canada were more enriched in (+)-oxychlordane than samples from Iceland (0.587). Fisk et al. determined $EF_+ > 0.500$ for heptachlor epoxide and oxychlordane in six seabird species and ringed seals from the Arctic [133,259]. In human adipose tissue of a male American $EF_+ > 0.500$ were determined [123]. In blubber of a polar bear (*Ursus maritimus*) the oxychlordane EF was also >0.5 [258].

In a Baltic cod (*Gadus morhua*) liver and a Baltic herring (*Clupea harengus*) sample the EFs of oxychlordane were >0.5 [123]. EFs of oxychlordane and heptachlor epoxide in Baltic herring ranged from 0.524 to 0.583 and 0.286 to 0.412, respectively [255]. Eggs of seagulls exhibited oxychlordane EFs of 0.600–0.697 ($n = 5$) and heptachlor epoxide EFs of 0.615–0.730 ($n = 2$) [59,245,257]. Depending on the species, EF_+ of oxychlordane were partly >0.5 or <0.5 [73,123,124,245,255].

EFs of heptachlor epoxide in blubber, liver and brain of two harbor seal samples were <0.167 in all tissues [245]. Strong depletion of (+)-heptachlor epoxide was also found in ringed seals, grey seals, and harbor seals from the Baltic (EFs = 0.090 to 0.333) [255] while those in blubber of ringed seals from the Canadian Arctic ranged from 0.430 to 0.480 [259]. In contrast, terrestrial animals (rabbit liver, deer liver) usually showed EFs >0.5 in liver (see above). This was also found in eggs of doves [245]. No significant differences in EFs of diverse chlordane compounds were found between M74 and non-M74 salmon [205]. Most of the chlordane compounds (*trans*-chlordane, oxychlordane, heptachlor epoxide, U82, MC5 and MC6) but also α-HCH and *o,p′*-DDT were racemic in both groups, indicating no differences. By contrast, starved occluded seals had more prominent EFs than unstarved occluded for *cis*- and *trans*-chlordane [205]. Despite some huge amount of data available, the understanding of the enantio-selective fate of chlordane-related compounds is not well-developed.

Results from plants are scarce. Lee et al. studied the *cis*- and *trans*-chlordane uptake by zucchini from air. Both compounds were virtually racemic in air. The compound profile in different plant compartments was different between air→plant [247] and soil→plant transfer [252] but the EFs were not markedly different. In further studies Mattina et al. have refined the chlordane transport into plants [260,261].

6.4.5 Endosulfan

Technical endosulfan is a non-systemic contact and stomach insecticide [262]. It consists of the diastereomers α- and β-endosulfan in the ratio 7:3. Both are nonchiral and have significantly different physicochemical properties (e.g. the melting points are 109 and 209°C, respectively [262]). The higher volatility of α-endosulfan leads to an enrichment in air and a depletion in rain, relatively to β-endosulfan [263]. Furthermore, there is a substantial interconversion of the β- into the α-enantiomer [263]. Despite these different properties, both stereoisomers have comparable LD_{50} values [262]. Unlike other cyclodiene pesticides, α-endosulfan is still in use in Africa, North America, and Europe [96,203] and is now one of the most abundant and ubiquitous organochlorine pesticide in the North-American atmosphere [203]. It is noteworthy as well, that molecular modeling and X-ray crystallography demonstrated that α-endosulfan is asymmetric (chiral) although the conformers providing chirality have not been resolved to date [263].

Compared to other cyclodiene pesticides such as chlordane, endosulfan is less stable and the metabolism is widely known [264]. Hydrolysis of the heterocyclic ring and elimination of sulfan/sulfate leaves the corresponding diol. The diol can lead to the hydroxyether or the lactone which both are chiral [96]. Lab incubations with anaerobic and aerobic soils demonstrated the non-racemic formation of both the hydroxyether (EF down to ~0.44) and more pronounced the lactone (EF up to ~0.7) [96]. Aerobic and anaerobic transformation led to inverse EFs for the lactone. Since anaerobic and aerobic conditions are both found in normal soil, their variation may account for site-specific and seasonal variations [96]. It has been noted that endosulfan I co-eluted with (−)-*cis*-chlordane on β-PMCD [231].

6.4.6 Compounds of technical toxaphene (CTT)

Toxaphene (Camphechlor) is one of the most widely applied pesticides. The technical product is obtained by the exhaustive chlorination of the chiral natural products

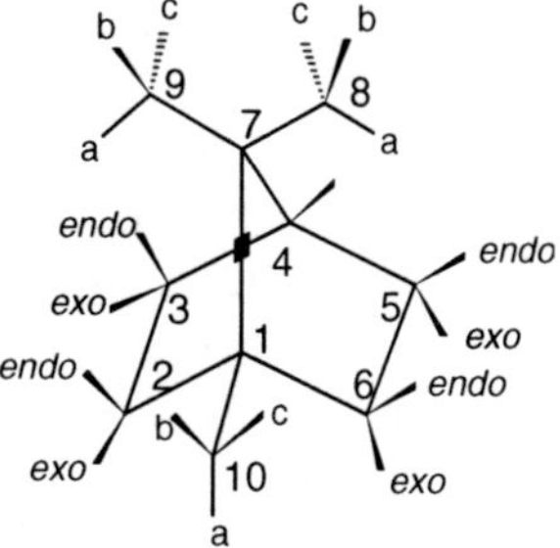

Fig. 6.13. Backbone and the respective carbon numbering of bornane, the major compound class found in toxaphene.

camphene and/or α-pinene to an average composition of $C_{10}H_{10}Cl_8$ [265]. This makes toxaphene different from any other organochlorine compound. Recently, Reddy et al. demonstrated this by the radiocarbon analysis [266]. Hydrocarbon rearrangement during synthesis leads to >90% polychlorinated 1,7,7-trimethyl-bicyclo[2.2.1]heptanes (chlorobornanes, Fig. 6.13); in addition, chlorocamphenes, chlorodihydrocamphenes, and chlorobornenes are formed in this process [265,267,268]. Theoretically, alone the chlorobornanes may exist in a variety of 16,128 pairs of enantiomers and only 511 achiral congeners [269]. According to that more than 97% of the the compounds of technical toxaphene (CTTs) are chiral and to date no non-chiral CTT has been identified.

The mirror image of a chlorobornane is obtained by exchange of the substituents at C2, C3, and C9 with those at C6, C5, and C8, respectively [269]. Thus, different chemical names (e.g. 2-*exo*,5,5,8,9,9,10-octachlorobornane is enantiomeric to 3,3,6-*exo*,8,8,9,10-octachlorbornane) have been ascribed to the enantiomers of chlorobornanes. Multiple substitutions on the primary carbons C8, C9, and C10 were suspected to cause hindered rotation. In fact recently, temperature dependent rotamers were described for a polychlorobornene [270]. For some components the population of these conformations were determined [271], and the rotational barrier has been calculated [272]. These data suggested that atropisomers do not exist in the case of the chlorobornanes discussed in this chapter. More details on the phenomenon of atropisomerism is presented in Section 6.4.1.1.

Technical products contain >1000 compounds [273] and most likely no individual compound with more than 3% of the total mass [274]. Insecticidal properties were assigned to the entire technical product. A baseline separation of the technical product is neither possible with the most powerful HRGC system nor with multidimensional GC [267]. This complexity would make impossible any attempts of enantioseparation of CTTs in samples. Fortunately, the toxaphene pattern is quickly changed in the environment due to extensive metabolization [267]. Depending on the potential of an organism for biotransformation, only a few recalcitrant CTTs are slowly transformed. In anaerobic media, the GC profile of CTTs is shifted towards lower chlorinated bornanes. The principal transformation pathway is reductive dechlorination primarily at carbons with geminal chlorine atoms [275]. Therefore, residues in sediment, soil, and sewage sludge are mainly recruited from hexa- and heptachlorobornanes. In contrast, biological samples, and particularly higher organisms of marine food chains accumulate a few hepta-, octa-, and nonachlorobornanes [267]. The selective production of important CTT standards has been achieved by the group of Parlar [276,277] and thereafter also by Nikiforov et al. [278]. Nevertheless, unlike all other organochlorine compounds discussed in this review, there are many more unknown than known CTTs to be detected in environmental samples. This complicates the analysis. The structures of the chlorobornanes described in this section are listed in Table 6.6.

In 1994, first enantioseparation was reported on handmade β-BSCD, and the commerically available BGB 172 column is still the most applied CSP for CTTs [64,73,118,266,279,280–285]. In the meantime, several CTTs were enantioseparated on randomly silylated β-TBDM (BGB 176, BGB Analytik) [68,89], randomly ethylated γ-PECD [286], and CB-β-PMCD (Chirasildex, Varian) but not on β-PMCD [69,85].

First enantioseparations of the technical product by GC and LC devices showed another peculiarity of toxaphene. Buser and Müller found a small levorotation of different technical toxaphene mixtures detected with chiroptical detector [118]. Furthermore, enantioselective gas chromatography in combination with MS/MS analysis of technical toxaphene mixtures using selected reaction monitoring (SRM) showed some compounds in a racemic composition while others were enantioenriched. This observation can be explained by the fact that the technical toxaphene synthesis starts with the chiral natural product camphene (or α-pinene) which undergoes skeleton rearrangement during the chlorination. Even more complex, the optical purity of natural camphene differs from country to country. While pine trees in North America contain racemic camphene, such trees in Europe may contain enantiopure camphene [118]. Recently, Nikiforov et al. demonstrated that the synthesis from pinene will lead to enantiopure or at least enantioenriched CTTs in the technical mixture whereas camphene leads to racemates [287]. A heptachlorobornane B7-1453 isolated from the technical product Melipax® (former German Democratic Republic) [288] showed an EF_{+} of 0.552–0.565 when determined by both GC/EI-MS and GC/ECNI-MS on β-BSCD and by GC/ECD on β-TBDM [61,289]. Another isolate from Melipax, B8-1412 showed only a weak, if any, deviation from the racemate [290]. Karlsson reported racemates for several components in Toxaphene® investigated by MDGC/HRMS [112]. Despite some open questions it appears that the US-product toxaphene which is the most applied product was racemic as were synthesized single CTTs [64,280,281]. Efforts in congener-specific toxaphene analysis, the prerequisite for enantioselective studies, were recently reviewed [132,267]. Pure enantiomers of ten environmentally relevant CTTs were obtained after enantioselective HPLC using a ChiraDex column (Merck, Germany) [291]. Although many CTTs are available as single standards, determination of the optical rotation of enantiomers has not been achieved. Moreover, hints on enantioselective crystallization of CTTs have been mentioned [292,293]. However, the breakthrough in the field is the enantioselective full synthesis by Nikiforov et al. which was just finished some months ago [287]. It should be noted that the CTTs discussed in this work possess three (B9-1025) to seven (e.g. B8-1413) asymmetric carbons [269]. Thus, transformation of one single enantiomer cannot yield

Table 6.6 Chemical names and abbreviations of CTTs discussed in this presentation

Systematic AV-code [337]	Chemical structure of enantiomer a (IUPAC name)	Chemical structure of enantiomer b	Parlar numbers [338]	Alternative acronyms	CAS-no.
B6-923	2-*exo*,3-*endo*,6-*exo*,8,9,10	2-*exo*,5-*endo*,6-*exo*-8,9,10	–	Hx-Sed [295]	70649-41-1
B7-515	2,2,5-*endo*,6-*exo*,8,9,10	2-*exo*,3-*endo*,6,6,8,9,10	P-32	Toxicant B [339]	51775-36-1
B7-1000	2-*endo*,3-*exo*,5-*endo*,6-*exo*,8,8,10	2-*exo*,3-*endo*,5-*exo*,6-*endo*,9,9,10	–	7-1 [340]	330663-51-9
B7-1001	2-*endo*,3-*exo*,5-*endo*,6-*exo*,8,9,10	2-*exo*,3-*endo*,5-*exo*,6-*endo*,8,9,10	–	Hp-Sed [295]	70649-42-2
B7-1453	2-*exo*, 3-*endo*,5-*exo*,9,9,10,10	3-*exo*, 5-*endo*,6-*exo*,8,8,10,10	–	TOX7 [288]	177334-52-0
B8-789	2,2,5,5,9,9,10,10	3,3,6,6,8,8,10,10	P-38	–	165820-15-5
B8-806	2,2,5-*endo*,6-*exo*,8,8,9,10	2-*exo*,3-*endo*,6,6,8,9,9,10	P-42	Toxicant A [339, 281]	187348-02-3
B8-809	2,2,5-*endo*,6-*exo*, 8,9,9,10	2-*exo*,3-*endo*,6,6,8,8,9,10			177695-50-0
B8-1412	2-*endo*,3-*exo*,5-*endo*,6-*exo*,8,8,9,10	2-*exo*,3-*endo*,5-*exo*,6-*endo*,8,9,9,10	–		205056-05-9
B8-1413	2-*endo*,3-*exo*,5-*endo*,6-*exo*,8,8,10,10	2-*exo*,3-*endo*,5-*exo*,6-*endo*,9,9,10,10	P-26	T2 [300], TOX8 [272]	142534-71-2
B8-1414	2-*endo*,3-*exo*,5-*endo*,6-*exo*,8,9,10,10	2-*exo*,3-*endo*,5-*exo*,6-*endo*,8,9,10,10	P-40		166021-27-8
B8-1945	2-*exo*,3-*endo*,5-*exo*,8,9,9,10,10	3-*exo*,5-*endo*,6-*exo*,8,8,9,10,10	P-41		165820-16-6
B8-2229	2-*exo*,5,5,8,9,9,10,10	3,3,6-*exo*,8,8,9,10,10	P-44		165820-17-7
B9-1025	2,2,5,5,8,9,9,10,10	3,3,6,6,8,8,9,10,10	P-62		154159-06-5
B9-1679	2-*endo*,3-*exo*,5-*endo*,6-*exo*,8,8,9,10,10	2-*exo*,3-*endo*,5-*exo*,6-*endo*,8,9,9,10,10	P-50	T12 [300], TOX9 [272]	66860-80-8

a racemic metabolite (without intermediate C–C cleavage). Reductive dechlorination is also thought to be the major transformation pathway of CTTs in biota. Thus, if one CTT is enantioselectively transformed into another CTT with one Cl less, this change of the EF should also be found in the metabolite.

6.4.6.1 EFs of CTTs in the environment

EFs in water and air are difficult to determine due to the complex, scarcely metabolized CTT-patterns found in these media. Even MDGC-MS was not able to provide exact data for each CTT [112]. Nonetheless, the EFs of the CTTs that could be investigated (B7-515, B8-1412, B8-1945, B8-1414) were identical in water and surface air from Lake Superior [112]. For B8-1414, the EF in surface water (0.541–0.548) almost matched that in air above water (0.531–0.545) which indicated volatilization of the compound from water into air in August [112], similarly to the method and effects described by Jantunen et al. [294].

Residues of toxaphene in (anaerobic) media are dominated by B6-923 or Hx-sed and B7-1001 (Hp-sed) [292,295]. In sediments from the Baltic Sea, B6-923 was poorly enantioresolved, and B7-1001 showed a slight if any deviation from the racemate [296]. Age-dated sediment core slices (Hanson Lake, Canada, early 1900s to 1990) contained racemic B6-923 but an excess was found for the second eluted B7-1001 enantiomer which varied with time (e.g. EF(1946) = 0.444, EF(1992) = 0.412) [297]. The results could not be interpreted in detail since these compounds are metabolites of different parent compounds whose transformation—if enantioselectively—may lead to preferential enantioenrichment with opposed optical rotation, and the initial concentration and kinetics of the transformation of the parent compounds may vary. Furthermore, the metabolites themselves might be enantioselectively degraded upon formation. In another study with sediments, the EF of B7-1001 was 0.474 [103]. EFs for B8-1412 (0.510), B8-1414 (0.512), and B8-1945 (0.502) were determined in sediments from Lake Superior [112]. Muir summarized data of his research group and found mostly racemic proportions of CTTs in sediment [284]. EFs of some CTTs in sediments from a marsh lands were between 0.474 and 0.524 [103].

Areas with particular high CTT contamination mainly exist in North America, and the CTT patterns in aquatic biota from these areas if often more linked with that found in microorganism-controlled media (soil, sediment, sewage sludge) and different to marine fish and marine mammals (see below). EF for eleven CTTs were determined in members of different trophic level of food webs of Lake Superior [112,284]. Most congeners were racemic in lake water and sediment but in biota, non-racemic proportions were observed for the most compounds. Already low trophic-level zooplankton showed non-racemic composition for several compounds. Surprisingly, even the major persistent congener in marine mammals, B9-1679, was non-racemic in zooplankton (EF = 0.440 ± 0.020), and thus more different from the racemate than in marine mammals (see below). Compared to the fish species the deviation from the racemate was the usual case with the exception of B9-1679 which was more racemic in fish than zooplankton. Significant alterations of the EFs of B7-515 and B8-1414, and a little less pronounced for B8-1412, B8-1945, and B8-806/9 were found in fish from Lake Superior [112]. Frequently, the EF was switched in fish relative to zooplankton.

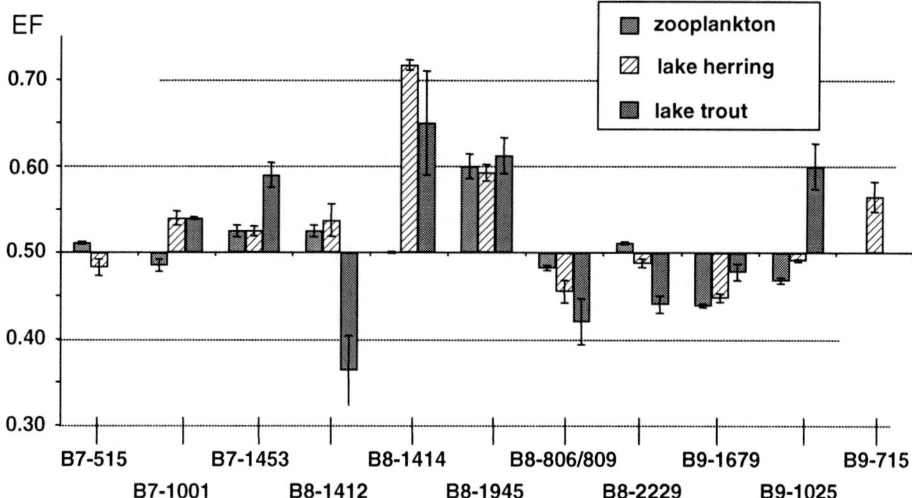

Fig. 6.14. EFs of eleven toxaphene compounds in zooplankton and two fish species from the same food chain [284]. The different EFs obtained for both different compounds and species illustrate the complexity of interpreting results found in environmental compartments.

Fig. 6.14 compiles some data determined in fish from Lake Superior samples [284]. Muir et al. also linked biomagnification factors with EFs and correlated EFs in different species. This allowed predicting that dietary transfer of the non-racemic congeners is occurring and that the CTT pattern obverved in lake trout is not entirely due to biotransformation in the trout [284]. A food web study in a marine marsh land indicated enantioselective depletion of the major abundant compounds in sediments (B6-923 and B7-1001) already on a very low trophic level (grass shrimp) [103]. EFs of B7-1001 were reversed in different species—grass shrimps had EFs >0.666 and mummichogs EFs of 0.333. To make the picture more complex, the EF in top predators at this site was between the low trophic samples indicating an uptake from different food sources and demonstrating the complexity of such ecosystems also in terms of interpretation of enantioselective data. The EF of the persistent B8-1413 was not affected irrespective of the trophic position [103]. The significant alteration of the EF of B7-1001 and B6-923 in small fish led to the conclusion that these metabolites are easily biotransformed [103]. For this reason an elimination study was performed [26]. These highly polluted fish were caught at the above mentioned site and released in uncontaminated water and fed with commercial toxaphene-free fish food [26]. B6-923 and B7-1001 were enantioselectively eliminated from the fish within 30 and 60 days, respectively [26]. For B6-923, the second eluted enantiomer was eliminated twice as fast, and the EF in fish was found to be in equilibrium between daily uptake and daily metabolism/excretion. Assuming a racemic uptake, determination of the EF in fish at the site along with the elimination rate of the enantiomers allowed calculations of the daily and annual throughput of the component with B6-923. This study was only possible with enantioselective analysis [26]. The shift in the EF of B6-923 during the elimination experiment suggested that B6-923 (and other chiral compounds) should be considered as a two-component mixture formed by the two B6-923-enantiomers. Two-component mixtures are characterized by non-ideal first-order kinetics (in contrast to single enantiomers) [26]. These results appeared to be promising, but different results were obtained when the study was repeated at lower water temperature [298]. Therefore, parameters such as water temperature obviously can affect the enantioselectivity in biota [298]. Similar effects of temperature were previously found for soils where the microbial communities even switched the EFs [299].

EFs of the two major CTTs in marine wildlife, B8-1413 and B9-1679 [272,300], were between 0.500 and 0.526 in marine mammals (blubber and brain), were 0.500 ± 0.026 in marine fish, 0.565/0.583 in monkey adipose tissue, and 0.524–0.565 in human milk, indicative of little or no enantioselectivity in their transformation [64,73,85,111,279,280,282]. However, other CTTs were significantly non-racemic in marine organisms. In seal blubber, the first eluting enantiomer of B8-2229 was significantly more abundant (EF = 0.722–0.772) [301]. The same was found for B8-1412 in seal blubber, human milk, and cod livers but this time the first eluting enantiomer was depleted [89,302]. In contrast the EF of B8-1412 was 0.474 in a polar bear (*Ursus maritimus*) and 0.688 in harbor porpoises (*Phocoena phocoena*) [85]. EFs of B7-1453 in cod and seal blubber from the Baltic Sea were 0.500 and 0.565, respectively [89,289]. EFs of B7-1000 in elephant seal, Weddell seal, and skua adipose tissue were 0.524, 0.333, and 0.230, respectively [303]. De Geus et al. found several significantly depleted CTTs in fish (hake liver) and dolphin (blubber) [111]. For B9-1025, the EF was 0.750–0.800, and in the case of B8-789 the EF was ~ 0.875 in hake liver and > 0.952 in dolphin blubber [111]. The second eluting enantiomer of B9-1025 (P-62) was also significantly depleted in blubber of Antarctic Weddell seals [85]. The transformation of B9-1025 into B8-2229 was suggested, and both compounds show EFs significantly deviating from the racemate [85]. By contrast, two potential metabolites of B9-1679 were usually much more non-racemic than the parent compound, which indicates that the EF of B8-1414 and B8-1412 is most likely the consequence of their own transformation or the additional formation from another parent compound. Thus, there is a small likeability left that the non-racemic proportions of B9-1679 in zooplankton (see above) are the result of the enantioselective transformation of a parent compound (here, decachlorobornane). A comprehensive study on the enantioselective transfer of toxaphene congeners from hen to eggs was recently performed by Hamed et al. [304]. After a feeding period of 38 days, changes in the EFs were obtained for eight individual CTTs that could be studied. Some congeners such as B8-2229 and B9-1025 showed significant changes in the EF, both in hen tissue as well as in egg yolk [304].

These findings, although not understood in details, led to the differentiation between persistent (EFs $\sim$ racemic) and partly degradable CTTs (significant enantioenrichment) [85]. Kallenborn and Heimstad distinguished 12 scenarios for enantioselective and non-enantioselective EFs of CTTs which included calculated stabilities, degree of chlorination, and levels in the environment [305]. A general classification to describe the environmental persistence of chiral toxaphene pollutants based on congener-specific elimination kinetics and susceptibility to biotransformation as measured by EFs is shown in Table 6.7.

Table 6.7 Classification of toxaphene congeners based on elimination kinetics and potential for biotransformation [298]

Classification	I	II	III	IV
Elimination processes[a]	Rapid, low activation barrier	Rapid, high activation barrier	Slow, low activation barrier	Slow, high activation barrier
Evidence of Enantioselectivity	No (racemic)	Yes (non-racemic)	No (racemic)	Yes (non-racemic)
Congener examples	Penta-2; B7-515	B6-923	B9-1679	B8-1414

[a]Temperature dependence of elimination is represented by the Arrhenius equation $k \alpha e^{-Ea/RT}$, where k = reaction rate constant, E_a = activation energy (barrier), R = universal gas constant and T [K].

6.4.7 Bromocyclene

Bromocyclene is another Diels-Alder condensation product utilized as insecticide which is synthesized as racemic mixture. Bromocyclene is used against ectoparasites in petcare and in sheep-herds. In industrialized countries, the main input into the environment is *via* waste water into sewage treatment plants (STPs) [306]. Incomplete elimination within STPs leads to contamination of surface waters. Bromocyclene in the discharge of STPs was found to be racemic. On the other hand some but not all fish (bream) in rivers show EFs significantly deviating from 0.5, which indicates biotransformation in the respective samples [306].

Though the main elimination step in STPs is probably sorption to sludge, transformation in STPs may occur in the anaerobic digesters of the STPs were the residence time is much higher than in the aeration basins. Additionally, dehalogenation reactions occur predominantly in anaerobic media in STPs. Thus, EFs deviating from 0.5 were determined in sludge samples after treatment in digesters. However, there are no data on the extend of transformation of bromocyclene in these digesters [307].

6.4.8 Modern pesticides

Modern pesticides are those which are currently used, they have been registered or authorized from 1960 to 2006. In most cases they are less lipophilic than the older organochlorines. They are used in all kinds of applications (insecticides, herbicides fungicides), whereas in the colder regions herbicides and fungicides predominate. However in opposite to the older organochlorines they are synthesized in a more targeted fashion and some care is taken that by-products are less relevant nowadays. A multitude of these compounds is chiral. Some of these are currently marketed predominately in their enantiopure forms (mecoprop, metalaxyl) while other compounds are still marketed as racemic mixtures (metolachlor).

6.4.8.1 Mecoprop

Mecoprop (IUPAC-name: (*R*)-2-(4-chloro-*o*-tolyloxy)propionic acid, trade name: MCPP, Fig. 6.15) is used as a herbicide in agriculture worldwide. This chiral compound has been

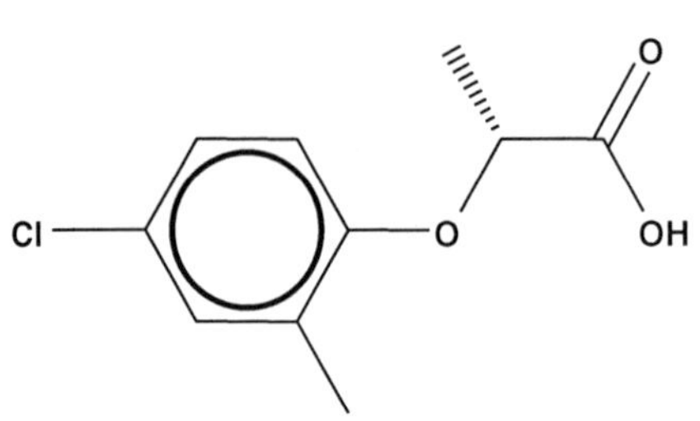

R- Mecoprop

Fig. 6.15. Structures of chiral musk fragrances.

marketed by the pesticide industry as enantiopure ingredient for several decades. Mecoprop is found regularly in surface water samples at 0.01–1 μg/L levels in most countries. However Bucheli et al. determined much more mecoprop in Swiss surface waters than what was modeled from use in agriculture [308]. Enantioselective GC/MS analysis revealed that the compound was not enantiopure in surface waters but held varying EFs. It was found out, that mecoprop is also used in (building) construction, e.g. to protect bitumous polymers used for the sealing of rooftops from destruction by plants growing on top of the roofs. Thus the respective sealings are protected by embedded MCPP. Opposite to the agricultural practice, the mecoprop in this application is not enantiopure but racemic. However the respective sealing and the polymer to which mecoprop is bonded is subject to photooxydation processes and thus degrades by and by. By this mechanism considerable amounts of MCPP are released with runoff waters due to rain [308].

Moreover, Bucheli et al. calculated the contributions of the two sources (agriculture and construction) for MCPP in Swiss surface waters [308]. It turned out that—though varying—indeed a major fraction (in average ~50%) of this herbicide stemmed from construction and not from agriculture. This finding had serious implication on the evaluation of the usage of this pesticide. Without the use of enantioselective determination it would have been extremely hard to trace the source of pollution in this case. Similar results were reported by Poiger et al. 2003 [309] for Swiss lakes. During April–September the mecoprop concentrations rose in these ecosystems, on the other hand R-mecoprop was analyzed with higher enantiomeric excess during these months. It was concluded, that during these months the application of enantiopure R-mecoprop took place in agriculture, while during the rest of the year racemic MCPP from rooftops equaled with those from agriculture. However due to different catchment areas the lakes received different patterns: while the lake Greifensee received less than 20% of its mecoprop load from agriculture, lake Baldeggersee received about 70% from agriculture.

In soil ecosystems phenoxyalkanoic acid herbicides were studied by Buser and Müller [310,311]. Not only did some enantiomers degrade faster than other ones, but also in some experiments selected enantiomers seemed to be converted into each other. For instance, R-MCPP was transformed into S-MCPP in soil experiments with a maximum of S-MCPP found after 5 days. This is making an overall assessment much more complex than thought of in earlier times. Similar results were described for other chiral phenoxyalkanoic acid herbicides such as 2-(2,4-dichlorophenoxy)propionic acid (DCPP). Obviously the transformation pathway goes *via* either a carbanionic or a enolic intermediate form at the previous asymmetric C-atom [311]. Moreover, the respective enantioselectivity is depended on pH values albeit the racemization rate was low [312]. Whether or not racemization occurs in other than specialized soil ecosystem is unknown at the moment.

6.4.8.2 Acetanilides

Several acetanilides (e.g. metalaxyl, metolachlor, alachlor) are chiral compounds. The stereoisomers of these compounds can be separated by GC as well as HPLC [313,314]. Enantioselective determination demonstrated that different soils transformed metalaxyl with different enantiopreference. While in a soil from moderate climate regimes the *R*-enantiomer was degraded faster than the *S*-enantiomer, the transformation of the tropical soil resulted in a quicker breakdown of the *S*-enantiomer [315]. It must thus be concluded that different enzymes and possibly different microbes were responsible for this transformation in the two different soils. Buerge et al. [312] demonstrated that enantiopreference of the transformation of metalaxyl was diverse for European soils, as well. These authors discussed pH dependence as well as different soil horizons. While the EF tended to higher values during transformation of racemic metalaxyl in the A_h horizon which is the topsoil, in the E horizon the racemic composition was constant, and the deepest horizon (B_s-90 cm) produced EFs about 0.33. In these studies the enantioselectivity (ES) of transformation was used (Eq. (11)):

$$\mathrm{ES} = \frac{k_R - k_S}{k_R + k_S} \tag{11}$$

with k being kinetic rate constants.

The ES values ranged from −0.5 at pH = 3 to +0.5 at pH = 7. The respective effect can also be triggered by manipulating the respective pH by adding carbonate or acid to the respective soil [312]. Enantioselective analysis was performed using a GC-column with a stationary phase consisting of octakis(bis-tert-butyldimetylsilyl)-γ-cyclodextrin. The respective acid metabolite was analyzed as ethyl-metalaxyl.

6.4.9 Environmentally relevant pharmaceuticals

Residues of pharmaceuticals have been demonstrated to be an issue in waste water treatment and drinking water production [316,317]. Currently, several pharmaceuticals are monitored in waste water treatment to determine how they can be eliminated from the water. Several of these compounds of environmental concern are chiral, and thus the enantiomeric composition at the diverse treatment steps can be used for the effectiveness of the biological treatment. Important compounds are ibuprofen and its transformation products as well as metoprolol.

Ibuprofen is used as a anti-inflammatory analgesic especially in case of reumatic disorders [309]. It is administered as a racemate but in the human body it is metabolized enantioselectively, thus in urine an EF of 0.95 (*S*-ibuprofen) is determined. A similar ee was determined in the waste-water plant inflow [309]. However in the effluent the EF is about 0.6. This indicates that the microbial community in the waste-water treatment plant degrades *S*-ibuprofen faster than the *R*-enantiomer, which is *vice versa* in the human body. In another experiment the same group incubated ibuprofen in water samples under controlled conditions. The transformation was fastest for the *S*-enantiomer especially if the incubation samples were illuminated and not put into the dark. Thus this experiment can be taken as indication not only bacteria but also algae are involved in enantioselective transformation processes [318].

6.4.10 Polycyclic musks

Polycyclic musks are marketed worldwide under different trade names (compare Table 6.8). In this chapter the abbreviations of the systematic names will be used predominantly. Musks are beyond the fragrances with the longest history in perfumery [319]. Historically, only the macrolide lactones excreted by the musk deer and musk cat have been used, but synthetic musk fragrances were produced since the twentieth century. At the beginning, nitromusks (i.e. derivatives of trinitrotoluene) were predominantly used [319]. However, these compounds were sometimes irritating to the skin, and were subject to photoreactions. Additionally some were discussed to be carcinogenic, and they were determined in huge amounts in waste water [320,321]. The next step was to switch to oxygenated hydrocarbons such as the polycyclic musks. Today, HHCB and to a somewhat lesser extent AHTN dominate the market in musk perfumery, though the next generation of macrocyclic musks is also marketed. They are considered to be still too expensive to be used in mass applications such as washing powders, shampoos, dishwashing detergents etc. though [319]. HHCB is sold with 1500 t/a and AHTN with 590 t/a in Europe (Balk and Ford, 1999 [322]). Other compounds such as ADBI, AHDI, ATII, and DPMI have one order of magnitude less market volume.

HHCB has two chiral centres, while AHTN has got one, thus these fragrances as well as other ones are chiral (Table 6.8, Fig. 6.16). Up to now, they are sold as racemates although only one enantiomer gives the desired sensoric effect, in fact, their mirror images are either inactive or exhibit an unwelcome sensoric effect [323]. Especially the polycyclic musk compounds are released with some thousand tons annually directly

Table 6.8 Overview on polycyclic aromatic musk fragrances that are currently used

Abbreviation	IUPAC name	Trade name
AHTN	7-Acetyl-1,1,3,4,4,6-hexamethyl-1,2,3,4-tetrahydronaphthalene	Tonalide
ADBI	1-[6-(1,1-Dimethylethyl)-2,3-dihydro-1,1-dimethyl-1H-inden-4-yl]-ethanone	Celestolide
AHDI	1-(2,3-Dihydro-1,1,2,3,3,6-hexamethyl-1H-inden-5-yl)ethanone	Phantolide
ATII	1-[2,3-Dihydro-1,1,2,6-tetramethyl-3-(1-methyl-ethyl)-1H-enden-5yl]ethanone	Traseolide
DPMI	1,2,3,5,6,7-Hexahydro-1,1,2,3,3-pentametyl-4H-inden-4one	Cashmeran
HHCB	1,3,4,6,7,8-Hexahydro-4,6,6,7,8,8-hexamethylcyclopenta-(g)-2-benzopyran	Galaxolide

4*R*,7*S*-HHCB (galaxolide)

4*R*,7*S*-HHCB-lactone (galaxolidone)

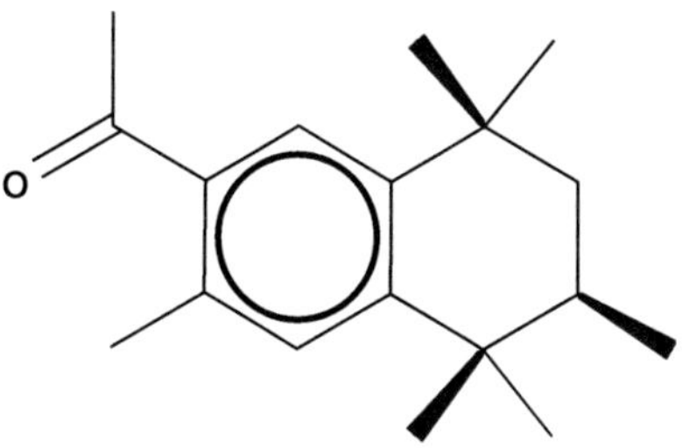

6*R*-AHTN (Tonalide)

Fig. 6.16. Structure of *R*-mecoprop.

to the waste water by washing/cleaning operations. Sewage treatment plants (STPs) have to cope with these compounds. However elimination rates range from 30 to 95% depending on the location and the construction of the respective plants [320,324,325]. Until 2004, it was not clear by what means the elimination took place: biotransformation, chemical transformation or sorption mechanisms. In some cases serious amounts of the transformation product HHCB-lactone were formed from the parent HHCB by oxidation processes [325,326]. Thus these compounds are released into the aquatic environment and reach even the North Sea [320,326,327]. Polycyclic musk compounds have been demonstrated to be subject to accumulation processes in fish [320,328]. New motivation on studies on polycyclic musks arose when Seinen et al. (1999) demonstrated that HHCB expressed some esterogenic effects [329].

Enantioselective analysis was first established by Franke et al. (1999) [330] for fish samples. Additionally, Gaterman et al. demonstrated that enantiomeric enrichment of polycyclic musks was taking place in fish [331]. Most interestingly the diverse fish species in the ecosystem held not only different concentrations, but also different EFs. This was probably due to different transformation efficiencies of the diverse fish species or due to different trophic levels, and thus different accumulation. However, significantly different enzymes were used by these fish to cope with these compounds. In another study Kallenborn et al. (2001) [329] detected the metabolite HHCB-lactone in fish samples from Norway. It remained unsolved whether this metabolite was formed by the fish or whether it accumulated HHCB-lactone from the water.

The same methodology of separation [330] of HHCB-lactone was then also used to discriminate between biological degradation/transformation and physicochemical processes, such as sorption to sludge, and chemical oxidation in waste water treatment [326] (Fig. 6.17). In most European STPs, it seems that sorption to sludge is the dominant elimination mechanism [325,326,332]. Transformation to HHCB-lactone is performed with enantiopreference and is thus determined to be a biological process [326]. However this transformation consumes about 10% of the incoming HHCB, but as the metabolite is more hydrophylic than the parent compound, the concentrations of HHCB and HHCB lactone are similar in the effluent [325,326]. Different enantiomeric patterns in the effluent of diverse STPs indicated, that different

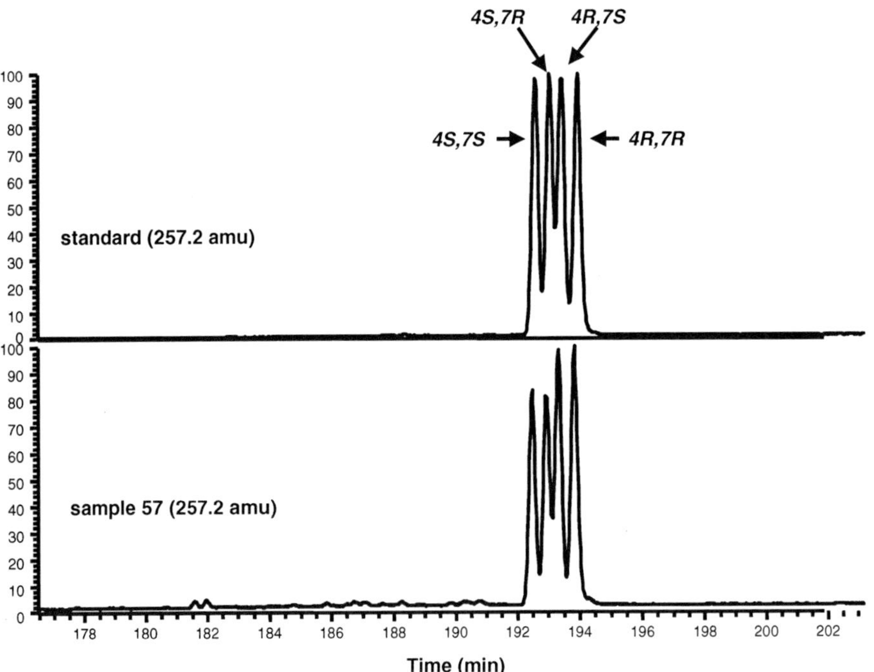

Fig. 6.17. Enantioseparation of HHCB-lactone in (top) a racemic standard in comparison to (bottom) an effluent sample from a sewage treatment plant.

biological processes predominate in the diverse plants [326]. (Fig. 6.16). In addition to these findings Berset et al. (2004) [333] determined EFs of HHCB, AHTN, AHDI, ATII DPMI and HHCB-lactone in sewage sludges, and effluent waters. Several of the reported values were significantly different from EF = 0.5, with 0.45 and 0.77, respectively being maximum values. Thus biodegradation was assumed to take place in the respective plants. Interestingly enough, the good separation, which was achieved in this work for HHCB-lactone in standard solutions could not be achieved for real-world samples. Enantioselective analysis may be used in future to further optimize STPs for the transformation of organic micropollutants.

6.5 CONCLUSIONS

After 15 years of research, GC enantioseparations at ultra trace concentrations have reached a high scientific level. Most of the work is currently published in top journals of the respective disciplines. Some outstanding and unique results have been obtained which certainly increased the scientific understanding of the environmental behaviour and fate of chiral pollutants. There is no doubt that this trend will continue.

While the first years were mainly restricted to researchers that used lab-made CSPs, today the overwhelming number of studies is performed with commercially available columns. In the meantime, more or less all volatile chiral pollutants have been successfully GC-enantioseparated. Despite a high variety available, a handful of CSPs cover ~90% of all applications, independent of the target compound. Analytical studies on basic enantioseparation techniques have become scarce in favour of environmental-oriented studies. However, these studies are performed with steadily improved equipment by using sophisticated techniques including GC/MS, GC/MSMS, MDGC and even GC × GC.

On the other hand, the relatively high number of publications in the field hides the fact that the author teams very often apply established methods for their research and there are lesser studies on testing and expanding the scope. QA is playing now a more and more important role since the researchers are aware of the pitfalls that can easily falsify the measurements and thus may lead to wrong conclusions. However, there is still a lack of intercalibration data in real samples or even validation for environmental samples.

The most complex point is still the interpretation of the data which aims to generalize the findings. To obtain knowledge from results we do have to know the meaning of measured EFs in a general content. The results presented in this article indicate that one should not be too quick in simplifying enantioselective data. For instance, the initial idea EFs in biota would correlate with the trophic position has been shown to be only valid under certain circumstances. Often each chiral compound appears to have its own enantioselective story in each matrix. Some spectacular findings in the initial phase had to be seen in relative terms within the recent years. One of the most spectacular results was that only (+)-α-HCH penetrates the blood-brain-barrier of seals [55]. However, recent studies indicate that this is not the case in other species. Furthermore, many other pollutants were almost racemic in brains of seals.

This and other examples clarify that the researchers in the first phase had golden hands when selecting samples for their research.

Two point source investigations provided interesting insights into the circulation and fate of chiral pollutants. At the moment when more studies in this field have been published, significant variations in EFs in the narrowest surrounding prevented the total understanding of the processes. When uptake and elimination studies appeared to give new understanding, indications that some "simple" factors—such as the surrounding water temperature—can have strong effects on the kinetics give cause for caution in the interpretation of EF data.

Nevertheless, studies and approaches such as using EF determinations in closed but complex systems such as sewage treatment plants is a promising tool to find out more about the fate of chemicals therein. Despite all uncertainties in EF data, the positive aspects prevail, and the determination of EFs in environmental samples is today by far more than the addition of a new parameter to monitoring data. The data compiled in this article may be helpful for stimulating new studies, eliciting so some hidden aspects of this acknowledged field of research.

REFERENCES

1 W.A. König, R. Krebber and P. Mischnick, Cyclodextrins as chiral stationary phases in capillary gas chromatography. Part V: Octakis(3-O-butyryl-2,6-di-O-pentyl)-γ-cyclodextrin. J. High Resol. Chromatogr., 12 (1989) 732–738.

2 R. Kallenborn, H. Hühnerfuss and W.A. König, Enantioselective metabolism of (+/−)-α-1,2,3,4,5,6-hexachlorocyclohexane in organs of the Eider duck. Angew. Chem. Int. Ed., 30 (1991) 320–321.

3 J. Faller, H. Hühnerfuss, W.A. König and P. Ludwig, Gas chromatographic separation of the enantiomers of marine organic pollutants. Distribution of α-HCH enantiomers in the North Sea. Mar. Poll. Bull., 22 (1991) 82–88.

4 L.H. Easson and E. Stedman, Studies on the relationship between chemical constitution and physiological action. V. Molecular dissymmetry and physiological activity. Biochem. J., 27 (1933) 1257–1266.

5 G.M. Ramos Tombo and D. Bellus, Chirality and crop protection. Angew. Chem. Int. Ed. Engl., 30 (1991) 1193–1215.

6 E.J. Ariens and J.J.S. van Rensen (Eds.), Stereoselectivity of Pesticides – Biological and Chemical Problems, Elsevier Science Publishers, Amsterdam, 1988.

7 S.J. Cristol, The structure of α-benzene hexachloride. J. Am. Chem. Soc., 71 (1949) 1894.

8 W. Vetter and W. Jun, Elucidation of a polychlorinated bipyrrole structure using enantioselective GC. Anal. Chem., 74 (2002) 4287–4289.

9 K.L.E. Kaiser, On the optical activity of polychlorinated biphenyls. Environ. Poll., 7 (1974) 93–101.

10 S. Safe, Polychlorinated biphenyls (PCBs) and polybrominated biphenyls (PBBs): biochemistry, toxicology, and mechanism of action. Crit. Rev. Toxicol., 13 (1984) 319–395.

11 Ananymous, Polychlorierte Biphenyle, DFG, VCH Verlag, 1988.

12 S.L. Schantz, J.J. Widholm and D.C. Rice, Effects of PCB exposure on neuropsychological function in children. Environ. Health Perspect., 11 (2003) 357–376.

13 H.-J. Lehmler, L.W. Robertson, A.G. Garrison and P.R.S. Kodovanti, Effects of PCB 84 enantiomers on [^{3}H]-phorbol ester binding in rat cerebellar granule cells and 45Ca^{2+}-uptake in rat cerebellum. Toxicol. Lett., 156 (2005) 391–400.

14 A. Miyazaki, T. Hotta, S. Marumo and M. Sakai, Synthesis, absolute stereochemistry, and biological activity of optically active cyclodiene insecticides. J. Agric. Food Chem., 26 (1978) 975–977.

15 A. Miyazaki, M. Sakai and S. Marumo, Synthesis and biological activity of optically active heptachlor, 2-chloroheptachlor, and 3-chloroheptachlor. J. Agric. Food Chem., 28 (1980) 1310–1311.
16 W.A. McBlain, The levo enantiomer of o,p′-DDT inhibits the binding of 17β-estradiol to the estrogen receptor. Life Science, 40 (1987) 215–222.
17 W. Vetter and V. Schurig, Enantioselective determination of chiral organochlorines in biota on modified cyclodextrins by gas chromatography. J. Chromatogr. A, 774 (1997) 143–175.
18 R. Kallenborn and H. Hühnerfuss, Chiral Environmental Pollutants. Trace Analysis and Ecotoxicology, Heidelberg, Springer, 2001.
19 W. Vetter, Enantioselective fate of chiral chlorinated hydrocarbons and their metabolites in environmental samples. Food Rev. Int., 71 (2001) 113–182.
20 W.J.M. Hegeman and R.W.P.M. Laane, Enantiomeric enrichment of chiral pesticides in the environment. Rev. Environ. Contam. Toxicol., 173 (2002) 85–116.
21 T. Müller and H. Kohler, Chirality of pollutants – effects on metabolism and fate. Appl. Microbiol. Biotechnol., 64 (2004) 300–316.
22 I. Ali and H.Y. Aboul-Enein, Chiral Pollutants: Distribution, Toxicity and Analysis by Chromatography and Capillary Electrophoresis, Wiley, UK, 2004.
23 A. Finizio, T.F. Bidleman and S.Y. Szeto, Emission of chiral pesticides from an agricultural soil in the Fraser Valley, British Columbia. Chemosphere, 36 (1998) 345–355.
24 B. Eitzer, W. Iannucci-Berger and M.I. Mattina, Volatilization of weathered chiral and achiral chlordane residues from soil. Environ. Sci. Technol., 37 (2003) 4887–4893.
25 M. Robson and Harrad. Chiral PCB signatures in air and soil: implications for atmospheric source apportionment. Environ. Sci. Technol., 38 (2004) 1662–1666.
26 W. Vetter, K.A. Maruya and K.L. Smalling, Interpreting non-racemic ratios of chiral organochlorines using naturally contaminated fish. Environ. Sci. Technol., 35 (2001) 4444–4448.
27 E. Gil-Av, B. Feibush and R. Charles-Sigler, Separation of enantiomers by gas liquid chromatography with an optically active stationary phase. Tetrahed. Lett., 7 (1966) 1009–1015.
28 E. Gil-Av, B. Feibush and R. Charles-Sigler, Separation of enantiomers by gas chromatography with optically active stationary phase, in: A.B. Littlewood (Ed.), Gas Chromatography, Institute of Petroleum, London, 1967, pp. 227–239.
29 H. Frank, G.J. Graeme and E. Bayer, Rapid gas chromatographic separation of amino acid enantiomers with a novel chiral stationary phase. J. Chromatogr. Sci., 15 (1977) 174–176.
30 I. Benecke and W.A. König, Isocyanate als universelle Reagentien bei der Derivatbildung für die gaschromatographische Enantiomerentrennung. Angew. Chem. Int. Ed. Engl., 21 (1982) 709.
31 V. Schurig, Resolution of a chiral olefin by complexation chromatography on an optically active rhodium(I) complex. Angew. Chem. Int. Ed. Engl., 16 (1977) 110.
32 T. Koscielski, D. Sybilska and J. Jurczak, J. Chromatogr., 261 (1983) 357–362.
33 Z. Juvancz, G. Alexander and J. Szejtli, Permethylated β-cyclodextrin as stationary phase in capillary gas chromatography. J. High Resol. Chromatogr., 10 (1987) 105–107.
34 A. Villiers Sur la fermentation de la fécule par l'action du ferment butyriqué. Compt Rendu Hebd Seances. Acad. Sci., 112 (1891) 536–538.
35 F. Schardinger, Über thermophile Bakterien aus verschiedenen Speisen und Milch, sowie über einige Umsetzungsproduke derselben in kohlenhydrathaltigen Nährlösungen, darunter krystallisierte Polysaccharide (Dextrine) aus Stärke. Z Unters Nahr u Genussm, 6 (1903) 865–880.
36 W.A. König, Gas Chromatographic Enantiomer Separation with Modified Cyclodextrins, Heidelberg, Hüthig, 1992.
37 G. Gattuso, S.A. Nepogodiev and J.F. Stoddart, Synthetic cyclic oligosaccharides. Chem. Rev., 98 (1998) 1919–1958.
38 J. Szejtli, Introduction and general overview of cyclodextrin chemistry. Chem. Rev., 98 (1998) 1743–1753.
39 V. Schurig, Enantiomer separation by gas chromatography on chiral stationary phases. J. Chromatogr. A, 666 (1994) 111–129.
40 V. Schurig and H.-P. Nowotny, Separation of enantiomers on diluted permethylated β-cyclodextrin by high-resolution gas chromatography. J. Chromatogr., 441 (1988) 155–163.

41 A. Dietrich, B. Maas and A. Mosandl, Diluted modified cyclodextrins as chiral stationary phases – influence of the polysiloxane solvent: heptakis(2,3-di-acetyl-6-O-tert-butyldimethylsilyl)-β-cyclodextrin. J. High Resol. Chromatogr., 18 (1995) 152–156.

42 D.W. Armstrong, W.Y. Li and J. Pitha, Reversing enantioselectivity in capillary gas chromatography with polar and nonpolar cyclodextrin derivative phases. Anal. Chem., 62 (1990) 214–217.

43 D.W. Armstrong, and H.L. Jin, New polar and non-polar cyclodextrin derivative gas chromatographic phases. J. Chromatogr., 502 (1990) 154–159.

44 W. Blum and R. Aichholz, Gas chromatographic enantiomer separation on tert-butyldimethylsilylated β-cyclodextrin diluted in PS-086. Simple method to prepare enantioselective glass capillary columns. J. High Resol. Chromatogr. 13 (1990) 515–518.

45 A. Dietrich, B. Maas, V. Karl, P. Kreis, D. Lehmann, B. Weber and A. Mosandl, Stereoisomeric flavor compounds. Part LV: Stereodifferentiation of some chiral volatiles on heptakis(2,3-di-O-acetyl-6-O-tert-butyldimethylsilyl)-β-cyclodextrin. J. High Resol. Chromatogr., 15 (1992) 176–179.

46 A. Dietrich, B. Maas, W. Messer, G. Bruche, V. Karl, A. Kaunziger and A. Mosandl, Stereoisomeric flavor compounds. Part LVIII: The use of heptakis(2,3-di-O-methyl-6-O-tert-butyldimethylsilyl)-β-cyclodextrin as a chiral stationary phase in flavor analysis. J. High Resol. Chromatogr., 15 (1992) 590–593.

47 B. Maas, A. Dietrich, T. Beck, S. Börner and A. Mosandl, Di-tert-butyldimethylsilylated cyclodextrins as chiral stationary phases: thermodynamic investigations. J. Microcol. Sep., 7 (1995) 65–73.

48 P. Fischer, R. Aichholz, U. Bölz, M. Juza and S. Krimmer, Permethyl-β-cyclodextrin – chemically bonded to polysiloxane: a chiral stationary phase with wider application range for enantiomer separation by capillary gas chromatography. Angew. Chem. Int. Ed. Engl., 19 (1990) 427–429.

49 V. Schurig, D. Schmalzing, U. Mühleck, M. Jung, M. Schleimer, P. Mussche, C. Duvekot and J.C. Buyten, Gas chromatographic enantiomer separation on polysiloxane-anchored permethyl-β-cyclodextrin (Chirasil-Dex). J. High Resol. Chromatogr., 13 (1990) 713–717.

50 V. Schurig, Z. Juvancz, G.J. Nicholson and D. Schmalzing, Separation of enantiomers on immobilized polysiloxane-anchored permethyl-β-cyclodextrin (CHIRASIL-DEX) supercritical fluid chromatography. J. High Resol. Chromatogr., 14 (1991) 58–62.

51 J. Dönnecke, W.A. König, O. Gyllenhaal, J. Vessman and C. Schultze, Enantiomer separation by capillary SFC and GC on immobilized oktakis(2,6-di-O-methyl-3-O-pentyl)-γ-cyclodextrin. J. High Resol. Chromatogr., 17 (1994) 779–784.

52 Z. Juvancz and J. Szejtli, The role of cyclodextrins in chiral selective chromatography. Trends Anal. Chem., 21 (2002) 379–388.

53 A. Mosandl, Enantioselective capillary gas chromatography and stable isotope ratio mass spectrometry in the authenticity control of flavors and essential oils. Food Rev. Int., 11 (1995) 597–664.

54 Z. Juvancz and P. Petersson, Enantioselective gas chromatography. J. Microcol. Sep., 8 (1996) 99–114.

55 K. Möller, C. Bretzke, H. Hühnerfuss, R. Kallenborn, J.N Kinkel, J. Kopf and G. Rimkus, The absolute configuration of (+)-α-1,2,3,4,5,6-hexachlorocyclohexane, and its permeation through the seal blood-brain barrier. Angew. Chem. Int. Ed., 33 (1994) 882–884.

56 J. Faller, H. Hühnerfuss, W.A. König, R. Krebber and P. Ludwig, Do marine bacteria degrade α-hexachlorocyclohexane stereoselectively. Environ. Sci. Technol., 25 (1991) 676–678.

57 H. Hühnerfuss, J. Faller, W.A. König and P. Ludwig, Gas chromatographic separation of the enantiomers of marine pollutants. 4. Fate of hexachlorocyclohexane isomers in the Baltic and the North Sea. Environ. Sci. Technol., 26 (1992) 2127–2133.

58 H. Hühnerfuss, R. Kallenborn, K. Möller, B. Pfaffenberger and G. Rimkus, Enzymatic enantioselective degradation of α-hexachlorocyclohexane in different tissues of marine and terrestric biota as determined by chiral gas chromatography. Proceedings of 15th International Symposium on Capillary Chromatography, Vol. 1, Riva del Garda, Italy, May 24–27, 1993, pp. 576–581.

59 H. Hühnerfuss, J. Faller, R. Kallenborn, W.A. König, P. Ludwig, B. Pfaffenberger, M. Oehme and G. Rimkus, Enantioselective and nonenantioselective degradation of organic pollutants in the marine ecosystem. Chirality, 5 (1993) 393–399.

60 S. Mössner, T.R. Spraker, P.R. Becker and K. Ballschmiter, Ratios of enantiomers of alpha-HCH and determination of alpha-, beta-, and gamma-HCH isomers in brain and other tissues of neonatal northern fur seals (Callorhinus ursinus). Chemosphere, 24 (1992) 1171–1180.
61 W. Vetter, U. Klobes, B. Luckas, G. Hottinger and G. Schmidt, Determination of (+/−) elution orders of chiral organochlorines by HPLC with a chiral detector and enantioselective gas chromatography. J. Assoc. Off. Anal. Chem., 81 (1998) 1245–1251.
62 E. Francotte, K. Grolimund and Z. Juvancz, Benzoylcellulose derivatives as chiral stationary phases for open tubular column chromatography. Chirality, 5 (1993) 232–237.
63 K. Otsuka, S. Honda, J. Kato, S. Terabe, K. Kimata and N. Tanaka, Effects of compositions of dimethyl-β-cyclodextrins on enantiomer separations by cyclodextrin modified capillary zone electrophoresis. J. Pharm. Biomed. Anal., 17 (1998) 1177–1190.
64 W. Vetter, U. Klobes, B. Luckas and G. Hottinger, Enantiomer separation of persistent compounds of technical toxaphene CTTs on t-butyldimethylsilylated β-cyclodextrin. Chromatographia, 45 (1997) 255–262.
65 M. Oehme, L. Müller and H. Karlsson, High-resolution chromatographic test for the characterisation of enantioselective separation of organochlorine compounds. Application to tert.-butyldimethylsilyl β-cyclodextrin. J. Chromatogr. A, 775 (1997) 275–285.
66 A. Jaus and M. Oehme, Benefits of partially alkylated cyclodextrins for the enantioselective separation of chiral polychlorinated compounds. Organohalogen Compd., 40 (1999) 387–390.
67 A. Jaus and M. Oehme, Enantioselective behaviour of ethylated γ-cyclodextrins as GC stationary phases for chlorinated pesticides and phase characterisation by HPLC. Chromatographia, 50 (1999) 299–304.
68 W. Vetter, U. Klobes, B. Luckas and G. Hottinger, Use of 6-O-tert.-butyldimethylsilylated β-cyclodextrins for the enantioseparation of chiral organochlorines. J. Chromatogr. A, 846 (1999) 375–381.
69 S. Ruppe, W. Vetter, B. Luckas and G. Hottinger, Application of well-defined β-cyclodextrins for the enantioseparation of compounds of technical toxaphene and further organochlorines. J. Microcol. Sep., 12 (2000) 541–548.
70 L.M.M. Jantunen and T.F Bidleman, Organochlorine pesticides and enantiomers of chiral pesticides in arctic ocean water. Arch. Environ. Contam. Toxicol., 35 (1998) 218–228.
71 G. Hottinger, Personal communication to W. Vetter, January and February 2005.
72 F. Kobor, K. Angermund and G. Schomburg, Molecular modelling experiments on chiral recognition in GC with specially derivatized cyclodextrins as selectors. J. High Resol. Chromatogr., 16 (1993) 299–311.
73 W. Vetter, U. Klobes, K. Hummert and B. Luckas, Gas chromatographic separation of chiral organochlorines on modified cyclodextrin phases and results of marine biota samples. J. High Resol. Chromatogr., 20 (1997) 85–93.
74 M.D. Müller, H.-R. Buser and C. Rappe, Enantioselective determination of various chlordane components and metabolites using chiral high-resolution gas chromatography with a β-cyclodextrin derivative of improved selectivity and electron-capture negative ion mass spectrometry detection. Chemosphere, 34 (1997) 2407–2417.
75 V. Schurig, H.-P. Nowotny and D. Schmalzing, Gas-chromatographic enantiomer separation of unfunctionalized cycloalkanes on permethylated β-cyclodextrin. Angew. Chem. Int. Ed., 28 (1989) 736–737.
76 V. Schurig, Personal communication to W. Vetter, March 2000.
77 S. Allenmark, A note concerning chromatography on mixed chiral stationary phases. Enantiomer, 4 (1999) 67–69.
78 H.-R. Buser and M.D. Müller, Enantioselective determination of chlordane components, metabolites, and photoconversion products in environmental samples using chiral high-resolution gas chromatography and mass spectrometry. Environ. Sci. Technol., 27 (1993) 1211–1220.
79 W. Vetter, K. Lehnert and G. Hottinger, Enantioseparation of chiral organochlorines on permethylated β- and γ-cyclodextrin, as well as 1:1 mixtures of them. J. Chromatogr. Sci., in press.

80 C. Bicchi, G. Artuffo, A. D'Amato, V. Manzin, A. Galli and M. Galli, Cyclodextrin derivatives for the GC separation of racemic mixtures of volatile compounds. Part VI: The influence of the diluting phase on the enantioselectivity of 2,6-di-O-methyl-3-O-pentyl-β-cyclodextrin. J. High Resol. Chromatogr., 16 (1993) 209–214.

81 S. Mössner and K. Ballschmiter, Separation of α-hexachlorocyclohexane (α-HCH) and pentachlorocyclohexene (PCCH) enantiomers on a cyclodextrin-phase (Cyclodex-B by HRGC/ECD.). Fresenius J. Anal. Chem., 348 (1994) 583–589.

82 H.-R. Buser and M.D. Müller, Isomer and enantioselective degradation of hexachlorocyclohexane isomers in sewage sludge under anaerobic conditions. Environ. Sci. Technol., 29 (1995) 664–672.

83 R.L. Falconer, T.F. Bidleman, D.J. Gregor, R. Semkin and C. Reixeira, Enantioselective breakdown of α-hexachlorocyclohexane in a small Arctic lake and its watershed. Environ. Sci. Technol., 29 (1995) 1297–1302.

84 E.M. Ulrich and R.A. Hites, Enantiomeric ratios of chlordane-related compounds in air near the Great Lakes. Environ. Sci. Technol., 32 (1998) 1870–1874.

85 W. Vetter and B. Luckas, Enantioselective fate of persistent and partly degradable toxaphene components in high trophic level biota. Chemosphere, 41 (2000) 499–506.

86 M. Oehme, R. Kallenborn, K. Wiberg and C. Rappe, Simultaneous enantioselective separation of chlordanes, a nonachlor compound, and o,p′-DDT in environmental samples using tandem capillary columns. J. High Resol. Chromatogr., 17 (1994) 583–588.

87 I.H. Hardt, C. Wolf, B. Gehrcke, D.H. Hochmuth, B. Pfaffenberger, H. Hühnerfuss and W.A. König, Gas chromatographic enantiomer separation of agrochemicals and polychlorinated biphenyls (PCBs) using modified cyclodextrins. J. High Resol. Chromatogr., 17 (1994) 859–864.

88 W.A. König, I.D. Hardt, B. Gehrcke, D.H. Hochmuth, H. Hühnerfuss, B. Pfaffenberger and G. Rimkus, Optisch aktive referenzsubstanzen für die umweltanalytik durch präparative enantioselektive gaschromatographie. Angew. Chem., 106 (1994) 2175–2177.

89 U. Klobes, W. Vetter, B. Luckas and G. Hottinger, Enantioseparation of compounds of technical toxaphene (CTTs) on 35% heptakis(6-O-tert.-butyldimethylsilyl-2,3-di-O-methyl)-β-cyclodextrin diluted in OV1701. Chromatographia, 47 (1998) 565–569.

90 R. Seemamahannop, A. Berthod, M. Maples, S. Kapila and D.W. Armstrong, Uptake and enantioselective elimination of chlordane compounds by common carp Cyprinus carpio, L. Chemosphere, 59 (2005) 493–500.

91 G. Koske, G. Leupold, D. Angerhöfer and H. Parlar, Multidimensional gas chromatographic enantiomer quantification of some chlorinated xenobiotics in cod liver and fish oils. Chemosphere, 39 (1999) 683–688.

92 C. Wong and A.W. Garrison, Enantiomer separation of polychlorinated biphenyl (PCB) atropisomers and polychlorinated biphenyl retention behavior on modified cyclodextrin capillary gas chromatography columns. J. Chromatogr. A, 866 (2000) 213–220.

93 A.W. Garrison, V.A. Nzengung, J.K. Avants, J.J. Ellington and N.L. Wolfe, Determining the environmental enantioselectivity of o,p′-DDT and o,p′-DDD. Organohalogen Compd., 31 (1997) 256–261.

94 A.W. Garrison, V.A. Nzengung, J.K. Avants, J.J. Ellington, W.J. Jones, D. Rennels and N.L. Wolfe, Phytodegradation of p,p′-DDT and the enantiomers of o,p′-DDT. Environ. Sci. Technol., 3 (2000) 1663–1670.

95 H. Karlsson, M. Oehme and G. Scherer, Isolation of the chlordane compounds U82, MC5, MC7, and MC8 from technical chlordane by HPLC including structure elucidation of U82 and determination of ECD and NICI-MS response factors. Environ. Sci. Technol., 33 (1999) 1353–1358.

96 E. Schneider and K. Ballschmiter, Transformation experiments with two chiral endosulfan metabolites by soil microorganisms – CHIRAL HRGC on lipophilic cyclodextrin derivatives. Fresenius J. Anal. Chem., 352 (1995) 756–762.

97 S. Meijer, C. Halsall, T. Harner, A. Peters, W. Ockenden, A. Johnston and K. Jones, Organochlorine pesticide residues in archived UK soil. Environ. Sci. Technol., 35 (2001) 1989–1995.

98 P. Hoekstra, B. Braune, C. Wong, M. Williamson, B. Elkin and D. Muir, Profile of persistent chlorinated contaminants, including selected chiral compounds, in wolverine (Gulo gulo) livers from the Canadian Arctic. Chemosphere, 53 (2003) 551–560.

99 K.C. Cundy and P.A. Crooks, Unexpected phenomenon in the high-performance liquid chromatographic analysis of racemic [^{14}C]-nicotine: separation of enantiomers in a totally achiral system. J. Chromatogr., 281 (1983) 17–23.

100 A. Kurganov, Separation of non-racemic mixtures of enantiomers. GIT Spezial., 1/97 (1997) 23–24.

101 B. Pfaffenberger, I. Hardt, H. Hühnerfuss, W.A. König, G. Rimkus, A. Glausch, V. Schurig and J. Hahn, Enantioselective degradation of α-hexachlorocyclohexane and cyclodiene insecticides in roe-deer liver samples from different regions in Germany. Chemosphere, 29 (1994) 1543–1554.

102 M. Schwinge, W. Vetter and B. Luckas, Enantioselective determination of atropisomeric PCBs after liquid chromatographic enrichment. Organohalogen Compd., 40 (1999) 405–408.

103 W. Vetter and K. Maruya, Congener and enantioselective analysis of toxaphene residues in sediment and biota representing several trophic levels of a contaminated estuarine wetland. Environ. Sci. Technol., 34 (2000) 1627–1635.

104 A. Götsch, E. Mariussen, R. von der Recke, D. Herzke, U. Berger and W. Vetter, Analytical strategies for successful enantioselective separation of atropisomeric PBB 132 and PBB 149 in environmental samples. J. Chromatogr. A, 1063 (2005) 193–199.

105 V.A. Nikiforov, V.G. Tribulovich and V.S. Karavan, Experience in isolation and identification of toxaphene congeners and prospects of congener-specific analysis of environmental samples. Organohalogen Compd. 26 (1995) 379–382.

106 R. Baycan-Keller, M. Oehme and B. Galliker, Optimization of individual column length of a tandem column system for the isomer and enantiomer selective separation of toxaphenes. Organohalogen Compd., 35 (1998) 229–233.

107 R. Baycan-Keller and M. Oehme, Optimization of tandem columns for the isomer and enantiomer selective separation of toxaphenes. J. Chromatogr. A, 837 (1999) 201–210.

108 A. Glausch, G.J. Nicholson, M. Fluck and V. Schurig, Separation of enantiomers of stable atropisomeric polychlorinated biphenyls (PCBs) by multidimensional gas chromatography on Chirasil-Dex. J. High Resol. Chromatogr., 17 (1994) 347–349.

109 V. Schurig and A. Glausch, Enantiomer separation of atropisomeric polychlorinated biphenyls (PCBs) by gas chromatography on Chirasil-Dex. Naturwissenschaften, 80 (1993) 468–469.

110 G.P. Blanch, A. Glausch, V. Schurig, R. Serrano and M.J. González, Quantification and determination of enantiomeric ratios of chiral PCB 95, PCB 132, and PCB 149 in shark liver samples (C. coelolepis) from the Atlantic Ocean. J. High Resol. Chromatogr., 19 (1996) 392–396.

111 H.-J. de Geus, R. Baycan-Keller, M. Oehme, J. de Boer and U.A.T. Brinkman, Enantiomer ratios of bornane congeners in biological samples using heart-cut gas chromatography with an enantioselective column. J. High Resol. Chromatogr., 21 (1998) 39–46.

112 H. Karlsson, D. Muir, W. Strachan, S. Backus, D. DeVault and D.M. Whittle, Enantiomer ratios of toxaphene in abiotic and biological samples from Lake Superior. Organohalogen Compd., 41 (1999) 597–600.

113 C. Wong, P. Hoekstra, H. Karlsson, S. Backus, S. Mabury and D. Muir, Enantiomer fractions of chiral organochlorine pesticides and polychlorinated biphenyls in standard and certified reference materials. Chemosphere, 49 (2002) 1339–1347.

114 L. Bordajandi, P. Korytár, J. de Boer and M. González, Enantiomeric separation of chiral polychlorinated biphenyls on β-cyclodextrin capillary columns by means of heart-cut multidimensional gas chromatography and comprehensive two-dimensional gas chromatography application to food samples. J. Sep. Sci., 28 (2005) 163–171.

115 J. Beens and U.A.T. Brinkman, Comprehensive two-dimensional gas chromatography – a powerful and versatile technique. The Analyst, 130 (2005) 123–127.

116 D. Ryan and P. Marriott, Comprehensive two-dimensional gas chromatography. Anal. Bioanal. Chem., 376 (2003) 295–297.

117 J. Dalluge, R.J.J. Vreuls, J. Beens and U.A.T. Brinkman, Optimization and characterization of comprehensive two-dimensional gas chromatography with time-of-flight mass spectrometric detection (GC × GC-TOF MS). J. Sep. Sci., 25 (2002) 201–214.
118 H.-R. Buser and M.D. Müller, Isomeric and enantiomeric composition of different commercial toxaphenes and chlorination products of (+)- and (−)- camphenes. J. Agric. Food Chem., 42 (1994) 393–400.
119 R. von der Recke, E. Mariussen, U. Berger, A. Götsch, D. Herzke and W. Vetter, Determination of the enantiomer ratio of PBB 149 by GC/NICI-tandem mass spectrometry in the selected reaction monitoring mode. Organohalogen Compd., 66 (2004) 213–218.
120 M.D. Müller and H.-R. Buser, Identification of the (+) and (−) enantiomers of chiral chlordane compounds using chiral high-performance liquid chromatography/chiroptical detection and chiral high-resolution gas chromatography and mass spectrometry. Anal. Chem., 66 (1994) 2155–2162.
121 S. Chu, A. Covaci and P. Schepens, Levels and chiral signatures of persistent organochlorine pollutants in human tissues from Belgium. Environ. Res., 93 (2003) 167–176.
122 B. Eitzer, M. Incorvia Mattina and W. Iannucci-Berger, Compositional and chiral profiles of weathered chlordane residues in soil. Environ. Toxicol. Chem., 20 (2001) 2198–2204.
123 H.-R. Buser and M.D. Müller, Enantiomer separation of chlordane components and metabolites using chiral high-resolution gas chromatography and detection by mass spectrometric techniques. Anal. Chem., 64 (1992) 3168–3175.
124 U. Klobes, W. Vetter, B. Luckas, K. Skírnisson and J. Plötz, Levels and enantiomeric ratios of (α-HCH, oxychlordane, and PCB 149 in blubber of harbour seals (Phoca vitulina) and grey seals (Halichoerus grypus) from Iceland and other species. Chemosphere, 37 (1998) 2501–2512.
125 W. Vetter and B. Luckas, Analytical artifacts during enantioselective determination of chiral organochlorines with GC/ECNI-MS. Organohalogen Compd., 35 (1998) 367–370.
126 A. Berthod, W. Li and D.W. Armstrong, Multiple enantioselective retention mechanisms on derivatized cyclodextrin as chromatographic chiral stationary phases. Anal. Chem., 64 (1992) 873–879.
127 B. Schwemmer, W.P. Cochrane and P.B. Polen, Oxychlordane, animal metabolite of chlordane: isolation and synthesis. Science, 169 (1970) 1087.
128 V.R. Meyer, Accuracy in the chromatographic determination of extreme enantiomeric ratios: a critical reflection. Chirality, 7 (1995) 567–571.
129 E. Benická, R. Novakowský, J. Hrouzek, J. Krupcik, P. Sandra and J. de Zeeuw, Multidimensional gas chromatographic separation of selected PCB atropisomers in technical formulations and sediments. J. High Resol. Chromatogr., 19 (1996) 95–98.
130 E.L. Eliel, S.H. Wilen and L.N. Mander, Stereochemistry of Organic Compounds, New York, Wiley Interscience, 1994.
131 V. Rautenstrauch, M. Lindström, B. Bourdin, J. Currie and E. Oliveros, Enantiomeric purities of (R)- and (S)-camphors from the chiral pool and high enantiomeric purities in general. Helv. Chim. Acta., 76 (1993) 607–615.
132 H.-J. de Geus, H. Besselink, A. Brouwer, J. Klungsøyr, B. McHugh, E. Nixon, G.G. Rimkus, P.G. Wester and J. de Boer, Environmental occurrence, analysis, and toxicology of toxaphene compounds. Environ. Health Persp., 107 (1999) 115–144.
133 A.T. Fisk, R.J. Norstrom, K.A. Hobson, J. Moisey and N.J. Karnovsky, Chlordane components and metabolites in six Arctic seabird species from the northwater polynya: relationships between enantioselective analysis of chlordane metabolites and trophic level. Organohalogen Compd., 40 (1999) 413–417.
134 K. Wiberg, R. Letcher, C. Sandau, R. Norstrom, M. Tysklind and T. Bidleman, Chiral analysis of organochlorines in the Arctic marine food chain: chiral biomagnification factors and relationships of enantiomers ratios, chemical residue and biological data. Organohalogen Compd., 35 (1998) 371–375.
135 W. Vetter, K. Hummert, B. Luckas and K. Skírnisson, Organochlorine residues in two seal species from western Iceland. Sci. Total Environ., 170 (1995) 159–164.

136 E. Ulrich, D. Helsel and W. Foreman, Complications with using ratios for environmental data comparing enantiomeric ratios ERs; and enantiomer fractions EFs. Chemosphere, 53 (2003) 531–538.
137 H.-R. Buser, M.D. Müller, T. Poiger and M. Balmer, Environmental behavior of the chiral acetamide pesticide metalaxyl: enantioselective degradation and chiral stability in soil. Environ. Sci. Technol., 36 (2002) 221–226.
138 M. Jung, D. Schmalzing and V. Schurig, Theoretical approach to the gas chromatographic separation of enantiomers on dissolved cyclodextrin derivatives. J. Chromatogr., 552 (1991) 43–57.
139 U. Beitler and B. Feibush, Interaction between asymmetric solutes and solvents. Diamides derived from L-Valine as stationary phases in gas-liquid partition chromatography. J. Chromatogr., 123 (1976) 149–166.
140 B. Koppenhöfer and E. Bayer, Chiral recognition in the resolution of enantiomers by GLC. Chromatographia, 19 (1984) 123–130.
141 W.A. König, D. Icheln, T. Runge, B. Pfaffenberger, P. Ludwig and H. Hühnerfuss, Gas chromatographic enantiomer separation of agrochemicals using modified cyclodextrines. J. High Resol. Chromatogr., 14 (1991) 530–536.
142 K. Watabe, R. Charles and E. Gil-Av, Temperature dependent inversion of elution sequence in the resolution of α-amino acid enantiomers on chiral diamide selectors. Angew. Chem. Int. Ed. Engl., 28 (1989) 192–194.
143 B. Koppenhoefer and B. Lin, Thermodynamic properties of enantiomers of underivatized diols *versus* the cyclic carbonates in gas chromatography on chirasil-val. J. Chromatogr., 481 (1989) 17–26.
144 L.S. Ettre, J.V. Hinshaw and L. Rohrschneider, Grundbegriffe und Gleichungen der Gaschromatographie, Heidelberg, Hüthig Verlag GmbH, 1996.
145 V.A. Davankov, Analytical chiral separation methods. Pure Appl. Chem., 69 (1997) 1469–1474.
146 W. Vetter, unpublished results, 2005.
147 G. Koske, G. Leupold and H. Parlar, Gas chromatographic enantiomer separation of some selected polycyclic xenobiotics using modified cyclodextrins. Fresenius Envir. Bull., 6 (1997) 489–493.
148 K. Breivik, A. Sweetman, J.M. Pacyna and K.C. Jones, Towards a global historical emission inventory for selected PCB congeners — a mass balance approach: 1 global production and consumption. Sci. Total Environ., 290 (2002) 181–198.
149 S. Jensen, Report of a new chemical hazard. New Sci., 32 (1966) 621.
150 G.M. Frame, A collaborative study of 209 PCB congeners and 6 Aroclors on 20 different HRGC columns: 2 semi-quantitative Aroclor congener distributions. Fresenius J. Anal. Chem., 357 (1997) 714–722.
151 K. Ballschmiter and M. Zell, Analysis of polychlorinated biphenyls (PCB) by glass capillary gas chromatography. Fresenius Z. Anal. Chem., 302 (1980) 20–31.
152 U. Berger, W. Vetter, A. Götsch and R. Kallenborn, Chromatographic enrichment and enantiomer separation of axially chiral polybrominated biphenyls in a technical mixture. J. Chromatogr. A, 973 (2002) 123–133.
153 P. Lloyd-Williams and E. Giralt, Atropisomerism, biphenyls and the Suzuki coupling: peptide antibiotics. Chem. Soc. Rev., 30 (2001) 145–157.
154 P. Haglund, Enantioselective separation of polychlorinated biphenyl atropisomers using chiral high-performance liquid chromatography. J. Chromatogr. A, 724 (1996) 219–228.
155 M. Püttmann, F. Oesch and L.W. Robertson, Characteristics of polychlorinated biphenyl (PCB) atropisomers. Chemosphere, 15 (1986) 2061–2064.
156 V. Schurig and S. Reich, Determination of the rotational barriers of atropisomeric polychlorinated biphenyls (PCBs) by a novel stopped-flow multidimensional gas chromatographic technique. Chirality, 10 (1998) 425–429.
157 V. Schurig, A. Glausch and M. Fluck, On the enantiomerization barrier of atropisomeric 2,2′,3,3′,4,6′-hexachlorobiphenyl (PCB 132). Tetrahedron Asymm., 6 (1995) 2161–2164.
158 M.T. Harju and P. Haglund, Determination of the rotational energy barriers of atropisomeric polychlorinated biphenyls. Fresenius J. Anal. Chem., 364 (1999) 219–223.

159 W.A. König, B. Gehrcke, T. Runge and C. Wolf, Gas chromatographic separation of atropisomeric alkylated and polychlorinated biphenyls using modified cyclodextrins. J. High Resol. Chromatogr., 16 (1993) 376–378.
160 W. Bürkle, H. Karfunkel and V. Schurig, Dynamic phenomena during enantiomer resolution by complexation gas chromatography. A kinetic study of enantiomerization. J. Chromatogr., 288 (1984) 1–14.
161 A. Mannschreck, H. Zinner and N. Pustet, The significance of the HPLC time scale: an example of interconvertible enantiomers. Chimia, 43 (1989) 165–166.
162 P. Haglund, Isolation and characterisation of polychlorinated biphenyl (PCB) atropisomers. Chemosphere, 32 (1996) 2133–2140.
163 A. Mannschreck, N. Pustet, L.W. Robertson, F. Oesch and M. Püttmann, Enantiomers of polychlorinated biphenyls. Semipreparative enrichment by liquid chromatography. Liebigs Ann. Chem., 1985 (1985) 2101–2103.
164 M. Püttmann, A. Mannschreck, F. Oesch and L.W. Robertson, Chiral effects in the induction of drug-metabolizing enzymes using synthetic atropisomers of polychlorinated biphenyls (PCBs). Biochem. Pharmacol., 38 (1989) 1345–1352.
165 P. Haglund and K. Wiberg, Determination of the gas chromatographic elution sequences of the (+)- and (−)-enantiomers of stable atropisomeric PCBs on Chirasil-Dex. J. High Resol. Chromatogr., 19 (1996) 373–376.
166 A. Glausch, G.P. Blanch and V. Schurig, Enantioselective analysis of chiral polychlorinated biphenyls in sediment samples by multidimensional gas chromatography-electron capture detection after steam distillation-solvent extraction and sulfur removal. J. Chromatogr. A, 723 (1996) 399–404.
167 J. Ren and S. Harrad, The relative contributions of primary and secondary sources of atmospheric PCB in Birmingham, UK. Organohalogen Compd., 40 (1999) 419–422.
168 C. Wong, A. Garrison and W. Foreman, Enantiomeric composition of chiral polychlorinated biphenyl atropisomers in aquatic bed sediment. Environ. Sci. Technol., 35 (2001) 33–39.
169 U. Pakdeesusuk, W. Jones, C. Lee, A. Garrison, W. O'Niell, D. Freedman, J. Coates and C. Wong, Changes in enantiomeric fractions during microbial reductive dechlorination of PCB132, PCB149, and Aroclor 1254 in Lake Hartwell sediment mirocosms. Environ. Sci. Technol., 37 (2003) 1100–1107.
170 A. Singer, C. Wong and D. Crowley, Differential enantioselective transformation of atropisomeric polychlorinated biphenyls by multiple bacterial strains with different inducing compounds. Appl. Environ. Microbiol., 68 (2002) 5756–5759.
171 S. Harrad, A. Jamshidi and M. Robson, Spatial variation in chiral signatures of PCBs in matched air and topsoil samples an an urban-rural transsect: implications for atmospheric source apportionment. Organohalogen Comp., 66 (2004) 460–466.
172 C. Wong, S. Mabury, D. Whittle, S. Backus, C. Teixeira, D. Devault and D. Muir, Organochlorine compounds in Lake Superior: chiral polychlorinated biphenyls and biotransformation in the aquatic food web. Environ. Sci. Technol., 38 (2004) 84–92.
173 C. Wong, A. Garrison, P. Smith and W. Foreman, Enantiomeric composition of chiral polychlorinated biphenyl atropisomers in aquatic and riparian biota. Environ. Sci. Technol., 35 (2001) 2448–2454.
174 A. Glausch, J. Hahn and V. Schurig, Enantioselective determination of chiral 2,2′,3,3′,4,6′-hexachlorobiphenyl (PCB 132) in human milk samples by multidimensional gas chromatography/electron capture detection and by mass spectrometry. Chemosphere, 30 (1995) 2079–2085.
175 C. Wong, F. Lau, M. Clark, S. Mabury and D. Muir. Rainbow trout (Oncorhynchus mykiss) can eliminate chiral organochlorine compounds enantioselectively. Environ. Sci. Technol., 36 (2002) 1257–1262.
176 H. Hühnerfuss, B. Pfaffenberger, B. Gehrcke, L. Karbe, W.A. König and O. Landgraff, Stereochemical effects of PCBs in the marine environment: seasonal variation of coplanar and atropisomeric PCBs in blue mussels (Mytilus edulis L.) of the German Bight. Mar. Poll. Bull., 30 (1994) 332–340.

177 O. Jiménez, B. Jiménez, L. Marsili and M.J. Gonzalez, Enantiomeric ratios of chiral polychlorinated biphenyls in striped cetaceans from the Mediterranean Sea. Organohalogen Compd., 40 (1999) 409–412.

178 S. Reich, B. Jiménez, L. Marsilli, L.M. Hernández, V. Schurig and M.J. González, Congener specific determination and enantiomeric ratios of chiral PCBs in striped dolphins (Stenella coeruleoalba) from the Mediterranean Sea. Organohalogen Compd., 35 (1997) 335–337.

179 P. Hoekstra, C. Wong, T. O'Hara, K. Solomon, S. Mabury and D. Muir, Enantiomer-Specific accumulation of PCB atropisomers in the bowhead whale balanea mysticetus. Environ. Sci. Technol., 36 (2002) 1419–1425.

180 E. Jakobsson and L. Asplund, New types of persistent halogenated compounds, in: J. Paasivirta (Ed.), The Handbook of Environmental Chemistry, Vol. 3, Part K, Springer-Verlag, Berlin, Heidelberg, 2001, pp. 97–126.

181 W. Vetter and B. Luckas, Theoretical aspects of polychlorinated bornanes and the composition of toxaphene in technical mixtures and environmental samples. Sci. Total Environ., 160/161 (1995) 505–510.

182 T. Nezel, F. Müller-Plathe, M.D. Müller and H.-R. Buser, Theoretical considerations about chiral PCBs and their methylthio and methylsulfonyl metabolites being possibly present as stable enantiomers. Chemosphere, 35 (1997) 1895–1906.

183 J. Döbler, N. Peters, C. Larssson, Å. Bergman, E. Geidel and H. Hühnerfuss, The absolute structures of separated PCB-methylsulfone enantiomers determined by vibrational dichroism and quantum chemical calculations. J. Molec. Struct., 586 (2002) 159–166.

184 T. Ellerichmann, Å. Bergman, S. Franke, H. Hühnerfuss, E. Jacobsson, W.A. König and C. Larsson, Gas chromatographic enantiomer separations of chiral PCB methyl sulfons and identification of selectively retained enantiomers in human liver. Fresenius Environ. Bull., 7 (1998) 244–257.

185 K. Wiberg, R. Letcher, C. Sandau, J. Duffe, R. Norstrom, P. Haglund and T.F. Bidleman, Enantioselective gas chromatography/mass spectrometry of methylsulfonyl PCBs with application to arctic marine mammals. Anal. Chem., 70 (1998) 3845–3852.

186 C. Larsson, K. Norström, I. Athanansiadis, A. Bignert, W. König and Å. Bergman, Enantiomeric specifity of methylsulfonyl-PCBs and distribution of bis 4-chlorophenyl; sulfone, PCB, and DDE methyl sulfones in grey seal tissues. Environ. Sci. Technol., 38 (2004) 4950–4955.

187 S. Chu, A. Covaci, K. Haraguchi, S. Voorpoels, K. van de Vijver, K. Das, J. Bouquegneau, W. de Coen, R. Blust and P. Schepens, Levels and enantiomeric signatures of methyl sulfonyl PCB and DDE metabolites in livers of harbor porpoises (Phocoena phocoena) from the Southern North Sea. Environ. Sci. Technol., 37 (2003) 4573–4578.

188 C. Larsson, T. Ellerichmann, S. Franke, M. Athanasiadou, H. Hühnerfuss and Å. Bergmann, Enantiomeric separation of chiral methylsulphonyl PCB congeners in liver and adipose tissue from rats dosed with A50. Organohalogen Compd., 40 (1999) 427–430.

189 C. Larsson, T. Ellerichmann, H. Hühnerfuss and Å. Bergman, Chiral PCB methyl sulfones in rat tissues after exposure to technical PCBs. Environ. Sci. Technol., 36 (2002) 2833–2838.

190 S.G. Chu, A. Covaci, K. van de Vijver, W. de Coen, R. Blust and P. Schepens, J. Environ. Monit., 5 (2003) 521–526.

191 J. de Boer, K. de Boer and J.P. Boon, Polybrominated biphenyls and diphenylethers, in: J. Paasivirta (Ed.), The Handbook of Environmental Chemistry, Vol. 3, Part K, Springer-Verlag, Berlin, Heidelberg, 2000, pp. 61–95.

192 A. Norström, K. Andersson and C. Rappe, Major components of some brominated aromatics used as flame retardants. Chemosphere, 4 (1976) 255–261.

193 G. Sundström, O. Hutzinger, S. Safe and V. Zitko, The synthesis and gas chromatographic properties of bromobiphenyls. Sci. Total Environ., 6 (1976) 15–29.

194 S.M. Getty, D.E. Rickert and A.L. Trapp, Polybrominated biphenyl (PBB) toxicosis: an environmental accident. CRC Crit. Rev. Environ. Control, 7 (1977) 309–323.

195 K. Kay, Polybrominated biphenyls (PBB) environmental contamination in Michigan, 1973–1976. Environ. Res., 13 (1977) 74–93.

196 T.N.K. Raju, The Nobel chronicles. 1948: Paul Hermann Müller (1899–1965). The Lancet, 353 (1999) 1196.
197 W. Perkow, Die Insektizide. 2. verb. u. überarb. Aufl. Heidelberg, Dr. Alfred, Hüthig Verlag, 1968.
198 E.C. Voldner and Y.F. Li, Global usage of selected persistent organochlorines. Sci. Total Environ., 160/161 (1995) 201–210.
199 Anonymous, World Health Organization. DDT and its derivatives: environmental aspects. Environ. Health Crit., (1989) 83.
200 H.-R. Buser and M.D. Müller, Isomer-selective and enantiomerselective determination of DDT and related compounds using chiral high-resolution gas chromatography/mass spectrometry and chiral high-performance liquid chromatography. Anal. Chem., 67 (1995) 2691–2698.
201 E.J. Aigner, A.D. Leone and R.L. Falconer, Concentration and enantiomeric ratios of organochlorine pesticides in soils from the U.S. corn belt. Environ. Sci. Technol., 32 (1998) 1162–1168.
202 R. Falconer, T.F. Bidleman and S.Y. Szeto, Chiral pesticides in soils of the Fraser Valley, British Columbia. J. Agric. Food Chem., 45 (1998) 1946–1951.
203 L. Shen, F. Wania, Y. Lei, C. Teixeira, D. Muir and T. Bidleman, Atmospheric distribution and long-range transport behavior of organochlorine pesticides in North America. Environ. Sci. Technol., 39 (2005) 409–420.
204 K. Wiberg, T. Harner, J. Wideman and T. Bidleman, Chiral analysis of organochlorine pesticides in Alabama soils. Chemosphere, 45 (2001) 843–848.
205 K. Wiberg, Å. Bergman, M. Olsson, A. Roos, G. Blomqvist and P. Haglund, Concentrations and enantiomer fractions of organochlorine compounds in Baltic species hit by reproductive impairment. Environ. Toxicol. Chem., 21 (2002) 2542–2551.
206 H.-R. Buser and M.D. Müller, Enantioselective analyses of persistent and modern pesticides. A step toward sustainable agriculture. Chimia, 51 (1997) 694–700.
207 L.L. McConnell, W.E. Cotham and T.F. Bidleman, Gas exchange of hexachlorocyclohexane in the Great Lakes. Environ. Sci. Technol., 27 (1993) 1304–1311.
208 H. Iwata, S. Tanabe, N. Sakai and R. Tatsukawa, Distribution of persistent organochlorines in the oceanic air and surface seawater and the role of ocean on their global transport and fate. Environ. Sci. Technol., 27 (1993) 1080–1098.
209 P. Ludwig, H. Hühnerfuss, W.A. König and W. Gunkel, Gas chromatographic separation of the enantiomers of marine pollutants. Part 3 Enantioselective degradation of α-HCH and γ-HCH by marine microorganisms. Marine Chem., 38 (1992) 12–23.
210 T.F. Bidleman and C.E. Olney, Chlorinated hydrocarbons in the Sargasso Sea atmosphere and surface water. Science, 183 (1973) 516–518.
211 A.H. Knap and K.S. Binkley, Chlorinated organic compounds in the troposphere over the western north Atlantic Ocean measured by aircraft. Atmos. Environ., 25A (1991) 1507–1516.
212 B.G. Loganathan and K. Kannan, Global organochlorine contamination trends an overview. Ambio, 23 (1994) 187–191.
213 D.A. Kurtz and E.L. Atlas, Distribution of hexachlorocyclohexanes in the Pacific Ocean basin, air and water, 1987, in: D.A. Kurtz (Ed.), Long Range Transport of Pesticides, Lewis, Chelsea, Michigan: ACS Symposium Series, 1990, pp. 143–160.
214 J. Schreitmüller and K. Ballschmiter, Air-water equilibrium of hexachlorocyclohexanes and chloromethoxybenzenes in the North and South Atlantic. Environ. Sci. Technol., 29 (1995) 207–215.
215 M. Oehme, J.-E. Haugen and M. Schlabach, Ambient air levels of persistent organochlorines in spring 1992 at Spitsbergen and the Norwegian mainland: comparison with 1984 results and quality control measures. Sci. Total. Environ., 160/161 (1995) 139–152.
216 W.E. Cotham Jr. and T.F. Bidleman, Estimating the atmospheric deposition of organochlorine contaminants to the Arctic. Chemosphere, 22 (1991) 165–188.
217 L. Jantunen, H. Kylin and T. Bidleman, Air–water gas exchange of α-hexachlorocyclohexane enantiomers in the South Atlantic Ocean and Antarctica. Deep-Sea Res. II, 51 (2004) 2661–2672.
218 D. Mackay and F. Wania, Transport of contaminants to the Arctic: partitioning, processes and models. Sci. Total Environ., 160/161 (1995) 25–38.

219 M. Hippelein and M.S. McLachlan, Soil/air partitioning of semivolatile organic compounds. 1. Method development and influence of physiscal-chemical properties. Environ. Sci. Technol., 32 (1998) 310–316.

220 M. Horstmann and M.S. McLachlan, Initial development of a solid-phase fugacity meter for semivolatile organic compounds. Environ. Sci. Technol., 26 (1992) 1643–1649.

221 M.D. Müller, M. Schlabach and M. Oehme, Fast and precise determination of α-hexachlorocyclohexane enantiomers in environmental samples using chiral HRGC. Environ. Sci. Technol., 26 (1992) 566–569.

222 R.L. Falconer, T.F. Bidleman and D.J. Gregor, Air–water gas exchange and evidence for metabolism of hexachlorocyclohexanes in Resolute Bay, N. W. T. Sci. Total Environ., 160/161 (1995) 65–74.

223 L.M. Jantunen and T. Bidleman, Air–water gas exchange of hexachlorocyclohexanes (HCHs) and the enantiomers of α-HCH in Arctic regions. J. Geophys. Res., 101 (1996) D22, 28837–28846.

224 J.J. Ridal, T.F. Bidleman, B.R. Kerman, M.E. Fox and W.M.J. Strachan, Enantiomers of α-hexachlorocyclohexane as tracers of air–water gas exchange in Lake Ontario. Environ. Sci. Technol., 31 (1997) 1940–1945.

225 B. Bethan, W. Dannecker, H. Gerwig, H. Hühnerfuss and M. Schulz, Seasonal dependence of the chiral composition of α-HCH in coastal deposition at the North Sea. Chemosphere, 44 (2001) 591–597.

226 K. Sundqvist, H. Wingfors, E. Brorström-Lundén and K. Wiberg, Air–sea gas exchange of HCHs and PCBs and enantiomers of α-HCH in the Kattegat Sea region. Environ. Poll., 128 (2003) 73–83.

227 T.F. Bidleman, Atmospheric transport of pesticides and exchange with soil. Air Water Soil Poll., 115 (1999) 115–166.

228 T.F. Bidleman, T. Harner, K. Wiberg, J.L. Wideman, K. Brice, K. Su, R.L. Falconer, E.J. Aigner, A.D. Leone, J.J. Ridal, B. Kerman, A. Finizio, H. Alegria, W.J. Parkhurst and S.Y. Szeto, Chiral pesticides as tracers of air-surface exchange. Environ. Poll., 102 (1998) 43–49.

229 T.F. Bidleman and R.L. Falconer, Enantiomer ratios for apportioning two sources of chiral compounds. Environ. Sci. Technol., 33 (1999) 2299–2301.

230 T. Harner, L. Jantunen, T. Bidleman and L. Barrie, Microbial degradation is a key elimination pathway of hexachlorocyclohexanes from the Arctic Ocean. Geophys. Res. Lett., 27 (2000) 1155–1158.

231 A. Leone, S. Amato and R. Falconer, Emission of chiral organochlorine pesticides from agricultural soils in the Cornbelt Region of the US. Environ. Sci. Technol., 35 (2001) 4592–4596.

232 L. Shen, C. Teixeira, D. Muir and T. Bidleman, Hexachlorocyclohexanes in the North American athmosphere. Environ. Sci. Technol., 38 (2004) 965–975.

233 S. Law, T. Bidleman, M. Martin and M. Ruby, Evidence of enantioselective degradation of α-hexachlorocyclohexane in groundwater. Environ. Sci Technol., 38 (2004) 1633–1638.

234 E. Ulrich, K. Willett, A. Caperell-Grant, R. Bigsby and R. Hites, Understanding enantioselective processes: a laboratory rat model for of α-hexachlorocyclohexane accumulation. Environ. Sci. Technol., 35 (2001) 1604–1609.

235 K. Wiberg, R.J. Letcher, C.D. Sandau, R.J. Norstrom, M. Tysklind and T.F. Bidleman, The enantioselective bioaccumulation of chiral chlordane and alpha-HCH contaminants in the polar bear food chain. Environ. Sci. Technol., 34 (2000) 2668–2674.

236 Anonymous, World Health Organization. Chlordane. Environ. Health Crit., 34 (1984) 1–82.

237 R.L. Lipnick and D.C.G. Muir, History of persistent, bioaccumulative and toxic chemicals, in: R.L. Lipnick, J.L.M. Hermens, K.C. Jones and D.C.G. Muir (Eds.), Persistent, Bioaccumulative, and Toxic Substances, Vol. I: Fate and Exposure. ACS Symposium Series No. 772, Washington DC, 2000, pp. 1–12.

238 M.A. Dearth and R.A. Hites, Complete analysis of technical chlordane using negative ionization mass spectrometry. Environ. Sci. Technol., 25 (1991) 245–254.

239 H. Parlar, K. Hustert, S. Gäb and F. Korte, Isolation, identification, and chromatographic characterization of some chlorinated C10 hydrocarbons in technical chlordane. J. Agric. Food Chem., 27 (1979) 278–283.

240 IARC, Occupational exposures in insecticide application, and some pesticides; chlordane and heptachlor, in: IARC Monographs on the Evaluation of Carcinogenic Risks to Humans, IARC, World Health Organization, Lyon, France, 1991, pp. 115–175.

241 D.C.G. Muir, R.J. Norstrom and M. Simon, Organochlorine contaminants in Arctic marine food chains: accumulation of specific PCB congeners and chlordane-related compounds. Environ. Sci. Technol., 22 (1988) 1071–1079.

242 A.A. Brimfield and J.C. Street, Mammalian biotransformation of chlordane: in vivo and primary hepatic comparisons. Ann. NY Acad. Sci., 320 (1979) 247–256.

243 M. Miyazaki, T. Yamagishi and M. Matsumoto, Isolation and structure elucidation of some components in technical grade clordane. Arch. Environ. Contam. Toxicol., 14 (1985) 475–483.

244 M.A. Dearth and R.A. Hites, Chlordane accumulation in people. Environ. Sci. Technol., 25 (1991) 1279–1285.

245 W.A. König, I.D. Hardt, B. Gehrcke, D.H. Hochmuth and H. Hühnerfuss, Optically active reference compounds for environmental analysis obtained by preparative enantioselective gas chromatography. Angew. Chem. Int. Ed. Engl., 33 (1994) 2085–2087.

246 K. Wiberg, L. Jantunen, T. Harner, J. Wideman, T. Bidleman, K. Brice, K. Su, R. Falconer, A. Leone, W. Parkhurst and H. Alegria, Chlordane enantiomers as source markers in ambient air. Organohalogen Compd., 33 (1997) 209–213.

247 T. Bidleman, L.M.M. Jantunen, P. Helm, E. Brorström-Lundén and S. Juntto, Chlordane enantiomers and temporal trends of chlordane isomers in Arctic Air. Environ. Sci. Technol., 36 (2002) 539–544.

248 T. Bidleman, F. Wong, C. Backe, A. Södergren, E. Brorström-Lundén, P. Helm and G. Stern, Chiral signatures of chlordanes indicate changing sources to the atmosphere over the past 30 years. Atmos. Environ., 38 (2002) 5963–5970.

249 A. Leone, E. Ulrich, C. Bodnar, R. Falconer and R. Hites, Organochlorine pesticide concentrations and enantiomer fractions for chlordane in indoor air from the US cornbelt. Atmos. Environ., 34 (2000) 4131–4138.

250 R.L. Falconer, A.D. Leone, C. Bodnar, K Wiberg, T.F. Bidleman, L.M. Jantunen, T. Harner, W.J. Parkhurst, H. Alegria, K. Brice and K. Su, Using enantiomeric ratios to determine sources of chlordane to ambient air. Organohalogen Compd., 35 (1998) 331–334.

251 T.F. Bidleman, L.M. Jantunen, K. Wiberg, T. Harner, K. Brice, K. Su, R.L. Falconer, A.D. Leone, E.J. Aigner and W.J. Parkhurst, Soil as a source of atmospheric heptachlor epoxide. Environ. Sci. Technol., 32 (1998) 1546–1548.

252 M. Incorvia Mattina, W. Iannucci-Berger, B. Eitzer, J. White, Rhizotron study of cucurbitaceae: transport of soil-bound chlordane and heavy metal contaminants differs with genera. Environ. Chem., 1 (2004) 86–89.

253 H. Karlsson, M. Oehme, S. Skopp and I.C. Burkow, Enantiomer ratios of chlordane congeners are gender specific in cod (*Gadus morhua*) from the Barents Sea. Environ. Sci. Technol., 34 (2000) 2126–2130.

254 G. Koske, G. Leupold, D. Angerhöfer and H. Parlar, Multidimensional gas chromatographic enantiomer quantification of some polycyclic xenobiotics in cod liver and fish oils. Organohalogen Compd., 35 (1999) 363–366.

255 K. Wiberg, M. Oehme, P. Haglund, H. Karlsson, M. Olsson and C. Rappe, Enantioselective analysis of organochlorine pesticides in herring and seal from the Swedish marine enviroment. Mar. Poll. Bull., 36 (1998) 345–353.

256 H.-R. Buser, M.D. Müller and C. Rappe, Enantioselective determination of chlordane components using chiral high-resolution gas chromatography-mass spectrometry with application to environmental samples. Environ. Sci. Technol., 26 (1992) 1533–1540.

257 H. Hühnerfuss, B. Pfaffenberger and G. Rimkus, Enantioselective transformation and accumulation of cyclodiene pesticides at different trophic levels of marine and terrestrial biota. Organohalogen Compd., 29 (1996) 88–93.

258 U. Klobes, W. Vetter, D. Glotz, B. Luckas, K. Skírnisson and P. Hersteinsson, Organochlorine levels and enantiomeric ratios in liver of polar foxes (Alopex lagopus) and liver and adipose tissue

of a polar bear (Ursus maritimus) sampled in Iceland. Int. J. Environ. Anal. Chem., 69 (1998) 67–81.
259 A.T. Fisk, M. Holst, K.A. Hobson, J. Duffe, J. Moise and R.J. Norstrom, Persistent organochlorine contaminants and enantiomeric signatures of chiral pollutants in ringed seals (Phoca hispida) collected on the east and west side of the Northwater Polynya, Canadian Arctic. Arch. Environ. Contam. Toxicol., 42 (2002) 118–126.
260 M. Incorvia Mattina, J. White, B. Eitzer and W. Iannucci-Berger, Cycling of weathered clordane residues in the environment: compositional and chiral profiles in contiguous soil, vegetation and air compartments. Environ. Toxicol. Chem., 21 (2002) 281–288.
261 M. Incorvia Mattina, B. Eitzer, W. Iannucci-Berger, W. Lee and J. White, Plant uptake and translocation of highly weathered soil-bound technical clordane residues: data from field and Rhizotron studies. Environ. Toxicol. Chem., 23 (2004) 2756–2762.
262 C.R. Worthing and R.J. Hance, The pesticide manual – a world compendium, 9th ed. Unwin Brothers Ltd, Old Woking, Surrey, UK: British Crop Protection Council, 1991.
263 W.F. Schmidt, S. Bilboulian, C.P. Rice, J.C. Fettinger, L.L. McConnell and C.J. Hapeman, Thermodynamic, spectroscopic, and computational evidence for the irreversible conversion of β- to α-endosulfan. J. Agric. Food Chem., 49 (2001) 5372–5376.
264 R. Martens, Degradation of endosulfan-8,9-^{14}C in soil under different conditions. Bull. Environ. Contam. Toxicol., 17 (1977) 438–446.
265 M.A. Saleh, Toxaphene: chemistry, biochemistry, toxicity and environmental fate. Rev. Environ. Contam. Toxicol., 118 (1991) 1–85.
266 C.M. Reddy, L. Xu, T.I. Eglinton, J.P. Boon and D.J. Faulkner, Radiocarbon content of synthetic and natural semi-volatile halogenated organic compounds. Environ. Poll., 120 (2002) 163–168.
267 W. Vetter and M. Oehme, New types of persistent halogenated compounds. in: J. Paasivirta (Ed.), The Handbook of Environmental Chemistry, Vol. 3, Part K, Springer-Verlag, Berlin, Heidelberg, 2000, pp. 237–287.
268 L. Kimmel, M. Coelhan, G. Leupold, W. Vetter and H. Parlar, FTIR spectroscopic characterization of chlorinated camphenes and bornenes in technical toxaphene. Environ. Sci. Technol., 34 (2000) 3041–3045.
269 W. Vetter, Theoretical aspects of the distribution of chlorinated bornanes including symmetrical aspects. Chemosphere, 26 (1993) 1079–1084.
270 H. Parlar, J. Burhenne, M. Coelhan and W. Vetter, Structure of the toxaphene compound 2,5-*endo*,6-*exo*,8,9,9,10,10-octachlorobornene-2: a temperature-dependent formation of two rotamers. Environ. Sci. Technol., 39 (2005) 1736–1740.
271 W. Vetter, U. Klobes, B. Krock, B. Luckas, D. Glotz and G. Scherer, Isolation, structure elucidation, and identification of a further major toxaphene compound in environmental samples. Environ. Sci. Technol., 31 (1997) 3023–3028.
272 W. Vetter, G. Scherer, M. Schlabach, B. Luckas and M. Oehme, An unequivocal ^{1}H NMR structural assignment of TOX8 and TOX9, the two most abundant toxaphene congeners in marine mammals. Fresenius J. Anal. Chem., 349 (1993) 552–558.
273 P. Korytár, L.L.P. van Stee, P.E.G. Leonards, J. de Boer and U.A.T.H. Brinkman, Attempt to unravel the composition of toxaphene by comprehensive two-dimensional gas chromatography with selective detection. J. Chromatogr. A, 994 (2003) 179–189.
274 W. Vetter, G. Gleixner, W. Armbruster, S. Ruppe, G. Stern and E. Braekevelt, Congener-specific concentrations and carbon stable isotope ratios ($\delta^{13}C$) of two technical toxaphene products (Toxaphene® and Melipax®). Chemosphere, 58 (2005) 235–241.
275 S. Ruppe, A. Neumann and W. Vetter, Anaerobic transformation of compounds of technical toxaphene. I Regiospecific reaction of chlorobornanes with geminal chlorine atoms. Environ. Toxicol. Chem., 22 (2003) 2614–2621.
276 J. Burhenne, D. Hainzl, L. Xu, B. Vieth, L. Alder and H. Parlar, Preparation and structure of high-chlorinated bornane derivatives for the quantification of toxaphene residues in environmental samples. Fresenius J. Anal. Chem., 346 (1993) 779–785.

277 D. Hainzl, J. Burhenne, H. Barlas and H. Parlar, Spectroscopic characterization of environmentally relevant C10-chloroterpenes from a photochemically modified toxaphene standard. Fresenius J. Anal. Chem., 351 (1995) 271–285.

278 V.A. Nikiforov, V.G. Tribulovich and V.S. Karavan, Experience in isolation and identification of toxaphene congeners and prospects of congener-specific analysis of environmental samples. Organohalogen Compd., 26 (1995) 379–382.

279 H.-R. Buser and M.D. Müller, Isomer- and enantiomer-selective analyses of toxaphene components using chiral high-resolution gas chromatography and detection by mass spectrometry/mass spectrometry. Environ. Sci. Technol., 28 (1994) 119–128.

280 R. Kallenborn, M. Oehme, W. Vetter and H. Parlar, Enantiomer selective separation of toxaphene congeners isolated from seal blubber and obtained by synthesis. Chemosphere, 28 (1993) 89–98.

281 H. Karlsson, M. Oehme and L. Müller, Enantioselective separation of toxaphene congeners. Results, problems and challenges. Organohalogen Compd., 28 (1996) 405–409.

282 L. Alder, R. Palavinskas and P. Andrews, Enantioselective determination of toxaphene components in fish, monkey adipose tissue from a feeding study and human milk. Organohalogen Compd., 28 (1996) 410–415.

283 W. Vetter, U. Klobes, B. Krock and B. Luckas, Congener specific separation of compounds of technical toxaphene (CTTs) on a non-polar CP-Sil 2 phase. J. Microcol. Separ., 9 (1997) 29–36.

284 D. Muir, Toxaphene in the food webs of Lake Superior and nearby lakes final report to U.S: EPA Great Lakes. National Program Office, 2001.

285 H. Karlsson, D. Muir, C. Teixeira, W. Strachan, S. Backus, D. DeVault, M. Whittle and C. Bronte, Toxaphene bioaccumulation in Lake Superior: insights from congener and enantiomer analysis. Organohalogen Compd., 47 (2000) 121–124.

286 A. Jaus and M. Oehme, Characterisation of partially ethylated γ-cyclodextrins, a well suited alternative for the enantioselective separation of toxaphenes by HRGC. Organohalogen Compd., 40 (1999) 423–425.

287 V. Nikiforov, A. Trukhin, F. Kruchkov, A. Kiprianova1, S. Miltsov and R. Kallenborn, Isolation of pure enantiomers of toxaphene congeners via hydrochlorination and chlorination of pinene and composition of Soviet polychloropinene. Organohalogen Compd., 66 (2004) 467–472.

288 W. Vetter, B. Krock, B. Luckas and G. Scherer, Structure elucidation of a main toxaphene heptachloro congener in marine organisms after isolation from Melipax. Chemosphere, 33 (1996) 1005–1019.

289 W. Vetter, B. Krock, U. Klobes and B. Luckas, Enantioselective analysis of a heptachlorobornane isolated from the technical product Melipax by gas chromatography/mass spectrometry. J. Agric. Food Chem., 45 (1997) 4866–4880.

290 U. Klobes, W. Vetter and B. Luckas, 1998, unpublished results.

291 W. Vetter and D. Kirchberg, Production of toxaphene enantiomers by enantioselective HPLC after isolation of the compounds from an anaerobically degraded technical mixture. Environ. Sci. Technol., 35 (2001), 960–965.

292 G. Fingerling, N. Hertkorn and H. Parlar, Formation and spectroscopic investigation of two hexachlorobornanes from six environmentally relevant toxaphene components by reductive dechlorination in soil under anaerobic conditions. Environ. Sci. Technol., 30 (1996) 2984–2992.

293 K.J. Palmer, R.Y. Wong, R.E. Lundin, S. Khalifa and J.E. Casida, Crystal and molecular structure of 2,2,5-*endo*,6-*exo*,8,9,10-heptachlorobornane, $C_{10}H_{11}Ci_7$, a toxic component of toxaphene insecticide. J. Amer. Chem. Soc., 97 (1975) 408–413.

294 L. Jantunen, T. Harner, T. Bidleman and J. Wideman, Sources and air-water of toxaphene to Lake Superior. Organohalogen Compd., 33 (1997) 285–289.

295 G.A. Stern, M.D. Loewen, B.M. Miskimmin, D.C.G. Muir and J.B. Westmore, Characterization of two major toxaphene compounds in treated lake sediment. Environ. Sci. Technol., 30 (1996) 2251–2258.

296 C. Rappe, P. Haglund, H.-R. Buser and M.D. Müller, Enantioselective determination of chiral chlorobornanes in sediments from the Baltic Sea. Organohalogen Compd., 31 (1997) 233–237.

297 W. Vetter, R. Bartha, G. Stern and G. Tomy, Enantioselective determination of two persistent chlorobornane congeners in sediment from a toxaphene treated Yukon lake. Environ. Toxicol. Chem., 18 (1999) 2775–2781.
298 K.A. Maruya, K.E. Smalling and W. Vetter, Temperature affects on the enantioselectivity of toxaphene elimination by fish. Environ. Sci. Technol., 39 (2005) 3999–4004.
299 D.L. Lewis, A.W. Garrison, K.E. Wommack, A. Whittemore, P. Steudler and J. Melillo, Influence of environmental changes on degradation of chiral pollutants in soils. Nature, 401 (1999) 898–901.
300 G.A. Stern, D.C.G. Muir, C.A. Ford, N.P. Grift, E. Dewailly, T.F. Bidleman and M.D. Walla, Isolation and identification of two major recalcitrant toxaphene congeners in aquatic biota. Environ. Sci. Technol., 26 (1992) 1838–1840.
301 W. Vetter, U. Klobes, B. Luckas and G. Hottinger, Enantioselective determination of toxaphene and other organochlorines on tert.-butyldimethylsilylated β-cyclodextrin. Organohalogen Compd., 33 (1997) 63–67.
302 U. Klobes, W. Vetter, B. Luckas and G. Hottinger, Enantioselective determination of 2-*endo*,3-*exo*,5-*endo*,6-*exo*,8,8,9,10-octachlorobornane (B8–1412) in environmental samples. Organohalogen Compd., 35 (1997) 359–362.
303 W. Vetter, E. Scholz, B. Luckas and K.A. Maruya, Structure of a persistent heptachlorobornane in toxaphene (B7–1000) agrees with molecular model predictions. J. Agric. Food Chem., 49 (2001) 759–765.
304 S. Hamed, G. Leupold, A. Ismail and H. Parlar, Enantioselective determination of chiral toxaphene congeners in laying hens and eggs using multidimensional high resolution gas chromatography. J. Agric. Food Chem., 53 (2005) 7156–7164.
305 R. Kallenborn and E.S. Heimstad, Theoretical modelling in combination with empirical investigations as a new tool for ecotoxicological evaluations of chiral pollutants. Chlorinated bornanes as an example. Organohalogen Compd., 40 (1999) 391–394.
306 B. Bethan, K. Bester, H. Hühnerfuss and G. Rimkus, Bromocyclen contamination of surface water, waste water and fish from northern Germany, and gas chromatographic chiral separation. Chemosphere, 34 (1996) 2271–2280.
307 K. Bester, Enantioselective degradation of bromocyclene in sewage plants. Organohalogen Compd., 66 (2004) 411–418.
308 T. Bucheli, S.R. Müller, A. Voegelin and R.P. Schwarzenbach, Bituminous roof sealing membranes as major sources of the herbicide R,S-Mecoprop in roof runoff waters potential contamination of groundwater and surface waters. Environ. Sci. Technol., 32 (1998) 3465–3471.
309 T. Poiger, H.R. Buser, M.D. Müller, M.E. Balmer and I.J. Buerge, Occurrence and fate of organic micropollutants in the environment: regional mass balances and source apportioning in surface waters based on laboratory incubation studies in soil and water, monitoring, and computer modeling. Chimia, 57 (2003) 492–498.
310 H.-R. Buser and M.D. Müller, Conversion of various phenoxyalkanoic acid herbicides in soil. 2. Elucidation of the enantiomerization process of chiral phenoxy acids from incubation in a D_2O/soil system. Environ. Sci. Technol., 31 (1997) 1960–1967.
311 M.D. Müller and H.-R. Buser, Conversion reactions of various phenoxyalkanoic acid herbicides in soil. 1. Enantiomerization and enantioselective degradation of the chiral 2-phenoxypropionic acid herbicides. Environ. Sci. Technol., 31 (1997) 1953–1959.
312 I.J. Buerge, T. Poiger, M.D. Müller and H.R. Buser, Enantioselective degradation of metalaxyl in soils: chiral preference changes with soil pH. Environ. Sci. Technol., 37 (2003) 2668–2674.
313 M.D. Müller and H.-R. Buser, Environmental behaviour of acetamide pesticide stereoisomers, 2. Stereo and enantioselective degradation in sewage sludge and soil. Environ. Sci. Technol., 29 (1995) 2031–2037.
314 H.-R. Buser and M.D. Müller, Environmental behaviour of acetamide pesticide stereoisomers. 1. Stereo- and enantioselective determination using chiral high resolution gas chromatography and chiral high performance liquid chromatography. Environ. Sci. Technol., 29 (1995) 2023–2030.

315 A. Monkiedje, M. Spiteller and K. Bester, Degradation behaviour and effects on microbial biomass of racemic mixture and enantiomeric forms of the fungicide metalaxyl in two soils. Environ. Sci. Technol., 37 (2003) 707–712.
316 T.H. Heberer, S. Butz and H.-J. Stan, Analysis of phenoxycarboxylic acids and other acidic compounds in tap, ground, surface and sewage water at the low ppt-level. Int. J. Environ. Anal. Chem., 58 (1995) 43–54.
317 R. Hirsch, T. Ternes, K. Haberer and K.L. Kratz, Occurrence of antibiotics in the aquatic environment. Sci. Total Environ., 225 (1999) 109–118.
318 H.R. Buser, T. Poiger and M.D. Müller, Occurrence and environmental behavior of the chiral pharmaceutical drug ibuprofen in surface waters and in wastewater. Environ. Sci. Technol., 33 (1999) 2529–2535.
319 M. Gautschi, J.A. Bajgrowicz and P. Kraft, Fragrance chemistry – milestones and perspectives. Chimia, 55 (2001) 379–387.
320 H.D. Eschke, J. Traud and H.J. Dibowski, Untersuchungen zum Vorkommen polycyclischer Moschusduftstoffe in verschiedenen Umweltkompartimenten- Nachweis und Analytik mit GC-MS in Oberflächen-, Abwässern und Fischen. UWSF-Z. Umweltchem. Ökotox., 6 (1994) 183–189.
321 H.D. Eschke, J. Traud and H.J. Dibowski, Analytik und Befunde künstlicher Nitromoschus-Substanzen in Oberflächen- und Abwässern sowie Fischen aus dem Einzugsgebiet der Ruhr. Vom Wasser, 83 (1994) 373–383.
322 F. Balk and R.A. Ford, Environmental risk assessment for the polycyclic musks AHTN and HHCB in the EU. 1. Fate and exposure assessment. Toxicol. Lett., 111 (1999) 57–79.
323 P. Kraft and G. Frater, Enantioselectivity of the musk odor sensation. Chirality, 13 (2001) 388–394.
324 S.L. Simonich, T.W. Federle, W.S. Eckhoff, A. Rottiers, S. Webb, D. Sabaliunas and W. De Wolf, Removal of fragrance materials during US and European wastewater treatment. Environ. Sci. Technol., 36 (2002) 2839–2847.
325 K. Bester, Retention characteristics and balance assessment for two polycyclic musk fragrances HHCB and AHTN in a typical German sewage treatment plant. Chemosphere, 57 (2004) 863–870.
326 K. Bester, Polycyclic musks in the Ruhr catchment area – transports, discharges of wastewater, and transformations of HHCB, AHTN, and HHCB-lactone. J. Environ. Monitor., 7 (2005) 43–51.
327 K. Bester, H. Hühnerfuss, W. Lange, G.G. Rimkus and N. Theobald, Results of non target screening of lipophilic organic pollutants in the German Bight. II Polycyclic musk fragrances. Water Res., 32 (1998) 1857–1863.
328 R. Gatermann, S. Biselli, H. Hühnerfuss, G.G. Rimkus, M. Hecker and L. Karbe, Synthetic musks in the environment. Part 1 Species-dependent bioaccumulation of polycyclic and nitro musk fragrances in freshwater fish and mussels. Arch. Environ. Contam. Toxicol., 42 (2002) 437–446.
329 W. Seinen, J.C. Lemmen, R.H.H. Pieters, E.M.J. Verbruggen and B. van der Burg, AHTN and HHCB show weak estrogenic – but no uterotrophic activity. Toxicol. Lett., 111 (1999) 161–168.
330 S. Franke, C. Meyer, N. Heinzel, R. Gatermann, H. Hühnerfuss, G. Rimkus, W.A. Konig and W. Francke, Enantiomeric composition of the polycyclic musks HHCB and AHTN in different aquatic species. Chirality, 11 (1999) 795–801.
331 R. Gatermann, S. Biselli, H. Hühnerfuss, G.G. Rimkus, S. Franke, M. Hecker, R. Kallenborn, L. Karbe and W.A. Konig, Synthetic musks in the environment. Part 2 Enantioselective transformation of the polycyclic musk fragrances HHCB, AHTN, AHDI, and ATII in freshwater fish. Arch. Environ. Contam. Toxicol., 42 (2002) 447–453.
332 E. Artola-Garicano, I. Borkent, J.L.M. Hermens and W.H.J. Vaes, Removal of two polycyclic musks in sewage treatment plants freely dissolved and total concentrations. Environ. Sci.. Technol., 37 (2003) 3111–3116.
333 J.D. Berset, T. Kupper, R. Etter and J. Tarradellas, Considerations about the enantioselective transformation of polycyclic musks in wastewater, treated wastewater and sewage sludge and analysis of their fate in a sequencing batch reactor plant. Chemosphere, 57 (2004) 987–996.
334 W.A. König, Enantioselective gas chromatography. Trends. Anal. Chem., 12 (1993) 130–137.

335 A.S. McNeish, T.F. Bidleman, R.T. Leah and M.S. Johnson, Enantiomers of methylhexachlorocyclohexane and hexachlorocyclohexane in fish, shellfish, and waters of the Mersey Estuary. Environ. Toxicol., 14 (1999) 397–403.

336 H. Karlsson, Isomer and enantiomer selective ultra trace analysis of chlordane in biota including isolation and structure elucidation of single congeners, Shaker Verlag, Aachen, Germany. Ph.D. Thesis, University of Basel, Switzerland, 1999.

337 P. Andrews and W. Vetter, A systematic nomenclature system for toxaphene congeners. Part 1: Chlorinated bornanes. Chemosphere, 31 (1995) 3879–3886.

338 H. Parlar, D. Angerhöfer, M. Coelhan and L. Kimmel, HRGC and HRGC-ECNI determination of toxaphene. Organohalogen Compd., 26 (1995) 357–362.

339 W.V. Turner, S. Khalifa and J.E. Casida, Toxaphene toxicant A. Mixture of 2,25-*endo*, 6-*exo*,8,8,9,10-octachlorobornane and 2,2,5-*endo*,6-*exo*,8,9,9,10-octachlorobornane. J. Agric. Food Chem., 23 (1975) 991–994.

340 W. Vetter, B. Krock and B. Luckas, Congener specific determination of compounds of technical toxaphene (CTTs) in different Antarctic seal species. Chromatographia, 44 (1997) 65–73.

© 2006 Elsevier B.V. All rights reserved.
Chiral Analysis
K.W. Busch and M.A. Busch, Eds.

CHAPTER 7

HPLC resolution using polysaccharide derivatives as CSP

Chiyo Yamamoto[1] and Yoshio Okamoto[2]

[1]*Department of Applied Chemistry, Graduate School of Engineering, Nagoya University, Japan*
[2]*EcoTopia Science Institute, Nagoya University, Furo-cho, Chikusa-ku, Nagoya 464-8603, Japan*

7.1 INTRODUCTION

The first baseline separation of enantiomers was realized by Davankov et al. in 1971 [1], and since that time, remarkable advances have been made in both the instrumentation for high-performance liquid chromatography (HPLC) and the development of chiral stationary phases (CSPs). Today, the resolution by HPLC using a CSP is one of the most useful techniques for estimating both enantiomer composition and obtaining pure enantiomers. More than a hundred CSPs are now commercially available and most chiral compounds can be resolved using these CSPs. Therefore, the key factor of this technique is how we choose a suitable CSP for the resolution of a target compound.

Fig. 7.1 indicates how organic chemists estimated enantiomer excess (ee) in the journal, "*Tetrahedron: Asymmetry*," during 1995–2003. The measurement of the optical rotation using a polarimeter, which is the classical method used for many years, has already lost its role. The three major methods are now nuclear magnetic resonance (NMR), gas chromatography (GC), and HPLC. The NMR method is gradually becoming less important probably because of its lower accuracy. The GC using a CSP is effective for volatile compounds of relatively low molecular weight and has been mostly used only for analytical purposes. On the other hand, the HPLC method using a CSP is suitable for compounds with a relatively high molecular weight. The compounds that we deal with will become larger and larger. This means that the resolution by HPLC will maintain its major role in the future.

7.2 CHIRAL STATIONARY PHASE (CSP) FOR HPLC

Fig. 7.2 shows some of the typical CSPs for HPLC. The CSPs can be classified into two types: one consists of optically active small molecules (**1–5**), and the other is based

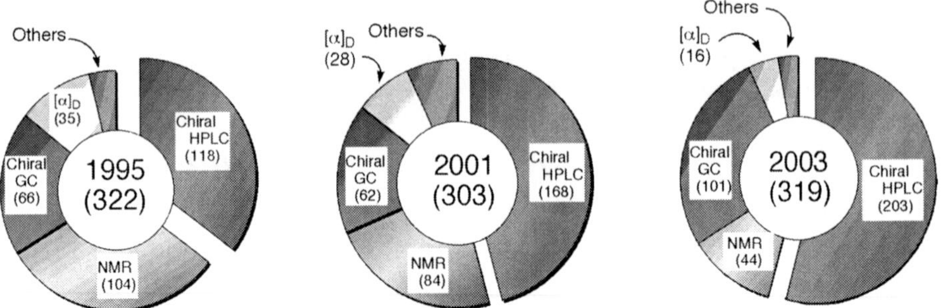

Fig. 7.1. Distribution of the methods for the determination of enantiomer composition appeared in *Tetrahedron Asymmetry* during 1995–2003.

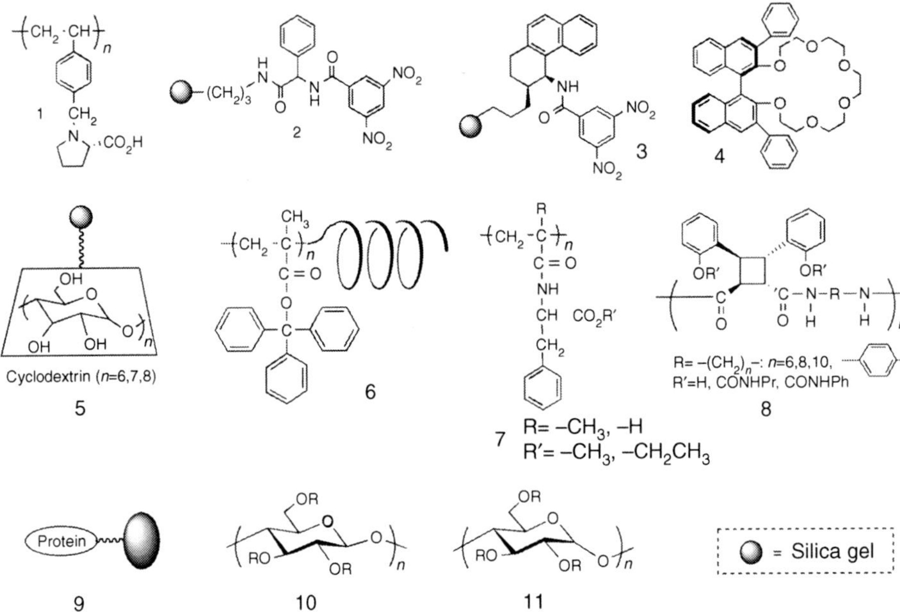

Fig. 7.2. Typical chiral stationary phases for HPLC.

on optically active polymers (**6–11**). The CSP **1** prepared by Davankov is capable of resolving α-amino acids in the presence of a copper ion [1,2]. CSP **2** is the first commercialized CSP found by Pirkle et al. [3]. He developed a large number of analogous CSPs including CSP **3** [4]. The crown ether containing CSP **4** shows a high chiral recognition ability to α-amino acids [5]. The cyclodextrin-bonded CSP **5** exhibits chiral recognition *via* host–guest complexation [6]. A small molecule as a chiral selector is usually bonded onto a support material such as silica gel. These CSPs have been called the "blush-type" after their shapes. On the other hand, a polymer can be utilized as a CSP by coating it on silica gel or by cross-linking as a gel. The chiral recognition ability of small-molecule CSPs significantly depends on the small molecules themselves. On the other hand, the resolution abilities of polymer-based CSPs are changeable depending on the higher order structure of the polymers, indicating that it is difficult to predict the chiral recognition ability of a polymer CSP from the monomer structure. For the high chiral recognition ability of a polymer CSP, one of the most important factors is for the polymer to have a regular higher order structure [7,8]. The one-handed helical poly(triphenylmethyl methacrylate) (**6**) has been prepared by the helix-sense-selective polymerization using a chiral anionic initiator. This is the first example of an optically active vinyl polymer, whose chirality is mainly caused by the helicity of the main chain [9]. The CSP **6** has been prepared by coating the polymer on silica surface [10]. This phase is especially useful for the resolution of chiral aromatic hydrocarbons that are difficult to resolve by other methods due to the lack of functionalities [7,8,10]. Poly(acrylamides) and poly(methacrylamides) (**7**) with chiral side groups were utilized as CSPs by Blaschke and coworkers [11,12]. These CSPs can resolve various drugs including thalidomide. Saigo prepared polyamides **8** from chiral cumarine derivatives and diamine [13]. The chiral recognition ability of **8** depends on the number of methylene groups, and an odd–even effect has been observed.

Besides the synthetic polymers, naturally occurring chiral polymers, such as proteins (**9**) and polysaccharides (**10**,**11**), can also be used as a chiral selector. Various proteins including albumins, glycoproteins, and enzymes have been evaluated as the CSPs for HPLC, and some of them have been commercialized. The loading capacity and stability of these CSPs are not high due to the existence of the limited number of chiral recognition sites and conformational instability [14]. Polysaccharides, particularly cellulose (**10**) and amylose (**11**), are the most readily available optically active and completely stereoregular polymers. While the chiral recognition ability of these polymers themselves are not sufficient as CSPs, they can be easily converted to esters and carbamates, which show much better chromatographic and enantioselective properties [7,8,15–19].

Fig. 7.3(a) shows the distribution of the CSPs during HPLC used for analyzing an enantiomer composition in the journal "*Tetrahedron: Asymmetry*" during 1995–2003. It is obvious that the polysaccharide-type CSPs are the most popular.

7.3 POLYSACCHARIDE-BASED CSPs

At present, more than ten cellulose- and amylose-based CSPs are commercially available. Their structures and suppliers are summarized in Fig. 7.4.

Fig. 7.5 shows the chromatogram of the HPLC resolution of Tröger base on amylose tris(3,5-dimethylphenylcarbamate). The (+)-isomer elutes first, followed by the (−)-isomer, and the complete baseline separation is achieved. The results of the separation can be evaluated by three parameters—capacity factors (k'), separation factor (α), and resolution factor (Rs)—defined as follows:

$$k'_1 = (t_1 - t_0)/t_0, \quad k'_2 = (t_2 - t_0)/t_0,$$

$$\alpha = (t_2 - t_0)/(t_1 - t_0) = k'_2/k'_1, \quad \mathrm{Rs} = 2(t_2 - t_1)/(w_1 + w_2)$$

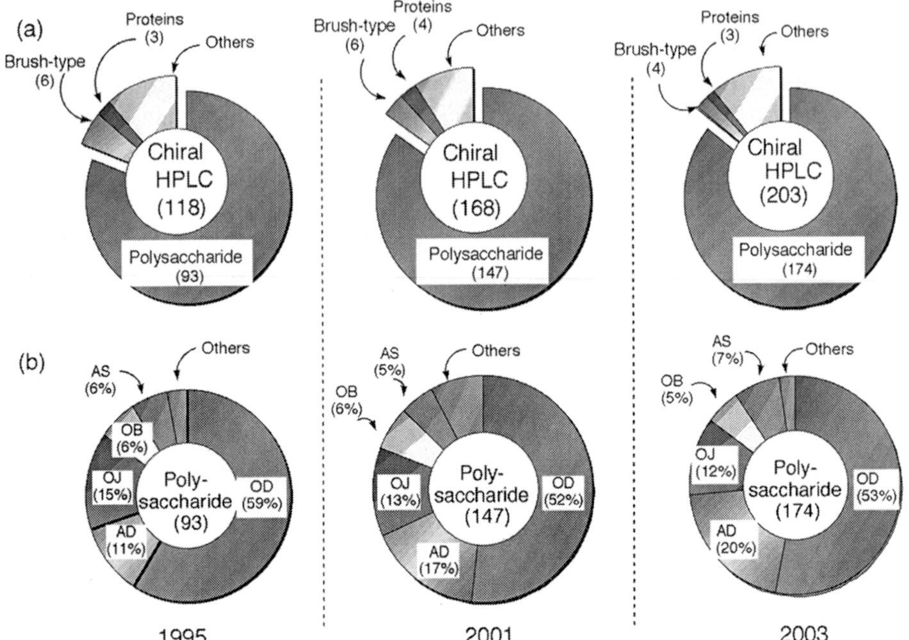

Fig. 7.3. Distribution of CSPs used in chiral HPLC reported in *Tetrahedron Asymmetry* during 1995–2003.

where t_1 and t_2 are the retention times of the enantiomers, t_0 is the dead time (retention time of a nonretained compound), and w_1 and w_2 are the bandwidths of the peaks. In a chromatographic separation, α is directly related to the chiral recognition ability of a CSP, and Rs is correlated to both the chiral recognition ability of a CSP and the column efficiency (theoretical plate number).

The cellulose- and amylose-based CSPs are classified into three types: cellulose esters, cellulose and amylose phenylcarbamates, and amylose benzylcarbamates.

7.3.1 Cellulose esters

Cellulose triacetate, one of the oldest CSPs, has been utilized in two different forms (**CTA-I** and Chiralcel **OA**). These two cellulose triacetates exhibit different chiral recognitions [20–22]. This is caused by different higher order structures depending on the preparation methods. **CTA-I** has been prepared by the heterogeneous acetylation of native microcrystalline cellulose (Avicel) in benzene, and has a crystallinity (form I) [23,24]. The derivative is ground and sieved into small particles to be used as a packing material. On the other hand, **OA** is prepared by coating **CTA-I** dissolved in a solvent on silica gel and does not maintain the same structure as that of **CTA-I**. On these two CSPs, the reversal of the elution order of the

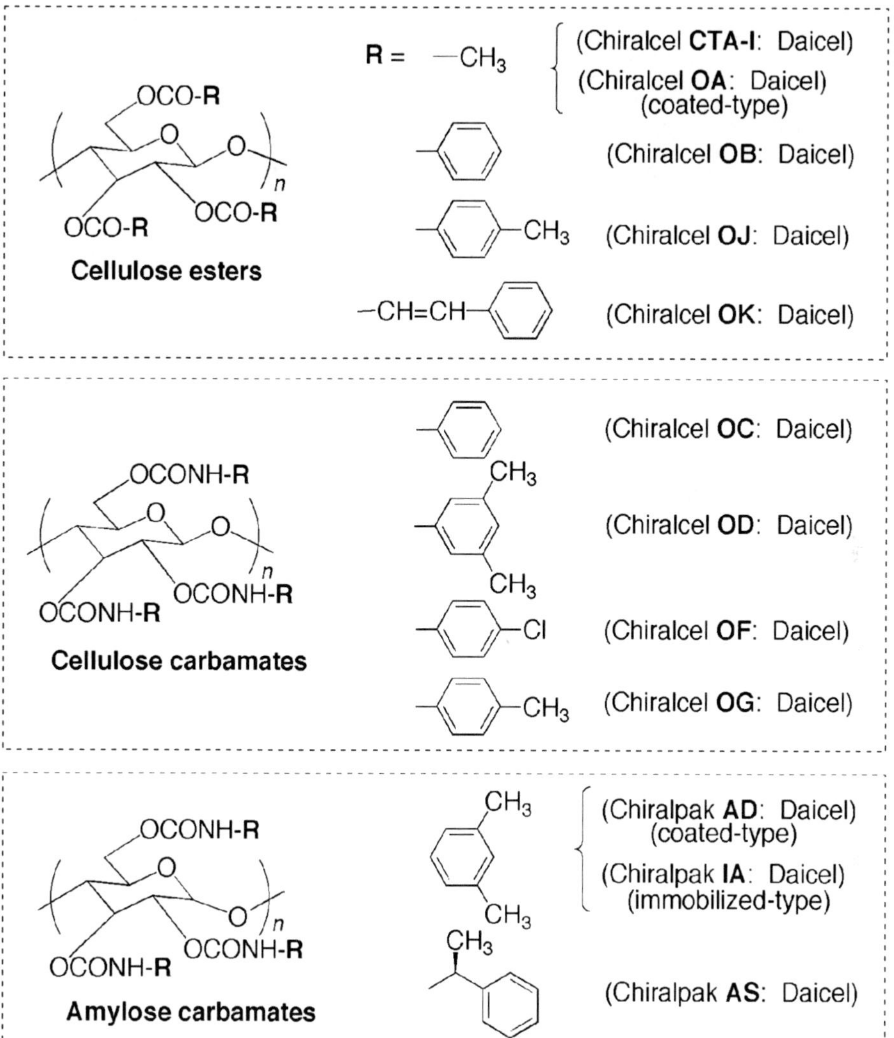

Fig. 7.4. Structures and suppliers of the commercially available polysaccharide-based CSPs. CTA-I is also available from Merck and Macherey-Nagel.

enantiomers has been observed for several racemates. **CTA-I** has a high loading capacity [25,26], and **OA** has a high column efficiency and durability. The resolutions of several stereochemically interesting racemates using **CTA-I** can be attained (Fig. 7.6) [27–34].

Besides the acetates, benzoate (**OB**) [35], 4-methylbenzoate (**OJ**) [36], and cinnamate (**OK**) [21] of cellulose coated on silica gel are also commercially available. The introduction of a methyl group at the *para*-position of **OB** results in a significant

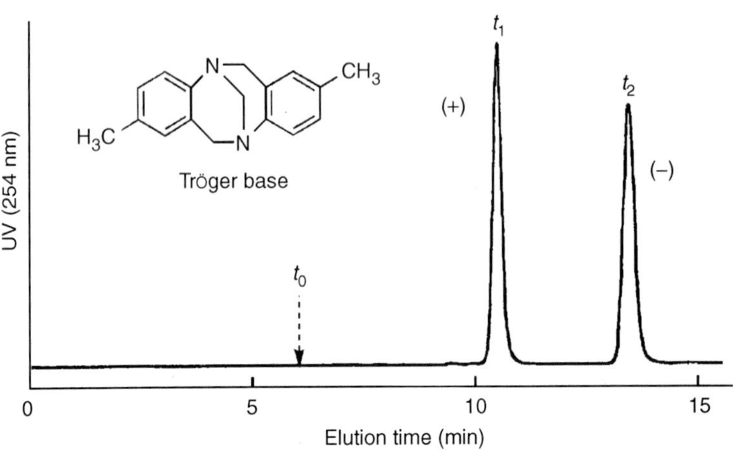

Fig. 7.5. Optical resolution of Tröger base on amylose tris(3,5-dimethylphenylcarbamate) column: 25 × 0.46 (i.d.) cm; eluent: hexane–2-propanol (90/10); flow rate: 0.5 ml/min.

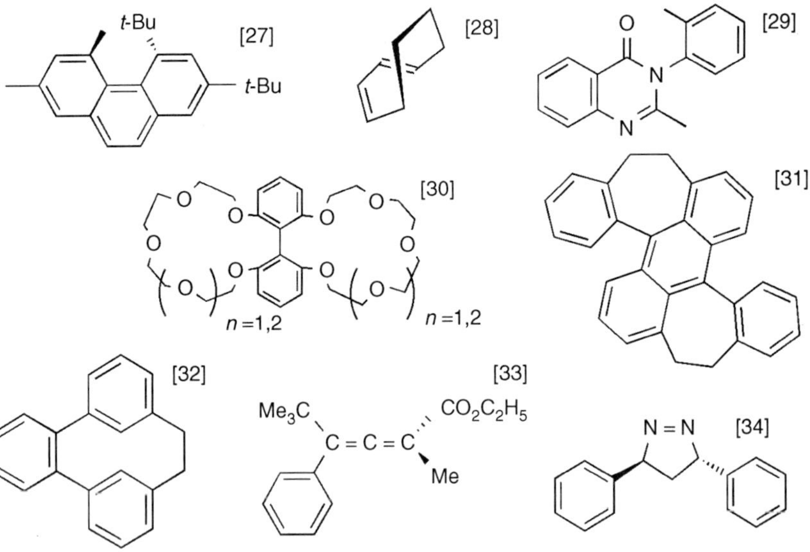

Fig. 7.6. Stereochemically interesting compounds resolved on **CTA-I**.

difference in the chiral recognition abilities of **OB** and **OJ**. For example, **OJ** appears to separate large-size molecules more efficiently than **OB** (Fig. 7.7) [22]. **OB** resolves non-aromatic chiral compounds (Fig. 7.8) [37]. **OK** shows a recognition similar to that of **OJ**. Amylose benzoates exhibit a much lower chiral recognition than the cellulose benzoates [36].

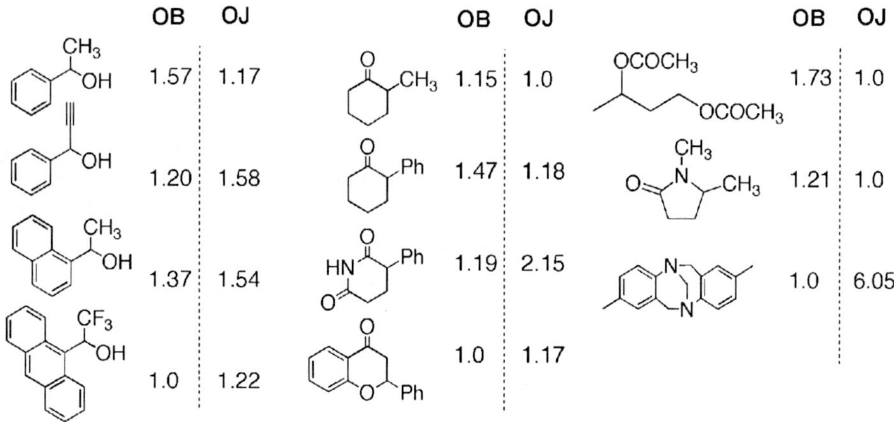

Fig. 7.7. Difference in chiral recognition ability (α) between **OB** and **OJ**.

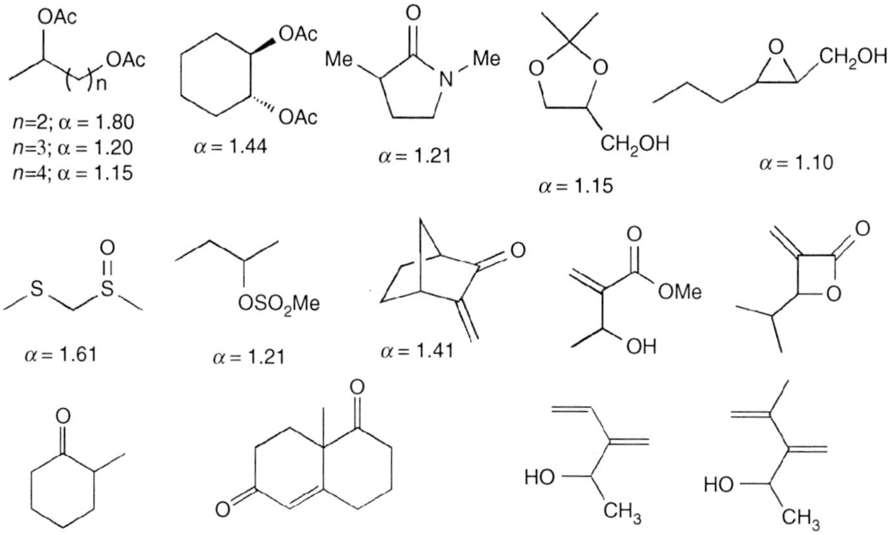

Fig. 7.8. Non-aromatic compounds resolved on **OB**.

7.3.2 Cellulose and amylose phenylcarbamates

Cellulose and amylose are easily converted to trisphenylcarbamate derivatives by reacting them with the corresponding phenyl isocyanates, and a variety of phenylcarbamates of cellulose and amylose have been prepared for use as the CSPs for HPLC [38–41]. Their chiral recognition abilities are significantly influenced by the substituents on the phenyl group. The introduction of an electron-donating alkyl group or an electron-withdrawing halogen at the *meta* and/or *para* position on the phenyl ring often improves the chiral recognition ability for many racemates [39]. Under a normal-phase condition using a hexane–alcohol mixture, polar racemates interact with the carbamate residue of CSPs *via* hydrogen bonding. Among the numerous phenylcarbamates, the tris(3,5-dimethylphenylcarbamate) derivatives of both the cellulose and the amylose exhibit an excellent resolving ability for a variety of racemates [39–41], and the corresponding CSPs, Chiralcel **OD** and Chiralpak **AD**, are commercially available. These two have been most frequently utilized for HPLC enantioseparation in many fields. For instance, **OD** resolves aromatic hydrocarbons, amines, carboxylic acids [42], alcohols, amino acid derivatives [43], and many drugs [44], including β-adrenergic blocking agents [45] with high α values. Figs. 7.9 [46–55] and 10 [54,56–64] show the chiral compounds resolved on **OD** and **AD**, respectively. On these two CSPs, the reverse elution order of enantiomers is often observed, suggesting that **OD** and **AD** are complementary in chiral recognition. In other words, many enantiomers unresolved on **OD** can be resolved on **AD**, and *vice versa*. Statistically, nearly 80% of the chiral compounds have been resolved by at least one of these two columns [15].

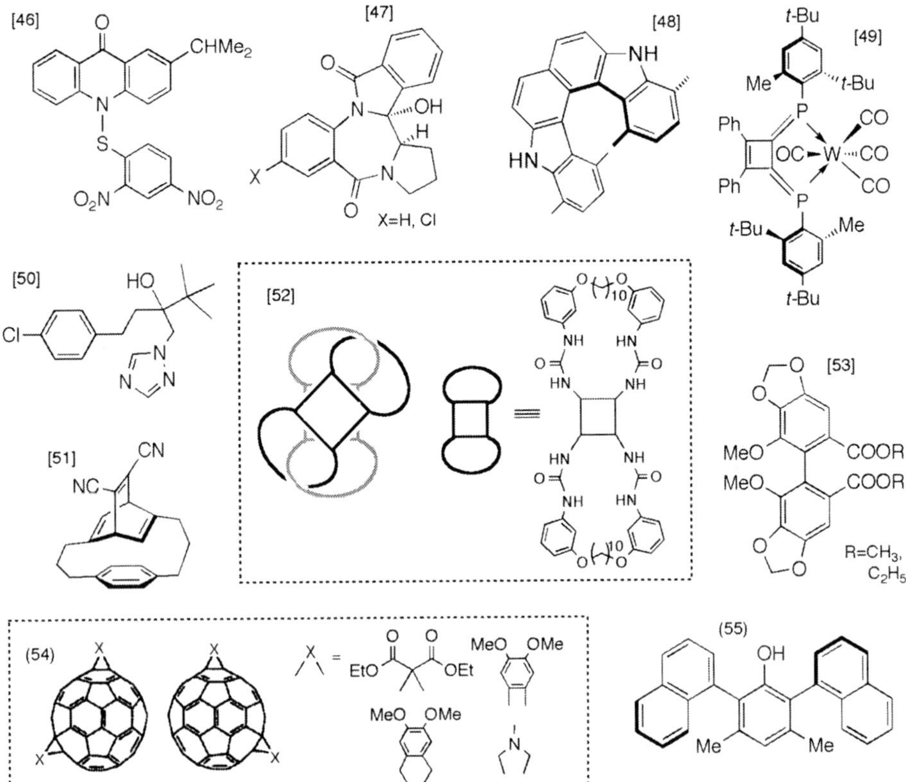

Fig. 7.9. Compounds resolved on cellulose tris(3,5-dimethylphenylcarbamate) (**OD**).

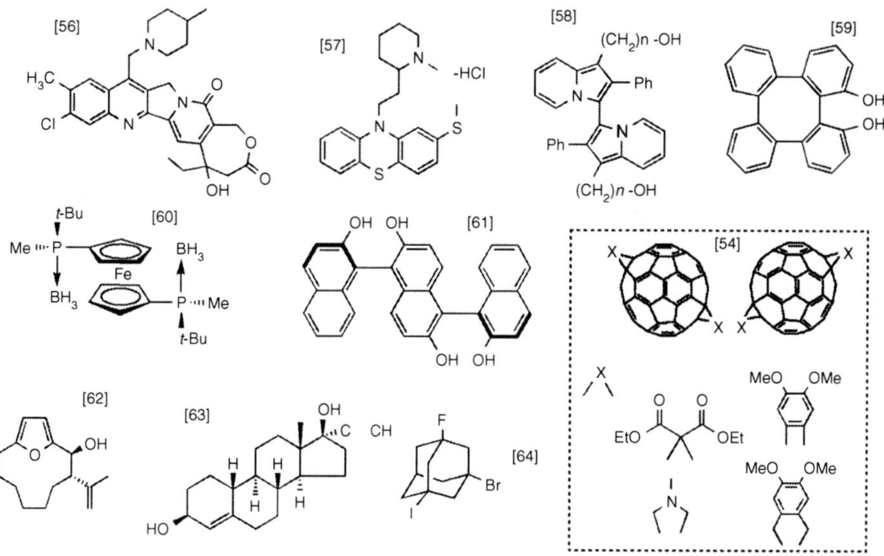

Fig. 7.10. Compounds resolved on amylose tris(3,5-dimethylphenylcarbamate) (**AD**).

Recently, a new CSP, Chiralpak **IA**, was commercialized. **IA** was developed in order to expand the variety of the eluents which can be used for **AD**. Except for **CTA-I**, all other polysaccharide-based CSPs are prepared by coating the polysaccharide derivatives on silica gel, and therefore, the solvents such as THF and chloroform, which dissolve or swell them, cannot be used as the main mobile phase. Since **IA** has been prepared by immobilizing amylose tris(3,5-dimethylphenylcarbamate) on a silica surface, most solvents can be used as the eluents. This enhances the possibility of the resolution of the racemates that cannot be resolved by the coated-type CSP. The addition of chloroform into an eluent can also increase the solubility of a sample in the eluents. This is particularly important for a preparative separation. Topologically, interesting catenanes and molecular knots [65–67], and the smallest chiral fullerene C_{76} [68] were successfully resolved using a hexane–chloroform–2-propanol mixture or a hexane–chloroform mixture (Fig. 7.11). For the polysaccharide derivatives, several immobilization methods have been reported [69–80].

7.3.3 Amylose benzylcarbamates

Amylose benzylcarbamates show a characteristic chiral recognition for many racemates different from those of the phenylcarbamates [81,82]. Among several benzylcarbamate derivatives, only 1-phenylethylcarbamate and 1-phenylpropylcarbamate exhibit a high chiral recognition, and other less or more bulky benzylcarbamates exhibit very low recognition abilities. In addition to the size, the chirality of the side chain also affects the chiral recognition; the (*S*)-1-phenylethylcarbamate of amylose (**AS**)

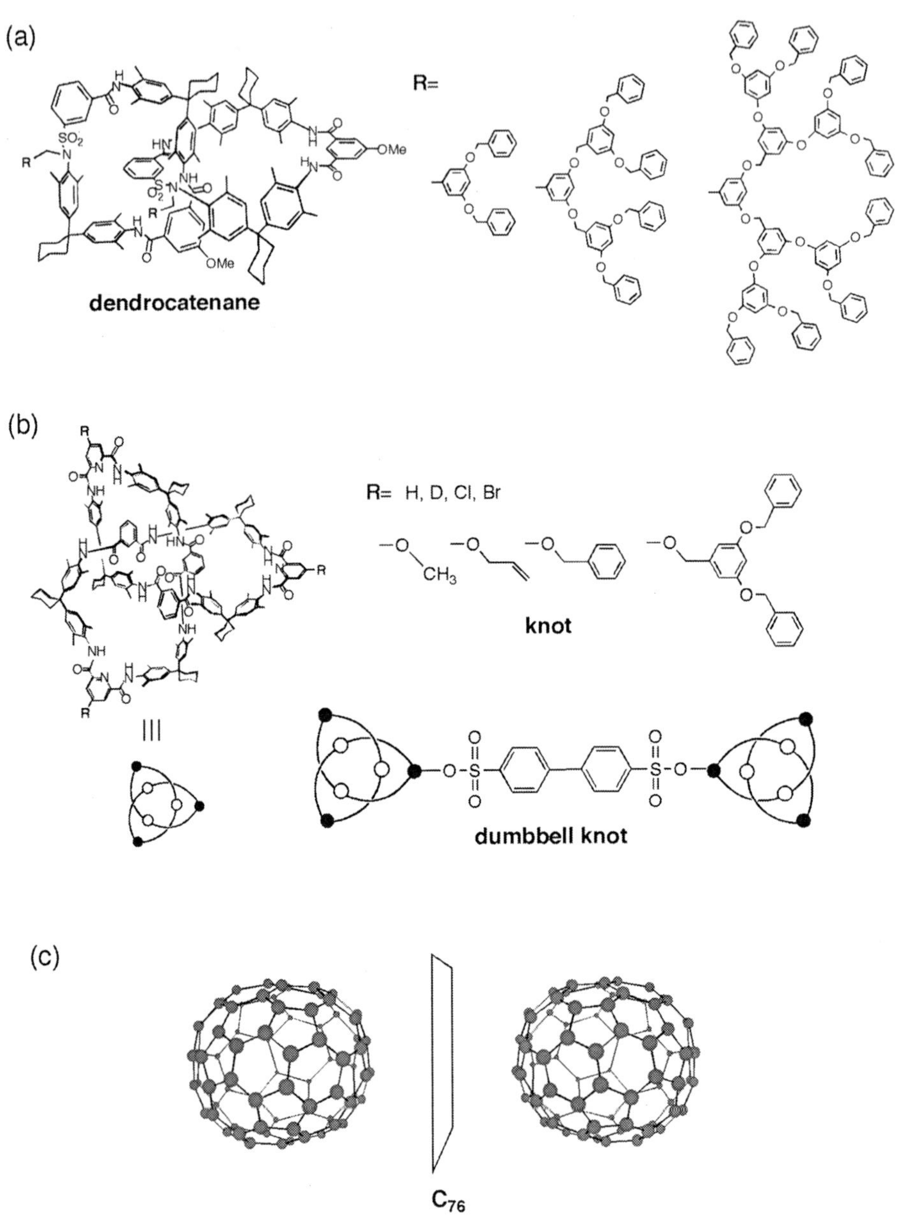

Fig. 7.11. Structures of dendrocatenanes (a), molecular knot, dumbbell knot (b), and C_{76} (c) resolved on chemically bonded-type CSP of **AD**.

shows a higher resolving ability than the (*R*) derivative and is commercially available. On the other hand, for the cellulose derivatives, the (*R*) derivative shows higher resolving abilities than the (*S*) derivative. **AS** is particularly useful for the resolution of β-lactams [83] and 3-hydroxy-2-cyclopentenone derivatives [84] (Fig. 7.12).

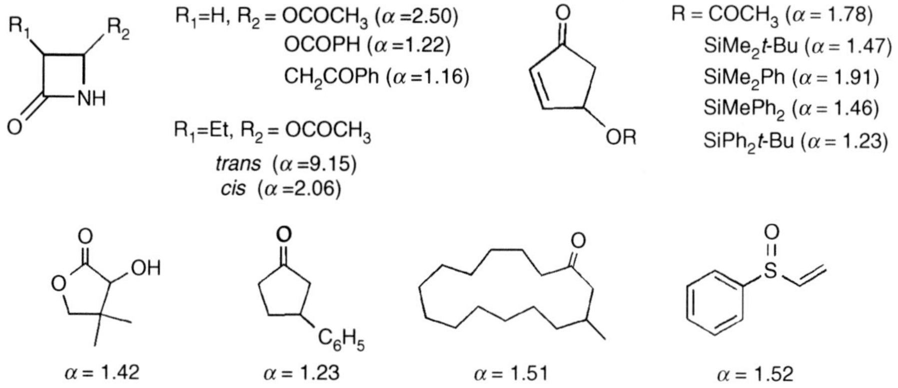

R_1=H, R_2 = $OCOCH_3$ (α =2.50)
OCOPH (α =1.22)
CH_2COPh (α =1.16)

R_1=Et, R_2 = $OCOCH_3$
trans (α =9.15)
cis (α =2.06)

R = $COCH_3$ (α = 1.78)
$SiMe_2t$-Bu (α = 1.47)
$SiMe_2Ph$ (α = 1.91)
$SiMePh_2$ (α = 1.46)
$SiPh_2t$-Bu (α = 1.23)

α = 1.42 α = 1.23 α = 1.51 α = 1.52

Fig. 7.12. Compound efficiently resolved on **AS**.

7.3.4 Other polysaccharide derivatives

Besides the above polysaccharide derivatives used for the commercialized CSPs, other carbamates of cellulose and amylose and the phenylcarbamates of other polysaccharides have been evaluated as CSPs for HPLC.

Cellulose tris(3,5-dichlorophenylcarbamate) shows a characteristic high resolving ability, and a very high enantioseparation factor (α = 112) was obtained for 2-(benzylsulfinyl)benzamide on this CSP using 2-propanol as the eluent (Fig. 7.13) [85]. However, this phase is soluble in the hexane containing 10–20% 2-propanol, which restricted its commercialization.

Alkylcarbamates of the polysaccharides, such as the methyl- and isopropyl-carbamates, show a very low chiral recognition. This may be because these alkyl groups are too small for the derivatives to maintain a regular higher order structure, which is important for a high chiral recognition. On the other hand, the cellulose and amylose cycloalkylcarbamates such as cyclohexyl and norbornylcarbamates (Fig. 7.14) show a high chiral recognition, which are comparable to those of **OD** and **AD**. In addition, the cycloalkylcarbamates can be used as the CSPs for thin-layer chromatography (TLC) due to the absence of a strong UV absorption above 220 nm [86,87]. A good correlation is observed between the α values on HPLC and TLC, which enables the rapid set up of the conditions for HPLC resolution.

As described above, the introduction of an alkyl group or halogen at the *meta* position on the phenyl ring improves the chiral recognition ability. Therefore, the 3,5-dimethyl- and 3,5-dichlorophenylcarbamates of other polysaccharides, such as chitin, chitosan, xylan, curdlan, dextran, galactosamine, and inulin, have been prepared in order to evaluate their chiral-recognition abilities as CSPs for HPLC (Fig. 7.15) [38,88–90]. The enantioselectivity and the elution order of enantiomers on these polysaccharides are significantly different, mainly due to the different sugar units. Among these derivatives, the 3,5-dimethylphenylcarbamates of xylan, chitosan, and chitin and the 3,5-dichlorophenylcarbamates of galactosamine and chitin showed

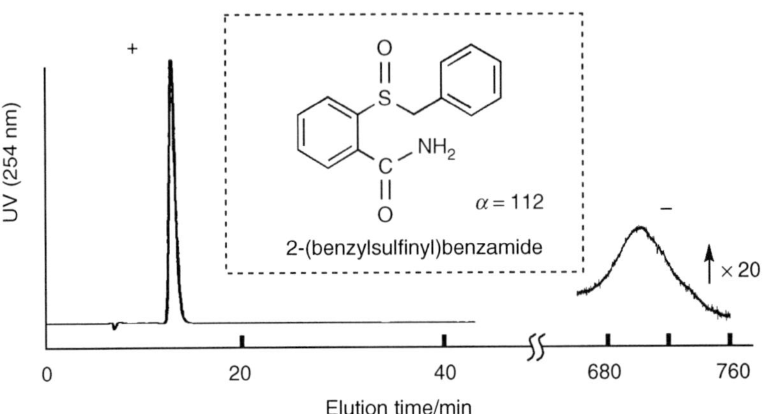

Fig. 7.13. Chromatogram of resolution of 2-(benzylsulfinyl)benzamide on cellulose tris(3,5-dichlorophenylcarbamate) using 2-propanol as an eluent. Chromatogram is reproduced, with permission, from Ref. [85] (Copyright 2000, Chemical Society of Japan).

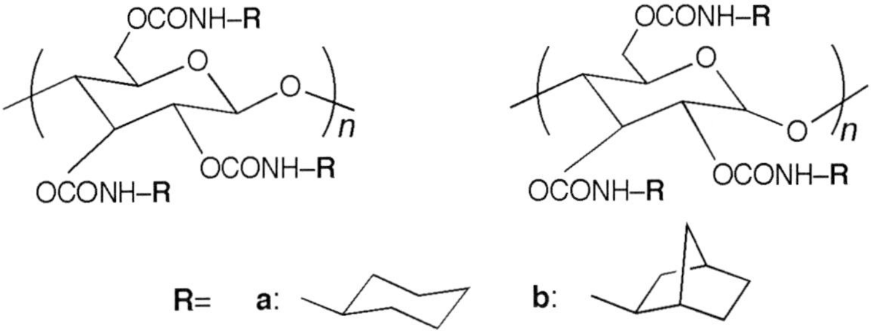

Fig. 7.14. Cyclohexyl (a) and norbornylcarbamates (b) of cellulose (left) and amylose (right).

relatively high chiral recognition abilities. Especially, the chitin phenylcarbamates are valuable due to their high durability for solvents. Due to the low solubility of the chitin derivatives on silica gel, various solvents can be utilized as the eluents. In addition, chitin phenylcarbamates are effective for the resolution of several acidic drugs such as the 2-arylpropionic acids including ketoprofen and ibuprofen (Fig. 7.16) [89].

7.4 SELECTION OF A SUITABLE COLUMN AND ELUENT FOR ENANTIOSEPARATION

For the efficient resolution of chiral compounds into enantiomers, it is very important to choose an appropriate CSP and an eluent for each compound. Polysaccharide-based CSPs have an excellent resolving ability for a variety of chiral compounds including

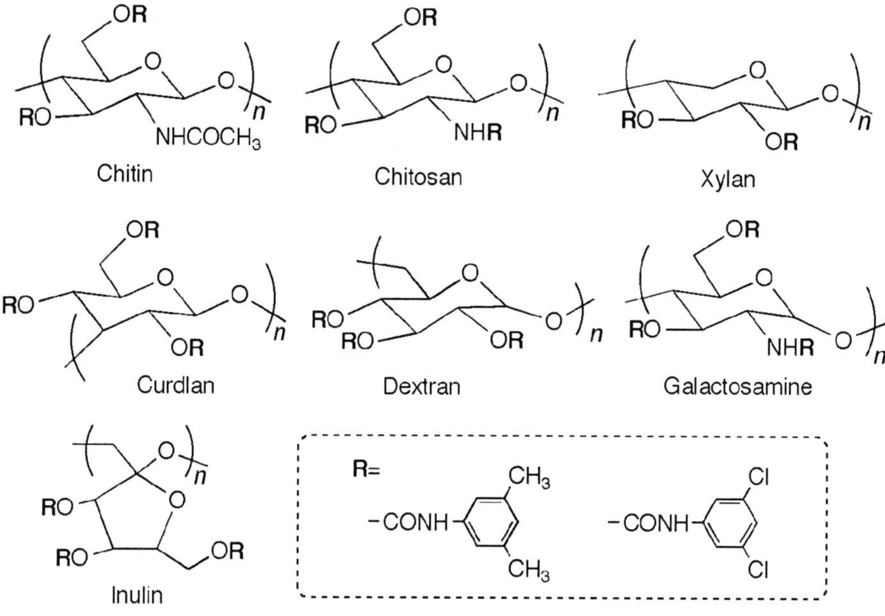

Fig. 7.15. Structures of phenylcarbamates of various polysaccharides.

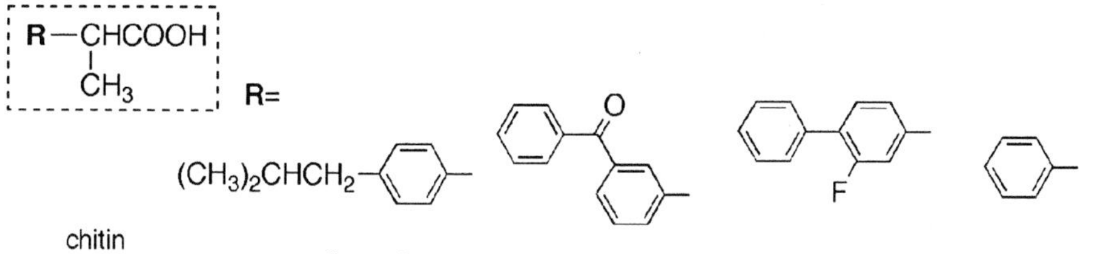

R—CH(CH$_3$)COOH R=

chitin phenylcarbamate	ibuprofen	ketoprofen	flurbiprofen	
	$(CH_3)_2CHCH_2-C_6H_4-$	3-benzoylphenyl	2-fluoro-4-biphenylyl	phenyl
3,5-$(CH_3)_2$	~ 1 (+)	1.21 (+)	1.08 (–)	1.41 (+)
3,5-$(Cl)_2$	1.11 (+)	1.72 (+)	1.10 (+)*	1.39 (+)

The signs in parentheses represent the optical rotation of the first-eluted enantiomer. Eluent: hexane–2-propanol–CF_3COOH (95/5/1, v/v/v). *Eluent: hexane–2-propanol–CF_3COOH (90/10/1, v/v/v).

Fig. 7.16. Resolution of 2-arylpropionic acids on chitin carbamates.

aromatic hydrocarbons, axially and planar dissymmetric compounds, metal-containing compounds, chiral sulfur or phosphorus compounds, cyano compounds, carbonyl compounds, amines, carboxylic acids, alcohols, amino acid derivatives, and ethers [15,18]. In addition, each polysaccharide-based CSP shows a characteristic and somehow complementary chiral resolution. On these CSPs, 80–90% of the chiral compounds seem to be resolved. Fig. 7.3(b) shows the distribution of the polysaccharide-type CSPs used in chiral HPLC. Among the CSPs, **OD**, **AD**, **OJ**, and **AS** show an especially high resolving ability for a variety of chiral compounds. Therefore, to resolve a racemate, it is recommended to first use these columns. As examples, the chromatograms for the resolution of several racemates bearing different functional groups are shown in Fig. 7.17 [42,43,45,91]. Chiralpak **AD** may be replaced by Chiralpak **IA**.

The resolution by HPLC is often significantly influenced by the eluents. For the polysaccharide-based CSPs, both normal-phase eluents, such as hexane containing an alcohol, and reversed-phase eluents [92], such as an alcohol or acetonitrile containing water, can be used. Phenylcarbamates of cellulose and amylose have a left-handed helical structure, in which the polar carbamate groups are located inside and the hydrophobic aromatic groups are placed at the outside of the polymer chain [93,94]. Therefore, for polar analytes, the resolution under a normal-phase condition may be recommended to enhance the hydrogen-bonding or dipole–dipole interaction between carbamate residues and enantiomers. The alcohols, which are often used as an additive of an eluent, influence the enantioselectivity, and the change in 2-propanol to ethanol shortens the elution time and improves the separation in some cases [95]. When the retention times are rather short, methyl *tert*-butyl ether is effective as an additive instead of alcohols. On the other hand, a reversed-phase condition is recommended for the resolution of non-polar analytes, which may interact with the polysaccharide derivatives through a hydrophobic interaction. Aqueous eluents are also valuable for investigating the pharmacokinetics, physiological, toxicological, and metabolic activities of drug enantiomers in living systems [96].

7.5 SEPARATION OF SEVERAL TYPICAL RACEMATES

7.5.1 Chiral acids

Fig. 7.18 shows the structures of chiral acids resolved on **OD** or **AD**. For the resolution of acidic compounds under normal phase conditions, it is highly recommended to add a small amount of a strong acid such as trifluoroacetic acid (ca. 0.5%) to an eluent in order to decrease the retention time and tailing of peaks [42,97,98]. An acidic mobile phase is also useful under reversed-phase conditions in order to suppress the dissociation of an analyte. An aqueous solution or buffer of pH 2 containing an organic modifier (alcohol or acetonitrile), such as $HClO_4$ aq. (pH 2)/CH_3CN (60/40) and 0.5 M $NaClO_4$–$HClO_4$ aq. (pH 2)/CH_3CN (60/40), often results in good resolution [92].

7.5.2 Chiral amines

In order to resolve an analyte with a basic primary, secondary, or tertiary amino group under normal phase conditions, the addition of a small amount of an amine, such as diethylamine or isopropylamine (ca. 0.1%), may bring about a better resolution without

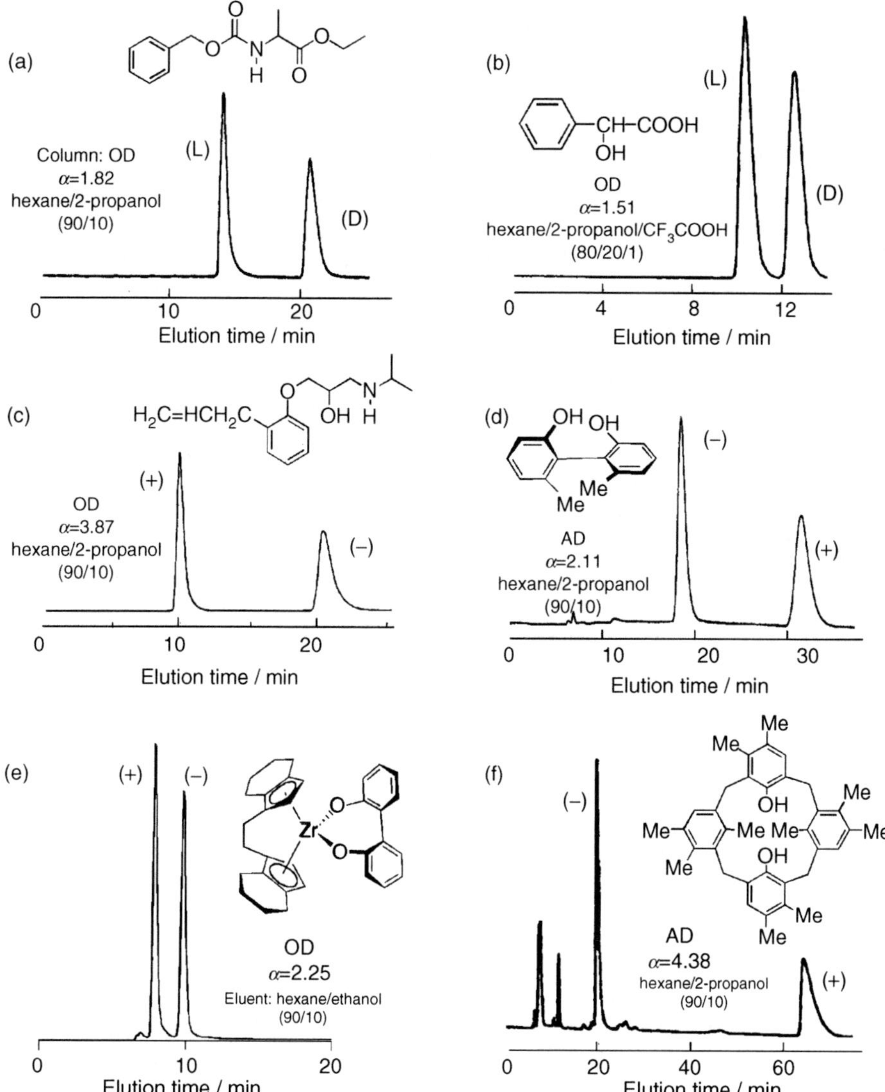

Fig. 7.17. Chromatograms of the resolution of (a) *N*-benzyloxycarbonyl alanine ethyl ester, (b) mandelic acid, (c) alprenolol, (d) 2,2′-dihydroxy-6,6′-dimethylbiphenyl, (e) the ansa-zirconocene derivative, and (f) the partially *O*-methylated Calix[4]arene. Column, 25 × 0.46 cm (i.d.); flow rate, 0.5 ml/min. Chromatograms (b), (c), and (e) are reproduced, with permission, from Ref. [42,45,91] (Copyright 1988, 1986, and 1996, Chemical Society of Japan), respectively.

the tailing of peaks [43,45]. For example, β-adrenergic blocking agents (β-blockers), which possess a general structure, $ArOCH_2CH(OH)CH_2NHCH(CH_3)$ (Ar = aromatic), have been resolved on **OD** with high α values using a hexane–2-propanol–diethylamine mixture as the eluent (Fig. 7.19) [45]. Under reversed-phase conditions, the silica gel

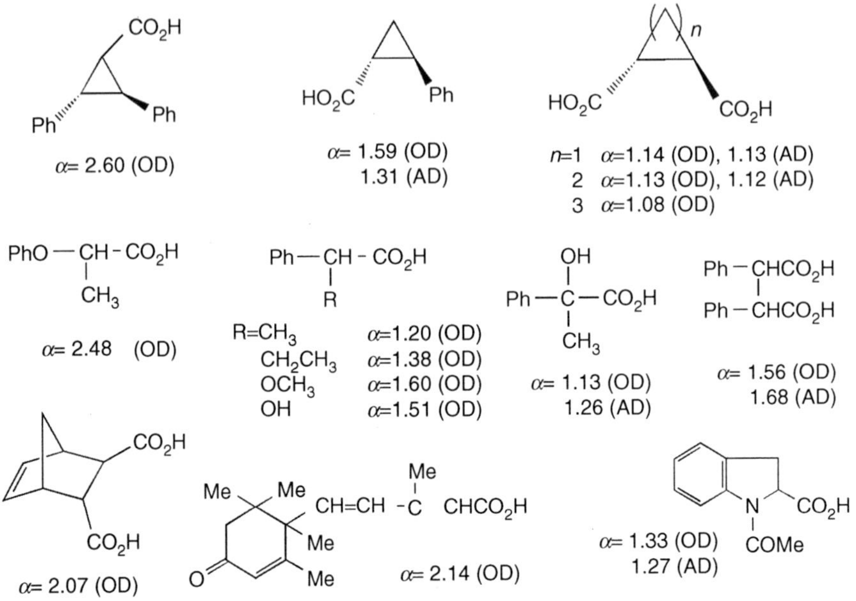

Fig. 7.18. Carboxylic acids resolved on **OD** or **AD**.

as a support material is unstable above pH 7, indicating that a basic eluent cannot be utilized. However, under neutral and acidic mobile phase conditions, positively charged basic compounds cannot efficiently interact with the CSP. Therefore, it is effective to add a considerable amount of anions, for instance, PF_6^-, BF_4^-, and ClO_4^-, to the mobile phase to form an ion pair. The use of 0.5 M $NaClO_4$ aq./CH_3CN (60/40) is recommended [92].

7.5.3 Chiral alcohols

Usually chiral alcohols can be directly resolved with hexane containing a small amount of an alcohol as the eluent, and the resolution of many alcohols on the polysaccharide-based CSPs have been reported. If one cannot directly resolve an alcohol, particularly an aliphatic alcohol, their resolution is often efficiently attained as the phenylcarbamate on **OD**. For example, the phenylcarbamates of 2-butanol and 2-pentanol are completely resolved on **OD** with a very high selectivity (Fig. 7.20) [99]. The derivatization of alcohols to benzoates is also effective for resolution. For chiral compounds having phenolic hydroxy groups, the addition of a small amount of an acid to the eluent is recommended to depress its dissociation.

7.5.4 Chiral aromatic hydrocarbons

For the resolutions of chiral aromatic compounds without polar substituents, both normal- and reversed-phase conditions are effective (Fig. 7.21). For example,

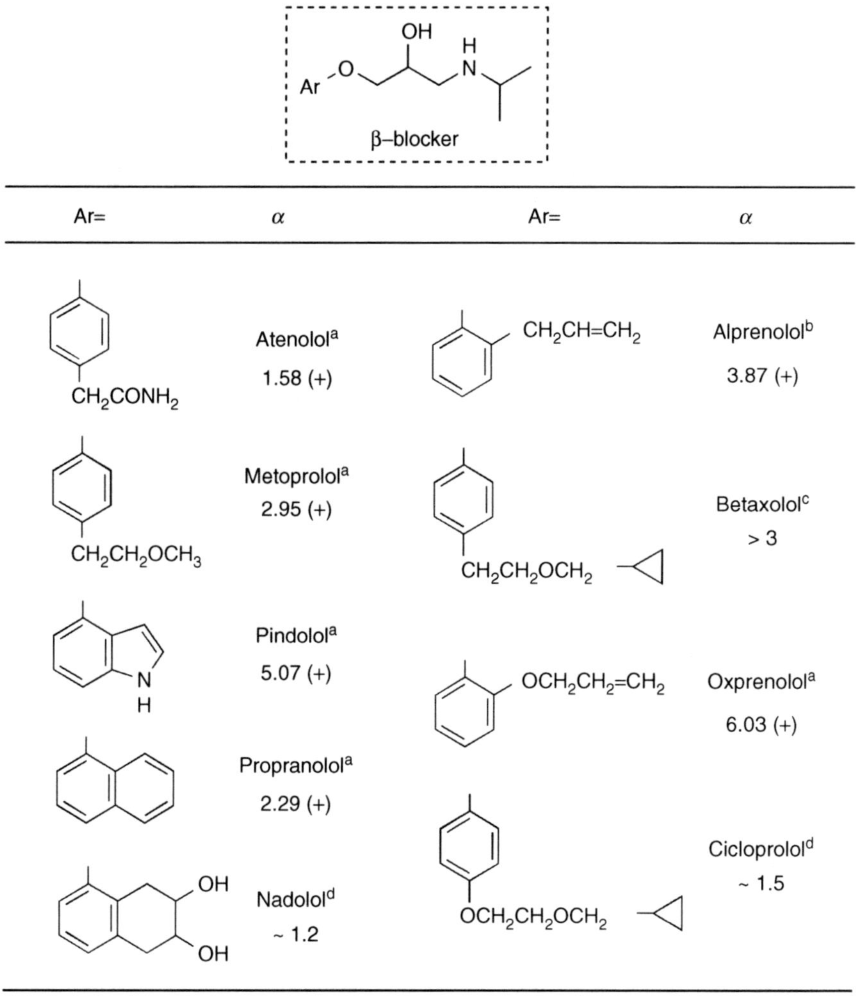

OH H N Ar O β–blocker

Ar=	α	Ar=	α
CH_2CONH_2	Atenolol[a] 1.58 (+)	$CH_2CH{=}CH_2$	Alprenolol[b] 3.87 (+)
$CH_2CH_2OCH_3$	Metoprolol[a] 2.95 (+)	$CH_2CH_2OCH_2$ — cyclopropyl	Betaxolol[c] > 3
N H	Pindolol[a] 5.07 (+)	$OCH_2CH_2{=}CH_2$	Oxprenolol[a] 6.03 (+)
	Propranolol[a] 2.29 (+)		
OH OH	Nadolol[d] ~ 1.2	$OCH_2CH_2OCH_2$ — cyclopropyl	Cicloprolol[d] ~ 1.5

[a]eluent, hexane-2-propanol-$HNEt_2$ (80:20:0.1). [b]eluent, hexane-2-propanol (90:10). [c]eluent, hexane-2-propanol-$HNEt_2$ (92:8:0.05). [d]eluent, hexane-2-propanol-ethnol-$HNEt_2$ (80:5:15:0.05).

Fig. 7.19. Resolution of β-blocker on **OD**.

chiral paracyclophanes (**12**) were resolved on **OD** using a hexane–2-propanol mixture as the eluent [100]. Doubly tethered biphenyl **13a** could be resolved on **OD** with a mixture of hexane–2-propanol. On the other hand, its tetraethyl derivative **13b** was not resolved with a mixture of hexane–alcohol and pure hexane, but it was separated on **OD** using iso-octane [101]. Compound **14** was also completely separated on **OD** using iso-octane [102,103]. Compound **15** was resolved on **AD** under the reversed-phase condition using an ethanol–water (9:1) mixture [104]. In an alcohol–water eluent,

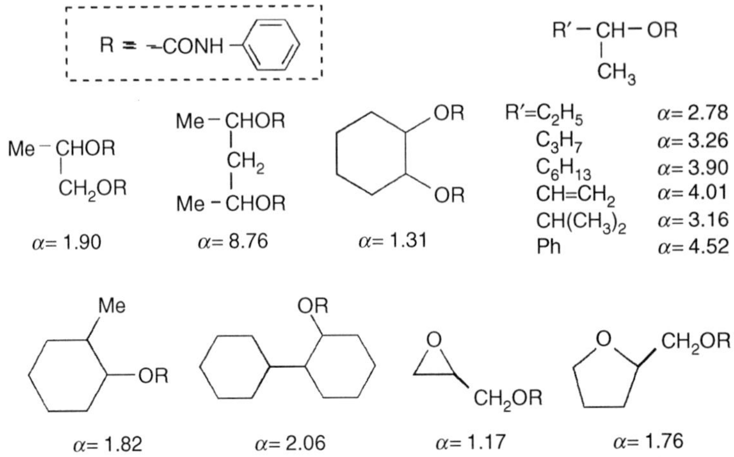

Fig. 7.20. Resolution of alcohols as phenylcarbamates.

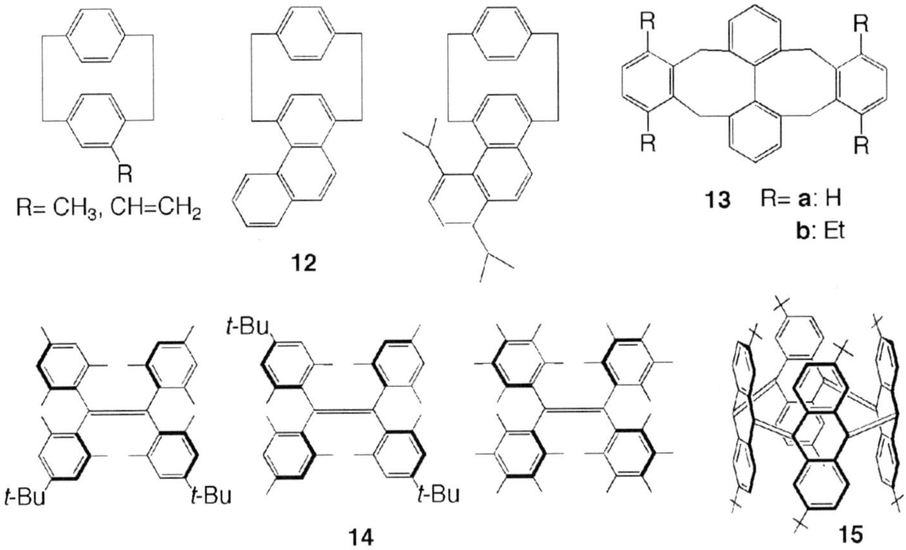

Fig. 7.21. Chiral aromatic hydrocarbons without polar substituents resolved on polysaccharide-based CSPs.

aromatic hydrocarbons interact with the polysaccharide derivatives through hydrophobic interactions like π–π stacking, and the addition of water to the alcohol delays the elution times of the enantiomers. For axially or planar chiral aromatic hydrocarbons, a one-handed helical poly(triphenylmethyl methacrylate) (**6**) also exhibits a high chiral recognition under reversed-phase conditions [7,8,10].

7.5.5 Non-aromatic chiral compounds

For the resolution on the polysaccharide-based CSPs, it is not necessary for chiral compounds to possess an aromatic moiety. The main interaction between the polysaccharide-based CSPs and polar analytes under a normal phase condition is a polar interaction represented by hydrogen bonding. As shown in Fig. 7.8, several non-aromatic compounds have been resolved on **OB** [37]. The detection of non-aromatic enantiomers by a UV detector is often difficult and a refractive index (RI) detector is more useful.

7.6 CONCLUSION

Chiral HPLC is a practically useful separation technique not only for determining optical purity, but also for obtaining optical isomers, and a large number of CSPs for HPLC have been commercialized. Polysaccharide-based CSPs are the most popular CSPs today, and can achieve the successful resolution of a wide range of chiral compounds particularly by optimizing the separation conditions.

A large-scale separation of enantiomers on polysaccharide-based CSPs has already been realized by using a simulated moving-bed (SMB) chromatographic system for the resolution on an industrial scale [105]. On the other hand, capillary liquid chromatography (CLC), capillary electrophoresis (CE), and capillary electrochromatography (CEC), which are the analytical methods using a capillary column, have been extensively investigated [106]. These techniques are useful for studying the pharmacokinetics, physiological, toxicological, and metabolic activities of drug enantiomers in living systems. A capillary column needs much lower amounts of a stationary phase and a mobile phase than a normal-size HPLC column, and therefore, these are environmentally friendly systems. The polysaccharide-based CSPs can also be utilized in supercritical fluid chromatography (SFC), which may enable the resolutions in a shorter time with a good separation [107,108].

The chiral recognition mechanism of the polysaccharide-based CSPs has been investigated on the basis of chromatographic methods, and spectroscopic and computational approaches at a molecular level [109–113]. The classification of the chiral recognition mechanism on the CSPs will enable the selection of suitable resolution conditions and the further development of more effective CSPs.

REFERENCES

1 S.V. Rogozhin and V.A. Davankov, Ligand chromatography on asymmetric complex-forming sorbents as a new method for resolution of racemates. J. Chem. Soc., Chem. Commun., 192 (1971) 490.

2 V.A. Davankov, Resolution of racemates by ligand exchanges chromatography. Adv. Chromatogr., 18 (1980) 139–195.

3 W.H. Pirkle, J.M. Finn, J.L. Schreiner and B.C. Hamper, A widely useful chiral stationary phase for the high-performance liquid chromatography separation of enantiomers. J. Am. Chem. Soc., 103 (1981) 3964–3966.

4 W.H. Pirkle and T.C. Pochapsky, Considerations of chiral recognition relevant to the liquid chromatographic separation of enantiomers. Chem. Rev., 89 (1989) 347–362.

5 T. Shinbo, T. Yamaguchi, K. Nishimura and M. Sugiura, Chromatographic separation of racemic amino acids by use of chiral crown ether-coated reversed-phase packings. J. Chromatogr., 405 (1987) 145–153.

6 S.C. Chang, G.L. Reid III, S. Chen, C.D. Chang and D.W. Armstrong, Evaluation of a new polar-organic high-performance liquid chromatographic mobile phase for cyclodextrin-bonded chiral stationary phases. Trends in Anal. Chem., 12 (1993) 144–153.

7 Y. Okamoto, E. Yashima and C. Yamamoto, Optically active polymers with chiral recognition ability, in: M.M. Green, R.J.M. Nolte and E.W. Meijer (Eds.), Materials Chiralty: Volume 24 of Topics in Stereochemistry, John Wiley & Sons Inc., New Jersey, 2003, pp. 157–208.

8 C. Yamamoto and Y. Okamoto, Optically active polymers for chiral separation. Bull. Chem. Soc. Jpn., 77 (2004) 227–257.

9 Y. Okamoto, K. Suzuki, K. Ohta, K. Hatada and H. Yuki, Optically active poly(triphenylmethyl methacrylate) with one-handed helical conformation. J. Am. Chem. Soc., 101 (1979) 4763–4765.

10 Y. Okamoto and K. Hatada, Resolution of enantiomers by HPLC optically active poly(triphenylmethyl methacrylate). J. Liq. Chromatogr., 9 (1986) 369–384.

11 G. Blaschke, Chromatographic resolution on racemates. Angew. Chem. Int. Ed. Engl., 19 (1980) 13–24.

12 G. Blaschke, H.P. Kraft, K. Fickentscher and F. Kohler, Chromatographic separation of racemic thalidomide and teratogenic activity of its enantiomers. Arzneim. Forsch., 29 (1979) 1640–1642.

13 K. Saigo, Synthesis and properties of polyamides having a cyclobutanedicarboxylic acid derivative as a component. Prog. Polym. Sci., 17 (1992) 35–86.

14 J. Haginaka, Protein-based chiral stationary phases for high-performance liquid chromatography enantioseparations. J. Chromatogr. A, 906 (2001) 253–273.

15 Y. Okamoto and Y. Kaida, Resolution by high-performance liquid chromatography using polysaccharide carbamates and benzoates as chiral stationary phases. J. Chromatogr. A, 666 (1994) 403–419.

16 E. Yashima and Y. Okamoto, Chiral discrimination on polysaccharides derivatives. Bull. Chem. Soc. Jpn., 68 (1995) 3289–3307.

17 Y. Okamoto and E. Yashima, Polysaccharide derivatives for chromatographic separation of enantiomers. Angew. Chem. Int. Ed., 37 (1998) 1020–1043.

18 E. Yashima, C. Yamamoto and Y. Okamoto, Polysaccharide-based chiral LC columns. Synlett, (1998) 344–360.

19 E. Yashima, Polysaccharide-based chiral stationary phases for high-performance liquid chromatographic enantioseparation. J. Chromatogr. A, 906 (2001) 105–125.

20 Y. Okamoto, M. Kawashima, K. Yamamoto and K. Hatada, Useful chiral packing materials for high-performance liquid chromatographic resolution. Cellulose triacetate and tribenzoate coated on macroporous silica gel. Chem. Lett., (1984) 739–742.

21 A. Ichida, T. Shibata, I. Okamoto, Y. Yuki, H. Namikoshi and Y. Toga, Resolution of enantiomers by HPLC on cellulose derivatives. Chromatographia, 19 (1984) 280–284.

22 K. Oguni, H. Oda and A. Ichida, Development of chiral stationary phases consisting of polysaccharide derivatives. J. Chromatogr. A, 694 (1995) 91–100.

23 G. Hesse and R. Hagel, Complete separation of a racemic mixture by elution chromatography on cellulose triacetate. Chromatographia, 6 (1973) 277–280.

24 G. Hesse and R. Hagel, Chromatographic resolution or racemates. Liebigs Ann. Chem., (1976) 966–1008.

25 E.J. Francotte, Preparative chromatographic separation of enantiomers. J. Chromatogr. A, 576 (1992) 1–45.

26 E.J. Francotte, Contribution of preparative chromatographic resolution to the investigation of chiral phenomena. J. Chromatogr. A, 666 (1994) 565–601.

27 H. Koller, K.-H. Rimböck and A. Mannschreck, High-pressure liquid chromatography on triacetylcellulose. J. Chromatogr. A, 282 (1983) 89–94.

28 R. Isaksson, J. Rochester, J. Sandström and L.-G. Wistrand, Resolution, circular dichroism spectrum, molecular structure, and absolute configuration of cis, trans-1,3-cyclooctadiene. J. Am. Chem. Soc., 107 (1985) 4074–4075.
29 A. Mannschreck, H. Koller, B. Stühler, M.A. Davies and J. Traber, The enantiomers of methaqualone and their unequal anticonvulsive activity. Eur. J. Med. Chem.-Chim. Ther., 19 (1984) 381–383.
30 G. Lindsten, O. Wennerström and R. Isaksson, Chiral biphenyl bis(crown ethers): synthesis and resolution. J. Org. Chem., 52 (1987) 547–554.
31 I. Agranat, M.R. Suissa, S. Cohen, R. Isaksson, J. Sandström, J. Dale and D. Grace, A novel titanium-induced aromatic dicarbonyl coupling. Synthesis of a chiral strained polynuclear aromatic hydrocarbon. J. Chem. Soc., Chem. Commun., (1987) 381–383.
32 M. Wittek, F. Vögtle, G. Stühler, A. Mannschreck, B.M. Lang and H. Irngartinger, New helical hydrocarbons. VIII. Enantiomer separation, circular dichroism, racemization, and x-ray analysis of benzo[2.2]metacyclophane. Chem. Ber., 116 (1983) 207–214.
33 N. Krause and G. Hnadke, Enantioseparation of allenes by liquid chromatography. Tetrahedron Lett., 32 (1991) 7225–7228.
34 M.P. Schneider and H. Bippi, Transfer of optical activity in the decomposition of (+)- and (−)-trans-3,5-diphenyl-1-pyraxoline: competing "biradical" and "cycloreversion" pathways. J. Am. Chem. Soc., 102 (1980) 7363–7365.
35 Y. Okamoto, M. Kawashima, K. Yamamoto and K. Hatada, Useful chiral packing materials for high-performance liquid chromatographic resolution. Cellulose triacetate and tribenzoate coated on macroporous silica gel. Chem. Lett., (1984) 739–742.
36 Y. Okamoto, R. Aburatani and K. Hatada, Cellulose tribenzoate derivatives as chiral stationary phases for high-performance liquid chromatography. J. Chromatogr., 389 (1987) 95–102.
37 T. Shibata, I. Okamoto and K. Ishii, Chromatographic optical resolution on polysaccharides and their derivatives. J. Liq. Chromatogr., 9 (1986) 313–340.
38 Y. Okamoto, M. Kawashima and K. Hatada, Useful chiral packing materials for high-performance liquid chromatographic resolution of enantiomers: phenylcarbamates of polysaccharides coated on silica gel. J. Am. Chem. Soc., 106 (1984) 5357–5359.
39 Y. Okamoto, M. Kawashima and K. Hatada, Controlled chiral recognition of cellulose trisphenylcarbamate derivatives supported on silica gel. J. Chromatogr., 363 (1986) 173–186.
40 Y. Okamoto, R. Aburatani, T. Fukumoto and K. Hatada, Useful chiral stationary phases for HPLC. Amylose tris(3,5-dimethylphenylcarbamate) and tris(3,5-dichlorophenylcarbamate) supported on silica gel. Chem. Lett., (1987) 1857–1860.
41 Y. Okamoto, R. Aburatani and K. Hatada, Chromatographic optical resolution on 3,5-disubstituted phenylcarbamates of cellulose and amylose. Bull. Chem. Soc. Jpn., 63 (1990) 955–957.
42 Y. Okamoto, R. Aburatani, Y. Kaida and K. Hatada, Direct optical resolution of carboxylic acids by chiral HPLC on tris(3,5-dimethylphenylcarbamate)s on cellulose and amylose. Chem. Lett., (1988) 1125–1128.
43 Y. Okamoto, Y. Kaida, R. Aburatani and K. Hatada, Optical resolution of amino acid derivatives by high-performance liquid chromatography on tris(phenylcarbamate)s of cellulose and amylose. J. Chromatogr., 477 (1989) 367–376.
44 Y. Okamoto, R. Aburatani, K. Hatano and K. Hatada, Optical resolution of racemic drugs by chiral HPLC on cellulose and amylose tris(phenylcarbamate) derivatives. J. Liq. Chromatogr., 11 (1988) 2147–2163.
45 Y. Okamoto, M. Kawashima, R. Aburatani, K. Hatada, T. Nishiyama and M. Masuda, Optical resolution of β-blockers by HPLC on cellulose triphenylcarbamate derivatives. Chem. Lett., (1986) 1237–1240.
46 M. Ben-David Blanca, C. Yamamoto, Y. Okamoto, S.E. Biali and D. Kost, Resolution and rotational barriers of quinolinone and acridone sulfenamide derivatives: demonstration of the S-N chiral axis. J. Org. Chem., 65 (2000) 8613–8620.

47 A.G. Griesbeck, W. Kramer and J. Lex, Diastereo-and enantioselective synthesis of pyrrolo[1,4]benzodiazepines through decarboxylative photocyclization. Angew. Chem. Int. Ed., 40 (2001) 577–579.

48 I. Pischel, S. Grimme, S. Kotila, M. Neiger, F.V.I. Pischel, S. Grimme, S. Kotila, M. Neiger and F. Vögtle, A configurationally stable pyrrolohelicene: experimental and theoretical structure-chiroptic relationships. Tetrahedron: Asymmetry, 7 (1996) 109–116.

49 M. Yoshifuji, Y. Ichikawa, K. Toyota, E. Kasashima and Y. Okamoto, Rotational isomers of a 1,2-diphenyl-3,4-diphosphinidenecyclobutene. Chem. Lett., (1997) 87–88.

50 E.N. Shapovalova, O.A. Shpigun, L.M. Nesterova and M.Y. Belov, Determination of the optical purity of fungicide of the triazole series. J. Anal. Chem., 59 (2004) 290–294.

51 W. Matsuda-Sentou and T. Shinmyozu, Resolution and chiroptical properties of [32](1,4)barrelenophanedicarbonitrile, Tetrahedron: Asymmetry, 12 (2001) 839–842.

52 A. Bogdan, M.O. Vysotsky, T. Ikai, Y. Okamoto and V. Böhmer, Rational synthesis of multicyclic bis[2]catenanes. Chem. Eur. J., 10 (2004) 3324–3330.

53 Y. Liu, W. Lao, Y. Zhang, S. Jiang and L. Chen, Direct optical resolution of the enantiomers of axially chiral compounds by high-performance liquid chromatography on cellulose tris(3,5-dimethylphenylcarbamate) stationary phase. Chromatographia, 52 (2000) 190–194.

54 Y. Nakamura, K. O-kawa, T. Nishimura, E. Yashima and J. Nishimura, Systematic enantiomeric separation of [60]fullerene bisadducts possessing an inherent chiral addition pattern. J. Org. Chem., 68 (2003) 3251–3257.

55 S. Saito, T. Kano, H. Muto, M. Nakadai and H. Yamamoto, Asymmetric coupling of phenols with arylleads. J. Am. Chem. Soc., 121 (1999) 8943–8944.

56 J.F. Goossens, C. Foulon, C. Bailly, D.C.H. Bigg, J.P. Bonte and C. Vaccher, Chiral resolution of enantiomers of homocamptothecin derivatives, antifumor topoisomerase I inhibitors, using high performance liquid chromatography on polysaccharide-based chiral stationary phases. Chromatographia, 59 (2004) 305–313.

57 F. Geiser and R. Shah, Enantioseparation of hydrochloride salts using carbon dioxide-based mobile phases with on-line polarimetric detection. Chirality, 16 (2004) 263–266.

58 F. Theil, H. Sonnenschein and T. Kreher, Lipase-catalysed resolution of 3,3′-bi-indolizines: the first preparative access to enantiomerically pure samples. Tetrahedron: Asymmetry, 7 (1996) 3365–3370.

59 J-F. Wen, W. Hong, K. Yuan, T.C.W. Mak and H.N.C. Wong, Synthesis, resolution, and applications of 1,16-dihydroxytetraphenylene as a novel building block in molecular recognition and assembly. J. Org. Chem., 68 (2003) 8918–8931.

60 H. Tsuruta and T. Imamoto, A new P-chiral bisphosphine, 1,1′-bis[(t-butyl)methylphosphino]-ferrocene, as an effective ligand in catalytic asymmetric hydrosilylation of simple ketones. Tetrahedron: Asymmetry, 10 (1999) 877–882.

61 T. Sugimura, S. Inoue and A. Tai, Selective preparation of optically pure (R,R)-1,1′:5′,1″-ternaphthalene-2,2′,6,2″-tetraol: a new higher homolog of BINOL. Tetrahedron Lett., 39 (1998) 6487–6490.

62 H.F. Kasai, M. Tsubuki, K. Takahashi, M. Shirao, Y. Matsumoto, T. Honda and Y. Seyama, Separation of stereoisomers of several furan derivatives by capillary gas chromatography-mass spectrometry, supercritical fluid chromatography, and liquid chromatography using chiral stationary phases. J. Chromatogr. A, 977 (2002) 125–134.

63 M. Kummer and G. Werner, Chiral resolution of enantiomeric steroids by high-performance liquid chromatography on amylose tris(3,5-dimethylphenylcarbamate) under reversed-phase conditions. J. Chromatogr. A, 825 (1998) 107–114.

64 P.R. Schreiner, A.A. Fokin, O. Lauenstein, Y. Okamoto, T. Wakita, C. Rinderspacher, G.H. Robinson, J.K. Vohs and C.F. Campana, Pseudotetrahedral polyhaloadamantanes as chirality probes: synthesis, separation, and absolute configuration. J. Am. Chem. Soc., 124 (2002) 13348–13349.

65 C. Reuter, G. Pawlittzki, U. Wörsdörfer, M. Plevoets, A. Mohry, T. Kubota, Y. Okamoto and F. Vögtle, Chiral dendrophanes, dendro[2]rotaxanes, and dendro[2]catenanes: synthesis and chiroptical phenomena. Eur. J. Org. Chem., (2000) 3059–3067.

66 J. Recker, W.M. Müller, U. Müller, T. Kubota, Y. Okamoto, M. Nieger and F. Vögtle, Dendronized molecular knot: selective synthesis of various generations enantiomer separation, circular dichroism. Chem. Eur. J., 8 (2002) 4434–4442.

67 O. Lukin, J. Recker, A. Böhmer, W.M. Müller, T. Kubota, Y. Okamoto, M. Nieger, R. Fröhlich and F. Vögtle, A topologically chiral molecular dumbbell. Angew. Chem. Int. Ed., 42 (2003) 442–445.

68 C. Yamamoto, T. Hayashi, Y. Okamoto, S. Ohkubo and T. Kato, Direct resolution of C_{76} enantiomers by HPLC using an amylose-based chiral stationary phase. Chem. Commun., (2001) 925–926.

69 Y. Okamoto, R. Aburatani, S. Miura and K. Hatada, Chiral stationary phases for HPLC: cellulose tris(3,5-dimethylphenylcarbamate) and tris(3,5-dichlorophenylcarbamate) chemically bonded to silica gel. J. Liq. Chromatogr., 10 (1987) 1613–1628.

70 E. Yashima, H. Fukaya and Y. Okamoto, 3,5-dimethylphenylcarbamates of cellulose and amylose regioselectively bonded to silica gel as chiral stationary phases for high-performance liquid chromatography. J. Chromatogr. A, 677 (1994) 11–19.

71 L. Oliveros, P. Lopez, C. Minguillón and P. Franco, Chiral chromatographic discrimination ability of a cellulose 3,5-dimethylphenylcarbamate/10-undecenoate mixed derivative fixed on several chromatographic matrixes. J. Liq. Chromatogr., 18 (1995) 1521–1532.

72 N. Enomoto, S. Furukawa, Y. Ogasawara, H. Akano, Y. Kawamura, E. Yashima and Y. Okamoto, Preparation of silica gel-bonded amylose through enzyme-catalyzed polymerization and chiral recognition ability of its phenylcarbamate derivative in HPLC. Anal. Chem., 68 (1996) 2798–2804.

73 P. Franco, A. Senso, C. Minguillón and L. Oliveros, 3,5-Dimethylphenylcarbamates of amylose, chitosan and cellulose bonded on silica gel. Comparison of their chiral recognition abilities as high-performance liquid chromatography chiral stationary phases. J. Chromatogr. A, 796 (1998) 265–272.

74 P. Franco, A. Senso, L. Oliveros and C. Minguillón, Covalently bonded polysaccharide derivatives as chiral stationary phases in high-performance liquid chromatography. J. Chromatogr. A, 906 (2001) 155–170.

75 T. Kubota, T. Kusano, C. Yamamoto, E. Yashima and Y. Okamoto, Cellulose 3,5-dimethylphenylcarbamate immobilized onto silica gel via copolymerization with a vinyl monomer and its chiral recognition ability as a chiral stationary phase for HPLC. Chem. Lett., (2001) 724–725.

76 E.R. Francotte, Enantioselective chromatography as a powerful alternative for the preparation of drug enantiomers. J. Chromatogr. A, 906 (2001) 379–397.

77 E.R. Francotte and D. Huynh, Immobilized halogenophenylcarbamate derivatives of cellulose as novel stationary phases for enantioselective drug analysis. J. Pharm. Biomed. Anal., 27 (2002) 421–429.

78 T. Kubota, C. Yamamoto and Y. Okamoto, Preparation of chiral stationary phase for HPLC based on immobilization of cellulose 3,5-dimethylphenylcarbamate derivatives on silica gel. Chirality, 15 (2003) 77–82.

79 T. Kubota, C. Yamamoto and Y. Okamoto, Preparation and chiral recognition ability of cellulose 3,5-dimethylphenylcarbamate immobilized on silica gel through radical polymerization. J. Polym. Sci. Part A, Polym. Chem., 41 (2003) 3703–3712.

80 T. Kubota, C. Yamamoto and Y. Okamoto, Phenylcarbamate derivatives of cellulose and amylose immobilized onto silica gel as chiral stationary phases for high-performance liquid chromatography. J. Polym. Sci. Part A, Polym. Chem., 42 (2004) 4704–4710.

81 Y. Okamoto, Y. Kaida, H. Hayashida and K. Hatada, Tris(1-phenylethylcarbamate)s of cellulose and amylose as useful chiral stationary phases for chromatographic optical resolution. Chem. Lett., (1990) 909–912.

82 Y. Kaida and Y. Okamoto, Optical resolution by high-performance liquid chromatography on benzylcarbamates of cellulose and amylose. J. Chromatogr., 641 (1993) 267–278.

83 Y. Kaida and Y. Okamoto, Optical resolution of β-lactams on 1-phenylethylcarbamates of cellulose and amylose. Chirality, 4 (1992) 122–124.

84 Y. Kaida and Y. Okamoto, Efficient optical resolution of 4-hydroxy-2-cyclopentenone derivatives by HPLC on 1-phenylethylcarbamates of cellulose and amylose. Chem. Lett., (1992) 85–88.
85 B. Chankvetadze, C. Yamamoto and Y. Okamoto, Extremely high enantiomer recognition in HPLC separation of racemic 2-(benzylsulfinyl)benzamide using cellulose tris(3,5-dichlorophenylcarbamate) as a chiral stationary phase. Chem. Lett., (2000) 1176–1177.
86 T. Kubota, C. Yamamoto and Y. Okamoto, Tris(cyclohexylcarbamate)s of cellulose amylose as potential chiral stationary phases for high-performance liquid chromatography and thin-layer chromatography. J. Am. Chem. Soc., 122 (2000) 4056–4059.
87 T. Kubota, C. Yamamoto and Y. Okamoto, Chromatographic enantioseparation by cycloalkylcarbamate derivatives of cellulose and amylose. Chirality, 14 (2002) 372–376.
88 Y. Okamoto, J. Noguchi and E. Yashima, Enantioseparation on 3,5-dichloro- and 3,5-dimethylphenycarbamates of polysaccharides as chiral stationary phases for high-performance liquid chromatography. React. Funct. Polym., 37 (1998) 183–188.
89 C. Yamamoto, T. Hayashi, Y. Okamoto and S. Kobayashi, Enantioseparation by using chitin phenylcarbamates as chiral stationary phases for high-performance liquid chromatography. Chem. Lett., (2000) 12–13.
90 C. Yamamoto, T. Hayashi and Y. Okamoto, High-performance liquid chromatographic enantioseparation using chitin carbamate derivatives as chiral stationary phases. J. Chromatogr. A, 1021 (2003) 86–91.
91 S. Habaue, H. Sakamoto and Y. Okamoto, Optical resolution of chiral ethylenebis(4,5,6,7-tetrahydro-1-indenyl)zirconium derivatives by high-performance liquid chromatography. Chem. Lett., (1996) 383–384.
92 K. Tachibana and A. Ohnishi, Reversed-phase liquid chromatographic separation of enantiomers on polysaccharide type chiral stationary phases. J. Chromatogr. A, 906 (2001) 127–154.
93 P. Zugenmaier, Structural investigations on cellulose derivatives. J. Appl. Polu. Sci.: Appl. Polym. Symp., 37 (1983) 223–238.
94 U. Vogt and P. Zugenmaier, Structural models for some liquid crystalline cellulose derivatives. Ber. Bunsenges. Phys. Chem., 89 (1985) 1217–1224.
95 J. Dingene, Polysaccharide phases in enantioseparation, Chapter 6, in: G. Subramanian (Ed.), A Practical Approach to Chiral Separations by Liquid Chromatography, VCH, New York, 1994.
96 A. Ishikawa and T. Shibata, Cellulosic chiral stationary phase under reversed-phase condition. J. Liq. Chromatogr., 16 (1993) 859–878.
97 Y. Okamoto, R. Aburatani and K. Hatada, Direct optical resolution of abscisic acid by high-performance liquid chromatography on cellulose tris(3,5-dimethylphenylcarbamate). J. Chromatogr., 448 (1988) 454–455.
98 Y. Okamoto, R. Aburatani, Y. Kaida, K. Hatada, N. Inotsume and M. Nakano, Direct chromatographic separation of 2-arylpropionic acid enantiomers using tris(3,5-dimethylphenylcarbamate)s of cellulose and amylose as chiral stationary phases. Chirality, 1 (1989) 239–242.
99 Y. Okamoto, Z.-K. Cao, R. Aburatani and K. Hatada, Optical resolution of alcohols as carbamates by HPLC on cellulose tris(phenylcarbamate) derivatives. Bull. Chem. Soc. Jpn., 60 (1987) 3999–4003.
100 H. Hopf, W. Grahn, D.G. Barrett, A. Gerdes, J. Hilmer, J. Hucher, Y. Okamoto and Y. Kaida, Optical resolution of [2.2]paracyclophanes by high-performance liquid chromatography on tris(3,5-dimethylphenylcarbamate) of cellulose and amylose. Chem. Ber., 123 (1990) 841–845.
101 L. Eshdat, E. Shabtai, A.S. Saleh, T. Sternfeld, M. Saito, Y. Okamoto and M. Rabinovitz, Flexibility vs rigidity of singly and doubly tethered biphenyls: structure, dynamic stereochemistry, and resolution of tribenzo[*a,c,f*]cyclooctane, tetrabenzo[*a,de,h,kl*]bicyclo[6.6.0]tetradecane, and their alkyl derivatives. J. Org. Chem., 64 (1999) 3532–3537.
102 K. Maeda, Y. Okamoto, N. Morlender, N. Haddad, I. Eventova, S.E. Biali and Z. Rappoport, Does the threshold enantiomerization route of crowded tetraarylethenes involve double bond rotation? J. Am. Chem. Soc., 117 (1995) 9686–9689.

103 K. Maeda, Y. Okamoto, O. Toledano, D. Becker, S.E. Biali and Z. Rappoport, Multiple buttressing interactions: enantiomerization barrier of tetrakis(pentamethylphenyl)ethane. J. Org. Chem., 59 (1994) 5473–5475.
104 R. Herges, M. Deichmann, T. Wakita and Y. Okamoto, Synthesis of a chiral tube. Angew. Chem. Int. Ed., 42 (2003) 1170–1172.
105 M. Schulte and J. Strube, Preparative enantioseparation by simulated moving bed chromatography. J. Chromatogr. A, 906 (2001) 399–416.
106 B. Chankvetadze and G. Blaschke, Enantioseparations in capillary electromigration techniques: recent developments and future trends. J. Chromatogr. A., 906 (2001) 309–363.
107 Y. Kaida and Y. Okamoto, Optical resolution by supercritical fluid chromatography using polysaccharide derivatives as chiral stationary phases. Bull. Chem. Soc. Jpn., 65 (1992) 2286–2288.
108 J.L. Bernal, L. Toribio, M.J. del Nozal, E.M. Nieto and M.I. Montequi, Separation of antifungal chiral drugs by SFC and HPLC: a comparative study. J. Biochem. Biophys. Methods, 54 (2002) 245–254.
109 E. Yashima, M. Yamada, Y. Kaida and Y. Okamoto, Computational studies on chiral discrimination mechanism of cellulose trisphenylcarbamate. J. Chromatogr. A, 694 (1995) 347–354.
110 E. Yashima, C. Yamamoto and Y. Okamoto, NMR studies of chiral discrimination relevant to the liquid chromatographic enantioseparation by a cellulose phenylcarbamate derivative. J. Am. Chem. Soc., 118 (1996) 4036–4048.
111 E. Yashima, M. Yamada, C. Yamamoto, M. Nakashima and Y. Okamoto, Chromatographic enantioseparation and chiral discrimination in NMR by trisphenylcarbamate derivatives of cellulose, amylose, oligosaccharides, and cyclodextrins. Enantiomer, 2 (1997) 225–240.
112 C. Yamamoto, E. Yashima and Y. Okamoto, Computational studies on chiral discrimination mechanism of phenylcarbamate derivatives of cellulose. Bull. Chem. Soc. Jpn., 72 (1999) 1815–1825.
113 C. Yamamoto, E. Yashima and Y. Okamoto, Structural analysis of amylose tris(3,5-dimethylphenylcarbamate) by NMR relevant to its chiral recognition mechanism in HPLC. J. Am. Chem. Soc., 124 (2002) 12583–12589.

© 2006 Elsevier B.V. All rights reserved.

CHAPTER 8

Chiral separations by capillary electrophoresis

A.M. Stalcup

Department of Chemistry, University of Cincinnati, P.O. Box 210172, Cincinnati, OH 45221-0172, USA

8.1 INTRODUCTION

Electrophoresis involves the separation of charged species under the influence of an electric field. An excellent review of the early history of electrophoretic methods is detailed by Vesterberg [1]. While the introduction of electrophoresis as a method for the characterization of proteins is generally attributed to Tiselius in the 1930s, the migration of ions in the presence of an applied electric field was noted as early as the mid- to late 1800s. The rapid development of capillary electrophoresis as an analytical tool for chiral as well as achiral separations has benefited tremendously from the extensive knowledge-base generated over decades in the application of classical electrophoretic methods (e.g. isoelectric focusing, SDS-PAGE on slab gels) to the separation of biopolymers. Similarly, it could be argued that advances in capillary electrophoresis have fueled, to some extent, the rapid development in microchip-based assays [2,3].

Capillary electrophoresis has been the subject of a number of excellent recent specialized texts [4–6]. The technique was developed largely in response to the relatively low efficiency, poor reproducibility and lack of automation inherent in classical electrophoretic methods. The remarkable flexibility inherent in capillary electrophoretic methods may be inferred by its applications ranging from the separation of simple ions (e.g. inorganic anions [7,8]) to the analysis of single-cell content [9] to the actual separation of cells [10]. While capillary electrophoresis offers considerable attractive analytical advantages (e.g. outstanding resolution, minimal amounts of chiral selector, nonhazardous waste), there are a number of limitations. For instance, sample loading is typically on the order of a few nanograms and while the separations may sometimes be scaled up and adapted to a classical electrophoretic platform [11], the reverse is far more common [12]. In addition, capillary electrophoresis is experimentally somewhat more challenging than high performance liquid chromatographic or gas chromatographic methods. The reasons for this will be discussed in subsequent sections.

It is perhaps in the area of chiral separations where capillary electrophoresis has demonstrated the greatest utility. Enantiomeric separations by capillary electrophoresis has been the subject of a number of excellent general reviews [13,14] as well as reviews of specific classes of chiral selectors [15–19]. Advantages of capillary electrophoresis for chiral separations include the small amounts of chiral selector required and the ability to rapidly change chiral selectors. Virtually any chiral molecule which is available enantiomerically pure can be used as a chiral selector in capillary electrophoresis. This flexibility has resulted in a bewildering plethora of reported methods for chiral separations which this chapter attempts to organize.

This chapter is intended for nonspecialists in the pharmaceutical and university communities likely more familiar with analytical chromatographic separation methods than electrophoretic methods. Hence, the chapter will begin by highlighting similarities/differences between capillary electrophoresis and common chromatographic methods. This will be followed by a brief summary of capillary electrophoretic procedures (e.g. sample injection) and hardware (e.g. columns and common detectors) before turning attention to the theory and general strategies behind chiral separations by capillary electrophoresis. The chapter will then provide a general review of particular classes of chiral selectors, emphasizing analogies in mechanisms with related chiral chromatographic phases that readers may have some familiarity with and analyte features that promote chiral recognition by individual classes of chiral selectors.

8.2 CAPILLARY ELECTROPHORESIS: THE EXPERIMENT

In capillary electrophoresis, the motive force driving the separation is the voltage applied across the column which is analogous to the flow rate in chromatography (e.g. higher flow rate or higher voltage, faster analysis). The migration of buffer ions in the capillary column in response to the applied field generates a current that in turn generates heat, or Joule heating [20], analogous to the pressure generated in a chromatographic column. Both pressure in chromatography and current in electrophoresis can be useful diagnostic tools for instrumental performance.

Buffers with higher ionic strength or ions with higher mobility (e.g. Na^+) generate more heat for a given applied voltage than buffers with lower ionic strength or lower ionic mobility (e.g. K^+, zwitterionic buffers). While pressure is largely ignored as an artifact in liquid chromatography, the heat accompanying the current in capillary electrophoresis cannot be ignored. If the heat arising from the high current (e.g. $\gtrsim$250 μamps for water-cooled capillaries) exceeds the heat dissipative capability of the system, the resultant thermal gradients within the capillary create convection which can diminish the overall efficiency of the separation. According to Ohm's Law ($V = iR$), plots of current *versus* applied voltage should be linear. These plots can be used to determine the highest voltage that can be used with a given buffer system.

While most capillary electrophoresis is performed using aqueous buffers, achiral and chiral separations can also be done using nonaqueous solvents such as methanol [21], acetonitrile [22], formamide [23] or mixtures of organic solvents [24]. The buffer constituents in these organic systems are usually organic acid/base pairs such acetic acid/triethylammonium acetate or formic acid/ammonium formate [25]. Tolerable current densities are generally lower than what can be used in aqueous systems because of the increased volatility and lower boiling points of many organic solvents relative to water.

Additional considerations with capillary electrophoresis not evident in chromatographic experiments include additional peak distortion sources and potential siphoning, if the column ends or reservoir levels are not at the same height. Peak tailing or fronting in chromatographic separations are usually associated with overlapped peaks, multimodal retention or column overload. For instance, analyte interactions with capillary walls (Fig. 8.1(a)) produce the capillary electrophoretic counterpart of tailing exhibited by amines arising from interactions with residual silanols on reversed phase silica-based columns. However, large differences in the electrophoretic mobility of the analyte and the background electrolyte counter-ion can also lead to peak distortion. Ion depletion in the analyte zone as a result of the mismatch in electrophoretic mobilities of the analyte ion and the background electrolyte counter-ion can cause an electric field discontinuity in the analyte zone. As a result of this electrodispersion, sawtooth analyte peaks can emerge which exhibit either fronting (Fig. 8.1(b)) or tailing [26]. The sample matrix can also have a profound effect on the sample peak shape in capillary electrophoresis [27] and will be discussed in more detail later.

8.2.1 Capillary electrophoresis: the experimental data

In chromatography, the experimental parameter measured is the retention time, t_r. However, the experimental value that is generally reported is k or k', the retention factor, determined from the retention time through

$$k' = \frac{t_r - t_o}{t_o}, \qquad (1)$$

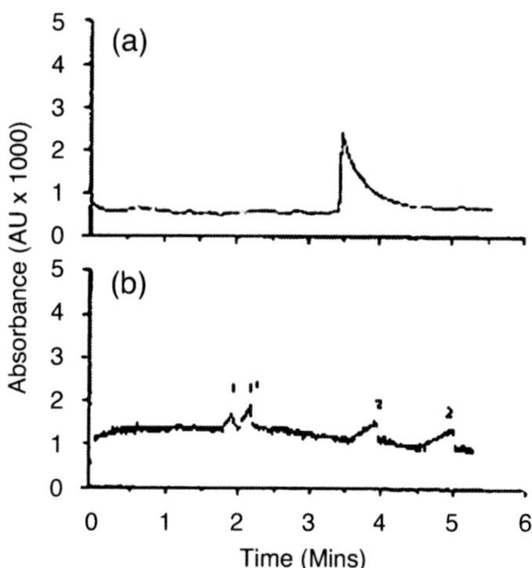

Fig. 8.1. Electropherograms showing peak distortion from (a) wall effects and (b) mobility mismatch.

where t_o is the retention time of an unretained solute. Essentially, the retention factor normalizes the data for differences in column length, mobile phase flowrate and extra column volumes, thereby allowing direct comparisons of retention data obtained in different systems.

In capillary electrophoresis, the experimental parameter measured is the migration time. However, the experimental value that is commonly reported is μ_{meas}, the apparent or measured electrophoretic mobility. It is calculated from the measured migration time through

$$\mu_{meas} = \frac{L_d/t_m}{V/L_t} = \frac{L_d L_t}{V t_m}, \tag{2}$$

where L_d is the length of the capillary to the detector window, L_t is the total length of the capillary and V is the applied voltage. The length to the detector is included because the migration time is measured at the detector, usually on-column or while the analyte is still physically within the capillary column instead of post-column as in liquid or gas chromatography. The total length of the capillary is included because the electric field is applied across the entire length of the column; the ratio of the applied voltage over the total length of the column is the electric field strength. Similarly, the migration time of a neutral analyte is used to calculate μ_{eof}, the electroosmotic mobility,

$$\mu_{eof} = \frac{L_d/t_{eof}}{V/L_t} = \frac{L_d L_t}{V t_{eof}} \tag{3}$$

an important parameter that will be discussed in more detail in the next section.

As in the case of k', μ_{meas} normalizes the data for differences in column length and applied voltages, thereby facilitating direct comparisons for results obtained in different systems. However, in contrast to k', which is unitless, μ_{meas} has units of cm^2/Vs and is typically in the range of 10^{-4} cm^2/Vs.

In the case of an analyte interacting with an auxiliary agent in the background electrolyte (e.g. CA, a chiral additive), the measured mobility of the analyte is given by

$$\mu_{meas} = \frac{\mu_f + \mu_c K[CA]}{1 + K[CA]}, \tag{4}$$

where μ_f is the mobility of the free analyte, μ_c is the mobility of the complex and K is the association constant for the complex [28]. Mobility data obtained at various concentrations of the additive can be conveniently used to determine the association or binding constants between the analyte and the additive [29,30].

Analogous to chromatographic selectivity, selectivity for CE can be determined from the migration time, after accounting for the contribution of electroosmotic flow (e.g. void volume in chromatography) to the migration time. Selectivity can be calculated from

$$\alpha = \frac{t_2 - t_{eof}}{t_1 - t_{eof}}, \tag{5}$$

where t_1 and t_2 are the migration times of the analytes and t_{eof} is the migration time of a neutral marker.

Calculation of resolution, R_S, is fairly straightforward and can be determined from

$$R_S = \frac{2(t_2 - t_1)}{(w_1 + w_2)}, \tag{6}$$

where w_1 and w_2 are the peak widths at the baseline. Theoretically, in the absence of any additional interactions between analytes and auxiliary agents in the background electrolyte, it can be shown that resolution is given by Ref. [31]

$$R_S = 0.177(\mu_1 - \mu_2)\left(\frac{V}{D(\mu_{ave} + \mu_{eof})}\right)^{\frac{1}{2}}, \tag{7}$$

where μ_1 and μ_2 are the measured mobilities of the two analytes, μ_{ave} is their average mobility, D is their average diffusion coefficient and V is the applied voltage. It can be seen from this relationship that the resolution increases with increasing voltage and is greatest when $\mu_{ave} \simeq \mu_{eof}$. The increasing resolution with increasing voltage arises because generally, in the absence of auxiliary agents which interact with the analyte, the most significant contribution to band broadening is longitudinal diffusion.

8.2.1.1 Relationship between μ_{meas} and μ_{ep}

The measured electrophoretic mobility, μ_{meas}, is a combination of the intrinsic electrophoretic mobility, μ_{ep}, and the electroosmotic flow, μ_{eof}

$$\mu_{meas} = \mu_{ep} + \mu_{eof}. \tag{8}$$

The intrinsic electrophoretic mobility, μ_{ep}, is a property of the analyte and can be expressed as

$$\mu_{ep} = \frac{q}{6\pi\eta r_s}, \tag{9}$$

where q is the charge on the analyte, η is the viscosity of the media and r_s is the hydrodynamic radius, related to the size of the analyte. Several points should be noted about the intrinsic electrophoretic mobility. First of all, the intrinsic electrophoretic mobility is zero for neutral analytes. Second, μ_{ep} can have positive or negative values. In addition, it should be noted that μ_{ep} decreases as the viscosity of the media increases and as the size of the analyte increases or charge on the analyte decreases.

The dependence of electrophoretic mobility on hydrodynamic radius, and not strictly molecular weight, in some cases may allow for the direct separation of cis/trans isomers [32] or diastereomeric derivatives of enantiomers. For instance, differences in internal hydrogen bonding interactions of diastereomeric Marfey's [33] derivatives of enantiomers (Fig. 8.2) can enable diastereomeric separations by capillary electrophoresis despite the identical charge and mass.

8.2.1.2 Relationship between μ_{meas} and μ_{eof}

The second contribution to μ_{meas}, the measured electrophoretic mobility, is μ_{eof}, the electroosmotic mobility. It may be helpful to consider the electroosmotic flow (typically, approximately a few nl/min) in comparison to the void volume in HPLC or GC.

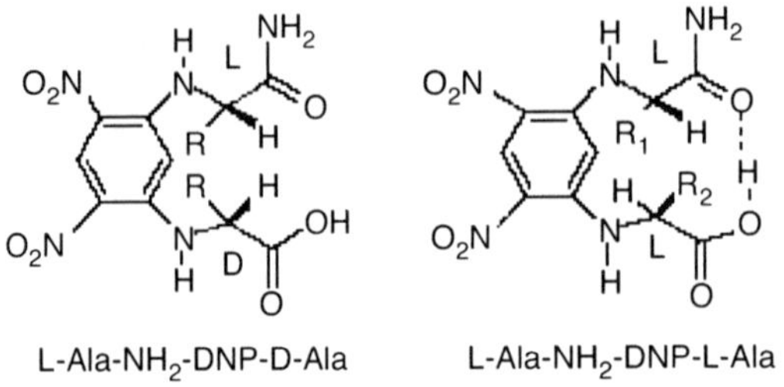

L-Ala-NH_2-DNP-D-Ala L-Ala-NH_2-DNP-L-Ala

Fig. 8.2. Structures of the Marfey's derivatives of D- and L-alanine. Adapted from Ref. [33].

In HPLC and GC, the void volume is the amount of mobile phase required to elute an analyte that does not interact with the stationary phase. In capillary electrophoresis, in the absence of any interactions with background electrolyte constituents, the electroosmotic flow marker indicates the migration time of an analyte which does not interact with the applied electric field.

An important distinction between chromatography and electrophoresis is that the chromatographic void volume is relatively invariant [34] for a given column and configuration. In contrast, the electroosmotic flow can be highly variable and even change directions within the same column if there is significant modification of the capillary wall by either sample or buffer components during the electrophoretic run. Hence, the electroosmotic flow can be a mortal enemy or a valuable ally. While total elimination of the electroosmotic flow may be highly desirable in some applications [35], the electroosmotic flow can also be exploited, if understood, to achieve the desired separations.

Understanding the electroosmotic flow is absolutely critical for obtaining robust, reproducible capillary electrophoretic separations and for understanding separation mechanisms. Because the electroosmotic flow has some profound consequences for the separations achieved in capillary electrophoresis, the next section will address this phenomenon at some length.

8.2.1.3 More about μ_{eof}

In the bare fused silica columns that are commonly used in capillary electrophoresis, the surface silanols are somewhat acidic and tend to be deprotonated at pH $\gtrsim 5$ although the extent of deprotonation is highly dependent upon the column pretreatment. The presence of the deprotonated surface silanols tends to attract cations (and their waters of hydration) from the bulk solution, resulting in the formation of an electrical double layer [36] comprised of a tightly held inner layer (the Helmholtz or Stern layer) and a diffuse outer layer (the Debye–Hückel or Gouy–Chapman layer). This diffuse layer has a thickness in the range of 5–10 Å at electrolyte concentrations of 0.1M and in the range of 50–100 nm for electrolyte concentrations of 1 mM [37]. It is this outermost diffuse layer that actually produces the "electroosmotic flow" or movement of bulk liquid in the capillary in the presence of an applied voltage.

In most routine applications, the voltage is applied such that the cations are attracted toward the detector end of the column (cathodic detection) and the anode is on the inlet side of the column. At high pH in a bare fused silica capillary, even anions, which have electrophoretic mobility in opposition to the electroosmotic flow, eventually reach the detector because of the magnitude of the electroosmotic flow.

The electroosmotic flow velocity (cm/s) is given by

$$v_{\text{eof}} = \frac{L_{\text{d}}}{t_{\text{eof}}} = \frac{\delta e}{\eta} E, \tag{10}$$

where L_{d} is the length of the column from the inlet to the detector, t_{eof} is the migration time of a neutral marker in seconds, η is the viscosity of the background electrolyte, δ is the electrical double layer thickness, e is the excess charge in solution per unit area and E is the electric field strength. The direct dependence of the electroosmotic flow on the thickness of the double layer implicates pH, and ionic strength as well as presence and type of buffers/modifiers as moderators of the magnitude and direction of the electroosmotic flow rate. In general, the electroosmotic flow rate in a bare fused silica column increases as,

- ionic strength of the solution decreases because of the increased thickness of the double layer;
- temperature increases because of the reduced background electrolyte viscosity;
- field strength goes up;
- pH increases because of the increasing ionization of the surface silanols (Fig. 8.3).

Buffer depletion, or the loss of buffering capacity in the background electrolyte over time, can arise from either migration of the buffer constituents or through

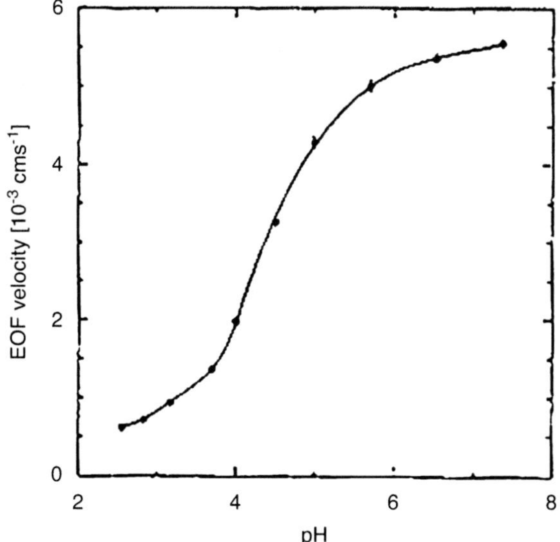

Fig. 8.3. Electroosmotic flow as a function of pH. Adapted from Ref. [38].

electrochemical reactions [39] at the electrodes which cause changes in the pH [40]. Buffer depletion can be minimized by using buffers near the pK_as of their acid or base constituents, using large buffer reservoirs or by exchanging the buffers in the reservoirs with increasing frequency as the run times get longer. Working at high ionic strength can also reduce buffer depletion but at a cost of high current densities.

Changes in organic composition can also influence the magnitude of the electroosmotic flow through changes in buffer viscosity. In the case of some buffer systems (e.g. phosphate) or organic modifiers (e.g. methanol), there is also thought to be some modification of the silica surface. Hence, most practitioners recommend "dedicating" a column to a specific buffer system.

Various strategies have been developed to deal with the challenges of μ_{eof} robustness including,

- standardized pre-run column conditioning (e.g. 0.1N NaOH or KOH wash for 2 min, followed by a 2-min water rinse, followed by a 2-min buffer rinse) to regenerate a consistent surface at the beginning of each run;
- covalently modifying the surface silanols [41] to engender [42], reduce or eliminate μ_{eof};
- coating the capillary walls (e.g., polymeric coatings [43], amine modifiers [44], sub-CMC concentrations of cationic surfactants [45], alternating polyelectrolyte layers [46]);
- operating at low [47] or high pH regions in which μ_{eof} is relatively insensitive to pH changes.

One of the primary experimental challenges of capillary electrophoresis is determining which particular strategy or combinations, thereof, will provide the desired robustness for a particular application. Hence, the pre-run wash conditions must be adapted to the surface conditions in the capillary. For instance, in the case of a siloxane-bonded coating on the inside surface of the capillary, a 0.1 N KOH wash would be inappropriate because the siloxane bond would not be stable to such treatment.

There are a variety of methods used to measure the electroosmotic flow [48]. Probably, the best approach is to use the migration time of a neutral marker (e.g. mesityl oxide or nitromethane). Given the potential changes in run-to-run electroosmotic flow, it is best if the neutral marker is added at low concentrations directly to the sample. The selection of a neutral marker becomes more problematic in electrokinetic chromatography. Electrokinetic chromatography, a variant of capillary electrophoresis that includes virtually all chiral separations by capillary electrophoresis, uses an additive that acts as a *pseudo*stationary phase. The presence of the additive can complicate selection of an electroosmotic flow marker because it may be difficult to identify a neutral candidate that does not interact with the additive [49].

Alternatively, the sample solvent peak is used to measure the electroosmotic flow. This is analogous to using the solvent or system peak in HPLC. While convenient and commonly used, this method can produce erroneous results. A sample solvent plug with a dramatically different composition than the bulk buffer can produce a discontinuity in the electric field strength and/or a localized change in the electroosmotic flow which can engender hydrodynamic flow [50].

Another method involves filling the capillary and one reservoir with a buffer of some composition, filling the other reservoir with buffer of identical but slightly less (or more) concentrated buffer, applying the run voltage and monitoring the current. Plots of current *versus* time reveal that the current changes gradually until the buffer in the column is totally exchanged [51]. The slope in the linear portion of the plot can be used to calculate the μ_{eof} [52].

8.3 THE COLUMN

In capillary electrophoresis, the polyimide-coated fused silica capillary columns used are identical to the capillary tubing used in capillary GC except that, in most applications, the interior of the fused silica tubing is bare. One of the biggest advantages of these uncoated capillary columns is their relatively low replacement costs.

Typical column dimensions are 25, 50, 75 or 100 μm internal diameters with lengths between 30 and 100 cm. The narrower diameter columns afford more efficient heat dissipation than larger bore columns but blockages and sensitivity can become problematic. However, as the column diameter increases, resolution often decreases. This loss in resolution is usually attributed to inefficient heat dissipation and radial thermal gradients within the column. The 50 and 75 μm I. D. columns are probably the most commonly used.

Many of the properties noted for surface silanols on silica-based chromatographic supports (e.g. protein adsorption, tailing of amines and hydrolytic instability) are relevant for fused silica surfaces as well. A variety of methods such as dynamic coating [53] or covalent linking of different groups on the silica surface has been developed [41] not only to address the electroosmotic flow issues mentioned previously but also to confer additional desired properties to the silica surface.

Capillary columns may also be filled with a gel matrix for biopolymer separations by sieving. While polyacrylamide gels, generated *in situ* [54] have been used, it is more common to use entangled polymer network solutions which can be readily replaced [55].

Electrochromatography is another variant of electrophoretic methods. It employs alternative separation platforms such as wall-modified capillaries [56], packed capillaries [57] or monolithic media [58,59] to introduce a true stationary phase. In these applications, the applied electric field serves primarily as a means of moving mobile phase and neutral analytes through the column although charged analytes may also be separated on the basis of charge to hydrodynamic radius. While wall-coated capillaries are used extensively in gas chromatography, the relatively low diffusion coefficients of analytes in liquid solution and the overall hydrolytic instability of siloxane bonds reduce the overall effectiveness of this platform for capillary electrophoresis. Incorporation of packing or a monolithic [60,61] chromatographic bed tends to be more common than wall-bonded stationary phases and increases the amount of sample that can be loaded onto the capillary column.

8.4 SAMPLE INJECTION

In contrast to most chromatographic methods, sample introduction in capillary electrophoresis does not use syringes. Instead, sample introduction is accomplished by placing the inlet end of the column in a sample vial and briefly either applying a finite pressure (hydrodynamic) or voltage (electrokinetic) across the column [62]. Hydrodynamic injection can be implemented either by raising the sample vial a specified height above the outlet vial or by applying a finite pressure on the sample vial or a finite vacuum on the outlet vial for a set period of time. Hydrodynamic injection requires additional instrumental hardware (e.g. vacuum, pressure supply or mechanical lift) and cannot be used with gel-filled capillaries but there is no analyte discrimination. Electrokinetic injection involves the brief application of an applied voltage which is generally lower than the run voltage. Electrokinetic injection requires no additional hardware, can be used with gel-filled capillaries and can be used for selective analyte enrichment.

While not intended as an exhaustive discussion of the impact of the sample matrix on peak shape, it should be noted that the sample matrix can have a more profound influence on sample peak shape than what is commonly encountered in HPLC. For instance, large differences in ionic strength between the sample matrix and the background electrolyte can actually induce hydrodynamic flow within the column [50] which can broaden sample peaks. Sample stacking [63], a method for enhancing capillary electrophoresis sensitivity, relies on using a sample matrix with a different ionic strength than the background electrolyte. For example, in the case of charged analytes, the sample matrix has a lower ionic strength than the background electrolyte. At the application of the run voltage, the charged analytes are accelerated through the low ionic strength sample matrix and focused at the boundary between the high and low ionic strength zones.

8.5 DETECTION

The most common type of detection used in capillary electrophoresis is on-column UV detection, likely arising from its ease of implementation, familiarity to HPLC users and versatility. Typically, a small detection window is created on the column by literally burning off the polyimide exterior coating with a lit match and washing off the resultant surface residue. For columns with interior surface modification, hot nitric acid washes are used to remove the polyimide column exterior coating. The region of the fused silica capillary tubing devoid of polyimide coating is extremely fragile and must be handled with care.

In contrast to high performance liquid chromatography where UV detection is commonly performed at 254 nm, capillary electrophoretic UV detection is routinely accomplished at 214 nm. In most on-column UV detection, the path length is the diameter of the capillary. UV absorption, A, can be described by the familiar Beer's Law relationship $A = \varepsilon bc$, where ε is the molar absorptivity, c is the concentration and b is the pathlength. Thus, using a wider diameter capillary increases the pathlength and hence sensitivity. However, as the path length increases, the contribution of any background absorbance by the buffer also increases and reduced efficiency in heat dissipation can adversely impact resolution. Thus, a compromise must be reached between the column efficiency of narrower columns and the higher sensitivity of the larger bore columns. Alternatively, columns have been used incorporating a so-called "Z-configuration" or a "bubble" which increases the pathlength [64]. However, the incorporation of these special detection cells reduces overall flexibility and their use is not commonplace.

An interesting difference between UV detection in HPLC and capillary electrophoresis is the impact of migration time on peak area. In HPLC, when the analytes exit the column, their velocity through the detector is dictated by the flow rate of the mobile phase. However, in capillary electrophoresis, when analytes have different migration times, their migration times through the detector are also different. Analytes with later migration times have slower transit through the detector with more opportunity for absorbance. Hence, for racemates migrating in the presence of a chiral additive, the second peak is larger. A correction factor incorporating the ratio of the migration times is sometimes used to correct for this effect.

One feature common to the use of chiral mobile phase additives in liquid chromatography and chiral additives in capillary electrophoresis is the challenge of detection in the presence of the chiral additive. While a number of UV-transparent chiral selectors are commonly used, an alternative strategy devised for UV detection in the presence of UV-absorbing chiral selectors is illustrated in Fig. 8.4 [65,66].

In this approach, the capillary is filled or partially filled with background electrolyte containing a chiral selector for which the electrophoretic mobility is directed towards the inlet reservoir. The outlet reservoir contains background electrolyte with no chiral selector. As the chiral selector moves towards the inlet end of the column, the enantiomers are separated while migrating through the zone containing the chiral selector. Once the separated enantiomers exit the zone containing the chiral selector,

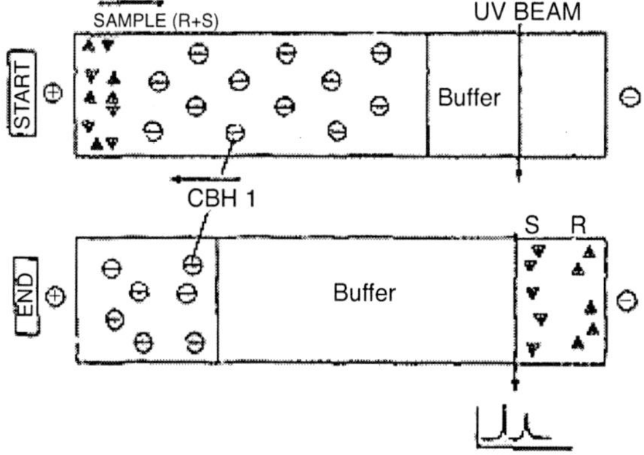

Fig. 8.4. Separation of enantiomers by capillary electrophoresis in free solution. Experimental conditions selected so that the chiral additive migrates in a direction opposite to that of the enantiomers, towards the injection end of the column. The UV-absorbing additive does not interfere with the UV detection of the enantiomers [65].

they have identical mobilities and remain separated as they pass through the detector window in the absence of chiral selector.

For analytes which lack a UV chromophore, a UV-absorbing additive or probe is added to the background electrolyte [67]. Successful indirect detection requires careful consideration of the relative mobilities of the probe and analyte as well as judicious choice of background electrolyte [68].

The second most common type of detection is laser-induced fluorescence which typically relies on the presence of a fluorophore on the analyte or the introduction of a fluorophore on the analyte through derivatization [69]. Nonphotometric-based detection alternatives include electrochemical and mass spectrometry. While detection based on voltammetry [70] and conductivity [71] have been reported, most electrochemical detection in capillary electrophoresis is accomplished using amperometric detection [72]. Amperometric detection, which requires electroactive analytes, is very sensitive ($\sim 10^{-7}$M) [73] and has a wide linear range. With the proper precautions, it offers a response that can be virtually independent of the particular electroactive species being detected for comparable electron processes. Most capillary electrophoretic interfacing to mass spectrometry is accomplished through electrospray ionization [74], using a liquid sheath interface [75] and requires volatile buffers.

8.6 CHIRAL SEPARATION THEORY IN CAPILLARY ELECTROPHORESIS

Virtually all chiral separations in capillary electrophoresis are accomplished by electrokinetic chromatography and will be our focus here. Electrokinetic chromatography is a variant of capillary electrophoresis which employs a background electrolyte containing an additive that acts as a *pseudo*stationary phase. In the case of enantiomeric separations, this additive is chiral and should be enantiomerically pure. Interactions between the chiral selector and a pair of enantiomers lead to the formation of two transient diastereomeric complexes. The conditions required for chiral separation, first derived by Wren and Rowe [28], can be written as

$$\Delta\mu_{1,2} = \frac{(\mu_{\mathrm{f}} - \mu_{\mathrm{c}})(K_2 - K_1)[\mathrm{CA}]}{(1 + K_1[\mathrm{CA}])(1 + K_2[\mathrm{CA}])}, \tag{11}$$

where $\Delta\mu_{1,2}$ is the difference in mobility of the two enantiomers; μ_{f} is the mobility of the free analyte and μ_{c} is the mobility of the complex; K_1 and K_2 are the association constants between the individual enantiomers and the chiral additive (CA). It should be noted that this derivation assumes that the mobilities of the two transient diastereomeric complexes formed by the enantiomers with the chiral additive are the same which, while likely a good first approximation, is not necessarily true [76].

According to this equation, three conditions must be satisfied to achieve chiral separation. First, there must be a chiral selector. Second, there must be a difference in the mobility of the free enantiomer and the transient diastereomeric complex. Thus, a neutral chiral selector cannot chirally resolve a pair of neutral enantiomers. Finally, there must be a difference in the association constants for the two enantiomers.

While it is common to encounter qualitative comparisons in overall effectiveness of chiral selectors in the literature, it should be noted that these comparisons are typically based on somewhat arbitrary experimental conditions derived in the absence of association constant information. However, Wren and Rowe predicted [28] that there was an optimum concentration of chiral additive given by

$$[\mathrm{CA}]_{\mathrm{optimum}} = \frac{1}{\sqrt{K_1 K_2}}. \tag{12}$$

This can be rationalized if one considers the measured mobility of the enantiomers in the presence of chiral additive as

$$\mu_{\mathrm{meas}} = f\mu_{\mathrm{c}} + (1 - f)\mu_{\mathrm{f}}, \tag{13}$$

where f represents the fraction of time the enantiomer spends in the complex and $(1 - f)$ represents the fraction of time the enantiomer spends in the uncomplexed state. Too little chiral additive and the apparent mobilities of both enantiomers approaches that of the free analyte. Too much chiral additive and the apparent mobilities of both enantiomers approach that of the complex. This can be seen in Fig. 8.5.

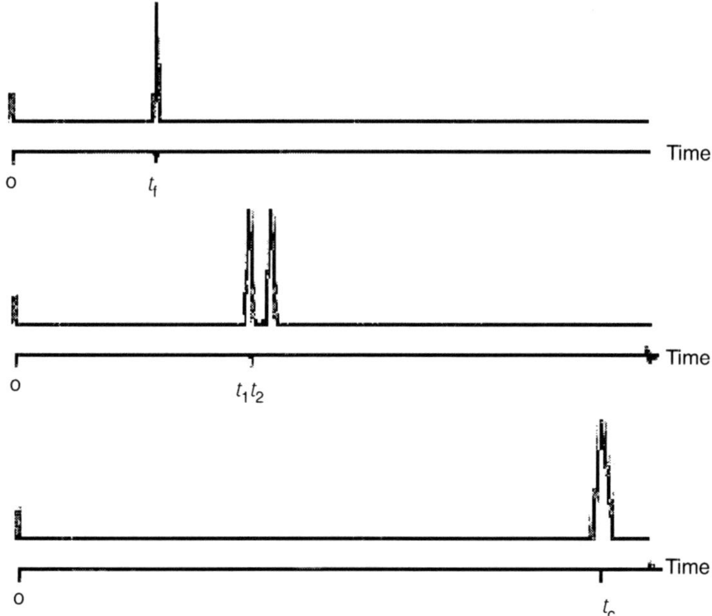

Fig. 8.5. Reconstructed electropherograms showing impact of increasing chiral additive concentration on the chiral resolution and migration time of an enantiomeric pair that is totally free (t_f), enantiomerically resolved (t_1,t_2) or totally complexed (t_c) for an enantiomeric pair that migrates before the chiral selector. Adapted from Ref. [77].

The interval between the migration time of the free analyte and the migration time of the fully complexed analyte is sometimes referred to as the "separation window" [78]. The bigger the difference between the electrophoretic mobility of the chiral selector and the uncomplexed enantiomer, the larger the potential separation window is.

8.6.1 Chiral separation strategies: neutral enantiomeric pairs

One of the earliest examples of a chiral separation of neutral analytes by capillary electrophoresis used a combination of a neutral chiral selector and an achiral ionic *pseudo*stationary phase [77]. In this report, chiral separation of neutral enantiomeric pairs (Fig. 8.6) was achieved using a neutral cyclodextrin with sodium dodecyl sulfate micelles (micellar electrokinetic chromatography (MEKC)) as an auxiliary achiral agent.

In this case, both enantiomers nonstereospecifically partitioned into or associated with the anionic micellar phase which migrated after the electroosmotic flow. When the enantiomers were free or associated with the micelle, they had identical mobilities. The extent to which they associated with the micelle was predicated on their stereospecific association with the cyclodextrin. It should be noted that this case also provides an excellent example of how complicated equilibria can become in electrophoresis applications. Examination of the various equilibria involved in this process, illustrated in Fig. 8.7, suggest that in this case, the chiral separation may not entirely require the complex between the cyclodextrin and the analyte to be neutral. Surfactants are known to form ternary complexes with small molecules and cyclodextrins [79]. Thus, co-inclusion of a surfactant molecule with the analyte could confer a negative charge to the complex.

Another example of electrolyte-neutral selector complexation occurs between borate and diols [80]. Indeed, a very clever variant of this approach was used by El Rassi et al. [81]. They exploited borate–diol complexation to obtain chiral separations using

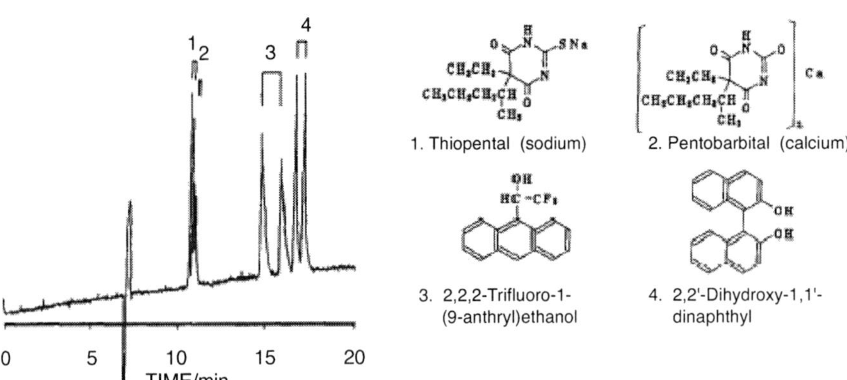

Fig. 8.6. Chiral separation of four enantiomers. Conditions: buffer, 0.02 M phosphate-borate buffer of pH 9.0 containing 0.05 M SDS and 30 mM γ-CD; capillary: 650 mm × 0.05 mm I.D. (effective length 500 mm); applied voltage, 20 kV; detection, 220 nm: temperature, ambient. Adapted from Ref. [77].

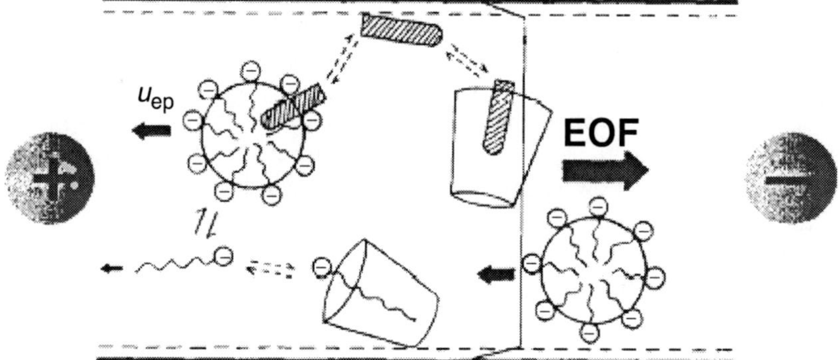

Fig. 8.7. Schematic illustrating the partitioning of a neutral analyte into a micelle and into a cyclodextrin. Adapted from Ref. [77].

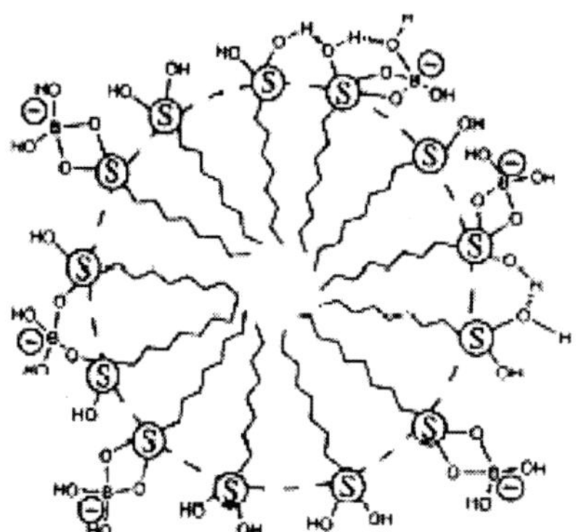

Fig. 8.8. Idealized structure of alkylglycoside-borate micelle showing borate complexation and hydrogen bonding on the outer surface of the micelle. Circled-S stands for the sugar head group of the surfactant possessing a cyclic sugar head group [82].

nonionic glycosidic-based surfactants that migrated anodically in the presence of borate buffers (Fig. 8.8.) [82]. The stereospecific interactions occurred between the chiral analytes and the glycosidic headgroups of the surfactant molecules.

A more common approach to chirally resolving neutral analytes is to use a charged chiral selector. A variety of selectors has proven useful for neutral analytes. Examples include macrocyclic antibiotics above or below their isoelectric points or ionizable functionalized cyclodextrins (e.g. sulfobutyl [83], sulfated [47] or aminated [84,85] cyclodextrins). In some cases, a combination of oppositely charged additives such as mixtures of cyclodextrin derivatives has proven efficacious for the separation of uncharged enantiomers [86].

The use of charged chiral selectors has evolved into two modes of operation. In the original or conventional mode, sometimes referred to as "normal polarity", the charged chiral selector may either migrate ahead of the electroosmotic flow marker or after but the analyte ultimately migrates through the detector window primarily under the influence of the electroosmotic flow. The separation window occurs between the migration time of the neutral analyte and the chiral selector.

The alternative mode, sometimes referred to as the carrier mode [87], has two key features—the electroosmotic flow is substantially suppressed and the electrophoretic mobility of the chiral selector is directed toward the detector end of the capillary. Hence, the primary mechanism for transport of the neutral analyte to the detector is through association with the chiral carrier. Suppression of the electroosmotic flow may be accomplished by using a carrier background electrolyte at relatively low pH (e.g. ~ pH 3 or lower) and/or coating the capillary with nonionic polymer or surfactant [47]. In the case of an anionic carrier, the polarity of the capillary electrophoresis system is reversed (i.e. anodic detection) relative to what is normally used, so this approach is sometimes referred to as *reversed polarity* [47,88]. Under these conditions, analytes which have the strongest affinity for the chiral selector migrate earliest in the electropherogram while analytes having lower affinity migrate later (Fig. 8.9.).

Chiral separations using carrier mode have considerable flexibility and offer a number of optimization strategies including reduction in chiral selector concentration or addition of auxiliary agents which can compete for the chiral selector. Because the analytes

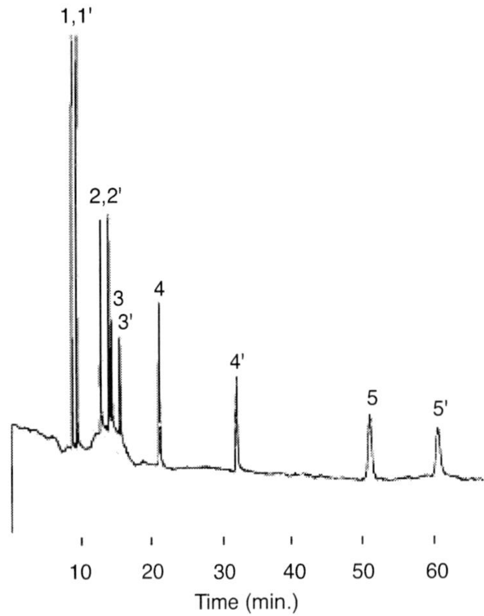

Fig. 8.9. Electropherogram showing the enantioseparations of aminoglutethimide (1,1′), 5-(4-methylphenyl)-5-phenylhydantoin (2,2′), 5-(4-hydroxyphenyl)-5-phenylhydantoin (3,3′), 5-cyclobutyl-5-phenylhydantoin (4,4′), and phensuximide (5,5′).Capillary electrophoresis (CE) conditions: 2% sulfated cyclodextrin; 10 mM phosphate buffer, pH 3.8; 15 kV. Under these conditions, aminoglutethimide is protonated whereas all the other analytes are neutral [47].

do not migrate to the detector in the absence of the chiral selector, the separation window, in essence, increases in the presence of competition from the auxiliary agents.

8.6.2 Chiral separation strategies: charged enantiomeric pairs

One of the earliest approaches to chirally resolving enantiomeric ions was to use a neutral chiral selector [89]. While neutral chiral selectors migrate with the electro-osmotic flow, the charged enantiomers may migrate either before or after the electro-osmotic flow. Generally, enantiomer migration before the neutral chiral selector is the most desirable approach because it affords the shortest analysis times.

Charged chiral selectors can be used for chiral separations of oppositely as well as similarly charged enantiomers. While both normal mode and carrier mode have been used, carrier mode has been particularly effective for the chiral separation of charged analytes. For instance, in the case of an anionic chiral carrier interacting with cationic enantiomers, the electrophoretic mobility of the free enantiomers is directed toward the inlet end of the capillary while the electrophoretic mobility of the chiral selector is directed toward the outlet. Disruption of the association between the anionic chiral carrier and the cationic enantiomers sets up a true counter-current process; effectively, the analytes see a "longer" column, thereby amplifying chiral recognition. An example of this is provided by the separation of cationic enantiomers using sulfated cyclodextrin in carrier mode. If the primary interaction between the cationic enantiomeric pair and sulfated cyclodextrin (SCD) is through inclusion complexation, disruption of this interaction with the addition of an organic solvent (e.g. methanol) can lead to increased separation. This enhancement in separation can be clearly seen in the case of the basic drug terbutaline [90], and arises because the solvent competes with the analytes for the sulfated cyclodextrin cavity (Fig. 8.10).

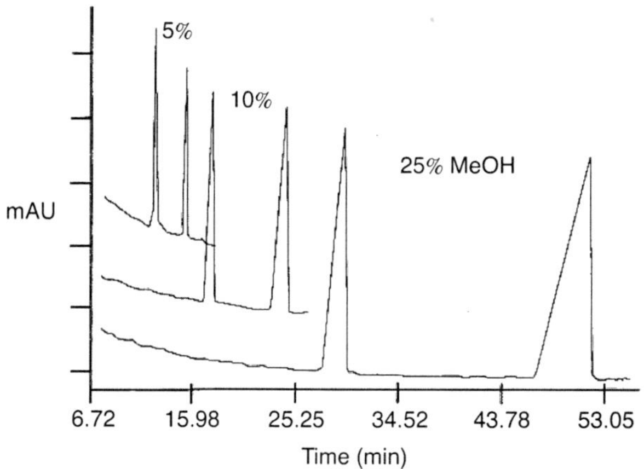

Fig. 8.10. Overlaid electropherograms of terbutaline chiral separation. SCD concentration was held constant at 2% in 25 mM phosphate buffer (pH 3): methanol concentrations 5, 10 and 25% (w/w); coated capillary; −10 kV; 20°C [90].

8.7 CHIRAL SELECTORS IN CAPILLARY ELECTROPHORESIS

Historically, many successful chiral selectors in high performance liquid chromatography were previously investigated as chiral mobile phase additives in thin layer chromatography [91,92] and high performance liquid chromatography [93]; thus, their introduction as chiral additives in capillary electrophoresis formed a natural extension. As a result, early practitioners in chiral separations by capillary electrophoresis benefited tremendously from these advances in chiral chromatography and rapid progress was made. Interestingly, since the advent of chiral separations by capillary electrophoresis, a number of new chiral stationary phases for liquid chromatography have evolved based on results obtained using unique chiral selectors in capillary electrophoresis, most notably the macrocyclic antibiotic phases [94] and the cinchona alkaloid-based phases [95–97].

As noted in the theoretical background for chiral separations by capillary electrophoresis, chiral separations require the formation of transient diastereomeric complexes between the chiral selector and the enantiomers to be separated. This transient complex is often the result of a primary interaction site that dominates the interaction between the enantiomer and the selector. This dominant interaction site could be inclusion complexation (e.g. cyclodextrin, some polysaccharides, crown ethers, macrocyclic antibiotics), covalent (e.g. ligand exchange) or electrostatic (e.g. ion pairing). This transient complex also manifests a number of secondary orientational interactions (e.g. hydrogen bonding, steric, dipole stacking or π–π interactions) that ultimately give rise to chiral recognition. The primary classes of chiral selectors are listed in Table 8.1. This list is by no means all inclusive and the following discussion of each class is merely intended to provide some general guidelines.

8.7.1 Chiral selectors based on cyclodextrins

By far and away, the most commonly used chiral selectors for capillary electrophoresis are the cyclodextrins (CD) [13]. These are macrocyclic compounds comprised of

Table 8.1 Primary classes of chiral additives used in capillary electrophoresis

Selector	Analyte structural features	Typical additive concentrations
Cyclodextrins (functionalized)	Hydrophobic and hydrogen bonding groups near stereogenic center	1–5%
Polysaccharide	Hydrophobic and hydrogen bonding groups near stereogenic center	1–10%
Crown ether	Primary amine near stereogenic center	5–10 mM
Ligand exchange	Two heteroatoms near stereogenic center	1–40 mM Cu^{2+} 1–80 mM CA
Chiral surfactants	Hydrophobic group near stereogenic center	5–160 mM
Protein	Hydrophobic and hydrogen bonding groups near stereogenic center	0.1–1 mM
Macrocyclic antibiotic	Hydrophobic and hydrogen bonding groups near stereogenic center	2–20 mM

Table 8.2 Summary of the properties of the most common native cyclodextrins

CD	No. of glucose units	MW	Cavity diameter (nm)	Water solubility (g/100 ml)
α	6	972	0.57	150 mM (14.5)
β	7	1135	0.78	17 mM (1.85)[a]
γ	8	1297	0.95	180 mM (23.2)

[a]The β-cyclodextrin has an order of magnitude lower solubility than either α- or γ-cyclodextrin. This reduced solubility is attributed to the more favorable internal hydrogen bonding through the hydroxyls lining the wider rim of the β-cyclodextrin than for α- or γ-cyclodextrin.

D-glucose, connected through α(1→4) linkages. Table 8.2 summarizes some of the properties of the most common native cyclodextrins. Their record as successful chiral selectors in gas and liquid chromatography coupled with their UV transparency has no doubt contributed to their success as chiral selectors in capillary electrophoresis.

For the most part, chiral recognition using cyclodextrin additives seems to rely on the formation of an inclusion complex as the primary interaction. Thus, as in chiral separations using cyclodextrin stationary phases [98], there needs to be a good "fit" between the cyclodextrin and the analytes [99]. For instance, analytes which favor inclusion with the β-cyclodextrin contain a hydrophobic group about the size of a naphthyl ring or a *t*-butyl group [100]. Chiral recognition also requires secondary interactions with the 2° hydroxyls which constitute the wider mouth of the cyclodextrin. Thus, the included portion of the molecule and hydrogen bonding groups must be in close proximity to the stereogenic center. Additional structural rigidity in the vicinity of the chiral center also facilitates chiral recognition.

To increase the applicability of neutral cyclodextrins to chiral separations by capillary electrophoresis, a number of acetylated [101], alkylated [30,102] or hydroxyalkylated [103,104] cyclodextrins have also been introduced. The substituents increase the depth of the hydrophobic pocket and increase steric bulk near the chiral mouth of the cyclodextrin. In the case of the hydroxypropyl cyclodextrin [105], the substituent potentially introduces another stereogenic center positioned slightly above the mouth of the cyclodextrin and has proven useful for analytes in which the included portion of the molecule is not directly adjacent to the hydrogen bonding moieties.

While the earliest work on cyclodextrins as chiral selectors involved the native cyclodextrins, the charged cyclodextrins [106] have dominated capillary electrophoretic applications since 1998 [109]. A variety of ionizable cyclodextrins (e.g. phosphated [108], carboxymethylated [109], aminated [110], and sulfobutylated [111]) have been employed as chiral selectors in capillary electrophoresis, although not all are commercially available. However, the sulfated cyclodextrins have thus far demonstrated the widest application range [112]. The high solubility and the pH-independent ionization of the sulfated cyclodextrins allow a large concentration and pH window for separation optimization which no doubt contributes to its wide application range. In addition, the presence of the sulfate esters introduces additional primary electrostatic interaction sites [113,114] which facilitate chiral separations of analytes that may be either too big or too small for effective inclusion complexation.

References pp. 267–275

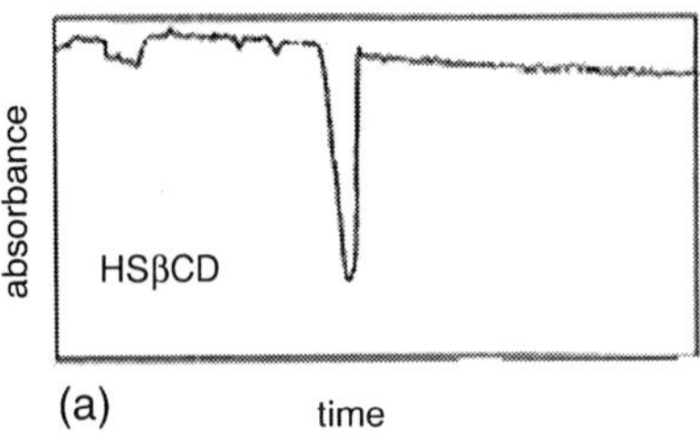

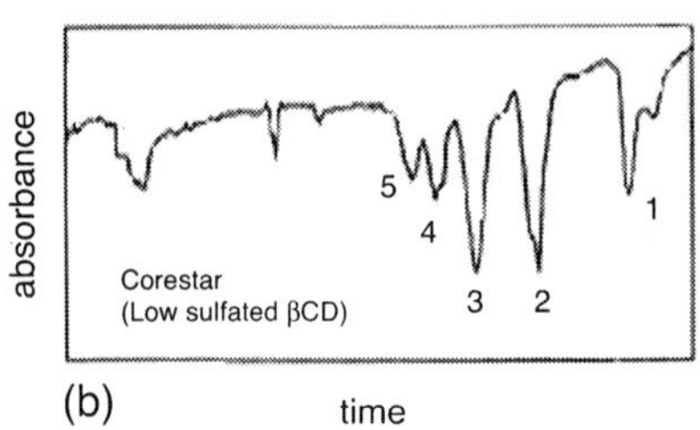

Fig. 8.11. Capillary electrophoretic analysis using UV indirect detection of a two commercially available derivatized cyclodextrins. (a) highly sulfated, HS-β-CD and (b) low sulfated-β-CD (Cerestar). Adapted from Ref. [115].

It is important to note that a significant number of the commercially available derivatized cyclodextrins are randomly substituted and hence, are complex mixtures of homologues and isomers because of the large number of reactive sites on the cyclodextrin (e.g. 21 OHs on the β-CD). The heterogeneity tends to be more of an issue for the lower degrees of substitution than for the higher degrees of substitution. For instance, Fig. 8.11 illustrates the capillary electrophoretic analysis of a couple of commercially available anionic cyclodextrins [115].

It may be argued that substitution on the randomly substituted cyclodextrins is not entirely random because of differences in the reactivity of the hydroxyls [116]. Nevertheless, as a result of the heterogeneity, the randomly substituted cyclodextrin concentrations in the background electrolyte are usually reported as weight percentages instead of molar concentrations. In response to the heterogeneity issue, Vigh et al. have developed families of single isomer cyclodextrin derivatives [117–119]. The randomly substituted cyclodextrins are generally less expensive than the single isomeric forms and often have higher aqueous solubility than either the native or single isomeric functionalized cyclodextrins. However, lot-to-lot reproducibility can be an issue for the randomly substituted cyclodextrins [120], especially for the cyclodextrins with low degrees of substitution.

8.7.2 Chiral selectors based on noncyclic polysaccharides

Because of their UV transparency, a variety of carbohydrates have been successfully used as chiral selectors in capillary electrophoresis. As in the case of randomly substituted cyclodextrins, polysaccharides are typically complex mixtures. Hence, concentrations are generally given in weight percentages (e.g. 1–10%). In general, oligosaccharides need to be at least 7–10 residues long to be effective as chiral selectors although some aminoglycosides comprised of 3–4 residues have found limited use [121]. The carbohydrate-based chiral selectors can be generally subdivided into neutral and charged polysaccharides.

The neutral noncyclic polysaccharides demonstrating the widest chiral application range are the dextrins [122] and maltodextrins [123,124]. Both are complex mixtures of linear polymers of D-glucose, bonded through $\alpha(1\rightarrow4)$ linkages. The maltodextrins, typically 3–10 residues long, and the dextrins, which are higher molecular weight (MW) oligomers, are hydrolysis products of amylose, a constituent of starch. The α(1→4) linkages impose a helical conformation on the chain which allows for inclusion complexation behavior analogous to cyclodextrins. Indeed, in the case of the maltodextrins, it was noted that chiral recognition required at least 7 glucose residues [125] which coincides with the number of residues in β-cyclodextrin. The opportunity for inclusion complexation with these selectors despite the absence of a rigid cavity might reduce some structural constraints of the analyte for successful chiral recognition imposed by cyclodextrin chiral selectors; however, it may also undermine the effectiveness of the primary inclusion interaction thereby limiting their enantioselectivity.

Ionizable polysaccharides offer a number of advantages including increased solubility and an increased separation window relative to the neutral polysaccharides. Of the naturally occurring ionic noncyclic polysaccharides, the anionic polysaccharides, incorporating sulfates and/or carboxylates, seem to be the most effective, particularly for basic drugs. Incorporation of the sulfate groups enables these chiral selectors to be used across a broad pH range. The most thoroughly investigated ionic noncyclic polysaccharides have been heparin [126], chondroitin sulfate [127] and dermatan sulfate [128]. The basic subunit of heparin is either a di-, tetra-, or hexasaccharide composed of uronic acid and glucosamine residues, linked through α(1→4) linkages. As a consequence of these linkages, heparin is thought to assume a helical conformation in solution [129] with inclusion as the primary interaction [126]. Dermatan sulfate, also known as chondroitin sulfate B, is particularly effective for basic compounds with hydroxyl-substituted aromatic rings [128,130] thus implicating a role for hydrogen bonding in the primary interaction with non-helical polysaccharides.

In addition, semi-synthetic ionizable polysaccharides produced by the introduction of sulfates (e.g. dextran sulfate [131], pentosan polysulfate [132]) or carboxymethyl groups (e.g. carboxymethyl cellulose, amylose or dextran [133]) have broadened the chiral recognition capability of noncyclic polysaccharides. Again, these anionic additives seem to be the most effective for the chiral separation of basic drugs.

8.7.3 Chiral selectors based on crown ethers

Chiral separations using chiral crown ethers in capillary electrophoresis usually employ (+)-(18-crown-6)-2,3,11,12-tetracarboxylic acid or its optical antipode as the chiral selector in the background electrolyte. As in the case of the chiral crown ether stationary phases for liquid chromatography [134,135], the separation mechanism is based on the primary formation of an inclusion complex between a protonated primary amine on the chiral analyte and the chiral crown ether (Fig. 8.12). Secondary interactions between the substituents on the stereogenic center of the molecule and the chiral crown ether contribute to the chiral recognition [136]. Thus, chiral recognition with this selector requires that the stereogenic center be in close proximity to the primary amine. Bulky substituents on the asymmetric center seem to enhance chiral recognition.

Chiral crown ethers offer a number of advantages including a well-understood chiral recognition mechanism and the availability of chiral selectors of either configuration.

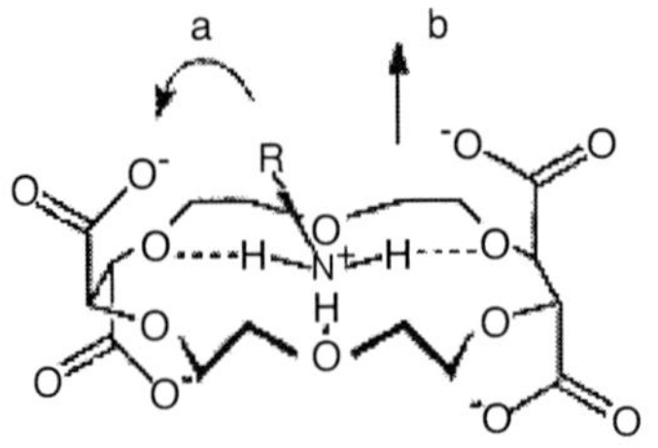

a. specific interaction

b. sterical or electrostatic repulsion

Fig. 8.12. Structure of the inclusion complex formed between a chiral crown ether selector and a primary amine [137].

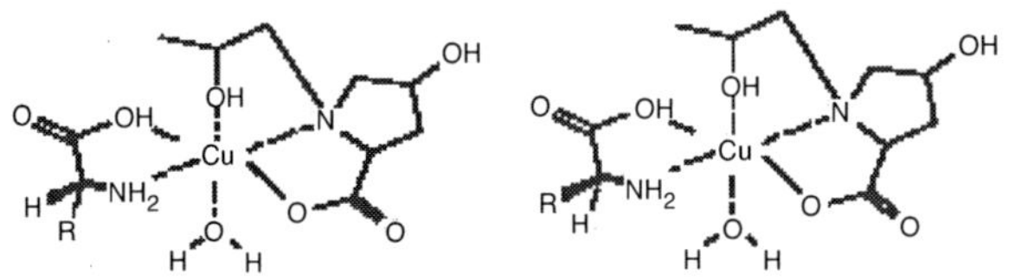

Fig. 8.13. Structures of the two diastereomeric complexes formed between the chiral selector, *N*-(2-hydroxypropyl)-L-4-hydroxyproline and D or L amino acids [139].

It should be noted that care must be taken in the selection of other background electrolyte constituents because alkali or alkaline-earth ions can compete for the polyether cavity, thereby diminishing overall chiral recognition. The pK_as of the four carboxylic acids on (18-crown-6)-2,3,11,12-tetracarboxylic acid are 2.13, 2.84, 4.29 and 4.88 [138]. Thus, pH is also an important experimental parameter.

8.7.4 Chiral selectors based on ligand exchange

Chiral separations by ligand exchange require the presence of a metal ion (e.g. Cu^{2+}) in the background electrolyte. They also require that both the chiral selector and chiral analytes have at least two hetero atoms (e.g. S, N or O) in fairly close proximity to their respective chiral centers and that can donate electron pairs to the metal ion to form a chelate. The formation of transient diastereomeric ternary mixed metal complex and the differences in their relative stability constants and/or hydrodynamic radii drive the separation. An example of two such diastereomeric complexes is illustrated in Fig. 8.13.

Amino acids, α-amino alcohols and α-hydroxy acids are all examples of analytes resolvable by a chiral ligand exchange mechanism. Indeed, the earliest report of a chiral separation in capillary electrophoresis was the enantioresolution of dansylated amino acids based on ligand exchange interactions using Cu^{2+} and histidine as the chiral selector [140].

In chiral ligand exchange electrokinetic chromatography, the chiral selector metal ion relative concentrations are either 1:1 or 2:1. In the case of 2:1 relative concentrations, the actual separation mechanism likely relies on competitive binding between the chiral analyte and one of the chiral selector ligands on the metal center to form the diastereomeric ternary mixed metal complexes. Often, the copper complexes are colored and the presence of the complexes in the background electrolyte results in a fairly high background signal. Given the reliance of the separation mechanism on chelation, it is not too surprising that the separations are also highly pH dependent. As in the case of chiral crown ether selectors, chiral ligand exchange in capillary electrophoresis has some significant advantages. First, the straightforward interaction mechanism typically allows for the prediction of likely separations based on the analyte structure. The well-understood interaction mechanism also allows for assignment of configuration in the absence of standards if enantiomerically pure standards are available for compounds related to the chiral analytes. In addition, the chiral selectors of either configuration are often commercially available.

8.7.5 Chiral selectors based on chiral surfactants

A variety of chiral surfactants developed from amino acids [141,142] and small sugars [19,82] have been used as micellar *pseudo*phases in capillary electrophoresis. As in achiral micellar-based separations, the micellar phase is characterized by a critical micelle concentration as well as an aggregation number, the average number of molecules in the aggregate. Examples are listed in Table 8.3.

Nonstereospecific hydrophobic interactions usually drive partitioning of the analytes into the micellar phase. Chiral recognition arises through secondary stereospecific interactions with the chiral polar head groups. The addition of organic modifiers (e.g. methanol) is sometimes required for highly hydrophobic analytes although the addition of the organic modifier can perturb micellization (e.g. shift the critical micelle concentration or the aggregation number). Warner et al. [145,146] developed "polymerized" chiral micelles derived from amino acids which tolerate a broad range of organic modifier concentrations and offer some advantages in capillary electrophoresis/mass spectrometry coupling [147].

Another class of chiral selectors based on solution aggregation is the bile salts [148]. Bile salts have a steroidal backbone with a hydrophilic side and a hydrophobic side. Their aggregates usually incorporate fewer molecules [149] than typical alkyl-based surfactants and aggregation is thought to occur as a two-step process [150]. As can be seen in Table 8.3, bile salts form very different kinds of aggregates than the more common spherical micelle. Indeed, they are known to form mixed micelles with cholesterol. Thus, it is not too surprising that analytes best resolved by micelles are somewhat plate-like (e.g. 1,1′-binaphthyl-2,2′-diyl hydrogen phosphate) [151].

Table 8.3 Structures and relevant physical properties of common chiral surfactants

Structure name of surfactant	CMC (mM)	Aggregation number	Schematic of micellar structure
$n = 7$ nonyl-β-D-glucopyranoside	25	~100–110 [143]	
octyl-β-D-maltopyranoside	24.3		Adapted from Ref. [82]
sodium cholate	13–15	7–12	Adapted from Ref. [144]

8.7.6 Chiral selectors based on proteins

Because many of the differences in enantiomeric bioactivity likely arise from stereospecific interaction with proteins, it is not too surprising that proteins can also be used as chiral selectors in capillary electrophoresis. Indeed, capillary electrophoretic investigations may provide considerable insight into chiral drug–protein interactions not accessible with HPLC data because of the absence of confounding contributions from linkers or the underlying solid support encountered in HPLC studies. In addition, drug–drug interactions may also be conveniently probed in well-designed capillary electrophoretic studies [152].

Typical protein concentrations are in the micromolar range. While the utility of proteins as chiral selectors is somewhat limited because of their interference in UV detection in capillary electrophoresis, partial fill methods have been used to circumvent this issue [153,154]. When using proteins for chiral selectors, the inner walls of the capillary are often coated (e.g. polyacrylamide [65,155] or polyvinyl alcohol [156]) to improve separation performance by reducing protein affinity for the capillary walls. The presence of ionizable moieties on the protein chiral selectors dictates that the magnitude as well as the direction of their intrinsic electrophoretic mobility is highly pH dependent.

A variety of proteins have been used as chiral selectors including various serum albumins [155], α_1-acid glycoprotein [157] and cellulase [65]. For some proteins, the drug-binding sites are well known [158] and can provide some guidance in the protein selection for a particular chiral separation. For instance, human serum albumin has a warfarin binding site and a ketoprofen binding site [152]. Because lipophilicity [158] is often the primary interaction with protein selectors, small amounts of organic modifiers are sometimes added to the background electrolyte to decrease analyte affinity for the protein. For some analytes in the presence of glycoproteins, chiral recognition has been shown to emanate primarily from the protein domain [157] while for other systems, the carbohydrate residues augment the separations achieved by the protein domain [159].

8.7.7 Chiral selectors based on macrocyclic antibiotics

The macrocyclic antibiotic class of chiral selectors is one of the most recently introduced and probably the most versatile of all of the immobilized chiral selectors in chiral HPLC. Although the strong UV absorption of this class of chiral selectors has somewhat inhibited their wide application in capillary electrophoresis (Fig. 8.14) [160], their excellent enantioselectivity allows for chiral separations to be obtained with low concentrations of selector (e.g. 1–5 mM for vancomycin). Alternatively, indirect detection methods may be used [161].

Examples of typical glycopeptide macrocyclic antibiotic chiral selectors are illustrated in Fig. 8.15. As can be seen in the figure, the glycopeptide macrocyclic antibiotics are comprised of peptide bonds, aromatic rings and sugar residues. The ansamycins, another family of macrocyclic antibiotics used as chiral selectors [162], do not incorporate carbohydrate residues. The presence of the peptide bonds in both the glycopeptides and the ansamycins renders these macrocyclic antibiotics somewhat

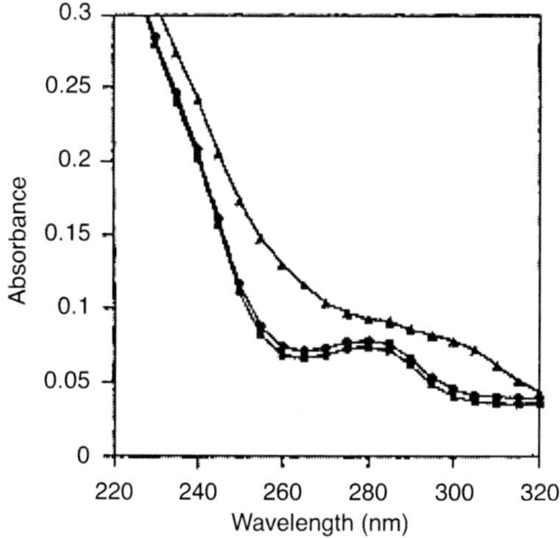

Fig. 8.14. UV spectra of ristocetin A (at a concentration of $7.2.10^{-3}$ mg/ml in 0.1 M phosphate buffer) at pH 4.0 (■). pH 7.0 (●) and pH 9.9 (▲) [160].

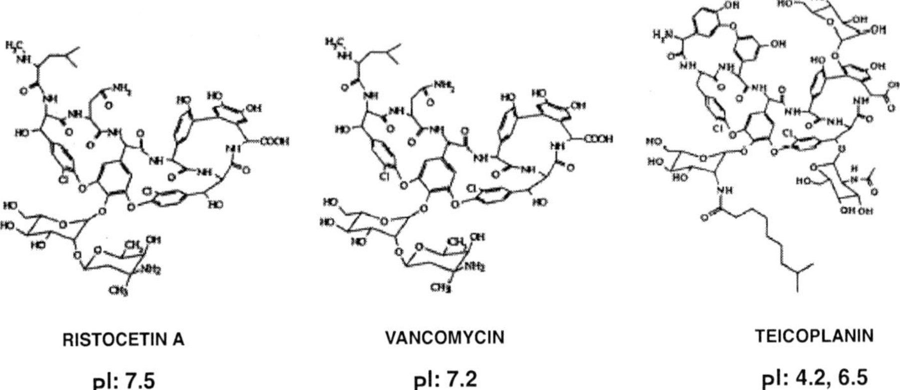

Fig. 8.15. Examples of three common macrocyclic antibiotic chiral selectors. Adapted from Ref. [15].

vulnerable to hydrolysis which is accelerated under high temperatures and basic pH. They incorporate numerous chiral centers as well as sites for electrostatic, dipole–dipole, hydrogen bonding and π-stacking interactions. Chiral centers are distributed throughout the macrocyclic antibiotic framework. Hence, there seems to be more flexibility with regard to analyte structural restrictions than encountered with some of the other chiral selectors (e.g. primary amine near stereogenic center for chiral crown ethers, noted previously). In addition, they provide cyclic moieties which can accommodate inclusion complexation [162].

As in the case of the protein chiral selectors, the inner walls of the capillary are often coated (e.g. hexadimethrine bromide [163]) when using glycopeptide chiral selectors to inhibit selector adsorption on the capillary walls. The presence of ionizable moieties in the macrocyclic antibiotics dictates that the magnitude and direction of their intrinsic electrophoretic mobility is highly dependent upon pH. This is illustrated in Fig. 8.16.

This dependence on pH must be taken into account when developing a separation method.

In the glycopeptide antibiotics, potential chiral recognition sites are found in both the polypeptide rings as well as the pendant sugar residues. Comparisons of chiral recognition obtained with either the native macrocyclic antibiotic or the aglycon basket have shown that for most compounds, the cyclic peptide regions are responsible for chiral recognition. However, examples can also be found in which the amino sugar plays a major but indirect role in chiral recognition [164] arising through sugar-promotion of self-aggregation of the chiral selector rather than through a direct stereospecific interaction. For other members of this class of chiral selectors, such as rifamycin B [165] and avoparcin [166], the tendency to self-aggregate is often parasitic to the chiral separation. Hence, organic modifiers (e.g. 30% 2-propanol) [162] are sometimes added to inhibit this self-association.

In general, macrocyclic antibiotic chiral selectors often offer complimentary chiral selectivity. For instance, the glycopeptide macrocyclic antibiotics tend to be

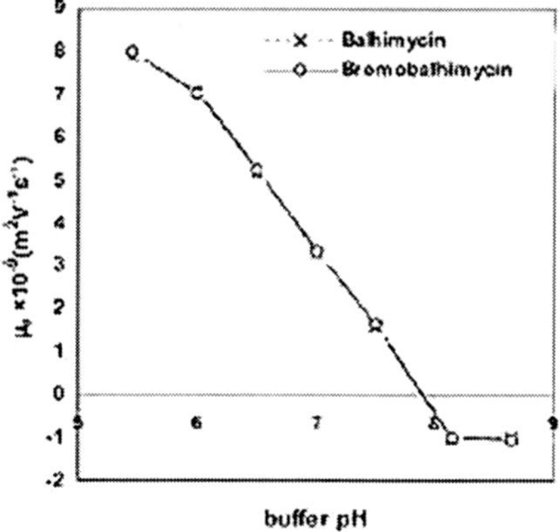

Fig. 8.16. Impact of pH on the electrophoretic mobilities of the glycopeptide macrocyclic antibiotics, balhimycin and bromobalhimycin [163].

most effective for acidic compounds [160,163,167]. In contrast, the ansamycin, rifamycin B, has good general enantioselectivity towards cationic compounds (e.g. amino alcohols) [161] while rifamycin SV is more effective at resolving anionic analytes (e.g. profens) [162].

8.8 CONCLUSIONS

Since its introduction in the early 1980s, capillary electrophoresis has emerged as one of the most effective and versatile analytical methods for chiral separations. While the assorted native and derivatized cyclodextrins remain the most commonly used chiral additives, there is a large variety of alternative chiral additives that have also been reported, thus providing the analyst with a wide range of choices for a particular application. The number of potential interaction sites for most chiral selectors obscures the prediction of likely selectors/conditions and migration order for a particular separation. However, awareness of the requirements for chiral recognition coupled with a basic understanding of capillary electrophoretic processes and familiarity with the vast published empirical data base greatly enhances the probability of success.

REFERENCES

1 O. Vesterberg, History of electrophoretic methods. J. Chromatogr., 480 (1989) 3–19.
2 J.P. Landers, Molecular diagnostics on electrophoretic microchips. Anal. Chem., 75 (2003) 2919–2927.
3 R.-L. Chien. Sample stacking revisited: a personal perspective. Electrophoresis, 24 (2003) 486–497.
4 C.F. Poole, Capillary-electromigration separation techniques, in: The Essence of Chromatography, Elsevier Science B.V., 2003, Amsterdam, The Netherlands, pp. 619–717.

5 P. Camilleri (Ed.), Capillary Electrophoresis: Theory and Practice, 2nd Ed. CRC Press, Boca Raton, FL, 1998.

6 J.P. Landers (Ed.), Handbook of Capillary Electrophoresis, 2nd Ed. CRC Press, Boca Raton, FL, 1998.

7 Y. Tanaka, N. Naruishi, H. Fukuya, J. Sakata, K. Saito and S. Wakida, Simultaneous determination of nitrite, nitrate, thiocyanate and uric acid in human saliva by capillary zone electrophoresis and its application to the study of daily variations. J. Chromatogr. A, 1051 (2004) 193–197.

8 T.M.H. Choy, L. Jia and C.W. Huie, Investigation of the effects of halide ions on indirect UV detection in capillary zone electrophoresis. J. Sep. Sci., 25 (2002) 333–341.

9 J.L. Zabzdyr and S.J. Lillard, New approaches to single-cell analysis by capillary electrophoresis. TRAC, 20 (2001) 467–476.

10 D.W. Armstrong, G. Schulte, J.M. Schneiderheinze and D.J. Westenberg, Separating microbes in the manner of molecules. 1. Capillary electrokinetic approaches. Anal. Chem., 71 (1999) 5465–5469.

11 R.M.C. Sutton, S.R. Gratz and A.M. Stalcup, The use of capillary electrophoresis as a method development tool for classical gel electrophoresis. Analyst, 123 (1998) 1477–1480.

12 A. Guttman, A.S. Cohen, D.N. Heiger and B.L. Karger, Analytical and micropreparative ultrahigh resolution of oligonucleotides by polyacrylamide gel high-performance capillary electrophoresis. Anal. Chem., 62 (1990) 137–141.

13 M. Blanco and I. Valverde, Choice of chiral selector for enantioseparation by capillary electrophoresis. TRAC, 22 (2003) 428–439.

14 B. Chankvetadze, Capillary Electrophoresis in Chiral Analysis, John Wiley & Sons, Ltd., Chichester, UK, 1997

15 C. Desiderio and S. Fanali, Chiral analysis by capillary electrophoresis using antibiotics as chiral selector. J. Chromatogr. A, 807 (1998) 37–56.

16 C.E. Evans and A.M. Stalcup, Comprehensive strategy for chiral separations using sulfated cyclodextrins in capillary electrophoresis. Chirality, 15 (2003) 709–723.

17 M.G. Schmid, N. Grobuschek, O. Lecnik and G. Gübitz, Chiral ligand-exchange capillary electrophoresis. J. Biochem. Biophys. Methods, 48 (2001) 143–154.

18 Y. Tanaka and S. Terabe, Recent advances in enantiomer separations by affinity capillary electrophoresis using proteins and peptides. J. Biochem. Biophys. Methods, 48 (2001) 103–116.

19 Z. El Rassi, Chiral glycosidic surfactants for enantiomeric separation in capillary electrophoresis. J. Chromatogr. A, 875 (2000) 207–233.

20 A.S. Rathore, Joule heating and determination of temperature in capillary electrophoresis and capillary electrochromatography columns. J. Chromatogr. A, 1037 (2004) 431–443.

21 A.M. Stalcup and K.H. Gahm, Quinine as a chiral additive in nonaqueous capillary zone electrophoresis. J. Microcolumn Sep., 8 (1996) 145–150.

22 J.L. Miller, D. Shea and M.G. Khaledi, Separation of acidic solutes by nonaqueous capillary electrophoresis in acetonitrile-based media. Combined effects of deprotonation and heteroconjugation. J. Chromatogr. A, 888 (2000) 251–266.

23 F. Wang and M.G. Khaledi, Non-aqueous capillary electrophoresis chiral separations with sulfated β-cyclodextrin. J. Chromatogr. B, 731 (1999) 187–197.

24 C. Czerwenka, M. Lämmerhofer and W. Lindner, Electrolyte and additive effects on enantiomer separation of peptides by nonaqueous ion-pair capillary electrophoresis using tert-butylcarbamoylquinine as chiral counterion. Electrophoresis, 23 (2002) 1887–1899.

25 L. Geiser, S. Cherkaoui and J.-L. Veuthey, Potential of formamide and N-methylformamide in nonaqueous capillary electrophoresis coupled to electrospray ionization mass spectrometry. Application to the analysis of β-blockers. J. Chromatogr. A, 979 (2002) 389–398.

26 D.J. Skanchy, G.-H. Xie, R.J. Tait, E. Luna, C. Demarest and J.F. Stobaugh, Application of sulfobutylether-β-cyclodextrin with specific degrees of substitution for the enantioseparation of pharmaceutical mixtures by capillary electrophoresis. Electrophoresis, 20 (1999) 2638–2649.

27 E.K.M. Andersson and I. Hägglund, Sample matrix influence on the choice of background electrolyte for the analysis of bases with capillary zone electrophoresis. J. Chromatogr. A, 979 (2002) 11–25.

28 Wren, S.A.C., Rowe, R.C., Theoretical aspects of chiral separation in capillary electrophoresis, I. Initial evaluation of a model. J. Chromatogr., 603 (1992) 235–241.
29 K.L. Rundlett and D.W. Armstrong, Methods for the determination of binding constants by capillary electrophoresis. Electrophoresis, 22 (2001) 1419–1427.
30 A. Amini, N. Merclin, S. Bastami and D. Westerlund, Determination of association constants between enantiomers of orciprenaline and methyl-β-cyclodextrin as chiral selector by capillary zone electrophoresis using a partial filling technique. Electrophoresis, 20 (1999) 180–188.
31 J.W. Jorgenson and K.D. Lukacs, Zone electrophoresis in open-tubular glass capillaries. Anal. Chem., 53 (1981) 1298–1302.
32 J.J.B. Nevado, A.M.C. Salcedo and G.C. Peñalvo, Simultaneous determination of cis- and trans-resveratrol in wines by capillary zone electrophoresis. Analyst, 124 (1999) 61–66.
33 R. Bhushan and H. Brückner, Marfey's reagent for chiral amino acid analysis: a review. Amino Acids, 27 (2004) 231–247.
34 A. Alvarez-Zepeda and D.E. Martire, Reversed-phase liquid chromatographic study of excess and absolute sorption isotherms of acetonitrile-water mixtures. J. Chromatogr., 550 (1991) 285–300.
35 G.M. Janini, G.M. Muschik and H.J. Issaq, Micellar electrokinetic chromatography in zero-electroosmotic flow environment. J. Chromatogr. B, 683 (1996) 29–35.
36 D.C. Harris, Quantitative Chemical Analysis, 5th Ed. W.H. Freeman and Company, NY, 1999, pp. 477.
37 R. Kuhn and S. Hoffstetter-Kuhn, Capillary electrophoresis: Principles and Practice, Springer-Verlag, Berlin, Germany, 1993, pp.25.
38 C. Chaiyasut, Y. Takatsu, S. Kitagawa and T. Tsuda, Estimation of the dissociation constants for functional groups on modified and unmodified silica gel supports from the relationship between electroosmotic flow velocity and pH. Electrophoresis, 22 (2001) 1267–1272.
39 J.N. Stuart, N.G. Hatcher, X. Zhang, R. Gillette and J.V. Sweedler, Spurious serotonin dimer formation using electrokinetic injection in capillary electrophoresis from small volume biological samples. Analyst, 130 (2005) 147–151.
40 M. Macka, P. Andersson and P.R. Haddad, Changes in electrolyte pH due to electrolysis during capillary zone electrophoresis. Anal. Chem., 70 (1998) 743–749.
41 X. Shao, Y. Shen, K. O'Neill and M.L. Lee, Capillary electrophoresis using diol-bonded fused-silica capillaries. J. Chromatogr. A, 830 (1999) 415–422.
42 W. Qin, H. Wei and S.F.Y. Li, 1,3-Dialkylimidazolium-based room-temperature ionic liquids as background electrolyte and coating material in aqueous capillary electrophoresis. J. Chromatogr. A, 985 (2003) 447–454.
43 A. Cifuentes, P. Canalejas, A. Ortega and J.C. Dıez-Masa, Treatments of fused-silica capillaries and their influence on the electrophoretic characteristics of these columns before and after coating. J. Chromatogr. A, 823 (1998) 561–571.
44 E.G. Yanes, S.R. Gratz and A.M. Stalcup, Tetraethylammonium tetrafluoroborate: a novel electrolyte with a unique role in the capillary electrophoretic separation of polyphenols found in grape seed extracts. Analyst, 125 (2000) 1919–1923.
45 G.M. Janini, K.C. Chan, J.A. Barnes, G.M. Muschik and H.J. Issaq. Separation of pyridinecarboxylic acid isomers and related compounds by capillary zone electrophoresis. Effect of cetyltrimethylammonium bromide on electroosmotic flow and resolution. J. Chromatogr. A, 653 (1993) 321–327.
46 T.W. Graul and J.B. Schlenoff, Capillaries modified by polyelectrolyte multilayers for electrophoretic separations. Anal. Chem., 71 (1999) 71, 4007–4013.
47 A.M. Stalcup and K.H. Gahm, Application of sulfated cyclodextrins to chiral separations by capillary zone electrophoresis. Anal. Chem., 68 (1996) 1360–1368.
48 J.L. Pittman, K.F. Schrumb and S.D. Gilman. On-line monitoring of electroosmotic flow for capillary electrophoretic separations. Analyst, 126 (2001) 1240–1247.
49 B.A. Williams and G. Vigh, Determination of accurate electroosmotic mobility and analyte effective mobility values in the presence of charged interacting agents in capillary electrophoresis. Anal. Chem., 69 (1997) 4445–4451.

50 J.L. Pittman, H.J. Gessner, K.A. Frederick, E.M. Raby, J.B. Batts and S.D. Gilman, Experimental studies of electroosmotic flow dynamics during sample stacking for capillary electrophoresis. Anal. Chem., 75 (2003) 3531–3538.

51 S. Arulanandam and D. Li. Determining ζ potential and surface conductance by monitoring the current in electro-osmotic flow. J. Col. Interf. Sci., 225 (2000) 421–428.

52 A. Sze, D. Erickson, L. Ren and D. Li, Zeta-potential measurement using the Smoluchowski equation and the slope of the current–time relationship in electroosmotic flow. J. Col. Interf. Sci., 261 (2003) 402–410.

53 G. Vanhoenacker, F. de l'Escaille, D. De Keukeleire and P. Sandra, Dynamic coating for fast and reproducible determination of basic drugs by capillary electrophoresis with diode-array detection and mass spectrometry. J. Chromatogr. B, 799 (2004) 323–330.

54 Y. Baba and M. Tsuhako, Gel-filled capillaries for nucleic acid separations in capillary electrophoresis. TRAC, 77 (1992) 280–287.

55 C. Heller, Capillary electrophoresis of proteins and nucleic acids in gels and entangled polymer solutions. J. Chromatogr. A, 698 (1995) 19–31.

56 E. Guihen and J.D. Glennon, Recent highlights in stationary phase design for open-tubular capillary electrochromatography. J. Chromatogr. A, 1044 (2004) 67–81.

57 J. Simal-Gándara, The place of capillary electrochromatography among separation techniques—a review. Crit. Rev. Anal. Chem., 34 (2004) 85–94.

58 A.-M. Siouffi, Silica gel-based monoliths prepared by the sol–gel method: facts and figures. J. Chromatogr. A, 1000 (2003) 801–818.

59 Z. Chen, H. Ozawa, K. Uchiyama and T. Hobo, Cyclodextrin-modified monolithic columns for resolving dansyl amino acid enantiomers and positional isomers by capillary electrochromatography. Electrophoresis, 24 (2003) 2550–2558.

60 D. Allen and Z. El Rassi, Capillary electrochromatography with monolithic silica column: I. Preparation of silica monoliths having surface-bound octadecyl moieties and their chromatographic characterization and applications to the separation of neutral and charged species. Electrophoresis, 24 (2003) 408–420.

61 B. Preinerstorfer, W. Bicker, W. Lindner and M. Lämmerhofer, Development of reactive thiol-modified monolithic capillaries and in-column surface functionalization by radical addition of a chromatographic ligand for capillary electrochromatography. J. Chromatogr. A, 1044 (2004) 187–199.

62 Z. Krivácsy, A. Gelencsér, J. Hlavay, G. Kiss and Z Sárvári, Electrokinetic injection in capillary electrophoresis and its application to the analysis of inorganic compounds. J. Chromatogr. A, 834 (1999) 21–44.

63 Z.K. Shihabi, Stacking in capillary zone electrophoresis. J. Chromatogr. A, 902 (2000) 107–117.

64 G. Hempel, Strategies to improve the sensitivity in capillary electrophoresis for the analysis of drugs in biological fluids. Electrophoresis, 21 (2000) 691–698.

65 L. Valtcheva, J. Mohammad, G. Pettersson and S. Hjertén, Chiral separation of β-blockers by high-performance capillary electrophoresis based on non-immobilized cellulase as enantioselective protein. J. Chromatogr., 638 (1993) 263–267.

66 T. Ward, I.C. Dann and A.P. Brown, Separation of enantiomers using vancomycin in a countercurrent process by suppression of electroosmosis. Chirality, 8 (1996) 77–83.

67 J.L. Beckers and P. Boček, Calibrationless quantitative analysis by indirect UV absorbance detection in capillary zone electrophoresis: the concept of the conversion factor. Electrophoresis, 25 (2004) 338–343.

68 P. Doble, M. Macka and P.R. Haddad. Design of background electrolytes for indirect detection of anions by capillary electrophoresis. TRAC, 19 (2000) 10–17.

69 L. Tao and R.T. Kennedy, Laser-induced fluorescence detection in microcolumn separations. TRAC, 17 (1998) 484–491.

70 P. Zakaria, M. Macka, G. Gerhardt and P.R. Haddad, Pulsed potentiometric detection in capillary electrophoresis using platinum electrodes. Analyst, 125 (2000) 1519–1523.

71 R.M. Guijt, C.J. Evenhuis, M. Macka and P.R. Haddad, Conductivity detection for conventional and miniaturised capillary electrophoresis systems. Electrophoresis, 25 (2004) 4032–4057.
72 L.A. Holland and A.M. Leigh, Amperometric and voltammetric detection for capillary electrophoresis. Electrophoresis, 23 (2002) 3649–3658.
73 T. Kappes, B. Galliker, M.A. Schwarz and P.C. Hauser, Portable capillary electrophoresis instrument with amperometric, potentiometric and conductometric detection. TRAC, 20 (2001) 133–139.
74 R.D. Smith, J.A. Olivares, N.T. Ngyuen and H.R. Udseth, Capillary zone electrophoresis mass-spectrometry using an electrospray ionization interface. Anal. Chem., 60 (1988) 436–441.
75 W.F. Smyth and P. Brooks, A critical evaluation of high performance liquid chromatography-electrospray ionisation-mass spectrometry and capillary electrophoresis-electrospray-mass spectrometry for the detection and determination of small molecules of significance in clinical and forensic science. Electrophoresis, 25 (2004) 1413–1446.
76 B. Chankvetadze, W. Lindner and G.K. E. Scriba, Enantiomer separations in capillary electrophoresis in the case of equal binding constants of the enantiomers with a chiral selector: commentary on the feasibility of the concept. Anal. Chem., 76 (2004) 4256–4260.
77 H. Nishi, T. Fukuyama and S. Terabe, Chiral separation by cyclodextrin-modified micellar electrokinetic chromatography. J. Chromatogr., 553 (1991) 503–516.
78 U. Pyell, Determination and regulation of the migration window in electrokinetic chromatography. J. Chromatogr. A, 1037 (2004) 479–490.
79 E. Schneiderman, A.M. Stalcup, B. Perly and E. Brooks. Binary and ternary complexes between nonionicsurfactant $C_{12}E_6$, benzoate and cyclodextrin. Part II. β-CD. J. Inclus. Phenom., 43 (2002) 43–50.
80 J. Palmer, S. Atkinson, W.Y. Yoshida, A.M. Stalcup and J.P. Landers, Charged chelate capillary electrophoresis of endogenous corticosteroids. Electrophoresis, 19 (1998) 3045 3051.
81 Y. Mechref, Z. El Rassi, Micellar electrokinetic capillary chromatography with in-situ charged micelles. VI. Evaluation of novel chiral micelles consisting of steroidalglycoside surfactant-borate complexes. J. Chromatogr. A, 724 (1996) 285–296.
82 J.T. Smith and Z. El Rassi, Micellar electrokinetic capillary chromatography with in situ charged micelles. IV. Influence of the nature of the alkylglycoside surfactant. J. Chromatogr. A, 685 (1994) 131–143.
83 R.J. Tait, D.O. Thompson, V.J. Stella and J.F. Stobaugh, Sulfobutyl ether b-cyclodextrin as a chiral discriminator for use with capillary electrophoresis. Anal. Chem., 66 (1994) 4013–4018.
84 Y. Tanaka and S. Terabe, Enantiomer separation of acidic racemates by capillary electrophoresis using cationic and amphoteric β-cyclodextrins as chiral selectors. J. Chromatogr. A, 781 (1997) 151–160.
85 J.L. Haynes III, S.A. Shamsi, F. O'Keefe, R. Darcey and I.M. Warner, Cationic β-cyclodextrin derivative for chiral separations. J. Chromatogr. A, 803 (1998) 261–271.
86 A.M. Abushoffa, M. Fillet, A.-C. Servais, P. Hubert and J. Crommen, Enhancement of selectivity and resolution in the enantioseparation of uncharged compounds using mixtures of oppositely charged cyclodextrins in capillary electrophoresis. Electrophoresis, 24 (2003) 343–350.
87 B. Chankvetadze, Separation selectivity in chiral capillary electrophoresis with charged selectors. J. Chromatogr. A, 792 (1997) 269–295.
88 G.M. Janini, G.M. Muschik and H.J. Issaq, Micellar electrokinetic chromatography in zero-electroosmotic flow environment. J. Chromatogr. B, 683 (1996) 29–35.
89 S. Fanali, Separation of optical isomers by capillary zone electrophoresis based on host-guest complexation with cyclodextrins. J. Chromatogr., 474 (1989) 441–446.
90 S.R. Gratz and A.M. Stalcup, Enantiomeric separations of terbutaline by CE with a sulfated β-cyclodextrin chiral selector: a quantitative binding study, Anal. Chem., 70 (1998) 5166–5171.
91 D.W. Armstrong, F.-Y. He and S.M. Han, Planar chromatographic separation of enantiomers and diastereomers with cyclodextrin mobile phase additives. J. Chromatogr., 448 (1988) 345–354.
92 N. Grinberg and S. Weinstein, Enantiomeric separation of Dns-amino acids by reversed-phase thin layer chromatography. J. Chromatogr., 303 (1984) 251–255.

93 J. Debowski, D. Sybilska and J. Jurczak, β-Cyclodextrin as a chiral component of the mobile phase for separation of mandelic acid into enantiomers in reversed-phase systems of high-performance liquid chromatography. J. Chromatogr., 237 (1982) 303–306.
94 K.H. Ekborg-Ott, X. Wang and D.W. Armstrong, Effect of selector coverage and mobile phase composition on enantiomeric separations with ristocetin A chiral stationary phases. Microchem. J., 62 (1999) 26–49.
95 V. Piette, M. Lammerhofer, W. Lindner and J. Crommen, Enantiomeric separation of N-protected amino acids by non-aqueous capillary electrophoresis using quinine or tert-butyl carbamoylated quinine as chiral additive. Chirality, 11 (1999) 622–630.
96 A.M. Stalcup and K.H. Gahm, Quinine as a chiral additive in nonaqueous capillary zone electrophoresis. J. Microcolumn Sep., 8 (1996) 145–50.
97 M. Lämmerhofer and W. Lindner, Quinine and quinidine derivatives as chiral selectors. I. Brush type chiral stationary phases for high-performance liquid chromatography based on cinchonan carbamates and their application as chiral anion exchangers. J. Chromatogr. A, 741 (19967 33–48.
98 D.W. Armstrong, Y.I. Han and S.M. Han, Liquid chromatographic resolution of enantiomers containing single aromatic rings with b-cyclodextrin bonded phases. Anal. Chim. Acta, 208 (1988) 275–281.
99 M.E. Amato, K.B. Lipkowitz, G.M. Lombardo and G.C. Pappalardo, High-field spectroscopic techniques combined with NMR molecular dynamics simulations for the study of the inclusion complexes of α- and β-cyclodextrins with the cognition activator 3-phenoxypyridine sulphate (CI-844). Magn. Reson. Chem., 36 (1998) 693–705.
100 S.C. Chang, L.R. Wang and D.W. Armstrong, Facile resolution of N-tert-butoxycarbonyl amino acids: the importance of enantiomeric purity in peptide synthesis. J. Liq. Chromatogr., 15 (1992) 1411–1429.
101 B. Chankvetadze, K. Lomsadze, N. Burjanadze, J. Breitkreutz, G. Pintore, M. Chessa, K. Bergander and G. Blaschke, Comparative enantioseparations with native β-cyclodextrin, randomly acetylated-β-cyclodextrin and heptakis-(2,3-di-O-acetyl)-β-cyclodextrin in capillary electrophoresis. Electrophoresis, 24, (2003)1083–1091.
102 H. Katayama, Y. Ishihama and N. Asakawa, Enantiomeric separation by capillary electrophoresis with an electroosmotic flow-controlled capillary. J. Chromatogr. A, 875 (2000) 315–322.
103 A. Guttman, S. Brunet and N. Cooke, Capillary electrophoresis separation of enantiomers by cyclodextrin array analysis. LC GC, 14 (1996) 32, 34 –36, 38, 40, 42.
104 N. Sidamonidze, F. Suss, W. Poppitz and G.K.E. Scriba, Influence of the amino acid sequence and nature of the cyclodextrin on the separation of small peptide enantiomers by capillary electrophoresis using α-, β-, and γ-cyclodextrin and the corresponding hydroxypropyl derivatives. J. Sep. Sci., 24 (2001) 777–783.
105 A. Guttman and N. Cooke, Practical aspects of chiral separations of pharmaceuticals by capillary electrophoresis. I. Separation optimization. J. Chromatogr. A, 680 (1994) 157–162.
106 T. de Boer, R.A. de Zeeuw, G.J. de Jong and K. Ensing, Recent innovations in the use of charged cyclodextrins in capillary electrophoresis for chiral separations in pharmaceutical analysis. Electrophoresis, 21 (2000) 3220–3239.
107 Z. Juvancz and J. Szejtli, The role of cyclodextrins in chiral selective chromatography. TRAC, 21(5), (2002) 379–388.
108 E.G. Yanes, S.R. Gratz, R.M. C. Sutton and A.M. Stalcup, Comparison of sulfated and phosphated β-cyclodextrin for chiral separations by capillary electrophoresis. Fres. J. Anal. Chem., 369 (2001) 412–417.
109 A. Van Eeckhaut, M.R. Detaevernier and Y. Michotte, Development of a validated capillary electrophoresis method for enantiomeric purity testing of dexchlorpheniramine maleate. J. Chromatogr. A, 958 (2002) 291–297.
110 J.L. Haynes III, S.A. Shamsi, F. O'Keefe, R. Darcey and I.M. Warner, Cationic β-cyclodextrin derivative for chiral separations. J. Chromatogr. A, 803 (1998) 261–271.
111 C. Desiderio and S. Fanali, Use of negatively charged sulfobutyl ether-β-cyclodextrin for enantiomeric separation by capillary electrophoresis. J. Chromatogr. A, 716 (1995) 183–196.

112 C.E. Evans and A.M. Stalcup, Comprehensive strategy for chiral separations using sulfated cyclodextrins in capillary electrophoresis. Chirality, 15 (2003) 709–723.

113 M. Wedig and U. Holzgrabe, Resolution of ephedrine derivatives by means of eutral and sulfated heptakis(2,3-di-O-acetyl)β-cyclodextrins using capillary lectrophoresis and nuclear magnetic resonance spectroscopy. Electrophoresis, 20 1999) 2698–2704.

114 C. Hellriegel, H. Handel, M. Wedig, S. Steinhauer, F. Sorgel, K. Albert and U. Holzgrabe, Study on the chiral recognition of the enantiomers of ephedrine derivatives with neutral and sulfated heptakis(2,3-O-diacetyl)-β-cyclodextrins using capillary electrophoresis, UV, nuclear magnetic resonance spectroscopy and mass spectrometry. J. Chromatogr. A, 914 (2001) 315–324.

115 F.-T.A. Chen, G. Shen and R.A. Evangelista, Characterization of highly sulfated cyclodextrins. J. Chromatogr. A, 924 (2001) 523–532.

116 K.H. Gahm, W.Y. Yoshida, W.P. Niemczura and A.M. Stalcup, Regioselective synthesis and characterization of naphthylethylcarbamoyl-ß-cyclodextrins. Carbohydr. Res., 248 (1993) 119–128.

117 J.B. Vincent, D.M. Kirby, T.V. Nguyen and G. Vigh, A family of single-isomer chiral resolving agents for capillary electrophoresis. 2. Hepta-6-sulfato-β-cyclodextrin. Anal. Chem., 69 (1997) 4419–4428.

118 S. Li and G. Vigh, Single-isomer sulfated α-cyclodextrins for capillary electrophoresis: hexakis(2,3-di-O-methyl-6-O-sulfo)-α-cyclodextrin, synthesis, analytical characterization, and initial screening tests. Electrophoresis, 25 (2004) 2657–2670.

119 H. Cai, T.V. Nguyen and G. Vigh, A family of single-isomer chiral resolving agents for capillary electrophoresis. 3. Heptakis(2,3-dimethyl-6-sulfato)-β-cyclodextrin. Anal. Chem., 70 (1998) 580–589.

120 U. Schmitt, M. Ertan and U. Holzgrabe, Chiral capillary electrophoresis: facts and fiction on the reproducibility of resolution with randomly substituted cyclodextrins. Electrophoresis, 25 (2004) 2801–2807.

121 H. Nishi, K. Nakamura, H. Nakai and T Sato, Enantiomer separation by capillary electrophoresis using DEAE-dextran and aminoglycosidic antibiotics. Chromatographia, 43, (1996) 426–430.

122 R. Gotti, R. Pomponio and V. Cavrini, Linear, neutral polysaccharides as chiral selectors in enantioresolution of basic drug racemates by capillary electrophoresis. Chromatographia, 52 (2000) 273–277.

123 A. D'Hulst and N. Verbeke, Chiral separation by capillary electrophoresis with oligosaccharides. J. Chromatogr., 608 (1992) 275–287.

124 T. Watanabe, K. Takahashi, M. Horiuchi, K. Kato, H. Nakazawa, T. Sugimoto and H. Kanazawa, Chiral separation and quantitation of pentazocine enantiomers in pharmaceuticals by capillary zone electrophoresis using maltodextrins. J. Pharm. Biomed. Anal., 21 (1999) 75–81.

125 H. Solnl, M. Stefansson, M.-L. Rlekkola and M.V. Novotny, Maltooligo-saccharides as chiral selectors for the separation of pharmaceuticals by capillary electrophoresis. Anal. Chem., 66 (1994) 3411–3484.

126 A.M. Stalcup and N.M. Agyei, Heparin: a novel chiral selector for capillary zone electrophoresis. Anal. Chem., 66 (1994) 3054–3059.

127 Y. Du, A. Taga, S. Suzuki, W. Liu and S. Honda, Effect of structure modification of chondroitin sulfate C on its enantioselectivity to basic drugs in capillary electrophoresis. J. Chromatogr. A, 947 (2002) 287–299.

128 R. Gotti, S. Furlanetto, V. Andrisano, V. Cavrini and S. Pinzauti, Design of experiments for capillary electrophoretic enantioresolution of salbutamol using dermatan sulfate. J. Chromatogr. A, 875 (2000) 411–422.

129 R.J. Linhardt, Heparin: structure and activity. J. Med. Chem., 46 (2003) 2551–2564.

130 R. Gotti, V. Cavrini, V. Andrisano and G. Mascellani, Dermatan sulfate as useful chiral selector in capillary electrophoresis. J. Chromatogr. A, 814 (1998) 205–211.

131 N.M. Agyei, K.H. Gahm and A.M. Stalcup, Chiral separations using heparin and dextran sulfate in capillary zone electrophoresis. Anal. Chim. Acta, 307 (1995) 185–191.

132 X. Wang, J.-T. Lee and D.W. Armstrong, Separation of enantiomers by capillary electrophoresis using pentosan polysulfate. Electrophoresis, 20 (1999) 162–170.

133 H. Nishi and Y. Kuwahara, Enantiomer separation by capillary electrophoresis utilizing carboxymethyl derivatives of polysaccharides as chiral selectors. J. Pharm. Biomed. Anal., 27 (2002) 577–585.

134 M.H. Hyun and Y.J. Cho, Preparation and application of a chiral stationary phase based on (+)-(18-crown-6)-2,3,11,12-tetracarboxylic acid without extra free aminopropyl groups on silica surface. J. Sep. Sci., 28 (2005) 31–38.

135 M.H. Hyun, Characterization of liquid chromatographic chiral separation on chiral crown ether stationary phases. J. Sep. Sci., 26 (2003) 242–250.

136 E. Bang, J.-W. Jung, W. Lee, D.W. Lee and W. Lee, Chiral recognition of (18-crown-6)-tetracarboxylic acid as a chiral selector determined by NMR spectroscopy. J. Chem. Soc., Perkin Trans. 2 (2001) 1685–1692.

137 K. Verleysen, J. Vandijck, M. Schelfaut and P. Sandra, Enantiomeric separations in capillary electrophoresis using 18-crown-6-tetracarboxylic acid ($18C6H_4$) as buffer additive. J. High Resolut. Chromatogr., 21 (1998) 323–331.

138 R. Kuhn, Enantiomeric separation by capillary electrophoresis using a crown ether as a chiral selector. Electrophoresis, 20 (1999) 2605–2613.

139 N. Grobuschek, M.G. Schmid, C. Tuscher, M. Ivanova and G. Gübitz, Chiral separation of β-methyl-amino acids by ligand exchange using capillary electrophoresis and HPLC. J. Pharma. Biomed. Anal., 27 (2002) 599–605.

140 E. Gassmann, J.E. Kuo and R.N. Zare, Electrokinetic separation of chiral compounds. Science, 230 (1985), 813–14.

141 M.E. Swartz, J.R. Mazzeo, E.R. Grover and P.R. Brown, Separation of amino acid enantiomers by micellar electrokinetic chromatography using synthetic chiral surfactants. Anal. Biochem., 231 (1995) 65–71.

142 J.R. Mazzeo, E.R. Grover, M.E. Swartz and J.S. Petersen, Novel chiral surfactant for the separation of enantiomers by micellar electrokinetic capillary chromatography. J. Chromatogr. A, 680 (1994) 125–135.

143 M. Aoudia and R. Zana, Aggregation behavior of sugar surfactants in aqueous solutions: effects of temperature and the addition of nonionic polymers. J. Col.Inter. Sci., 206 (1998) 158–167.

144 H. Kawamura, Y. Murata, T. Yamaguchi, H. Igimi, M. Tanaka, G. Sugihara and J.P. Kratohvil, Spin-label studies of bile salt micelles. J. Phys. Chem., 93 (1989) 3321–3326.

145 S.A. Shamsi, J. Macossay and I.M. Warner, Improved chiral separations using a polymerized dipeptide anionic chiral surfactant in electrokinetic chromatography: separations of basic, acidic, and neutral racemates. Anal. Chem., 69 (1997) 2980–2987.

146 K.A. Agnew-Heard, M.S. Peña, S.A. Shamsi and I.M. Warner, Studies of polymerized sodium N-undecylenyl-L-valinate in chiral micellar electrokinetic capillary chromatography of neutral, acidic, and basic compounds. Anal. Chem., 69 (1997) 958–964.

147 C. Akbay, S.A. A. Rizvi and S.A. Shamsi, Simultaneous enantioseparation and tandem UV-MS detection of eight β-blockers in micellar electrokinetic chromatography using a chiral molecular micelle. Anal. Chem., in press.

148 R.O. Cole, M.J. Sepaniak and W.L. Hinze, Optimization of binaphthyl enantiomer separations by capillary zone electrophoresis using mobile phases containing bile salts and organic solvent. J. High Resolut. Chromatogr., 13 (1990) 579–582.

149 R. Ninomiya, K. Matsuoka and Y. Moroi, Micelle formation of sodium chenodeoxycholate and solubilization into the micelles: comparison with other unconjugated bile salts. Biochim. Biophys. Acta, 1634 (2003) 116–125.

150 D.G. Oakenfull and L.R. Fisher, The role of hydrogen bonding in the formation of bile salt micelles. J. Phys. Chem., 81 (1977) 1838–1841.

151 Szökő, J. Gyimesi, Z. Szakács and M. Tarnai, Equilibrium binding model of bile salt-mediated chiral micellar electrokinetic chromatography. Electrophoresis, 20 (1999). 2754–2760.

152 X. Zhu, Y. Ding, B. Lin, A. Jakob and B. Koppenhoefer, Study of enantioselective interactions between chiral drugs and serum albumin by capillary electrophoresis. Electrophoresis, 20 (1999) 1869–1877.

153 J.J. Martínez-Pla, Y. Martín-Biosca, S. Sagrado, R.M. Villanueva-Camañas and M.J. Medina-Hernández, Fast enantiomeric separation of propranolol by affinity capillary electrophoresis using human serum albumin as chiral selector: application to quality control of pharmaceuticals. Anal. Chim. Acta, 507 (2004) 171–178.
154 H. Xu, X. Yu and H. Chen, Enantiomeric separation of basic drugs with partially filled serum albumin as chiral selector in capillary electrophoresis. Anal Sci., 20 (2004) 1409–1413.
155 X.-X. Zhang, F. Hong, W.-B. Chang, Y.-X. Ci and Y.-H. Ye, Enantiomeric separation of promethazine and d,l-α-amino-β-[4-(1,2-dihydro-2-oxo-quinoline)] propionic acid drugs by capillary zone electrophoresis using albumin as chiral selectors. Anal. Chim. Acta, 392 (1999) 175–181.
156 E. DeLorenzi, G. Massolini, M. Quaglia, C. Galbusera and G. Caccialanza, Evaluation of qual egg white riboflavin binding protein as a chiral selector in capillary electrophoresis by applying a modified partial filling technique. Electrophoresis, 20 (1999) 2739–2748.
157 Y. Sadakane, H. Matsunaga, K. Nakagomi, Y. Hatanaka and J. Haginaka, Protein domain of chicken α_1-acid glycoprotein is responsible for chiral recognition. Biochem. Biophys. Res. Comm., 295 (2002) 587–590.
158 F. Kilár and B. Visegrády, Mapping of stereoselective recognition sites on human serum transferrin by capillary electrophoresis and molecular modeling. Electrophoresis, 23 (2002) 964–971.
159 H. Matsunaga and J. Haginaka, Separations of basic drugs by capillary electrophoresis using ovoglycoprotein as a chiral selector: comparison of chiral resolution ability of ovoglycoprotein and completely deglycosylated ovoglycoprotein. Electrophoresis, 22 (2001) 3251–3256.
160 D.W. Armstrong, M.P. Gasper and K.L. Rundlett, Highly enantioselective capillary electrophoretic separations with dilute solutions of the macrocyclic antibiotic ristocetin A. J. Chromatogr. A., 689 (1995) 285–304.
161 D.W. Armstrong, K. Rundlett and G.L. Reid III, Use of a macrocyclic antibiotic, rifamycin B, and indirect detection for the resolution of racemic amino alcohols by CE. Anal. Chem., 66 (1994) 1690–1695.
162 T.J. Ward, C. Dann III and A. Blaylock, Enantiomeric resolution using the macrocyclic antibiotics rifamycin B and rifamycin SV as chiral selectors for capillary electrophoresis. J. Chromatogr. A., 715 (1995) 337–344.
163 Z. Jiang, J. Kang, D. Bischoff, B. Bister, R.D. Süssmuth and V. Schurig, Evaluation of balhimycin as a chiral selector for enantioresolution by capillary electrophoresis. Electrophoresis, 25 (2004) 2687–2692.
164 J. Kang, D. Bischoff, Z. Jiang, B. Bister, R.D. Su1ssmuth and V. Schurig, A mechanistic study of enantiomeric separation with vancomycin and balhimycin as chiral selectors by capillary electrophoresis. Dimerization and enantioselectivity. Anal. Chem., 76 (2004) 2387–2392.
165 D.W. Armstrong, J. Schneiderheinze, U. Nair, L.J. Magid and P.D. Butler, Self-association of Rifamycin B: possible effects on molecular recognition. J. Phys. Chem. B, 103 (1999) 4338–4341.
166 K.H. Ekborg-Ott, G.A. Zientara, J.M. Schneiderheinze, K.H. Gahm and D.W. Armstrong, Avoparcin, a new macrocyclic antibiotic chiral run buffer additive for capillary electrophoresis. Electrophoresis, 20 (1999) 2438–2457.
167 K.L. Rundlett, M.P. Gasper, E.Y. Zhou and D.W. Armstrong, Capillary electrophoretic enantiomeric separations using the glycopeptide antibiotic, teicoplanin. Chirality, 8 (1996) 88–107.

© 2006 Elsevier B.V. All rights reserved.
Chiral Analysis
K.W. Busch and M.A. Busch, Eds.

CHAPTER 9

Chiral separations in microfluidic devices

Detlev Belder

Institut für Analytische Chemie, Universität Regensburg 93040, Regensburg, Germany

9.1 INTRODUCTION

In recent years, capillary electrophoresis (CE) has become an established technique for chiral separation of enantiomers, especially for pharmaceutical compounds. Chiral analysis has become one of the main application areas of CE and is often used as fast alternative to high-performance liquid chromatography (HPLC).

Chiral CE is nowadays regarded as a mature technique which is also reflected by the numerous reviews which appeared over the last years [1–17] describing the fundamentals and numerous applications in this field.

Microchip electrophoresis (MCE) which can be regarded as a higher miniaturized variation of CE has been introduced about ten years ago by Manz's group. Electrophoretic separations are performed in intersected channels apparent in the so-called microfluidic chips [18–21]. MCE has been applied for electrophoretic separations for a variety of biochemical and chemical analyses [22–25]. The main field of application is to date the analysis of biomolecules such as oligonucleotides, proteins and peptides. DNA separations on microchips have meanwhile become a rather mature technique which can compete with CE and classical gel electrophoresis [26]. This is also reflected by the introduction of commercial products such as Agilent 2100 Bioanalyser, Shimadzu MCE 2010 and Hitachi SV 1100 and SV 1210, which appear to aim mainly on this application area.

Chiral separations on micromachined electrophoretic devices have been reported only recently [27–37]. In this chapter, I want to give an introduction into the current status of MCE as a new technique for chiral analysis.

Motivations for performing chiral separations in MCE as a higher miniaturized format compared to classical CE are (i) improved separation performance; (ii) reduced instrument size and reagent consumption; (iii) enhanced separation speed and sample throughput and (iv) realization of new integrated devices.

The reduction of size of a potential instrument has also been an issue mentioned in the first article on chiral MCE by Hutt et al. [27], who explored the feasibility to use

References pp. 292–295

microfabricated electrophoresis devices to analyze for extinct or extant life signs in extraterrestrial environments.

One of the main motivations of our group to develop chiral separations in microfluidic devices was the development of methods for chiral high-throughput screening (HTS). While traditionally the main challenge in chiral analysis was to achieve chiral separation, the significant improvement in sample throughput in chiral analysis is only a recent demand [38,39]. One of the reasons for the demand of such chiral HTS systems is the recent introduction of combinatorial techniques for the development of asymmetric catalysts, such as directed evolution of enzymes [40,41]. Using these methods, thousands of potential catalysts can be generated per day using time-saving parallel synthesis. Testing of these large libraries for enantioselectivity is, however, a big task for the analytical chemist. Accordingly, in analogy to parallel synthesis of the catalysts, multiplexing the analysis step is an obvious way to improve the sample throughput. Using electrophoretic techniques this can be easily done by chiral capillary array electrophoresis (CAE) using adapted commercially available CAE systems [30], which have originally been developed for DNA sequencing for the human genome project. One of the most promising features of MCE in this context is its great potential for high-throughput analysis due to the ease of multiplexing the separation channels [42–45] combined with the possibility for high-speed separations. Furthermore, as chemical reactors can be integrated in such a device the design of a micro-total analysis system (μ-TAS) [46] as an integrated tool for the fast development of asymmetric catalysts is feasible as recently shown [68]. In this context, the reduction of the size of a potential instrument is only of secondary importance.

9.2 METHODOLOGY AND INSTRUMENTATION

The separation principles in chiral MCE are basically the same as in classical CE. Therefore, most applications which can be realized in classical CE can also be transferred to MCE. The basics of chiral CE have been discussed in a previous chapter.

The main difference between classical CE and MCE is the format of the separation device, and consequently the experimental setup. While conventional CE is performed in fused silica (FS) capillaries, MCE is accomplished on planar devices with microfabricated channels. An attractive feature using microfabricated devices is the possibility of generating very complex structures which can be utilized for on-chip reactions, labeling and separations.

The most commonly used design consists of two crossed channels with a longer separation channel intersected by a second channel for sample introduction. At the end of each channel are access holes which serve as reservoirs for buffer, buffer waste, sample and sample waste. While the typical length of a CE-capillary is about 30–50 cm, the separation channel in microfluidic devices is only several centimeter long. Such microchips are typically made of two fused parts. While one plate carries the actual microstructure (often the bottom plate) another plate with the holes is used as a cover plate. A schematic drawing of such a microchip is shown in Fig. 9.1.

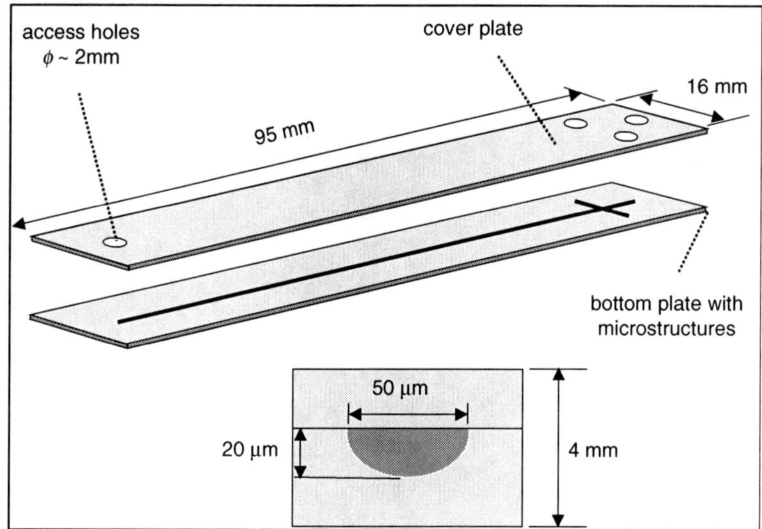

Fig. 9.1. Schematic drawing of a microfluidic chip with typical dimensions for MCE.

While in classical CE, FS is nearly exclusively used as a column material, quite diverse materials are used for microchip fabrication.

Glass and especially FS are very attractive materials for microfluidic chips due to their excellent optical properties, high insulating properties, and due to the inertness towards a variety of different solvents. There are, however, some manufacturing issues associated with glass microchips, which makes the use of alternate materials like different polymers very attractive [47].

An advantage of polymer-based microfluidic chips is the wide choice in microfabrication techniques like injection molding [48], laser ablation [49], imprinting [50] or hot embossing [51]. For such microfluidic devices, inexpensive mass production is possible which makes them very attractive as disposable devices for commercial applications. Furthermore, channels with high aspect ratios are much easier to produce compared to glass devices, which is often advantageous for optical detection.

9.2.1 Injection

One of the main difference between classical CE and MCE is the injection process. While both electrokinetic and hydrodynamic injection schemes are utilized in classical CE, electrokinetic injection is nearly exclusively used in MCE. This is due to the demand of injecting only very narrow sample zones to achieve sufficient resolution for demanding applications like chiral separations. In MCE, this can elegantly be performed as the size of the injection plug is defined by the size of the channel intersection and can further be manipulated by electric potentials applied at the reservoirs [52,53]. This injection process which allows the injection of very narrow sample plugs is one of the main methodic differences between classical CE and MCE.

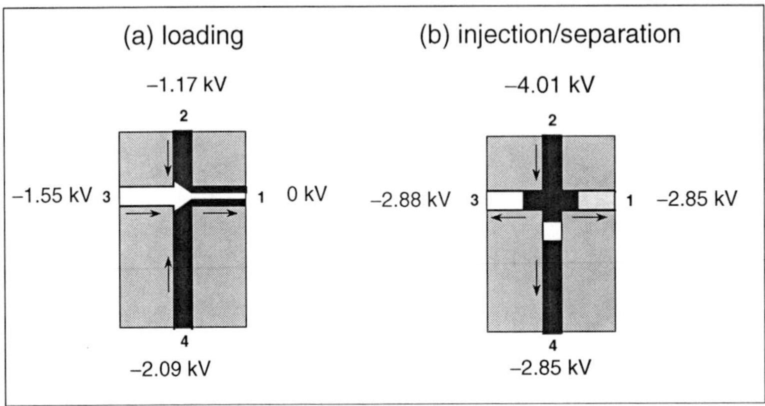

Fig. 9.2. Schematic drawing of a pinched injection process.

The complex fluid handling is performed by directed manipulation of the applied voltages at the reservoirs due to electroosmosis and electrophoresis. The pinched injection mode [54], where very narrow sample zones can be created, has become common practice in MCE especially for chiral separations. For this purpose pinching potentials are applied at the buffer inlet and outlet, inducing buffer flows to sample outlet. This prevents sample leakage into the separation channel and by that band broadening during injection. In the following separation phase (t_0 in Fig. 9.3), counter voltages are applied at the sample inlet and outlet reservoirs which purges the injection channel and inhibits sample leakage and injection-induced tailing.

A schematic drawing of such a pinched injection with exemplarily applied potentials and resulting flows is shown in Fig. 9.2.

In this pinched injection mode, the analyte is drawn by electroosmosis and electrophoresis to the intersection of the channels. In order to ensure that a representative sample is injected, relatively long injection times are required for the analytes to fill the intersection. For this purpose the analyte is usually dissolved in the chiral electrolyte. In the case of very high enantioselectivities, where the enantiomers differ significantly in effective electrophoretic mobilities it can be difficult to inject a representative sample within a reasonable time frame. This can lead to a systematic error in the determination of the enantiomer ratios. The injection time necessary for generating a representative analyte plug is often longer than the separation time and therefore limits the analytical speed.

In Fig. 9.3, a schematic drawing of an MCE experiment including the injection and the following process is shown.

For repetitive injections the gated injection method [55,56] is advantageous, where a continuous analyte stream is filling the sample cross, while an electrolyte stream prevents sample leakage into the separation channel. For injection the analyte flow is deflected into the separation channel for a short time. The gated injection scheme is schematically shown in Fig. 9.4. This injection method is however rarely used in chiral MCE, may be due to the difficulty of dispensing very narrow sample zones.

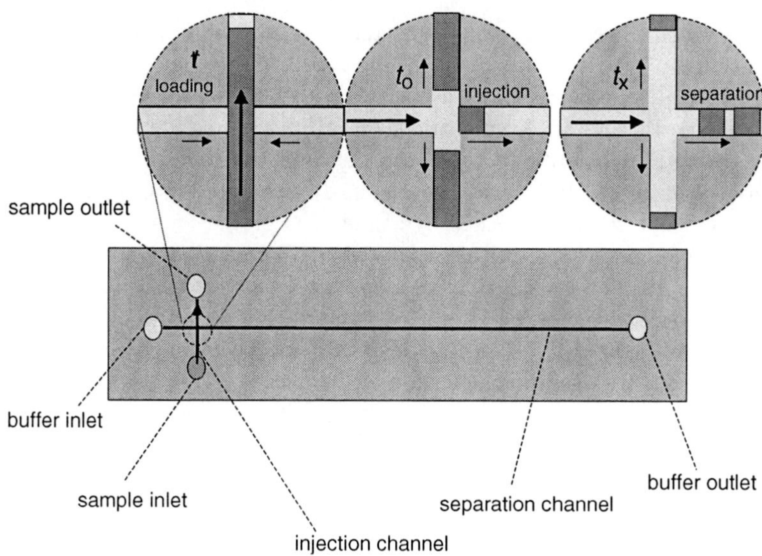

Fig. 9.3. Schematic drawing of the injection–separation process in microchip electrophoresis (MCE) applying pinched injection.

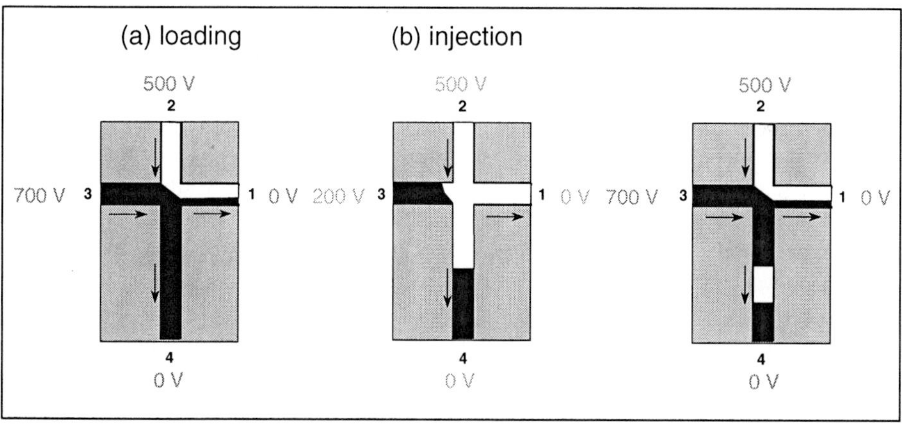

Fig. 9.4. Schematic drawing of a gated injection process.

Promising alternatives for narrow sample injectors, reducing also the complexity of voltage control, have recently been introduced and discussed by Zhang and Manz [57].

As mentioned above, the possibility of generating very sharp sample zones is one of the main reasons that demanding separations like chiral separations can be accomplished on tiny chips in a considerably shorter time compared to classical CE with conventional injection. That very fast chiral separations can also be performed in classical CE if very narrow injection zones are generated was shown by Thompson et al. [58]. In this study, they coupled microdialysis online with fluorescence labeling and optically gated CE to determine D- and L-aspartate in tissue samples. Optical-gating injection is based on the photodecomposition of a fluorescent compound under irradiation. The gating laser beam is focused onto the capillary/channel, which causes photo bleaching of the fluorophore which is continuously introduced into the capillary. When the bleaching beam is briefly blocked by a shutter, a plug of intact fluorophore is injected onto the column which can be separated and detected downstream with fluorescence detection. Using β-cyclodextrin as an additive they were able to separate D,-L aspartate in 3 s over a separation length of 15 mm. This shows that fast chiral separations can also be realized in classical capillaries if the injection of very narrow sample zones is combined with sensitive detection techniques.

9.2.2 Detection

As discussed in Section 9.2.1, the injection of very narrow sample zones in MCE enables very fast separations in tiny devices utilizing high field strengths. A drawback of injecting very sharp sample zones is, however, that the detection sensitivity usually also decreases with decreasing amount of analyte injected. The development of appropriate detection methods providing sensitivity for a wide variety of analytes is accordingly a challenging task in MCE. Detection is currently however one of the weak points in MCE. While the highly miniaturized and flat format of the separation device has several advantages with regard to separation performance, size and chemical integration, this compact format is often disadvantageous for the implementation of powerful detection methods.

The most commonly used detection technique in MCE is fluorescence detection due to its high sensitivity. Besides high sensivity detection, another reason for the popularity of fluorescence detection in MCE is that this technique can easily be implemented in the experimental setup. A common setup for MCE with epifluorescence, where emission and excitation light are spatially separated by a dichroic mirror, is shown in Fig. 9.5.

While in most reports on chiral MCE with fluorescence detection a monochromatic laser has been used as a light source [27–30], the use of a lamp-based system with flexible variable wavelength detection was also reported [32]. This appears to be advantageous in chiral MCE as a wide variety of different fluorescent tags with different required excitation wavelengths can be utilized. The possibility to use a wide variety of diverse fluorescent tags simplifies method development in chiral analysis, as the chemical nature of the derivatizing agent will affect the enantiomeric separation.

Fluorescein isothiocyanate (FITC) has been the most popular reagent for fluorescence labeling in chiral MCE. This compound is suitable for labeling amine functionalities, fluorescence can be excited with the 488-nm line of popular Argon-ion lasers. Tagging with FITC was applied for chiral separation of amino acids by Hutt et al. [27], Wang et al. [31] and Rodriguez et al. [29]. For the separation of various chiral amines FITC isomer I was used [30,31] according to reaction Scheme 1.

Fluorescence labeling of amines with FITC results in fluorescent compounds which are anionic at high pH. The fact that the derivatives are ionic and by that exhibit an electrophoretic mobility simplifies method development for chiral separations. In this case, chiral resolution can be obtained by using non-charged cyclodextrins and respective derivatives which are available in a wide variety. The addition of surfactants or charged host molecules is possible but not necessary as for uncharged analytes.

Wallenborg et al. [28] used 4-fluoro-7-nitrobenzofurazane for fluorescence labeling of amine functionalities for chiral MCE of amphetamine and related compounds. This fluorophore generates uncharged derivatives which can also be excited at 488 nm by an

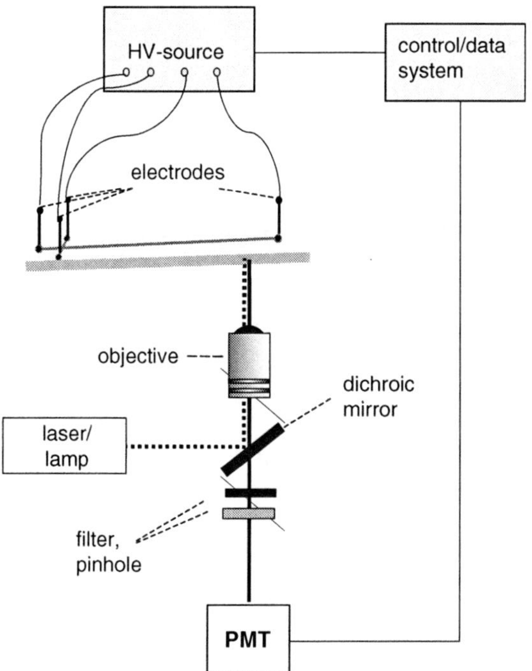

Fig. 9.5. Basic setup for microchip electrophoresis (MCE) with epifluorescence detection.

argon-ion laser. Electrophoretic mobility of the analytes was promoted by applying a micellar system using sodium dodecylsulfate (SDS) in combination with uncharged γ-cyclodextrin as chiral selector.

A drawback of fluorescence detection is, however, that most of the analytes have to be derivatized prior to analysis, which is not only inconvenient and sometimes troublesome but also affects the chiral recognition by selectors such as cyclodextrins. If the compounds to be analyzed exhibit natural fluorescence at the excitation wavelength troublesome labeling can be omitted. This was shown by Il Cho et al. for the chiral separation of gemifloxacin applying a He/Cd-laser operated at 325 nm [35] and recently Belder et al. [67] for small aromatics using a deep UV-laser at 266 nm.

Amperometric detection was introduced by Schwarz and Hauser [33] in chiral MCE for the separation of basic drugs. Detection was carried out with a two electrode amperometric detector, eliminating the need for individual counter and reference electrodes. Rather good detection sensitivities were obtained for catecholamines which were in the order of 10^{-7} M. Because the electrochemical oxidation of catecholamines is pH-dependent, higher detection sensitivities could be obtained at higher pH values. Direct electrochemical oxidation of ephedrine and pseudoephedrine was achieved at a gold electrode at a high pH of 12.6. As the basic compounds are uncharged at this pH a charged cyclodextrin derivative had to be used for chiral resolution. It was observed that the detection sensitivity is also affected by the type and concentration of chiral and achiral additives applied for achieving resolution. As mentioned by the authors, electrochemical detection has the advantage that derivatization can be omitted. On the other hand, the freedom of buffer choice is limited as conditions have to be arrived which satisfy the requirements for both separation and detection. Another important aspect is that electrochemical detection lacks universality.

While conductivity detection is regarded as rather insensitive in zone electrophoresis it can be very attractive in isotachophoresis due to the online focusing effect during analyses. Chiral isotachophoresis in microfluidic devices with on-column conductivity detection was successfully demonstrated by Ölvecka et al. [34] for the chiral separation of tryptophan enantiomers. For this purpose they used a coupled separation channel configuration.

The most dominant detection technique in conventional CE is UV detection due to its versatility and ease of operation. UV detection has however only rarely been applied to MCE [59–61] due to geometrical constraints and especially because of the limited optical path lengths. Furthermore rather expensive quartz chips have to be used for high UV transparency. Only recently UV detection was used in chiral MCE applying the commercial instrument MCE 2010 from Shimadzu [62]. In this contribution, fast chiral separation of a wide variety of basic and acidic compounds could be realized without derivatization. Using highly sulfated cyclodextrins as chiral selectors baseline separation of 18 compounds could be achieved in less than a minute, as shown in Table 9.1. The fastest separation was obtained in 2 s which was at the time the world record for the fastest chiral separation.

The MCE 2010 instrument from Shimadzu uses a linear imaging detector. Accordingly the obtained data format using this instrument is different from that

Table 9.1 Electrophoretic data for on-chip chiral separation of 18 drugs using different cyclodextrins as additives, required time t and space x for baseline separation ($R \sim 1.5$)

Compound	Selector	t (s)	x (mm)	Compound	Selector	t (s)	x (mm)
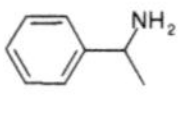	HS-γ-CD	39	11.2	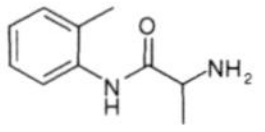	HS-α-CD	7	4.8
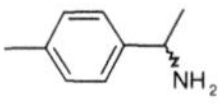	HS-α-CD	15	13.6	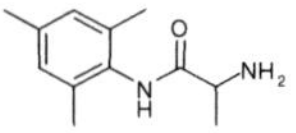	HS-γ-CD	11	11.1
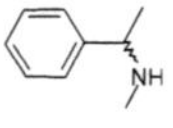	HS-β-CD	13	15.4	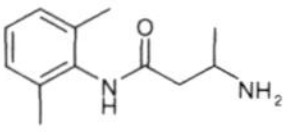	HS-γ-CD	9	5.9
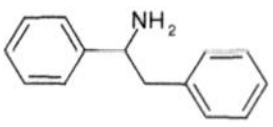	HS-γ-CD	4	6.0	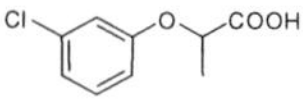	HS-α-CD	27	16.1
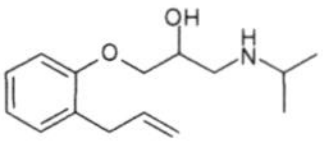	HS-α-CD	20	21.6	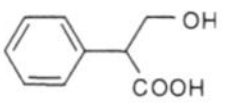	HS-β-CD	22	10.8
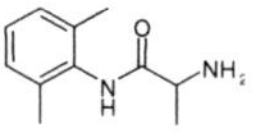	HS-γ-CD	4	4.7	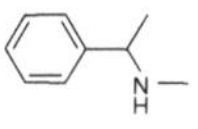	HS-γ-CD	58	18.1
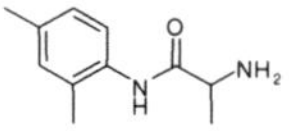	HS-α-CD	31	11.3	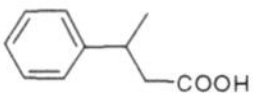	HS-γ-CD	31	13.2
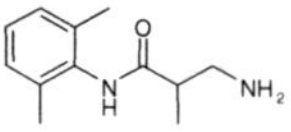	HS-γ-CD	23	15.8	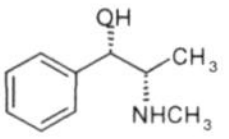	HS-β-CD	10	15.8
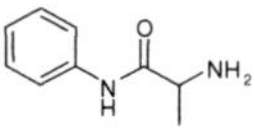	HS-γ-CD	11	8.5	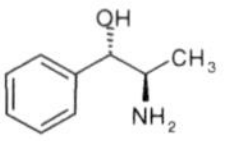	HS-γ-CD	2	2

typically known from traditional devices, where the detection is realized at a fixed point at the end of the separation channel. The linear imaging detector enables to observe a real-time migration pattern.

Such migration patterns recorded at different times during the separation process is shown in Fig. 9.6 for the chiral separation of the antiarrhytmicum tocainide. The upper image in Fig. 9.6 shows the sample zone filling the intersection of the separation channel and the injection channel. The device was switched to separation state by switching the potentials according to the voltage program described in the experimental section. After 2 s already partial separation can be observed. Complete baseline separation is observed after 4 s utilizing a separation length of 4.7 mm. Utilizing the whole separation length of 25 mm a resolution of $R = 7$ is obtained in 24 s.

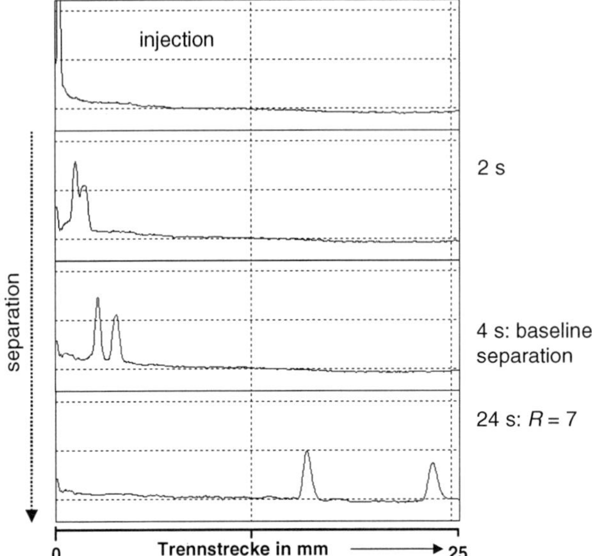

Fig. 9.6. Real-time migration patterns for the chiral separation of tocainide. Electrolyte: 5% HS-γ-CD, 25 mM triethylammonium phosphate buffer pH 2.5; detection: UV at 200 nm. Other conditions are given in the procedure section.

Successful separation of a mixture of 3 chiral drugs could be performed in a single run in less than 11 s utilizing a separation length of only 12 mm.

Although these results show that common UV detection is possible for chiral MCE the detection sensitivity has still to be improved as rather high sample concentrations (about 2 mg/ml) were used.

Only recently online chemiluminescence (CL) detection was used by Liu et al. in chiral MCE [36]. In this case no light source is required which makes instrument configuration very easy. They used peroxalate-peroxide as CL system for the detection of dansyl species and were able to detect D/L-DNS-phenylalanine at a concentration of 5 μM after chiral separation using hydroxypropyl β-CD.

9.2.3 Chiral electrolytes

As enantiomers have identical electrophoretic mobilities they can only be separated in a chiral environment. To enable the differentiation of enantiomers by electrophoresis, chiral additives are dissolved in the electrolyte as the so-called pseudo-stationary phases. When two enantiomers bind to the dissolved selector to a different extent this can result in different effective mobilities of the enantiomers which enables chiral resolution in CE. As the chiral selector is dissolved in the electrolyte the chiral selectivity of the system can easily be altered by changing the buffer constitution. This is done by using different types of chiral selectors at different concentrations, by varying the pH value and also by the addition of promoting non-chiral compounds. Development of a suitable method is rather easy and rapid since the replacement of the electrolyte is done automatically in modern CE systems.

As the separation principles in chiral MCE and CE are similar, analogous approaches for enabling chiral recognition are used in chip electrophoresis. Often the method can simply be transferred from classical CE to chip devices.

As in classical CE, the most commonly used chiral selectors in chip electrophoresis are cyclodextrins and their derivatives which are commercially available in a wide variety. Ionic analytes can be separated in rather simple electrolyte systems employing non-ionic CD derivatives as additives. The non-ionic cyclodextrin (CD) derivative hydroxypropyl-γ-CD was used for the separation of negatively charged FITC-labeled amines [30,32] with fluorescence detection and for the separation of DNS-phenylalanine with CL detection [34]. The native α-cyclodextrin was used in chiral isotachophoresis for the separation of basic tryptophan enantiomers [34]. By using highly sulfated ionic cyclodextrins, 19 acidic or basic compounds could be separated in seconds employing UV detection [62].

SDS can be used as a non-chiral additive to enable the separation of uncharged analytes, as they can interact with the charged micelles. This technique is widely used in classical CE and is called micellar electrokinetic chromatography (MEKC or MECC). SDS is also used in combination with uncharged cyclodextrins to enable chiral separation of non-ionic compounds and also for promoting resolution of charged analytes. Such SDS–cyclodextrin systems have also been applied in MCE for the separation of 4-fluoro-7-nitrobenzofuran derivatives of amphetamine and analogs [28]. In this study, best results were obtained using a highly sulfated γ-cyclodextrin in combination with a low concentration of SDS. A combination of SDS with cyclodextrins was also used for promoting the chiral separation of FITC-labeled amino acids [27,31]. Schwarz and Hauser [33] used the non-chiral crown ether 18-crown-6 for promoting the separation of noradrenaline enantiomers in combination with carboxymethylated cyclodextrins as chiral selector.

Chiral crown ether, (−)-(18-crown-6)-tetracarboxylic acid ($18C_6H_4$), was used by Il Cho et al. [35] as an effective chiral selector for resolving gemifloxacin in sodium containing media. Adding the chelating agent, ethylenediaminetetraacetic acid (EDTA) to the run buffer was found to greatly improve the separation efficiencies and peak shapes.

9.2.4 Enhancing resolution and performance

As discussed above, very fast chiral separations can be performed employing highly miniaturized devices which makes MCE very attractive for HTS systems. Furthermore due to the compact format, portable hand-held devices for chiral separations should be feasible, which should allow to bring such analysis systems to remote sites, e.g. for drug analysis in-field.

The compact format of the separation device can however be disadvantageous for very demanding chiral separations with rather low selectivities. While in classical CE the column length is flexible and rather long separation capillaries are usually

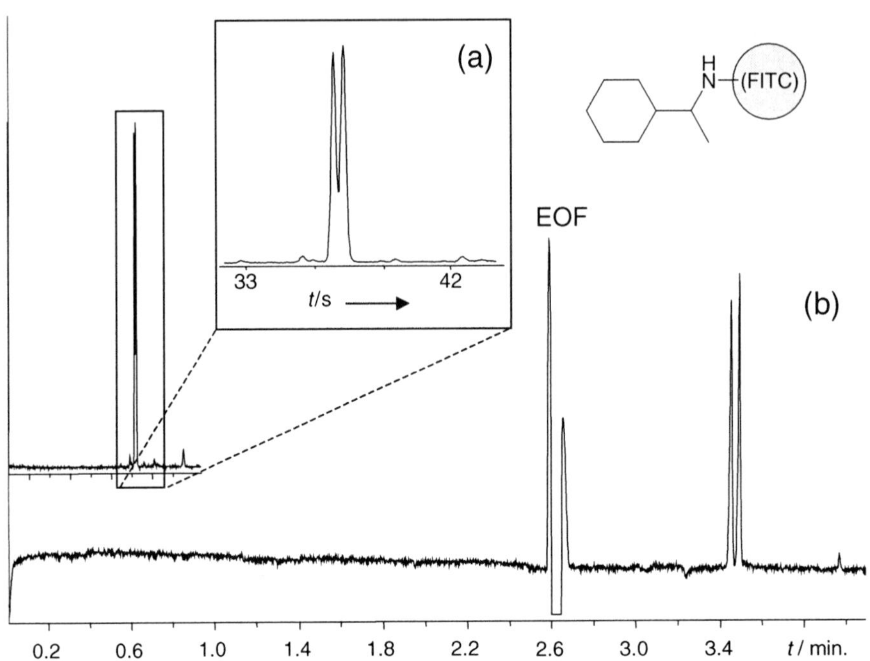

Fig. 9.7. Comparison of chiral separations obtained in microchip electrophoresis (MCE) (a) and in classical capillary electrophoresis (CE) (b).

used (~30–50 cm), the separation channel in MCE has a fixed length of only several centimeters. As discussed in Section 9.2.1, separation in such miniature dimension is possible due to the ability to inject very narrow sample zones.

However, in practice the resolution achievable in MCE devices is often lower compared to that obtainable in classical CE utilizing considerably longer separation capillaries. This is shown in Fig. 9.7, where the chiral separation of an FITC-labeled amine obtained at typical conditions in MCE and in classical CE is compared.

For both the experiments the same electrolyte was used, while the column length was limited to 7 cm in MCE, a column of 40 cm effective length was used in CE. The separation time in MCE is about 40 s, considerably lower compared to 3.5 min in classical CE. The resolution in MCE is however inadequate for a reliable determination of the enantiomeric ratio.

In order to obtain sufficient resolution in chiral MCE, different strategies have been used:

(i) The most obvious way to improve chiral separation is to enhance the enantioselectivity of the system as much as possible.

Il Cho et al. [35] also found that chiral separation in MCE was poorer to that obtained in classical CE. They were able to obtain sufficient resolution by increasing the amount of chiral selector. This approach is however very limited once the optimal electrolyte constitution is reached.

Highly sulfated cyclodextrins (HS-CD) have proven to exhibit extraordinarily high enantioselectivities in classical CE [63]. Accordingly, these selectors are very useful to realize chiral separations in very short columns on chips. HS-CDs have been used by Wallenborg et al. [28] to realize chiral separations of 4-fluoro-7-nitrobenzofurazane in combination with a folded channel. Ludwig et al. showed that by the use of HS-CDs chiral separation of a variety of compounds is possible utilizing only several millimeter long separation channels, see Fig. 9.5 [62].

(ii) By the use of folded separation channels, the column length can be extended without enlarging the compact footprint of the device, as shown in Fig. 9.8. This approach was used for the very first chiral MCE separation by Hutt et al. [27]. They found that a folded 19-cm-long channel gave superior enantiomeric resolution compared to a 5-cm straight channel geometry. A similar approach was used by Wallenborg et al. [28] who found that a short 6-cm-long channel did not provide sufficient separation length to obtain chiral separation. Improved chiral resolution was observed in a longer U-folded channel and best resolution was obtained in the longest S-folded channel with about 170 mm length. A similar channel geometry with 103-mm column length was used by Il Cho et al. for chiral separation on a chip [35]. However, the need to incorporate turns, in order to realize long channels on a compact microchip, adds a geometric contribution to analyte dispersion [64,65].

(iii) The use of coated channels can also improve chiral separations on microfluidic devices considerably. The use of internal coatings to improve separation performance by the suppression of both analyte wall interaction and electroosmosis is common practice in classical CE but is so far rarely used in MCE. The impact of channel coating with poly(vinyl alcohol) on a chiral separation in MCE is shown in Fig. 9.9 for the separation of an FITC-labeled amine [66].

While in an uncoated glass channel only incomplete separation with an R value of 0.9 is obtained, baseline separation is observed in a coated device. The resolution was enhanced by a factor of more than 2. Using a coated device, baseline separation was even possible using half the separation length. Using such a PVA-coated glass microchip, reliable determination of high enantiomeric ratios was possible, as shown in Fig. 9.10. About 1% of the minor enantiomer could be reliably determined in a coated

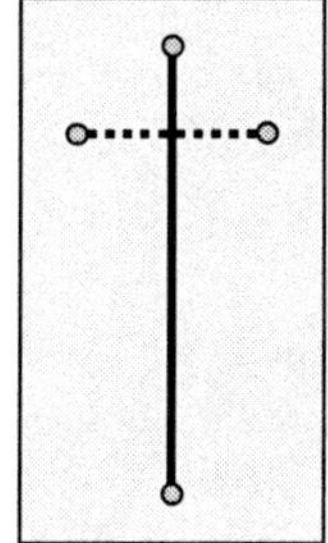
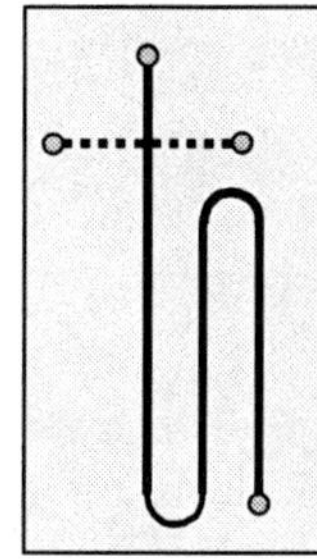
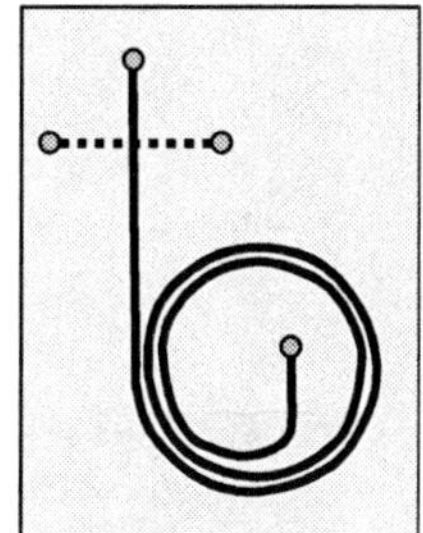

Fig. 9.8. Channel layouts enabling long separation channels on a small device.

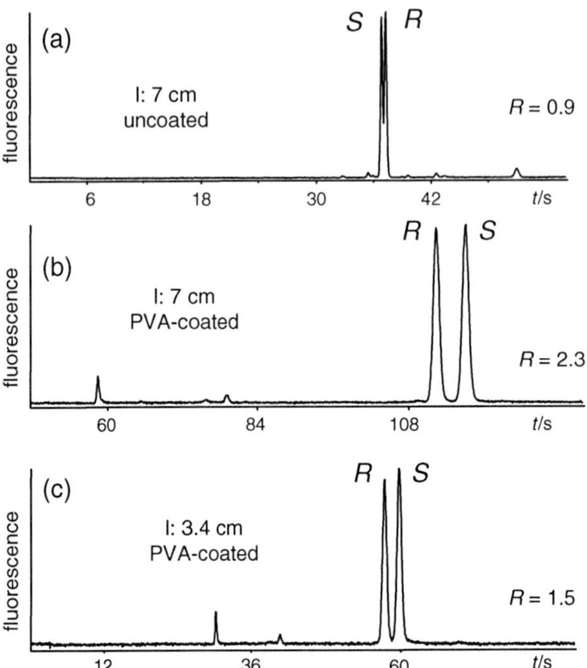

Fig. 9.9. Influence of PVA-channel coating on chiral resolution of FITC-labeled (*R*)-(−) and (*S*)-(+)-1-cyclohexylethylamine in MCE. The effective separation lengths were 7 cm for the uncoated channel (a) and 7 cm (b) and 3.4 cm (c) for the PVA-coated microchip. Buffer: 40 mM CHES, 6.25 mM HP-γ-CD, pH 9.2.

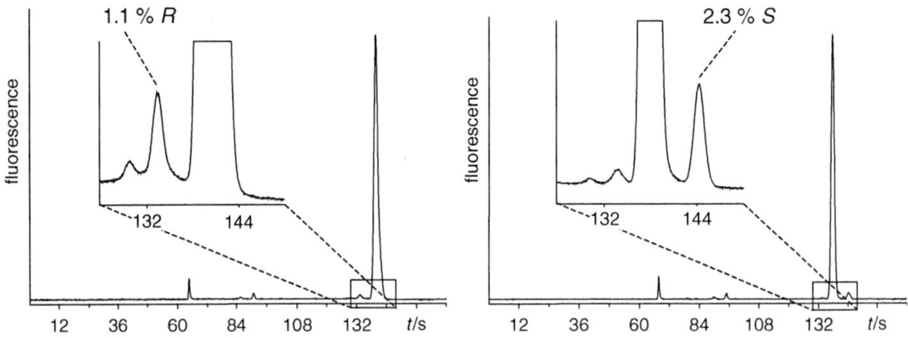

Fig. 9.10. Quantitative determination of small enantiomeric impurities in FITC-labeled (*R*)-(−) and (*S*)-(+)-1-cyclohexylethylamine.

chip while this was not possible using an uncoated devices. This was however not the determination limit but the maximum optical purity of the compounds available.

Using such PVA-coated glass microchips, reliable determination of the enantiomeric ratios of FITC-labeled cyclohexyethylamine samples was possible within a wide range

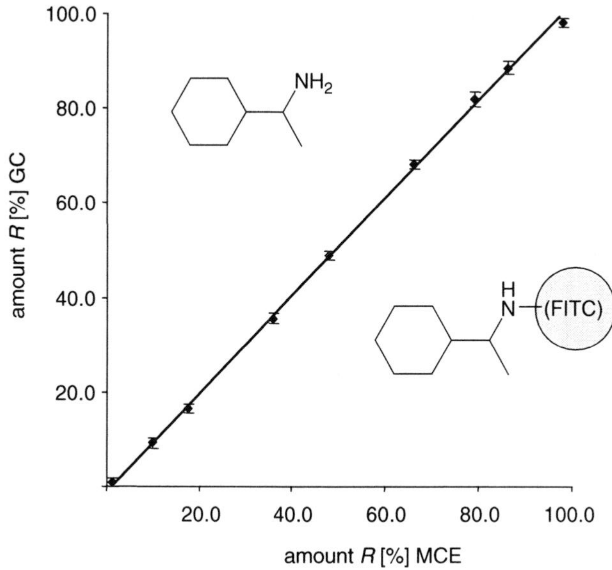

Fig. 9.11. Determination of enantiomer ratios with GC or MCE. Sample: mixtures of FITC-labeled (*R*)-(−) and (*S*)-(+)-1-cyclohexylethylamine. Error bars represent the standard deviation for three repetitive experiments. Sample concentration: 2 μM for MCE and 6.7 mM for GC analyses.

from about 1 to 98% amount of the *R*-enantiomer. A rather good average relative standard deviation (RSD) of about 1.9% was found for the multiple ee determinations. The results obtained with MCE were compared with those using gas chromatography (GC). For this purpose, the samples were splitted prior to the derivatization step as the unlabeled compounds were separated in GC equipped with flame ionization detection (FID) detection while the FITC-labeled compounds were employed for MCE with fluorescence detection. A linear correlation of the determined enantiomer ratios was found, as shown in Fig. 9.11, where the amount of the *R*-enantiomer determined with GC is plotted *versus* the amount obtained with MCE. The excellent R^2 value (0.9993) obtained shows that chiral MCE is a reliable method for the determination of the enantiomeric purity. While the GC analysis took about 19 min, the MCE separation could be performed in about 2 min.

Using highly selective chiral selectors and high field strength it is even possible to obtain chiral separations in less than a second. This was recently demonstrated for the separation of DNS-amino acids [67]. This is shown in Fig. 9.12 for DNS-tryptophan which could be separated in about 720 ms which is currently the fastest chiral separation reported.

Such fast separations could be realized in MCE using high field strength up to 2600 V/cm and short separation length of several millimeters. Baseline separation of a more complex mixture of three chiral compounds could be achieved in 3.3 s. These results show that MCE has great potential for chiral HTS and for real-time process control and for multidimensional separations.

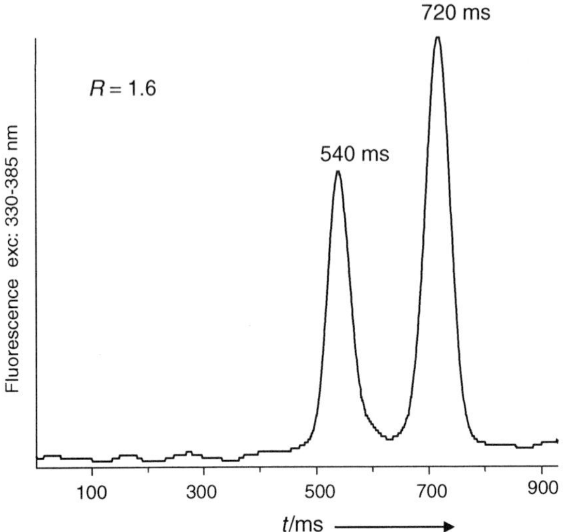

Fig. 9.12. Subsecond chiral separation of DNS-tryptophan. Electrolyte: 2% HS-γ-CD, 25 mM triethylammonium phosphate buffer pH 2.5.

9.3 OUTLOOK

In this chapter, an introduction and an overview of the possibilities of MCE for chiral separation are presented. Especially, the capabilities of chiral MCE to perform chiral separations with unsurpassed speed show that MCE has great potential for chiral HTS and also for real-time process control. With the availability of mature commercial instrumentation, chiral MCE can become a fast alternative to HPLC and also CE, especially for applications where improving speed is the main challenge. In this context, the ease of multiplexing is a strong feature of MCE. Furthermore, the potential integration of enantioselective reactions on chips could lead to integrated devices for the fast development of enantioselective catalysts.

REFERENCES

1 T.J. Ward, Chiral separations. Anal. Chem., 74 (2002) 2863–2872.
2 H. Wan and L.G. Blomberg, Chiral separation of amino acids and peptides by capillary electrophoresis. J. Chromatogr. A, 875 (2000) 43–88.
3 A. Amini, Recent developments in chiral capillary electrophoresis and applications of this technique to pharmaceutical and biomedical analysis. Electrophoresis, 22 (2001) 3107–3130.
4 G. Gübitz and M.G. Schmid, Recent progress in chiral separation principles in capillary electrophoresis. Electrophoresis, 21 (2000) 4112–4135.
5 T. de Boer, R.A. de Zeeuw, G.J. de Jong and K. Ensing, Recent innovations in the use of charged cyclodextrins in capillary electrophoresis for chiral separations in pharmaceutical analysis. Electrophoresis, 21 (2000) 3220–3239.
6 K.L. Rundlett and D.W. Armstrong, Methods for the determination of binding constants by capillary electrophoresis. Electrophoresis, 22 (2001) 1419–1427.

7 A. Rizzi, Fundamental aspects of chiral separations by capillary electrophoresis. Electrophoresis, 22 (2001) 3079–3106.

8 R. Vespalec and P. Bocek, Chiral separations in capillary electrophoresis. Chem. Rev., 100 (2000) 3715–3753.

9 G. Gübitz and M.G. Schmid, Chiral separation by chromatographic and electromigration techniques. A review. Biopharm. Drug. Dispos., 22 (2001) 291–336.

10 G. Gübitz and M.G. Schmid, Recent progress in chiral separation principles in capillary electrophoresis. Electrophoresis, 21 (2000) 4112–4135.

11 B. Chankvetadze and G. Blascke, Enantioseparations in capillary electromigration techniques: recent developments and future trends. J. Chromatogr. A, 906 (2001) 309–363.

12 B. Chankvetadze, Enantioseparation of chiral drugs and current status of electromigration techniques in this field. J. Sep. Sci., 24 (2001) 691–705.

13 U. Schmitt, S.K. Branch and U. Holzgrabe, Chiral separations by cyclodextrin-modified capillary electrophoresis - determination of the enantiomeric excess. J. Sep. Sci., 25 (2002) 959–974.

14 N.M. Maier, P. Franco and W. Lindner, Separation of enantiomers: needs, challenges, perspectives. J. Chromatogr., 906 (2001) 3–33.

15 S. Fanali, Enantioselective determination by capillary electrophoresis with cyclodextrins as chiral selectors. J. Chromatogr. A, 875 (2000) 89–122.

16 K. Otsuka and S. Terabe, Enantiomer separation of drugs by micellar electrokinetic chromatography using chiral surfactants. J. Chromatogr. A, 875 (2000) 163–178.

17 B. Chankvetadze, Enantiomer migration order in chiral capillary electrophoresis. Electrophoresis, 23 (2002) 4022–4035.

18 A. Manz, D.J. Harrison, E.M.J. Verpoorte, J.C. Fettinger, A. Paulus, H. Ludi and H.M. Widmer, Planar chips technology for miniaturization and integration of separation techniques into monitoring systems – capillary electrophoresis on a chip. J. Chromatogr., 593 (1992) 253–258.

19 D.J. Harrison, K. Fluri, K. Seiler, Z.H. Fan, C.S. Effenhauser and A. Manz, Micromachining a miniaturized capillary electrophoresis-based chemical-analysis system on a chip. Science, 261 (1993) 895–897.

20 D.J. Harrison, A. Manz, Z.H. Fan, H. Ludi and H.M. Widmer, Capillary electrophoresis and sample injection systems integrated on a planar glass chip. Anal. Chem., 64 (1992) 1926–1932.

21 C.S. Effenhauser, A. Manz and H.M. Widmer, Glass chips for high-speed capillary electrophoresis separations with submicrometer plate heights. Anal. Chem., 65 (1993) 2637–2642.

22 D. Schmalzing, L. Koutny, A. Adourian, P. Belgrader, P. Matsudaira, and D. Ehrlich, DNA typing in thirty seconds with a microfabricated device. Proc. Natl. Acad. Sci. USA., 94 (1997) 10273–10278.

23 S.C. Jacobson, C.T. Culbertson, J.E. Daler and J.M. Ramsey, Microchip structures for submillisecond electrophoresis. Anal. Chem., 70 (1998) 3476–3480.

24 S.R. Liu, H.J. Ren, Q.F. Gao, D.J. Roach, R.T. Loder, T.M. Armstrong, Q.L. Mao, I. Blaga, D.L. Barker and S.B. Jovanovich, Automated parallel DNA sequencing on multiple channel microchips. Proc. Natl. Acad. Sci. USA., 97 (2000) 5369–5374.

25 S.R. Wallenborg and C.G. Bailey, Separation and detection of explosives on a microchip using micellar electrokinetic chromatography and indirect laser-induced fluorescence. Anal. Chem., 72 (2000) 1872–1878.

26 L.H. Zhang, F.Q. Dang and Y. Baba, Microchip electrophoresis-based separation of DNA. J. Pharm. Biomed. Anal., 30 (2003) 1645–1654.

27 L.D. Hutt, D.P. Glavin, J.L. Bada and R.A. Mathies, Microfabricated capillary electrophoresis amino acid chirality analyzer for extraterrestrial exploration. Anal. Chem., 71 (1999) 4000–4006.

28 S.R. Wallenborg, I.S. Lurie, D.W. Arnold and C.G. Bailey, On-chip chiral and achiral separation of amphetamine and related compounds labeled with 4-fluoro-7-nitrobenzofurazane. Electrophoresis, 21 (2000) 3257–3263.

29 I. Rodriguez, L.J. Jin and S.F.Y. Li, High-speed chiral separations on microchip electrophoresis devices. Electrophoresis, 21 (2000) 211–219.

30 M.T. Reetz, K.M. Kühling, A. Deege, H. Hinrichs and D. Belder, Super-high-throughput screening of enantioselective catalysts by using capillary array electrophoresis. Angew. Chem. Int. Ed., 39 (2000) 3891–3893.

31 H. Wang, Z.P. Dai, L. Wang, J.L. Bai and B.C. Lin, Enantiomer separation of amino acids on microchip-based electrophoresis. Chinese J. Anal. Chem., 30 (2002) 665–669.

32 D. Belder, A. Deege, M. Maass and M. Ludwig, Design and performance of a microchip electrophoresis instrument with sensitive variable-wavelength fluorescence detection. Electrophoresis, 23 (2002) 2355–2361.

33 M.A. Schwarz and P.C. Hauser, Rapid chiral on-chip separation with simplified amperometric detection. J. Chromatogr. A, 928 (2001) 225–232.

34 E. Ölvecka, M. Masar, D. Kaniansky, M. Jöhnck and B. Stanislawski, Isotachophoresis separations of enantiomers on a planar chip with coupled separation channels. Electrophoresis, 22 (2001) 3347–3353.

35 S. Il Cho, K.N. Lee, Y.K. Kim, J. Jang and H. Chung, Chiral separation of gemifloxacin in sodium-containing media using chiral crown ether as a chiral selector by capillary and microchip electrophoresis. Electrophoresis, 23 (2002) 972–977.

36 B.F. Liu, M. Ozaki, Y. Utsumi, T. Hattori and S. Terabe, Chemiluminescence detection for a microchip capillary electrophoresis system fabricated in poly(dimethylsiloxane). Anal. Chem., 75 (2003) 36–41.

37 D. Belder and M. Ludwig, Microchip electrophoresis for chiral separations. Electrophoresis, 24 (2003) 2422–2430.

38 M.G. Finn, Emerging methods for the rapid determination of enantiomeric excess. Chirality, 14 (2002) 534–540.

39 M.T. Reetz, Combinatorial and evolution-based methods in the creation of enantioselective catalysts. Angew. Chem. Int. Ed., 40 (2001) 284–310.

40 M.T. Reetz, A. Zonta, K. Schimossek, K. Liebeton and K.E. Jaeger, Creation of enantioselective biocatalysts for organic chemistry by in vitro evolution. Angew. Chem. Int. Ed., 36 (1997) 2830–2832.

41 M.T. Reetz and K.E. Jaeger, Enantioselective enzymes for organic synthesis created by directed evolution. Chem. Eur. J., 6 (2000) 407–412.

42 A.T. Woolley, G.F. Sensabaugh and R.A. Mathies, High-speed DNA genotyping using microfabricated capillary array electrophoresis chips. Anal. Chem., 69 (1997) 2181–2186.

43 Y.N. Shi, P.C. Simpson, J.R. Scherer, D. Wexler, C. Skibola, M.T. Smith and R.A. Mathies, Radial capillary array electrophoresis microplate and scanner for high-performance nucleic acid analysis. Anal. Chem., 71 (1999) 5354–5361.

44 J.W. Simpson, M.C. Ruiz-Martinez, G.T. Mulhern, J. Berka, D.R. Latimer, J.A. Ball, J.M. Rothberg and G.T. Went, A transmission imaging spectrograph and microfabricated channel system for DNA analysis. Electrophoresis, 21 (2000) 135–149.

45 C.-X. Zhang and A. Manz, Narrow sample channel injectors for capillary electrophoresis on microchips. Anal. Chem., 73 (2001) 2656–2662.

46 A. Van den Berg and T.S.J. Lammerink, Micro total analysis systems: microfluidic aspects, integration concept and applications, in: H. Becker and A. Manz (Eds.), Microsystems Technology in Chemistry and Life Science, Vol. 194, Topics in Current Chemistry, 1998, pp. 21–49.

47 S.A. Soper, A.C. Henry, B. Vaidya, M. Galloway, M. Wabuyele and R.L. McCarley, Surface modification of polymer-based microfluidic devices. Anal. Chim. Acta, 470 (2002) 87–99.

48 R.M. McCormick, R.J. Nelson, M.G. Alonso-Amigo, J. Benvegnu and H.H. Hooper, Microchannel electrophoretic separations of DNA in injection-molded plastic substrates. Anal. Chem., 69 (1997) 2626–2630.

49 M.A. Roberts, J.S. Rossier, P. Bercier and H. Girault, UV laser machined polymer substrates for the development of microdiagnostic systems. Anal. Chem., 69 (1997) 2035–2042.

50 L. Martynova, L.E. Locascio, M. Gaitan, G.W. Kramer, R.G. Christensen and W.A. MacCrehan, Fabrication of plastic microfluid channels by imprinting methods. Anal. Chem., 69 (1997) 4783–4789.

51 H. Becker and W. Deitz, Microfluidic devices for μ-TAS applications fabricated by polymer hot embossing. Proc. SPIE, 3515 (1998) 177.

52 L.L. Shultz-Lockyear, C.L. Colyer, Z.H. Fan, K.I. Roy and D.J. Harrison, Effects of injector geometry and sample matrix on injection and sample loading in integrated capillary electrophoresis devices. Electrophoresis, 20 (1999) 529–538.

53 J.P. Alarie, S.C. Jacobson, C.T. Culbertson and J.M. Ramsey, Effects of the electric field distribution on microchip valving performance. Electrophoresis, 21 (2000) 100–106.

54 S.C. Jacobson, R. Hergenroder, L.B. Koutny, R.J. Warmack and J.M. Ramsey, Microchip capillary electrophoresis with an integrated postcolumn reactor. Anal. Chem., 66 (1994) 1107–1113.

55 Y.J. Liu, R.S. Foote, S.C. Jacobson, R.S. Ramsey and J.M. Ramsey, Electrophoretic separation of proteins on a microchip with noncovalent, postcolumn labeling. Anal. Chem., 72 (2000) 4608–4613.

56 S.C. Jacobson, L.B. Koutny, R. Hergenröder, A.W. Moore, and J.M. Ramsey, Microchip capillary electrophoresis with an integrated postcolumn reactor. Anal. Chem., 66 (1994) 3472–3476.

57 C.X. Zhang and A. Manz, Narrow sample channel injectors for capillary electrophoresis on microchips. Anal. Chem., 73 (2001) 2656–2662.

58 J.E. Thompson, T.W. Vickroy and R.T. Kennedy, Rapid determination of aspartate enantiomers in tissue samples by microdialysis coupled on-line with capillary electrophoresis. Anal. Chem., 71 (1999) 2379–2384.

59 K.B. Mogensen, N.J. Petersen, J. Hubner and J.P. Kutter, Monolithic integration of optical waveguides for absorbance detection in microfabricated electrophoresis devices. Electrophoresis, 22 (2001) 3930–3938.

60 H. Salimi-Moosavi, Y.T. Jiang, L. Lester, G. McKinnon and D.J. Harrison, A multireflection cell for enhanced absorbance detection in microchip-based capillary electrophoresis devices. Electrophoresis, 21 (2000) 1291–1299.

61 H. Nakanishi, T. Nishimoto, A. Arai, H. Abe, M. Kanai, Y. Fujiyama and T. Yoshida, Fabrication of quartz microchips with optical slit and development of a linear imaging UV detector for microchip electrophoresis systems. Electrophoresis, 22 (2001) 230–234.

62 M. Ludwig, F. Kohler and D. Belder, High-speed chiral separations on a microchip with UV detection. Electrophoresis, 24 (2003) 3233–3238.

63 C. Perrin, Y. Vander Heyden, M. Maftouh and D.L. Massart, Rapid screening for chiral separations by short-end injection capillary electrophoresis using highly sulfated cyclodextrins as chiral selectors. Electrophoresis, 22 (2001) 3203–3215.

64 C.T. Culbertson, S.C. Jacobson and J.M. Ramsey, Microchip devices for high-efficiency separations. Anal. Chem., 72 (2000) 5814–5819.

65 B.M. Paegel, L.D. Hutt, P.C. Simpson and R.A. Mathies, Turn geometry for minimizing band broadening in microfabricated capillary electrophoresis channels. Anal. Chem., 72 (2000) 3030–3037.

66 D. Belder, A. Deege, F. Kohler and M. Ludwig, Poly(vinyl alcohol)-coated microfluidic devices for high-performance microchip electrophoresis. Electrophoresis, 23 (2002) 3567–3573.

67 N. Piehl, M. Ludwig and D. Belder, Subsecond chiral separations on a microchip. Electrophoresis, 25 (2004) 2463–2466.

68 D. Belder, M. Ludwig, L.-W. Wang and M.T. Reetz, Enantioselective catalysis and analysis on a chip. Angew. Chem. Int. Ed., 45 (2006) 2463–2466.

PART III

© 2006 Elsevier B.V. All rights reserved.
Chiral Analysis
K.W. Busch and M.A. Busch, Eds.

CHAPTER 10

Instrumental aspects of chiroptical detection

Kenneth W. Busch, Marianna A. Busch and Carlos Calleja-Amador

Department of Chemistry and Biochemistry, Baylor University, One Bear Place #97348, Waco, Texas 76798, USA

Enantiomers are mirror-image isomers with identical chemical and physical properties in an achiral environment. To distinguish between a pair of enantiomers, one must either place them in a chiral environment or use a probe that is inherently chiral in nature (like a polarized light beam). In the former case, enantiomeric discrimination can be carried out chemically by means of chiral auxiliary agents. A chiral auxiliary agent is a homochiral form of a chiral molecule that can interact with the two enantiomeric forms of a chiral pair to produce diastereomeric pairs with different chemical and physical properties that can then be distinguished from one another.

Alternatively, enantiomeric discrimination can also be carried out physically with a chiral probe using a polarized light beam. Although the phenomenon of "polarization of light" has been known since its discovery by Christian Huygens in 1690,[1] and is used extensively today in chemistry and other areas, many research scientists with an interest in enantiomeric discrimination may be unfamiliar with the various practical aspects of the polarization of light as it relates to chiroptical instrumentation.

This chapter will be divided into two parts which covers two basic instrumental topics relevant to modern chiroptical instruments. Part A will discuss the properties and production of polarized light and endeavor to provide a basic introduction to the topic that can serve both as a tutorial and background for later chapters. Because most readers will be interested in the practical aspects of the topic, the mathematical aspects will be kept to a minimum. Readers wishing a more detailed, in-depth discussion of light polarization should consult more advanced texts devoted to this subject [1–4].

Modern chiroptical instrumentation depends critically on the ability to detect very weak signals. As a result, Part B of this chapter will discuss some aspects of signal handling that allow very weak signals to be detected in the presence of noise. Modulation followed by synchronous demodulation provides a powerful means of locking-in on a signal of interest while simultaneously rejecting other signal components

---

[1] C. Huygens, Traité de la lumiè, Van der Aa, Leyden, 1690.

that are not of interest. As a result, the chapter will conclude with a discussion of phase-sensitive detection as a means of signal-to-noise enhancement in signal recovery.

PART A—THE POLARIZATION OF LIGHT

10.1 LIGHT AS A WAVE

Many of the properties of light as it interacts with matter can be explained by assuming that light is an electromagnetic wave propagating in space, as shown in Fig. 10.1. The wave consists of an oscillating electric vector along with an oscillating magnetic vector at right angles to the electric vector. Such a wave is referred to as a transverse electromagnetic wave because the oscillating vectors are orthogonal to the direction of propagation. In most of our discussions, only the electric vector will be considered.[2]

Sinusoidal wave motion of the type illustrated in Fig. 10.1 can be generated by a phasor rotating counterclockwise at a particular frequency, as shown in Fig. 10.2. As shown in the figure, sine wave 1 (the projection of the phasor on the y-axis) is generated as phasor 1, whose magnitude is A, rotates counterclockwise with an angular frequency of ω radians per second. As the phasor rotates, its y amplitude is given by

$$y = A\sin(\omega t - \phi), \tag{1}$$

where ϕ is the phase angle of the waveform. In Fig. 10.2, ϕ equals zero for phasor 1 and -45° for phasor 2. Since the phasors are rotating counterclockwise, phasor 1 is said to

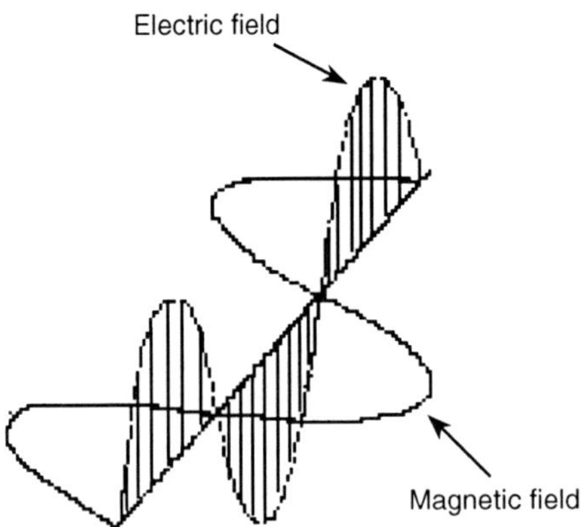

Fig. 10.1. Light as a transverse electromagnetic wave.

[2] Since both vectors are always proportional to one another, specifying one is tantamount to specifying the other. From a chemical standpoint, the electric vector is generally considered in preference to the magnetic vector by convention because it is the vector that is involved with inducing dipoles in the matter during wave propagation.

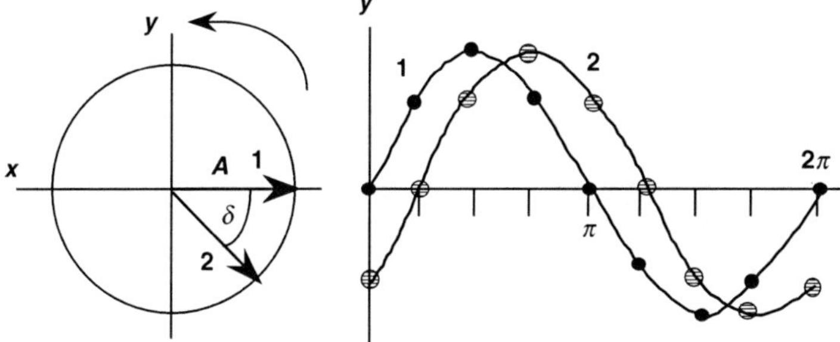

Fig. 10.2. Generation of a sine wave with a rotating phasor.

lead phasor 2 by 45°. For some discussions, it is helpful to express ϕ in terms of a length x, where x is given by

$$x = \left(\frac{\phi}{2\pi}\right)\lambda \tag{2}$$

Substituting Eq. (2) into Eq. (1) gives

$$y = A\sin(\omega t - kx), \tag{3}$$

where k is the propagation number given by $2\pi/\lambda$.

The vector represented by Eq. (1) can also be represented in complex notation as

$$y = Ae^{i(\omega t - kx)} = Ae^{i\omega t} \cdot e^{-i\delta} \tag{4}$$

where δ is equal to kx. Equation (4) is useful when studying the combined influence of several waves where the resultant amplitude depends on the amplitudes and phase angles of the individual waves, but not on the frequency of the wave, which does not vary. In the complex plane, $Ae^{-i\delta} = A(x + iy) = A\,(\sin\delta + i\cos\delta)$. In this notation, the real part of the expression is the instantaneous magnitude of the electric vector.

10.2 TYPES OF POLARIZED LIGHT

Light that is emitted from an ordinary light source (i.e. a non-laser source) can be considered to arise from a collection of independent radiators. A given independent radiator from a collection of radiators will emit a wave train that varies in its phase angle as well as the plane of the electric vector from wave trains emitted by other radiators in the collection. If we consider only the electric vector of the light wave over a period of time, the plane of the electric vector will vary randomly, and the composite of all of the independent wave trains will be a light beam where the electric vector is more or less randomly oriented about the direction of propagation, as shown in Fig. 10.3(a).[3]

[3] The convention of portraying unpolarized light as a many-pointed star is not strictly true. Indeed, most light is neither completely polarized nor completely unpolarized. See Ref. [2].

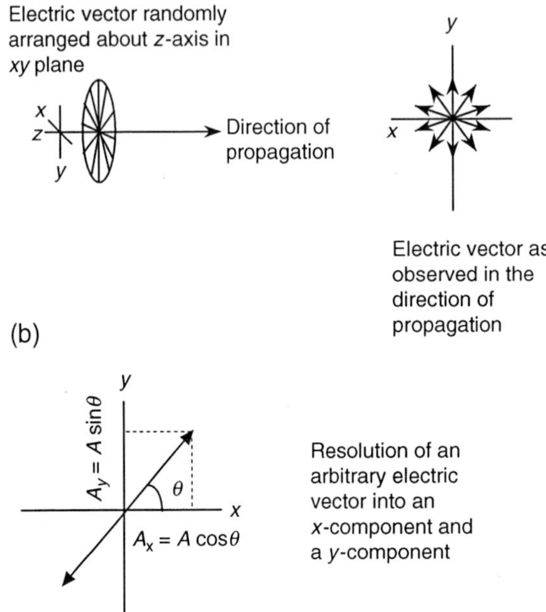

Fig. 10.3. (a) Electric vector is randomly arranged about the z-axis in unpolarized light; (b) Resolution of an arbitrarily oriented electric vector into an x-component and a y-component.

Such a beam is referred to as being unpolarized.[4] Fig. 10.3(b) shows how an electric vector with an arbitrary orientation with respect to the x-axis can be resolved into an x-component and a y-component. If θ is allowed to vary randomly with time, all possible angles will be observed. The net effect with an unpolarized beam will be that $\langle A_x \rangle = \langle A_y \rangle$, where the brackets refer to the time average value rather than the instantaneous value. Such a beam can be envisioned as being composed of two orthogonal vectors of equal magnitude.

Fig. 10.4 shows a convenient way of representing the two orthogonal vectors. Electric vectors in the x-plane are coming out of the plane of the paper and are referred to as s-polarized and are represented in figures as a series of dots along the direction of propagation. Electric vectors in the y-plane are in the plane of the paper and are referred to as p-polarized and are represented as a series of equally spaced short lines. Unpolarized light, which is made up of equal amounts of x-polarization and y-polarization, is represented as a combination of both lines and dots.

In contrast to unpolarized light, polarized light can be said to have a preference for a particular transverse plane (or a preference for a particular handedness, in case of

---

[4] In an unpolarized beam, the orientation of the electric vector is envisioned to vary randomly over time intervals on the order of 10^{-8} s. See Ref. [2].

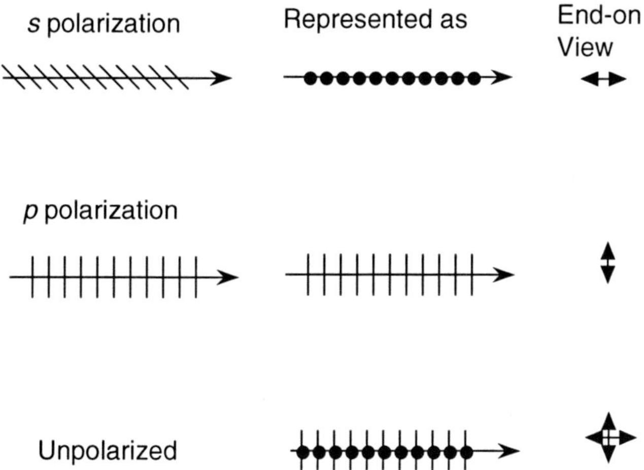

Fig. 10.4. Diagrammatic representation of *s*- and *p*-polarizations.

circularly polarized light). Polarized light can be classified into three general categories with the first two as special cases of the third:

- linearly polarized light,
- circularly polarized light,
- elliptically polarized light.

As we will see later, any form of polarization can be converted into another form by use of a polarization-form converter.

10.2.1 Linearly polarized light

If the electric vectors of a propagating beam have preference for a particular plane, as shown in Fig. 10.5, the light is referred to as being linearly polarized. Linear polarization can be envisioned mathematically as being generated by two counter-rotating phasors of equal length, one rotating counterclockwise and the other rotating clockwise at the same frequency, as shown in Fig. 10.6. As the phasors rotate, the y-projection of their vector sum produces a linearly polarized resultant in the y-direction.

10.2.2 Circularly polarized light

In the previous section, we envisioned how linearly polarized light could be visualized as being generated mathematically by looking at the y-component of two counter-rotating phasors of equal length, rotating at the same frequency. In the case of circularly polarized light, consider two linearly polarized wave trains that are orthogonal to each other, as shown in Fig. 10.7. These wave trains can be represented by two complex expressions of the form $Ae^{-i\delta}$, where the y-component is written as $A_y e^{-\delta_y}$ and the

Linearly Polarized Light

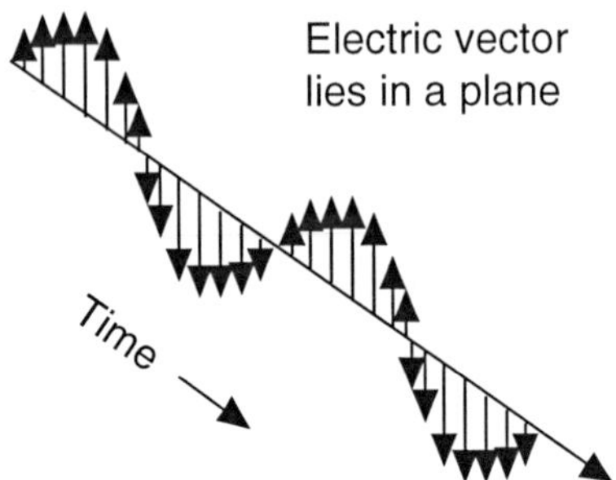

Fig. 10.5. Linearly polarized light.

x-component is written as $A_x e^{-i\delta_x}$. Circularly polarized light will result if $A_y = A_x$, and the phase difference between the x- and y-components is 90° (see Appendix C). If $(\delta_y - \delta_x) = \gamma = +90°$, the beam is said to be right-circularly polarized.[5] With a right-circularly polarized beam, an observer looking directly at the beam from some distance would observe an electric vector of constant magnitude whose tip would trace out a circle in a clockwise sense.[6] Viewed at an angle to the direction of propagation, the tip of the electric vector would rotate as it moves along the propagation direction and trace out a right-handed helix.[7] If $\gamma = -90°$, the beam would be left-circularly polarized and would trace out a left-handed helix when viewed at an angle to the direction of propagation.[8] Fig. 10.8 shows the pattern traced by the electric vector for left-circularly polarized light.

10.2.3 Elliptically polarized light

Consider again two linearly polarized wave trains that are orthogonal to each other. Elliptically polarized light will result when there is a phase shift between the x- and y-components and $A_x \neq A_y$. Elliptically polarized light is characterized by two parameters, as shown in Fig.10.9: α, which is the azimuthal angle,[9] and the ellipticity, which is

---

[5] The vertical component leads the horizontal component by 90°.

[6] Since linearly polarized light can be generated mathematically by two counter-rotating phasors, it is clear that linearly polarized light can be thought of as being composed of equal amounts of left- and right-circularly polarized light.

[7] A right-handed helix is similar to the thread of a typical machine screw. If a right-handed helix is moved toward an observer through a fixed transverse plane, the intersection of the helix with the plane will trace out a circle in the clockwise direction.

[8] An observer looking directly at the beam from some distance would observe an electric vector of constant magnitude whose tip would trace out a circle in a counterclockwise sense.

[9] The azimuthal angle is the angle between the major axis of the ellipse and the x-(horizontal) axis.

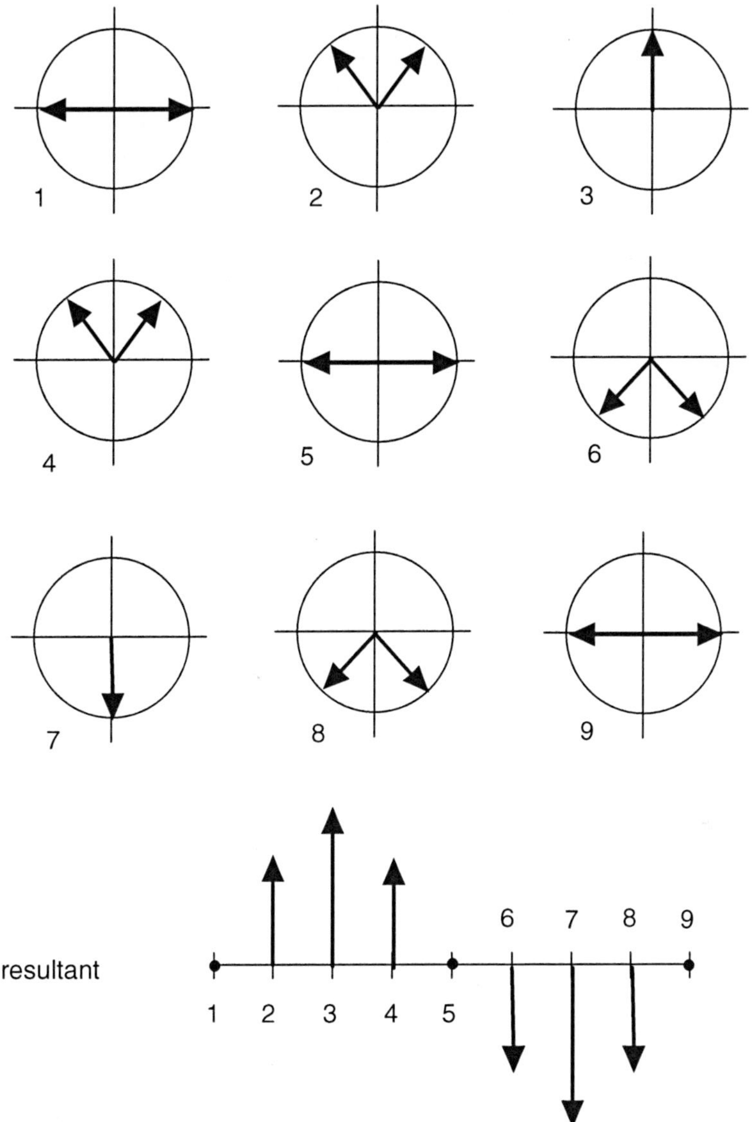

Fig. 10.6. Generation of linearly polarized light by two counter-rotating vectors of equal magnitude.

the ratio of the minor ellipse semi-axis (b) to the major ellipse semi-axis (a). Linear polarization and circular polarization are two special cases of elliptical polarization, where linear polarization corresponds to an ellipticity of 0 and circular polarization corresponds to an ellipticity of 1. Fig. 10.10 shows the sectional patterns produced by the electric vector of a wave train over a suitable time period as observed by an observer looking back at the light source from some distance.

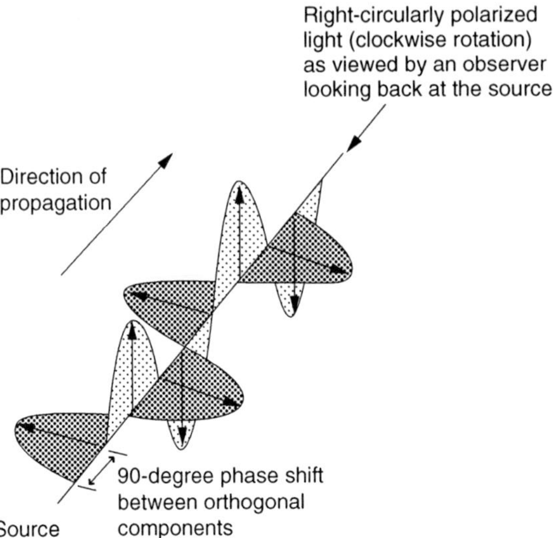

Fig. 10.7. Production of right-circularly polarized light for two orthogonal waves of equal amplitude but phase-shifted by 90°.

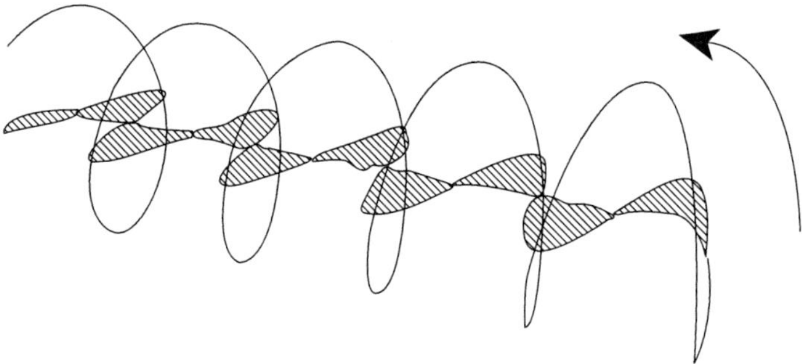

Fig. 10.8. Left-circularly polarized light forms a helix rotating counterclockwise when viewed looking back along the z-axis at the source.

10.3 PRODUCTION OF LINEARLY POLARIZED LIGHT

Having defined the types of polarized light, we will now discuss how polarized light can be produced. Polarization is generally accomplished by means of an optical device known as a polarizer.[10] In this section, we will discuss the different types of polarizers that are available and their performance parameters.

Polarizers divide an incident beam into two orthogonally linearly polarized beams that can be separated spatially by some means. The basic strategy behind a polarizer is to

[10] A polarizer is a purely optical device that can convert a beam of unpolarized light into one that is appreciably polarized in some form. Linear polarizers produce linearly polarized light, circular polarizers produce circularly polarized light, etc.

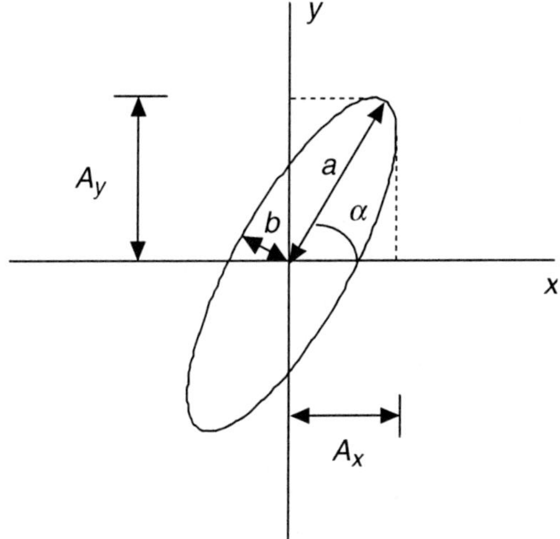

Fig. 10.9. Elliptically polarized light.

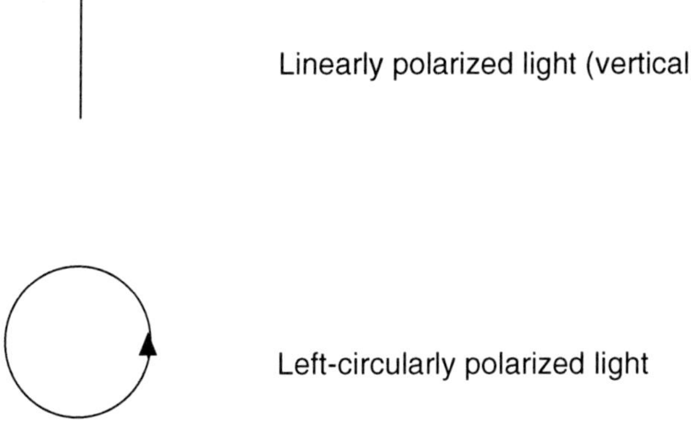

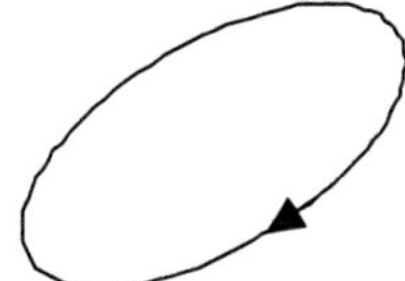

Fig. 10.10. Sectional patterns for linearly polarized light, left-circularly polarized light, and right-elliptically polarized light.

References p. 341

transmit one polarized beam and absorb, reflect, or divert the other. Four optical phenomena can be used to achieve these goals:

- scattering,
- reflection,
- birefringence,
- dichroism.

Before discussing the different types of polarizers available, it is worthwhile to consider some of the operational characteristics that impact the selection of an appropriate polarizer. Polarizers are characterized by a variety of performance factors:

- Useful wavelength range,
- Extinction ratio—the ratio of the light transmitted by two crossed polarizers to that transmitted when the polarization axes are parallel (see Appendix B),
- Angular aperture—the extent of beam divergence that can be tolerated,
- Spatial aperture—the cross-sectional area of the beam that can be transmitted,
- Effect of dispersion[11] on the performance,
- Transmission efficiency—how much light is lost,
- Power-handling ability
  - Is the unwanted polarization component dumped without injury to the device or laboratory personnel?
  - Can any cement used in fabricating the device withstand the anticipated power levels?
- Expense—many of the best polarizers are costly because they are made from naturally occurring crystals whose supply is limited.

10.3.1 Polarization by scattering

When light passes through a collection of small particles, photons of the light beam can undergo elastic collisions with the particles, thereby altering the directions of the scattered photons. Such elastic scattering is referred to as Rayleigh scattering, which is responsible for the blue color of the sky in the daytime. Rayleigh scattering is:

- proportional to the intensity of the incident beam,
- proportional to the number of scattering centers per unit volume,
- proportional to λ^{-4}, where λ is the wavelength of the incident light—short wavelength light is preferentially scattered.

Fig. 10.11 illustrates the situation with Rayleigh scattering. The figure shows a tank of water to which a few drops of milk have been added to make the liquid slightly turbid. An unpolarized light beam emitted from a source at the left can be considered as being composed equally of wave trains vibrating in the yz and xz planes. Fig. 10.11 shows that

---

[11] Dispersion is the change in refractive index with wavelength.

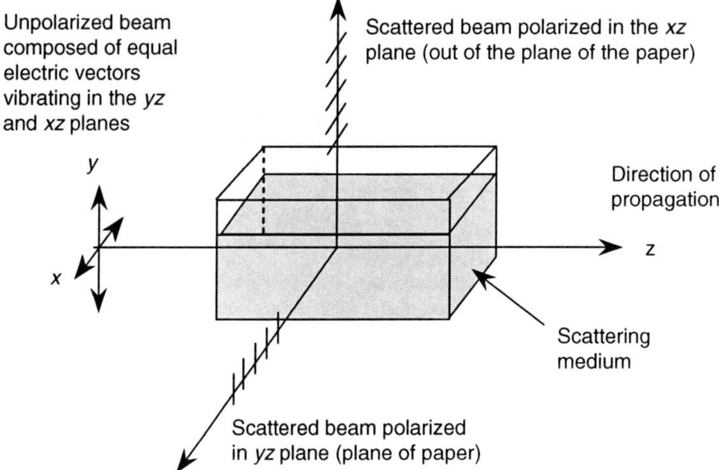

Fig. 10.11. Polarization of a beam of light scattered by small particles suspended in a liquid medium.

light scattered in the horizontal direction will be polarized in the *yz* plane, while light scattered in the vertical direction will be polarized in the *xz* plane. This polarization phenomenon is a direct consequence of the transverse nature of light.[12] While scattering does produce polarization, it is rarely used to polarize light in the optical region and so will not be discussed further in this chapter.[13]

10.3.2 Polarization by reflection

In 1809, Etienne Malus, a student of Fourier, published an article on the polarization of light by reflection. Consider an unpolarized beam that is incident on an optical surface at some angle to the normal. Such a beam can be considered as being composed of equal amounts of *s*- and *p*-polarized components. As shown in Fig. 10.12, part of the beam will be reflected from the surface and part will be transmitted into the medium according to the Snell's law.[14] According to Fresnel's equations, the reflectance for the *s*- and *p*-components will vary with the angle of incidence of the beam:

$$R_p = \frac{\tan^2(i - \phi)}{\tan^2(i + \phi)} \tag{5}$$

$$R_s = \frac{\sin^2(i - \phi)}{\sin^2(i + \phi)}, \tag{6}$$

[12] If light scattered in the horizontal direction were polarized in the *xz* plane, the wave would be a longitudinal wave (like a sound wave) rather than a transverse wave.

[13] Compton scattering has been used to polarize X-rays and λ rays.

[14] See Appendix A for a brief discussion of pertinent optical concepts.

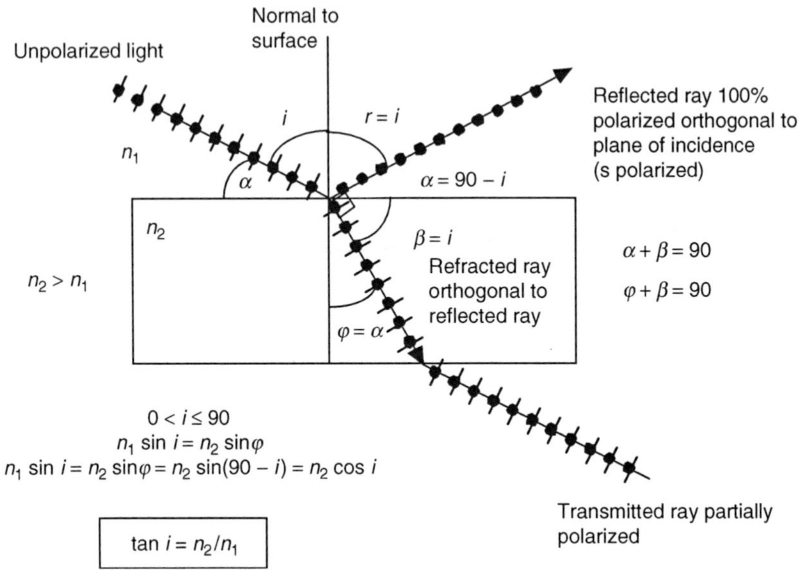

Fig. 10.12. Polarization by reflection when light strikes a dielectric material at Brewster's angle.

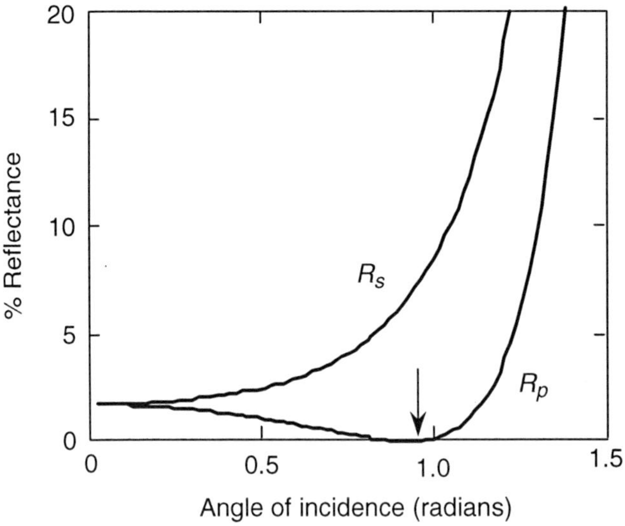

Fig. 10.13. Plot of Fresnel's equations for *s*- and *p*-components of light. At Brewster's angle, the % reflectance for the *p*-component becomes zero.

where R is the reflectance (i.e. the ratio of the reflected intensity to the incident intensity), i is the angle of incidence, and ϕ is the angle of refraction given by Snell's law. Fig. 10.13 shows the plots of Eqs. (5) and (6) as a function of the angle of incidence for a material with a refractive index of 1.3. As indicated by the arrow, the plot for R_p goes

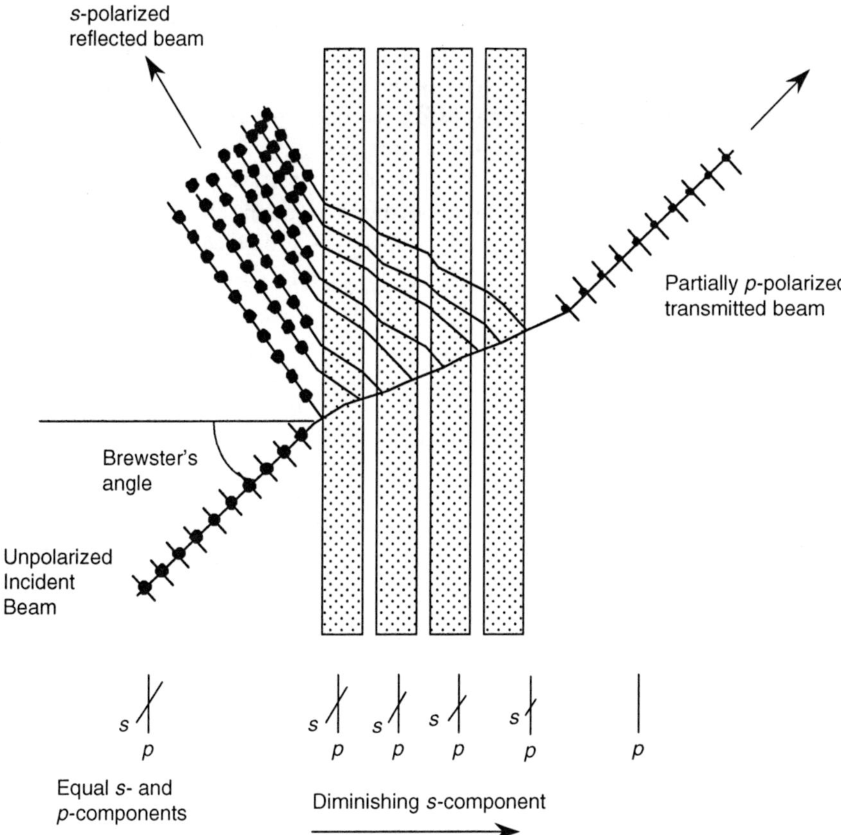

Fig. 10.14. Polarization with a stacked-plate polarizer.

to zero at a particular angle of incidence known as Brewster's angle [5]. Fig. 10.12 shows that Brewster's angle is given by

$$\tan i_{\mathrm{B}} = \frac{n_1}{n_2}, \tag{7}$$

where i_{B} is Brewster's angle and n_1 and n_2 are the refractive indices of the media.

For a material with a refractive index of 1.3, Brewster's angle is 52°, assuming n_1 is 1.0. Fig. 10.12 also shows that at Brewster's angle, the reflected ray and the refracted ray are orthogonal. At Brewster's angle, 100% of the reflected ray is *s*-polarized.

In practice, polarizers based on reflection are generally composed of a series of plates that make up what is referred to as a pile. Such polarizers are often known as stacked-plate polarizers. Fig. 10.14 shows a schematic diagram of a hypothetical stacked-plate polarizer where the incident beam is at Brewster's angle. For convenience, most stacked-plate polarizers are operated in the transmission mode rather than using the reflected beam. Various designs for stacked-plate polarizers have been proposed over the years [6].

Because the transmitted beam is not completely polarized, effective stacked-plate polarizers are generally composed of six or more plates.

Stacked-plate polarizers are particularly useful in the vacuum ultraviolet and the infrared beyond 2.3 μm [7]. Extinction ratios on the order of 10^{-2} can be achieved by careful design [8]. While spatial aperature is not a problem, angular aperture and wavelength dispersion can cause problems.[15] To be effective, the incident radiation must be at Brewster's angle, limiting stacked-plate polarizers to essentially collimated beams. Although power-handling ability is good, transmission efficiency in the vacuum ultraviolet may not exceed 40%.[16] The physical size of stacked-plate polarizers can be an impediment in applications where space is limited.

10.3.3 Polarization by birefringence

In the last quarter of the seventeenth century, the Danish scientist Rasmus Bartholin observed that when objects were viewed through Iceland spar (a naturally occurring form of calcite found in Iceland), the image appeared to be doubled. Today, this phenomenon whereby a beam of light is split into two beams is referred to as birefringence or double refraction. Birefringence is observed in all anisotropic crystals including quartz, calcite, and tourmaline.[17] Fig. 10.15 shows the double image produced when an object is viewed through a calcite crystal.[18] If the crystal is rotated in the direction shown by the arrow in the figure (i.e. the plane of the page), the hatched letters will rotate around the solid letters, tracing the path of the circle shown in the figure. The rays that form the stationary image are referred to as ordinary rays or O-rays (because they are refracted in the normal manner); those that form the image that precesses about the solid letters as the crystal is rotated behave in an extraordinary fashion and are, therefore, referred to as extraordinary rays or E-rays. The O-rays and the E-rays are both linearly polarized orthogonally to one another. Birefringent materials can be divided into two classes: positively birefringent materials and negatively birefringent materials, depending on the magnitude of the refractive indices for the O- and E-rays. In negatively birefringent materials, like calcite, $n_o > n_e$. For positively birefringent materials, the opposite is true.

[15] Wavelength dispersion (i.e. the change of refractive index with wavelength) in the material will cause Brewster's angle, given by Eq. (7) to vary.

[16] While this may seem like a severe disadvantage, in the vacuum UV, alternative polarizers are difficult to come by.

[17] Crystalline solids are made up of periodic structures (unit cells) that form a crystal lattice. If the crystal lattice is such that the atoms that compose it are uniformly spaced along three orthogonal axes, the crystal is said to be isotropic. On the other hand, if the atomic spacing along one of the axes is different from those along the other axes, the crystal is aniostropic. Crystalline anisotropy leads to optical anisotropy, which, in turn, means that the refractive index of the material varies with the direction of the light beam through the crystal.

[18] Calcite is a uniaxial, rhombohedral crystal in the hexagonal crystallographic system.

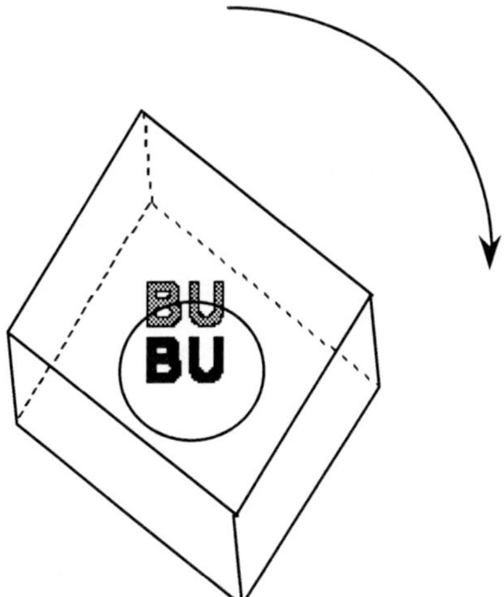

Fig. 10.15. Double image produced by birefringence in a calcite crystal. As the crystal is rotated, the hatched letters move in a circle around the solid letters.

Birefringence prism polarizers made from anisotropic crystals, like calcite, are especially useful in the ultraviolet region of the spectrum.[19] Over the years, various prism configurations have been designed to act as polarizers. These include the

- Ahrens polarizer
- Glazebrook polarizer
- Wallaston polarizer
- Rochon polarizer
- Glan–Foucault polarizer
- Glan–Taylor polarizer
- Glan–Thompson polarizer

Ahrens polarizer. Fig. 10.16 shows a schematic diagram of an Ahrens polarizer, which consists of three prisms of calcite whose optic axes are all parallel, as indicated by the dash/dotted lines. The three prisms that make up the Ahrens polarizer are cemented together with Canada balsam,[20] which acts as an optical coupling material as well as a cement to hold the prisms together. The central wedge-shaped prism presents a relatively large angle of incidence (~ 75°) to the incoming beam because the length-to-width ratio of the polarizer is 1.9. As a result, when unpolarized light strikes the entrance face of the polarizer as shown in the figure, the O-rays undergo total internal reflection at the internal prism interface[21] and are absorbed by black paint that coats the lateral faces of the polarizer. The E-rays, on the other hand, which, for calcite, have a lower

[19] The transmission range for calcite is from 240 to 1800 nm.

[20] Canada balsam is an oleoresin exuded from the balsam fir tree. It has a refractive index of ~ 1.55, which is less than that of calcite, and is transparent down to about 330 nm.

[21] The O-rays have a higher refractive index, which results in a smaller critical angle. If this angle is exceeded, total internal reflection will occur. See Appendix A for a discussion of total internal reflection.

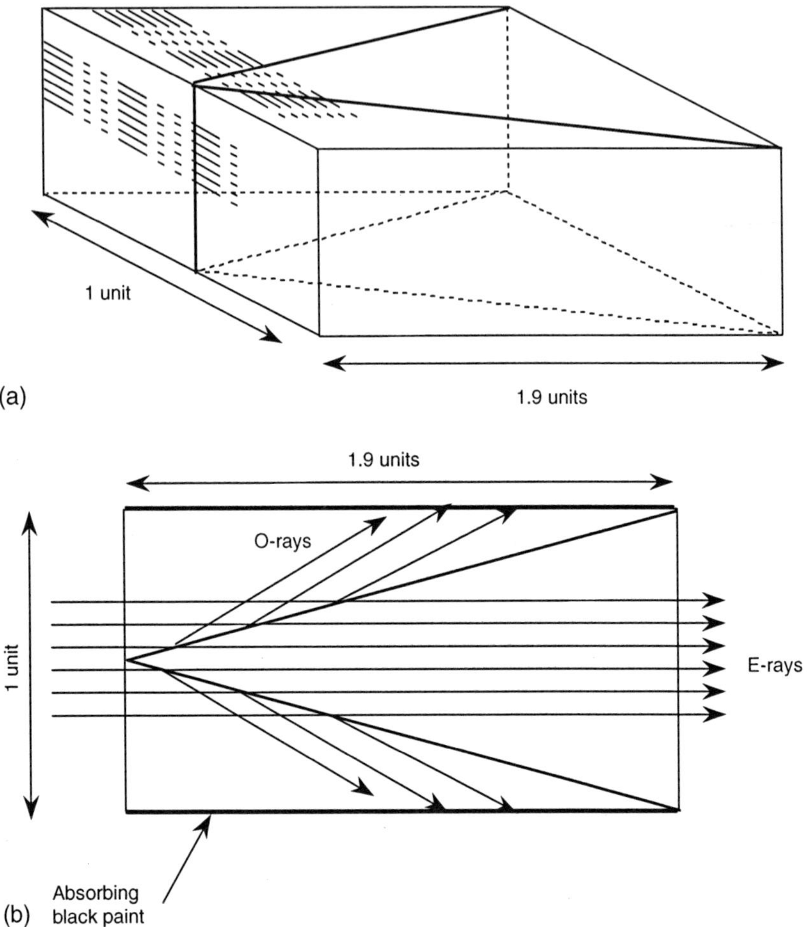

Fig. 10.16. The Ahrens polarizer. (a) Perspective view. Dashed-dot lines show orientation of optic axis. (b) Top view showing directions of E- and O-rays.

refractive index (and therefore a larger critical angle) than the O-rays because of birefringence, pass through the prism interface undeflected. Although its use is limited to the visible region because of the Canada balsam, the Ahrens polarizer does have some advantages:

- large linear aperture,
- large acceptance angle ($\sim 20^\circ$),
- high extinction ratio,
- entrance face orthogonal to incident beam,
- visible wavelength range.

Because black paint is used to absorb the unwanted O-rays, this polarizer is not recommended for use with high-power laser sources, which generally require that the unwanted beam be dumped externally to the polarizer.

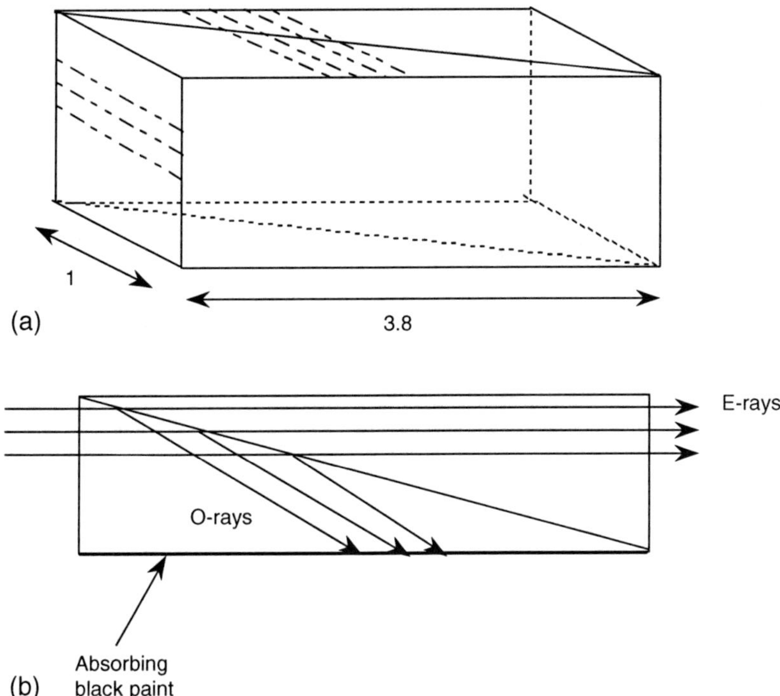

Fig. 10.17. The Glazebrook polarizer. (a) Perspective view. Dashed-dot lines show orientation of optic axis. (b) Top view showing directions of E- and O-rays.

Glazebrook polarizer. The Glazebrook polarizer, shown in Fig. 10.17, is a simpler variation of the Ahrens polarizer and is composed of two prisms cemented together with Canada balsam. Comparing Fig. 10.16 with Fig. 10.17, one sees that the Glazebrook polarizer is essentially one-half of an Ahrens polarizer. As a result, the length-to-width ratio for the Glazebrook polarizer is 3.8. Like the Ahrens polarizer, the E-rays are transmmitted and the O-rays undergo total internal reflection and are absorbed by a layer of black paint.

Wollaston polarizer. Fig. 10.18 shows a schematic diagram of a Wollaston polarizer, which consists of two birefringent prisms cemented together with Canada balsam. As shown in the figure, the optical axes of the two prisms are orthogonal to one another with the optic axis of the entrance-face prism oriented horizontally. With this polarizer, the O-ray experiences a decrease in the refractive index upon entering the second prism and is refracted away from the prism interface normal. The E-ray experiences an increase in the refractive index on entering the second prism and is refracted toward the prism interface normal. As a result, the incident beam is divided into two orthogonally polarized beams that emerge from the prism with a deviation angle ϕ, which depends on the wavelength and the wedge angle θ. The device is especially useful if both polarized beams are required. Because Canada balsam is used as a cement, the Wollaston polarizers are generally restricted to the visible region and limited to low-power levels.

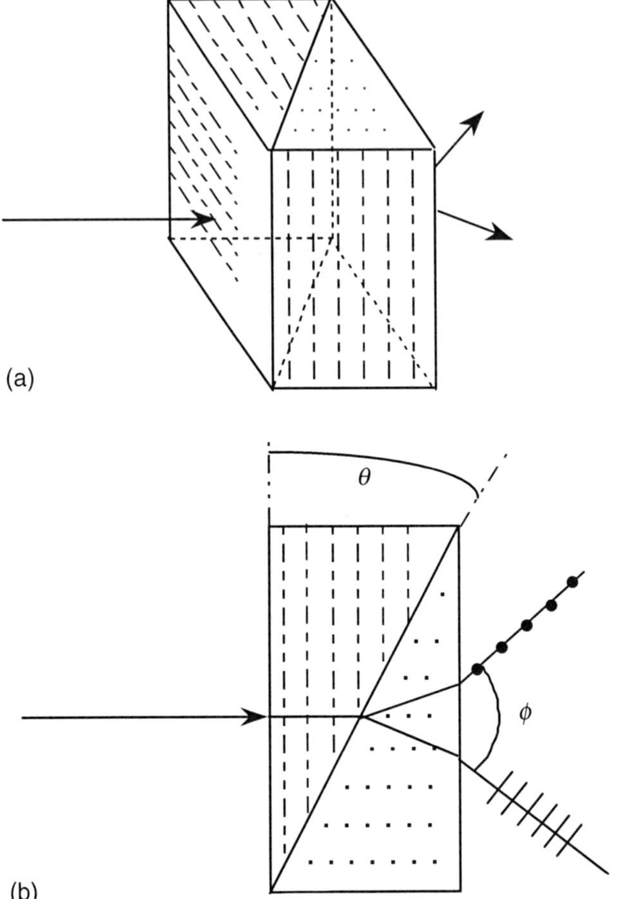

Fig. 10.18. The Wollaston polarizer: θ is the prism angle; ϕ is the deviation between the two emerging beams. (a) Perspective view. Dashed-dot lines show orientation of optic axis. (b) Top view showing directions of emerging rays.

Rochon polarizer. The Rochon polarizer, shown in Fig. 10.19, is similar to the Wollaston polarizer, being made up of two birefringent prisms cemented together with Canada balsam. With this polarizer, the O-ray proceeds through the prism interface undeflected, while the E-ray experiences a decrease in refractive index, causing it to be refracted away from the prism interface normal. The deviation angle ϕ between the two polarized beams depends on θ and the wavelength of the light.

Glan polarizers. Glan polarizers are some of the more popular birefringent polarizers available commercially today. There are three variations of Glan polarizers, which consist of two birefringent prisms joined together in various configurations:

- the Glan–Foucault polarizer (Fig. 10.20),
- the Glan–Taylor polarizer (Fig. 10.21),
- the Glan–Thompson polarizer.

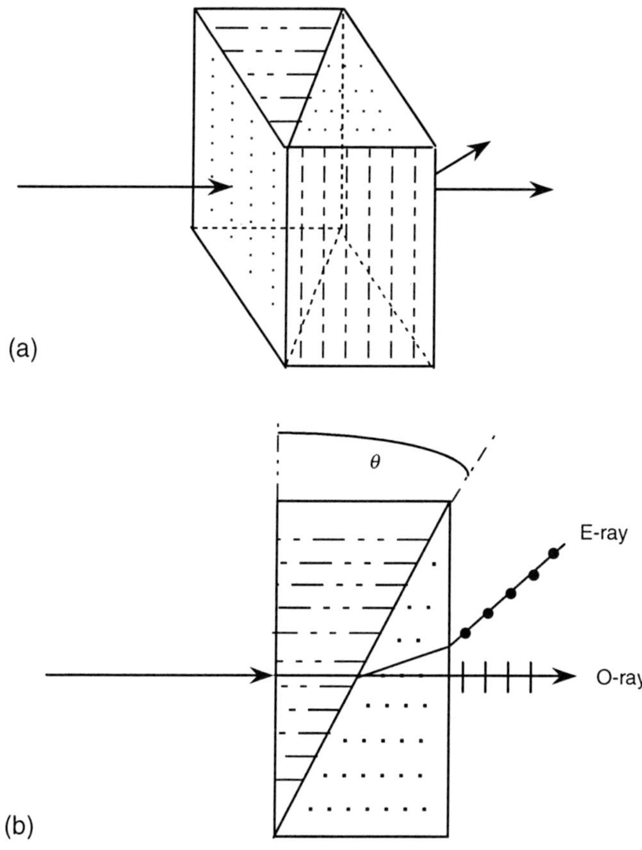

Fig. 10.19. The Rochon polarizer: θ is the prism angle. (a) Perspective view. Dashed-dot lines show orientation of optic axis. (b) Top view showing directions of E- and O-rays.

All Glan polarizers are based on a total-internal reflection principle, where one polarization component (the O-ray) is transmitted by the polarizer and the other is rejected by internal reflection (see Appendix A). In the Glan–Foucault configuration, which was designed specifically to transmit UV light, no coupling cement is used, and the two prisms are separated by a small air gap. In this configuration, the E-ray is transmitted and the O-ray is reflected internally. Because of the air gap, there are two air–calcite interfaces that result in substantial reflection losses for the E-ray, resulting in only modest transmission through the polarizer.

In 1948, Archard and Taylor [9], proposed a modification to the Glan–Foucault configuration that substantially reduced the reflection losses caused by the air gap. For polarization studies in the UV, the Glan–Taylor arrangement is arguably the best birefringent polarizer available commercially today. Because it is based on a total-internal reflection principle, which only works over a small range of angles, the acceptance angle for Glan–Taylor polarizers is about 7°.

In the Glan–Thompson configuration, the two prisms are cemented together. Use of a coupling-cement permits a full acceptance angle cone of 15°. If Canada balsam is used,

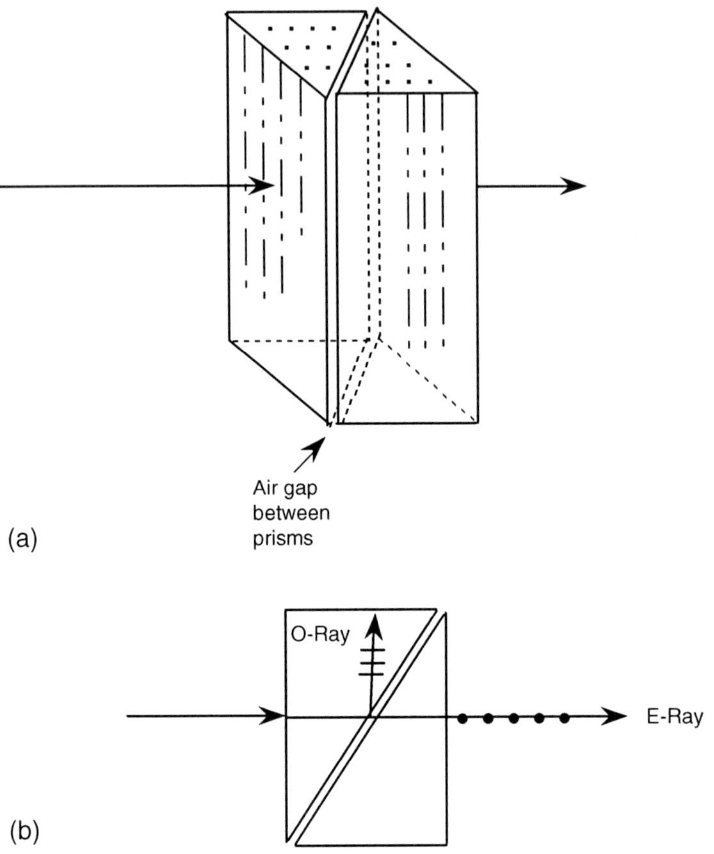

Fig. 10.20. The Glan–Foucault polarizer. (a) Perspective view. Dashed-dot lines show orientation of optic axis. (b) Top view showing directions of E- and O-rays.

this limits the short-wavelength transmission limit to about 330 nm. It is possible to lower the short-wavelength transmission to about 250 nm by using gedamine as a cement instead of Canada balsam.[22]

Table 10.1 compares the performance parameters of the Wollaston polarizer with those of the Glan–Thompson and Glan–Taylor polarizers.

10.3.4 Polarization by wire grids

Fig. 10.22 shows a finely spaced array of parallel wires that can function as a polarizer. Consider an unpolarized beam of light made up of two orthogonal components A_x and A_y, as shown in Fig. 10.22. When the beam encounters the wire grid, the vertical electric

[22] Gedamine is a solution of urea formaldehyde in butyl alcohol with a refractive index of 1.52. The solution forms a resin when the alcohol evaporates.

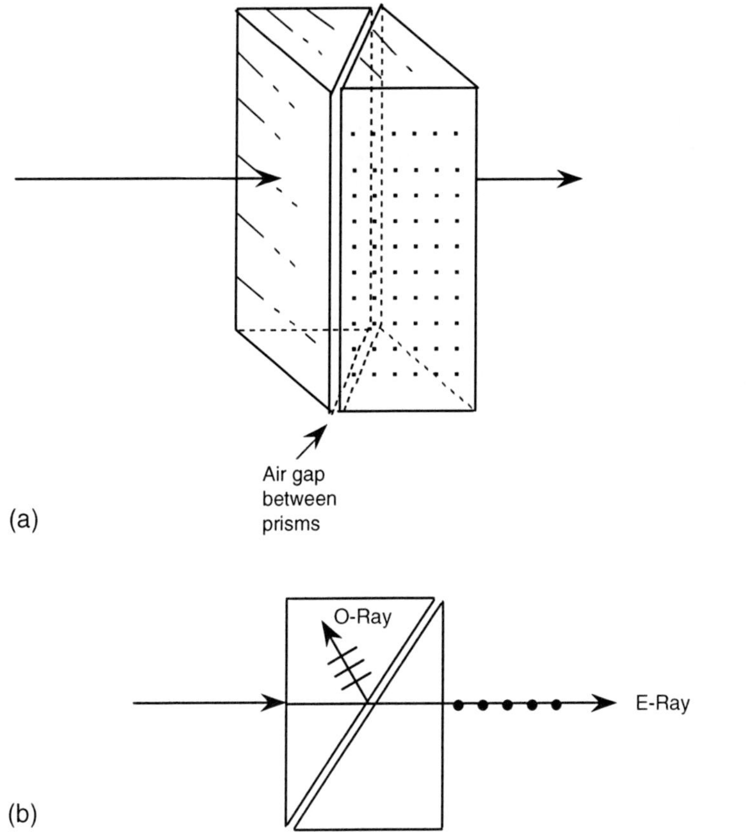

Fig. 10.21. The Glan–Taylor polarizer. (a) Perspective view. Dashed-dot lines show orientation of optic axis. (b) Top view showing directions of E- and O-rays.

Table 10.1 Comparison of prism polarizers

Prism parameter	Wollaston	Glan–Taylor	Glan–Thompson
Material	Calcite	Calcite	Calcite
Wavelength range (nm)	350–2500	220–2500	350–2500
Transmittance (%)	90	88	>90
Extinction ratio	10^{-5}	$<10^{-5}$	$<10^{-5}$
Linear aperture (mm)	10	10	10
Angular aperture	17–22°	8°	16°
Cement	Yes	No	Yes

vector A_y, which is parallel with the wires, will induce alternating currents in the wires that will lead to dissipation by Joule heating (I^2R, where I is the current induced in the wire and R is its resistance). As a result, the vertical-electric vector will be absorbed by the grid. By contrast, the horizontal-electric vector A_x is not shorted out by the wire grid and is transmitted. Because the spacing between the wires must be small with respect to

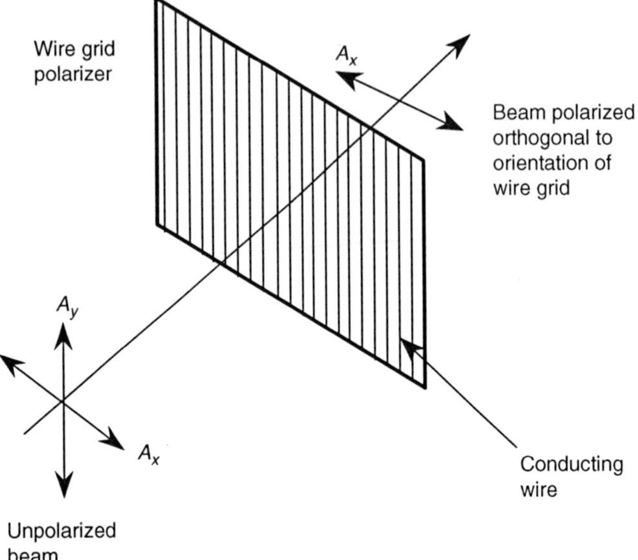

Fig. 10.22. The wire-grid polarizer: A_x and A_y are the electric vectors of the unpolarized light, where $A_x = A_y$.

the wavelength of the incident light, wire polarizers are only feasible for long wavelength radiation like the infrared.[23]

10.3.5 Polarization by dichroism

Polarization by dichroism is similar, in some respects, to polarization with wire grids in that polarization by dichroism involves preferential absorption of one component of unpolarized light and transmission of the other. Dichroism is a property of certain materials that have the ability to absorb light vibrating in a particular plane more strongly than light vibrating in an orthogonal plane.

The phenomenon of dichroism was discovered in 1815 by the French scientist Jean Biot while studying the optical properties of tourmaline [10]. Biot observed that the color of a tourmaline crystal varied when viewed through a slowly rotating linear polarizer. While certain single crystals, like tourmaline, exhibit dichroism, their use as polarizers is limited by high-cost and poor-performance parameters. Perhaps the most serious limitation is the lack of availability of large single crystals with which to make polarizers with sufficient apertures. In spite of efforts to grow single crystals of dichroic materials in the laboratory, the results achieved to date have been limited [11].

To circumvent the problem caused by a general lack of suitable large single crystals of dichroic materials, Edwin Land reasoned that it should be possible to duplicate the

[23] Wire grid polarization can easily be demonstrated with microwaves, where an ordinary oven rack can be used.

effect of a single crystal with a statistical array of microcrystals of the appropriate type [12]. If an array of microcrystals could, by some means, be suspended in a transparent matrix and aligned so that their absorption axes were roughly parallel, a sheet of this material would behave like a thin single crystal of dichroic material. Using crystals of quinine iodosulfate [13,14][24] suspended in cellulose acetate, Land was able to fabricate the first dichroic sheet polarizer (termed the J-sheet polarizer by the Polaroid Corporation®).

Today, J-sheet microcrystalline polarizers have largely been superseded by molecular-sheet polarizers. Molecular-sheet polarizers are made by orienting the transition-dipole moments of molecules in a plastic sheet so that they are all aligned in a particular direction. In this way, light waves with their electric vectors aligned with the transition-dipole moments of the absorbing molecules will be absorbed. Light waves with their electric vectors orthogonal to the transition-dipole moments of the aligned molecules will not be absorbed. In H-type sheet polarizers, sheets of poly(vinyl alcohol) (PVA) are stretched prior to being stained with molecular iodine. The procedure orients the iodine molecules so that they align parallel to the PVA polymer axis (which has been aligned in a particular direction by stretching).

In K-type sheet polarizers, an oriented PVA film is heated in the presence of a catalyst (typically HCl), where it undergoes dehydration. After dehydration, the film becomes highly dichroic. K-type sheets have their transmission axis perpendicular to the direction of stretch. Because they are not doped (like H-sheet polarizers), K-sheet polarizers are more tolerant of higher temperatures.

Sheet polarizers have the following advantages:

- low cost,
- large acceptance angle ($\sim 30°$),
- large linear aperture (5 cm or larger).

While sheet polarizers are available in the near-ultraviolet region, their extinction coefficients compared with prism polarizers is poor. In addition, most sheet polarizers that employ glass protective plates will not transmit light below about 330 nm. Finally, sheet polarizers are not recommended for high-power applications because the unwanted polarization component is actually absorbed by the polarizing film where it is ultimately converted to heat, which must be dissipated.

10.4 PRODUCTION OF CIRCULARLY POLARIZED LIGHT

Generation of circularly polarized light is important in chiroptical instruments like circular dichroism spectrometers. Production of circularly polarized light can be envisioned as a two-step process. First, unpolarized light is linearly polarized and then the linearly polarized beam is subsequently decomposed into circularly polarized light by

[24] Quinine iodosulfate ($C_{80}H_{194}I_6N_8O_{20}S_3$), a derivative of quinine from the bark of the cinchona tree, is a crystalline alkaloid material with a polarizing material 5 times that of tourmaline. It is sometimes called Herpathite after its discoverer William Herapath, an English physician.

References p. 341

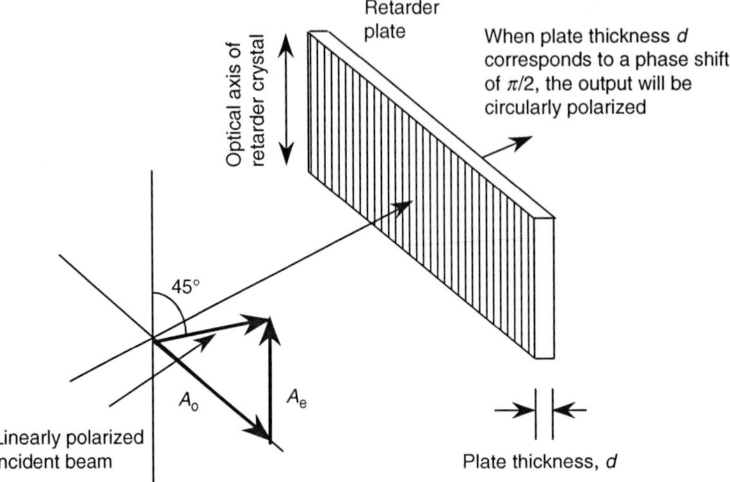

Fig. 10.23. Birefringent retarder plate: A_e and A_o are the electric vectors that make up the incident linearly polarized light whose vibration plane is at a 45° angle to the optical axis of the retarder plate. Because the vibration plane is at 45°, $A_e = A_o$.

means of a polarization-form converter known as a retarder.[25] Retarders perform two basic functions: they resolve an incident linearly polarized beam into two orthogonal beams and then retard one of the beams relative to the other so that the two beams emerge from the retarder with a phase difference δ (which for circularly polarized light is 90° or $\pi/2$).

Fig. 10.23 illustrates the process to produce circularly polarized light. As shown in the figure, linearly polarized light strikes the calcite retarder so that there is a 45° angle between the vibration plane of the polarized light and the optic axis (the fast axis[26]) of the retarder. This causes the linearly polarized beam to be decomposed into two orthogonal components—an O-component (A_o) and an E-component (A_e)—with equal magnitudes. If the retarder is made of a negative crystal like calcite, where $n_o > n_e$, the E-ray, whose vibrations are parallel to the optic axis of the retarder, will travel faster than the O-ray. Both orthogonal beams traverse the retarder along the same ray path, but with different optical path lengths.[27] For the O-ray, the optical path length is given by $n_o d$, where d is the thickness of the retarder plate. For the E-ray, the optical path length will be $n_e d$. Comparing both rays, the optical path length difference, Δ, between the rays when they emerge from the retarder will be

$$\Delta = (n_o - n_e)d \tag{8}$$

[25] Retarders are also known as phase shifters or wave plates.

[26] For a negative crystal like calcite, the fast axis is parallel to the optic axis. For a postive crystal, the reverse is true (i.e. the slow axis is parallel to the optic axis).

[27] The optical path of a ray is the product of the refractive index for the ray times the distance the ray travels in the medium.

This optical path length difference, Δ, can be converted into a corresponding phase difference, δ (in radians), by multiplying by $2\pi/\lambda$ (see Eq. (2)) to give

$$\delta = (2\pi/\lambda)d(n_o - n_e) \tag{9}$$

Eq. (9) shows that the phase difference produced by the retarder for a given wavelength depends on the refractive index difference between the O- and E-rays and the thickness of the retarder plate. Since a phase-shift of 2π does not alter the waves, a multiple of 2π can be added to the left-hand side of Eq. (9) to give plate thicknesses that are realistic:

$$m(2\pi) + \delta = (2\pi/\lambda)\, d\, (n_o - n_e) \tag{10}$$

where m is an interger known as the order. When $m=0$, Eq. (10) reduces to Eq. (11). Wave plates where $m=0$ are known as zero-order plates, while those with a postive value for m are known as multiple-order retarders.

Using calcite as an example, Eq. (10) can be used to get some idea of the plate thickness needed to produce circularly polarized light. For calcite at 589 nm, n_o is 1.658 and n_e is 1.486. To produce circularly polarized light, a phase difference of 90° ($\pi/2$) between the O- and E-rays is needed.[28] For a zero-order retarder, Eq. (9) gives a value of 0.856 μm for d. This is far too thin to be practical to fabricate and to use. By contrast, a multiple-order ($m=100$, for example) quarter-wave plate can be made from calcite with a plate thickness of $d=0.343$ mm. Such a plate must be ground to the precise thickness and optically polished. Since, in calcite, the E-ray travels faster than the O-ray, a quarter-wave plate of calcite will produce left-circularly polarized light.

Since zero-order wave plates have a larger acceptance angle than multiple-order wave plates, other materials besides calcite are often used as retarders. Mica, a birefringent mineral substance, offers a number of advantages as a retarder. At 589 nm, (n_o–n_e) is on the order of 0.005. Using this value in Eq. (9) gives a zero-order thickness for mica quarter-wave plates of 0.059 mm. Since mica is made up of thin sheets, thicknesses of this magnitude can easily be achieved. Of considerable significance to its use as a retarder, the refractive indices for the O- and E-rays in mica do not vary appreciably with wavelength. This means that mica can be used over a wide range of wavelengths (400–700 nm) with satisfactory performance.[29]

10.5 PRODUCTION OF ELLIPTICALLY POLARIZED LIGHT

Elliptically polarized light is produced basically in the same manner as circularly polarized light with linearly polarized light incident on a retarder, as shown previously in Fig. 10.23. Instead of orienting the vibration plane of the incident linearly polarized light

[28] Since one wavelength corresponds to a phasor rotation of 360° or 2π, a phase shift of 90° corresponds to $\lambda/4$. As a result, a retarder that produces a 90° phase shift is referred to as a quarter-wave plate. By analogy, a half-wave plate would produce a phase shift of 180°.

[29] By comparison, for quartz, only a 14 nm range can be tolerated for a 90 ± 1° phase shift. This means that a quartz wave plate must be selected for the wavelength of anticipated use.

at 45° with respect to the optic axis of the retarder (as was done to produce circularly polarized light), another angle is used. At this other angle, $A_o \neq A_e$, and the result is elliptically polarized light.

10.6 PHOTOELASTIC MODULATORS

Certain modern chiroptical instruments, like circular dichroism spectrometers, measure the absorbance difference of a sample for left- and right-circularly polarized light. Because this difference is quite small, AC measurements are used to extract the desired information (the small absorbance difference) from other larger potential instrumental variations like source drift, which would interfere with measurements on a longer time scale. To accomplish this goal, a high-speed alternating source of left- and right-circularly polarized light is needed. High-speed generation of an alternating source of left- and right-circularly polarized light can be accomplished with a device known as a photoelastic modulator or PEM.

PEMs are based on the photoelastic effect whereby birefringence is induced in a transparent solid material like silica by application of a stress.[30] Fig. 10.24(a) shows a schematic diagram of a PEM, where a rectangular bar of a suitable material like silica is coupled to a piezoelectric transducer. PEMs can be thought of as variable retarders. When an electrical signal is applied to the piezoelectric transducer, the silica bar vibrates at a resonant frequency determined by the length of the bar. The piezoelectric transducer, which is driven by an electric control circuit that controls the amplitude of the vibration, is tuned to the resonant frequency of the optical element (say 50 kHz). The birefringence induced in the rectangular bar by the vibration of the piezoelectric transducer varies with time and is a maximum in the center of the optical bar.

Fig. 10.24(b) illustrates the effect of the PEM on a linearly polarized beam that is incident at 45° to the modulator axis (horizontal plane). This 45° incident beam can be considered to be composed of a vertical oscillating electric vector (E_y) and a horizontal oscillating electric vector (E_x) of equal magnitude. As the piezoelectric tranducer applies a stress to the optical bar, birefringence is induced, and the bar behaves like a variable retarder. The amount of retardation, $\Delta(t)$, produced is a function of time and is given by

$$\Delta(t) = d[n_x(t) - n_y(t)], \tag{11}$$

where $\Delta(t)$ is the optical path difference as a function of time, d is the thickness of the optical bar, $n_x(t)$ is the instantaneous value of the refractive index in the horizontal plane, and $n_y(t)$ is the instaneous value of the refractive index in the vertical plane. In terms of a phase difference, δ, Eq. (11) can be written as

$$\delta(t) = (2\pi/\lambda)d[n_x(t) - n_y(t)], \tag{12}$$

where λ is the wavelength of the linearly polarized incident beam.

---

[30] The birefringence induced in the optical element by the applied stress is proportional to the resulting strain produced.

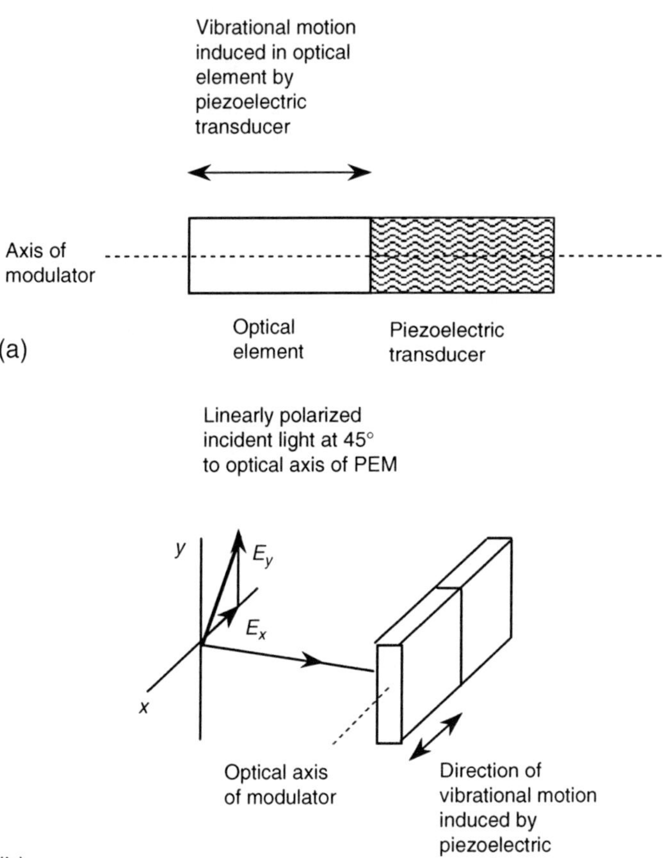

Fig. 10.24. Photoelastic modulator. (a) Axis of modulator; (b) Linearly polarized light incident at 45° to optical axis of modulator: E_x and E_y are the electric vectors that make up the incident linearly polarized light. Because the vibration plane is at a 45° angle, $E_x = E_y$.

We are now in a position to consider the retardation produced over a complete cycle of the piezoelectric oscillator. When the optical bar is not stressed, no birefringence is produced, and $n_x = n_y$. At this point in the cycle, $\delta(0) = 0$, and the incident linearly polarized light at 45° passes through the modulator unaffected (see point a in Fig. 10.25(a)). When the optical bar is stretched by the piezoelectric transducer, the refractive index in the vertical plane, n_y, decreases relative to that in the horizontal plane, n_x. This means that electric vector of the vertical component will lead the electric vector of the horizontal component. According to Eq. (12), this will produce a positive phase shift that will increase until the difference between n_x and n_y reaches a maximum (when the optical bar is stretched to the maximum by the piezoelectric transducer—see point b in Fig. 10.25(a)). By adjusting the magnitude of the control signal applied to the piezoelectric modulator, the maximum positive phase shift can be adjusted so that it is equal to $+90°$. In this condition, the PEM will behave as a quarter-wave plate and right-circularly polarized light will emerge from the modulator.

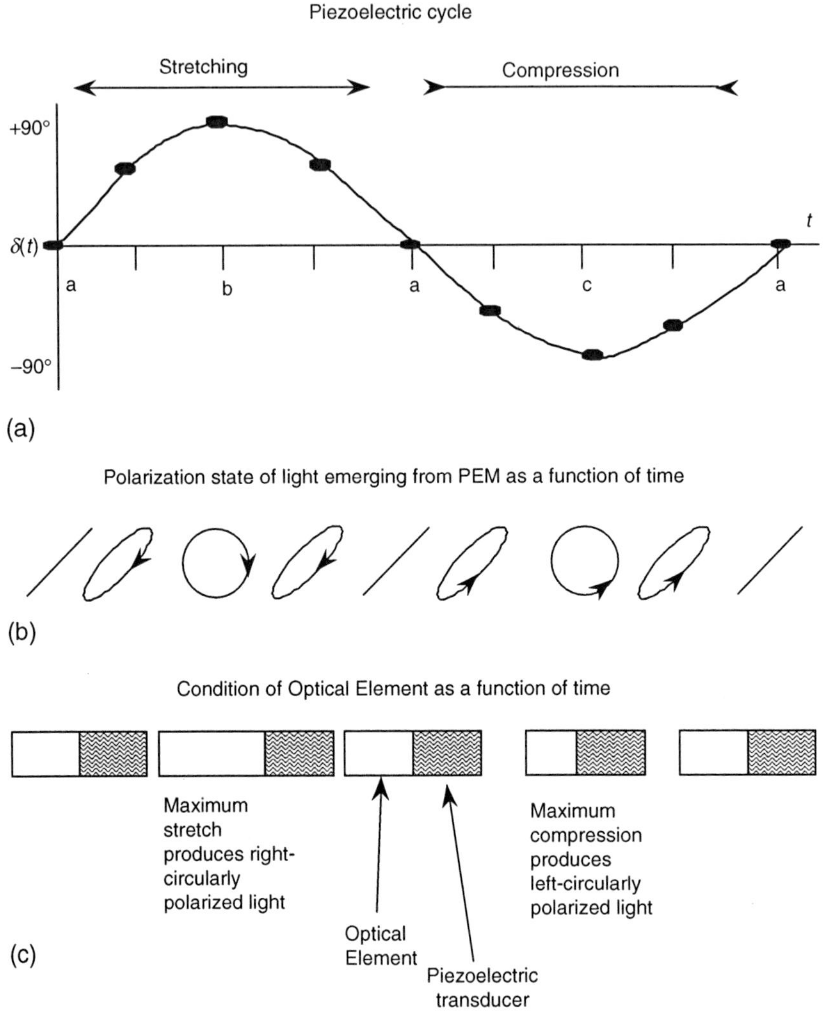

Fig. 10.25. (a) Piezoelectric cycle; (b) Polarization state of emerging light; (c) Condition of optical element as a function of time.

When the optical bar is compressed by the piezolectric transducer, the refractive index of the horizontal component, n_x, decreases relative to the refractive index in the vertical plane, n_y. In this condition, the electric vector of the horizontal component will lead the electric vector of the vertical component. According to Eq. (12), this will produce a negative phase shift that will increase until the difference between n_x and n_y once again reaches a maximum[31] (when the optical bar is compressed to the maximum by the piezoelectric transducer—see point c in Fig. 10.25(a)). With the control voltage set to

[31] Since $n_y > n_x$, the difference will be negative.

the proper level, the maximum phase shift in this condition will be -90°. A phase shift of -90° will produce left-circularly polarized light.

As shown in Fig. 10.25(b), as the piezoelectric transducer goes through a complete cycle, the output of the modulator will cycle between linearly polarized light, elliptically polarized light, and circularly polarized light. At the maximum stretching condition of the cycle, right-circularly polarized light will emerge from the modulator. At the maximum compression condition of the piezoelectric cycle, left-circularly polarized light will be produced.

PART B—SIGNAL HANDLING

10.7 PHASE-SENSITIVE DETECTION

Modern chemical instrumentation is dependent on the ability to detect very weak signals in the presence of noise. One way to accomplish this task is to use a form of demodulation known as phase-sensitive detection or synchronous demodulation. To understand this concept, it is worthwhile to review some aspects of amplitude modulation.

Amplitude modulation. Amplitude modulation occurs when the amplitude of a high-frequency signal (the carrier) is varied or modulated by a lower frequency signal containing the information to be transmitted. This process is represented schematically in Fig. 10.26, where the information (low-frequency cosine wave) is encoded in the amplitude of the carrier. Mathematically, the process is analogous to multiplication of the two waveforms. Consider two cosine waves that can be represented by

$$A_c \cos \omega_c t \text{ for the carrier} \tag{13}$$

$$A_s \cos \omega_s t \text{ for the signal,} \tag{14}$$

where A_c and A_s[32] are the amplitudes of the carrier and signal, respectively, and ω_c and ω_s are the frequencies of the carrier and signal, respectively, where $\omega_c \gg \omega_s$. When Eq. (13) and (14) are multiplied together, the result is:

$$A_c \cos \omega_c t \times A_s \cos \omega_s t = \frac{A_c A_s}{2}\left\{\cos(\omega_c + \omega_s) + \cos(\omega_c - \omega_s)\right\}, \tag{15}$$

where the sum and difference terms are two new frequencies referred to as sidebands. These sidebands contain the low-frequency information, which has now been shifted out to the region around ω_c.

Fig. 10.27 shows a computer-generated plot of two cosine waves and their product. It is clear that the information that was once present in the lower frequency signal is now encoded in the variation of the amplitude of the carrier wave with time. Shifting the information from a low-frequency region to a higher frequency regime can be beneficial

[32] A_s must be less than A_c to avoid overmodulation.

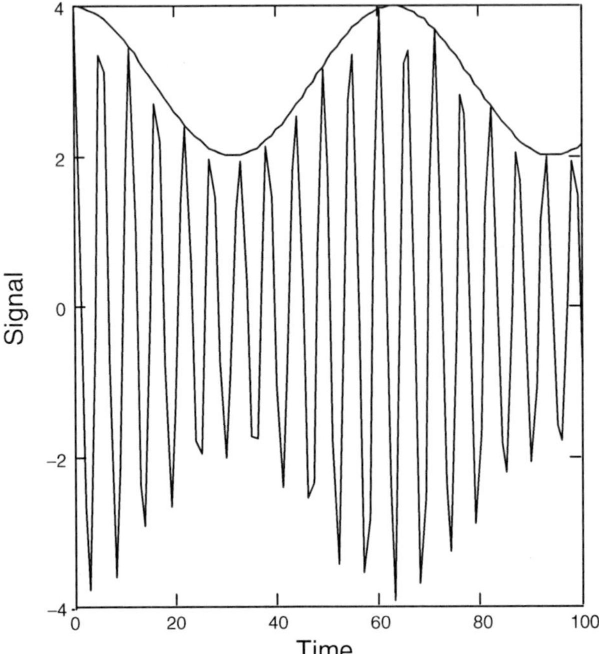

Fig. 10.26. An amplitude-modulated waveform where the information (low-frequency cosine wave) is encoded in the amplitude of the higher frequency carrier.

in avoiding the effects of low-frequency $1/f$ noise, which is sometimes referred to as drift, as shown in Fig. 10.28.

Demodulation. To recover the original signal, the process must be reversed by demodulation. Demodulation of a modulated signal can be accomplished by two basic procedures known as

- envelope detection,
- synchronous demodulation.

Envelope detection. Fig. 10.29 illustrates the use of envelope detection as a means of demodulation. The process is similar to half-wave rectification. As shown in the figure, the modulated signal is passed through a diode, where the negative portion of the carrier wave is removed. The diode is then followed by a low-pass RC filter, which removes the high-frequency spikes, restoring the original low-frequency signal (i.e. the information). While envelope detection is a simple means of signal recovery, it is not particularly efficient because half of the signal (the negative portion of the carrier) is rejected. Furthermore, envelope detection is not as powerful as synchronous detection when it comes to noise rejection.

Synchronous demodulation. Synchronous detection or synchronous demodulation is analogous to full-wave rectification. Mathematically, synchronous demodulation can be accomplished by multiplying the modulated waveform by the original carrier wave, as

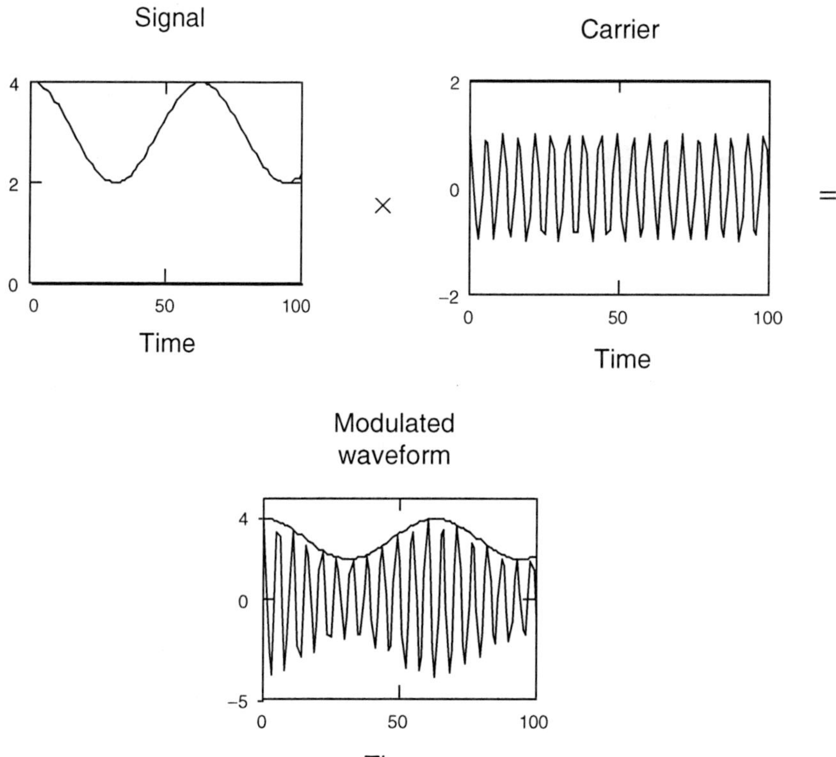

Fig. 10.27. Modulated waveform generated by multiplying the signal waveform by the carrier waveform.

shown schematically in Fig. 10.30. Once again the demodulated signal is passed through a low-pass RC filter, whose time constant determines the noise level of the output.

Fig. 10.31 shows a computer-generated plot of the carrier waveform and the modulated waveform on the same time axis. Notice that the two waveforms have postive and negative components. In this plot, it is clear that both the frequency and the phase of the two waveforms are the same (they are "in-sync"). If we multiply these two waveforms together, the demodulated waveform that results will be entirely positive because positive portions of the carrier are always multiplied by positive portions of the modulated wave and negative peaks in the carrier are always multiplied by negative peaks in the modulated waveform. Fig. 10.32 shows the results obtained when the two waveforms are actually multiplied together by the computer. Notice that in this plot, all the spikes are positive and the information is still contained in the envelope of the demodulated waveform (the low-frequency cosine wave). When this high-frequency waveform is passed through a low-pass RC filter, the high-frequency spikes are removed, and the original signal is recovered. The horizontal line in Fig. 10.32 shows the DC level of the demodulated waveform, which is removed by the low-pass filter.

Fig. 10.33 shows a computer-generated plot of the result obtained if the carrier and the modulated waveforms have the same frequency, but are shifted in phase by 90°.

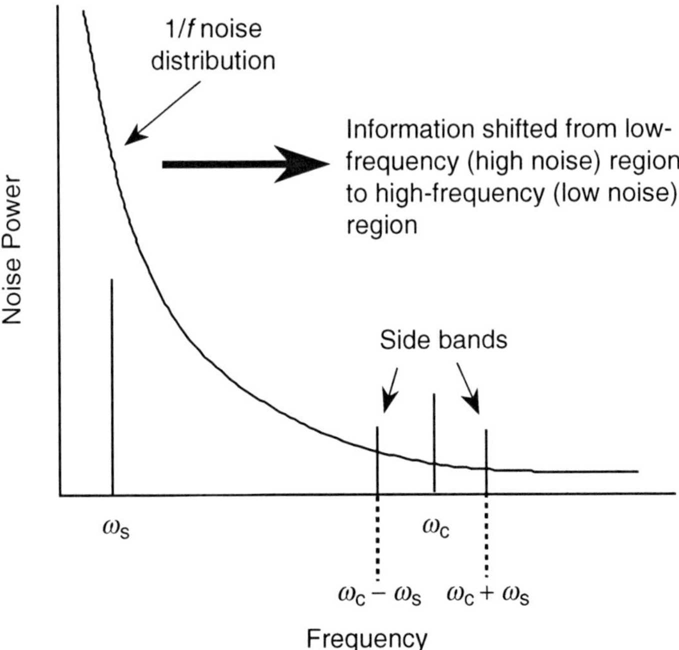

Fig. 10.28. Plot of noise power *versus* frequency for a system dominated by $1/f$ noise. Modulation of an information signal with frequency ω_s shifts the information to a lower noise region around the frequency of the carrier ω_c. The two sidebands are symmetrically located on either side of ω_c.

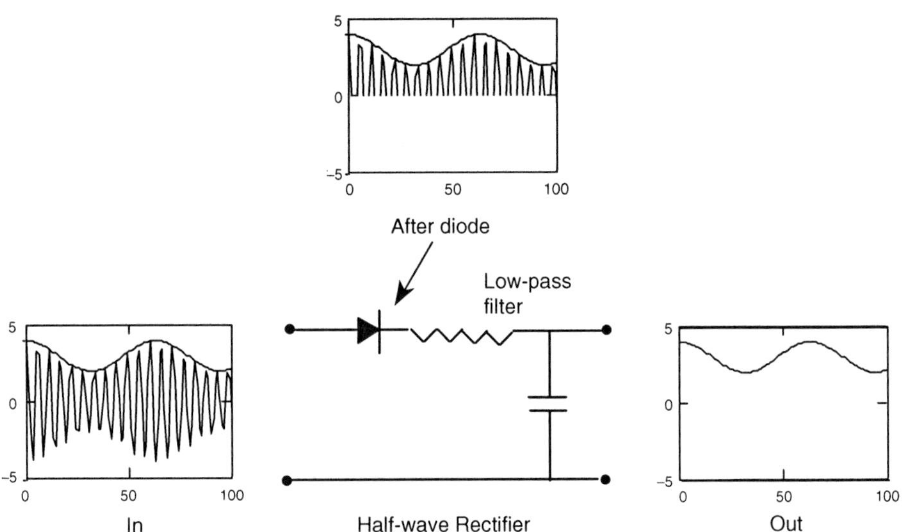

Fig. 10.29. Demodulation by envelope detection using a half-wave rectifier as a demodulator.

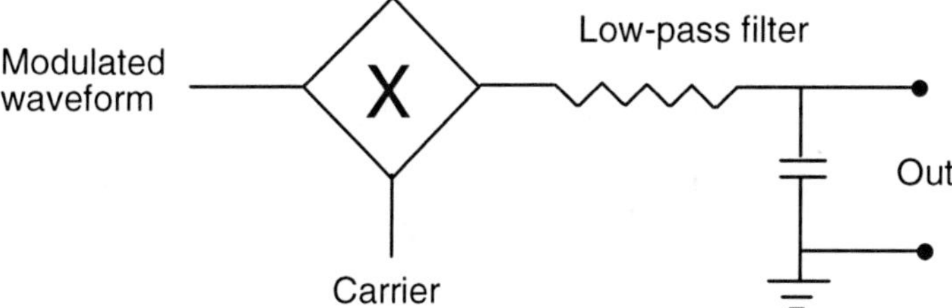

Fig. 10.30. Synchronous demodulation can be accomplished by multiplying the modulated waveform by the carrier. This amounts to full-wave rectification (all the negative peaks are converted to positive peaks). The time constant of the low-pass *RC* filter determines the noise level of the output.

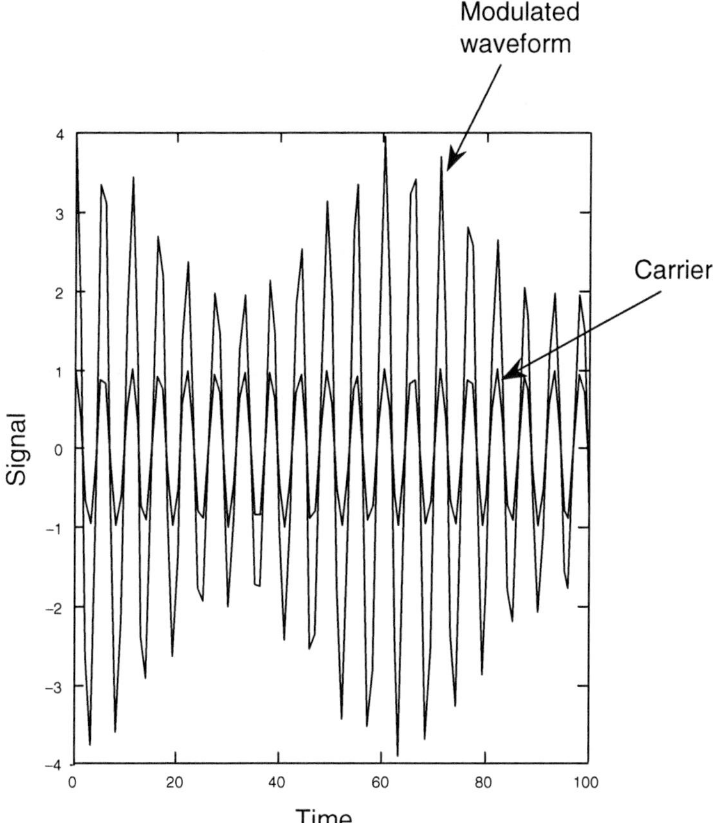

Fig. 10.31. Computer-generated plot showing the modulated waveform and the carrier waveform on the same time axis. Both waveforms are synchronized (same frequency and phase).

Comparing this plot to the one in Fig. 10.32, it is seen that a phase shift of 90° results in a product whose waveform has both positive and negative components. When this waveform is passed through a low-pass filter, the output will tend to be zero, because there is no DC level associated with this waveform. As a result, in order for a signal to

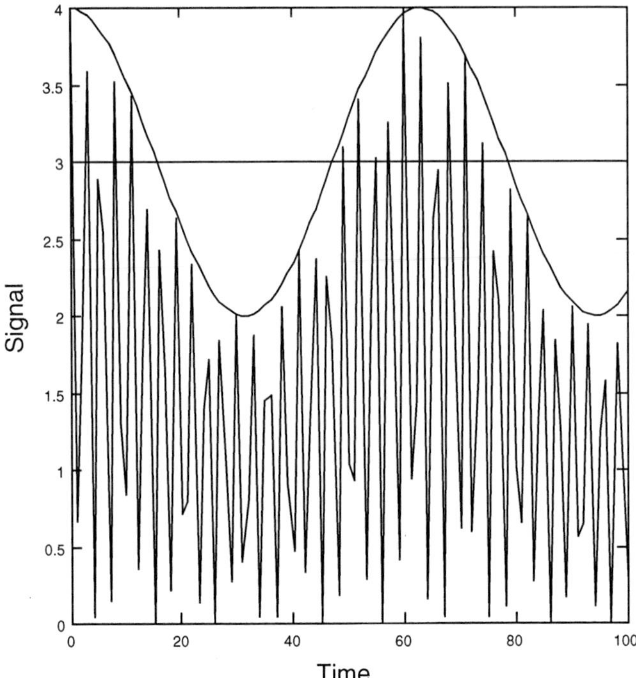

Fig. 10.32. Computer-generated plot of the result obtained when the modulated waveform is multiplied by the carrier with the same frequency and phase. The horizontal line in the figure shows the positive DC level of the resulting demodulated signal, which is removed by the low-pass RC filter.

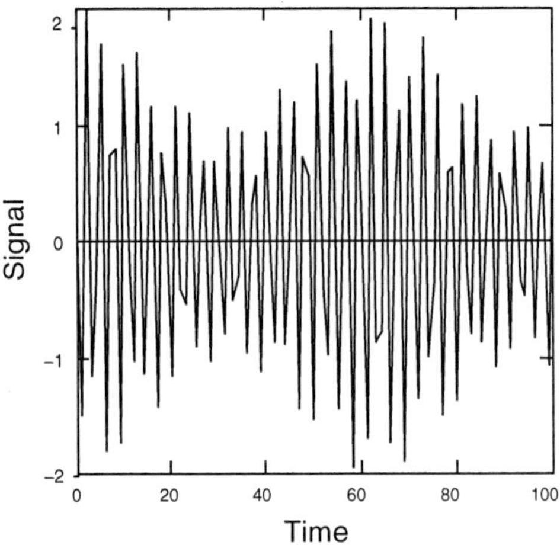

Fig. 10.33. Computer-generated plot of the result obtained if the modulated signal is shifted by 90° from the carrier waveform. Both signals still have the same frequency. The horizontal line in the figure shows that the demodulated signal in this case has a DC level of zero. The demodulated signal is entirely high-frequency AC components that are rejected by the low-pass RC filter.

produce an output from the synchronous demodulator, it must have a DC level associated with the demodulated waveform. To produce the maximum output, the modulated signal and the carrier must have the same frequency and a definite *phase* relationship (ideally the same phase). For this reason, synchronous demodulation is sometimes referred to as phase-sensitive detection.

So even if, by chance, a noise component has the same frequency as the carrier wave, the output will tend to zero because one of the characteristics of noise is a random phase. This means the phase of a noise component will vary randomly from positive to negative phase angles. As it does, the output from the phase-sensitive detector will vary from positive to negative. Averaged over a suitable time period by the low-pass filter, the noise component will average to zero.

Benefits of modulation and synchonous demodulation. Fig. 10.34(a) shows a hypothetical signal produced by a light source that is subject to drift. For convenience, the drift, which is low frequency, is represented as a slowly varying cosine wave. In techniques like circular dichroism (CD), where the small absorbance difference between left- and right-circularly polarized light is to be measured, the slow change in source intensity (I°)[33] with time can be much larger than the small change in I due to the absorbance difference for the two forms of circularly polarized light. This means that slow, sequential measurements of the absorbance for left- and right-circularly polarized light will be problematic in CD work because they occur over the same time interval as the source drift.

This situation can be ameliorated by using AC signal processing with a high-speed lock-in amplifier.[34] Suppose that the time between the two absorbance measurements for left- and right-circularly polarized light can be shortened dramatically. Now the small intensity change (ΔI) resulting from switching between left- and right-circularly polarized light will appear as a high-frequency waveform superimposed on the source drift, as shown in Fig. 10.34(b). By using AC signal processing, like a lock-in amplifier, the slow source drift will easily be removed from the signal by the high-pass RC filter in the input of the lock-in amplifier, as shown in Fig. 10.34(c).

In this way, only the AC-component of the signal will be processed by the lock-in amplifier. Moreover, if a PEM is used to sequentially produce left- and right-circularly polarized light (at 50 kHz), the control signal used to control the piezoelectric transducer can be used as a reference signal (i.e. the carrier) in synchronous demodulation because the left- and right-circularly polarized light will be generated in phase with the control signal. The DC output of the phase-detector will be proportional to the intensity difference between left- and right-circularly polarized light. For samples that are not optically active, this difference will be zero. When this is true, no AC waveform will be produced by the piezolectric modulator, and the output of the lock-in amplifier will be zero.

Finally, phase-sensitive detection is especially powerful in reducing noise introduced in the transmission channel of the instrument. The transmission channel is that part of the

---

[33] Since absorbance is defined as log (I°/I), a change in either the numerator or the denominator will change the absorbance.

[34] A lock-in amplifier is an AC amplifier that provides for phase-sensitive detection of signals.

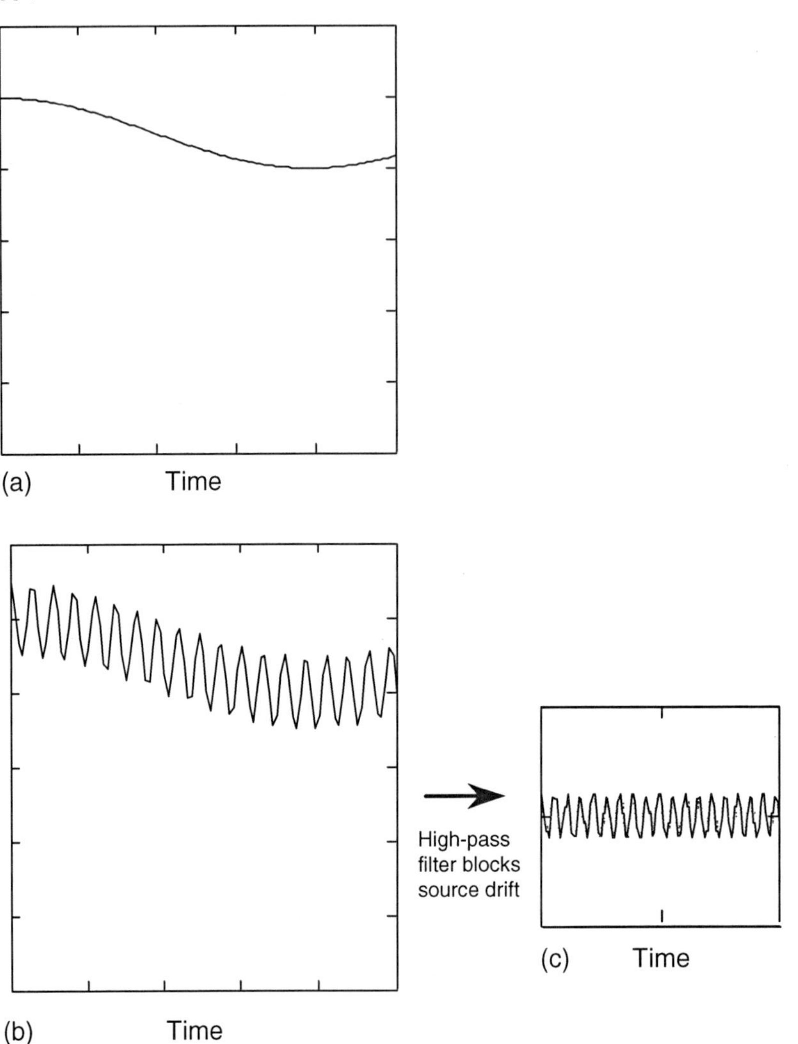

Fig. 10.34. (a) Detector signal as a function of time for a source whose intensity drifts; (b) Detector signal *versus* time showing a high-frequency modulated intensity superimposed on the intensity drift of the source; (c) Detector signal *versus* time after passing through the high-pass filter at the input of the lock-in amplifier. The slow time-varying drift component has been removed by the filter.

signal-processing channel that falls after the modulator and before the demodulator. Noise introduced in this portion of the channel has not been modulated.[35] As a result, it will not have the requisite frequency and phase stability to pass through the demodulation process. A major source of noise in the transmission channel often arises in the detector of the spectrometer. In the infrared region of the spectrum, in particular,

[35] It should be realized that any noise source that passes through the modulator will not be rejected by phase-sensitive detection because it has been encoded along with the signal of interest.

spectrometers are often detector-noise limited [15]. Phase-sensitive detection is a powerful tool in reducing detector noise, and is, therefore, extremely important in infrared spectrometric measurements.

APPENDIX A. BASIC OPTICS

1A. Refractive index

When light propagates through a transparent dielectric medium[36] like glass, the electric vector of the light wave interacts with the electron clouds around the atoms to induce oscillating dipoles in the medium that, in turn, re-emit the radiation. This virtual absorption and re-emission process retards the progress of the wave in the medium, thereby reducing its speed relative to the speed of light in a vacuum. The refractive index, n, of the medium is defined as

$$n = c/\upsilon, \tag{A1}$$

where c is the speed of light in a vacuum and υ is the speed of the light in the medium. In general, the denser the medium, the slower the light propagates and the larger the refractive index. The refractive index is also a function of the frequency of the radiation and the change in refractive index with wavelength is referred to as dispersion.

2A. Snell's law

When light traveling in one medium strikes the interface of another medium, its direction of travel is altered by the phenomenon of refraction. Fig. A10.1 shows the refraction of light as it travels from a medium of lower density (medium 1) to one of higher density (medium 2). As the wavefronts enter medium 2, they slow down and begin to stack up, causing the wavelength to decrease in the more dense medium. At the same time, the direction of the wavefronts in medium 2 is altered. Fig. A10.1 also shows the derivation of Snell's law from the wavelengths of the light in the two media. Distances A and B must correspond to an integral number (m) of wavelengths in the respective media. Solving the equations for A and B for m and equating the results, leads to Snell's law, given by

$$n_1 \sin i = n_2 \sin \phi, \tag{A2}$$

where n_1 and n_2 are the refractive indices, i is the angle of incidence, and ϕ is the angle of refraction.

[36] The dielectric medium is considered to be an array of atomic nuclei with their associated bound electrons. Since the atomic nuclei are relatively massive compared with the mass of an electron, only the bound electrons are forced into oscillation by a high-frequency periodic electromagnetic field.

References p. 341

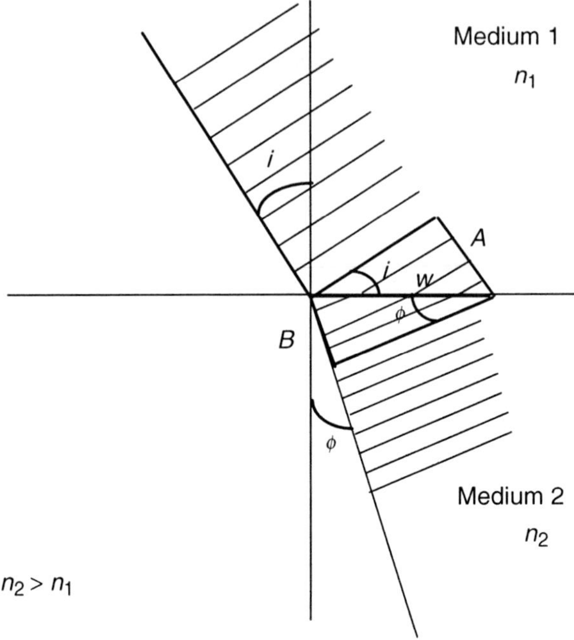

$A = m\lambda_1 = w \sin i$

$B = m\lambda_2 = w \sin \phi$

$m = (w \sin i)\lambda_1 = (w \sin \phi)\ \lambda_2$

$n_1 \sin i = n_2 \sin \phi$

$\lambda_1 = c/n_1\nu$

$\lambda_2 = c/n_2\nu$

Fig. A10.1. Derivation of Snell's law: n_1 and n_2 are the refractive indices of medium 1 and 2, respectively; w is the hypotenuse of the two triangles shown in bold lines; i is the angle of incidence, ϕ is the angle of refraction; m is an integer; c is the speed of light in a vacuum; ν is the frequency of the light. In order for the wavefronts to be preserved, the number of wavelengths in length A must be equal to the number of wavelengths in length B.

Fig. A10.2(a) shows the path of a light beam incident on a block of transparent dielectric material. At each interface, light is reflected and refracted. Fig. A10.2(b) shows the effect when light is incident on a trigonal prism. As the angle i increases, the ray emerging from prism begins to approach a value of $\phi = 90^\circ$ with respect to the normal to the second surface. When this point is reached, the ray no longer emerges from the prism, but is reflected to the right, as shown in Fig. A10.2(b). The angle of incidence for which this occurs is known as the critical angle and is given by

$$\sin i_c = \frac{n_1}{n_2}, \tag{A3}$$

where i_c is the critical angle.

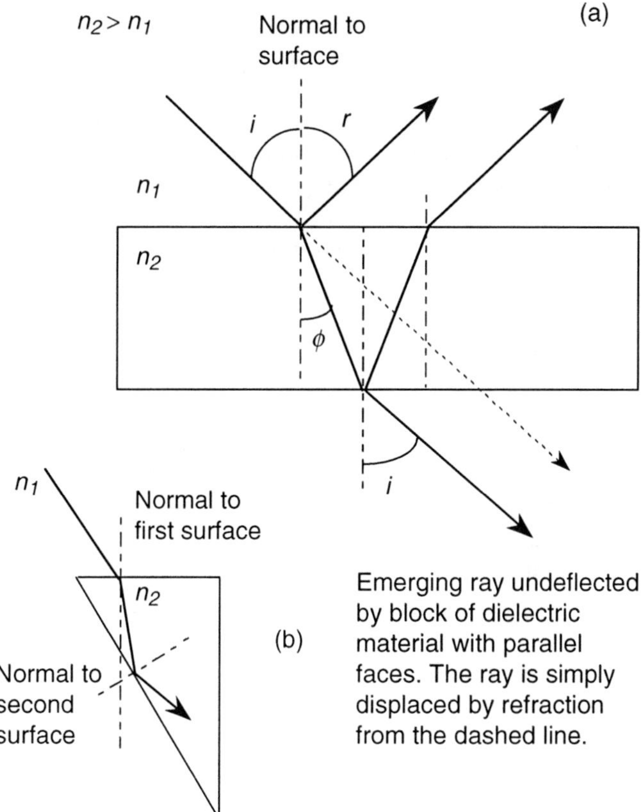

Fig. A10.2. Reflection and refraction: (a) from a block of dielectric material with parallel faces—n_1 and n_2 are the refractive indices of the two media; i is the angle of incidence; r is the angle of reflection; ϕ is the angle of refraction. (b) Total internal reflection from the interior surface of a prism when the critical angle is exceeded.

APPENDIX B. THE LAW OF MALUS

Polarizers are frequently used in pairs where the first polarizer is followed by a second, known as the analyzer. Unpolarized light incident on the polarizer will be converted into plane polarized light vibrating in a particular plane. When the polarization plane of the analyzer is coincident with that of the polarizer (the polarizers are said to be in a parallel alignment), the maximum intensity will be passed by the pair. If the polarization plane of the analyzer is rotated by 90°, the two polarization planes will be orthogonal (or crossed), and an observer will notice that virtually no light will be transmitted by the pair of polarizers. If the analyzer is rotated through an additional 90° in the same direction, the polarization planes of both polarizers will again be parallel, and the maximum intensity will again be passed by the pair. Finally, as the analyzer is rotated again by 90° in the same direction (270° total), the polarizers will be crossed and the transmitted light will again be extinguished.

Consider a polarizer–analyzer combination, where the polarization plane of the polarizer is oriented in the vertical direction. Let the polarization plane of the analyzer be set at some arbitrary angle θ to that of the polarizer. Let A be the amplitude of the light beam transmitted by the polarizer. For any angle θ, A can be resolved into two components—one parallel to the plane of the analyzer (A_p) and the other orthogonal to the plane of the analyzer (A_o). Since the orthogonal component is rejected by the analyzer (its electric vector is crossed with the plane of the analyzer), only A_p will be transmitted by the analyzer. The magnitude of A_p is given by

$$A_\mathrm{p} = A\cos\theta \tag{B1}$$

Since the intensity of a light beam is proportional to the square of the amplitude, the intensity of the light transmitted by the analyzer (assuming a perfect analyzer with no losses) will be given by

$$I = A_\mathrm{p}^2 = A^2\cos^2\theta = I^0\cos^2\theta, \tag{B2}$$

where I^0 is the intensity of the light beam incident on the analyzer and I is the intensity of the light transmitted by the analyzer. The cosine-squared relationship between the transmitted intensity and the angle θ between the two polarization planes was discovered in 1808 by Etienne-Louis Malus, a French scientist, and is known as Malus' Law [16].

APPENDIX C. LISSAJOUS FIGURES AND FORMS OF POLARIZATION

As mentioned at the beginning of this chapter, light can be considered to be composed of an oscillating electric vector and an oscillating orthogonal magnetic vector. In a vacuum, the propagation of the magnetic and electric fields is in phase, but the amplitude of the electric field is larger by a factor of c, the speed of light in a vacuum.[37] Because of this difference in amplitude, changes in the electric field are easier to measure. As a result, the phenomenon of polarization is generally studied considering only the electric component of light. From this point of view, the orientation of the electric field vector is taken as the orientation of the polarization of the radiation.

An electric field vector of light, E, with any arbitrary orientation can be resolved into two orthogonal components, as shown in Fig. C10.1(a). If the z-axis is taken as the direction of propagation, these two orthogonal electric vectors can be represented as E_x and E_y. The former can be visualized to lie in the plane perpendicular to the paper, and the latter can be visualized to be in the plane of the paper.

Whenever two orthogonal oscillating waveforms given by

$$\begin{aligned} x &= E_x\cos(\omega t + \delta_x) \\ y &= E_y\cos(\omega t + \delta_y) \end{aligned} \tag{C1}$$

[37] The amplitude of the magnetic vector B_o will be given by E_o/c, where E_o is the amplitude of the electric vector and c is the speed of light. If E_o has units of $\mathrm{Vm^{-1}}$ and c is in $\mathrm{ms^{-1}}$, B_o will have units of $\mathrm{Vsm^{-2}}$ or Tesla.

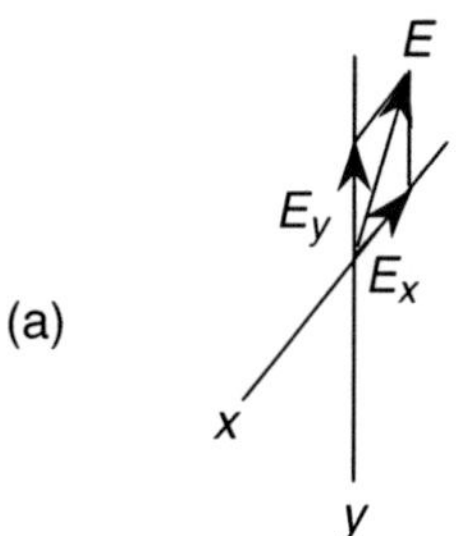

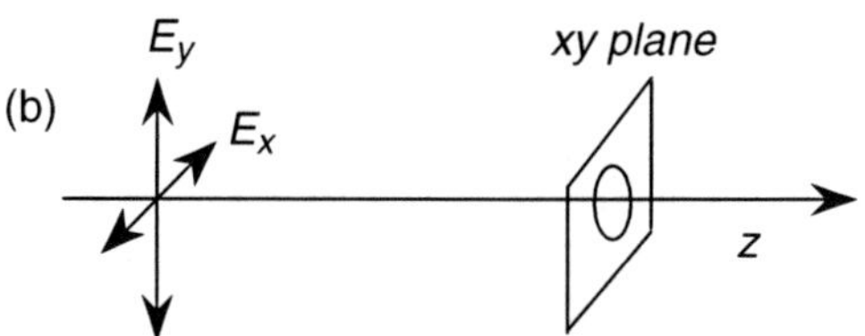

Fig. C10.1. (a) An arbitrary electric vector E resolved into E_x and E_y components. (b) The orthogonal E_x and E_y vectors generate a Lissajous pattern in the xy plane. The z-axis is the direction of propagation.

are added together, the combined effect of the vertical and horizontal oscillations traces a two-dimensional figure, known as a Lissajous figure,[38] in the xy plane as shown in Fig. C10.1(b). The shape of the Lissajous figure produced depends on the amplitudes E_x and E_y, their relative frequencies ω, and their relative phase angles δ_x and δ_y. When the frequencies of both waveforms are the same, the two-dimensional pattern produced takes the general form of an ellipse. An ellipse is a conic section produced when a plane intersects with a right circular cone at some angle. The limiting forms of an ellipse are the circle and a line segment.

In considering the polarization state of light, it is the phase relationship of δ_y with respect to δ_x that is important. This phase difference is termed Δ and can be taken as $\delta_y - \delta_x$. As Δ varies from 0 to 2π, different forms of polarization occur. Circular polarization and linear polarization, which are special limiting cases of elliptical polarization, occur when Δ takes on specific values.

As Δ varies, the Lissajous figure produced can have the shapes shown in Fig. C10.2. Fig. C10.2 shows that when Δ is 0 or π, the light is linearly polarized. If Δ is $\pi/2$ or $3\pi/2$, the light is circularly polarized, and for all other angles, the light is elliptically polarized.

For two orthogonal waveforms given by Eq. (C1), having the same frequency ω, it is possible to combine the two equations so as to eliminate t [17]. When this is done, the resulting equation is

$$\frac{x^2}{E_x^2} - 2\frac{x}{E_x}\frac{y}{E_y}\cos\Delta + \frac{y^2}{E_y^2} = \sin^2\Delta, \tag{C2}$$

[38] Lissajous figures are readily observed by applying sinusoidal waveforms to the vertical and horizontal inputs of an oscilloscope.

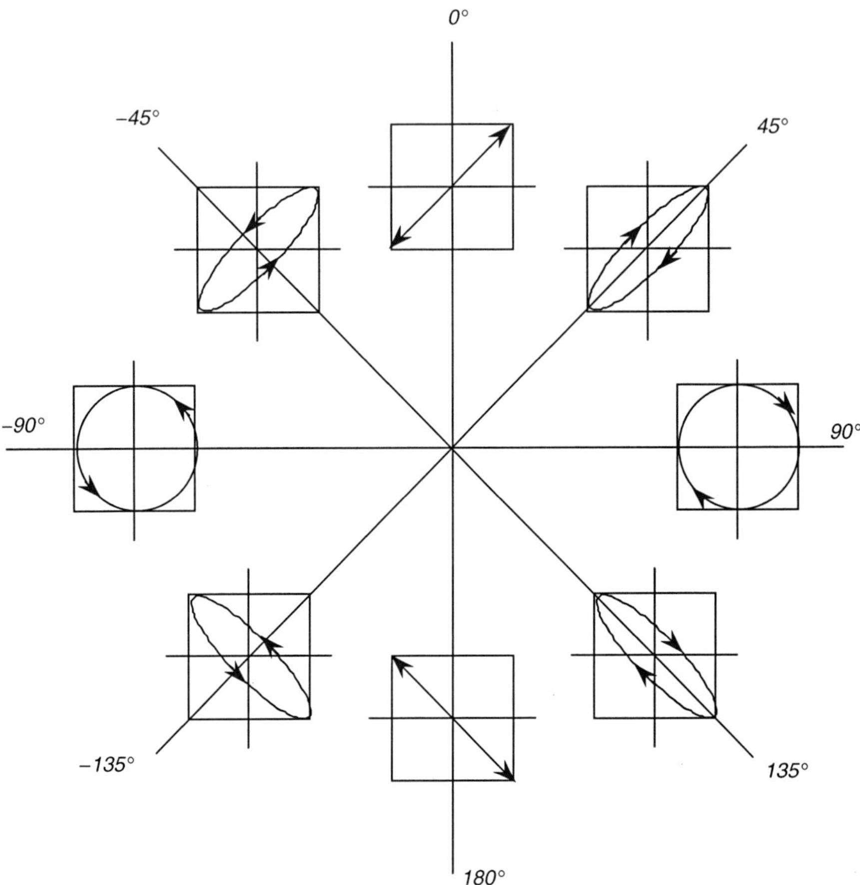

Fig. C10.2. Lissajous patterns produced for various values of Δ.

where x is the horizontal component of the Lissajous figure, y is the vertical component, and E_x and E_y are the amplitudes of the respective waveforms. Eq. (C2) can be used to determine the polarization forms produced by different values of Δ.

Linear polarization. Consider the situation where $E_x = E_y$ and Δ is 0 or 180°. When Δ is 0°, cos Δ is 1 and sin Δ is 0. Thus, when Δ is 0 and $E_x = E_y$, Eq. (C2) reduces to

$$\begin{aligned} x^2 - 2xy + y^2 &= 0 \\ (x - y)^2 &= 0 \\ x - y &= 0 \\ y &= x \end{aligned} \tag{C3}$$

Equation (C3) is the equation for a straight line with a slope of + 1.

When Δ is 180° and $E_x = E_y$, cos Δ is −1, sin Δ is 0, and Eq. (C2) reduces to

$$\begin{aligned} x^2 + 2xy + y^2 &= 0 \\ (x - y)^2 &= 0 \\ x + y &= 0 \\ y &= -x \end{aligned} \tag{C4}$$

Again, a linear relationship is observed only this time the slope is −1.

Circularly polarized light. When $E_x = E_y$ and Δ is equal to $\pm n\pi/2$, where n is an integer greater than zero, cos $\Delta = 0$ and sin $\Delta = \pm 1$. In this case, Eq. (C2) reduces to

$$x^2 + y^2 = E^2, \tag{C5}$$

where $E = E_x = E_y$. Eq. (C3) is the general equation for a circle with a radius of E.

When $\Delta = 90°$, the y-component (i.e. vertical component) leads the x-component (horizontal component) by 90°. When this is true, the Lissajous figure for the circle is traced in a clockwise fashion, and the light is right-circularly polarized. When $\Delta = -90°$, the x-component leads the y-component by 90°. This is equivalent to $\delta_y = -90°$. When the y-component lags the x-component by 90°, the Lissajous figure for the circle is traced in a counterclockwise fashion, and the light is left-circularly polarized.

REFERENCES

1 F.A. Jenkins and H.E. White, Fundamentals of Optics, 4th Ed. McGraw-Hill, New York, 1976.
2 W.A. Shurcliff, Polarized Light, Harvard University Press, Cambridge, MA, 1966.
3 D.S. Kliger, J.W. Lewis and C.E. Randall, Polarized Light in Optics and Spectroscopy, Academic Press, Boston, 1990.
4 D. Goldstein, Polarized Light, 2nd Ed. Marcel Dekker, New York, 2003.
5 D. Brewster, On the laws which regulate the polarization of light by reflection from transparent bodies. Phil. Trans., 105 (1815), 125.
6 A.S. Makas and W.A. Shurcliff, New arrangement of silver chloride polarizer for the infrared. J. Opt. Soc. Amer., 45 (1955), 998.
7 Ref. [3], p. 36.
8 Ref. [3], p. 37.
9 J.F. Archard and A.M. Taylor, Improved Glan-Foucault prism. J. Sci. Instr., 25 (1948), 407.
10 J.R. Biot, Sur un mode particulier de polarisation qui s'observe dans la tourmaline. Bull. Soc. Philomath. Paris, 6 (1815), 26.
11 Ref. [2], p. 59.
12 E.H. Land, Some aspects of the development of sheet polarizers. J. Opt. Soc. Amer., 41 (1951) 957.
13 W.B. Herapath, On the optical properties of a newly discovered salt of quinine which crystalline substance possesses the power of polarizing a ray of light, like tourmaline, and at certain angles of rotation of depolarizing it, like selenite. Phil. Mag., 3(4) (1852) 161.
14 W.B. Herapath, Further researches into the properties of the sulfate of iodo-quinone or herapathite. Phil. Mag., 9 (1855) 366.
15 K.W. Busch and M.A. Busch, Multielement Detection Systems for Spectrochemical Analysis, Wiley, New York, 1990.
16 E. Malus, Sur une propriété de la lumière réfléchie, Mémoires de Physique et de Chimie de la Société d'Arcueil, 2 (1809) 143–158.
17 Ref. [1], p. 253.

© 2006 Elsevier B.V. All rights reserved.
Chiral Analysis
K.W. Busch and M.A. Busch, Eds.

CHAPTER 11

Micro-scale polarimetry

Darryl J. Bornhop and Stephen Dotson

Department of Chemistry, Vanderbilt University, 7330 Stevenson Center, Nashville, TN 37235, USA

Optical activity is a unique characteristic of chiral molecules that can be used to predict their utility in a variety of fields including optics, nutrition, biotechnology and medicine. Determination of chirality is particularly important in the development of new drugs [1], and production of safe pharmaceuticals. With the worldwide sales of single-enantiomer drugs increasing substantially in recent years, it is necessary to quantify enantiomeric purity and thus detect chiral molecules [2]. It is now known that one enantiomer can produce a desired biological response while the other can have harmful side effects [3]. Combinatorial chemical methods and the implementation of parallel synthesis have further increased the pace of research in drug discovery [4–8], demanding improved analysis methods. Also, a number of enzymatic processes that are vital in the development of chiral drugs contain optically active moieties. Improvements in the enzyme specificity and activity are being obtained by "directed evolution" methodologies [8–10]. Determination of the purity of the chiral molecules obtained from high-throughput (HT) methods is vital to the success in finding new and better therapies. Yet successful HT implementation is dependent on the capability to perform polarimetry on the nanoliter volume scale. This chapter will outline existing micro-scale polarimetry methods and progress toward the HT micro-scale screening of compounds for optical activity.

One application of conventional polarimetry could benefit from smaller volumes and a higher throughput is enantioselective enzyme production monitoring. While enzymes are useful tools, many lack selectivity, and produce poor yields or low enantiomeric excess (ee). In directed evolution of enantioselective enzymes, thousands of mutants are created to find an enzyme that produces superior selectivity for a desired reaction [8–10]. Enantioselectivity needs to be analyzed in a HT fashion for directed evolution to be practical. While other techniques show promise, with progress having been made toward HT screening [11–16], there is still a need for an HT polarimeter. For polarimetry to be practiced in a HT format, the required sample volume needs to be significantly reduced and the optical configuration substantially simplified.

There are a couple of fascinating techniques for HT ee determinations, including a color test using chirality-dependent doped films of liquid crystals [17], fluorescent reporting using a DNA microarray [12], an assay employing antibodies that is an analog of competitive enzyme immunoassay [18], and a technique termed EMDee which is an enzymatic method for determining ee [11]. Here, we focus on the development of a low volume (< 1 μl) polarimeter.

11.1 CONVENTIONAL MICRO-SCALE POLARIMETRY

Conventional polarimetry typically requires relatively large sample volumes and lacks the ability to be multiplexed. Yet several groups have improved the performance and reduced the required sample volume to do polarimetry. In one of the first successful attempts to miniaturize polarimetry, Bobbitt and Yeung [19] performed microbore liquid chromatography with polarimetric detection to obtain optical activity. Fig. 11.1 shows a block diagram of the experimental setup, which consists of a 1 μl flow cell with a 1 cm path length interfaced to a liquid chromatography system [19]. A focused argon ion laser served as the illumination source. A Faraday rotator (air-based) was used to provide optical rotation at a frequency of 1 kHz. A detector was connected to a lock-in amplifier that was connected to a digital voltmeter and a computer, which was utilized for data storage [20]. The system was able to measure both optically and non-optically active eluents. The direct detection of optical activity produced the detection limits of 15 μdeg and 11 ng of fructose [19].

As an added benefit of this complicated optical design that gives high sensitivity, non-optically active samples could be determined by indirect polarimetry. In this approach, the mobile phase is optically active and the non-optically active samples are detected by the absence of a polarimetric signal. Fig. 11.2 demonstrates the detection with two different eluents. Fig. 11.2(a) shows an indirect polarimetry measurement, demonstrating that the system was able to detect decane, tetradecane, and hexadecane. In Fig. 11.2(b), the racemic eluent, interestingly small peaks for C10, C14, and C16 were

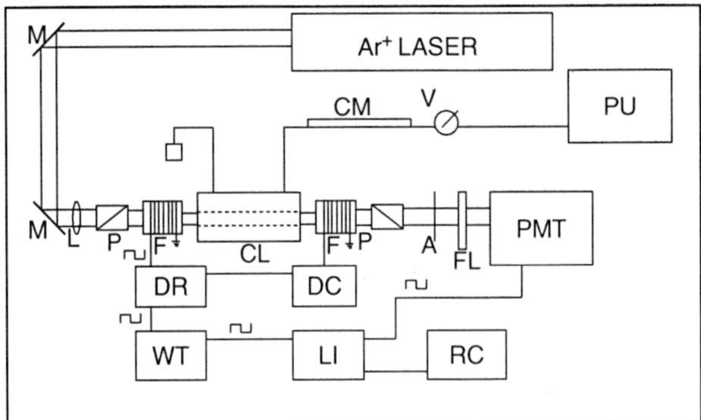

Fig. 11.1. Experimental arrangement for optical activity detector, M, Mirrors; P, Glan prism; F, Faraday rotator; CL, flow cell; A, aperture; FL, filter; PMT, photomultiplier tube; DR, driver; DC, power supply; WT, wave generator; LI, lock-in amplifier; RC, recorder; PU, pump; V, injection valve; CM, column; L, collimation lens. (Reproduced by permission of American Chemical Society.)

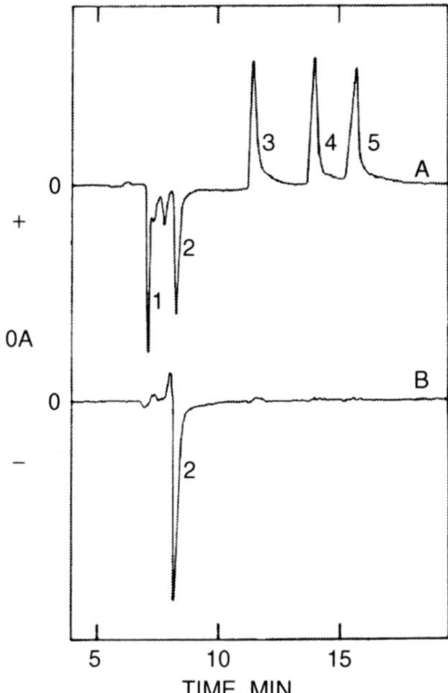

Fig. 11.2. Chromatograms of the mixture in the two eluents: (a) (−)-2-methyl-1-butanol, (b) (±)-2-methyl-1-butanol; (1) injection peak, (2) 2-octanol, (3) decane, (4) tetradecane, (5) hexadecane. (Reproduced by permission of American Chemical Society.)

still observed. This was attributed to the refractive index change that occurred as the sample elutes. The possibility of obtaining sensitivities of 12 ng makes indirect polarimetry promising [19], but high instrumentation costs and optical complexity limit this method for HT applications.

11.2 RI COMPENSATION

In order to improve signal-to-noise ratio and eliminate possible false-positive readings for commercial polarimetric measurements in HPLC, Maystre et al. [21] improved the sensitivity of conventional commercial polarimetric measurements in HPLC by using a refractive index (RI) equalizer to remove pseudo-rotation (false-positive rotations), caused by changes in RI. RI compensation was accomplished by applying a mixer and RI detector to the system (Fig. 11.3). The RI detector (positioned after a mixer) feeds a signal to a servocontroller unit, which introduces nonperturbing low or high RI liquid into the eluents before the mixer to correct any RI changes. This system requires that the additive does not react with the samples or mobile phase. For example, using 40% MeOH ($n = 1.3400$) as a mobile phase with the addition of 60% MeOH ($n = 1.3420$) will raise the RI to help correct for a decrease in RI. The eluents mixed with the RI compensating liquid proceed to a polarimeter for measurement [21].

Fig. 11.4 shows the effect of RI perturbations of polarimetric measurements of 2-phenylethanol (achiral). The open circles represent the signal measured by the

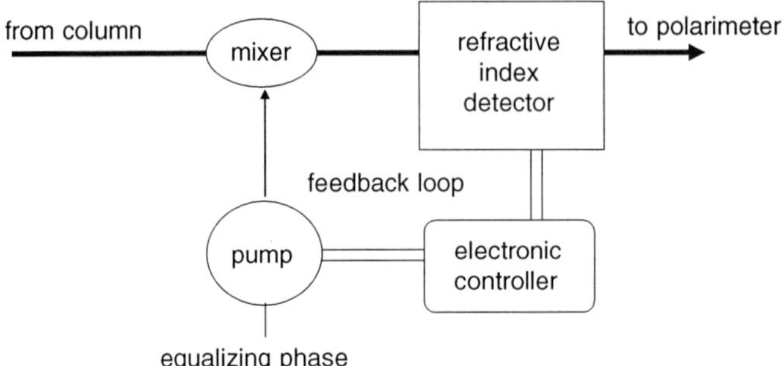

Fig. 11.3. Principle of operation of the RI equalizer. The feedback loop maintains the RI of the eluents at a constant value by mixing them with a variable amount of an optically denser phase. (Reproduced by permission of American Chemical Society.)

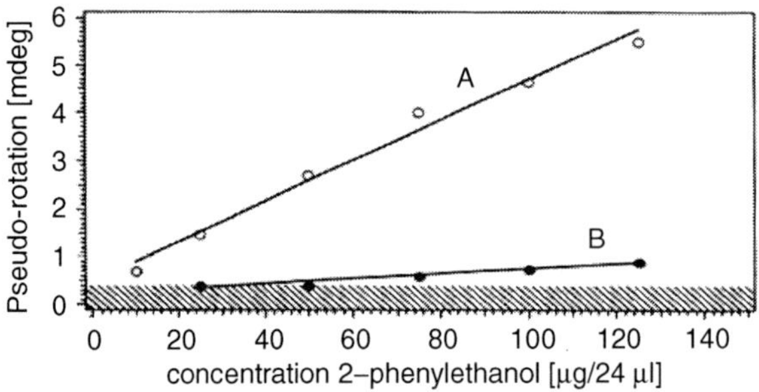

Fig. 11.4. Response of the polarimeter to an achiral compound 2-phenylethanol. Open circles (line A) were measured with the eluents flowing directly through the polarimeter. Closed circles (line B) were measured with the active RI compensation. (Reproduced by permission of American Chemical Society.)

polarimeter without RI compensation. Same concentrations of sample measured with RI compensation are shown as closed circles. The shaded portion corresponds to the noise level of the polarimeter. The data shows that the pseudo-rotation is not entirely removed but greatly reduced. Enhanced accuracy for determining the ee of (*R*,*S*)-1-phenylethanol was also demonstrated by eliminating pseudorotation. The prototype instrument had a dead volume of 3 μl and a detection volume of 1 μl. The system related the volume of the equalizer and the volume of the peak by $V_e \leq V_p/20$. Thus the suppression of RI peaks as small as 60 μl was possible [21].

11.3 MOLASSES ANALYSIS

Lloyd et al. [22] produced a polarimeter with a diode laser for chromatography applications. The polarimeter includes a lens, polarizer, a calibrator to standardize rotation, a modulator (allowing a 1° change in plane of polarization at a frequency of 0.5–2.5 kHz), an 8-μl flow cell, an analyzing polarizer, and a photodetector

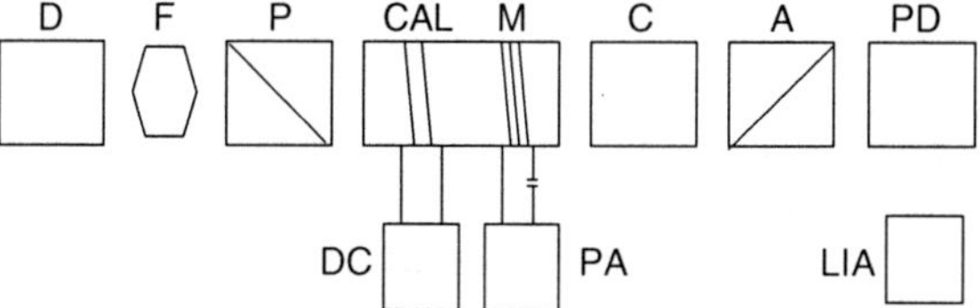

Fig. 11.5. Optical rotation detector block diagram; D, diode laser; F, lens; P, polarizer; CAL, calibrator; M, modulator; C, cell; A, analyzer; PD, photodiode; DC, DC power supply; PA, audio frequency power amplifier; LIA, lock-in amplifier. (Reproduced by permission of American Chemical Society.)

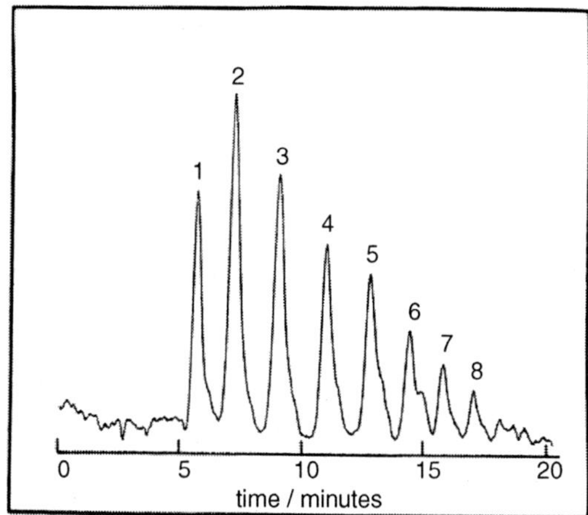

Fig. 11.6. Chromatogram of glucose syrup with optical rotation detection; 1, glucose; 2, maltose; 3, maltotriose and higher sugars, with numbers indicating degree of polymerization. (Reproduced by permission of American Chemical Society.)

(Fig. 11.5). The silicon photodectector is connected to a lock-in amplifier for signal recovery. Coupled with a UV detector, the enantiomeric purity of HPLC elutants was quantified without chiral separation.

Mixtures of D- and L-tryptophan in various proportions have been analyzed and enantiomeric purity of 1% for 50 μg injections was observed. The illumination source was an IR laser diode (820 nm) allowing determination of opaque samples, for example molasses, to be analyzed. The eluent used for separations was not chiral, thus no signal for optical activity was generated from the eluent. However, the authors noted that large changes in RI could defocus the laser, causing scattering, which may result in false optical activity readings. The specially designed flow cells used in these experiments (a tapered bore and a cylindrical bore cell were used) helped to reduce such effects (compared to conventional z-type straight-bore cells). The results from the analysis of a 20 μl injection of 2% syrup are shown in Fig. 11.6. The peaks indicate glucose, maltose, maltotriose, and higher oligomers. This method was reported to produce microdegree sensitivities [22], which is superior to most commercial polarimeters.

11.4 ADVANCED LASER POLARIMETER

The most sensitive solution-phase commercial polarimetric detector is a laser-based instrument produced by PDR-Chiral Inc. and shown in Fig. 11.7 [23]. In the PDR-Chiral Advanced Laser Polarimeter, a highly polarized and well-focused laser beam is passed through specially designed flow cells (for HPLC, process control, etc.), with volumes as small as 18 μl. The observed rotation is directly proportional to the net value of optical activity present in the flow cell. The system consists of a diode laser, a polarizer, Faraday rotator, flow cell, an analyzer plate, and a photodetector. The polarization state is modulated by the Faraday rotator and then detected by the photodetector. Thus, this commercial instrument is similar to that designed by Bobbit and Yeung discussed earlier, but has a 20-fold larger sample volume and twofold less sensitive detection limits [24].

The Advanced Laser Polarimeter has also been utilized to monitor kinetics for reactions where optical activity changes as the reaction proceeds. The glucose-oxidase catalyzed reaction of D-β-glucose and oxygen to D-gluconic acid and H_2O_2 (decrease in optical activity) was monitored. The reaction was observed for 50 min and at two different glucose oxidase concentrations. As shown in Fig. 11.8, the increase in glucose oxidase concentration leads to an increase in the reaction rate. This instrument can be used for multiple applications with a variety of flow cells available. Currrently the smallest volume flow cell available is 18 μl (model A-18, length 50 mm, bore 0.69 mm).

This commercial product has mature software and is very powerful, but it is expensive and complicated. Rotational sensitivities of 25 μdeg for a sample with a $\Delta\alpha =$ 10 (deg × ml)/(g × dm) and a limit of detection of 5 μg/ml (25 μM; MW = 200) for a 5 cm interaction length can be obtained [24]. Multiplexing this system for high-throughput screening appears unlikely.

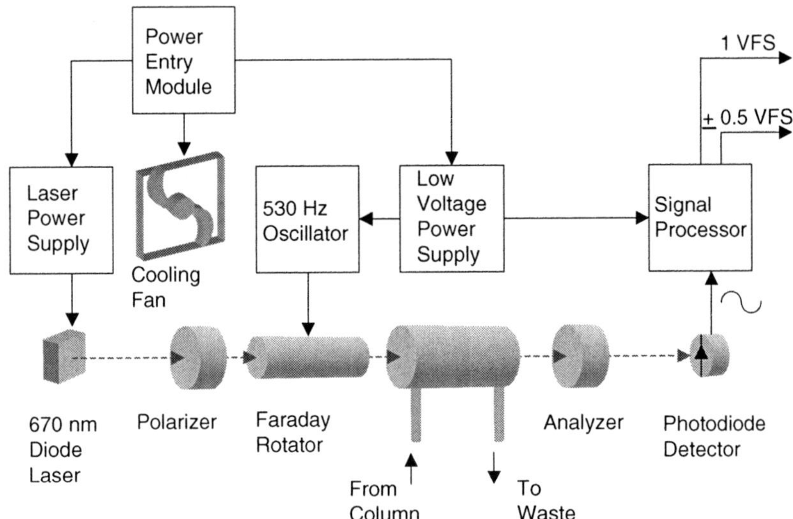

Fig. 11.7. Diagram of PDR-Chiral Advanced Laser Polarimeter. (Reproduced by permission of PDR-Chiral.)

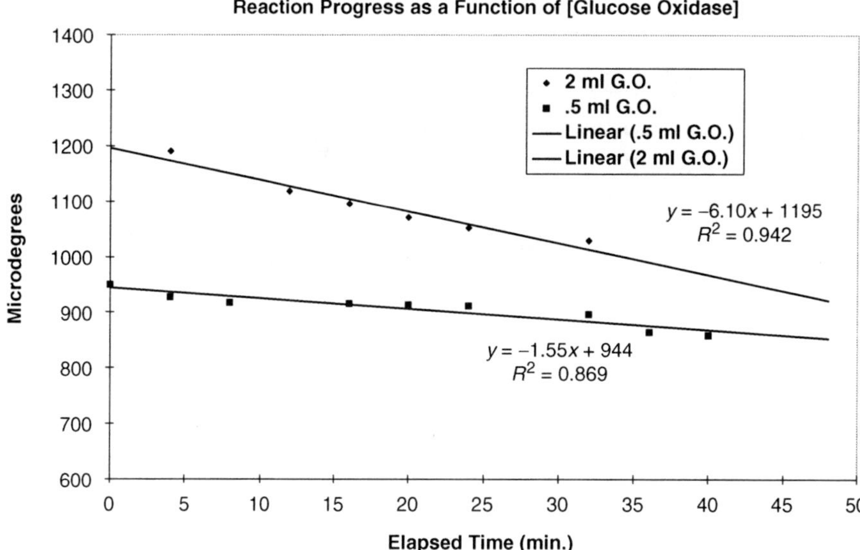

Fig. 11.8. Polarimetric response during glucose conversion. (Reproduced by permission of PDR-Chiral.)

11.5 IMAGING POLARIMETER

Polarimetry can be a powerful technique but the throughput has traditionally been poor. There have been two recent articles focused on developing polarimeters with significantly increased throughput [25,26]. Gibbs et al. [25] have developed an imaging polarimeter (Fig. 11.9) capable of measuring optical activity in several channels simultaneously. This system utilizes a 575 W diffused theatrical spotlight. The wavelength was filtered to pass emission near the sodium d-line. Illumination was then polarized prior to traversing an array of wells. The light then passed an analyzing polarizer before being imaged with a camera with a telecentric lens. The cell contains 37 channels, each with a path length of 12 cm and a volume of approximately 3.5 ml. Results showed that $\pm 0.08^\circ$ rotation could be quantified in this relatively large volume configuration.

In an extension of this basic optical train, it was also demonstrated that dextrose solutions of 0.5° and 1.0° optical rotations could be discriminated from water. Here, denser plates (384- and 1536-well plates) with volumes in the microliter range, were used with substantial reduction in system sensitivity performance. In another experiment, an enantioselective enzyme reaction was performed on a 96-well plate using the imaging polarimeter. Sucrose invertase converted sucrose [$\alpha = +66.5^\circ$] to dextrose [$\alpha = +52.7^\circ$] and fructose [$\alpha = -92.4^\circ$] and the reduction of optical rotation was monitored as a function of time. Fig. 11.10 shows a false color image (reproduced in gray scale) of a portion of the well plate at three different time points.

When a 384-well plate is used in imaging polarimetry, the system has an effective volume of approximately 75–100 μl per well. This technique does represent a good first step toward a functional small volume, multiplexed polarimeter, yet when conventional

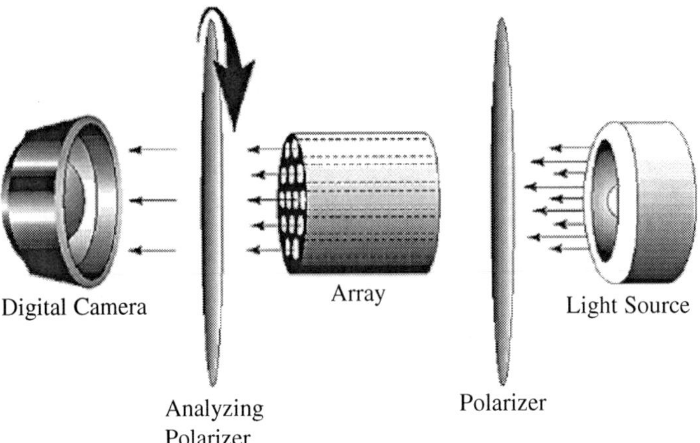

Fig. 11.9. Imaging polarimeter. (Reproduced by permission of American Chemical Society.)

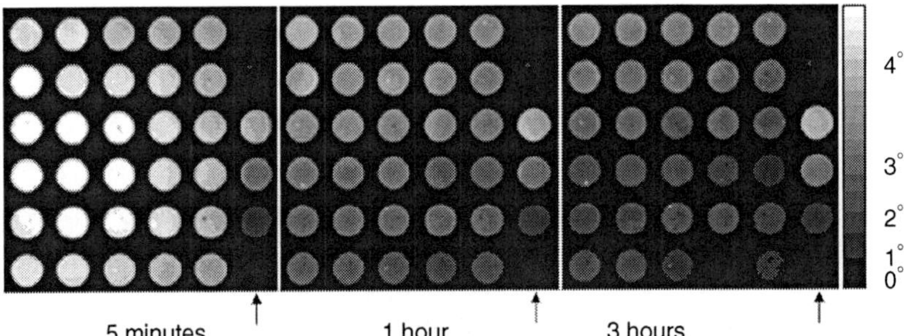

Fig. 11.10. Time series with a standard marked by an arrow. The brightness of the image is related to fit rotation values by the legend guide. (Reproduced by permission of American Chemical Society.)

well-plates are used detection limits fall significantly (It was reported that a 0.5° rotation could be detected in a 384-well plate.). The authors suggest that a better light source (laser) and higher crossed-polarizer extinction coefficients would be necessary to improve the polarimetric resolution or facilitate determination with short path length, standard multiwell plates [25].

11.6 AUTOMATED POLARIMETER

Schonfeld and Bornscheuer increased the throughput of a commercial micro-polarimeter by utilizing a pipetting robot (Fig. 11.11) to enable the assay of α-amino acid racemase activity [26] with a model enzyme, glutamate racemase, from *Lactobacillus fermintii*. Optical rotation was determined using a POLARmonitor (IBZ-Messtechnik, Hannover, Germany), utilizing a 40 μl sample volume cuvette. The overall automated system required a minimum of 350 μl. The system was automated by using a liquid handling system and a peristaltic pump. In a 96-well plate, 300 μl of purified enzyme samples were

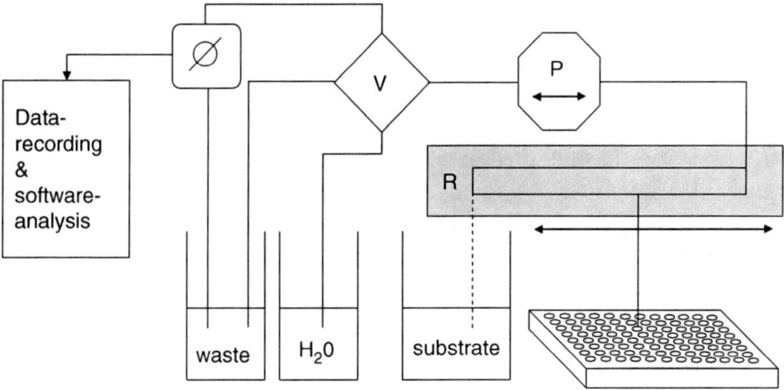

Fig. 11.11. Medium throughput polarimeter; R, liquid handler; P, peristaltic pump; V, value; Ø, POLARmonitor. (Reproduced by permission of American Chemical Society.)

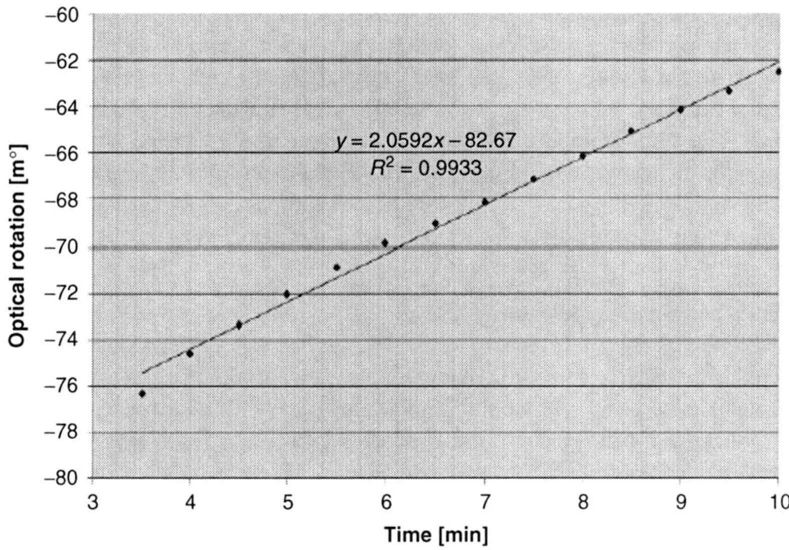

Fig. 11.12. Linear range used for the calculation of specific enzyme activity. (Reproduced by permission of American Chemical Society.)

tape-sealed and then 50 μl of substrate was added to the sample by the liquid handling system and mixed by pipetting. The mixture was then pumped into the POLARmonitor cuvette and monitored for 9 min. The cuvette was washed for 2 min with distilled water between runs.

Racemase activity could be measured in a 96-well plate using purified enzymes. Fig. 11.12 shows the linear range used for the calculation of enzyme rate from one of the reactions. The system was programmed to take 11 min for each run requiring 17 h for the entire 96-well plate; however, the authors suggest that the analysis time could be reduced and throughput increased to approximately three 96-well plates per day. The detection limit of the system is defined by the POLARmonitor, which has a 0.001°

resolution for a 100-mm cell. The specific activity of the glutamate racemase was determined to be 111.4 mdeg/min, which corresponded to 45.7 μmol · min^{-1} · mg^{-1} of purified enzyme. The automation by application of a liquid handling system increased the throughput of the system; however, only one sample may be analyzed at a time. The limitations of this approach include modest sensitivity and moderate throughput [26].

These improved polarimetric measurement methods still require sample volumes ranging from 1 μl to 1 ml and have a path length dependency that inherently limits further miniaturization. To realize the potential of polarimetry for HT screening, an alternative approach to optical activity measurements is necessary. Below, we discuss a recently developed polarimetric measurement method with nanoliter volumes and high sensitivity.

11.7 BACKSCATTER TECHNIQUE

Recently it has been demonstrated that backscattering interferometry could be used for polarimetry [27]. The backscatter detector utilizes a simple optical train based on interferometry, where light interacts in constructive and destructive interference. A coherent light source illuminates a capillary and part of the light is reflected from the surface of the capillary, while some of the light traverses the capillary multiple times before escaping. The light from the surface and the light that traversed the solution interact forming a fringe pattern. Furthermore, backscattering interferometry has a unique multipass [28,29] optical train providing the potential to perform optical activity determinations with ultra low volumes and at high sensitivity.

11.7.1 Capillary polarimetric detector (CPD)

The capillary polarimetry detector (CPD) has a simple optical train (Fig. 11.13) based on a 4-mW polarized He/Ne laser, a Polarcor™ polarizing plate with an extinction ratio of 1:10,000 (to further purify the polarization state of the beam), a fused silica capillary tube, and a transducer. With CPD, a polarized laser is directed onto an unmodified polyimide-coated capillary, illuminating the entire width of the tube. The result of this light–tube interaction was a fan of scattered light in 360° with well-defined light and dark

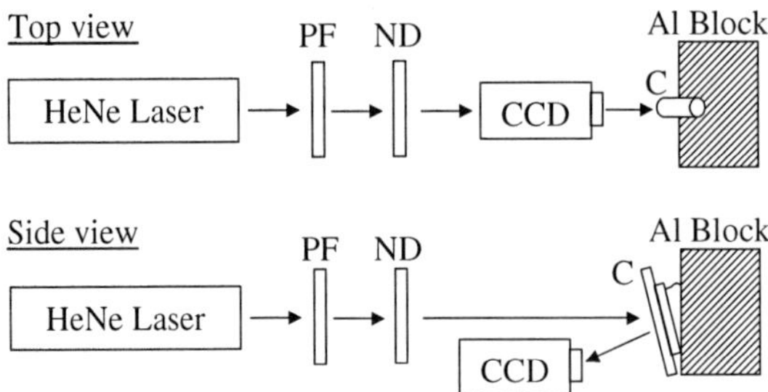

Fig. 11.13. Block diagram for the capillary polarimeter. PF is a polarizing filter, ND is a neutral density filter, C is a fused silica capillary, Al block is a mounting block made of aluminum, and CCD is a camera. (Reproduced by permission of American Chemical Society.)

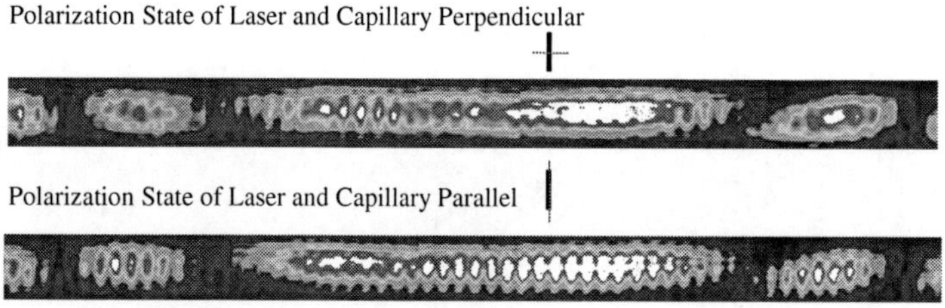

Polarization State of Laser and Capillary Perpendicular

Polarization State of Laser and Capillary Parallel

Fig. 11.14. Fringe pattern polarization-state dependency. (Reproduced by permission of American Chemical Society.)

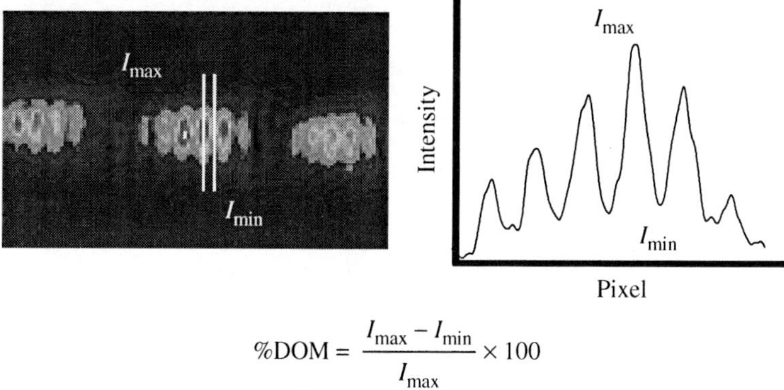

$$\%\text{DOM} = \frac{I_{max} - I_{min}}{I_{max}} \times 100$$

Fig. 11.15. CPD fringe and DOM sensing. (Reproduced by permission of Applied Spectroscopy.)

spots, or interference fringes detectable in the backscatter position. When the plane of polarization from the laser is parallel to the capillary central axis, both low-frequency and high-frequency fringes are observed. Upon rotation of the polarization plane to a perpendicular orientation the high-frequency fringes are attenuated. A charge-coupled device (CCD) camera and laser beam analyzer (LBA) are used to capture images of the fringes and the resulting changes in adjacent high-frequency fringes caused by optical rotation. This intensity change was determined to be quantitative relative to optical activity or plane of polarization (Fig. 11.14). By following changes in the depth of modulation (DOM), nanoscale polarimetry measurements were possible [27].

A cross section was taken from a fringe pattern (Fig. 11.15) and the resulting profile provided I_{max} (the intensity of a selected high-frequency fringe) and I_{min} (the intensity of the adjacent null). It was found that DOM was directly related to the optical activity of a molecule and that changes in the optical activity signal could be detected at the level $10^{-4\circ}$ in nanoliter volumes [27]. The System's response to *R*-mandelic acid is shown in Fig. 11.16 indicating a concentration dependence that appears nonlinear. Closer observation suggests two linear sections that could be explained by aggregation commonly seen at higher concentrations of this solute in aqueous solution [27]. Using a capillary with a 250-μm inner diameter and a 350-μm outer diameter and a beam diameter of 0.6 mm, the probe volume was 30 nl, 20 times small than previous dimensions. It is possible to further reduce the detection volume by using a smaller diameter capillary.

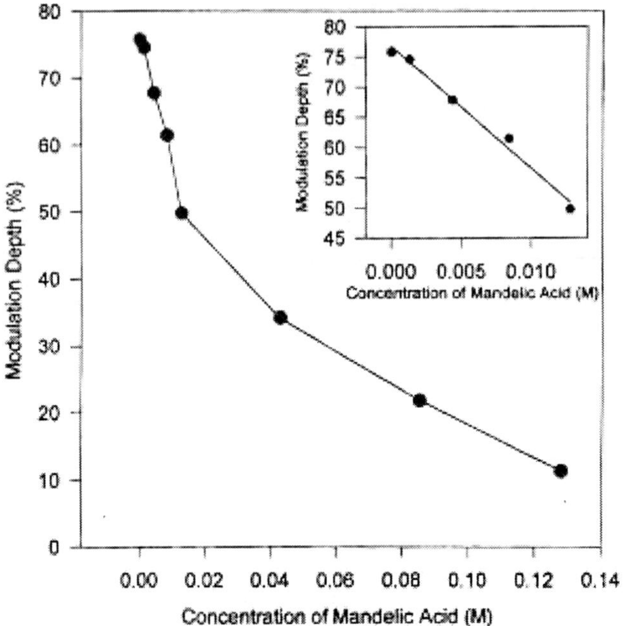

Fig. 11.16. DOM of *R*-mandelic acid. Insert shows first linear region from 0 to 0.012 M. The two linear sections suggest solute aggregation at higher concentrations. (Reproduced by permission of American Chemical Society.)

Further characterization of the CPD showed it follows common polarimetric response to Malus' Law [30], which states that the intensity of light is proportional to the $\cos^2$ of the angle of the plane of polarization to the principle plane of the illumination source, Eq. (1) [30].

$$I = I_0 \times \cos^2(\theta) \tag{1}$$

To test Malus' law response, a half-wave plate, which rotates the plane of polarization 2° for every 1° change, was mounted in a rotation stage and placed after the polarizer in the experimental setup shown in Fig. 11.13. Fig. 11.17 shows that the DOM signal follows Malus' law just as the intensity measurements do in a typical polarimeter [31]. For CPD to have broad applicability it will be necessary to perform DOM measurements in real-time. Furthermore, sampling by an LBA or DOM is an off-line process using a limited number of data points from the fringes. To circumvent these limitations, a new method of fringe interrogation is needed.

11.7.2 CPD: analysis by Fourier transform

Fourier transform methods allow the analysis of complex waveforms in terms of their sinusoidal components [32]. Fourier analysis transforms a waveform into its spectral components and has been utilized in mass spectrometry, infrared spectrometry, and nuclear magnetic resonance. Fourier analysis techniques and a CCD are used in fringe

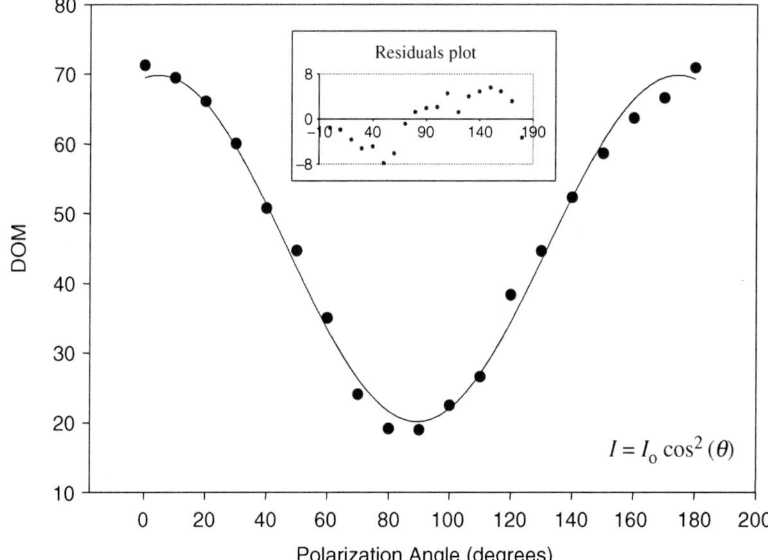

Fig. 11.17. DOM for the rotation of the plane polarization. (Reproduced by permission of Electrophoresis.)

interrogation and to improve measurements of the polarization signal obtained by the CPD. As before, a half-wave plate rotates the plane of polarization of the laser beam by 180° to demonstrate the Fourier signal *versus* the polarization angle follows Malus' law. Fig. 11.18 shows the Fourier spectra of the polarimetric signal for CPD at two different polarization angles. This plot demonstrates that a single peak (frequency) is produced by CPD even when the polarization state is changed. Fourier analysis of the fringe pattern can then be used to extract a parameter related to the polarimetric measurement. Indeed, the amplitude of the peak in the Fourier spectrum correlates with the strength or magnitude of the polarimetric signal. By performing the Fourier transform on fringe profiles generated at different polarization angles, a response curve can be generated. Fig. 11.19 shows this, illustrating the response to the rotation of the plane of polarization follows Malus' law. Using a fast Fourier transform (FFT) with CPD should facilitate near real-time measurements of optical activity, enabling the quantization of a transient event as seen in flow injection analysis (FIA) or HPLC. An improvement of 2.65 in signal-to-noise ratio and a detection limit of 0.64 mM for enzyme catalysis reactions were achieved with an FT using the off-line DOM method [31].

11.7.3 CPD: absolute optical activity measurements

While relative polarimetry response is important to quantify, some circumstances require the determination of absolute optical rotation power. The CPD was configured to determine if such measurements would be possible [28]. A second polarizer was mounted in a micro-calibrated rotation stage (Newport) and served as the "analyzer." By rotating

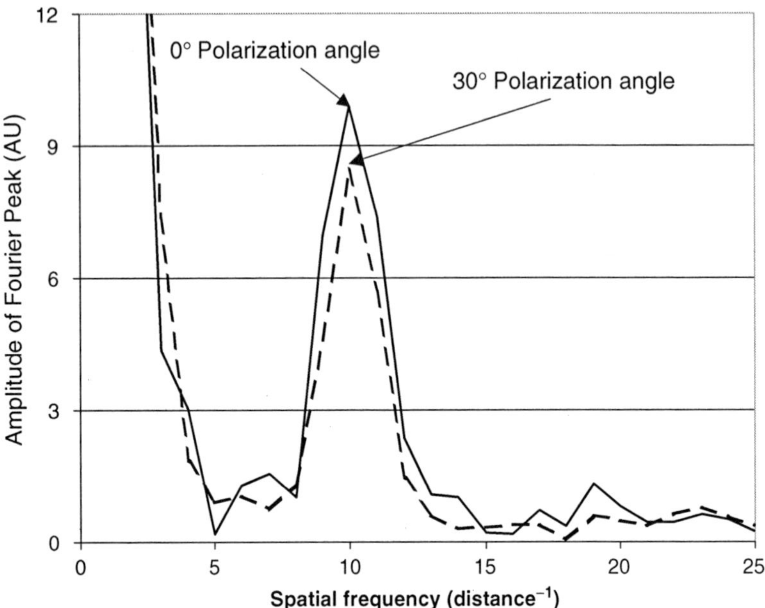

Fig. 11.18. Example of Fourier analysis results showing a Fourier peak at about 10 (distance^{-1}) whose amplitude correlates with the desired polarimetric signal. (Reproduced by permission of Electrophoresis.)

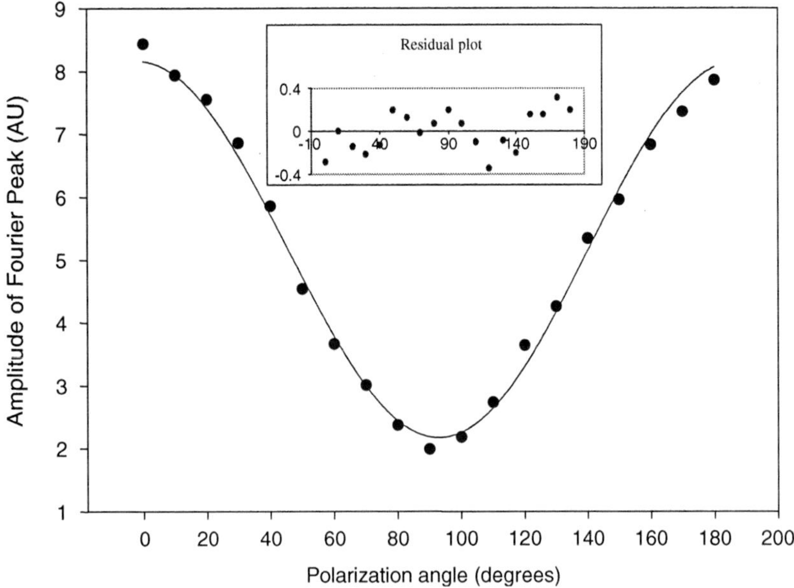

Fig. 11.19. Amplitude of the Fourier peak as a function of polarization angle, showing a behavior consistent with Malus' law. (Reproduced by permission of Electrophoresis.)

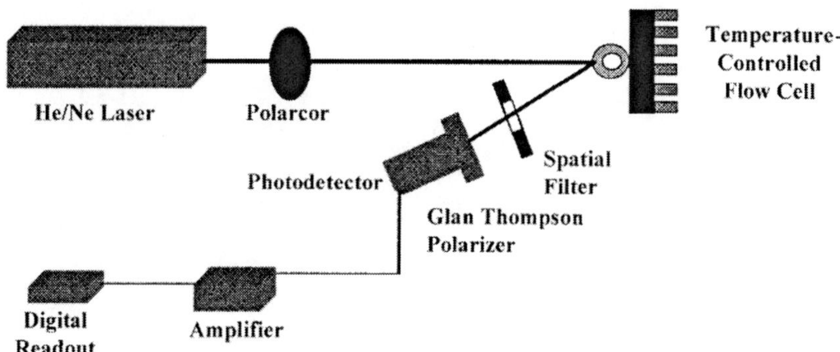

Fig. 11.20. Block diagram of the modified CPD micropolarimeter. (Reproduced by permission of Applied Spectroscopy.)

the analyzer, the maximum intensity provides the observed rotation as sensed with a single element photodetector. Fig. 11.20 shows the optical train. Using this system, it is possible to distinguish between *R* and *S* enantiomers (see Fig. 11.21(a)–(c)) [28], a first for nanoliter volume polarimetry. Fig. 11.21(a) shows that solutions of glycerol (a change in RI but not optically active) produce no change in rotation, which is a unique aspect of the CPD system [21]. Concentration limits of detection of 14×10^{-6} M for *R*-mandelic acid and 13×10^{-6} M for *S*-mandelic acid (an order of magnitude from the PDR-Chiral instrument) were quantifiable within a detection volume of 40 nl formed from an unmodified capillary with an inner diameter of 250 μm and a 0.8 mm diameter laser beam. The resolution performance of the detector is $4 \times 10^{-4\circ}$. In other words, the micropolarimeter provides picogram sensitivity (85 pg *R*-mandelic acid/79 pg *S*-mandelic acid) in sample volumes of 40 nl compared to the 50 μl for the smallest commercial cell, a decrease of approximately 3 orders of magnitude in volume [28].

The capillary-scale polarimeter has been constructed using a diode laser as an illumination source [33], and has been used for the detection of D-glucose [34], the examination of a kinetics study of D-β-hydroxybutyrate [35], and for polarimetric analysis in flowing streams [36]. Further advancement of this technique has led to an on-chip embodiment.

11.8 CONCLUSION

It has become increasingly important to measure optical activity, but obtaining the sensitivity necessary to measure rotation in small volumes has always presented a problem. The ability to minimize effects of RI changes, reduction of the amount of sample required, and performance of multiple measurements have presented problems for polarimetry for many years. However, there have been improvements made in polarimetry methodologies. Table 11.1 provides a comparison of the different optical activity measurements discussed.

Bobbitt and Yeung developed a successful small-volume polarimetric technique, which modulated the polarization of the illumination source. By locking into this

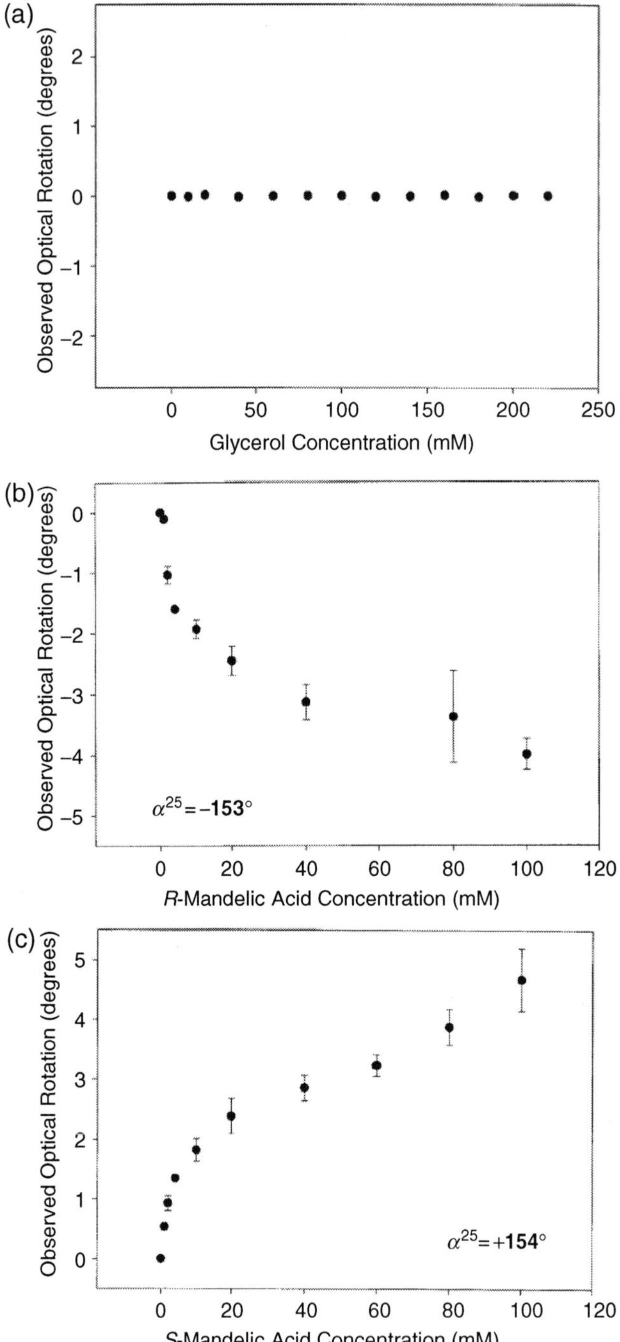

Fig. 11.21. (a) Response of the modified CPD micropolarimeter to glycerol; (b) response to *R*-mandelic acid; and (c) response to *S*-mandelic acid. (Reproduced by permission of Applied Spectroscopy.)

Table 11.1. Summary of various techniques

	Sample volume	Rotation detection limit	System note
Gibbs et al. [25]	75 μl	0.08°	Array polarimeter
Schonfeld and Bornscheuer [26]	40 μl	1 μ deg	Robotic pipettor
PDR-Chiral [23]	18 μl	25 μ deg	Modulation
Lloyd et al. [22]	8 μl	100 μ deg	820-nm light source
Bobbit and Yeung [19]	1 μl	15 μ deg	Modulation
CPD [27]	30 nl	100 μ deg	Capillary scale
CPD: absolute [31]	40 nl	400 μ deg	Capillary scale

modulation frequency, detection of 15 μ deg was possible. Utilizing a similar technique, Lloyd, Goodall, and Scrivener produced a similar instrument but utilized an 820-nm diode laser as an illumination source. This allowed for the analysis of molasses samples that are opaque at visible wavelengths and accomplished sensitivities in the microdegree range. PDR-Chiral's Advance Laser Polarimeter also modulates the polarization source. This is the most sensitive commercial instrument at 25 μ deg sensitivity with sample volumes as small as 18 μl.

Psuedorotation, a false-positive rotation caused by a change in the refractive index, can present a problem for conventional polarimeters. Even Bobbitt and Yeung discussed a false polarization signal caused by RI change. Maystre et al. [21] addressed this issue by means of an RI equalizer that introduced nonperturbing low or high RI liquid to correct any RI changes. Compensation for peaks as small as 60 μl was accomplished.

The throughput of polarimetry has typically been poor. Schonfeld and Bornscheuer [26] created an automated polarimeter by utilizing a liquid handling system and a peristaltic pump. The POLARmonitor used in the experiment has a detection limit of 1 millideg. In this system, 96 samples in a standard well plate could be analyzed in an automated fashion. This represents an advancement in throughput, yet it is still limited to one measurement at a time. Gibbs et al. [25] developed an imaging polarimeter that has the ability to measure an array or even multiwell plates. This provides a means of multiple optical activity measurements at once. However, sensitivity was only 0.08°.

The capillary polarimetric detector (CPD) has shown for the first time that optical activity measurements can be accomplished in nanoliter volumes. The simplistic design, consisting of a HeNe laser, polarizer, capillary, and detector, produces a unique analytical instrument. The changes in the polarization state of the illumination source were monitored by employing depth of modulation (DOM) analysis of the backscattered interferometric fringe pattern. In order to test this system, increasing concentrations of *R*-mandelic acid were introduced into the CPD. The DOM changed with the increasing concentrations of optically active solution. This technique achieved 100 μ deg sensitivities in a 30 μl probe volume.

The addition of an analyzer polarizer produced an instrument closely related to conventional polarimetry, where the optical activity is measured by the intensity of the light. In this configuration, the analyzer plate was rotated to determine the observed rotation. Solutions of glycerol, *R*- and *S*-mandelic acid in increasing concentrations were introduced into the CPD. As expected, the glycerol produced no observable rotation, while the *R*- and the *S*-mandelic acid produced opposite rotations, demonstrating that enantiomers can be distinguished a 30 μl volume.

The capillary polarimetric detector represents the first polarimeter that allows optical activity determinations to be performed in sample volumes down to the tens of nanoliters. It is also important to note that the sensitivity is not greatly sacrificed by the decrease in volume, which has been a previous limitation in polarimetry methodologies. The system is insensitive to RI changes, which can also be a hindrance for some polarimeters. Measurements of optically active molecules on a small scale are now possible, providing opportunities for analyses to be performed that were previously impossible due to sample size. An array of capillaries would provide a unique multiplexed nanoliter polarimeter.

REFERENCES

1 A.M. Rouhi, Taking a measure of chiral riches – researchers respond to high demand for ways to measure enantioenrichment quickly. Chem. Eng. News, 80 (2002) 51–57.

2 A.M. Rouhi, Chiral roundup – as pharmaceutical companies face bleak prospects, their suppliers diligently tend the fertile fields of chiral chemistry in varied ways. Chem. Eng. News, 80 (2002) 43–50.

3 A.M. Rouhi, Chirality at work. Chem. Eng. News, 81 (2003) 56–61.

4 S. Borman, Combinatorial chemistry. Chem. Eng. News, 80 (2002) 43–45.

5 S.J. Fox, M.A. Yund and S.F. Jones, Assay innovations vital to improving HTS. Drug Disc. & Dev., (2000) 40–43.

6 J. Kuhlmann, Drug research: from the idea to the product. Int. J. Clin. Pharmacol. Ther., (1997) 541–552.

7 S. Resetar and E. Eiseman, Anticipating technological change: combinatorial chemistry and the environment. EPA MR-1394.0-EPA, (2001).

8 M.T. Reetz, Combinatorial and evolution-based methods in the creation of enantioselective catalysts. Angew. Chem. Int. Ed., 40 (2001) 284–310.

9 O. May, P.T. Nguyen and F.H. Arnold, Inverting enantioselectivity by directed evolution of hydantoinase for improved production of L-methionine. Nat. Biotechnol., 18 (2000) 317–320.

10 F.H. Arnold, Design by directed evolution. Acc. Chem. Res., 31 (1998) 125–131.

11 P. Abato and C.T. Seto, EMDee: an enzymatic method for determining enantiomeric excess. J. Am. Chem. Soc., 123 (2001) 9206–9207.

12 G.A. Korbel, G. Lalic and M.D. Shair, Reaction microarrays: a method for rapidly determining the enantiomeric excess of thousands of samples. J. Am. Chem. Soc., 123 (2001) 361–362.

13 N. Cohen, S. Abramov, Y. Dror and A. Freeman, In vitro enzyme evolution: the screening challenge of isolating the one in a million. Trends Biotechnol., 19 (2001) 507–510.

14 A.M. Rouhi, Chiral chemistry. Chem. Eng. News, 82 (2004) 47–62.

15 M.G. Finn, Emerging methods for the rapid determination of enantiomeric excess. Chirality, 14 (2002) 534–540.

16 M. Tsukamoto and H.B. Kagan, Recent advances in the measurement of enantiomeric excesses. Adv. Synth. Catal., 344 (2002) 453–463.

17 R.A. van Delden and B.L. Feringa, Color indicators of molecular chirality based on doped liquid crystals. Angew. Chem. Int. Ed., 113 (2001) 3298–3300.

18 O. Hofstetter and H. Hofstetter, Antibodies as chiral selectors for the determination of enantioenrichment. Enantiomer, 6 (2001) 153–158.

19 D.R. Bobbitt and E.S. Yeung, Direct and indirect polarimetry for detection in microbore liquid-chromatography. Anal. Chem., 56 (1984) 1577–1581.

20 E.S. Yeung, L.E. Steenhoek, S.D. Woodruff and J.C. Kuo, Detector based on optical-activity for high-performance liquid-chromatographic detection of trace organics. Anal. Chem., 52 (1980) 1399–1402.
21 F. Maystre, A.E. Bruno, C. Kuhner and H.M. Widmer, Enhanced polarimetric detection in HPLC using a refractive-index equalizer. Anal. Chem., 66 (1994) 2882–2887.
22 D.K. Lloyd, D.M. Goodall and H. Scrivener, Diode-laser-based optical-rotation detector for high-performance liquid-chromatography and online polarimetric analysis. Anal. Chem., 61(1989) 1238–1243.
23 www.pdr-chiral.com.
24 PDR-Chiral Inc., Application Manual for Chiral Detection, 1999, pp. 1–50.
25 P.R. Gibbs, C.S. Uehara, P.T. Nguyen and R.C. Willson, Imaging polarimetry for high throughput chiral screening. Biotechnol. Prog., 19 (2003) 1329–1334.
26 D.L. Schonfeld and U.T. Bornscheuer, Polarimetric assay for the medium-throughput determination of alpha-amino acid racemase activity. Anal. Chem., 76 (2004) 1184–1188.
27 D.J. Bornhop and J. Hankins, Polarimetry in capillary dimensions. Anal. Chem., 68 (1996) 1677–1684.
28 K. Swinney, J. Nodorft and D.J. Bornhop, Nanoliter volume polarimetry. Appl. Spectrosc., 56 (2002) 134–138.
29 D.J. Bornhop, Microvolume index of refraction determinations by interferometric backscatter. Appl. Opt., 34 (1995) 3234–3239.
30 R.W. Ditchburn, Light, 3rd Ed. Academic Press, New York, 1976.
31 D.A. Markov, K. Swinney, K. Norville, D. Lu and D.J. Bornhop, A Fourier analysis approach for capillary polarimetry. Electrophoresis, 23 (2002) 809–812.
32 W.C. Elmore and M.A. Heald, Physics of Waves, Dover, New York, 1985.
33 K. Swinney, D. Markov, J. Hankins and D.J. Bornhop, Micro-interferometric backscatter detection using a diode laser. Anal. Chim. Acta, 400 (1999) 265–280.
34 K. Swinney, J. Hankins, D.J. Bornhop, Laser-based capillary polarimeter. J. Cap. Elec. and Microchip Tech., 6 (1999) 93–96.
35 K. Swinney and D.J. Bornhop, D-beta-Hydroxybutyrate reaction kinetics studied in nanoliter volumes using a capillary polarimeter. Appl. Spectrosc., 54 (2000) 1485–1489.
36 K. Swinney, J. Nodorft and D.J. Bornhop, Capillary-scale polarimetry for flowing streams. Analyst, 126 (2001) 673–675.

© 2006 Elsevier B.V. All rights reserved.

CHAPTER 12

Chiral analysis by regression modeling of spectral data

Kenneth W. Busch and Marianna A. Busch

Department of Chemistry & Biochemistry, Baylor University, One Bear Place #97348, Waco, Texas 76798, USA

12.1 INTRODUCTION

Asymmetric synthesis by means of enantioselective catalysis continues to be a major thrust in modern organic chemistry [1–3]. This research, in turn, has spurred the development of new analytical methods for the determination of enantiomeric purity. Enantiomers are mirror-image isomers of a molecule whose physical and chemical properties are indistinguishable in an achiral environment. The phenomenon of enantiomeric discrimination (ED), which permits a pair of optical isomers to be distinguished chemically or physically, is an active area of research from a fundamental as well as applied standpoint [4]. Experimental discrimination of enantiomers is carried out conventionally by means of chiral auxiliary agents such as chiral shift reagents (as in nuclear magnetic resonance (NMR)), and chiral complexing agents and chiral solvents (as in chiral chromatography) [5]. This diastereomeric discrimination arises when a given enantiomer of the chiral auxiliary interacts with two enantiomeric forms of a compound to produce diastereomeric pairs with different chemical and physical properties.

Indeed, the need for improved strategies for the assessment of enantiomeric composition of unknown samples is a topic of current interest. Reetz has reviewed the need for advanced technology for the determination of enantiomeric excess in large combinatorial libraries [6]. Finn, in a review of emerging high-throughput screening methods, concludes that spectroscopic methods show the most promise for rapid determination of enantiomeric purity [7,8].

The need for improved strategies for the assessment of enantiomeric purity arises from increased pressure on the pharmaceutical industry by government agencies for documentation on the pharmacological effects of individual enantiomers, and the simultaneous demand in drug development for the determination of enantiomeric excess in large combinatorial libraries as demand for single-enantiomer drugs continues to increase. In the area of drug metabolism, for example, it is well known that the pharmacological activity of a given molecule may be highly dependent on the enantiomeric form of the chiral drug molecule. When drug molecules interact with biological systems (which are themselves optically active), one enantiomer may be metabolized at a different rate or by a different pathway from the other. For example, differences in binding to plasma proteins and differences in binding at the active sites of pharmacological action may be different for the two enantiomers [9]. This may be particularly true for drugs whose metabolic pathway involves high-affinity, low-capacity enzymes such as the human cytochromes P450, 2C18, and 2D6 [10]. In the case where all the pharmacological activity may reside in a single enantiomer, the other enantiomer may be inactive, produce a qualitatively different effect, or may even be toxic [2].

To optimize the therapeutic value of enantiomeric drugs and avoid unintended adverse effects, methods for establishing enantiomeric purity of pharmaceutical products are urgently needed. Such methods are needed even for single-enantiomer drugs because racemization can occur *in vitro* as well as *in vivo*. Potential methods of chiral analysis should facilitate the routine analysis of racemates, pure enantiomers, and any intermediates in the manufacturing process. In pharmacological research and drug discovery, they are also needed for investigating the potential of an enantiopure drug to undergo inversion during storage and metabolism as well as for high-throughput screening of combinatorial libraries for enantiomeric excess (ee). While a variety of methods for determining ee have been developed, spectroscopic methods are considered to be the best choice for high-throughput screening [7]. Regardless of the method employed, any form of chiral analysis ultimately depends on an understanding of the factors involved in ED.

In this chapter, we will explore the role of chemometric methods, such as multivariate regression analysis, as a means of extracting information about the enantiomeric composition of samples from ordinary (achiral) spectral data. Before we discuss the details of chiral analysis by the regression modeling of spectral data, a brief discussion on multivariate calibration will be provided. For a more comprehensive discussion of this topic, the reader should consult more advanced works in this area [11–17].

12.2 INTRODUCTION TO MULTIVARIATE CALIBRATION

With the advent of modern computer-controlled analytical instrumentation, the amount of information potentially available as instrumental output is phenomenal and continues to increase exponentially. In the past, where one might monitor the progress of a reaction spectroscopically at a single wavelength with a monochromator-based instrument, today it is routine to measure a spectral range of wavelengths as a function of time with a polychromator equipped with a solid-state array detector. This increase in the amount of informational content available from modern analytical instruments has led to the development of a new discipline known as chemometrics, which, among other things, seeks to provide mathematical and statistical techniques to extract the maximum amount of information from the wealth of data now available from modern instrumentation. Moreover, the interface of the digital computer to analytical instrumentation in the chemistry laboratory has initiated a paradigm shift from univariate conceptualization to multivariate conceptualization.

In spite of the ready availability of powerful computers and multivariate statistical techniques, the change from univariate thinking to multivariate thinking has been slow. Consider a typical UV–visible absorption spectrum of a chromophore-containing organic molecule in a solution. In the current literature, the spectrum of this compound is frequently reported by stating the molar absorptivity of the compound at the wavelength of maximum absorption. This univariate approach reduces the entire spectrum down to a single number and completely overlooks all the ro-vibrational information present in the envelope of the absorption band.

In looking at the world around us, very few systems of interest are univariate. As chemists, we have been trained from our earliest experimental experience to handle this problem in the laboratory by varying one variable while keeping all the others constant. Thus, we study the effect of temperature on the volume of a gas while keeping the pressure and the number of moles of gas constant. Some disciplines, like medicine and psychology, do not have the luxury of a univariate approach. With human beings as subjects, it is not possble to test a drug on a hundred identical patients. Indeed, it is in behavioral science disciplines like psychology where some of the earliest attempts were made to model and predict outcomes based on multivariate data [18].

A major concept in analytical chemistry is the notion of calibration. In general, most instrumental methods rely on prior calibration to produce a reliable result. If a sample contains components that alter the instrument response to an analyte, these substances are referred to as interferents and must be removed or sequestered to avoid erroneous analytical results. This leads to the concept of specificity in analytical chemistry, which refers to the ability of the instrumental system to focus on a particular analyte in the presence of other sample concomitants. Instrumental techniques with high specificities are desirable because they reduce the need for analytical separations prior to an analysis.

Suppose, however, that another approach besides separation could be used that would still provide reliable analytical results in the presence of potentially interfering concomitants without the need for separations. Such an approach would be mathematical in nature rather than physical. With such an approach, we would have to turn our conventional thinking around. Instead of trying to isolate the analyte from potential interferents, we would calibrate the instrument with a set a samples that contain all the possible interferents at all their typical concentrations. To do this, our measurements would need to be multivariate in nature. In the spectral realm, we would use entire spectral regions rather than monitor a single wavelength. In this scenario, the regression analysis would focus on those variables that correlate with the parameter of interest while, at the same time, ignoring those variables that do not correlate with the parameter of interest. A basic assumption that must be met if this mathematical approach is to work is that the measured multivariate data contains information about the parameter of interest. If this is not true, no mathematical technique, no matter how sophisticated, will be able to produce a mathematical model for the parameter of interest. In other words, somewhere in the multivariate data, variations must exist that can be correlated with the parameter of interest.

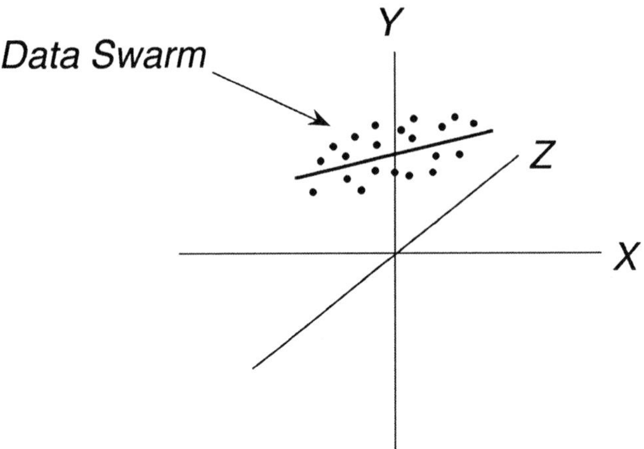

Fig. 12.1. Hypothetical data swarm plotted on an XYZ coordinate system. The line shows the first principal component drawn in the direction of the maximum variance in the data.

In linear multivariate regression modeling, the goal is to develop a regression model (i.e. a mathematical expression) that relates the dependent variable (the parameter that we wish to predict) with the independent variables (the variables that we measure). In linear regression modeling, this regression model takes the form of a linear equation:

$$\hat{Y} = b_0 + b_1X_1 + b_2X_2 + b_3X_3 + \cdots + b_nX_n, \tag{1}$$

where $\hat{Y}$ (Y-hat) is the predicted parameter of interest, X_i are the measured variables, and the b_n are the regression coefficients. The process of regression modeling involves the determination of an appropriate set of regression coefficients that will allow the prediction of the parameter of interest from the measured variables. To see how this can be done we will first consider a data transformation known as principal component analysis (PCA).

12.3 INTRODUCTION TO PRINCIPAL COMPONENT ANALYSIS

Consider a set of data with p variables. To represent this data will require a p-dimensional coordinate system or p-dimensional variable space. So that we can visualize the process graphically, let us consider a data set with 3 variables ($p = 3$). A fundamental assumption in multivariate analysis is that directions with maximum variance are associated with information content in the data.[1] Fig. 12.1 shows a hypothetical set of data plotted on a three-dimensional coordinate system. The data as plotted on the 3 coordinates can be thought of as a swarm of points. Suppose that by some means we are able to draw a line through this swarm in the direction of the

[1] In information theory, the information content of a signal is related to the signal variability. Signals that do not vary can not transmit information.

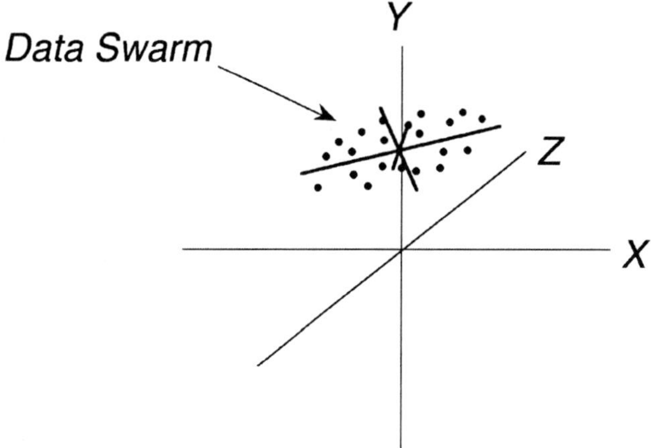

Fig. 12.2. Hypothetical data swarm plotted on an XYZ coordinate system. The lines show the first three principal components that comprise a new orthogonal coordinate system that can be used to represent the data.

maximum variance, such a line might resemble that in Fig. 12.1. In PCA, this line is referred to as the first principal component (PC1). Suppose we repeat the process and draw a second line through the data swarm in the direction of the next largest variance so that the second line (the second principal component (PC2)) is orthogonal to the first line. We can repeat this process successively with each new line orthogonal to the previous ones and representing less of the variance in the data.

Fig. 12.2 shows three principal components drawn through the hypothetical data swarm. The intersection of the three lines occurs at a point that represents the average value of all the data points in the swarm. PC1, PC2, and PC3 now represent a set of variance-scaled eigenvectors that provide a new orthogonal coordinate system on which data can be represented. The coordinates of the data on the new coordinate system are referred to as scores.

The contributions of the original variables X_1, X_2, and X_3 to the individual principal components are referred to as loadings. The higher the loading, the more important the given variable is to that PC. In mathematical terms, PC1 can be represented as

$$\mathrm{PC1} = p_{11}X_1 + p_{21}X_2 + p_{31}X_3 + \cdots + p_{p1}X_p, \tag{2}$$

where the given principal component is a linear combination of the original **p** vectors that make up the original variable space. The coefficients (p_{ij}) of the linear combination shown in Eq. (2) are the loadings. When the linear combinations for each principal component are written, the loadings make up a matrix known as a **P** matrix. The **P** matrix is a transformation matrix between the original variable space and the new space spanned by the PCs.

So far, all that we have done is to determine a new coordinate system on which to represent the data. That being said, one might wonder what of any real benefit has been accomplished. There are two potential benefits that accrue from this procedure. The first

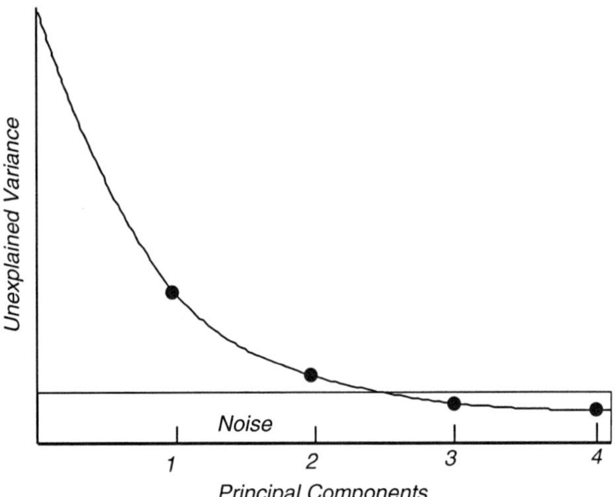

Fig. 12.3. Plot of unexplained variance as a function of the number of principal components (PCs) in the model. Higher PCs explain less and less of the unexplained variance and eventually only model noise.

is the removal of colinearity in the data. Colinearity arises when a variable is a function of one or more other variables that make up the original variable space. When the original data is represented in terms of the new coordinate system any colinearity that might have existed is removed because the principal components are all orthogonal to one another (i.e. mathematically independent).

The second potential benefit is a reduction in the dimensionality of the coordinate space needed to represent the data. When we go about determining a set of principal components, we can continue to determine new orthogonal vectors up to a maximum given by the smaller of

- n, where n is the number of variables in the original data ($X_1, X_2, \ldots, X_n$) or
- $K-1$, where K is the number of samples in the data swarm.

In the example that we used, there were three variables, so the maximum number of PCs that we could generate is three.[2] However, it should be realized that each new PC explains or accounts for a smaller amount of variance in the data than the previous one. At some point, new PCs do not provide substantially new information about the data and, in fact, become associated with noise. As a result, it may not be necessary to use all the possible PCs that are available to represent the data. If we plot the unexplained variance in the data as a function of the number of PCs determined, we get an ideal plot that may look something like that shown in Fig. 12.3. In this case, 4 PCs are possible, but only 2 are needed to represent the data (PC1 and PC2). The others (PC3 and PC4) simply account for noise in the data and are not necessary. By eliminating PC3 and PC4, we have effectively reduced the variable space from 4 to 2, making it easier to visualize the data.

[2] This assumes that $K > 3$, which is the case in Fig. 12.1.

By plotting the data in terms of the new coordinate system made up of PCs, hidden relationships that were formerly obscured in the original coordinate system can often be revealed. This can be illustrated chemically by performing a principal component analysis on some fundamental properties of some chemical elements. Table 12.1 shows 10 physical properties associated with 25 elements. From this set of data, 9 PCs are possible. Looking at the 250 entries in the table, it is hard to grasp any relationship between the elements. Fig. 12.4 shows a scores plot of the data for the first two principal components. PC1 explains 40% of the variance and PC2 explains 27% of the variance in the data.

The two-dimensional scores plot in Fig. 12.4 reveals some interesting relationships among the elements that would probably be diffult, if not impossible, to discern by looking at the original data in Table 12.1. Looking at the plot, we see that the data fall into four groups. In quadrant II, we see the halogens arranged diagonally in the order of increasing atomic number. In quadrant III, we find the rare gases arranged diagonally in the order of increasing atomic number. Near the origin of the scores plot, we find the alkali metals arranged more or less diagonally in the order of increasing atomic number. Finally, in quadrants I and IV, we find elements with 2+ oxidation states. Magnesium, calcium, and strontium are arranged diagonally in the order of increasing atomic number; however, as might be expected chemically, beryllium appears distinctly different from the other alkaline earth metals. Finally, in quadrants I and IV, we see a small group of metals with 2+ oxidation states (Mn, Cu, Ni, Co, Fe, and Pb). Notice that Pb, while having a 2+ oxidation state is separated from the others, which are all transition metals.

Fig. 12.5 shows a plot of the loadings for the first two PCs. From this plot, we can conclude that each of the ten variables played a significant role in the PCA model (none of the variables fall near the origin of the plot so all contribute to the model).

12.3.1 Principal component analysis revisited

In this section, we will take a more detailed look at PCA. Consider a set of spectra for a series of unknowns. Each spectrum consists of the absorbance (or intensity) at p wavelengths. In other words, there are p variables in the set. If there are n samples in the set, the data can be represented as an n (rows) $\times$ p (columns) array of numbers, or a data matrix that can be symbolized as **X**. The question that we wish to consider is whether this data matrix can be decomposed into a structure part and a noise part? In mathematical terms, can **X** be described by

$$\mathbf{X} = \mathbf{T}\mathbf{P}^{\mathrm{T}} + \mathbf{E}, \tag{3}$$

where the structure part is given by $\mathbf{T}\mathbf{P}^{\mathrm{T}}$ and the noise is given by the matrix **E**. In Eq. (3), **T** is the scores matrix and $\mathbf{P}^{\mathrm{T}}$ is the transpose of the loadings matrix.[3]

[3] For an introductory discussion of matrix algebra, see Appendix A at the end of the chapter.

Table 12.1 Physical properties of some elements

Element	MP[a]	BP[b]	Density[c]	IP[d]	EA[e]	HCap[f]	Z[g]	OxNum[h]	AtRad[i]	ENeg[j]
Ar	8.37E+01	8.74E+01	1.70E+00	1.52E+03	−3.60E−01	2.08E+01	18	0	0.95	0.0
Be	1.55E+03	3.24E+03	1.80E+03	8.99E+02	−2.50E+00	1.64E+01	4	2	1.11	1.5
Br	2.66E+02	3.32E+02	3.10E+03	1.14E+03	3.36E+00	3.61E+01	35	−1	1.14	2.8
Ca	1.12E+03	1.76E+03	1.54E+03	5.99E+02	−1.62E+00	2.59E+01	20	2	1.97	1.0
Cl	1.72E+02	2.39E+02	3.20E+00	1.25E+03	3.61E+00	3.39E+01	17	−1	0.99	3.0
Co	1.77E+03	3.17E+03	8.90E+03	7.58E+02	9.36E−01	2.48E+01	27	3	1.25	1.8
Cs	3.02E+02	9.60E+02	1.87E+03	3.77E+02	4.72E−01	3.22E+01	55	1	2.65	0.7
Cu	1.36E+03	2.87E+03	8.93E+03	7.45E+00	1.28E+00	2.44E+01	29	2	1.28	1.9
F	5.35E+01	8.50E+01	1.70E+00	1.68E+03	3.34E+00	3.13E+01	9	−1	0.71	4.0
Fe	1.81E+03	3.30E+03	7.87E+03	7.59E+02	5.82E−01	2.51E+01	26	3	1.24	1.8
He	9.00E−01	4.20E+00	2.00E−01	2.37E+03	−2.20E−01	2.08E+01	2	0	0.31	0.0
I	3.87E+02	4.57E+02	4.94E+03	1.01E+03	3.06E+00	3.69E+01	53	−1	1.33	2.5
K	3.37E+02	1.05E+03	8.60E+02	4.19E+02	5.01E−01	2.96E+01	19	1	2.27	0.8
Kr	1.17E+02	1.21E+02	3.50E+00	1.35E+03	−4.00E−01	2.08E+01	36	0	1.10	0.0
Li	4.52E+02	1.59E+03	5.34E+02	5.20E+02	6.20E−01	2.49E+01	3	1	1.52	1.0
Mg	9.24E+02	1.38E+03	1.74E+03	7.38E+02	−2.40E+00	2.49E+01	12	2	1.60	1.2
Mn	1.52E+03	2.37E+03	7.44E+03	7.17E+02	m	2.63E+01	25	2	1.37	1.5
Na	3.71E+02	1.17E+03	9.70E+02	4.96E+02	5.48E−01	2.82E+01	11	1	1.86	0.9
Ne	2.45E+01	2.72E+01	8.00E−01	2.08E+03	−3.00E−01	2.08E+01	10	0	0.65	0.0
Ni	1.73E+03	3.01E+03	8.90E+03	7.57E+02	1.28E+00	2.61E+01	28	2	1.25	1.8
Pb	6.00E+02	2.02E+03	1.13E+03	7.15E+02	1.05E+00	2.67E+01	82	2	1.75	1.8
Rb	3.12E+02	9.61E+02	1.53E+03	4.03E+02	4.86E−01	3.11E+01	37	1	2.48	0.8
Sr	1.04E+03	1.66E+03	2.60E+03	5.50E+02	−1.74E+00	2.64E+01	38	2	2.15	1.0
Xe	1.61E+02	1.66E+02	5.50E+00	1.17E+03	−4.20E−01	2.08E+01	54	0	1.30	0.0
Zn	6.93E+02	1.18E+03	7.14E+03	9.06E+02	m	2.54E+01	30	2	1.33	1.6

[a]Melting point (K).
[b]Boiling point (K).
[c]Density (Kgm^{-3}).
[d]Ionization potential ($KJmol^{-1}$).
[e]Electron affinity $\times 10^{-2}$ ($KJmol^{-1}$).
[f]Heat capacity ($JK^{-1}mol^{-1}$).
[g]Atomic number.
[h]Oxidation number.
[i]Atomic radius (Å).
[j]Electronegativity.

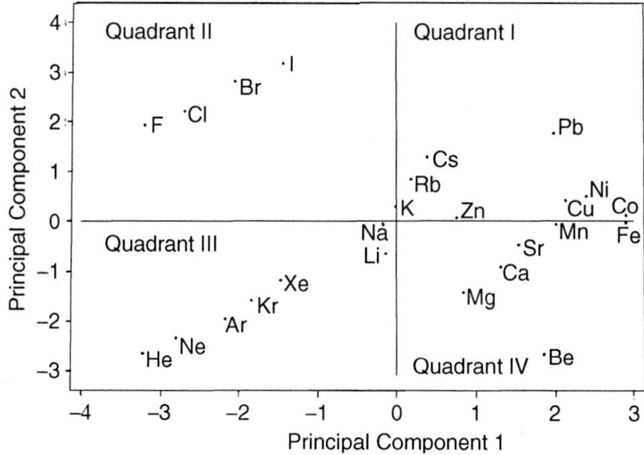

Fig. 12.4. Scores plot of principal component (PC) 2 *versus* principal component (PC) 1 from a principal component analysis (PCA) of the data in Table 12.1 for 25 elements.

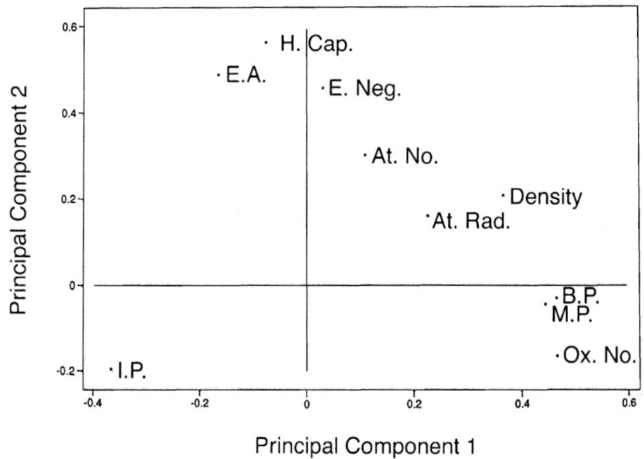

Fig. 12.5. Loadings plot of principal component 2 *versus* principal component 1 from a principal component analysis (PCA) of the data in Table 12.1 for 25 elements.

To see how this can be done, we will use a technique developed in 1966 by Herman Wold [19] known as non-linear iterative projections by alternating least squares (NIPALS).[4] Fig. 12.6 shows the steps in the algorithm. The NIPALS algorithm begins by mean centering the data matrix **X** by subtracting the average spectrum. Starting with an index of 1, we select a column in $\mathbf{X}_1$ to serve as a trial value for $\mathbf{t}_1$. A good choice for this trial $\mathbf{t}_1$ vector is the largest column in **X**. We then calculate $\mathbf{p}_1 = \mathbf{X}_1^T\mathbf{t}_1/|\mathbf{X}_1^T\mathbf{t}_1|$, where the denominator $|\mathbf{X}_1^T\mathbf{t}_1|$ represents the length of the vector (See Appendix A). Using $\mathbf{p}_1$,

[4] This is not the only way to do principal component analysis.

Steps in the NIPALS Algorithm

1. Start with a column vector in **X**. A good choice is the largest column. This will be our initial trial $\mathbf{t}_1$.
2. Calculate the loading vector, $\mathbf{p}_1$. From $\mathbf{p}_1$, calculate a new $\mathbf{t}_1$.
3. Continue until $\mathbf{t}_1$ converges. Calculate $\mathbf{t}_1\mathbf{p}_1^T$ with the final results for $\mathbf{t}_1$ and $\mathbf{p}_1$. $\mathbf{t}_1\mathbf{p}_1^T$ is the first principal component.
4. Update **X**. $\mathbf{X}_2 = \mathbf{X}_1 - \mathbf{t}_1\mathbf{p}_1^T$
   $\mathbf{X}_3 = \mathbf{X}_2 - \mathbf{t}_2\mathbf{p}_2^T = \mathbf{X}_1 - \mathbf{t}_1\mathbf{p}_1^T - \mathbf{t}_2\mathbf{p}_2^T$
5. Continue until $\mathbf{X}_A$ where A is the maximum number of PCs for the model. Higher values of $\mathbf{X}_A$ represent noise.

Fig. 12.6. The NIPALS algorithm.

we then calculate a new estimate for $\mathbf{t}_1$ by letting $\mathbf{t}_1' = \mathbf{X}_1\mathbf{p}_1$. If $|\mathbf{t}_1' - \mathbf{t}_1|$ is less than some small residual, we proceed to step 4 in Fig. 12.6. Otherwise, we return to step 2 and use $\mathbf{t}_1'$ as our estimate in calculating a new $\mathbf{p}_1'$. Eventually, our estimate of $\mathbf{t}_1$ converges and we can calculate $\mathbf{t}_1\mathbf{p}_1^T$, which is a matrix that represents the first principal component.[5] We then subtract the first principal component from $\mathbf{X}_1$ to get $\mathbf{X}_2$. We then repeat the entire process again with $\mathbf{X}_2$ to get the second principal component and so on until we arrive at $\mathbf{X}_A$, where A is the maximum allowed number of PCs for the model. Since not all the allowed PCs are necessary, the higher PCs are associated with the noise.

12.3.2 Multivariate regression

So far, we have discussed how a data set can be decomposed into a structure part and noise part. We have seen that by plotting the data in terms of a new coordinate system made up of principal components hidden relationships can be revealed that were formerly obscured in the original coordinate system. In this section, we will discuss how calibration can be performed in a multivariate fashion. Multivariate calibration involves relating two sets of data, **X** and **y**,[6] where **X** is a data matrix made up of the independent variables (i.e. the measured variables) and **y** is a vector made up of the dependent variables (i.e. the quantity that we are trying to predict). Fig. 12.7(a) shows schematically how these two data sets are combined to produce a multivariate regression model. The goal of multivariate regression modeling is prediction as shown schematically in Fig. 12.7(b).

12.3.2.1. Multiple linear regression

We are all familiar with the idea of solving a set of equations for *n* unknowns. To do this, we must have *n* independent equations (there must be no colinearity in the set of the equations). Suppose, for example, that **X** is a data matrix composed of spectral data for

[5] The product $\mathbf{t}_i\mathbf{p}_i^T$ is an outer product. Here, $\mathbf{p}_i$ is a column vector so its transpose is a row vector. The product is a matrix of rank one with the same dimensions as **X** so that subtraction is possible. See Appendix A for details.

[6] The predicted set of data can also be a matrix, but for our discussion, we will assume that the predicted quantities are a vector.

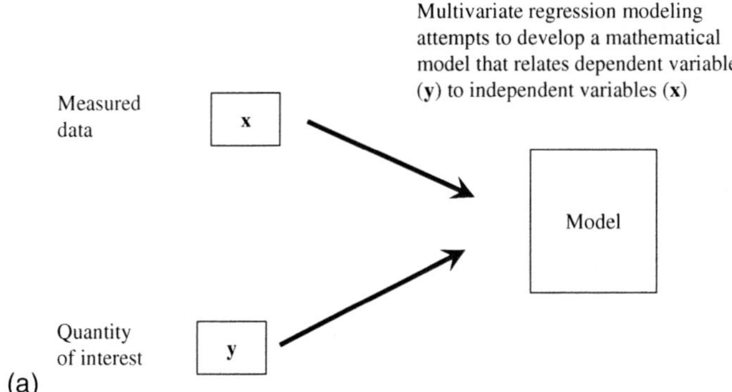

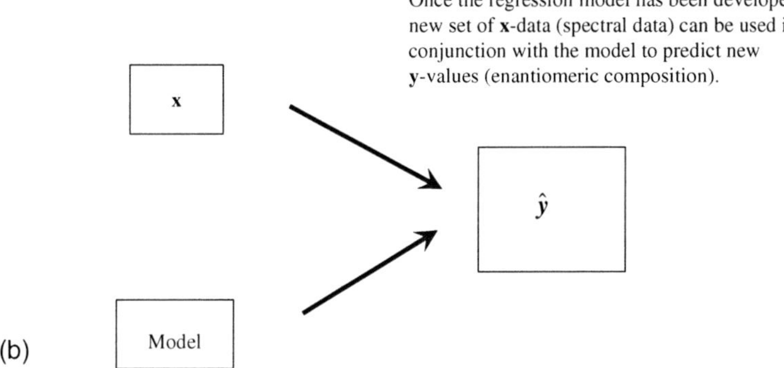

Fig. 12.7. Regression modeling. (a) Developing the model from the measured data and the quantity of interest; (b) using the measured data and the model to predict the quantity of interest.

a representative set of n samples.[7] This is the independent data. Suppose also that **y** is a column vector composed of the known values of the dependent variable (the quantity that we wish to predict) for the n samples. We can relate **X** to **y** by the following equation:

$$\mathbf{y} = \mathbf{X}\mathbf{b}, \tag{4}$$

where **b** is a column vector known as the regression vector that we wish to determine. Using matrix algebra, we can determine **b** by solving Eq. (4) as follows:

$$\mathbf{X}^{\mathrm{T}}\mathbf{y} = \mathbf{X}^{\mathrm{T}}\mathbf{X}\mathbf{b} \tag{5}$$

$$(\mathbf{X}^{\mathrm{T}}\mathbf{X})^{-1}\mathbf{X}^{\mathrm{T}}\mathbf{y} = (\mathbf{X}^{\mathrm{T}}\mathbf{X})^{-1}\mathbf{X}^{\mathrm{T}}\mathbf{X}\mathbf{b} = \mathbf{b} \tag{6}$$

[7] In general, **X** will not be a square matrix because there are usually more wavelengths in a spectrum than samples. The product of $\mathbf{X}^{\mathrm{T}}$ and **X** will be a square matrix, however.

MLR becomes problematic and may fail when

- there is colinearity in **X**
- the **X** data are noisy
- there are more variables than samples (the matrix is rectangular rather than square)
- interferences exist between variables in **X**

Fig. 12.8. Limitations of multiple linear regression modeling.

The regression vector **b** in Eq. (6) is made up of regression coefficients that represent the multivariate regression model. The procedure shown above is known as classical multiple linear regression and depends critically on the existence of $(\mathbf{X}^T\mathbf{X})^{-1}$. If colinearity in the **X** data exists, the inverse of $\mathbf{X}^T\mathbf{X}$ may not exist.[8] Fig. 12.8 shows some situations where multiple linear regression may fail.

12.3.2.2 Principal component regression

To get around the colinearity problem, we can perform a principal component analysis on the **X** data matrix. This will eliminate any colinearity because the scores vectors in PCA are clearly all orthogonal to each other. Therefore, instead of using the original **X** data matrix, we can use the scores matrix **T**. So, instead of $\mathbf{y} = \mathbf{Xb}$, we write

$$\mathbf{y} = \mathbf{Tb} \tag{7}$$

$$\mathbf{T}^T\mathbf{y} = \mathbf{T}^T\mathbf{Tb} \tag{8}$$

$$(\mathbf{T}^T\mathbf{T})^{-1}\mathbf{T}^T\mathbf{y} = (\mathbf{T}^T\mathbf{T})^{-1}(\mathbf{T}^T\mathbf{T})\mathbf{b} = \mathbf{b} \tag{9}$$

This should work because all the columns in **T** are orthogonal. The above procedure is known as principal component regression (PCR). While PCR gets around colinearity problems in the **X** data, it is not necessarily the most effective means of multivariate regression modeling.

One of the problems associated with PCR is the fact that only the **X** data are used to determine the PCs. In determining the principal components, we obtained a set of variance-scaled orthogonal vectors based on the variance of the **X** data alone. Suppose, however, that the directions of the maximum variances in **X** data are not optimally related to the **y** data that we wish to predict. To get around this problem, we can use another multivariate approach known as partial least squares regression or PLS regression.

12.3.2.3 Partial least squares regression

PLS-regression is similar to PCR in that the data decomposition results in a new orthogonal coordinate system based on PLS components. For space reasons, we will not discuss the details of how the PLS components are determined. For a discussion of this

[8] Colinearity in **X** is said to exist if a column in **X** can be written as an approximate or exact linear combination of other columns.

procedure, the reader is referred to more advanced texts [20]. The PLS algorithm is especially powerful for regression because both the **X** and **y** data are actively involved in the construction of the new basis set. In this way, the PLS algorithm focuses on those aspects of the **X** data that are most important in predicting **y**. The PLS method attempts to balance the two objectives by seeking a basis set of PLS components that explains both the variation in **X** data while, at the same time, taking into account the **y** data that we wish to predict.

12.3.2.4 Multivariate regression modeling

Multivariate regression modeling is a two-phase process. In the first, or calibration phase, a set of calibration samples is assembled. In assembling this calibration set of samples, it is important that the samples selected for the set be representative of the typical samples likely to be encountered. In general, the more the samples in the set, the better the regression model.[9] An important aspect of multivariate modeling is that *real* samples are used in the calibration phase rather than artificially prepared standards as are frequently used in univariate calibration.

To illustrate this point, suppose that we wish to develop a spectroscopic method to determine the fat content in beef. In selecting a calibration set of samples for this purpose, we would choose a large number of naturally occuring beef samples with different fat levels.[10] To make the model, we will need to analyze these samples independently to determine the actual fat levels. So, we need a reliable, accurate reference method of analysis in order to determine the **y** data for the model. This reference method may be relatively labor intensive to avoid possible interferences from other substances in the samples. Once the **y** values (fat content, in this case) have been established independently, we can collect the spectral data (over some appropriate spectral region) for each sample. This data (one spectrum per sample) will form the **X** matrix. As shown schematically in Fig. 12.7(a), these two data sets can then be related by means of PLS regression to give a regression vector made up of $p+1$ regression coefficients,[11] where p is the number of variables (wavelengths, in this case) in the **X** (spectral) data.

These regression coefficients are the mathematical model. Once we know what these values are, we can use them in conjunction with new spectral data to predict the y values for any given unknown sample:

$$\hat{y} = b_0 + \mathbf{xb}, \tag{10}$$

[9] This is especially true if the samples chosen represent distinctly different sample types. Statisticians refer to this situation by saying that the samples span the sample space. What this means is that every possible type of variation likely to occur with the samples is included in the calibration stage of regression modeling.

[10] It is important to emphasize that the meat samples with different fat contents must be naturally occuring. Under no circumstances would it be acceptable to take a lean sample of beef and prepare different samples from it by adding fat. Such samples would not be representative of naturally occuring samples, which may contain other concomitants as the fat level increases.

[11] The additional regression coefficient b_0 is calculated from the mean data.

where y-hat is the predicted value for y for the given sample based on a spectrum (a row vector with p variables) and the regression vector (a column vector with p regression coefficients) determined in the calibration phase.

The second phase of regression modeling is the validation phase. In this phase of regression modeling, we establish whether the values predicted by the model are accurate (in other words, is the model any good). The best way to establish this is with another set of real samples known as the validation set. Once again, these samples must be analyzed independently using an accurate reference method to determine the analyte of interest. Again, spectral data are collected for each unknown sample over the same spectral range used to develop the model. This spectral data is now used with the model (the regression vector **b** in Eq. (10)) to predict the y values for each sample in the validation set. By comparing the values predicted by the model with those determined by the reference analytical method, we can determine whether the accuracy of the regression model is sufficient for the particular task.

Regression modeling of this type is really an ongoing process. While chemometric techniques are very adept at making predictions based on very small differences in **X** data, they are not good at extrapolation. That is why it is so important to select an appropriate calibration set of samples. Down the road when using the model, if a particular sample is encountered that is different in some way from those used in the calibration phase to make the model, the model will not be prepared to handle that sample.[12]

Unfortunately, it is not uncommon for a reliable model that has been in use successfully for some time to suddenly breakdown and begin predicting erroneous results. This can usually be traced to some new set of conditions that was not included in the original model.[13] To overcome these problems, regression modeling must continue over time and periodic checking of the results with known samples is necessary. This means that periodically, new samples need to be analyzed by the reference analytical method and these new samples need to be added to those used in the initial model. With this expanded set of calibration samples (the original set and the new set), an updated model can be prepared by repeating the calibration phase with the new expanded set of samples. By periodically updating the calibration model with new samples, one continues to build a robust model that can account for samples with ever more variability, because the model has been trained with a wide variety of samples.

Finally, in the early stages of developing a new model, it is worthwhile to maintain a healthy dose of skepticism about the model until it has been shown to be reliable over time. Because chemometric techniques are so sensitive to small variations in the **X** data, it is not always clear exactly what the regression model is keying on, and it is possible to be easily fooled into thinking that the model is based on one set of

[12] We can think of regression modeling as a training process whereby we train the computer to recognize different samples based on their spectral profile. As with any training process, when new situations arise, the previous training may be inadequate.

[13] For example, if a model was developed with material from a particular supplier and the supplier modifies its process or the source of its raw materials, a previous regression model made with the old material will probably not work well.

criteria when it is really modeling something quite different that is related to the measured quantity.[14]

12.4 USE OF CHEMOMETRICS IN DETERMINING ENANTIOMERIC COMPOSITION

Chiral analysis by the regression modeling of spectral data involves the combination of three areas: spectroscopy, chemometrics, and guest–host chemistry. Before discussing the combination of the three areas, a brief discussion of cyclodextrins as chiral auxiliaries will be provided.

12.4.1 Use of a chiral auxiliary

ED is conventionally accomplished in the laboratory when a pair of enantiomers interacts with a chiral auxiliary to break the mirror symmetry of the enantiomeric pair by forming diastereomeric products, as shown in Scheme I.

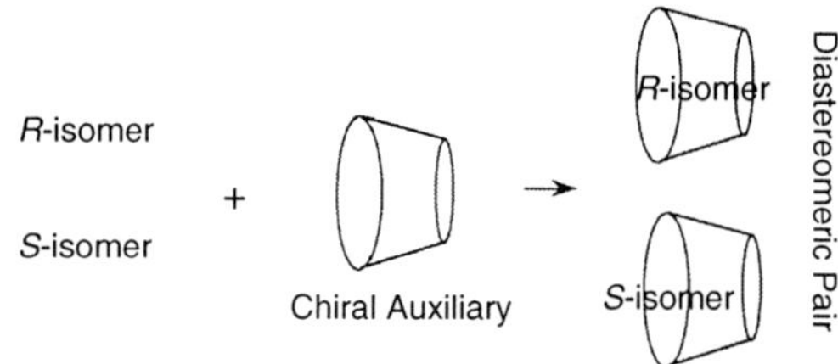

Cyclodextrins have been used widely as chiral auxiliaries in chromatography and electrophoresis because of their ability to form transient, non-covalent inclusion complexes (guest–host complexes) that provide a proven means for ED [21].

Cyclodextrins (CDs) are cyclic toroidal-shaped homochiral oligosaccharides produced by enzymatic degradation of starch that readily form transient, non-covalent complexes with suitable guest molecules [22]. While CDs can contain from 6 to 12 (+)-glucopyranose units bonded through α-(1,4) linkages so that all the glucose units are in a C-1 (D) chair conformation, the most common native CDs are α-CD (6 glucopyranose units), β-CD (7 glucopyranose units), and γ-CD (8 glucopyranose units). Fig. 12.9 shows the structures of α-, β-, and γ-CD. The molecules are homochiral because all the glucopyranose units are in a C-1 (D) chair conformation. Table 12.2 gives some properties of the common native CDs.

Any given CD, which has the shape of a truncated cone, has a hydrophobic interior cavity (produced by C-H groups) and a hydrophilic mouth (produced by the secondary hydroxyl groups on the C-2 and C-3 carbons of the glucose molecules). In contrast to

[14] For example, spectral features in the near-infrared (NIR) region are often dependent on temperature, salt content, and pH. Variations in parameters like these may correlate under certain circumstances with the parameter of interest. Such a model may work until this hidden correlation breaks down.

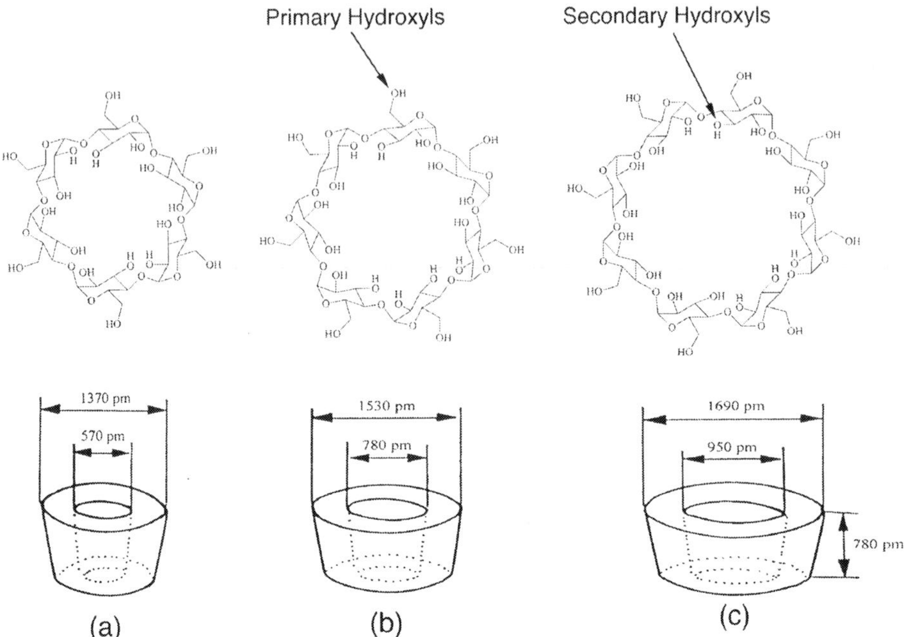

Fig. 12.9. Structures and sizes of cyclodextrins. (a) α-cyclodextrin; (b) β-cyclodextrin; (c) γ-cyclodextrin. J. Szejtli, Chapter 2, Figs. 2 and 5, in: J. Szejtli and T. Osa (Eds.), Comprehensive Supramolecular Chemistry, Vol. 3 Cyclodextrins, Pergamon, Oxford, 1996, p. 7, 12.

Table 12.2 Some properties of α-CD, β-CD, and γ-CD[a]

Property	α-CD	β-CD	γ-CD
Number of gluco-pyranose units	6	7	8
Cavity diameter (Å, approximately)	4.7–5.3	6.0–6.5	7.5–8.3
Cavity height (Å, approximately)	7.9	7.9	7.9
Cavity volume (Å^3, approximately)	174	262	427
pK[a]	12.3	12.2	12.1

[a]Source: J. Szejtli, Chapter 2, Table 4, in: J. Szejtli and T. Osa (Eds.), Comprehensive Supramolecular Chemistry, Vol. 3 Cyclodextrins, Pergamon, Oxford, 1996, p. 17.

the secondary hydroxyl groups on the mouth of the cavity, which are relatively rigid, the primary hydroxyl groups (on the C-6 carbon of the glucose molecule) on the truncated end of the cone can rotate to block the cavity.

In aqueous media, transient non-covalent guest–host complexes form with CDs as a result of induced or permanent dipole–dipole (van der Waals' type) interactions, hydrogen-bonding interactions, π–π interactions, and the hydrophobic properties of the guest molecule. To be effective in forming diastereomeric adducts, it is generally believed that three requirements must be met:

1. the chiral analyte must form an inclusion complex with the cyclodextrin;

2. the inclusion complex between the guest molecule and the cyclodextrin cavity must be rigid to avoid movement that would result in the same average environment for both enantiomers; and
3. the chiral center of the guest molecule must be near the mouth of the cavity and interact with it.

Indeed, interaction of the primary hydroxyl groups on the C-2 and C-3 carbons at the mouth of the cavity appears to be especially important in the formation of diastereomeric adducts. Molecular modeling studies [23] of the inclusion complexes between β-CD and (*R*)- and (*S*)-propranolol (a β-adrenergic blocker used as a cardiac depressant) have revealed that important differences exist between the enantiomers with regard to the secondary amine group. In particular, these studies have shown that hydrogen bonding between the nitrogen and the C-2 and C-3 hydroxyls is more favorable for the (*R*)-propranolol than for the (*S*)-isomer.

While advances in understanding basic supramolecular chemistry are fundamental to understanding the underlying principles of chiral discrimination, the situation with chiral discrimination using cyclodextrins is far from clear. In a recent article on the elucidation of the principles of chiral recognition, Topiol and Sabio [24] discuss the distinction between contact points and interaction points, the role of chiral centers, and the use of distance-matrix approaches in understanding chiral discrimination. In computational studies with complexes between Fluvastatin and γ-CD, they conclude that the γ-CD host provides a restraining effect that reduces the dimensionality of the system. On the other hand, in the studies of chiral discrimination of supramolecular complexes of enantiomeric binaphthyl derivatives with cyclodextrins [25], the high enantioselectivity observed was interpreted in terms of a positive entropic contribution due to extensive desolvation of the guest molecule and the oxygens on the wider side of the CD cavity. Li and Purdy [26], on the other hand, contend that the interaction of functional groups around the chiral center with the mouth of the cavity and/or that of another CD is essential for chiral discrimination. Their theory agrees with the results obtained by Bortolus et al. [27] in their studies of chiral discrimination of camphorquinone enantiomers with cyclodextrins.

12.5 CHIRAL ANALYSIS BY MULTIVARIATE REGRESSION MODELING OF SPECTRAL DATA

In 1996, MacDonald and Hieftje [28] studied the use of β-cyclodextrin for the enantiomeric differentiation of α-pinene enantiomers by near-infrared (NIR) spectroscopy over the spectral range of 1100–2498 nm. They found that the (1*R*)-(+) and (1*S*)-(−) enantiomers of α-pinene could be differentiated by means of principal component analysis of the first-derivative NIR spectral data. While the study by MacDonald and Hieftje demonstrated the feasibility of enantiodifferentiation of cyclodextrin inclusion complexes with NIR spectroscopy, the scope of the study was limited to complexes with α-pinene and β-cyclodextrin, solubility constraints limited the work to a study of thin films deposited on quartz plates, and no regression studies were done to determine whether enantiomeric purity could be determined by NIR.

In mid-2000, our group began an investigation, supported by the Robert A. Welch Foundation, into the use of regression modeling of NIR spectral data to see if regression models could be made that would predict the enantiomeric composition of unknown samples from NIR spectral data. NIR absorption bands typically arise from overtone vibrations and combination bands involving CH, OH, and NH groups, among others. Indeed, NIR studies by Politi et al. [29] on the inclusion complexes between cyclodextrins and aromatic compounds had shown that the oscillator strengths of the C–H overtone bands of the guest compounds were increased by two- to five-fold upon forming complexes. Moreover, previous studies of NIR spectra of α-, β-, and γ-cyclodextrin had shown that these compounds had bands at ~1180, 1430, 1510, and 1670 nm, which were assigned to the first and second overtone stretching vibrations and combination bands of C–H and O–H [30,31].

While NIR spectroscopy should be able to provide important information about the inclusion complexes of cyclodextrins with chiral molecules, and inclusion complex formation should result in diastereomeric adducts with different spectral properties, in contrast to the work of Tran [32], our attempts to develop regression models with NIR spectral data were disappointing. In our initial NIR studies, solubility limitations with native cyclodextrins were encountered. In these studies, it was found that the NIR bands for water (the solvent, 1940, 1450, and 1190 nm) overwhelmed any small spectral changes in this region that may have been due to hydrogen bonding in the guest–host complex (because of its limited concentration). While we were able to produce what looked like reasonable regression models from NIR spectral data in the calibration phase, we found that these models did not provide satisfactory prediction results in the validation phase of regression modeling. As a result, we began to look at using UV–visible absorption spectral data.

In ED by spectroscopic means, it is conventionally thought that successful discrimination requires sharp spectral features that are resolvable instrumentally. Indeed, in the area of chemical shift reagents in NMR, for example, a great deal of effort has gone into developing suitable reagents that can provide adequate shift to be resolvable instrumentally [33]. By contrast, the UV (electronic) region of the spectrum has been long neglected for ED because conventional wisdom maintains that the broad, somewhat featureless bands, which often overlap, do not contain much useful information and that measurements need to be made at the band maximum and be resolvable instrumentally. However, because chemometric techniques are so adept at extracting useful information from seemingly insignificant differences in the data, it is precisely in this type of application where their use can be of most benefit.

Although conventional wisdom maintains that UV–visible absorption spectra do not contain much useful information beyond the molar absorptivity at the absorption band maximum, actually the envelope of an absorption band, which is made up of vibrational transitions from the ground to the excited state, contains information about the molecule that may not be apparent from simple visual inspection. However, as we have seen with the example of the chemometric analysis of data from the periodic table, just because information does not appear to be present from simple visual inspection does not mean that it is not there.

Early in 2001, we began to explore the role of multivariate regression modeling of UV–visible absorption spectral data as a means for the rapid determination of the enantiomeric composition of unknown samples using ordinary (achiral) spectroscopic methods. The ultimate goal of this research was the development of rapid, reliable, and robust methods for assessing enantiomeric purity by ordinary inexpensive spectroscopic methods. The combination of chemometric (regression) techniques with guest–host chemistry and ordinary spectroscopic methods was a unique feature of this research. In 2002, we organized a full-day symposium on chiral analysis for FACSS 2002,[15] where we presented an article[16] on the results that we obtained from regression modeling of UV–visble spectral data obtained with chiral analytes in the presence of cyclodextrin host molecules.

12.5.1 Initial studies with amino acids

Early in 2003, Busch et al. published a communication in the *Journal of the American Chemical Society* dealing with using chemometric analysis of UV–visible spectral data for the determination of the enantiomeric composition of unknown samples of amino acids [34]. In this brief report, the authors showed how the enantiomeric composition of 2-phenylglycine and other amino acid samples could be determined by means of ordinary absorption spectrophotometry using β-cyclodextrin as a chiral auxiliary. It had been discovered during the course of this study that the absorption spectrum of solutions containing a fixed concentration of 2-phenylglycine and a fixed concentration of β-CD varied slightly as the enantiomeric composition of the 2-phenylglycine was varied. Fig. 12.10 shows the spectral data from 250 to 500 nm for a series of nine aqueous solutions (pH 12) containing 30 mM β-CD and 15 mM 2-phenylglycine. While the total concentration of 2-phenylglycine was fixed, the enantiomeric composition of the 2-phenylglycine used to prepare the solutions was varied from mole fraction 0.50 to 0.90 of (*R*)-2-phenylglycine, using different amounts of the commercially available enantiomers.

As shown in Fig. 12.10, the spectral changes observed did not seem to follow any particular order with changes in enantiomeric composition and the spectra of solutions with different enantiomeric compositions often crossed one another.[17] While these spectral changes were small, and might have been dismissed as having little real analytical utility, they seemed ideal for multivariate regression modeling.[18] Using a partial least squares algorithm (PLS-1), the Busch group demonstrated that multivariate

[15] 29th Annual Meeeting of the Federation of Analytical Chemistry & Spectroscopy Societies, October, 2002, Providence, RI.

[16] Paper 370, Book of Abstracts, FACSS 2002.

[17] It is clear from the figure that the spectral data do not vary in a simple manner with respect to enantiomeric composition of the samples (i.e. the spectra are not simply offset in a uniform fashion from one another). For this reason, multivariate techniques that use entire spectral regions (rather than a univariate approach using a single wavelength) were necessary.

[18] We had become accustomed to dealing with small spectral changes by previous work in the NIR region.

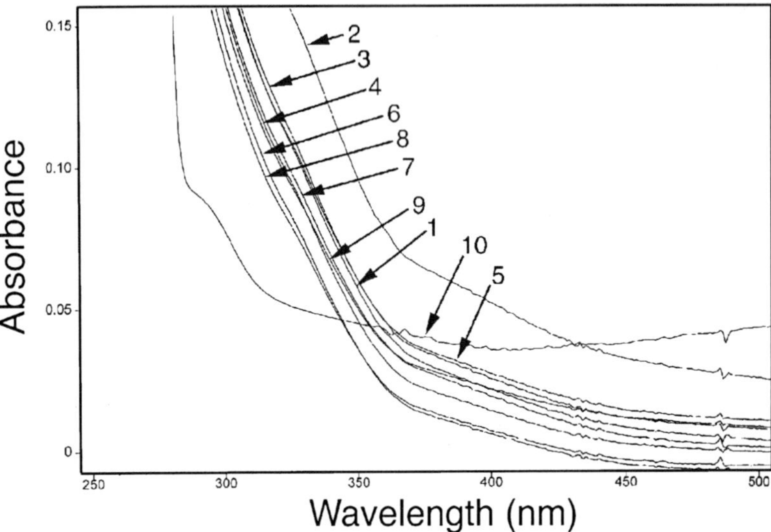

Fig. 12.10. Absorption spectra of solutions (pH 12) containing 30 mM β-CD and 15 mM ϕ-Gly of various enantiomeric composition (1–9). Mole fraction of (*R*)-2-phenylglycine: **1**, 0.460; **2**, 0.500; **3**, 0.566; **4**, 0.600; **5**, 0.634; **6**, 0.700; **7**, 0.800; **8**, 0.854; **9**, 0.900. Spectrum of 30 mM β-CD (10). (Reprinted from Ref. [34])

regression models could be developed from this spectral data that could predict the enantiomeric composition of the phenylglycine samples to within ±3% over a range of mole fractions from 0.5 to 0.9 of (*R*)-2-phenylglycine.

12.5.2 Studies with pharmaceutical compounds

In a subsequent study [35], the technique was used to determine the enantiomeric composition of samples of ibuprofen and norephedrine using regression models based on UV-spectral data of cyclodextrin guest–host complexes. In this study, it was found that the root-mean-square percent relative error (RMS%RE) for determining the mole fraction of (*S*)-(+)-ibuprofen depended on the particular cyclodextrin used: 2.7% (α-CD), 0.8% (β-CD), and 0.7% (γ-CD). Studies conducted with norephedrine gave RMS%RE values in the mole fraction of (1*S*, 2*R*)-norephedrine of: 5.9% (α-CD), 3.4% (β-CD), and 3.3% (γ-CD).

To illustrate the procedure in more detail, the results obtained for ibuprofen will be used an as an example. Fig. 12.11 shows the spectra obtained over the spectral region from 278 to 291 nm for a series of solutions containing a fixed amount of β-CD (7.5 mM) and a fixed amount of ibuprofen (7.5 mM) where the enantiomeric composition of the guest molecule was varied from mole fraction 0.55 to 0.95 of the *S*-isomer. These spectra were taken with a Hewlett-Packard photodiode array (Model 8455) spectrophotometer with a 1-cm quartz cell. Since both the *S* form and the racemic form of ibuprofen were available commercially, solutions with known concentrations and enantiomeric composition could be prepared directly by weight without need for additional analysis.

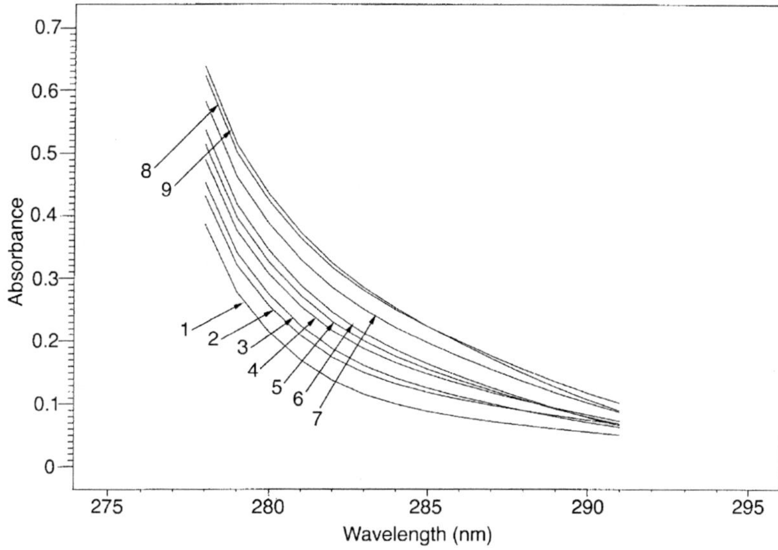

Fig. 12.11. Absorption spectra of solutions (pH 11.1) containing 7.5 mM β-CD and 7.5 mM ibuprofen of various enantiomeric composition (1–9). Mole fraction of (*S*)-ibuprofen: **1**, 0.55; **2**, 0.60; **3**, 0.65; **4**, 0.70; **5**, 0.75; **6**, 0.80; **7**, 0.85; **8**, 0.90; **9**, 0.95. (Reprinted from Ref. [35])

Fig. 12.12 shows a summary of a typical PLS regression model generated from the spectral data in Fig. 12.11 and the known enantiomeric compositions of the laboratory-prepared samples. This regression model was prepared using a commercially available chemometrics program.[19] Fig. 12.12(a) shows a PLS scores plot of the first PLS component (PLS_1, abscissa) *versus* the second PLS component (PLS_2, ordinate). The numbers in the plot are the sample numbers. It can be seen from the plot that the sample numbers increase from left to right with sample 1 on the left (mole fraction (*S*)-ibuprofen 0.55) through sample 9 on the right (mole fraction (*S*)-ibuprofen 0.95), indicating that PLS_1 is related to the enantiomeric composition of the samples.

According to the model, PLS_1 explains 98% of the enantiomeric composition and PLS_2 explains 2%. This is illustrated in Fig. 12.12(c), which shows a plot of residual *y* variance as a function of the number of PLS components in the model. From the graph, it can be seen that only two PLS components are actually needed to represent the data.[20]

Fig. 12.12(b) shows a plot of the regression coefficients for the model as a function of the wavelengths of the spectral data. It can be seen from the plot that the regression coefficients are positive for wavelengths less than 285 nm, zero at 285 nm, and negative for wavelengths greater than 285 nm. In this example, the regression coefficients do not

[19] The Unscrambler from CAMO, ASA, Oslo, Norway.

[20] In PLS modeling, the goal is to use a minimum number of components that explain most of the desired information. This is beneficial because it reduces the dimensionality of model. Use of more components than are actually needed can result in a model that is overfitted. This is undesirable because it means that some of the noise has actually been included in the model. Overfitting generally results in models with poor predictive ability.

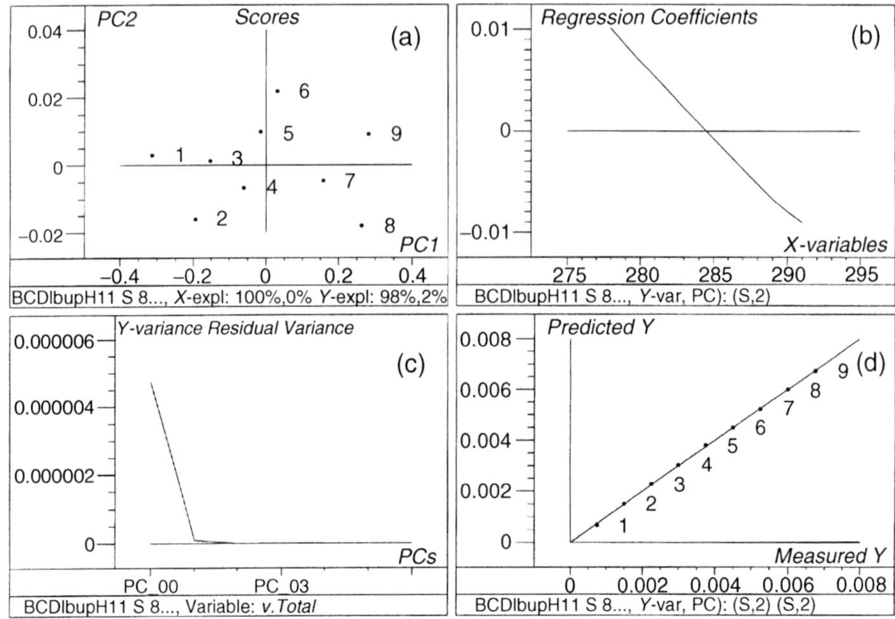

Fig. 12.12. Summary of regression results for ibuprofen: (a) scores plot; (b) regression coefficients as a function of wavelength; (c) residual variance as a function of the number of PLS components; (d) plot of concentration of (*S*)-ibuprofen predicted by the model *versus* the known values. (Reprinted from Ref. [35])

show much structure with wavelength; however, in many models with other analytes, the regression coefficients may vary in a complex manner with wavelength.

Finally, Fig. 12.12(d) shows a plot of the values predicted by the model *versus* the known values for the laboratory-prepared samples.[21] In a plot of this type, an ideal model would have a slope of 1, and intercept of 0, and a correlation coefficient of 1. While the model shown in Fig. 12.12 looks quite good, the real test of any regression model is its ability to correctly predict future unknown samples.

Table 12.3 gives the results of a validation study conducted with the model developed for ibuprofen. In this study, nine new samples of ibuprofen (7.5 mM) were prepared with different enantiomeric compositions from those used to prepare the original model.[22] The spectra of these samples[23] were taken with the spectrophotometer. Table 12.3 gives the results obtained with the validation samples for solutions containing α-, β-, and

[21] It should be stressed that the plot in Fig. 12.12(d) is not a calibration curve! There are no calibration curves in multivariate regression modeling. The model is the set of regression coefficients that are used to *calculate* the predicted values from new spectral data.

[22] The new samples had the same total concentration (7.5 mM) as the solutions used to make the original model and covered the same range of mole fractions of (S)-ibuprofen. They did not have the same enantiomeric compositions as the samples used to make the original model.

[23] These samples also contained 7.5 mM cyclodextrin.

Table 12.3 Relative errors obtained for ibuprofen with an independently prepared test set for the three CDs studied (Ref. [35])

	α-CD		β-CD		γ-CD	
Actual mole fraction	Predicted mole fraction	% Relative error	Predicted mole fraction	% Relative error	Predicted mole fraction	% Relative error
0.580	0.556	−4.1	0.580	0	0.579	−0.2
0.636	0.632	−0.6	0.634	−0.3	0.635	−0.2
0.690	0.678	−1.7	0.694	0.6	0.688	−0.3
0.715	0.706	−1	0.712	−0.4	0.716	0.1
0.780	0.777	−0.4	0.773	−0.9	0.786	0.8
0.840	0.891	6.1	0.834	−0.7	0.835	−0.6
0.867	0.893	3.0	0.878	1.3	0.881	1.6
0.915	0.915	0	0.906	−1	0.915	0
0.940	0.947	0.7	0.931	−1	0.945	0.5
RMS%RE		2.7		0.8		0.7

γ-cyclodextrin. It can be seen from the table that the RMS%RE obtained varies with the particular cyclodextrin used to make the model. This suggests that the quality of the model depends on the nature and extent of the guest–host inclusion complex that forms between the chiral analyte and the host.

12.5.3 Studies with modified cyclodextrins

In addition to naturally occuring CDs (or native CDs), a large number of chemically modified CDs have been synthesized in which the hydroxy groups of the CD have been substituted with a variety of groups. Indeed, Easton and Lincoln have reviewed the chiral discriminating ability of naturally occuring cyclodextrins (CDs) [36]. They concluded that complexes formed with naturally occuring CDs have limited enantio-selectivity because there is often limited interaction between the chiral centers of the cyclodextrin and those of the guest molecule. They further concluded that modified CDs often show improved stereoselectivity because they increase the extent of interaction between guest–host chiral centers by additional secondary bonding interactions.

To study the influence of modified CDs on chiral analysis by regression modeling of spectral data, a study [37] was conducted with four guest molecules of pharmaceutical interest (ephedrine, norephedrine, norepinephrine-L-bitartrate, and tryptophan methyl ester). The modified CDs used were: methyl-β-CD, α-, β-, and γ-carboxymethyl CDs, and α-, β-, and γ-hydroxypropyl CDs. As in previous studies, PLS-1 regression models were developed by correlating the known enantiomeric composition of laboratory-prepared samples with ordinary UV absorption spectral data.

Fig. 12.13 summarizes the results obtained in this study for the four chiral guest molecules and the seven different cyclodextrins. From the figure, it can be seen that the predictive ability as measured by the RMS%RE for the validation study was highly dependent on the particular combination of guest–host pair studied. For

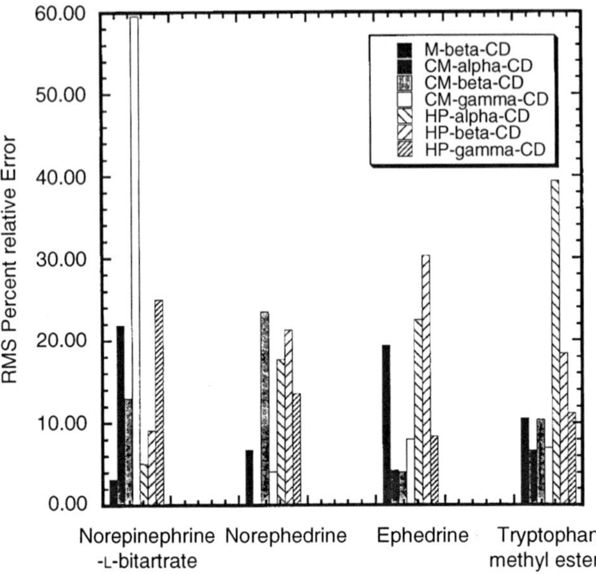

Fig. 12.13. Root mean square percent relative error (RMS%RE) obtained from regression models for norepinephrine-L-bitartrate, norephedrine, ephedrine, and tryptophan methyl ester with various chiral host molecules. (Reprinted from Ref. [37])

norepinephrine-L-bitartrate the best model in terms of predictive ability (i.e. the smallest RMS%RE) was obtained when methyl-β-CD was used as the host molecule (3.2% RMS%RE). The next best model employed hydroxypropyl-α-CD (5.1% RMS%RE). In the case of norephedrine, the best model was obtained with carboxymethyl-γ-CD (4.1% RMS%RE). For ephedrine, two hosts gave similar predictive abilities (carboxymethyl-α-CD, 4.3% RMS%RE; carboxymethyl-β-CD, 4.1% RMS%RE). Finally, for tryptophan methyl ester, two hosts again gave similar predictive abilities (carboxymethyl-α-CD, 6.7% RMS%RE; carboxymethyl-γ-CD, 7.0% RMS%RE).

If the results of this study with modified cyclodextrins are compared with previous results obtained with native cyclodextrins [35] for norephedrine, the native cyclodextrins are found to give somewhat better models (3.4% RMS%RE with β-CD and 3.3% RMS%RE with γ-CD). None of the models prepared in this study with modified CDs gave results that were close to those obtained for ibuprofen when native CDs were used (0.8% RMS%RE with β-CD and 0.7% RMS%RE with γ-CD).

12.5.4 Studies using flourescence

In an effort to improve the sensitivity of the technique with respect to the amount of chiral compound needed to develop a model, the use of fluorescence measurements was studied [38]. Fluorescence spectrometry offers several potential advantages when compared with absorption spectrophotometry. Fluorescence measurements, for example, are often more sensitive than corresponding measurements made by absorption spectrophotometry. This is due to the fact that fluorescence emission is detected against a low background isolated from the excitation source. By comparison, absorption measurements must distinguish the difference between two large light levels (the incident and transmitted light levels). As the detection limit is approached, these two light levels approach one another in magnitude and distinguishing a difference becomes difficult. In addition, since fluorescence is essentially an emission technique, it can be somewhat easier to implement in certain situations, because it is not necessary to monitor both the incident intensity and the transmitted intensity as required for absorption measurements.

While purely instrumental factors may influence the sensitivity of detection of an analyte in conventional fluorometry, it should be realized that with chiral analysis by multivariate regression modeling of spectral data, the question of sensitivity is not based solely on instrumental considerations. The spectral information that is actually used in regression modeling is encoded in the shape of the emission band envelope (and the extent to which this shape varies with enantiomeric composition) and not with detecting the intensity of the band maximum over a small background (i.e. the conventional detection limit) as would be done in conventional univariate fluorometry.

To determine whether fluorescence spectroscopy did indeed offer advantages compared with previous chiral analyses obtained with absorption spectroscopy, the possibility of determining the enantiomeric composition of phenylalanine samples by multivariate regression modeling of fluorescence spectral data was investigated. The goal of these studies was to determine whether satisfactory regression models could be made at lower concentration levels of chiral analyte and cyclodextrin than were used previously in absorption studies.

Fig. 12.14(a) shows the fluorescence spectra from 200 to 500 nm obtained for nine samples of phenylalanine (3.75 mM) of different enantiomeric composition in the presence of 7.50 mM β-CD. The samples were all excited at 257 nm. Fig. 12.14(b) shows an expanded plot of the fluorescence spectra of the nine samples over the wavelength range from 310 to 375 nm. Fig. 12.14 shows that definite spectral changes do indeed occur as the enantiomeric composition of the phenylalanine is varied (while keeping the total concentrations of phenylalanine and cyclodextrin fixed). The observed spectral changes are small, however, and occur in solutions containing mixtures of complexed and uncomplexed guest and host molecules. In circumstances such as these, multivariate regression modeling is frequently useful because it can focus on those spectral features that correlate with the parameter of interest (in this case, enantiomeric composition), while ignoring those spectral features that do not correlate.

Fig. 12.15 shows the mean-centered emission spectra from 249 to 444 nm for the nine samples shown in Fig. 12.14. It is clear from Fig. 12.15 that the mean-centered spectra beyond 303 nm correlate with the enantiomeric composition of the samples. In the spectral region from about 309 to 393 nm, the spectra are more or less uniformly displaced vertically with enantiomeric composition. Samples 1–4 with mol fractions of D-phenylalanine less than 0.5 have negative spectral signatures while samples 6–9 with mole fractions of D-phenylalanine greater than 0.5 have positive spectral signatures. Sample 5, which is the racemic mixture, is essentially at zero on the ordinate of Fig. 12.15.

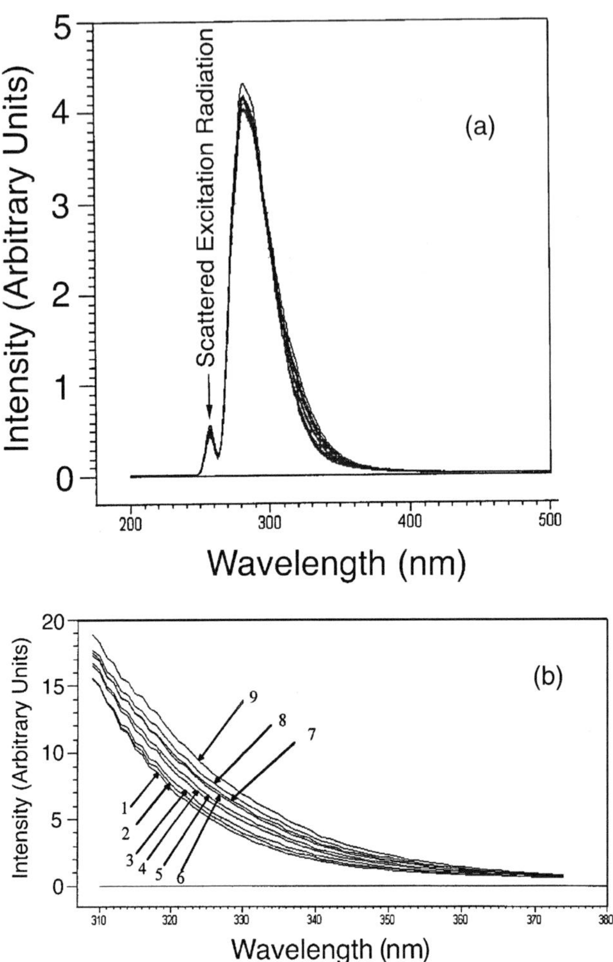

Fig. 12.14. (a) Fluorescence emission spectra from 250 to 500 nm of nine samples containing 7.50 mM β-CD and 3.75 mM phenylalanine of various enantiomeric compositions. Excitation at 257 nm. (b) Expanded view of fluorescence emission spectra from 310 to 375 nm of solutions containing 7.50 mM β-CD and 3.75 mM phenylalanine of various enantiomeric compositions (1–9). Mole fraction of D-phenylalanine: **1**, 0.100; **2**, 0.200; **3**, 0.340; **4**, 0.400; **5**, 0.500; **6**, 0.600; **7**, 0.700; **8**, 0.800; **9**, 0.900. Excitation at 257 nm. (Reprinted from Ref. [38])

Table 12.4 gives the prediction results obtained for both enantiomers of phenylalanine (0.469 mM) in the presence of β- and γ-CD (0.938 mM). Each model was validated with an independently prepared set of nine validation samples whose enantiomeric compositions were different from the calibration samples used initially to develop the regression model. When this regression model was used one week later on two occasions to predict the enantiomeric composition of new validation samples from new spectral data, the results obtained were reproducible and comparable to those obtained in the intial study. Indeed, the average RMS%RE in the mole fraction of D-phenylalanine

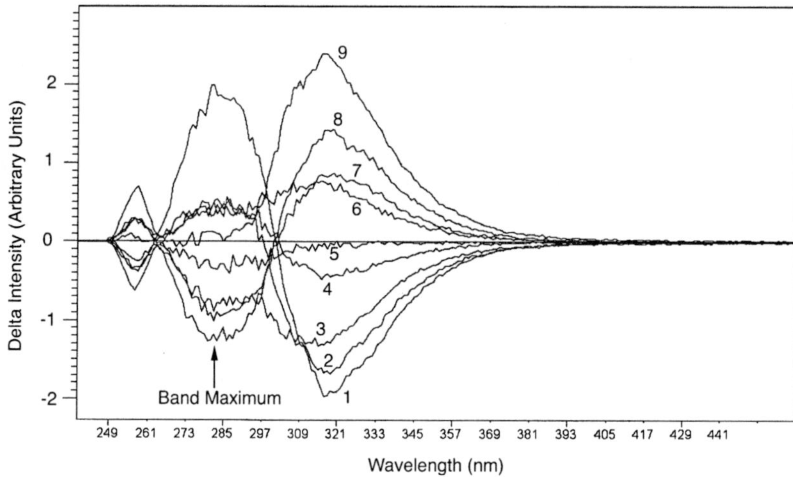

Fig. 12.15. Mean-centered spectra from 249 to 444 nm of nine samples containing 7.50 mM β-CD and 3.75 mM phenylalanine of various enantiomeric composition (1–9). Mole fraction of D-phenylalanine: **1**, 0.100; **2**, 0.200; **3**, 0.340; **4**, 0.400; **5**, 0.500; **6**, 0.600; **7**, 0.700; **8**, 0.800; **9**, 0.900. (Reprinted from Ref. [38])

from the two repeat experiments where β-CD was the host was 5.0%, while the two repeat experiments with γ-CD gave an average RMS%RE in the mole fraction of D-phenylalanine of 4.5%.

This study showed that the fluorescence spectra of phenylalanine–cyclodextrin inclusion complexes vary in certain ED regions with the enantiomeric composition of the phenylalanine when the concentrations of the phenylalanine and cyclodextrin are maintained constant. Correlation of the fluorescence spectral data in these ED regions with enantiomeric composition of the samples by means of PLS-1 regression gave regression models with good predictive ability for independently prepared validation sets of samples. Regression models made when γ-CD was used as the host gave similar prediction results to those obtained with models where β-CD was the host molecule. However, compared with similar studies done in absorption, fluorescence data were found to be more sensitive and the spectral differences observed as a function of enantiomeric composition were more uniformly spaced, making regression modeling more reliable. As a result, good regression models could be made at lower concentrations than were possible previously when absorption measurements were used.

12.6 CONCLUSIONS

This chapter has shown that chiral analysis by regression modeling of spectral data is feasible when cyclodextrins are used as chiral auxiliaries and the regression models produced can predict the enantiomeric composition of unknown samples to within about ±5%. While the use of multivariate regression modeling of spectroscopic data for chiral

Table 12.4 Relative errors obtained with the regression models made with the fluorescence spectral data obtained with solutions containing 0.469 mM L- or D-phenylalanine and 0.938 mM cyclodextrin

		β-CD				γ-CD			
Actual mole fraction D	Actual mole fraction L	Predicted mole fraction D	% Relative error for D	Predicted mole fraction L	% Relative error for L	Predicted mole fraction D	% Relative error for D	Predicted mole fraction L	% Relative error for L
0.264	0.736	0.266	0.8	0.734	−0.3	0.278	5.3	0.722	−1.9
0.328	0.672	0.333	2	0.667	−0.7	0.325	−0.9	0.675	0.4
0.452	0.548	0.443	−2	0.557	2	0.441	−2.4	0.559	2.0
0.548	0.452	0.546	−0.4	0.454	0.4	0.555	1	0.445	−2
0.620	0.380	0.624	0.6	0.376	−1	0.603	−2.7	0.397	4.5
0.715	0.285	0.730	2.1	0.270	−5.3	0.700	−2.1	0.300	5.3
0.752	0.248	0.762	1.3	0.238	−4.0	0.770	2.4	0.230	−7.3
0.844	0.156	0.842	−0.2	0.158	1	0.858	1.7	0.142	−9.0
0.892	0.108	0.894	0.2	0.106	−2	0.900	0.9	0.100	−7
RMS%RE			1.3		2.5		2.5		5.2

analysis is promising as an emerging potential analytical technology, in its present stage of development, it cannot necessarily compete with more mature existing methods of chiral analysis. It should be stressed, however, that progress in analytical chemistry depends on the study of new potential techniques that have potential analytical utility, even if the actual analytical applications lie in the future. Clearly more work will be needed to extend the technique to real samples like biological fluids and pharmaceuticals so that the full analytical potential of the concept can be realized.

APPENDIX A. MATRIX ALGEBRA

Introductory concepts. This section will introduce some topics dealing with linear equations. Eq. (A1) below is an example of a linear equation.

$$y = a_{11}x_1 + a_{12}x_2 + \ldots + a_{1n}x_n \tag{A1}$$

In this equation, the dependent variable y is a linear combination or weighted sum of the n independent variables x and the coefficients a_{ij} are the weighting factors.

Systems of linear equations can be represented as arrays known as matrices. Thus, the following system of equations can be represented as

$$\begin{aligned} y_1 &= a_{11}x_1 + a_{12}x_2 + a_{13}x_3 \\ y_2 &= a_{21}x_1 + a_{22}x_2 + a_{23}x_3 \\ y_3 &= a_{31}x_1 + a_{32}x_2 + a_{33}x_3 \end{aligned} \tag{A2}$$

or as

$$\mathbf{y} = \mathbf{A}\mathbf{x}, \tag{A3}$$

where $\mathbf{y}$ and $\mathbf{x}$ are 1×3 column vectors and $\mathbf{A}$ is a 3×3 matrix.

In this formalism, only the weighting coefficients shown in Eq. (A2) are used in the matrix $\mathbf{A}$.

$$\mathbf{A} = \begin{matrix} a_{11} & a_{12} & a_{13} \\ a_{21} & a_{22} & a_{23} \\ a_{31} & a_{32} & a_{33} \end{matrix} \tag{A4}$$

Matrices are generally indicated by uppercase bold capital letters. The matrix in Eq. (A4) is referred to as a 3×3 matrix, indicating that it has three rows and three columns. Vectors, which have only one variable, are written as lowercase bold capital letters as shown in Eq. (A5).

$$\mathbf{y} = \begin{matrix} y_{11} \\ y_{21} \\ y_{31} \end{matrix} \tag{A5}$$

Eq. (A5) shows a 3×1 column vector.

Matrix transposition. The transpose of a matrix is an operation that exchanges rows for columns and is symbolized as $\mathbf{X}^{\mathrm{T}}$. Thus, if $\mathbf{X}$ is an $i \times k$ matrix represented by the rectangle in Eq. (A6),

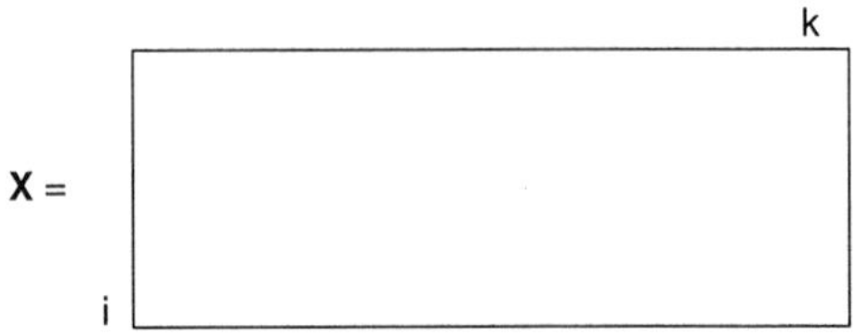

(A6)

its transpose is a $k \times i$ matrix represented as shown in Eq. (A7).

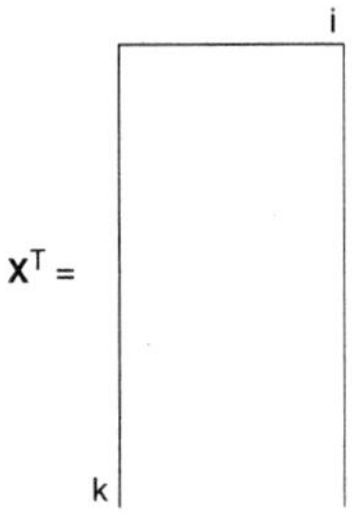

(A7)

Matrix addition. Matrix addition or subtraction is represented by

$$\mathbf{Y} = \mathbf{X} + \mathbf{Z} \tag{A8}$$

and is accomplished by adding (or subtracting) each term element by element. To add or subtract matrices, they must be compatible in terms of dimensions (they must have the same number of row and columns).

Matrix multiplication. Matrix multiplication is represented algebraically by

$$\mathbf{T} = \mathbf{UV}, \tag{A9}$$

where $\mathbf{T}$ is the product of two compatible matrices $\mathbf{U}$ and $\mathbf{V}$. In this notation, $\mathbf{V}$ is said to be pre-multiplied by $\mathbf{U}$ or $\mathbf{U}$ is post-multiplied by $\mathbf{V}$. When multiplying two matrices or vectors, they must have compatible dimensions and rows are multiplied into columns as shown in Eqs. (A10) and (A11).

$$\mathbf{T} = \underset{\mathbf{U}}{\begin{bmatrix} a & b \\ c & d \end{bmatrix}} \underset{\mathbf{V}}{\begin{bmatrix} e & f \\ g & h \end{bmatrix}} = \begin{bmatrix} w & x \\ y & z \end{bmatrix} \tag{A10}$$

$$\begin{aligned} w &= ae + bg \\ x &= af + bh \\ y &= ce + dg \\ z &= cf + dh. \end{aligned} \tag{A11}$$

Matrix division. Matrix division is represented algebraically by

$$\mathbf{T} = \mathbf{UV}^{-1} \tag{A12}$$

where $\mathbf{V}^{-1}$ is the inverse of the matrix $\mathbf{V}$. The inverse of a matrix $\mathbf{V}$ (if it exists) is $\mathbf{V}^{-1}$, such that $\mathbf{V}\mathbf{V}^{-1} = \mathbf{V}^{-1}\mathbf{V} = \mathbf{I}$, where $\mathbf{I}$ is the identity matrix given by

$$\mathbf{I} = \begin{matrix} 1 & 0 & 0 \\ 0 & 1 & 0 \\ 0 & 0 & 1 \end{matrix} \tag{A13}$$

for a 3×3 situation. In order to be invertible, a matrix must be of full rank. That means that all the columns (or rows) are linearly independent.

Length of a vector. Consider a 3×1 vector $\mathbf{x}$.

$$\mathbf{x} = \begin{matrix} a_{11} \\ a_{21} \\ a_{31}. \end{matrix} \tag{A14}$$

If we think of a vector in three-dimensional space as an arrow, the length or magnitude of the vector is given by

$$|\mathbf{x}| = \sqrt{a_{11}^2 + a_{21}^2 + a_{31}^2} \tag{A15}$$

where the vertical brackets are used to symbolize the length of the vector. In matrix algebra notation, Eq. (A15) is symbolized as

$$|\mathbf{x}| = (\mathbf{x}^{\mathrm{T}}\mathbf{x})^{1/2}. \tag{A16}$$

Outer product. The outer product of two vectors $\mathbf{t}$ and $\mathbf{p}$ is given by $\mathbf{tp}^{\mathrm{T}}$. While $\mathbf{t}$ and $\mathbf{p}$ are both vectors, their outer product is a matrix as shown by Eq. (A17).

$$\underset{3\times 1}{\begin{vmatrix} 1 \\ 4 \\ 2 \end{vmatrix}} \quad \underset{1\times 4}{\begin{vmatrix} 2 & 3 & 0 & 1 \end{vmatrix}} = \underset{3\times 4}{\begin{vmatrix} 2 & 3 & 0 & 1 \\ 8 & 12 & 0 & 4 \\ 4 & 6 & 0 & 2 \end{vmatrix}} \tag{A17}$$

Since the rows of the 3×4 matrix in Eq. (A17) are linear combinations of each other (i.e. they are not independent), the outer product is a matrix with a rank of one.

Mean centering. Mean centering involves subtracting the average value from the data. Consider the matrix

$$\begin{matrix} 1 & 4 \\ 3 & 0 \\ 2 & 5 \end{matrix} \tag{A18}$$

The average values of the first and second columns are 2 and 3, respectively. If we subtract the average values from the matrix in Eq. (A18), we get

$$\begin{matrix} -1 & 1 \\ 1 & -3 \\ 0 & 2 \end{matrix} \tag{A19}$$

which is the mean-centered matrix.

REFERENCES

1 E.N. Jacobsen, A. Pfaltz and H. Yamamoto (Eds.), Comprehensive Asymmetric Catalysis, Vols. I to III, Springer-Verlag, Berlin, 1999.
2 T.P. Yoon and E.N. Jacobsen, Privileged chiral catalysts. Science, 299 (2003) 1691.
3 S.C. Stinson, Chiral chemistry. Chem. Eng. News, 79 (20) (2001) 45.
4 N. Maier, P. Franco and W.J. Lindner, Separation of enantiomers: needs, challenges, perspectives. Chromatog. A, 906 (2001) 3–33.
5 G.R. Sullivan, Chiral lanthanide shift reagents. Top. Stereochem., 10 (1978) 287.
6 M. Reetz, Combinatorial and evolution-based methods in the creation of enantioselective catalysts. Angew. Chem. Int. Ed., 40 (2001) 284.
7 M. Finn, Emerging methods for the rapid determination of enantiomeric excess. Chirality, 14 (2002) 534.
8 M. Finn, Chapter 4, this book.
9 M. Islam, J. Mahdi and I. Bowen, Pharmacological importance of stereochemical resolution of enantiomeric drugs. Drug Safety, 17 (1997) 149–165.
10 J. Caldwell, Stereochemical determinants of the nature and consequences of drug metabolism. J. Chromatog. A, 694 (1995) 39–48.
11 H. Martens and T. Naes, Multivariate Calibration, John Wiley, New York, 1989.
12 M. Otto, Chemometrics, Wiley-VCH, Weinheim, Germany, 1999.
13 R. Kramer, Chemometric Techniques for Quantitative Analysis, Marcel Dekker, New York, 1998.
14 E.R. Malinowski, Factor Analysis in Chemistry, 2nd Ed. Wiley Interscience, New York, 1991.
15 K.R. Beebe, R.J. Pell and M.B. Beasholtz, Chemometrics – A Practical Guide, Wiley-Interscience, New York, 1998.
16 D.L. Massart, B.G.M. Vandeginste, L.M.C. Buydens, S. DeJong, P.J. Lewi and J. Smeyers-Verbeke, Handbook of Chemometrics and Qualimetrics: Part A, Elsevier, Amsterdam, 1997.
17 B.G.M. Vandeginste, D.L. Massart, L.M.C. Buydens, S. DeJong, P.J. Lewi and J. Smeyers-Verbeke, Handbook of Chemometrics and Qualimetrics: Part B, Elsevier, Amsterdam, 1998.
18 K. McNeil, I. Newman and F.J. Kelly, Testing Research Hypotheses with the General Linear Model, Southern Illinois Univ. Press, Carbondale, 1996.
19 H. Wold, Estimation of principal components and related models by iterative least squares, in: P.R. Krishnaiah (Ed.), Multivariate Analysis, Academic Press, New York, 1966.
20 K. Esbensen, Multivariate Data Analysis–In Practice, 5th Ed. CAMO Process AS, Oslo, Norway, 2002.
21 E. Schneiderman and A. Stalcup, Cyclodextrins: a versatile tool in separation science. J. Chromatog. B, 745 (2000) 83–102.
22 S. Li and W. Purdy, Cyclodextrins and their application in analytical chemistry. Chem. Rev., 92 (1992) 1457–1470.
23 D.W. Armstrong, T.J. Ward, R.D. Armstrong and T.E. Beesley, Separation of drug stereoisomers by the formation of beta-cyclodextrin inclusion complexes. Science, 232 (1986) 1132.
24 S. Topiol and M. Sabio, Elucidation of chiral recognition principles. Enantiomer, 1 (1996) 251–265.
25 K. Kano, Y. Kato and M. Kodera, Mechanism for chiral recognition of binaphthyl derivatives by cyclodextrins. J. Chem. Soc. Perkin Trans., 2 (1996) 1211.
26 S. Li and W. Purdy, Circular dichroism, ultraviolet, and proton nuclear magnetic resonance spectroscopic studies of the chiral recognition mechanism of beta-cyclodextrin. Anal. Chem., 64 (1992) 1405.
27 P. Bortolus, G. Marconi, S. Monti and B. Mayer, Chiral discrimination of camphorquinone enantiomers by cyclodextrins: a spectroscopic and photophysical study. J. Phys. Chem., 106 (2002) 1686–1694.
28 S.A. MacDonald and G.M. Hieftje, Use of shift reagents to determine enantiomers by near-infrared analysis. Appl. Spectrosc., 50 (1996) 1161.
29 M.J. Politi, C.D. Tran and G-H. Gao, Near-infrared spectroscopic investigation of inclusion complexes between cyclodextrins and aromatic compounds. J. Phys. Chem., 99 (1995) 14137.
30 L.G. Weyer, Appl. Spectrosc. Rev., 21 (1985) 1.

31 C. Manzanares, I.V.M. Blunt and J. Peng, Vibrational spectroscopy of nonequivalent carbon-hydrogen bonds in liquid cis- and trans-3-hexene. Spectrochim. Acta, 49A (1993) 1139–1152.
32 C.D. Tran, V.I. Grishko and D. Oliveira, Determination of enantiomeric compositions of amino acids by near-infrared spectrometry through complexation with carbohydrates. Anal. Chem., 75 (2003) 6455.
33 T. Wenzel, Trends Org. Chem., 8 (2000) 51–64.
34 K.W. Busch, I.M. Swamidoss, S.O. Fakayode and M.A. Busch, Determination of the enantiomeric composition of guest molecules by chemometric analysis of the UV-visible spectra of cyclodextrin guest-host complexes. J. Am. Chem. Soc., 125 (2003) 1690–1691.
35 K.W. Busch, I.M. Swamidoss, S.O. Fakayode and M.A. Busch, Determination of the enantiomeric composition of some molecules of pharmaceutical interest by chemometric analysis of the UV spectra of cyclodextrin guest-host complexes. Anal. Chim. Acta, 525 (2004) 53–62.
36 C.J. Easton and S.F. Lincoln, Chiral discrimination by modified cyclodextrins. Chem. Soc. Rev., 25(3) (1996) 163–170.
37 S.O. Fakayode, I.M. Swamidoss, M.A. Busch and K.W. Busch, Determination of the enantiomeric composition of some molecules of pharmaceutical interest by chemometric analysis of the UV spectra of guest-host complexes formed with modified cyclodextrins. Talanta, 65 (2004) 837–844.
38 S.O. Fakayode, M.A. Busch, D. Bellert and K.W. Busch, Determination of the enantiomeric composition of phenylalanine samples by chemometric analysis of the fluorescence spectra of cyclodextrin guest-host complexes. The Analyst, 130 (2005) 233–241.

© 2006 Elsevier B.V. All rights reserved.

CHAPTER 13

Electronic circular dichroism for chiral analysis

Jacek Gawronski and Pawel Skowronek

Department of Chemistry, A. Mickiewicz University, Poznan, Poland

13.1 INTRODUCTION—THE EXCITON COUPLING MECHANISM

Circular dichroism (CD), a spectroscopic technique based on differential absorption of left- and right-handed circularly polarized light, is ideally disposed to analyze molecular structure, composition and interactions of chiral systems. Apart from X-ray diffraction analysis it is the only method capable of absolute stereostructure assignment, (e.g. for the determination of absolute configuration). An advantage of the CD method is its independence of the physical form of the substance analyzed. While high-resolution X-ray diffraction experiment heavily relies on the availability of good quality monocrystal, CD measurements can be done with samples in solution, gas phase, solid dispersions (CD of randomly oriented molecules), films, gels, liquid crystals and even on monocrystals (CD of oriented systems).

The CD technique is based on registering CD spectra, representing the wavelength dependence of $\Delta\varepsilon$, where,

$$\Delta\varepsilon = \varepsilon_{\mathrm{L}} - \varepsilon_{\mathrm{R}}$$

ε_{L} and ε_{R} are molar absorption coefficients of left- and right-handed circularly polarized light, respectively. $\Delta\varepsilon$ is subject to the Beer–Lambert law, i.e.

$$\Delta\varepsilon = \frac{\Delta A}{c \cdot 1} \text{ (in mol}^{-1}\text{dm}^{3}\text{ cm}^{-1}\text{ units)}$$

where ΔA is directly measured difference of absorbance of left- and right-polarized light. In cases where concentration (c) is unknown it is common to present CD data in another unit, ellipticity, θ (in millideg)

$$\theta\,(\text{millideg}) = 32{,}982 \cdot \Delta A$$

One can also use molar ellipticity $[M]$, defined as

$$[M] = 3298 \cdot \Delta\varepsilon$$

Like electronic absorption spectrophotometers, commercial dichrographs usually register the spectra in the wavelength range 180–800 nm, with optional extensions to either side, i.e. 170 or 1000 nm.

CD spectroscopy, now forty odd years old, underwent significant development not only in the technical aspects but also in the theoretical treatment of the experimental data. There are several books [1–8] and book chapters [9–15] on the subject available which address the theory, applications, methodology and historical background.

The pioneering work of Djerassi et al. [16,17] on the optical activity of ketones resulted in the formulation of a well-established and widely used empirical *octant rule* for the n–π^* transition of saturated ketones (Figs. 13.1 and 13.2).

The rule correlates the sign of the Cotton effect associated with the n–π^* transition with sum of contributions of substituents in the octants of alternating signs—positive and negative. In this way the octant rule provides a correlation of chiroptical property of the substance—the sign of the Cotton effect—with structural property, i.e. configuration (absolute or relative) or conformation of the molecule. The most comprehensive and in-depth presentation of the experimental and theoretical development, applications and limitations of the octant rule is given in a book by Lightner and Gurst [8], see also Ref. [19].

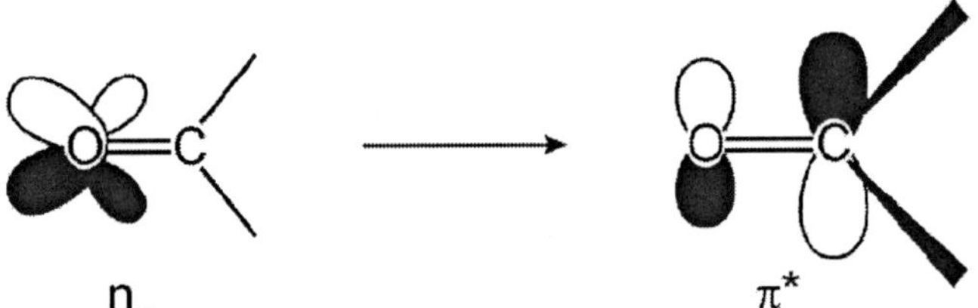

Fig. 13.1. The carbonyl valence orbitals participating in the n–π^* transition. Adapted from Ref. [18].

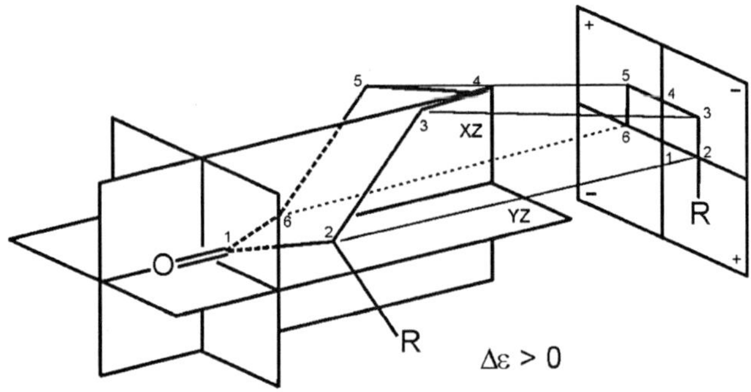

Fig. 13.2. Octant projection in the octant rule and signs of the four rear octants. Application to substituted cyclohexanone. Adapted from Ref. [8].

The n–π^* transition in carbonyls is an example of transition which, to a first approximation, has a magnetic dipole transition moment (mdtm) but no electric dipole transition moment (edtm). This transition is electric dipole forbidden because there is no linear motion of charge along the symmetry-defined axis (the C=O bond). Because of its forbidden character (low ε_{max}) the n–π^* transition Cotton effects are easy to measure, since the $\Delta\varepsilon/\varepsilon$ ratio is comparatively large (high signal to noise ratio in the measurement). The transition aquires electronic character by a mechanism known as dynamic coupling of the mdtm (n–π^*) with edtms in groups (chromophores) in the chiral environment of the C=O group [6].

In the pioneering years of CD spectroscopy many sector rules have been proposed on the basis of accumulated experimental data. These sector rules were used to analyze the stereostructure (absolute configuration, conformation) of new molecules from their CD spectra but they usually did not have a theoretical background for understanding the molecular basis of chiroptical phenomena. The obvious gap between the theory and applications of CD spectroscopy, stimulated efforts of many research groups, S.F. Mason in the U.K. and G. Snatzke in Germany, were among others, who ultimately developed the CD method as a standard tool for stereochemical analysis.

In a general sense, the Rosenfeld equation determines the sign and magnitude of the rotational strength, R

$$R = \mathrm{Im}\{(\mathrm{o}|\mu|\mathrm{a}) \cdot (\mathrm{a}|m|\mathrm{o})\},$$

where Im denotes the imaginary part, μ is the edtm of the transition from state o to a and m is the mdtm of the reverse transition. A complete and correct application of the CD measurement for solving any stereochemical problem is based on the *ab initio* computation of the CD spectrum (i.e. rotational strength, R for a number of transitions) of a molecule of assumed absolute stereochemistry and its confrontation with the experimental CD spectrum. This is a yes/no experiment: if the spectra match, the assumed stereostructure is correct, if they do not, the stereostructure is of mirror image. A general procedure is given in Scheme 13.1.

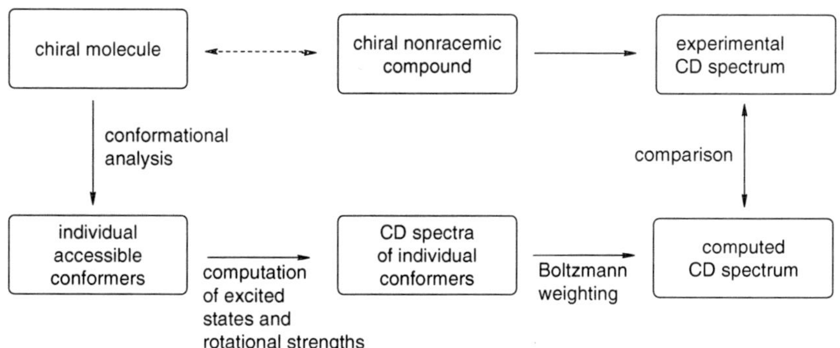

Scheme 1. Computational protocol for the determination of stereostructure of a chiral molecule from the CD spectra.

The success of *ab initio* computations is dependent on the level of calculations applied; the use of the DFT method with a large basis set is becoming more and more popular. There are several commercial programs available for computing electronic transitions and CD spectra, these include Gaussian 03 [20], DALTON [21] and TURBOMOLE [22], among others. By way of example, we show the results of quantum chemical computation of the CD spectrum of palmarumycin CP_3, a secondary fungal metabolite [23]. The molecule of palmarumycin CP_3 with the assumed 2*S*,4a*S*,8*R*,8a*S* absolute configuration exists as a mixture of two conformers, (A) and (B) (from AM1 calculation), for which the individual CD spectra were calculated using the BDZDO/MCDPD method (Fig. 13.3 (a)). After Boltzmann weighing for conformer population the averaged CD spectrum (b) matched well with the experimental one (c). Thus, natural palmarumycin CP_3 was found to have the initially assumed 2*S*,4a*S*,8*R*,8a*S* configuration (Fig. 13.3).

The dynamic coupling mechanism is applicable to coulombic coupling of the edtms μ_1 and μ_2 in two separate, non-coplanar chromophores. This coupled oscillators mechanism, developed theoretically by Kuhn [24] and Kirkwood [25] and later by Mason [26,27], found widespread use in organic stereochemistry due to its two basic advantages: (i) its theoretical foundations are easily translated to stereochemical models, (ii) many chiral molecules are, or can be made, two or more chromophoric entities.

Due to the seminal work of Harada and Nakanishi [4,28] the *exciton coupling* mechanism is now the most frequently used non-empirical method of CD spectroscopy, known as exciton coupled CD (ECCD) or *exciton chirality* method [29]. In a two-chromophoric system one chromophore provides the edtm while the second

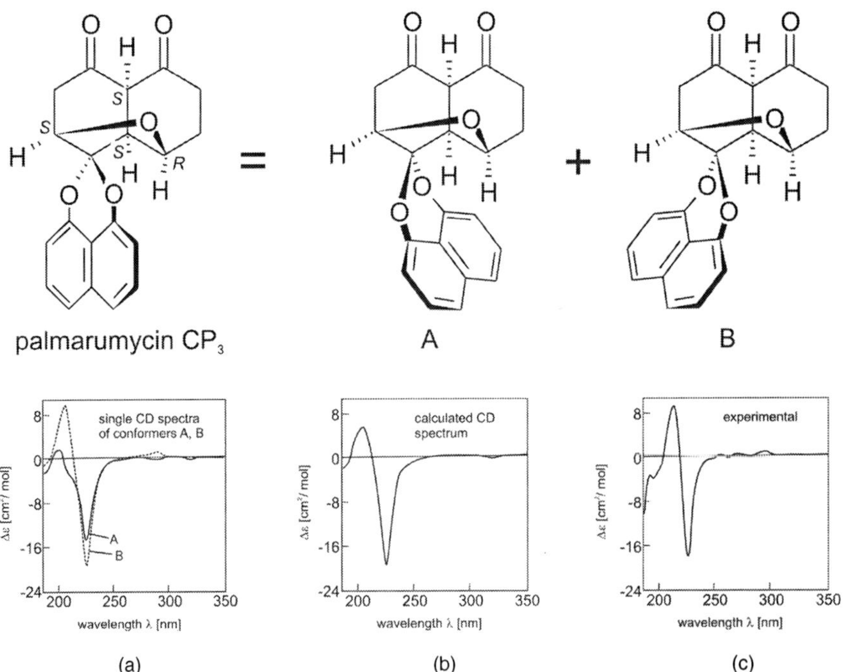

Fig. 13.3. Confrontation of computed (a,b) and experimental (c) CD spectra of palmarumycin CP_3. Adapted from Ref. [23].

chromophore provides an external origin for the mdtm produced by the edtm of the first chromophore and *vice versa*. When the two chromophores are identical (degenerate μ_1–μ_2 mechanism) the CD spectrum is represented by a pair of Cotton effects with opposite signs and comparable areas (*exciton couplet*). The two Cotton effects arise from the transitions to dimer excited states E^{α} and E^{β} formed by the symmetric and antisymmetric combination of the two excited configurations. The observable effect in the UV/VIS spectrum due to splitting of the states is either band broadening, or in extreme cases, band splitting. In the CD spectrum the sequence of signs of Cotton effects from the lower to the higher energy reveals the helicity of the two edtms system: either positive to negative for *P* helicity ($\omega > 0$) or negative to positive for *M* helicity ($\omega < 0$), Fig. 13.4.

Exciton chirality method is applicable to both homochromophoric (degenerate case) and heterochromophoric (non-degenerate case) systems, having two or more chromophores. It can be used to analyze the following chiral properties: absolute configuration, conformation and chiral interactions (see following sections of this chapter). Fig. 13.5 shows various applications of exciton chirality method, from simple

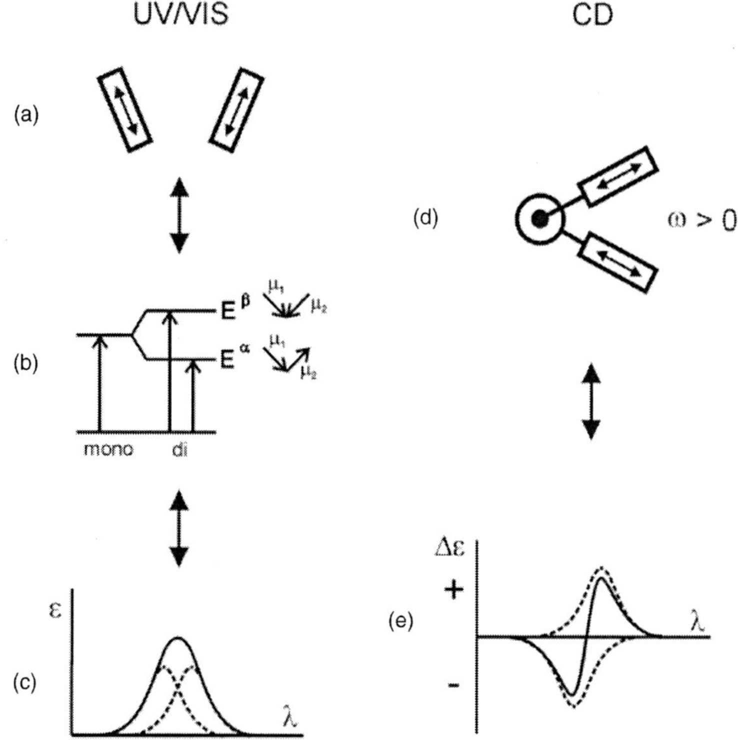

Fig. 13.4. Origin of the exciton splitting: oblique orientation of the two chromophores (a, rectangular boxes) and the two edtms (a, doubleheaded arrows) leads to splitting or band broadening of the allowed transition in the UV/VIS spectrum of the bichromophore (b, c). If the two edtms are non-coplanar (d, *P*-helicity) the corresponding CD curve (e) shows positive (lower energy) to negative (higher energy) exciton-split Cotton effect.

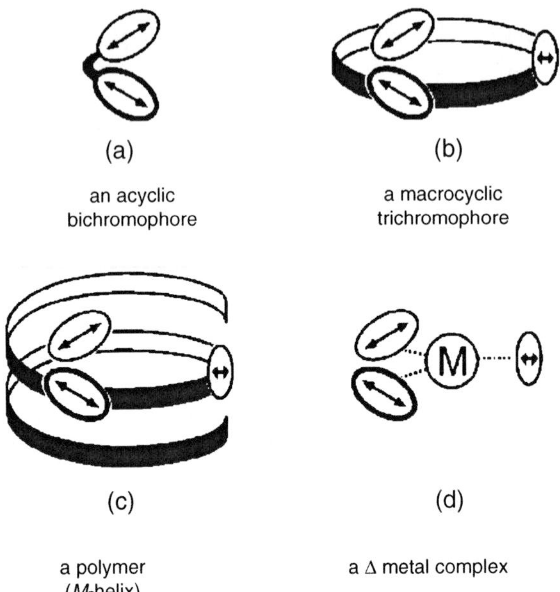

Fig. 13.5. All these chiral chromophoric assemblies give a negative exciton Cotton effect.

use of bichromophoric coupling for absolute configuration determination (a), through determination of configuration or conformation of a macrocycle (b) or conformation of a polymer (c), to absolute configuration determination of a metal (M) complex with chromophoric ligands (d).

Examples of cases (a) and (c) will be discussed in the following sections. Examples of a chiral macrocycle (b) with pyromellitic diimide chromophores (**1**) and a metal complex (d) with bipy or phen chromophores (**2**) are shown below:

1

(−) exciton Cotton effect
for 240 nm transition [30]

2

(−) exciton Cotton effect
for 270 nm transition [31]

Strictly speaking, electronic transitions in these two examples of chromophoric assemblies (b) and (d) belong to symmetry group D_3 and electronic transitions are either of A_2 type (lower energy, polarized along the C_3 axis) or E type (higher energy, doubly degenerate, polarized in the average molecular plane perpendicular to the C_3 axis). Neverthless, a negative bisignate Cotton effect is observed in the cases (b) and (d), as it is observed in the cases (a) and (c). Among advantages of the exciton chirality method are the following:

(1) the method is non-empirical, i.e. no assumptions or comparative examples are needed, nor in many cases computations are necessary.
(2) there are numerous possibilities to introduce necessary chromophores to non-chromophoric chiral molecules (see, the following section).
(3) very little sample is needed, depending on chromophores used. Because of high $\Delta\varepsilon/\varepsilon$ ratio, usually in the range 10^{-3}, low sample concentrations can be used with modern CD instruments. While microscale structural determinations are now routine [32], nanogram-scale derivatization procedures of hydroxy groups, for example with 2-naphthoate or 2-anthroate chromophores [33], have been introduced for highly sensitive HPLC/MS/CD detection [34].

Much of the popularity of applications of the exciton chirality method is due to simplicity of obtaining reliable structural data from experimental CD spectra. In the most frequently used *geometrical analysis* one needs to consider the orientation (helicity) of the two interacting edtms, resulting in a helical displacement of charge, which in turn interacts differently with left and right circularly polarized light. For a degenerate coupled oscillator system the rotational strength, $R^{\pm}$, of the two exciton CD bands is given by the equation:

$$R^{\pm} = \pm \frac{E\mu^2 r_{\mathrm{AB}}}{4\hbar} \sin\alpha \sin\beta \sin\omega,$$

where E is the transition energy, μ is the edtm, r_{AB} is the vector from A origin to the B origin and the angles α, β and ω are defined in Fig. 13.6.

In the situation shown in Fig. 13.6 ($\omega < 0$) the exciton Cotton effect is negative since R at lower energy E^- is negative and R at higher energy E^+ is positive. An example, corresponding to the case in Fig. 13.6 and typical for many analytical stereochemical applications is a diequatorial orientation of the two benzoyl chromophores attached

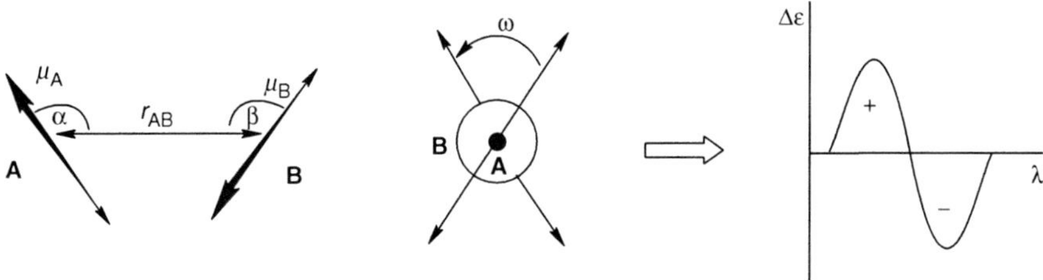

Fig. 13.6. Illustration of the geometry of the two interacting edtms in chromophores A and B. α and β are the angles between μ_{A} or μ_{B} and r_{AB} vectors, respectively; ω is the torsion angle involving the three vectors. The resulting exciton Cotton effect (right) is negative.

to the cyclohexane skeleton. This arrangement produces exciton Cotton effect of the sign reflecting the sign of the torsion angle X-C-C-Y (Fig. 13.7).

However it must be noted that the energies of the two bands, E^+ and E^-, are also dependent on the angles α, β and ω according to equation

$$E^{\pm} = E \pm \frac{\mu^2}{r_{\mathrm{AB}}^3}(\sin\alpha\sin\beta\cos\omega + 2\cos\alpha\cos\beta)$$

In certain cases the energies of the two CD bands are inverted, i.e. when the geometrical factor in Eq. (3) has a negative value. In a series of 1,1′-binaphthyl derivatives of the same absolute configuration the sign of the exciton Cotton effect due to the coupling of the ^1B_b edtms will change, when the dihedral angle between the naphthalene rings is larger than ca. 110° [35,36]. In practice, the dihedral angle is less than 110° and therefore 1,1′-binaphthyl derivatives show a positive Cotton effect for *S* configuration (Fig. 13.8).

Exciton chirality method is an ideal way to analyze absolute configurations of atropisomeric biaryls and it has been successfully applied to numerous compounds, including 1,1′-binaphthalene-2,2′-diol-based oligomers [37].

The case of 1,2-diequatorially substituted cyclohexane derivatives needs additional comment. In order to obtain absolute configuration from the analysis of exciton Cotton

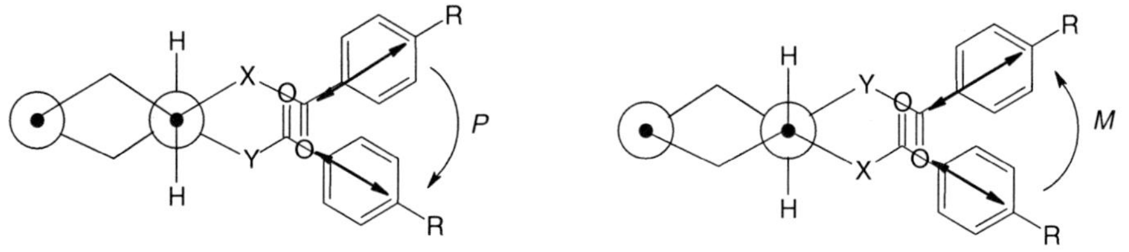

angle X-C-C-Y > 0
positive exciton Cotton effect

angle X-C-C-Y < 0
negative exciton Cotton effect

X, Y = O, NH

Fig. 13.7. Relation between benzoate or benzamide orientation and the sign of the exciton Cotton effect.

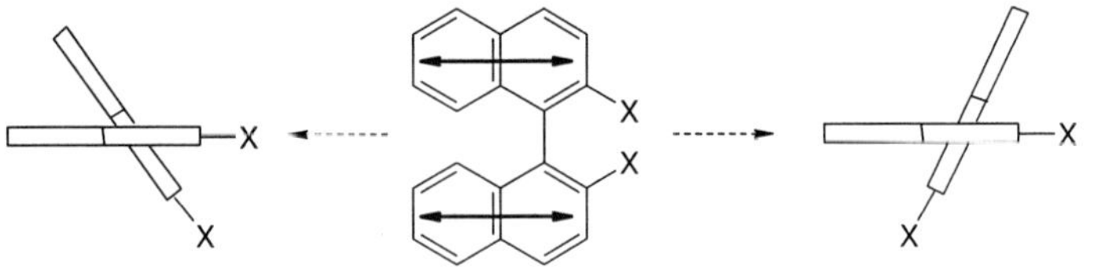

S configuration, *P* helicity
(+) exciton Cotton effect at ca. 230 nm,
e.g. for X = OH
$\Delta\varepsilon$ = +210 (234 nm)
−150 (223 nm)

S configuration, *P* helicity
predicted (−) exciton Cotton effect
at ca. 230 nm

Fig. 13.8. Conformational dependence of 1,1′-binaphthyl derivative exciton Cotton effect.

effect it is important to realize that, although in cyclohexane vicinal equatorial bonds are at the torsion angle ca. $+60$ or $-60°$, the edtms of the chromophores are not. For example, two derivatives of (*S*,*S*)-1,2-diaminocyclohexane (**3**,**4**) with long polyene imine chains and having a positive N-C-C-N torsion angle, show negative exciton Cotton effects, in clear opposition to intuitive (geometrical) analysis.

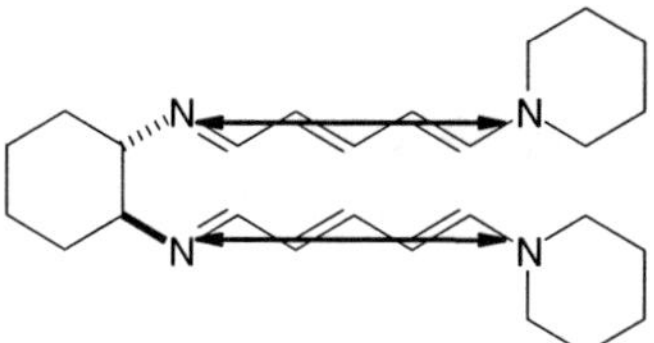

$\Delta\varepsilon$ −232 (546 nm)
+ 231 (475 nm) (as biscation) [38]

3

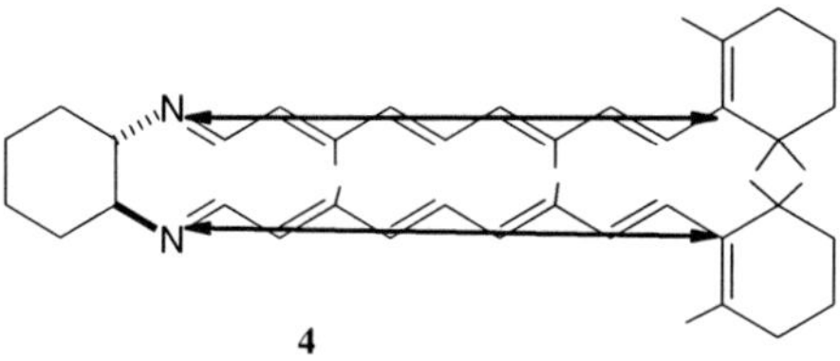

−34 (382 nm)
+34 (324 nm) [39]

4

In these chromophoric systems angle ω is approaching 0°, making analytical use of the CD data in a straightforward manner questionable.

C_3-symmetrical cyclotriguaiacylene (**5**) is another example of difficulties involved in geometrical analysis of CD data. Its exciton-coupled Cotton effects in the range 330–230 nm change sign upon ionization of the phenolic group. By semiempirical calculation this change could be explained as due to the rotation of the edtms in the benzene chromophores upon ionization [40].

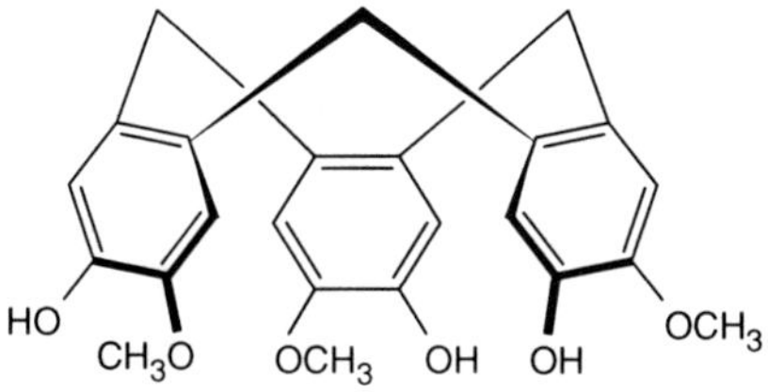

HO CH_3O OCH_3 OH OH OCH_3

5

By and large, analysis of the CD spectra using semiempirical coupled oscillator computations is the most convenient and reliable way to obtain structural data. Whereas *ab initio* computational methods may not be best suited for large molecules, application of semiempirical DeVoe polarizability model [41] can be applied to analyze CD spectra due to exciton coupling of any molecule, including polymers and biopolymers, especially proteins and polynucleotides.

In this model, recently reviewed by Rosini et al. [42] and Salvadori et al. [43], the electronic transitions are considered as oscillators. The induced dipole moment in a given oscillator has two sources: the electromagnetic field of the incident light and the field produced by the induced dipole moments of all other oscillators. Without going into detailed mathematical analysis the expression for the CD ($\Delta\varepsilon$) as a function of frequency (ν, cm^{-1}) is given by equation [42]

$$\Delta\varepsilon = 0.0073\pi^2 N\nu^2 \sum_{i>j} \mathbf{e}_i \times \mathbf{e}_j \cdot \mathbf{R}_{ij} \mathrm{ImA}_{ij},$$

where $\mathbf{e}_i$ and $\mathbf{e}_j$ are the unit vectors along induced dipole moments i and j, $\mathbf{R}_{ij}$ is the vector from i to j and ImA_{ij} is the imaginary part of the element A_{ij}. A is the inverse of matrix B, where

$$B_{ij} = (\delta_{ij}/\alpha_i) + G_{ij}$$

and G_{ij} is the point-dipole interaction term. Complex polarizability term $\alpha(\nu)$ is defined by equation:

$$\alpha(\nu) = R_i(\nu) + iI_i(\nu),$$

where $I_i(\nu)$ is obtainable from the absorption band of model compounds and $R_i(\nu)$ can be calculated from $I_i(\nu)$ by means of a Kronig–Kramers transform.

In practice, one first gets the optimized structure of the analyzed molecule from molecular modeling and then calculates the CD spectrum for assumed absolute configuration in a selected spectral range where the chromophores absorb, using the experimental data for absorption band intensity (ε_{max}), λ_{max} and $\Delta\lambda$ (bandwidth at half height).

Dipole strength (D) calculated from equation:

$$D = 91.8 \times 10^{-40} \varepsilon_{max} \cdot \Delta\lambda/\lambda_{max}$$

provides the data to calculate $I(\nu)$ and $R(\nu)$ and then to compute the CD spectrum as $\Delta\varepsilon(\nu)$. The computed CD spectrum is then confronted with the experimental one to see if the assumed absolute configuration was correct. As an example we show the CD spectra of 1,1′-binaphthalene derivative (**6**), experimental and computed by the DeVoe model (Fig. 13.9).

The assumed absolute configuration for computing the CD spectrum of (**6**) was *S* (*P*). The CD spectrum was computed at first-order approximation (i.e. only the effect of external field plus the field due to the other oscillator perturbed by the external field was taken into computation) and it showed generally satisfactory correspondence of the signs of the Cotton effects. A much better, nearly quantitative agreement was obtained with all-order calculation; this means that the measured CD spectrum corresponds to *S* absolute configuration of (**6**) [42].

The above examples illustrate also the general notion that both configuration and conformation of a molecule have an effect on its chiroptical properties.

In rare cases where Cotton effects due to two different electronic transitions are generated within the same chromophore, both absolute configuration and conformation can be determined from single CD measurement. For example, derivatives of 9,10-dihydrophenanthrene (**7**) display two bands in the UV spectrum, at ca. 270 nm (A)

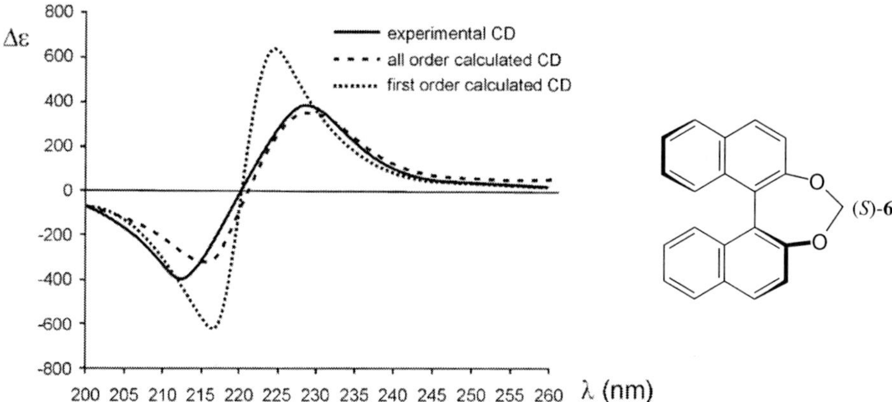

Fig. 13.9. Experimental and computed CD spectra of (*S*)-(+)-4,5-dihydro-3H-dinaphtho[2,1-c:1′,2′-*e*]oxepine (**6**). Adapted from Ref. [42].

and at ca. 220 nm (B). The sign of the Cotton effect associated with A band is determined by configuration of the chiral center (*allylic chirality*) and is independent of conformation, whereas the sign of band B Cotton effect is due to biaryl conformation (exciton coupling mechanism) [44]. Allylic chirality mechanism is based on the coupling of the edtm (band A) with the induced dipole of the adjacent C–X bond and is applicable to numerous cases of π-electron chromophores with allylic substituents (**8**).

B A B X X

P helicity

7 **8**

In the following sections main themes of the analytical applications of CD spectroscopy will be discussed and illustrated with selected examples from the recent literature. Emphasis will be put on the use of the exciton coupling method, as the most practical non-empirical approach.

13.2 ABSOLUTE CONFIGURATION DETERMINATION

Exciton coupling method is a very convenient, non-empirical and unequivocal way to determine absolute configuration of organic molecules, including biomolecules [45,46]. It should be realized, however, that chiral molecule having two chromophores with electric-dipole allowed transitions (usually of π–π^* type) gives exciton Cotton

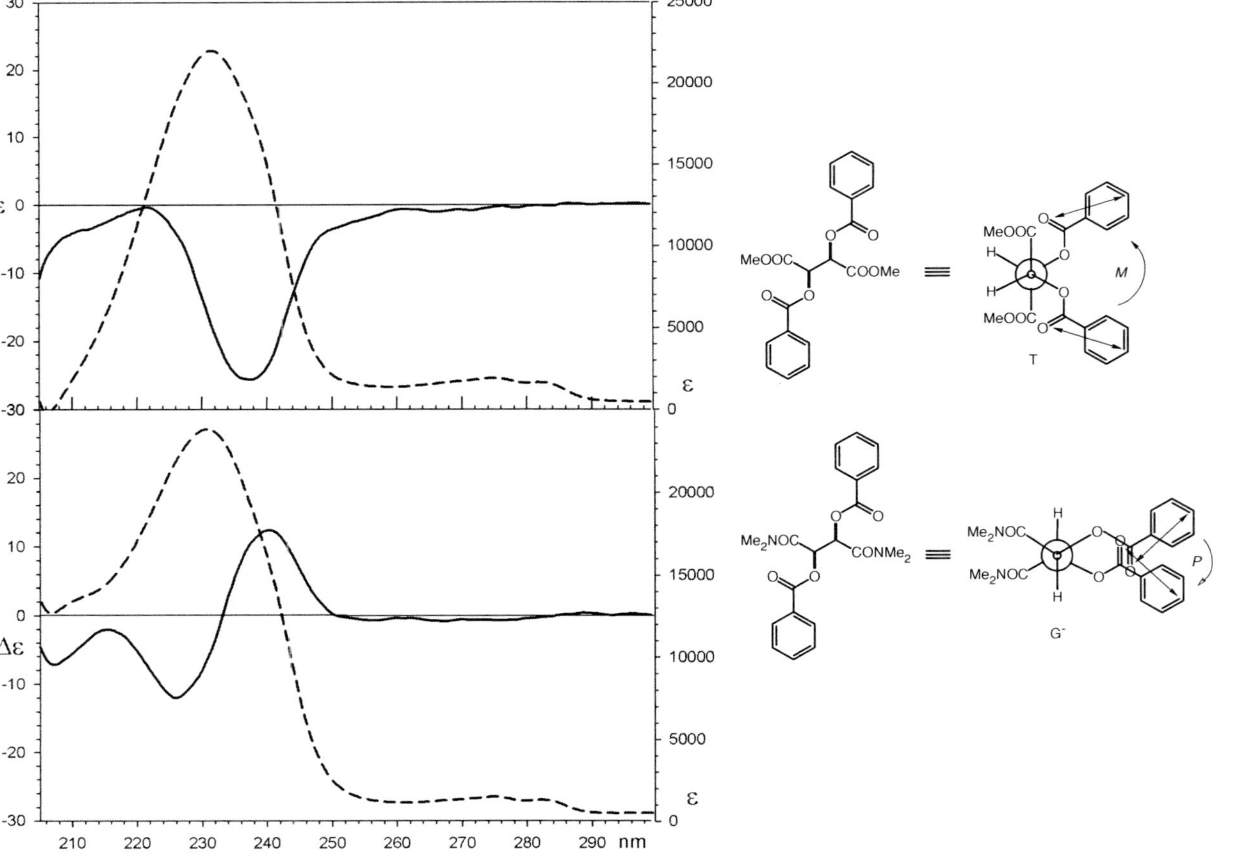

Fig. 13.10. ECCD spectra (solid lines), UV spectra (broken lines) and conformations of two derivatives of (*R*,*R*)-tartaric acid. Adapted from Ref. [47].

effect(s) reflecting the helicity (*P* or *M*) of the edtm. Therefore ECCD does not provide directly absolute configuration at the chiral atom; this has to be deduced from helicity of the edtm taking into account the conformation of the molecule, i.e.

$$\text{helicity} = \text{absolute configuration} \times \text{conformation}$$

The necessary data about conformation of the molecule are routinely obtained from computational modeling or from ancillary experimental data, e.g. from X-ray diffraction analysis or NMR spectra.

An example of dependency of exciton Cotton effect on both absolute configuration and conformation is shown in Fig. 13.10.

The two *O,O′*-dibenzoyl derivatives of dimethyl (*R,R*)-tartrate and *N,N,N′,N′*-tetramethyl-(*R,R*)-tartramide show opposite-sign exciton Cotton effects due to the benzoate 230 nm transition (in the case of the diester derivative only the lower-energy part of the exciton couplet is observed). The difference is of conformational origin; in the case of the diester conformer T with planar carbon framework and a negative helicity of the O-C-C-O bond chain is dominant. The tetraalkyldiamide derivative is characterized by a bent (G^-) carbon framework and a positive helicity of the O-C-C-O bond system (*P* helicity of the edtms), as confirmed by X-ray and NMR data [48].

13.2.1 Internal chromophores

In the following section, we show three examples of analysis of the absolute configuration of chiral bichromophoric molecules from their CD spectra. Paracyclophanes are the molecules with limited conformational freedom and the analysis of their CD spectra provides directly the absolute configuration, as shown in Fig. 13.11.

For an example of DFT/RPA computational analysis of the CD spectra of a cyclophane with two different chromophores—azulene and benzene—see Ref. [50].

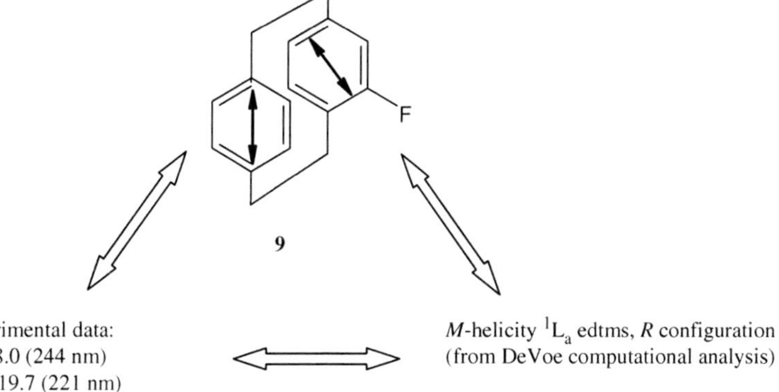

Fig. 13.11. Determination of the absolute configuration of (**9**) by confrontation of the experimental CD spectrum with the DeVoe computational analysis [49].

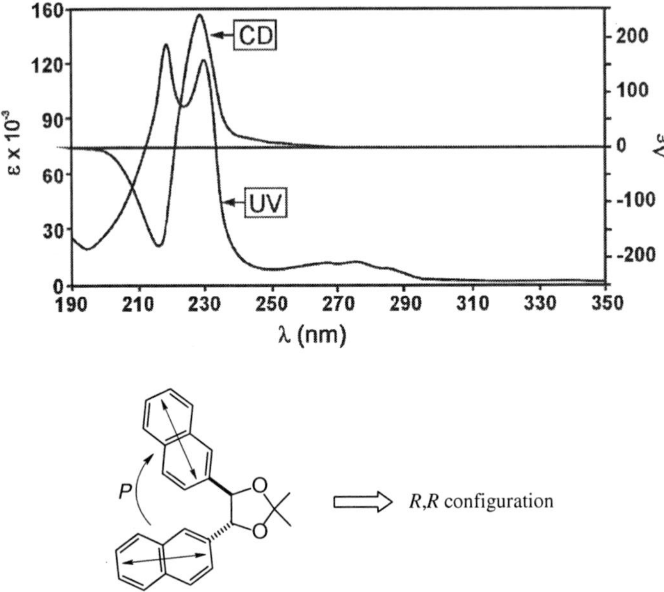

Fig. 13.12. Determination of the absolute configuration of (**10**) by geometrical analysis of the helicity of edtms. Adapted from Ref. [51].

Absolute configuration of 1,2-dinaphthylethane-1,2-diol obtained by asymmetric dihydroxylation was determined by analysis of the CD spectrum of its isopropylidene acetal (**10**), which is conformationally less flexible. Due to exciton interactions the allowed naphthalene ^1B_b transition around 225 nm splits into two bands, clearly visible in both the UV and the CD spectra (Fig. 13.12).

The CD spectrum shows a strong positive exciton couplet with the amplitude $A = +421$ which corresponds to *R,R* configuration of the diol. It is of interest to note that in this case conformational analysis of molecule (**10**) is not necessary since *P* helicity of the edtms is obtained independently of the rotation of the naphthyl groups [51].

Computational analysis (B3LYP/TZP) of both the conformation and rotational strengths of the electronic transitions was necessary to determine the absolute configuration of thalidomide (**11**) and its biotransformation products (Fig. 13.13).

Here, the short-wavelength (below 250 nm) part of the CD spectrum results from coupling of the edtms of the phthalimide chromophore [53] with edtm of the glutarimide. The experimental spectrum was found to match reasonably well the computed one for *S* absolute configuration [52].

13.2.2 External chromophores

In many cases in order to obtain exciton coupled CD spectra one, two or more chromophores are introduced into the parent molecule. Chromophoric derivatization

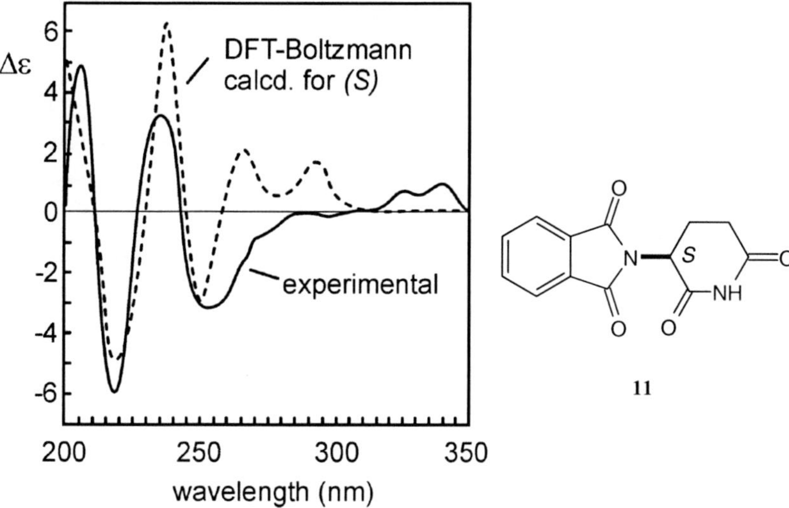

Fig. 13.13. Comparison of experimental and DFT calculated/Boltzmann averaged CD spectra of (*S*)-thalidomide (**11**). Adapted from Ref. [52].

plays an important role in ECCD since in this way one can deliberately introduce chromophores conforming to the following characteristics:

(1) in the case of monochromophoric molecules added chromophore enhances the effect of exciton coupling if λ_{max} of the allowed transitions in both chromophores match (nearly-degenerate case).
(2) introduction of two red-shifted (λ_{max}) chromophores allows to observe exciton Cotton effect essentially undisturbed by the electronic transitions in the parent chromophore.
(3) higher ε_{max} of added chromophore(s) means less sample required for the CD measurement.
(4) if possible, chromophoric derivatization should be carried out *in situ* (no necessity for isolation of the chromophoric derivative).

A detailed review of chromophores for stereochemical analysis by ECCD has been recently published [54].

By far the most popular way to make chromophoric derivatives is by acylation of hydroxy, amino or sulfanyl groups with aromatic acid derivatives, such as chlorides, fluorides or imidazolyl derivatives. These derivatives include benzoates and 4-substituted benzoates, originally introduced by Harada and Nakanishi [28], 2-naphthoates [55,56], 2-anthroates [33,57] as well as (substituted) cinnamates. For example, dibenzoate exciton Cotton effects in combination with ^{1}H NMR data were used to characterize configurations of di-*O*-benzoyl-*sn*-glycerols [58]. The usefulness of some of the chromophoric derivatives is further enhanced by their strong fluorescence which allows for easy microscale manipulations. In a model study, as little as 40 ng of (2*S*)-1,2-propanediol could be anthroylated with 2-anthroylimidazole, then purified by HPLC and used for CD/UV study [33].

Table 13.1 UV and CD data of various (2*R*,3*R*)-2,3-butanediol derivatives

Cl–C₆H₄–CO–O–	MeO–C₆H₄–CO–O–	MeO–C₆H₄–CH=CH–CO–O–	2-naphthoyl–O–	2-anthroyl–O–
ε_{240} 33,900[a,b]	ε_{255} 36,100[a,c]	ε_{307} 45,000[d]	ε_{232} 108,000[e]	ε_{258} 147,000[e]
235 nm (+4)	250 nm (+10)	278 nm (+14)	229 nm (+99)	253 nm (+170)
250 nm (−6)	268 nm (−7)	317 nm (−22)	242 nm (−116)	274 nm (−152)
$A = -10$	$A = -17$	$A = -36$	$A = -215$	$A = -322$

[a]UV and CD data were taken from Ref. [59].
[b]UV and CD spectra were measured in dioxane.
[c]UV and CD spectra were measured in ethanol.
[d]UV and CD spectra were measured in methylcyclohexane.
[e]UV and CD spectra were measured in acetonitrile.

A comparison of the CD and UV data for various derivatives of (3*R*,3*R*)-2,3-butanediol is given in Table 13.1 [33].

It is seen that the ratio of amplitude *A* of the exciton Cotton effect to ε_{max} of the compound dramatically increases with strongly absorbing chromophores; the increase is 7-fold between the 4-chlorobenzoate and the 2-anthroate derivatives. Intensely absorbing chromophores are particularly useful in cases where exciton Cotton effects are difficult to measure due to conformational averaging (in acyclic molecules) or due to long interchromophoric distance.

The latter problem can be solved efficiently with the use of porphyrin-based chromophores, such as 5-(4-carboxyphenyl)-10,15,20-triphenylporphyrin (TPP). This chromophore has extremely intense, red-shifted Soret band at 414 nm, ε_{max} 350,000 and is capable of producing large exciton Cotton effects, for example in derivatives of steroids [60]. A Cotton effect with the amplitude −10 could still be observed with a derivative of brevetoxin having appended two TPP chromophores at a distance 40–50 Å [61] (Fig. 13.14).

Selective derivatization of polyhydroxy compounds, such as carbohydrates, with two different chromophoric derivatives provides exciton-coupled CD spectra which are characteristic ("fingerprint type") for each stereochemical pattern of polyol structure [62]. To this end Nakanishi et al. used 9-anthroyl derivative for the primary hydroxy group and 4-methoxycinnamoyl derivatives for secondary hydroxyls of acyclic polyols. Both degenerate (cinnamate/cinnamate) and non-degenerate (cinnamate/anthroate) interactions of participating conformers contribute to the Cotton effects pattern, characteristic of a given absolute and relative configuration. These fingerprint spectra allow to analyze unknown structures simply by comparing the CD spectra of the sample and the model structures (for an application to absolute configuration determination of two bacteriohopanoids with acyclic pentol side chains see Ref. [63]).

The amplitude (*A*) of the exciton Cotton effect of a polychromophoric derivative may be treated as the sum of amplitudes of the exciton Cotton effects due to all pairs of interacting chromophores (additivity rule [64]). Thus for a trichromophoric molecule

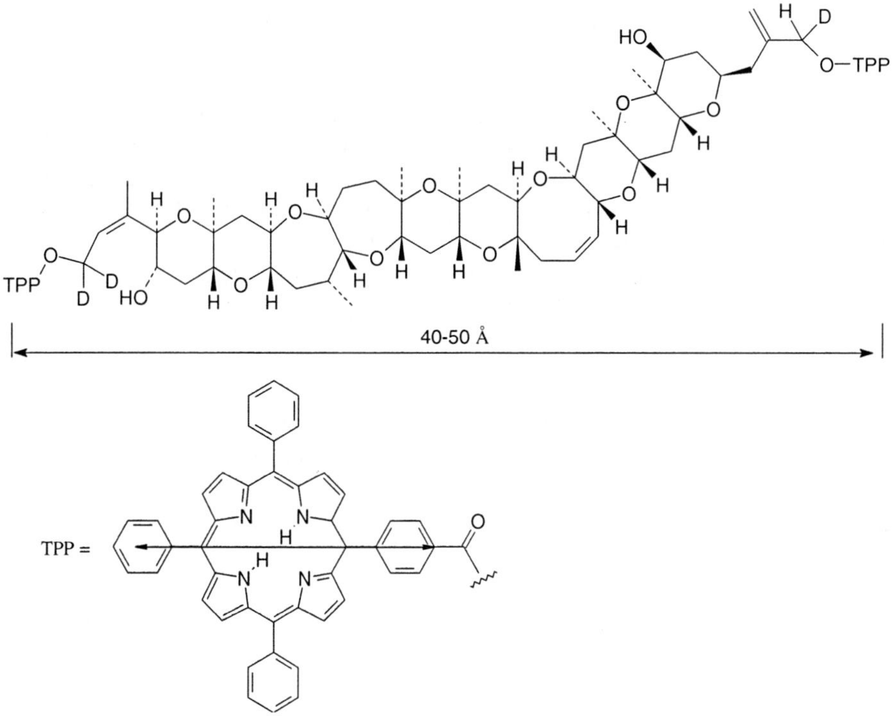

Fig. 13.14. Structure of brevetoxin-bridged porphyrin dimer.

there are three pairs of interacting chromophores, for a tetrachromophoric molecule there are six such pairs:

$$A_{1,2,3} = A_{1,2} + A_{2,3} + A_{1,3}$$
$$A_{1,2,3,4} = A_{1,2} + A_{2,3} + A_{3,4} + A_{1,3} + A_{2,4} + A_{1,4}$$

On the basis of the additivity rule one can analyze absolute configuration at a specific chiral center of a polyol. For example, the configuration at C2 of chiral 1,2,4-triol is related to the sign of a differential Cotton effect, as in Scheme 13.2 [65].

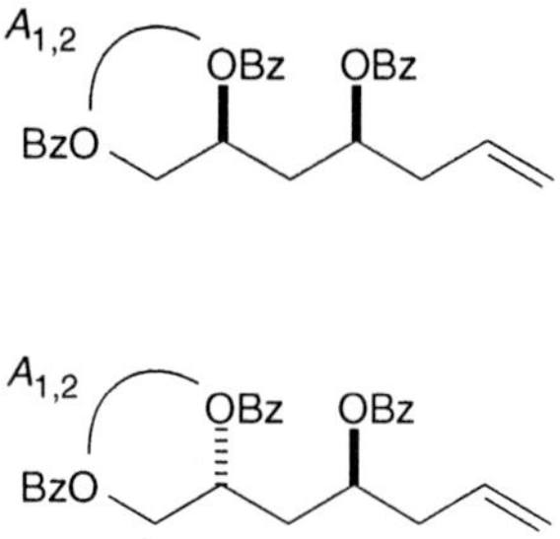

$A_{2,4}$ = ca. 0 (known)
$A_{1,2,4}$ = +10.4 (measured)
$A_{1,2} = A_{1,2,4} - A_{2,4}$ = +10, hence 2*R* configuration

$A_{2,4}$ = +21.2 (known)
$A_{1,2,4}$ = +14.0 (measured)
$A_{1,2} = A_{1,2,4} - A_{2,4}$ = 7, hence 2*S* configuration

Scheme 13.2

This empirical method is based on the assumption that $A_{1,4}$ is very small, thus $A_{1,2} \approx A_{1,2,4} - A_{2,4}$. For a related analysis of absolute configuration of a dimeric sphingolipid (−)-rhizochalin see Ref. [66].

While aromatic acyl derivatives are the most popular as chromophores for alcohols, for the amine group a number of structurally different chromophoric derivatives have been devised [54]. In addition to benzamide type derivatives, primary amino group can be transformed to phthalimide [53,67], 2,3-naphthalimide [68], 1,8-naphthalimide [69], *s*-triazine [70] and aromatic imine derivatives [71].

In contrast to benzamide derivatives of primary amines, analysis of the CD spectra of benzamide derivatives of secondary amines may present a problem due to complicated conformational equilibria [72,73]. Tertiary amines can be derivatized as quaternary ammonium salts bearing a biphenyl chromophore [74].

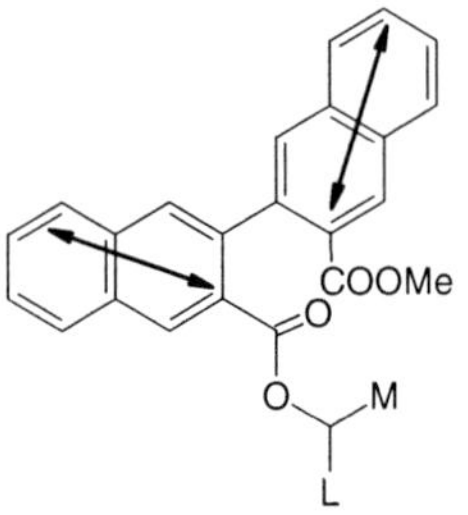

COOMe O O M L

12

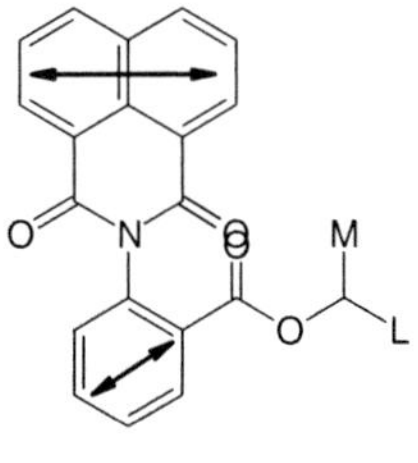

O N O M O L

13

X Y M N Cu N L N

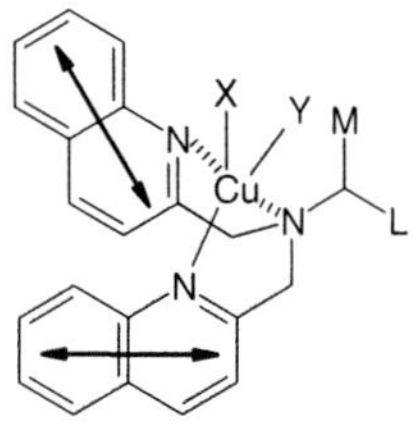

14

For chromophoric derivatization of the carboxy group esterification reaction with 2-naphthol [75] or 9-anthryldiazomethane [76] has been used. Very recently derivatization of the alkene group with the styrene chromophore using cross methathesis reaction has been proposed [77].

Finally, it should be added that the analysis of absolute configuration with the use of ECCD is not limited to molecules having two or more functional groups or chromophores. In the case of molecules having just a hydroxy or amino group at a chiral center the absolute configuration can be determined by analysis of the exciton Cotton effect of a bichromophore, such as **12–14**.

All these derivatives (**12**–**14**) exist in solution as a mixture of diastereomeric *P* and *M* conformers in equilibrium. The diastereomer ratio is determined by the difference in size of L and M groups attached to the chiral atom and it is reflected in the measured CD spectrum. Derivatives (**12**) [78] and (**13**) [79] are obtained by acylation of secondary alcohol whereas (**14**) [80] is a derivative of primary amine prepared by its alkylation with 2-bromomethylquinoline and complexation with a Cu(II) salt.

13.3 CONFORMATIONAL ANALYSIS

ECCD is a well-suited method for conformational analysis of chiral molecules [81]. For a molecule of known absolute configuration the preferred conformation can be obtained from the sign of the exciton Cotton effect, as in the example below (Fig. 13.15).

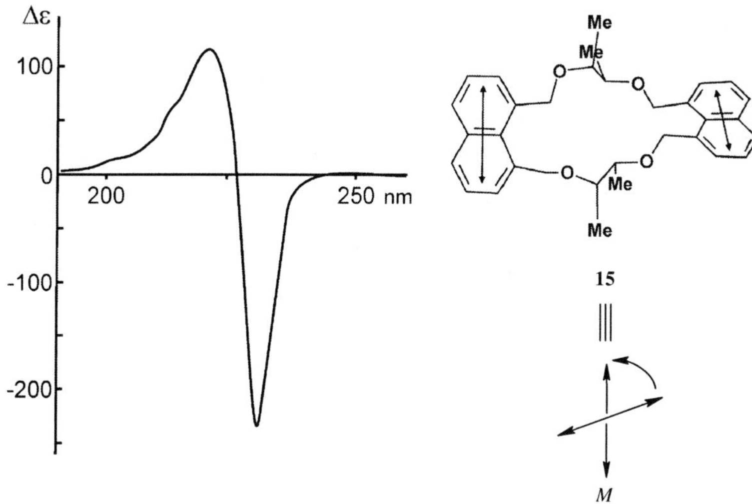

Fig. 13.15. CD spectrum of (**15**) in acetonitrile solution (adapted from Ref. [82]) and helicity of edtms.

The conformation of the chiral macrocycle (**15**), follows from the strong negative exciton Cotton effect due to ^1B_b naphthalene electronic transition and is further confirmed by the results of AM1 computational analysis and X-ray diffraction data [82].

In this section we will mainly consider the use of ECCD for the analysis of stereostructure of chiral molecules having helical π-electron chromophores, as well as chiral polymers and biopolymers (proteins, nucleic acids). All these structures can be classified as *P* or *M* helical, without resorting to configurational descriptors at the chiral atoms (*R* or *S*).

13.3.1 Atropisomers

Various types of resolved atropisomers have been characterized by CD spectra. They include such entities as stable chiral enols, e.g. (**16**) [83] and atropisomers of diaryl bicyclononanones, e.g. (**17**) [84]. In the latter case a negative exciton Cotton effect due to ^1B_b (in naphthalene) and ^1L_a (in 1,4-disubstituted benzene) transitions is assigned to the *M*-helical atropisomer. In the case 1,2-bis(*N*-benzoyl-*N*-methylamino)benzene (**18**) chiral non-racemic conformers were obtained by spontaneous crystallization and could be analyzed by CD spectra only in the crystals, as KBr pellet [85].

Atropisomeric biaryls are a class of helical molecules particularly well-suited for conformational analysis by ECCD. Due to nearly orthogonal arrangement of two (or more) aryl rings electronic exchange between the aryl rings is small, so they can be treated as coupled oscillators and their CD spectra analyzed according to the exciton chirality model. However, there is no simple relation between helicity (the sign of the dihedral angle between the aryl rings) and the sign of the exciton Cotton effect. This is because both the dihedral angle and the structure of the aromatic/heterocyclic rings affect the mutual position of the edtms in the two chromophores. As mentioned earlier, in 1,1′-binaphthyl derivatives the reversal of sign of the exciton Cotton effect

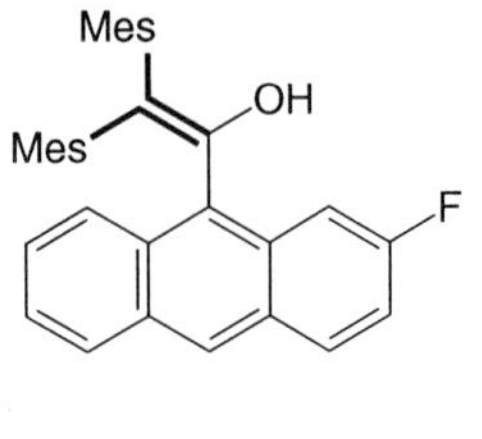

Mes
OH
Mes
F

(*P*)-**16** (Mes = 2,4,6-trimethylbenzene)

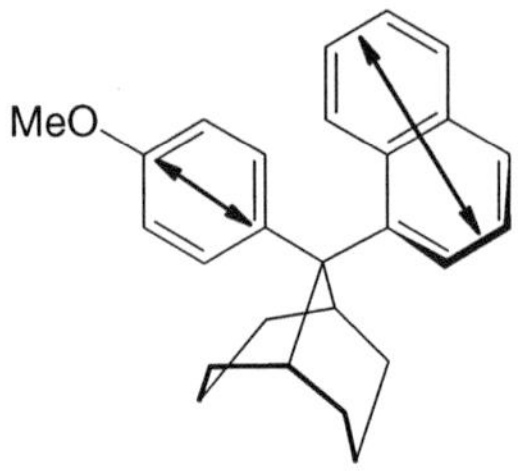

MeO

(*M*)-**17**

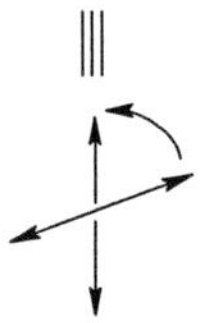

M

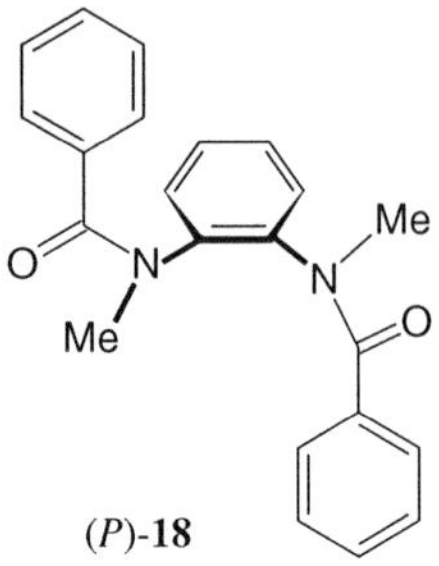

O
N
Me
Me
N
O

(*P*)-**18**

due to ^1B_b electronic transition takes place when the interchromophoric dihedral angle crosses the value ca.110°.

There are numerous reports on the application of the exciton chirality method to the analysis of conformation of biaryls, in many instances supported by the results of computation of the CD spectra. Examples include a 4,4′-biphenanthryl derivative (**19**) [86], a 1,1′-binaphthyl lactone (**20**) [87] and a naphthyltetrahydroisoquinoline alkaloid (**21**) [88].

Two chiral olefins, (*E*) and (*Z*)-1,1′,2,2′,3,3′,4,4′-octahydro-4,4′-biphenanthrylidenes

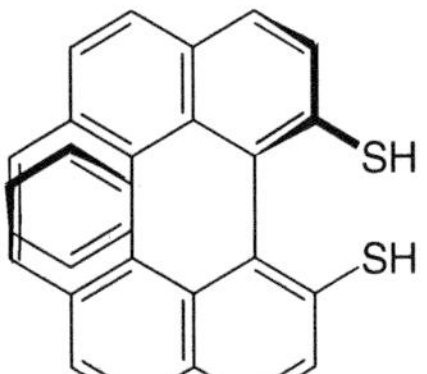

SH
SH

(*M*)-**19**
A = -285
(261 nm band)

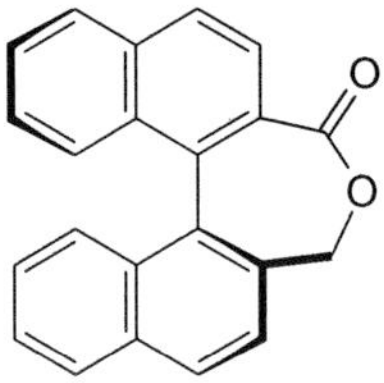

O
O

(*P*)-**20**
A = -3890
(225 nm band)

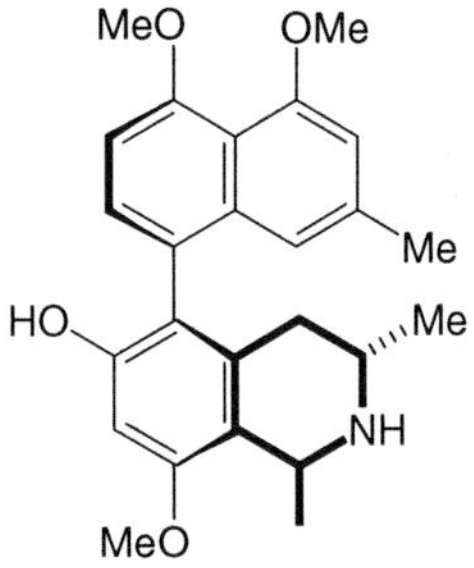

MeO
OMe
Me
HO
Me
NH
MeO

(*P*)-**21**
A = -340
(230 nm band)

(**22**) and (**23**), present a challenging test for the application of exciton approach to analyze their absolute stereochemistry. They are composed of two naphthalene chromophores connected by the central double bond and could be considered as fully conjugated (monochromophoric). The UV spectra, however, show bands resembling those of the naphthalene chromophores, in particular strong ^1B_b bands in the range 250–200 nm; these bands are associated with strong positive exciton Cotton effects ($A = +211.5$ for (**22**) and $+429$ for (**23**)), as shown in Fig. 13.16.

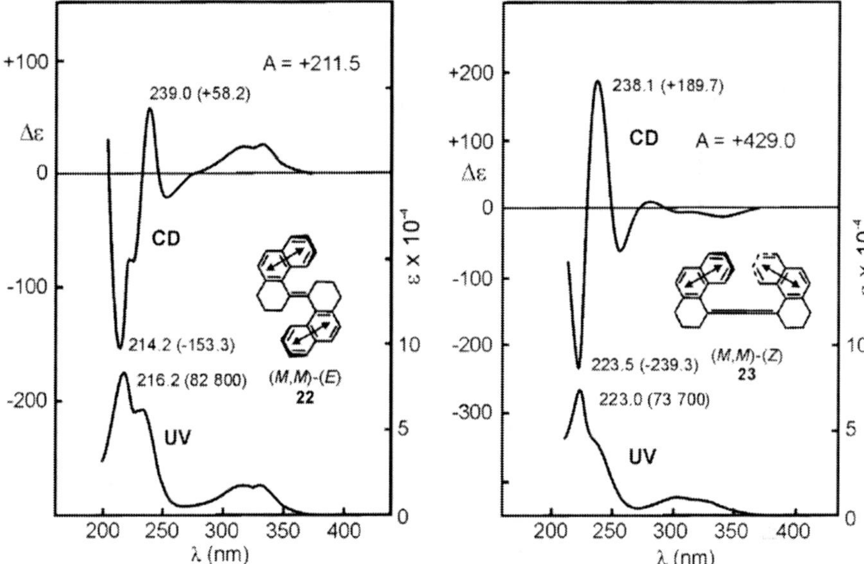

Fig. 13.16. CD and UV spectra of (**22**) (in methanol) and (**23**) (in hexane). Adapted from Ref. [89].

The absolute stereochemistry of (**22**) and (**23**) was determined by calculation of the UV and CD spectra by a semiempirical SCF-CI-DV MO method. The spectra computed for *M*,*M* helicity of the molecules matched well with the experimental data, thus establishing the sense of helicity of chiral olefins (**22**) and (**23**). In addition, the computation revealed that these molecules, although constituting a border case for application of the exciton chirality method, can still be analyzed as coupled oscillators and that the main features of the CD spectra arise from exciton coupling between the naphthalene chromophores. This is because due to steric reasons the central C=C bond is highly twisted, breaking full conjugation between the naphthalene chromophores [89].

For related applications of CD spectroscopy to the analysis of isomerisation of molecular rotors and motors see Ref. [90].

However, the analysis of stereostructure of helical, *fully* delocalized conjugated monochromophores by the exciton chirality method is not justified. Absolute stereostructure of molecules belonging to the broad group of helicenes in which helicity of the π-electron system is due to steric constrains, is most correctly done by confrontation of the computed and experimental CD spectra. Take for example *M*-hexahelicene (**24**), a typical C_2-helical chromophore (Fig. 13.17).

In C_2-symmetric chromophores the electronic excitations can be classified as A or B type, according to irreducible representations of the point group. A-type transitions are polarized along the *z*-axis, which is a two-fold symmetry axis and both edtm and mdtm are parallel to *z*. B-type transitions are orthogonal to *z*, i.e. they are polarized in the *xy*-plane. The well-established Wagniere's C_2-rule [91,92] states that A-type transitions have positive rotatory strength in *M*-helical chromophores (antiparallel edtm and mdtm) whereas B-type transitions have negative rotatory strength. To identify transition type in the electronic spectra it is customary to carry out computations of the

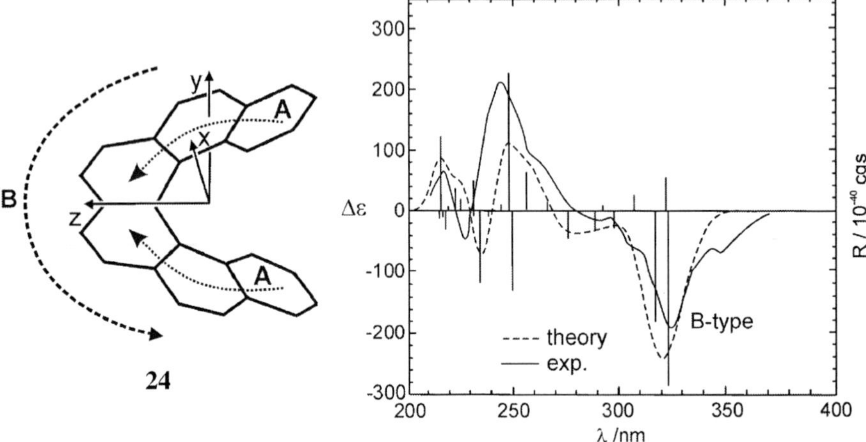

Fig. 13.17. Comparison of computed and experimental CD spectra of (*M*)-hexahelicene (**24**). The sticks indicate the positions and rotatory strengths of the calculated states. All calculated excitation energies are blue-shifted by 0.45 eV. Adapted from Ref. [91].

excited states. In the right panel of Fig. 13.17, the experimental and time-dependent DFT (TDDFT) computed spectra of (**24**) are compared, showing that B-type transitions prevail within the lower energy CD band (negative Cotton effect at ca. 320 nm) whereas the positive Cotton effect at ca. 250 nm is the superposition of more intense A-type (positive) transition and less intense B-type (negative) transition [91]. The C_2-rule appears to be applicable to the analysis of stereostructure of a wide variety of helical π-conjugated chromophores, such as substituted helicenes [93,94], pyrrolohelicene [95], zinc bilinone derivatives (**25**) [96], chiral pentamethine dye (**26**) [97] and hypericin derivative (**27**) [98]. In derivative (**25**) the preferred helicity is induced by an appended chiral OR* substituent, whereas in the case of (**27**) the stable diastereomers could be separated chromatographically. Except (**27**), other chromophores display lowest energy transition of B-type; therefore the sign of the lowest energy Cotton effect correlates directly with the sense of helicity of the C_2-symmetry chromophore. In the case of (**27**), according to semiempirical calculations, the lowest energy transition is of A-type and thus for *P* helicity shown the sign of the lowest energy Cotton effect is negative.

25 **26** **27**

13.3.2 Chiral polymers

Conformational analysis of chiral oligomers and polymers is one of the most promising applications of CD spectroscopy. Since many functional polymers are composed of chromophoric-type monomers, CD spectra of polymers, at least in cases where the monomers are achiral, is due to secondary polymer structure, i.e. due to formation of either *P* or *M* helices. Chromophores in helical chain arrangement are able to interact according to the exciton model, thus making possible the analysis of stereostructure of the polymer according to the exciton chirality method.

There are several possibilities for the induction of helicity in polymers:

(1) diastereoisomeric structures are formed by appending or adding a chiral auxiliary to the polymer chain. This is usually done by making the monomer chiral prior to polymerization or by using a chiral component in the polymerization process. The diastereomeric *P* and *M* helical polymers may or may not be in dynamic equilibrium under ambient temperature.
(2) enantiomeric but non-racemic polymer helices are obtained by asymmetric polymerization of achiral monomers or by resolution of racemic polymer. These *P* and *M* helices are not in equilibrium.
(3) optical activity is induced in racemic polymers by non-covalent interaction with the chiral molecules. This interaction shifts the equilibrium towards one of the helical structures. For the discussion on this subject see the section on induced CD.

There are numerous examples for inherently chiral polymers or polymers with an appended chiral auxiliaries whose structural analysis with the aid of CD spectroscopy suggests the presence (or absence) of a helical conformation with dominating one-handed screw sense. In the case of main chain π-conjugated polymers (conducting polymers) exciton coupling mechanism for the generation of CD spectra may not be operative, nevertheless changes in the sign and intensity of Cotton effects are indicative of conformation (i.e. secondary structure) alterations [99]. Examples of polymers belonging to this group are polyacetylenes (PA) with chiral appended XR* groups (**28**) [100–108].

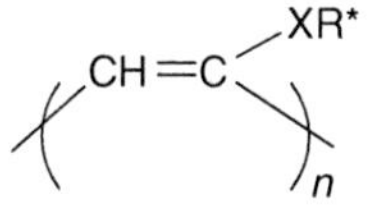

$$\left(CH{=}C(XR^*) \right)_n$$

28

It should be noted that CD spectra of these polymers can reveal inversion of helix sense (*P* or *M*), detected as a change of magnitude or sign of the Cotton effect. This has been observed with the change of temperature of polymer solution or with the change of the solvent used (Fig. 13.18) [104,105].

Changes in the CD spectra are also observed in copolymers of chiral and achiral phenylacetylenes indicating that the prevailing helix sense of the chiral–achiral random

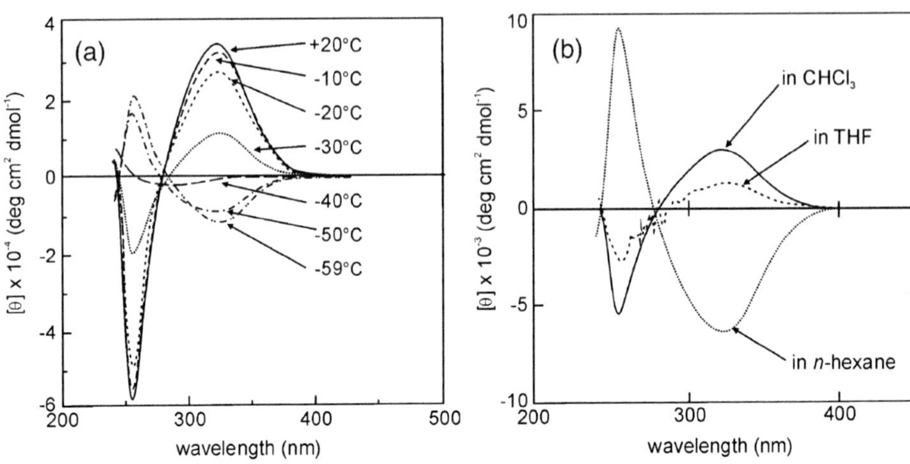

Fig. 13.18. (a) CD spectra of (**28**) (XR* = COO…) at various temperatures in $CHCl_3$ ($c = 6.0 \times 10^{-4}$ mol/l). (b) CD spectra of (**28**) (XR* = COO… and COO…, ratio 1:9) at 20°C in various solvents. Adapted from Ref. [104].

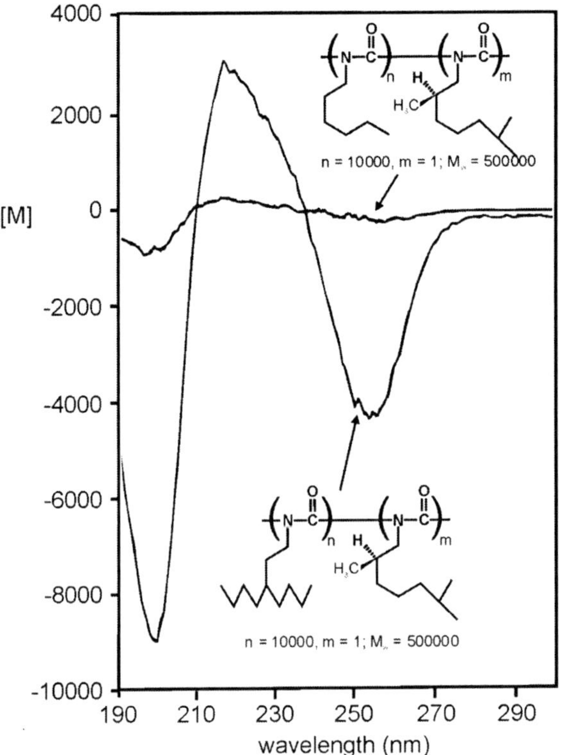

Fig. 13.19. Structure and circular dichroism spectra for the copolymers of (*R*)-2,6-dimethylheptyl isocyanate with 2-butylhexyl isocyanate and with *n*-hexyl isocyanate. Adapted from Ref. [113].

copolymers is determined by a subtle interaction between the chiral and achiral side chains [106].

Polyisocyanates (**29**) are stiff helical main chain polymers with a high degree of cooperativity along the polymer chains. Consequently they are composed of long, alternating sequences of *M* and *P* helices, separated by the helix reversal points. Due to low helix inversion barriers helix inversion occurs readily in solution and this process can be visualized as the movement of helix reversal points along the polymer backbone [106]. These polymers are extremely sensitive to very small chiral influences; for example if the pendant groups differ from their mirror images only by the substitution of a deuterium for a hydrogen, measurable optical activity is obtained [109]. Similarly, substantial optical activity is seen in copolymers of chiral and achiral units having very low proportions of the chiral units [110] ("*sergeant and soldiers principle*" [110–112]). CD spectroscopy can reveal the preferred chain helicity of polyisocyanate copolymer with high dilution of the chiral component; in the example shown in Fig. 13.19 this dilution is as high as 1:10,000 [113].

Polyisocyanides (**30**) have a rather well-defined 4_1 helical conformation (four repeat units per one helical turn) which is stabilized by sterically demanding substituents R*.

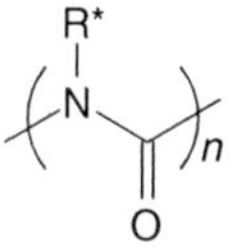

R* N n O

29

C n N X R*

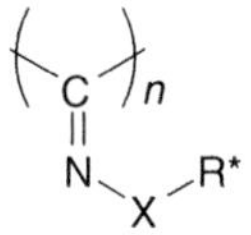

30

When X is 4-carboxyphenyl substituent, polyisocyanide displays a strong Cotton effect at 360 nm, the sign of which has been correlated with the helicity of the polymer chain [114]. CD spectra were instrumental for the detection of induction and memory of macromolecular helicity in a chiral polyisocyanide [115]. A method of determining the helicity of polyisocyanide chain by the induced exciton Cotton effects within the Soret band of pendant porphyrin chromophores has been proposed [116].

Another example of helical polymers with chromophoric main chain are polysilanes (**31**). In such polymers the UV absorption between 320 and 400 nm (depending on the chain substituents) is due to σ–σ* electronic transitions in the silicon chain. It is recognized that the absorption originates from segment-like domains consisting of 10–20 silicon atoms. These segments of *P* or *M* helicity are diastereoisomeric and not isostructural. They respectively give positive and negative Cotton effects which are separated by an energy gap and hence produce a CD spectrum of an apparent bisignate type (Fig. 13.20) [117].

Single-screw-sense rigid rod-like helical polysilane with a long alkyl ((**31**), $R = C_{10}H_{21}$) or triethereal moiety ((**31**), $R = Et(OCH_2CH_2)_3(CH_2)_3$) is characterized by an intense, narrow single Cotton effect at 323 nm [118]. With appropriate substitution pattern ((**31**), R = (*S*)-Et(MeHSi(CH$_2$)$_2$), R* = (*R*)-*i*Pr(CH$_2$)$_3$(Me)HSi(CH$_2$)$_2$) it is possible to induce helix–helix transition by changing the temperature: single negative Cotton effect (λ_{max} 325 nm) is observed at +25°C in isooctane solution, whereas at −60°C single positive

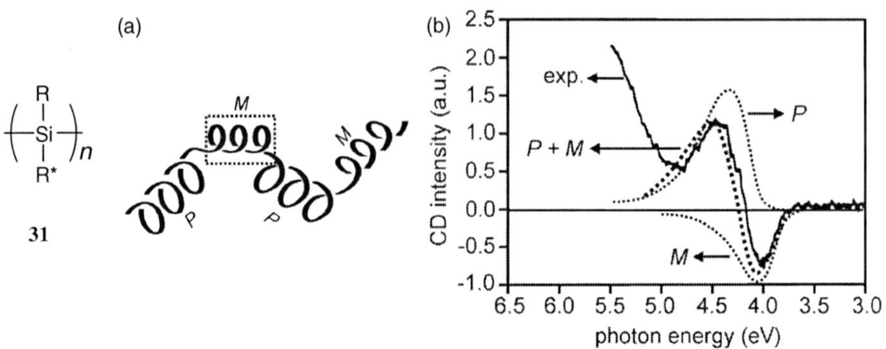

Fig. 13.20. (a) Segment-like *P* and *M* domains of polysilanes. (b) CD spectra of (**31**) (R = Me, R* = $CH_2SiH(Me)Et$). Bold solid curves represent the experimental CD spectra, bold dotted curves are the simulated CD spectra, and dotted curves stand for the deconvoluted CD spectra, and *P* and *M* mean the helicity of Si sequence. Adapted from Ref. [117].

Cotton effect appears at λ_{max} 320 nm. This is interpreted as almost complete switching from *M* to *P* helix upon cooling the solution of a chiral polysilane [119].

Changes in the CD spectra of chiral π-conjugated oligomers and polymers may also be due to solvent-induced aggregation [120]. Self-assembly of folded *m*-phenylene-ethynylene oligomers into helical columns has been recently suggested on the basis of solvent and temperature dependent CD spectra [121]. In the case of chirally *N*-substituted poly-3,6-carbazoles (**32**) both polymer chain helicity and supramolecular structure based on interchain π–π stacking are contributing to the solvent-dependent CD spectra [122].

32 **33**

Polythiophenes (PT) substituted with chiral alkyl groups in 3-position (**33**) maintain predominantly a planar structure in good solvents; therefore they do not exhibit significant CD spectra in the region of the main chain π–π^* transitions. However in poor solvents they form aggregates in which chiral side chains control ordered structure of the aggregate and the resulting strong CD spectra can be analyzed in terms of exciton coupling between individual polythiophene chains [123,124]. Such a situation is observed also in polymers containing arylene–vinylene units [125]. Furthermore, strong bisignate Cotton effects of aggregates of regioregular chiral polythiophenes can be switched off and on through electron transfer (doping with Cu(II) and undoping with triethylenetetramine (TETA)), Fig. 13.21.

This demonstrates the utility of the CD spectroscopy in uncovering supramolecular chirality of PT polymers.

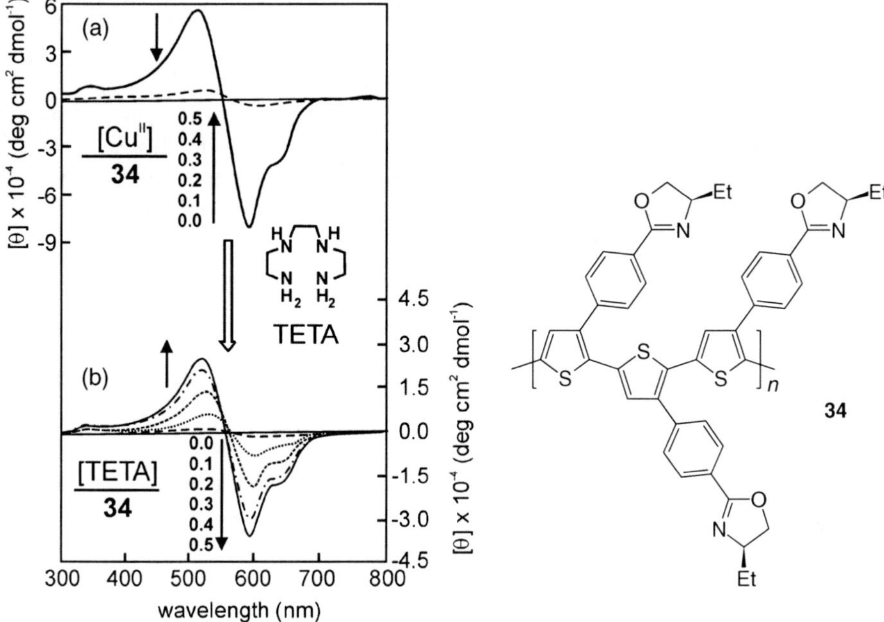

Fig. 13.21. CD changes of (**34**) upon addition of $Cu(OTf)_2$ to the polymer solution in a chloroform—acetonitrile mixture (1:1) (a) and a further addition of TETA to the polymer solution with $Cu(OTf)_2$ (b) Adapted from Ref. [126].

Optically active poly(aryl carbonates) of the formula (**35**) are of special type as their main chain is composed of conformationally locked 1,1′-binaphthyl units. In such a case CD spectrum is composed of contributions due to exciton interaction within individual 1,1′-binaphthyl components plus contributions due to exciton interactions between naphthalene chromophores from neighbouring 1,1′-binaphthyl units. Consequently the CD spectra of the monomer and polymer are very different. By comparing the calculated and experimental spectra for a decamer of (**35**) it was concluded that it forms a stable 4_1-helix structure [127].

35

Polymers of substituted ethylene, such as polystyrenes (PS, (**36**)), polyacrylates (PA, (**37**)) and polymethacrylates (PMA (**38**)) do not possess chromophoric groups in the main chain, however their chirality can be analyzed by means of CD spectroscopy with the aid of pendant reporter groups. Aromatic reporter groups are of particular interest. Cyclopolymers of chiral 4-vinylbenzoates with styrene ((**36**), R = Ph or 4-$C_6H_4COOR^*$) have been analyzed by exciton chirality method using the pendant vicinal benzoate chromophoric groups [128]. A terphenyl chromophore ((**36**), R = 2,5-bis(4′-alkoxyphenyl)phenyl) has also been used as a reporter group [129].

$-(CH_2-CH(R))_n-$ **36** $-(CH_2-CH(C{=}O)XR)_n-$ **37** $-(CH_2-CMe(C{=}O)XR)_n-$ **38**

The CD spectrum of helical poly(triphenylmethyl methacrylate) (PTrMA, (**38**), RX = $OCPh_3$) is used as a standard for a perfect, one-handed helical PMA in the analysis of helical conformations of PA (**37**) with similar reporter groups, obtained by asymmetric anionic polymerization [130].

Photoresponsive and reporter azobenzene group in the side chain of chiral acrylates and methacrylates is of considerable interest. This reporter group absorbs in the visible region (around 400 nm) and the observed CD couplets are due to interactions of the azobenzene π–π^* edtms, reflecting the ordered helicity of the polymer main chain. Fig. 13.22 shows an example of the CD spectra of chiral polymer film bearing substituted azobenzene chromophores.

The azobenzene group can act as photoresponsive switch group. CD spectra show that photochemical *trans* → *cis* isomerization of the azo group may shift the equilibrium between the diastereoisomeric *P* and *M* helical main chains, not only in homopolymers but also in copolymers [132].

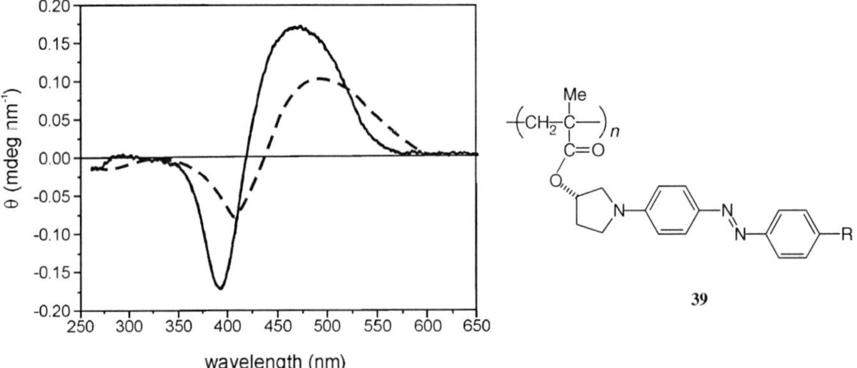

Fig. 13.22. CD spectra of native thin films on fused silica of (**39**), R = CN (120 nm thick, full line) and (**39**), R = NO_2 (140 nm thick, dashed line). Adapted from Ref. [131].

13.3.3 Protein secondary structure

Circular dichroism spectroscopy is an efficient tool for analyzing the secondary structure of proteins in solution and in a films [133–139]. In comparison to the X-ray crystallography and NMR technique, CD spectroscopy has few advantages. The principal one is speed and convenience; good quality spectra can be obtained within minutes. Also the amount of required material is small 100–500 μg, but acceptable spectra can be recorded with as little as 10 μg. The CD technique is non-destructive so it is possible to recover protein from solution and use it for other experiments.

The very important feature of CD analysis of protein is that CD studies can be performed over variety of conditions like temperature, pH, and additives. It has to be pointed that, contrary to X-ray, CD measurements are performed in a quasi-natural environment and enable to observe the dynamic processes such as folding, aggregation or binding. The main disadvantage of CD technique is that only proportion of secondary structure motifs can be obtained, without exact spatial arrangement of those motifs (although some information about tertiary structure can be obtained) [134,135,140–142]. Also generally no distinction between similar secondary structure motifs can be made, for example between α-helix and 3_{10} helix.

The CD spectra of proteins are usually divided into three regions, based on the group of chromophores which are active in the given range (Figure 13.23).

In the first region (below ~250 nm) electronic transitions due to the amide group are observed. This region is used for secondary structure analysis. UV and CD bands between ~250–300 nm are due to aromatic aminoacids side chains (Fig. 13.24). Each of the aromatic aminoacids has a characteristic band shape. Tryptophan has a maximum at 290 nm with vibronic structure (290–305 nm), tyrosine maximum is at 275 and 282, phenylalanine shows absorption band with vibronic structure between 255–270 nm. Cotton effects due to aromatic side chains may be used to estimate tertiary or quaternary structure of protein [143–145].

Contribution of the aromatic side chains to the CD spectra may lead to unusual spectra [146,147] and incorrect results of secondary structure analysis, especially to underestimation of α-helices contribution [148].

In the third region above 300 nm absorption due to ions or molecules which bind to peptide chain can be observed.

Main contribution to the absorption of peptides in far UV region (down from 250 nm) has the amide chromophore. Its electronic structure was investigated using *ab initio*

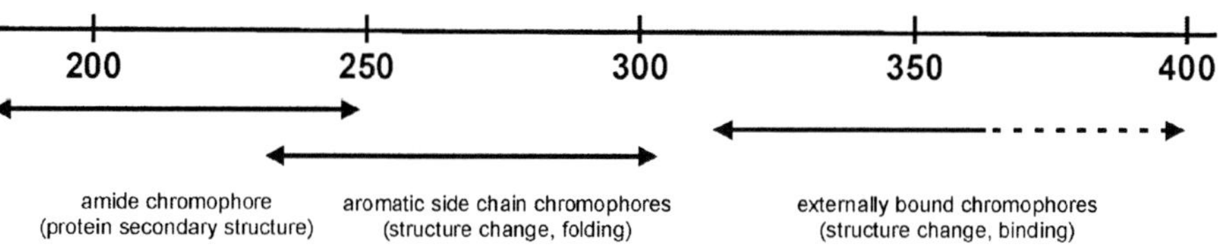

Fig. 13.23. Regions in CD spectra of proteins.

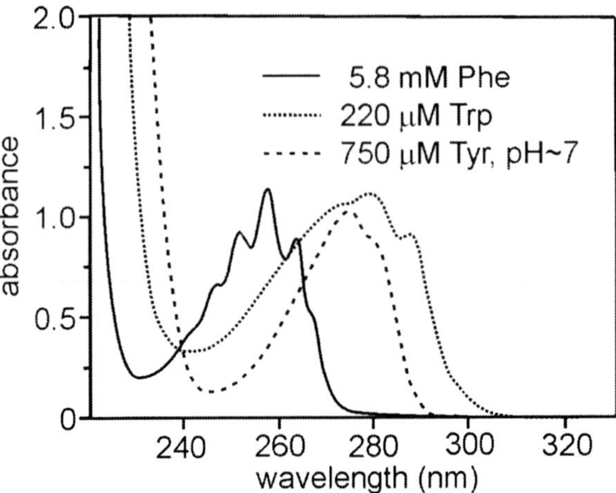

Fig. 13.24. UV spectra of aromatic aminoacids. Adapted from Ref. [6].

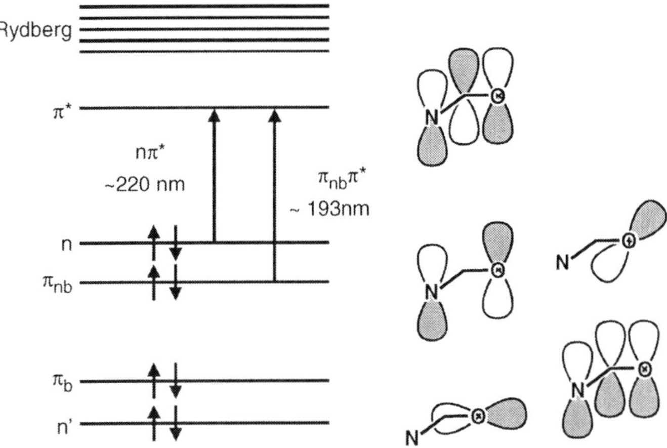

Fig. 13.25. Energy levels and transitions in amide chromophore. Adapted from Ref. [154].

calculations [149–152] of monoamides combined with polarized reflection spectroscopy [153]. Energy levels and transitions of the amide chromophore are shown in Fig. 13.25.

A weak band at 220 nm is described as forbidden $n\pi^*$ transition. A strong band at around 190 nm can be assigned to electronically allowed $\pi\pi^*$ transition. The next calculated electronic transition occurs at around 130–140 nm and corresponds to the $n'\pi^*$ and $\pi_b\pi^*$ transitions. For commercial CD spectrophotometers accessible range is down to 180 nm due to limited light source power and design. Using synchrotron radiation [155,156] or vacuum UV-CD spectrophotometer [157] CD spectra down to 160 nm can be recorded but the assignment of CD bands below 180 nm is still uncertain.

Calculation of CD spectra. CD spectra of amide chromophore in peptides arise due to nπ^* transition in a single amide group and due to exciton interaction between $\pi\pi^*$ transitions of the neighboring amide groups. High-level *ab initio* calculation of CD spectra are possible only for small rigid molecules [158].

To calculate CD spectra of a protein, a matrix method [135,149,154,159] derived from DeVoe polarizability model [41] is used. This method utilizes classical electrostatic interaction between chromophores to determine transition energies and rotational strength in large molecules. Proteins are described as consisting of M chromophoric groups with independently defined electronic properties. This method was used to calculate CD spectra for peptides with known X-ray diffraction determined structure [135,149,154] with quite a good agreement. Calculations of the CD spectra were used to investigate protein folding [160].

Similar approach represents extended Fano–DeVoe polarizability theory which uses different type of interaction energy between the chromophores [161,162].

Analysis of CD spectra. Polypeptide chain of protein is linked by a secondary amide group (except proline which forms a tertiary amide and accounts for about 5% of protein residues). Secondary structure of protein strongly affects the arrangement of the amide group leading to different electrostatic (exciton) interaction between electronic transition moments in case of $\pi\pi^*$ transitions and modification of chiral surrounding for nπ^* transitions. This consequently leads to different CD spectra.

Spatial orientation of the peptide backbone is described by two dihedral angles ϕ and φ. The most common motifs of α-helices and β-sheets are characterized by the angles −60, −45 and −120, +120 respectively. Another secondary structure is β-turn which reverses the direction of chain propagation. In some proline-rich proteins (example collagen) a poly(Pro)II type structure P_2 (ϕ, φ: −78, +150) is present. Secondary structures which do not belong to the classes mentioned above are usually described as unordered or random coil conformation.

For an α-helical protein a negative CD band due to nπ^* transition is observed near 220 nm. A $\pi\pi^*$ transition shows up as a couplet with a negative band at around 208 nm and a positive band near 192 nm. The intensity of this couplet is reduced for short helices [163] whereas in the case of the band at 220 nm such reduction is not observed. β-Sheet is represented by a negative band at 216 nm and positive band around 197 nm (Fig. 13.26).

Estimation of secondary structure of peptides is based on comparison of the CD spectrum of a protein with unknown secondary structure with the CD spectra of peptides of known secondary structure. Secondary structures of reference peptides were determined from their crystal structures available in Protein Data Bank [135,164]. Approximation of CD spectrum (C_λ) of a particular protein could be described as a linear combination of secondary structure component spectra $B_{K\lambda}$

$$C_\lambda = \sum f_K B_{K\lambda}$$

where f_K is a fraction of secondary structure K, $B_{K\lambda}$ is obtained from polypeptides of specific conformation [135,134]. Due to similarity of their CD spectra α-, 3_{10}- and π-helices are combined into one helical fraction (there are however attempts

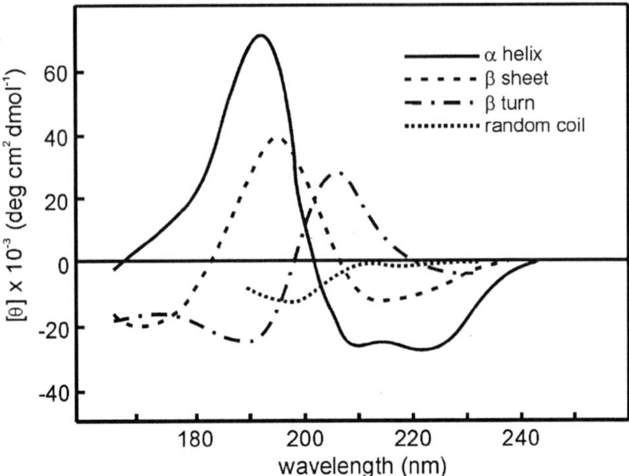

Fig. 13.26. CD spectra of protein with various types of secondary structure. Adapted from Ref. [133].

to distinguish between these three helices [165]). Similarly, β-turns and S-bends are combined to turns fraction. Correct estimation of secondary structure strongly depends on the reference set used. Recently 42-protein reference set was created combining three sets developed independly by three research groups [164]. For a better analysis of denaturated proteins (consisting mainly of unordered structure) five reference CD spectra of denaturated protein were included in the reference set [166].

In the past the most frequently used method of fitting the CD spectrum was multilinear regression [134]. Developed later Nonconstrained Least-square Analysis gives good estimation of α-helices. This method can be used to evaluate change in helical content due to interaction with the solvent, ligand or due to membrane formation. The Consistent Least-square Analysis improves the estimation of β-sheet and β-turns content by constraining the sum of the fractional weight to 1. All these methods may be used to analyze protein secondary structure even when only a limited wavelength range (200–240 nm) CD data are available, with only slight loss of accuracy.

One of the presently used methods is Ridge Regression Analysis (CONTIN) [134,164,167,168]. In this method, the contribution of each reference spectrum is kept small unless it contributes to the good agreement between the calculated best fit curve and the real CD spectrum. In another method (SELCON) [134,164,168], the analyzed CD spectrum of the protein is included in the reference set with initial guess of the structure of protein whose CD spectrum mostly resembles that of analyzed protein. In the next steps calculated conformation of secondary structure replaces the previous one and the protein with the least fitting CD spectrum is deleted from the basis set. The whole procedure is repeated until self-consistency is achieved. This method may overestimate the amount of α-helices and may underestimate the amount of β-sheet structure. This error may arise from different rotational strengths due to long β-sheets and short β-sheets (the latter are represented in globular protein, used as a reference set).

CDSSTR method [164,169] requires only eight reference CD spectra for successful analysis. The reference spectra are randomly selected from a large reference set. The composition of secondary structure is obtained using self-consistent approach. Although the results of analysis depend on the choice of reference spectra, performance of this method is very good [135,164]. These methods are included in the CDPro software available at the website http://lamar.colostate.edu/~sreeram/CDPro.

Another approach for secondary structure analysis is based on neural networks [170–172]. K2D [170] program utilizes data between 200 and 240 nm as an input. It gives best estimate of β-sheet contents but it does not evaluate the content of β-turns. Calculation of protein structure with the above methods is possible on a DICHROWEB [173] online server at http://www.cryst.bbk.ac.uk/cdweb.

A very important parameter for exact secondary structure assignment is the concentration of protein. In cases when concentration is unknown (for example when investigating insoluble proteins like amyloid or prion aggregates in films) secondary structure can be estimated using dissymmetry factors [174] or normalization procedure [175].

Another important aspect of protein secondary structure analysis is the contribution of aromatic side chain to the measured CD spectrum between 200 and 240 nm (see Fig. 13.24). Methods to estimate this contribution to the CD spectra and corrections in the secondary structure analysis were presented [176–178].

CD spectra of proteins can also be used for the analysis of their tertiary structure. Using designed reference set it is possible to estimate the number of α-helical and β-strand segments in proteins [134,141,142,179] as an extension to the existing secondary structure fitting programs or neural network analysis [180].

Conformational changes of protein. Circular dichroism is a convenient method to investigate changes in secondary structure and parameters which induce these changes. Recently there is interest in prion proteins which cause Bovine Spongiform Encephalopathy (BSE) and Creutzfeld–Jacob disease. Mechanism of the disease evolution involves conformational change of protein rich in α-helices to β-sheet [181,182]. It was shown that β-rich isoform of mouse prion protein (α-MoPrP) is thermodynamically more stable than α-helical isoform. Conformational transition between isoforms requires overcoming the large energy barrier due to unfolding to unordered transition state. It was found than under partially denaturating acidic conditions prion protein avoids α-helical isoform formation and folds directly to thermodynamically more stable β-rich isoform [183] (Fig. 13.27).

Amyloid β-peptide (Aβ), a key substance in Alzheimer's disease is characterized by its abnormal folding into neurotoxic aggregates. CD analysis suggests that in buffers free of organic solvents Aβ adopts unordered conformation [184,185]. Transition to the β-sheet containing fibrils causes aggregation on cell membranes. This change may be promoted by binding with Zn cation [186]. It has to be noted that both above cases could not be investigated by NMR or X-ray crystallography due to low solubility of proteins and impossibility of obtaining good quality crystals.

Similar conformational change from random-coil to β-sheet and the following aggregation of natural [187] or spider [188] silk protein was observed.

Folding of proteins is an important subject which can be easily followed by CD spectroscopy [189]. Folding of the cytochrome C protein and identification of folding

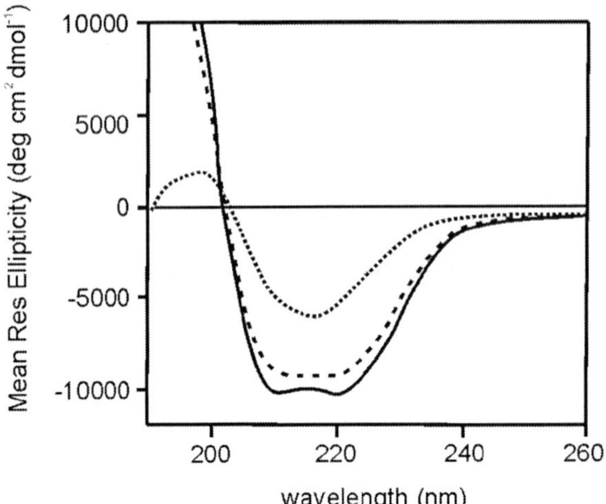

Fig. 13.27. Changes in the CD spectrum of α-MoPrP (solid line) after dilution with 10 M urea solution (dashed line) and after five-week incubation (dotted line). Adapted from Ref. [183].

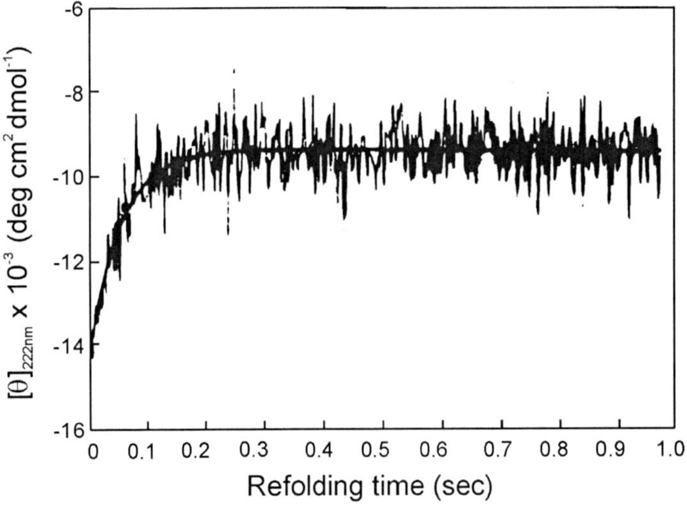

Fig. 13.28. Kinetic refolding curves of β-lactoglobulin in 0.01 M PBS buffer (pH 2.0) monitored by CD at 222 nm. Adapted from Ref. [192].

intermediates was investigated using time-resolved CD [190] and stopped flow CD [191]. Stopped flow CD technique enabled also investigation of refolding bovine β-lactoglobulin. As it was found by CD, kinetic intermediate is rich in α-helices whereas native conformation is mainly β structure with only one α-helix [192]. Fig. 13.28 shows changes of the CD signal measured at 222 nm during refolding of bovine β-lactoglobulin.

By detailed CD investigation of the 14 different globular proteins it was proposed that substantial part of the protein secondary structure is formed at the earliest stages of folding [193].

CD spectroscopy was applied to prove insulin chemical and conformational stability during encapsulation in poly(lactide-co-glycolide) [194], glyceryl monooleate–water gel [195] and sodium dodecyl sulfate [196] for future drug delivery. Changes in secondary structure followed by CD spectra were used to investigate the ability of protein to bind metal ions [197–199], drugs [200] or lipid membranes [201].

Changes in electronic ligand properties (induced CD observed above 300 nm) were used to investigate interaction between retinoic acid-β-lactoglobulin [202] or curcumin–serum albumin [203] or quercetin–human serum albumin complexes [204].

Membrane proteins differ from soluble albumins used to create reference sets in secondary structure analysis programs. Crystal structures for that class of proteins proved difficult to determine. Till now only 13 crystal structure of membrane proteins are determined [205]. Membrane proteins have somewhat different spectral characteristics (changes of band peaks and, to lesser extent, relative band areas) as shown by synchrotron radiation CD [200]. In membrane proteins two distinct types of α-helices are observed. α-Helix in soluble domain and α_T-helix for transmembrane helices have different spectral characteristics due to different environment and due to the fact that the average length of α_T-helix is about twice the length of α-helix in soluble proteins. CD spectra of the other components correlated well with those for soluble protein [206]. Recently reference set based on membrane protein is included in analyzing programs to improve secondary structure assignment [205,207].

Another class of peptides, i.e. *synthetic β-peptides* were also investigated by CD spectroscopy. It is known that β-peptides can fold into 3_{14}-helix, 12/10-helix, hairpin turn and parallel sheet [208–210]. The strong Cotton effect at 198 nm and the opposite-sign one at 215 nm are assigned to 3_{14} helix. Although much effort was devoted to derive conformation of β-peptides from CD spectra [211] a deeper understanding of the relationship between the conformation and CD spectra is required to calculate the latter. Attempts of the matrix calculation of CD spectra of β-peptides give unsatisfactory results [212]. It was found that fine changes in peptide geometry have significant effect on the CD spectrum.

13.3.4 Nucleic acids

UV and CD spectra of nucleic acids are dominated by absorption and differential absorption due to purine and pyrimidine bases. Phosphate groups have no absorption in the range accessible for standard CD spectrophotometers, i.e. above 180 nm. Sugars (ribose in the case of RNA and deoxyribose in the case of DNA) contribute to the spectra at about 190 nm, and below.

Electronic properties of nucleic acid bases. UV and CD spectra of purines and pyrimidines are dominated by π–π^* transitions (Fig. 13.29). Due to complexity of transitions pattern their exact arrangement and energies are not fully recognized [6,213–217]. The bases are planar and all transition moments lie in the plane of the

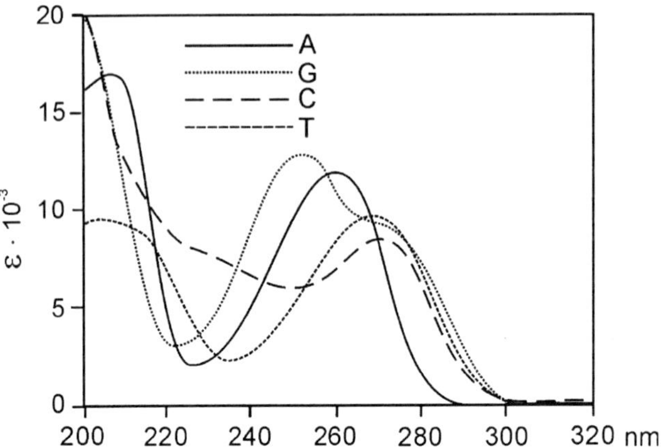

Fig. 13.29. UV spectra of DNA nucleotides: A-deoxyadenosine 5′-monophosphate, G-deoxyguanosine 5′-monophosphate, C-deoxycytidine 5′-monophosphate, T-deoxythymidine 5′-monophosphate. Adapted from Ref. [6].

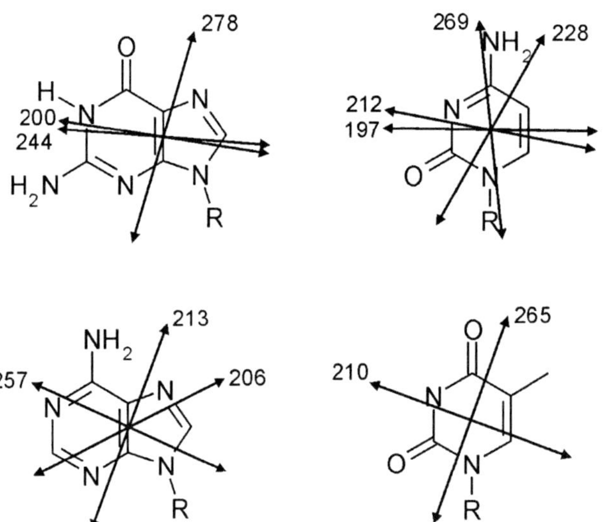

Fig. 13.30. Directions of electric dipole transition moments in nucleic acid bases.

molecule. In Fig. 13.30 are presented the π–π^* transition moments directions for nucleic acid bases.

Bases are achiral molecules and are not optically active. However, in the case of isolated nucleotides, they are chiral because of appended ribose sugar and induced CD signal can be recorded. Magnitude of these Cotton effects is of the order of 2 (Fig. 13.31).

Spectral characteristic of different DNA structure. The CD spectra of DNAs are mainly due to the excitonic interaction between transition dipoles in nucleic acid bases.

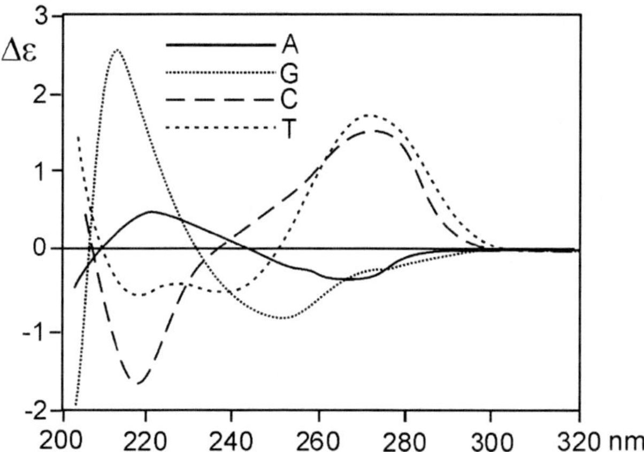

Fig. 13.31. Induced CD spectra of nucleotides: A-deoxyadenosine 5′-monophosphate, G-deoxyguanosine 5′-monophosphate, C-deoxycytidine 5′-monophosphate, T-deoxythymidine 5′-monophosphate. Adapted from Ref. [6].

The Cotton effects appear in a narrow spectral range (200–300 nm) and in principle it is not possible to assign a particular band to a particular base–base interaction.

In solution the most common structure of DNA is in the form of B double-stranded helix. This polymorph is characterized by a positive Cotton effect at 275 nm and a negative Cotton effect centered at about 240 nm with zero crossing at about 258 nm. At shorter wavelengths at 190 nm a strong positive CD band is observed. The structure of the DNA under certain conditions (80% ethanol or 2,2,2,-trifluoroethanol) can be forced to adopt A-form. This form is characterized by a strong positive CD band at about 260 nm and a fairly intense positive CD band at 220 nm and with a positive Cotton effect at 190 nm. This structure is also adopted by the naturally occurring RNA which shows identical spectra (Fig. 13.32).

Left-handed Z-form of DNA is easily formed by poly[d(G–C)]$_2$ (a transition from B-form to Z-form occurs after increasing the salt concentration). Its CD spectrum is a quasi-mirror of the spectrum of a B-form and has a negative CD band at 290 nm and the positive one at 260 nm. This form can be easily distinguished from A-form, since the latter has also a positive Cotton effect at 260 and a negative CD signal at around 200 nm.

The best example of parallel triplex is poly(dA.dT.dT) [219,220]. Its CD spectrum shows positive Cotton effects at near 280 and 260 nm, negative at 250 and 210 nm and positive ones at 190 and 177 nm. The negative band at 210–215 nm is believed to be an indication of the triplex formation.

Guanosine-rich oligodeoxyribonucleotides are known to adopt a quadruplex structure [218,221]. Similar structure was proposed for isoguanosino oligonucleotides [222]. Oligomers d(GGGA)$_5$ and d(GGA)$_7$ in the presence of LiCl form homoduplexes characterized by strong CD bands at 265, 250 and 215 nm (positive, negative, positive signs, respectively). CD spectra of d(GGGA)$_5$ and d(GGA)$_7$ tetraplexes are very similar to the spectra of duplexes and the magnitude of observed Cotton effects is approximately twice of these observed for duplexes. It is suggested that a strong

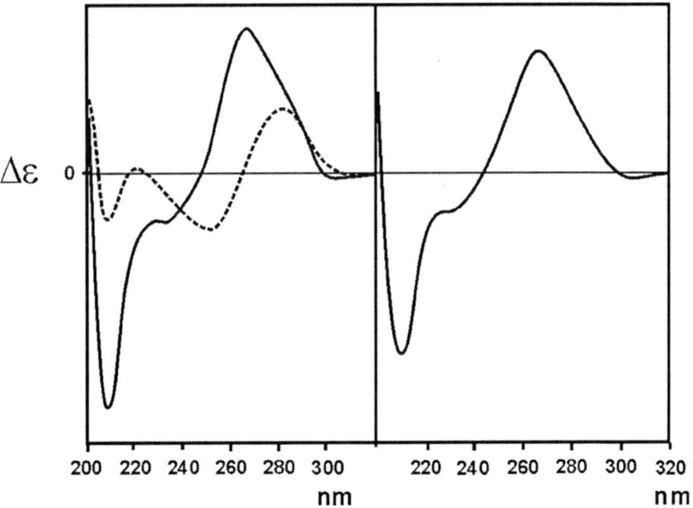

Fig. 13.32. The CD spectra of B (dashed) and A forms (solid) of DNA (left side) and CD spectrum of RNA (right side) of the same sequence. Adapted from Ref. [218].

Cotton effect at 260 nm observed for tetraplex in fact comes from intrastrand guanine–guanine interaction as it is also visible in B and A forms of anti-parallel duplex of d(C_4G_4) [218]. Although it is generally accepted that the parallel tetramers show positive band at 265 nm and a negative band at 240 nm while folded quadruplexes (monomers and dimers) show positive CD band at 295 nm and a negative CD band at 265 nm, an investigation of several guanine-rich oligonucleotides suggests that the CD spectra are very sensitive to small changes in quadruplex structure and the simple correlation between biophysical properties of tetraplex (UV melting, CD, calorimetry, NMR) and structure cannot be made [223].

Application of CD to analysis of DNA secondary structure and observation of conformational changes. CD spectroscopy is an useful tool to determine whether changes made in the nucleotide structure cause significant changes to the overall structure of the DNA. However, in case of the DNA, there are no such developed tools to analyze its structure as those available for proteins. Contrary to proteins the DNA is assumed to exist only in one particular conformation for the whole molecule. Analysis of the DNA conformation by CD usually is based on comparison of the recorded CD spectrum of DNA to that of the CD signal pattern characteristic for a particular conformation. An example of such analytical approach is the analysis of influence of base modification or sugar–phosphate chain modification on the stability of DNA duplexes as shown below.

Fluorescent DNA bases received recently much attention as fluorescent markers for detecting nucleic acids and monitoring changes in their structure. 1,3-Diaza-2-oxophenothiazine (tC) was found to be a very good replacement of cytosine base. From CD measurements of tC-PNA-DNA was concluded that tC does not seem to distort significantly duplex structure [224,225]. Similarly, fluorescein-labeled thymidine also does not perturb DNA B-form conformation [226].

The effect of 8-amino-2′-deoxyguanosine, a guanosine "analog" found in DNA damaged by 2-nitropropane, on the DNA structure was investigated to estimate distortion of helix made by this modified-base. CD studies indicated that this analog did not alter the structure of right-handed B-DNA [227]. Also in the case of 2′-deoxyguanosine carrying polycyclic aromatic hydrocarbon (9-methylanthracene) covalently bound to the exocyclic nitrogen [228] as well as in the case of 4′-alkylated thymidines [229] CD spectroscopy confirmed that B structure of polynucleotide was not significantly altered.

Conformationally restricted oligodeoxynucleotide analogues, where sugar moiety was replaced by a rigid modified sugar [230] or a rigid non-sugar structure [231], are of interest in antisense technology. An ideal antisense inhibitor should display high affinity and selectivity towards target sequence. CD analysis of tricyclo-DNA indicate the formation of a very stable Watson–Crick base pairing system in A-type helix [231]. Similary, Locked Nucleic Acid (LNA) containing 2′-*O*,4-C-methylene bridge which locks nucleotide in C3′-*endo* furanose conformation forms LNA–DNA duplexes of B-type having characteristic features in the CD spectra [230].

More advanced approach using CD/UV spectroscopy at different temperatures and analyzing data by multivariate curve resolution alternating least-squares method (MCR-ALS) enabled to identify the coexistence of three different conformers: monomeric dumbell-like structure, a dimeric four-stranded conformer and a random coil (unordered) conformer in the solution of cyclic oligonucleotide d(pTGCTCGGCT) [232]. Similar methodological approach using the UV, fluorescence and CD data was used to study the base-pairing schemes between mutagenic 2-aminopurine and cytosine or thymine [233] and transition from right-handed B conformation to left-handed Z conformation of poly(dG-dC) · poly(dG-dC) and poly(A) · poly(U) double stranded polynucleotides [234].

Analysis of the CD spectra recorded during the transition of DNA from B to Z form by locally linearized model [235], usually used to determine protein secondary structure, indicate the existence of two distinct intermediates [236]. CD spectroscopy was used to follow transition from B to Z form induced by polyamines [237] or a polyamine–platinum complex [238].

Triplex structure was found in pyrimidine motif DNA and RNA [239]. Creation of triplex structure is one important approach to nucleic acid recognition. The formation of such intramolecular helix structures in single stranded DNA and RNA *in vivo* is suggested. Oligodeoxyxylonucleotides built from 1-(β-D-2′deoxy-*threo*-pentafuranosyl) thymine were shown by CD to form triple helices with complementary purine RNA and DNA [240]. Formation of triplex is sometimes induced by binding a drug. Berenil, a diarylamidine derivative, induces formation of antiparallel triplex structure [241].

Quadruplex DNA attracts much attention as such guanose-rich sequences are found in telomers. In the case of adenine–guanine, depending on the nucleotide sequence of the oligomer a variety of strands of structures such as tetraplexes, homoduplexes and single-stranded conformers were detected [221].

To understand how tetraplex structure of telomere is joined to a double-stranded B-helix a novel DNA construct was described [242]. The structure containing both double helix and tetraplex fragments forms spontaneously and is thermodynamically stable. Fig. 13.33 shows CD spectra of (5′-AGGGTTAGGGTTAGGGTTAGGG-3′)

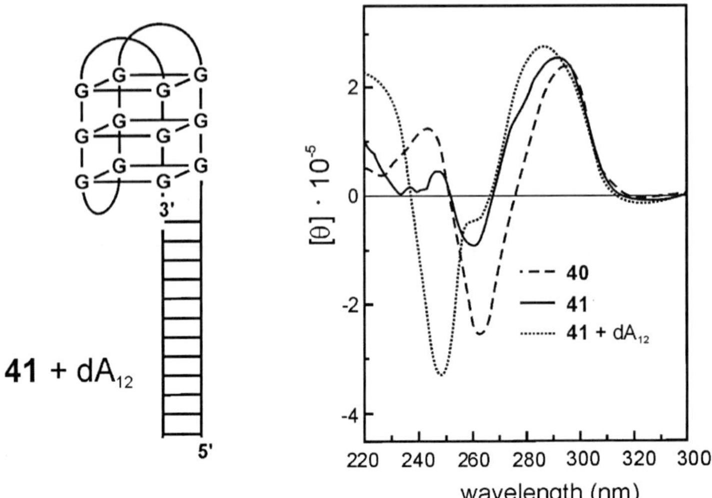

Fig. 13.33. Schematic representation of possible structure of (**41**) + dA_{12} and CD spectra of oligonucleotides (**40**) (dashed line), (**41**) (solid line), (**41**) + dA_{12} (dotted line). Adapted from Ref. [242].

sequence (**40**) which is known to form antiparalel quadruplex and its modification (**41**) after attachment of dT_{12} tail to 5′ side. Addition of complementary dA_{12} results in the CD spectrum ((**41**) + A_{12}) which is consistent with a hybrid structure containing a folded quadruplex linked to a $dT_{12} \cdot dA_{12}$ duplex.

Binding of single molecules absorbing in the UV region outside the DNA absorption to DNA polynucleotide can be observed by induced CD signals due to the interaction of such a molecule with chiral environment of DNA helix. It is known that the intercalators with long axis polarized transitions oriented parallel to DNA bases will induce negative CD signal in random sequence DNA [6,243]. The intercalators with long axis which pokes out into DNA grooves exhibit positive induced CD signal for long-axis polarized transitions.

Good example of the binding system is the interaction of porphyrin with DNA, studied by UV- and CD-spectroscopy. The interaction strongly depends on the nature of the inserted metal ion. Porphyrins and metallo-porphyrins having no axial ligands show two possible ways of intercalation. External binding, in the minor groove is characterized by negative-sign induced CD signals in the Soret region, whereas major groove porphyrins give positive signal in the same region. Metalloporphyrins which have axial ligand when binding to DNA do not intercalate due to steric repulsion of ligands. They either exhibit a positive induced CD signal or are CD silent in the Soret region [244,245].

13.4 ANALYSIS OF CHIRAL INTERACTIONS BY INDUCED CIRCULAR DICHROISM

In this section, we will discuss the use of CD spectroscopy in the analysis of chiral interactions between molecular structures not connected by a covalent bond.

Table 13.2 CD response in self-association (**A**) and host–guest interactions (**B**, **C**)

Case	Host	Guest	CD
A	Chiral, non-racemic (self-association)	None	CD change
B	Chiral, non-racemic	Chiral, non-racemic	CD change
	Chiral, non-racemic (e.g. *cyclodextrins*)	Achiral	Guest ICD (*extrinsic*)
C	Achiral or chiral, racemic (e.g. *calixarenes, porphyrins, bilirubins, polyanilines*)	Chiral, non-racemic	Host ICD (*intrinsic*)

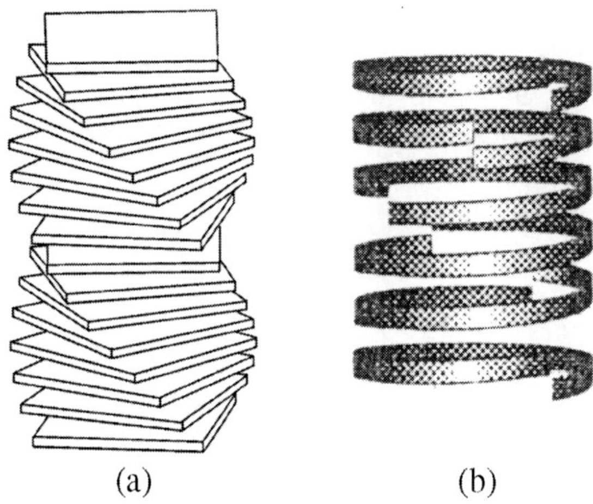

Fig. 13.34. Columnar self-association of chiral planar molecules (card-pack type) (a) and of helical molecules (b).

Different variants of such interactions can be envisaged but a common feature of such interactions is the generation of induced CD (ICD), i.e. the CD spectrum of the system is significantly different from the sum of CD spectra of the components.

The variants of chiral interactions that can be analyzed by CD spectroscopy are listed in Table 13.2.

They include self-association events (case **A**), resulting in a change of the CD spectrum, as well as an interaction of chiral host with chiral or achiral guest (case **B**) and racemic or achiral host with chiral guest (case **C**). The latter two cases of interaction may lead to generation of ICD, often of exciton coupling type. In general, induction of CD involves molecular interactions causing structural changes and/or chiral coupling of edtms. Detection and measurement of ICD is currently of high analytical interest, in connection with applications to supramolecular chemistry, chiral discrimination, chiral nanostructure organization and it has been recently reviewed [246,247].

13.4.1 Self-association

Self-association (**A**) of chiral planar molecules can take a columnar arrangement, with a twist around column axis due to chiral substitution of the molecules. The two (equivalent) arrangements are shown in Fig. 13.34.

Case (a) is represented by elongated molecules, like carotenoids [248], perylene dyes [249] and camptothecin derivative [250] as well as by disc-shaped molecules [251], this includes guanosine derivatives, forming hydrogen-bonded "G-quartets" [252]. A characteristic feature of these systems is a very weak CD activity of such compounds under conditions preventing association and a strong optical activity under conditions (solvent, concentration, temperature) favoring self-association. For example, carotenoid (6′*R*)-capsantholon (**42**) is nearly CD silent in ethanol solution but a very high exciton-type Cotton effect is observed upon dilution with water, due to formation of a card pack aggregate (H-type), with a negative tilt angle between longitudinally polarized edtms (Fig. 13.35) [248].

Case (b), Fig. 13.34, refers to helical molecules, like (**43**) [253] and oligo(phenylene ethylene)s coupled to 1.1′-binaphthalene [254]. These compounds show significantly changed CD spectra due to the increased concentration and/or the use of poor solvents.

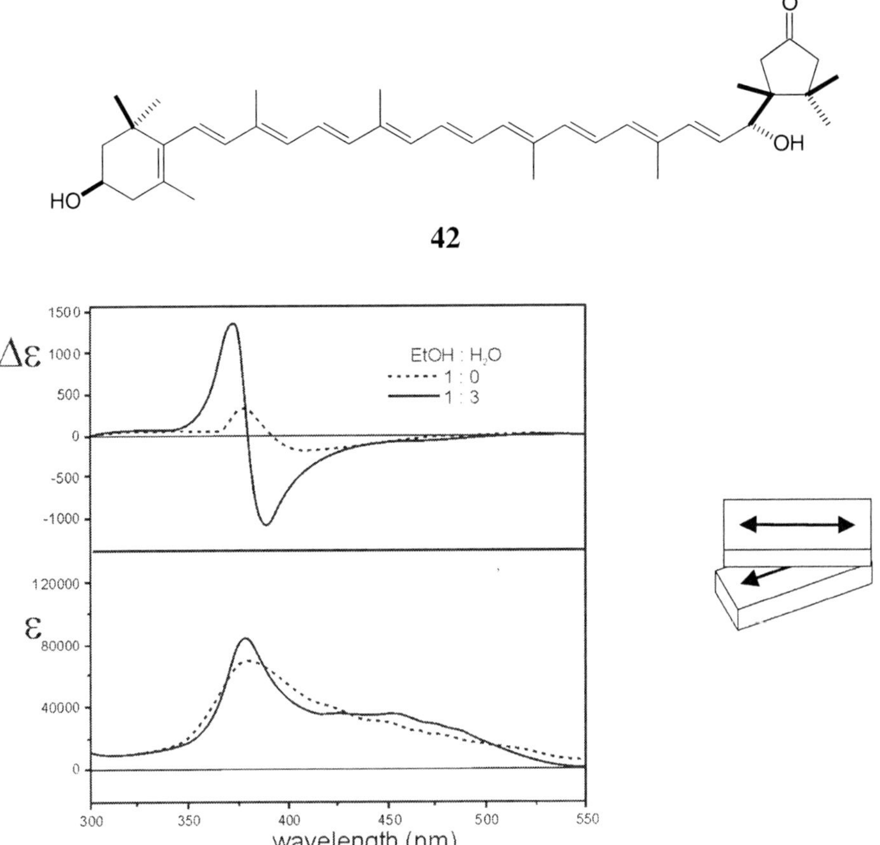

Fig. 13.35. CD (upper part) and UV/VIS (lower part) spectra of (**42**) in ethanol (- - -) and ethanol–water (1:3) (—). Adapted from Ref. [248].

A similar analysis of CD data can be applied to assemblies of molecules in organogels, where fibers of low-molecular weight organogelators are held together by non-covalent forces, such as π–π stacking, hydrogen bonding and solvophobic interactions [255–257]. It should be pointed out that oriented systems such as gels, films, membranes and liquid crystals are affected by macroscopic anisotropy [258] and the CD spectra often include components arising from linear dichroism (LD) [255,259]. It is therefore of importance to include in the analysis both the CD and the LD spectra, measured under identical conditions. If the LD spectrum is silent in the wavelength range of interest, the analysis of the CD spectrum can be carried out safely. For example CD spectrum of azobenzene-substituted cholesterol (**44**) in solution is very weak whereas a strong exciton-type Cotton effect due to the coupling of π–π^* transitions of the azobenzene chromophores is seen in a gel; the LD spectra are in both cases very weak. Therefore the sign of the observed Cotton effect is associated with the helical sense of the gel fibers (Fig. 13.36) [255].

$OC_{12}H_{25}$ $OC_{12}H_{25}$ $OC_{12}H_{25}$ $OC_{12}H_{25}$

O O O O

43

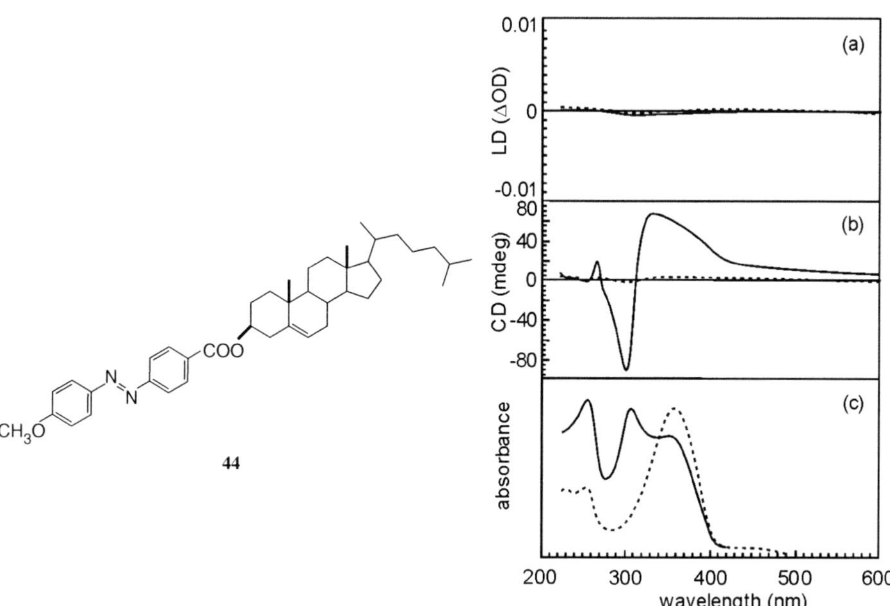

Fig. 13.36. (a) LD spectra, (b) CD spectra and (c) absorption spectra of (**44**) in 1-butanol solution at 60°C (—) and in the gel state at 25°C (—). Adapted from Ref. [255].

In discotic columnar liquid crystals the chirality (and associated CD spectra) of individual molecules may be reinforced not only by the formation of columns (stacking of the molecules) but also by an additional effect of helical ordering of the columns [260].

13.4.2 Host–guest association

Host–guest association between chiral cyclophane [261] or crown host [262] and chiral guest molecule generally results in a change of the CD spectrum due to exciton type interaction between edtms of the host and the guest. This can be employed to study chiral recognition (chiral discrimination) on the molecular level. Association between a chiral host and an achiral guest (**B**, Table 13.2) is best illustrated with the much studied in the past thirty years complexation of cyclodextrins [263]. Cyclodextrins are transparent in the absorption spectra down to 210 nm, therefore their complexes with achiral, chromophoric guests can be conveniently analyzed with the aid of CD spectroscopy. Induced extrinsic CD spectra (usually weak) reveal the mode of association of the guest within the cavity of cyclodextrin. According to Harata's rule induced Cotton effect of a chromophore located inside the cyclodextrin cavity is positive if its edtm is parallel to the principal axis of the cyclodextrin; if edtm is in the plane perpendicular to the principal axis the induced Cotton effect is negative [264]. This is illustrated in Fig. 13.37 with the induced Cotton effects due to ^1L_b and ^1L_a transitions of phenol in β-cyclodextrin [265].

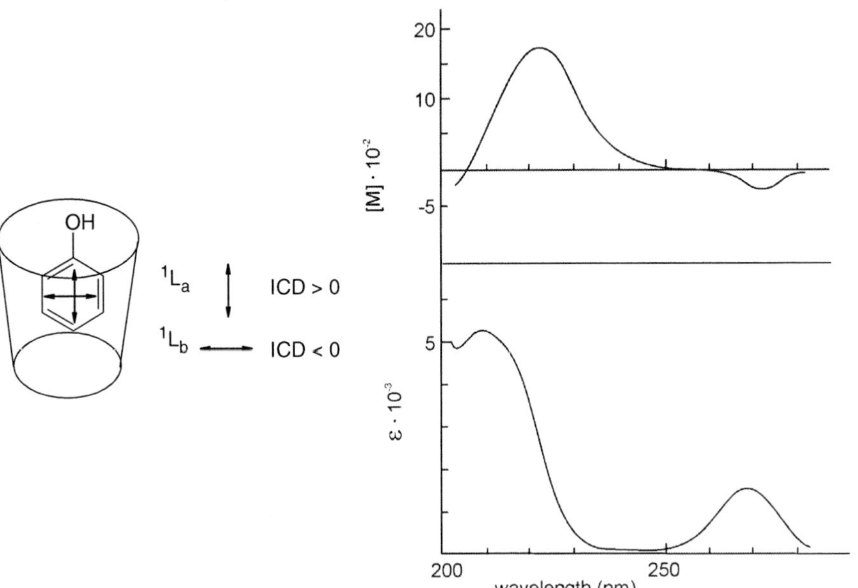

Fig. 13.37. ICD (upper) and UV (lower) spectra of the inclusion complex of phenol (1.49×10^{-4} M) in β-cyclodextrin (1.26×10^{-2}M) at pH = 6.2. Adapted from Ref. [265].

Kodaka has shown that induced Cotton effects due to chromophoric guest located outside the cavity of a chiral macrocycle have reversed signs [266]. For example, the induced Cotton effects due to β-naphthol transitions are of opposite sign in its complexes with α- and β-cyclodextrin, since in the former complex the β-naphthol molecule is not completely included in the small α-cyclodextrin moiety [267]. In either case the ICD of a macrocycle—chromophoric guest can be explained by a dipole–dipole interaction between the edtm of a guest molecule and the induced dipoles of the bonds comprising the chiral host macrocycle.

Much stronger exciton-type Cotton effects are obtained in the case of bichromophoric guest molecules, existing as racemic *P* and *M* conformers. Complexation of molecules such as diarylmethanes [268,269] or oligosilanes [270] with cyclodextrins leads to conformer discrimination of the guest and hence to a strong ICD. A further case of this type is the inclusion of two guest molecules in the large cavity of γ-cyclodextrin. A twisted sandwich structure of pentamethine cyanine dye dimer in the cavity of γ-cyclodextrin was proposed on the basis of the observed strong exciton Cotton effect [271].

By and large, ICD of cyclodextrin complexes is useful in the analysis of their structures. It has also been used to study complexation properties of cyclodextrins bearing additional sensing substituent, such as phenolphthalein [272], methyl red [273], azobenzene [274] or a dansyl group [275].

Calix[*n*]arenes (**45**) ($n=4$, 6 and 8) represent case **C** (Table 13.2). In contrast to cyclodextrins they are achiral and have strong π–π^* transitions due to substituted benzene chromophores. They can be made chiral by suitable substitution and when chiral, they show clear exciton coupling bands in the CD spectra [276]. ICD of achiral water-soluble calix[n]arene (**45**) ($R = SO_3H$) is observed in inclusion complexes with chiral ammonium guests. The sign of the exciton Cotton effect due to the coupling of ^1L_a transitions around 210 nm is related to the displacement of the benzene rings from time-averaged symmetrical arrangement [277].

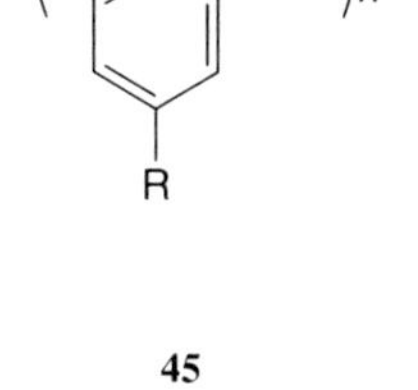

$R = (CH_2)_{10}CH_3$

45 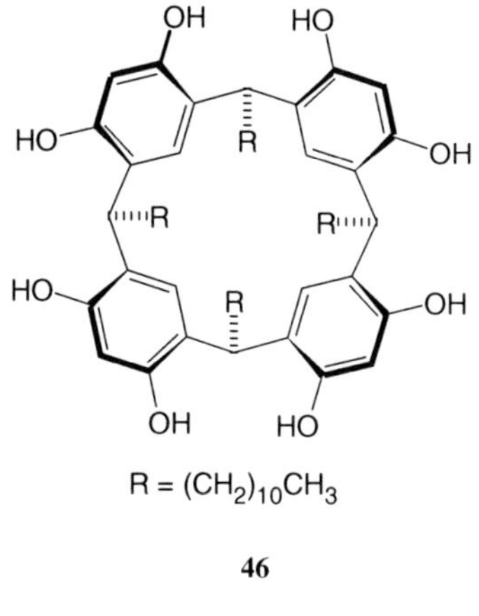 **46**

By contrast, hydrophobic calix[4]resorcarene (**46**) forms readily host–guest complexes with chiral alcohols, diols and polyols in non-polar solvents. Exciton type

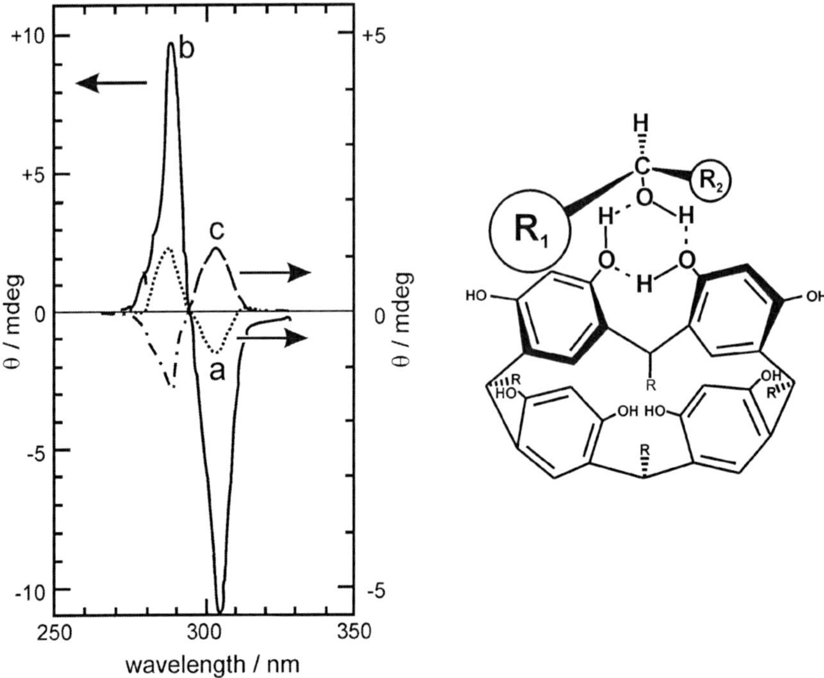

Fig. 13.38. ICD spectra of $CHCl_3$ solutions of resorcarene (**46**) (1.0 mM) with (−)-menthol (a), borneol (b) or cholesterol (c) (100 mM) and a model of conformational change upon complexation. Adapted from Ref. [278].

Cotton effects around 295 nm are induced upon complexation and they originate by chirality transfer from the guest to the host caused by hydrogen bonding and CH–π interactions, Fig. 13.38 [278].

These Cotton effects can be used for the analysis of absolute stereostructure of the guests in solution [279] and in the solid state [280].

Porphyrins (**47**) are a class of achiral host chromophores which have been used widely for the generation of ICD. Their distinct properties can be summarized as follows:

(1) they have an intense absorption band in the visible, $\lambda_{max} > 400$ nm (Soret band) which makes identification of their induced Cotton effects straightforward.
(2) high intensity of the Soret band ($\varepsilon = 300{,}000$–$400{,}000$) assures high-intensity of the exciton Cotton effects due to porphyrin/porphyrin-coupling.
(3) porphyrins have well-recognized stacking abilities—a property useful for the development of ICD.
(4) various metal complexes of porphyrins are readily available and can be used in the ICD studies.

Due to the highly symmetrical structure (D_{4h}) of porphyrin, the electronic transition corresponding to the Soret band is degenerate; theoretical interpretation of the

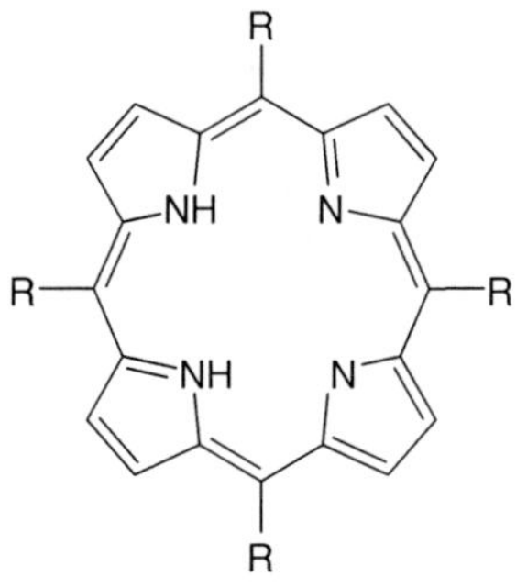

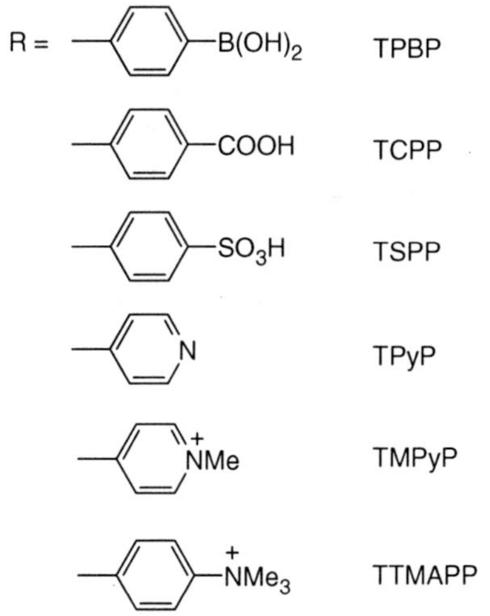

R = —C6H4—B(OH)2 TPBP
—C6H4—COOH TCPP
—C6H4—SO3H TSPP
—C5H4N TPyP
—C5H4N+Me TMPyP
—C6H4—N+Me3 TTMAPP

47

electronic spectra is a matter of current debate [281–283]. Nevertheless, any chiral arrangement of two or more porphyrin molecules will produce significant Cotton effects, even if the two molecules are located at a large distance. For analytical purposes various *meso*-tetrakis substituted porphyrins were introduced (see R in (**47**)). Zinc complexes of TCPP were used to study the interaction of metalloporphyrins with the monoclonal antibody [284] and tetraphenylboronic acid appended porphyrin (TPBP) was applied to study configurations of sugars [285]. "Double decker" cerium (IV) bis[tetrakis (4-pyridyl)-porphyrinate (TPyP) complex gives induced exciton-coupled Cotton effects in the presence of optically active dicarboxylic acids [286].

Dimeric porphyrin hosts ("tweezers") in which two porphyrin rings are connected by a short chain have been developed by the groups of Nakanishi–Berova [287,288] and Inoue–Borovkov [289,290] as sensitive tools for the analysis of guests stereostructure. Bis(zinc porphyrin) tweezer (**48**) is used for the determination of absolute configuration of amines, amino acids, amino alcohols and α-chiral carboxylic acids [287,288], as shown in Fig. 13.39.

Bis(zinc or magnesium porphyrin) (**49**) host in which porphyrins are connected by a short ethane bridge, are sensitive probes for absolute configuration of the guest molecules, such as monoamines (M = Zn) [289] and monoalcohols (M = Mg) [290].

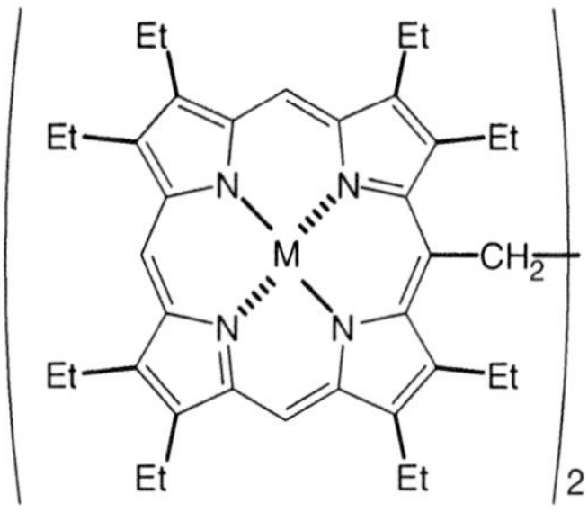

49

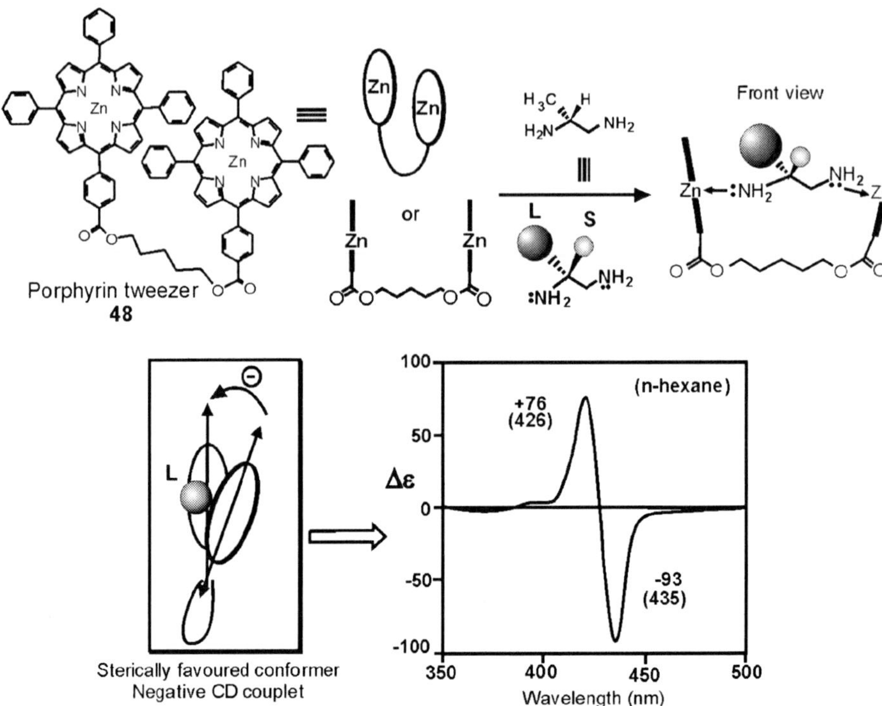

Fig. 13.39. Formation of the adduct between porphyrin tweezer (**48**) and (*R*)-1,2-diaminopropane and induction of exciton Cotton effect due to sterically favored conformer. Adapted from Ref. [287].

A further development of chiral sensing ability using zinc porphyrin dimers was due to the introduction of a biphenyl linker which enables strong induced Cotton effects with chiral diamine guests [291]. In all these metalloporphyrins guest binding to host metal center is essential for the generation of ICD.

Bilirubin (**50**), the yellow pigment of jaundice, is a bichromophoric tetrapyrrole, adopting either of the two enantiomeric folded conformations. It is therefore racemic (case **C**, Table 13.2). The folded conformers are stabilized by a network of intramolecular hydrogen bonds, involving the carboxylic groups and the two dipyrrinone chromophores (Fig. 13.40).

Whereas bilirubin is CD inactive, suitable chiral guest molecules may displace the equilibrium from the racemate toward either of the helical (*M*) or (*P*) forms (deracemization), generating exciton type Cotton effect within bilirubin long-wavelength (ca. 420 nm) transition. Lightner and co-workers studied extensively ICD of bilirubins and related pigments [293], under the influence of chiral amines [294], sulfoxides [295] and cyclodextrins [296]. Bilirubin also binds enantioselectively to bovine β-lactoglobulin [297]. It has been established that negative exciton Cotton effect is associated with the conformer of *M* helicity, estimated amplitude of the Cotton effect reaching $|A| = 500$ [298]. ICD of bilirubin provides an evidence that its molecules in solution are not planar; should the planar, extended, achiral structure be dominant, the ICD spectra would be weak.

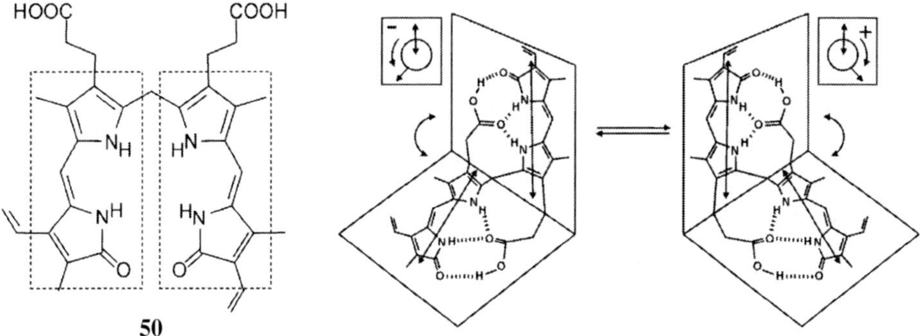

Fig. 13.40. Bilirubin-IXα (**50**) and its conformational structures of left (*M*) and right (*P*) helicity. Dashed lines are hydrogen bonds, and double-headed arrows represent the dipyrrinone long-wavelength edtms. Adapted from Ref. [292].

It should be noted that ICD spectra of diastereoisomeric bilirubin complexes may be quite different and therefore it may not be correct to determine the ratio of diastereoisomers solely from the ICD spectra; this is particularly true, if the $\Delta\varepsilon$ values of induced Cotton effects are low [299].

Achiral functional polymers which form enantiomeric helices and can bind chiral guest molecules represent another example of case **C** (Table 13.2). ICD is therefore a method to analyze absolute stereostructure of such polymers. Polyaniline (PANI, (**51**)), a conducting polymer, can be obtained in preferentially homohelical, charged form either by electropolymerization in the presence of a chiral acid, such as camphorsulfonic acid [300,301], carboxylic or phosphoric acids [302] or postsynthetically, by protonation of polyaniline emeraldine base with aminoacids [303]. Interestingly, a copolymer of aniline and *o*-toluidine (PANMA) has opposite helicity to that of polyaniline, obtained under identical conditions (DDQ oxidation). This is demonstrated by the ICD spectra, shown in Fig. 13.41 [304].

A water-soluble polyaniline, poly(2-methoxyaniline-5-sulfonic acid) (**52**), shows ICD spectrum when obtained by the electropolymerization in the presence of optically active 1-phenylethylamine [305].

HO_3S — NH — n — OMe

51 **52**

ICD spectra due to preferred *P* or *M* helical form of poly[(4-carboxyphenyl) acetylene] (**53**) are obtained in the presence of excess chiral amine. Bisignate Cotton effects are observed in the wavelength range 400–450 nm and the sign sequence of these Cotton effects could be correlated with the structure and absolute configuration of the

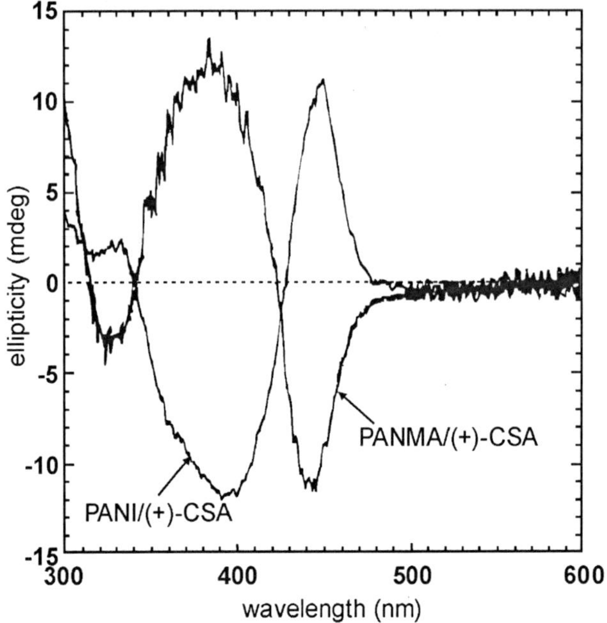

Fig. 13.41. ICD spectra of PANI/(+)-CSA and PANMA/(+)-CSA (aniline to *o*-toluidine ratio 1:1) in *m*-cresol. Adapted from Ref. [304].

guest amine [306]. A similar observation of ICD was reported for other functional poly(phenyl acetylene) hosts, bearing an amino group (**54**), guest molecules chiral acids) [307], boronate group (**55**), guest molecules diols, amino alcohols, hydroxy acids) [308] or a pendent crown ether (**56**), guest molecules amino acids) [309]. An example of such ICD spectrum is shown in Fig. 13.42 [309].

Amino acid salt added to (**56**) forms a hydrogen-bonded complex with the crown ether moiety, transferring its chirality to the helical sense of the polymer chain. In a broader sense, the CD spectra of all these functional poly(phenylacetylene) hosts can be used for determining the chirality of the guest molecules.

13.5 MISCELLANEOUS ANALYTICAL APPLICATIONS

Electronic CD spectroscopy offers now more methods of chiral structural analysis of interest for chemists. CD spectroscopy finds increasingly analytical applications to the forensic, chemical, pharmaceutical and food science [310,311]. For example, CD spectroscopy is used for detection of amphetamine enantiomers [312] or to determine drug levels in human serum [313]. Exciton coupling method was applied to analyze the structure of mitochondrial cytochrome b [314]. In chemical analysis CD spectroscopy has been used to follow by titration the formation of a [2]pseudorotaxane [315] or the binding of dinitrobenzoyl leucine to cinchona alkaloids [316]. CD spectra revealed the

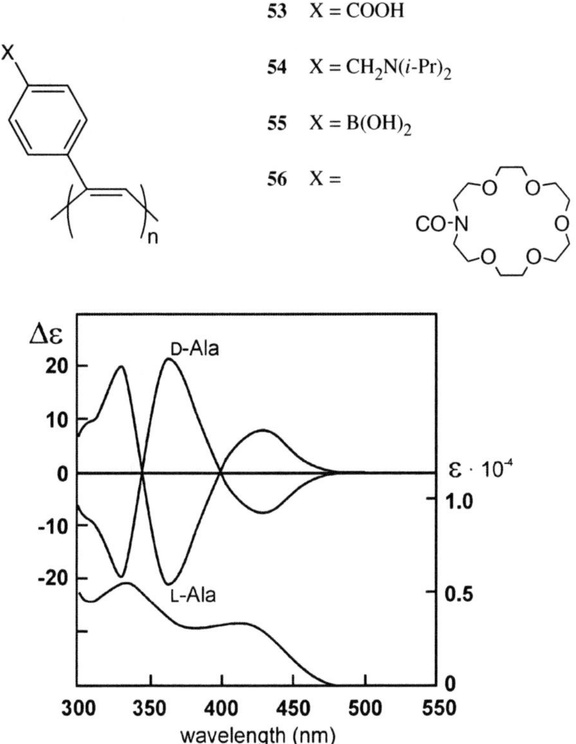

Fig. 13.42. ICD and UV spectra of (**56**) in the presence of alanine perchlorate in acetonitrile solution at 25°C.

product of [2+2] cycloaddition formed by sample irradiation during the CD/UV measurement [317].

In the following, we will briefly review analytical applications of CD spectroscopy which are related to the progress of instrument design.

On-line circular dichroism detection in HPLC combines high sensitivity of the CD technique with low sample load in analytical LC [318–321]. HPLC-CD/UV combined detection is now commercially available [322] and it allows to analyze mixtures of chiral compounds in two modes: qualitatively and quantitatively. With the use of achiral HPLC non-racemic compounds can be characterized by the CD/UV spectra. Moreover, their enantiomeric purity can be determined by comparing the CD/UV data of the sample with these of a reference sample (see below). HPLC on a chiral phase coupled to CD/UV detection allows to separate racemic mixtures of stereoisomers and their structural characterization. This technique is particularly powerful when combined with HPLC–NMR and HPLC–MS/MS techniques for structural analysis of extracts of bioactive compounds, without the need of isolation and purification of the sample [323]. A chiral derivatizing agent, TMBC (2-*tert*-butyl-2-methyl-1,3-benzodioxole-4-carboxylic acid), which is a chiral derivative of benzoic acid, has been introduced for HPLC-CD structural analysis of non-chromophoric diols, amino alcohols and related compounds with the use of exciton chirality method [324].

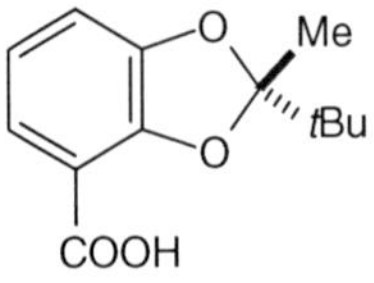

(*S*)-TMBC

On the technical side, CD detection in LC may have potential drawbacks. Overlapping peaks of enantiomers due to insufficient resolution on a chiral column occur, but they can be deconvoluted [325]. The effect of stop-flow HPLC method on the accuracy of the CD spectra obtained by on-line recording has been discussed [326].

There are many reports on the application of HPLC-CD to the analysis of drugs, with the use of either achiral [327,328] or chiral LC technique [329–331]. The latter technique is also of use in the analysis of drugs in plasma [332] and in food analysis [333].

Rapid and convenient determination of enantiomeric excess is in demand and in the process of development [334]. Simultaneous measurement of the CD and UV spectra with modern dichrographs allows to determine enantiomeric excess of a sample [335] in a way similar to the determination of optical purity by measurement of specific rotation. We propose to call the measured value *chiroptical purity* (c.p.)

$$\text{c.p.} = \frac{\Delta\varepsilon_{\text{obs}}}{\Delta\varepsilon_{\text{max}}},$$

where $\Delta\varepsilon_{\text{obs}}$ is the measured Cotton effect on the sample and $\Delta\varepsilon_{\text{max}}$ is the reference magnitude of the Cotton effect of (presumably) enantiomerically pure sample. Like in the conventional optical purity measurements, equivalence of chiroptical purity to enantiomeric excess is subject to certain restrictions, for example linear dependency of CD on sample concentration. However, much lower sample concentration needed for CD measurements makes the occurrence of aggregate artifacts in CD determinations much less likely to occur, compared to polarimetric measurements. Moreover, for the determination of c.p. knowledge of sample concentration is not required as it can be normalized against sample absorption (ε) by the anisotropy factor g

$$g = \frac{\Delta\varepsilon}{\varepsilon} = \frac{\Delta A}{A}$$

Since ΔA and A are obtained directly from CD/UV measurement, chiroptical purity can be defined as

$$\text{c.p.} = \frac{g_{\text{obs}}}{g_{\text{max}}}$$

This makes c.p. determination very convenient when coupled to achiral resolution of stereoisomers by HPLC [336–338], since no sample separation is needed.

HPLC–CD/UV technique is used for high-throughput screening of combinatorial libraries of enantioselective catalysts, by determining e.e. of the product within a very short time [339,340].

By some estimates, HPLC-CD detection offers two orders of magnitude higher sensitivity than a conventional optical rotation detection [341]. This is because numerous chiral compounds (e.g. certain 1,1′-binaphthyls) are characterized by low $[\alpha]_D$ values whereas they show strong (exciton-type) Cotton effects. The limit of detection in the analysis of drug substance in clinical plasma extracts using achiral HPLC–CD/UV system is at the nanogram range [342]. A method of amplification of sensitivity of c.p. measurement by derivatizing weakly CD active/UV inactive compound (e.g. chiral aliphatic alcohols) with racemic chromophoric derivatizing agent, based on 1,1′-binaphthyl skeleton, has been proposed [343]. In cases where the *g* value is very small (10^{-4}), reliable results can be obtained if proper procedure of the measurement (base line stability) is maintained [344].

An assay has shown that for the quality control of chiral drugs using achiral HPLC–CD/UV detection high accuracy can be obtained for samples showing *g* values in the range 10^{-1}–10^{-4}, also when the c.p. values were in the extremely demanding ranges, $\leq 1\%$ or $\geq 99\%$ [345].

It is expected that analytical applications of CD will gain broad perspectives with techniques greatly expanding the scope and sensitivity of the measurements. Impressive gains in sensitivity have been obtained by fluorescence detection of CD (FDCD) [346,347]. These measurements require fluorescent chromophores and have been applied to ECCD [348] and to the determination of enantiomeric excess [349].

CD measurements in the vacuum UV region are greatly enhanced with the use of synchrotron radiation as the light source (SRCD) which provides higher intensity radiation at low wavelengths (160 nm). Such measurements found applications in secondary structure analysis of proteins [350].

Finally, time-resolved CD spectroscopy (TRCD) offers possibilities to measure, with nano- and pico-second time resolution, the conformational changes of proteins upon light excitation [351–353].

13.6 ACKNOWLEDGMENT

This work was supported by the Ministry of Education and Science, grant no. PBZ-KBN-126/T09/10.

REFERENCES

1 P. Salvadori and F. Ciardelli (Eds.), Fundamental Aspects and Recent Developments in Optical Rotatory Dispersion and Circular Dichroism, Heyden & Sons, New York, 1973.
2 S.F. Mason (Ed.), Optical Activity and Chiral Discrimination, D. Reidel, Dordrecht, 1978.
3 E. Charney, The Molecular Basis of Optical Activity: Optical Rotatory Dispersion and Circular Dichroism, J. Wiley & Sons, New York, 1979.

4 N. Harada and K. Nakanishi, Circular Dichroism Spectroscopy—Exciton Coupling in Organic Stereochemistry, University Science Books, Mill Valley, 1983.
5 G.D. Fasman (Ed.), Circular Dichroism and the Conformational Analysis of Biomolecules, Plenum, New York, 1996.
6 A. Rodger and B. Nordén, Circular Dichroism and Linear Dichroism, Oxford University Press, Oxford, 1997.
7 N. Berova, K. Nakanishi and R.W. Woody (Eds.), Circular Dichroism—Principles and Applications, 2nd Ed. J. Wiley-VCH, New York, 2000.
8 D.A. Lightner and J.E. Gurst, Organic Conformational Analysis and Stereochemistry from Circular Dichroism Spectroscopy, J. Wiley-VCH, New York, 2000.
9 M. Le Grand and M.J. Rougier, Application of optical activity to stereochemical determinations, in: H.B. Kagan and G. Thieme (Eds.), Stereochemistry, Fundamentals and Methods, Vol. 2, Determination of Configurations by Dipole Moments, CD or ORD, G. Thieme, Stuttgart, 1977.
10 E.R. Morris, Chiroptical methods, in: S.B. Ross-Murphy (Ed.), Physical Technical Study of Food Biopolymers, Blackie, Glasgow, 1994.
11 N. Purdie and H.G. Brittain (Eds.), Analytical Applications of Circular Dichroism, Technical Instrumentation in Analytical Chemistry, Vol. 14, Elsevier, Amsterdam, 1994.
12 J. Gawronski, Determination of absolute and relative configuration by chiroptical methods, Vol. E21a, stereoselective synthesis, in: G. Helmchen, R.W. Hoffmann, J. Mulzer and E. Schaumann (Eds.), Houben - Weyl Methods of Organic Chemistry, 4th Ed. G. Thieme, Stuttgart, 1995.
13 J.B. Lambert, H.F. Shurvell, D.A. Lightner and R.G. Cooks, Organic Structural Spectroscopy, Part III: Electronic Absorption and Chiroptical Spectroscopy, Prentice Hall, Upper Saddle River, 1998.
14 P. Salvadori, L. Di Bari and C. Rosini, Electronic circular dichroism—fundamentals, methods and applications, in: W.J. Longh and I.W. Wainer (Eds.), Chirality in Natural and Applied Science, Blackwell, Oxford, 2002.
15 P. Salvadori, C. Bertucci and C. Rosini, Circular dichroism spectroscopy in the analysis of chiral drugs, in: H.Y. Aboul-Enein, I.W. Wainer (Eds.), Chemical Analysis, Vol. 142, Impact of Stereochemistry on Drug Developments and Use, J. Wiley & Sons, 1997.
16 C. Djerassi, Optical Rotatory Dispersion and its Applications to Organic Chemistry, McGraw Hill, New York, 1960.
17 W. Moffitt, W.B. Woodward, A. Moscowitz, W. Klyne and C. Djerassi, J. Am. Chem. Soc., 83 (1961) 4013.
18 A. Rodger and M. G. Moloney, J. Chem. Soc. Perkin Trans., 2 (1991) 919.
19 D.A. Lightner, Chapter 10 in Ref. [7].
20 M.J. Frisch, G.W. Trucks, H.B. Schlegel, G.E. Scuseria, M.A. Robb, J.R. Cheeseman, V.G. Zakrzewski, J.A. Montgomery Jr., R.E. Stratmann, J.C. Burant, S. Dapprich, J.M. Millam, A.D. Daniels, K.N. Kudin, M.C. Strain, O. Farkas, J. Tomasi, V. Barone, M. Cossi, R. Cammi, B. Mennucci, C. Pomelli, C. Adamo, S. Clifford, J. Ochterski, G.A. Petersson, P.Y. Ayala, Q. Cui, K. Morokuma, D.K. Malick, A.D. Rabuck, K. Raghavachari, J.B. Foresman, J. Cioslowski, J.V. Ortiz, B.B. Stefanov, G. Liu, A. Liashenko, P. Piskorz, I. Komaromi, R. Gomperts, R.L. Martin, D.J. Fox, T. Keith, M.A. Al-Laham, C.Y. Peng, A. Nanayakkara, C. Gonzalez, M. Challacombe, P.M.W. Gill, B.G. Johnson, W. Chen, M.W. Wong, J.L. Andres, M. Head-Gordon, E.S. Replogle and J.A. Pople, Gaussian98 and Gaussian03; Gaussian Inc., Pittsburgh, PA, USA.
21 T. Helgaker, H.J. Aa. Jensen, P. Joergensen, J. Olsen, K. Ruud, H. Aagren, A.A. Auer, K.L. Bak, V. Bakken, O. Christiansen, S. Coriani, P. Dahle, E.K. Dalskov, T. Enevoldsen, B. Fernandez, C. Haettig, K. Hald, A. Halkier, H. Heiberg, H. Hettema, D. Jonsson, S. Kirpekar, R. Kobayashi, H. Koch, K.V. Mikkelsen, P. Norman, M.J. Packer, T.B. Pedersen, T.A. Ruden, A. Sanchez, T. Saue, S.P.A. Sauer, B. Schimmelpfennig, K.O. Sylvester-Hvid, P.R. Taylor and O. Vahtras, DALTON, Release 1.2; 2001.
22 R. Ahlrichs, M. Bar, H.-P. Baron, R. Bauernschmitt, S. Bocker, M. Ehrig, K. Eichkorn, S. Elliott, F. Furche, F. Haase, M. Haser, H. Horn, C. Hattig, C. Huber, U. Huniar, M. Kattannek, A. Kohn, C. Kolmes, M. Kollwitz, K. May, C. Ochsenfeld, H. Öhm, A. Schafer, U. Schneider, O. Treutler,

M.V. Arnim, F. Weigend, P. Weis and H. Weiss, TURBOMOLE, Version 5.6; Universität Karlsruhe, 2002.
23 G. Bringmann, S. Busemann, K. Krohn and K. Beckmann, Tetrahedron, 53 (1997) 1655.
24 W. Kuhn, Trans. Faraday Soc., 26 (1930) 293.
25 J. G. Kirkwood, J. Chem. Soc., 5 (1937) 479.
26 S.F. Mason, General Models for Optical Activity, Chapter 1, Ref. [2].
27 S.F. Mason, Quart. Rev., 17 (1962) 20; S.F. Mason, Proc. Chem. Soc., (1962) 362; S.F. Mason and G.W. Vane, J. Chem. Soc. B, (1966) 370.
28 N. Harada and K. Nakanishi, J. Am. Chem. Soc., 91 (1969) 3989; N. Harada and K. Nakanishi, Acc. Chem. Res., 5 (1972) 257.
29 N. Berova and K. Nakanishi, Exciton Chirality Method: Principles and Applications, Chapter 12, Ref. [7].
30 J. Gawronski, M. Brzostowska, K. Gawronska, J. Koput, U. Rychlewska, P. Skowronek and B. Nordén, Chem. Eur. J., 8 (2002) 2484.
31 S.G. Telfer, N. Tajima and R. Kuroda, J. Am. Chem. Soc., 126 (2004) 1408.
32 M. Chang, H.V. Meyers, K. Nakanishi, M. Ojika, J.H. Park, M.H. Park, R. Takeda, J.T. Vazquez and W.T. Wiesler, Pure Appl. Chem., 61 (1989) 1193.
33 I. Akritopoulou-Zanze, K. Nakanishi, H. Stepowska, B. Grzeszczyk, A. Zamojski and N. Berova, Chirality, 9 (1997) 699.
34 N. Zhao, J.-S. Guo, L.-C. Lo, N. Berova, K. Nakanishi, G.T. Haupert, M. Warrack and A.A. Tymiak, Chem. Commun., (1997) 43.
35 S.F. Mason, R.H. Seal and D.R. Roberts, Tetrahedron, 30 (1974) 1671.
36 L. Di Bari, G. Pescitelli and P. Salvadori, J. Am. Chem. Soc., 121 (1999) 7998.
37 M.-K. Ng, H.-F. Chow, T.-L. Chan and T.C.W. Mak, Tetrahedron Lett., 37 (1996) 2979.
38 D. Gargiulo, F. Derguini, N. Berova, K. Nakanishi and N. Harada, J. Am. Chem. Soc., 113 (1991) 7046; N. Berova, D. Gargiulo, F. Derguini, K. Nakanishi and N. Harada, J. Am. Chem. Soc., 115 (1993) 4769.
39 V. Buss, K. Kolster and B. Görs, Tetrahedron: Asymmetry, 4 (1993) 1.
40 A. Collet and G. Gottarelli, J. Am. Chem. Soc., 104 (1982) 7383.
41 H. DeVoe, J. Chem. Phys., 41 (1964) 393; H. DeVoe, J. Chem. Phys., 43 (1965) 3199.
42 S. Superchi, E. Giorgio and C. Rosini, Chirality, 16 (2004) 422.
43 C. Rosini, M. Zandomeneghi and P. Salvadori, Tetrahedron: Asymmetry, 4 (1993) 545.
44 J. Gawronski, P. Grycz, M. Kwit and U. Rychlewska, Chem. Eur. J., 8 (2002) 4210.
45 S.G. Allenmark, Nat. Prod. Rep., 17 (2000) 145.
46 D.A. Lightner, Tech. Instrum. Anal. Chem., 14 (1994) 131.
47 J. Gawronski, K. Gawronska and U. Rychlewska, Tetrahedron Lett., 30 (1989) 6071.
48 J. Gawronski, K. Gawronska, P. Skowronek, U. Rychlewska, B. Warżajtis, J. Rychlewski, M. Hoffman and A. Szarecka, Tetrahedron, 53 (1997) 6113.
49 C. Rosini, R. Ruzziconi, S. Superchi, F. Fringuelli and O. Piermonti, Tetrahedron: Asymmetry, 9 (1995) 55.
50 S. Grimme, W. Mennicke, F. Vögtle and M. Nieger, J. Chem. Soc., Perkin 2 (1999) 521.
51 C. Rosini, S. Scamuzzi, G, Ucello-Barretta and P. Salvadori, J. Org. Chem., 59 (1994) 7395.
52 M. Meyring, J. Mühlbacher, K. Messer, N. Kastner-Pustet, G. Bringmann, A. Mannschreck and G. Blaschke, Anal. Chem., 74 (2002) 3726.
53 J. Gawronski, F. Kazmierczak, K. Gawronska, U. Rychlewska, B. Nordén and A. Holmén, J. Am. Chem. Soc., 120 (1998) 12083.
54 J. Gawronski and P. Skowronek, Curr. Org. Chem., 8 (2004) 65.
55 A. Kawamura, N. Berova, K. Nakanishi, B. Voigt and G. Adam, Tetrahedron, 53 (1997) 11961.
56 M. Hartl and H.-U. Humpf, J. Org. Chem., 66 (2001) 3678.
57 B. Weckerle, P. Schreier and H.-U. Humpf, J. Org. Chem., 66 (2001) 8160.
58 H. Uzawa, Y. Nishida, H. Ohrui and H. Meguro, J. Org. Chem., 55 (1990) 116.
59 S. Shapiro, Enantiomer, 1 (1996) 151.

60 S. Matile, N. Berova, K. Nakanishi, S. Novkova, I. Philipova and B. Blagoev, J. Am. Chem. Soc., 117 (1995) 7021.
61 S. Matile, N. Berova, K. Nakanishi, J. Fleischhauer and R.W. Woody, J. Am. Chem. Soc., 118 (1996) 5198.
62 W.T. Wiesler and K. Nakanishi, J. Am. Chem. Soc., 111 (1989) 9205; W.T. Wiesler and K. Nakanishi, J. Am. Chem. Soc., 112 (1990) 5574.
63 N. Zhao, N. Berova, K. Nakanishi, M. Rohmer, P. Mougenot and U.J. Jürgens, Tetrahedron, 52 (1996) 2777.
64 H. Liu and K. Nakanishi, J. Am. Chem. Soc., 104 (1982) 1178.
65 Y. Mori, Y. Kohchi, M. Suzuki and H. Furukawa, Tetrahedron Lett., 33 (1992) 2029; Y. Mori and H. Furukawa, Tetrahedron, 51 (1995) 6725.
66 T.F. Molinski, T.N. Makarieva and V.A. Stonik, Angew. Chem. Int. Ed., 39 (2000) 4076; G.M. Nicholas and T.F. Molinski, J. Am. Chem. Soc., 122 (2000) 4011.
67 F. Kazmierczak, K. Gawronska, U. Rychlewska and J. Gawronski, Tetrahedron: Asymmetry, 5 (1994) 527.
68 V. Dirsch, J. Frederico, N. Zhao, G. Cai, Y. Chen, S. Vunnam, J. Odingo, H. Pu, K. Nakanishi, N. Berova, D. Liotta, A. Bielawska and Y. Hannun, Tetrahedron Lett., 28 (1995) 4959.
69 J. Gawronski, K. Gawronska, P. Skowronek and A. Holmén, J. Org. Chem., 64 (1999) 234.
70 A. Iuliano, I. Voir and P. Salvadori, J. Org. Chem., 64 (1999) 5754.
71 N. Berova and K. Nakanishi, Chapter 12 in Ref. [7].
72 J.D. Chisholm, J. Golik, B. Krishnan, J.A. Matson and D.L. Van Vranken, J. Am. Chem. Soc., 121 (1999) 3801.
73 J. Gawronski, H. Kołbon and M. Kwit, Enantiomer, 7 (2002) 85.
74 N. Zhao, N. Kumar, K. Neuenschwander, K. Nakanishi and N. Berova, J. Am. Chem. Soc., 117 (1995) 7844.
75 M. Hartl and H.-U. Humpf, Tetrahedron: Asymmetry, 11 (2000) 1741.
76 K. Hör, O. Gimple, P. Schreier and H.-U. Humpf, J. Org. Chem., 63 (1998) 322.
77 K. Tanaka, K. Nakanishi and N. Berova, J. Am. Chem. Soc., 125 (2003) 10802.
78 S. Hosoi, M. Kamiya and T. Ohta, Org. Lett., 3 (2001) 3659.
79 J. Gawronski, M. Kwit and K. Gawronska, Org. Lett., 4 (2002) 4185.
80 S. Zahn and J.W. Canary, Org. Lett., 1 (1999) 861; A.E. Holmes, S. Zahn and J.W. Canary, Chirality, 14 (2002) 471; J. Zhang, A.E. Holmes, A. Sharma, N.R. Brooks, R.S. Rarig, J. Zubieta and J. W. Canary, Chirality, 15 (2003) 180.
81 J. Sandström, Chirality, 7 (1995) 181.
82 J. Grochowski, B. Rys, P. Serda and U. Wagner, Tetrahedron: Asymmetry, 6 (1995) 2059.
83 E. Rochlin and Z. Rappoport, J. Org. Chem., 68 (2003) 216.
84 D. Casarini, C. Rosini, S. Grilli, L. Lunazzi and A. Mazzanti, J. Org. Chem., 68 (2003) 1815.
85 I. Azumaya, I. Okamoto, S. Nakayama, A. Tanatani, K. Yamaguchi, K. Shudo and H. Kagechika, Tetrahedron, 55 (1999) 11237.
86 G. Gottarelli, G. Proni, G.P. Spada, D. Fabbri, S. Giadiali and C. Rosini, J. Org. Chem., 61 (1996) 2013.
87 G. Bringmann, T. Hartung, O. Kröcher, K.-P. Gulden, J. Lange and H. Burzlaff, Tetrahedron, 50 (1994) 2831.
88 G. Bringmann, K.-P. Gulden, H. Busse, J. Fleischhauer, B. Kramer and E. Zobel, Tetrahedron, 49 (1993) 3305; G. Bringmann, K.-P. Gulden, Y.F. Hallock, K.P. Manfredi, J.H. Cardellina, M.H. Boyd, B. Kramer and J. Fleischhauer, Tetrahedron, 50 (1994) 7807.
89 N. Harada, A. Saito, N. Koumura, H. Uda, B. De Lange, W.F. Jager, H. Wynberg and B.L. Feringa, J. Am. Chem. Soc., 119 (1997) 7241.
90 N. Koumura, E.M. Geertsema, A. Meetsma and B.L. Feringa, J. Am. Chem. Soc., 122 (2000) 12005; N. Koumura, E.M. Geertsema, M.B. van Gelder, A. Meetsma and B.L. Feringa, J. Am. Chem. Soc., 124 (2002) 5037; R.A. van Delden, N. Koumura, A. Schoevaars, A. Meetsma and B.L. Feringa, Org. Biomol. Chem., 1 (2003) 33.

91 F. Furche, R. Ahlrichs, C. Wachsmann, E. Weber, A. Sobanski, F. Vögtle and S. Grimme, J. Am. Chem. Soc., 122 (2000) 1717.
92 G. Wagniere and W. Hug, Tetrahedron Lett., 11 (1970) 4765; W. Hug and G. Wagniere, Tetrahedron, 28 (1972) 1241.
93 C. Wachsmann, E. Weber, M. Czugler and W. Seichter, Eur. J. Org. Chem., (2003) 2863.
94 Y. Ogawa, M. Toyama, M. Karikomi, K. Seki, K. Haga and T. Uyehara, Tetrahedron Lett., 44 (2003) 2167.
95 I. Pischel, S. Grimme, S. Kotila, M. Nieger and F. Vögtle, Tetrahedron: Asymmetry, 7 (1996) 109.
96 T. Mizutani, S. Yagi, T. Morinaga, T. Nomura, T. Takagishi, S. Kitagawa and H. Ogoshi, J. Am. Chem. Soc., 121 (1999) 754.
97 G. Nuding, E. Zimmermann and V. Buss, Tetrahedron Lett., 42 (2001) 2649.
98 R. Altmann, C. Etzlstorfer and H. Falk, Monatsh. Chem., 128 (1997) 785.
99 E. Yashima and Y. Okamoto, Chapter 18 in Ref. [7].
100 Q. Sun and B.Z. Tang, Polymer Prepr., 40 (1999) 560.
101 T. Masuda, H. Nakako, R. Nomura and M. Tabata, Polymer Prepr., 40 (1999) 521; H. Nakako, R. Nomura, M. Tabata and T. Masuda, Macromolecules, 32 (1999) 2861.
102 R. Nomura, S. Nishiura, J. Tabei, F. Sanda and T. Masuda, Macromolecules, 36 (2003) 5076.
103 G. Gao, F. Sanda and T. Masuda, Macromolecules, 36 (2003) 3932.
104 R. Nomura, H. Nakako and T. Masuda J. Mol. Cat. A, 190 (2002) 197.
105 K.K.L. Cheuk , J.W.Y. Lam, J. Chen, L.M. Lai and B.Z. Tang, Macromolecules, 36 (2003) 5947.
106 K. Morino, K. Maeda, Y. Okamoto, E. Yashima and T. Sato, Chem. Eur. J., 8 (2002) 5112.
107 T. Aoki, M. Kokai, K. Shinohara and E. Oikawa, Chem. Lett., (1993) 2009.
108 E. Yashima, K. Maeda and O. Sato, J. Am. Chem. Soc., 123 (2001) 8159.
109 S. Lifson, C. Andreola, N.C. Peterson and M.M. Green, J. Am. Chem. Soc., 111 (1989) 8850.
110 M.M. Green, M.P. Reidy, R.J. Johnson, G. Darling, D.J.O'Leary and G. Wilson, J. Am. Chem. Soc., 111 (1989) 6452.
111 M.M. Green, N.C. Peterson, T. Sato, A. Teramoto, R. Cook and S. Lifson, Science, 268 (1995) 1860.
112 M.M. Green, Chapter 17 in Ref. [7].
113 S.K. Jha, K.-S. Cheon, M.M. Green and J.V. Selinger, J. Am. Chem. Soc., 121 (1999) 1665.
114 N. Hida, F. Takei, K. Onitsuka, K. Shiga, S. Asaoka, T. Iyoda and S. Takahashi, Angew. Chem. Int. Ed., 42 (2003) 4349.
115 M. Ishikawa, K. Maeda, Y. Mitsutsuji and E. Yashima, J. Am. Chem. Soc., 126 (2004) 732.
116 F. Takei, H. Hayashi, K. Onitsuka, N. Kobayashi and S. Takahashi, Angew. Chem. Int. Ed., 40 (2001) 4092.
117 M. Fujiki, J. Am. Chem. Soc., 116 (1994) 11976.
118 M. Fujiki, S. Toyoda, C.-H. Yuan and H. Takigawa, Chirality, 10 (1998) 667; M. Fujiki, J. Am. Chem. Soc., 116 (1994) 6017.
119 M. Fujiki, J.R. Koe, M. Motonaga, H. Nakashima, K. Terao and A. Teramoto, J. Am. Chem. Soc., 123 (2001) 6253.
120 E. Peters, R.A.J. Janssen, S.C.J. Meskers and E.W. Meijer, Polymer Prepr., 40 (1999) 519.
121 L. Brunsveld, E.W. Meijer, R.B. Prince and J.S. Moore, J. Am. Chem. Soc., 123 (2001) 7978.
122 Z.-B. Zhang, M. Motonaga, M. Fujiki and C.E. McKenna, Macromolecules, 36 (2003) 6956.
123 B.M.W. Langeveld-Voss, M.M. Bouman, M.P.T. Christiaans, R.A.J. Janssen and E.W. Meijer, Polymer Prepr., 37 (1996) 499; B.M.W. Langeveld-Voss, D. Beljonne, Z. Shuai, R.A.J. Janssen, S.C. Meskers, E.W. Meijer and J.-L. Bredas, Adv. Mater., 10 (1998) 1343.
124 H. Goto, Y. Okamoto and E. Yashima, Macromolecules, 35 (2002) 4590.
125 E. Peeters, M.P.T. Christiaans, R.A.J. Janssen, H.F.M. Schoo, H.P.J.M. Dekkers and E.W. Meijer, J. Am. Chem. Soc., 119 (1997) 9909; J.J.L.M. Cornelissen, E. Peeters, R.A.J. Janssen and E.W. Meijer, Acta Polym., 49 (1998) 471.
126 H. Goto and E. Yashima, J. Am. Chem. Soc., 124 (2002) 7943.
127 T. Takata, Y. Furusho, K. Murakawa, T. Endo, H. Matsuoka, T. Hirasa, J. Matsuo and M. Sisido, J. Am. Chem. Soc., 120 (1998) 4530.

128 O. Haba, Y. Morimoto, T. Uesaka, K. Yokota and T. Kakuchi, Macromolecules, 28 (1995) 6378; T. Kakuchi, O. Haba, E. Hamaya, T. Naka, T. Uesaka and K. Yokota, Macromolecules, 29 (1996) 3807; T. Kakuchi, O. Haba, T. Uesaka, M. Obata, Y. Morimoto and K. Yokota, Macromolecules, 29 (1996) 3812; T. Kakuchi, O. Haba, T. Uesaka, Y. Yamaguchi, M. Obata, Y. Morimoto and K. Yokota, Macromol. Chem. Phys., 197 (1996) 2931; T. Uesaka, Y. Sugiura, M. Obata, A. Narumi, K. Yokota and T. Kakuchi, Polym. J., 31 (1999) 342.
129 Z. Yu, X. Wan, H. Zhang, X. Chen and Q. Zhou, Chem. Commun., (2003) 974.
130 S. Habaue, T. Tanaka and Y. Okamoto, Macromolecules, 28 (1995) 5973; T. Tanaka, S. Habaue and Y. Okamoto, Polym. J., 27 (1995) 1202.
131 L. Angiolini, R. Bozio, L. Giorgini, D. Pedron, G. Turco and A. Dauru, Chem. Eur. J., 8 (2002) 4241.
132 S. Mayer and R. Zentel, Macromol. Chem. Phys., 199 (1998) 1675.
133 S.M. Kelly and N.C. Price, Curr. Protein Pept. Sci., 1 (2000) 349.
134 N.J. Greenfield, Anal. Biochem., 235 (1996) 1.
135 N. Sreerama and R.W. Woody, Computation and Analysis of Protein Circular Dichroism Spectra, in: L. Brand and M.L. Johnson (Eds.), Methods in Enzymology, Elsevier Inc., Vol. 383, 2004, p. 318.
136 A. Bierzyński, Acta Biochim. Pol., 48 (2001) 1091.
137 J.T. Pelton and L.R. McLean, Anal. Biochem., 277 (2000) 167.
138 N.J. Greenfield, Trends Anal. Chem., 18 (1999) 236.
139 P. Wallimann, R.J. Kennedy and D.S. Kemp, Angew. Chem. Int. Ed., 38 (1999) 1290.
140 S.Y. Venyaminov and K.S. Vassilenko, Anal. Biochem., 222 (1994) 176.
141 N. Sreerama, S.Yu. Venyaminov and R.W. Woody, Anal. Biochem., 299 (2001) 271.
142 N. Sreerama, S.Yu. Venyaminov and R.W. Woody, Protein Sci., 8 (1999) 370.
143 D. Andersson, U. Carlsson and P.-O. Freskgård, Eur. J. Biochem., 268 (2001) 1118.
144 R. Li, Y. Nagai and M. Nagai, Chirality, 12 (2000) 216.
145 O.K. Gasymov, A.R. Abduragimov, T.N. Yusifov, and B.J. Glasgow, Anal. Biochem., 318 (2003) 300.
146 R.W. Woody, Biopolymers, 17 (1978) 1451.
147 P.-O. Freskgård, L.G. Martensson, P. Jonasson, B.H. Johnsson and U. Carlsson, Biochemistry, 33 (1994) 14281.
148 S. Bhattacharjee, G. Tóth, S. Lovas and J.D. Hirst, J. Phys. Chem. B, 107 (2003) 8682.
149 J.D. Hirst, K. Colella and A.T.B. Gilbert, J. Phys. Chem. B, 107 (2003) 11813.
150 R.W. Woody, J. Chem. Phys., 49 (1968) 4797.
151 N.A. Besley and J.D. Hirst, J. Phys. Chem. A, 102 (1998) 10791.
152 N.A. Besley and J.D. Hirst, J. Mol. Struct. (Theochem), 506 (2000) 161.
153 R.W. Woody and A. Koslowski, Biophys. Chem., 101–102 (2002) 535.
154 D.M. Rogers and J.D. Hirst, Chirality, 16 (2004) 234.
155 J.G. Lees and B.A. Wallace, Spectroscopy, 16 (2002) 121.
156 B.A. Wallace, J. Synchrotron Rad., 7 (2000) 289.
157 W.C. Johnson Jr. and I. Tinoco Jr., J. Am. Chem. Soc., 94 (1972) 4389.
158 N.A. Besley, M.-J. Brienne and J.D. Hirst, J. Phys. Chem. A, 104 (2000) 12371.
159 J.D. Hirst, J. Phys. Chem., 109 (1998) 782.
160 J.D. Hirst, S. Bhattacharjee and A.V. Onufriev, Faraday Discuss., 122 (2002) 253.
161 H. Ito, Bull. Chem. Soc. Jpn., 76 (2003) 59.
162 H. Ito, Y. Arakawa and Y.J. I'Haya, J. Chem. Phys., 98 (1993) 8835.
163 C.K. Larive, S.M. Lunte, M. Zhong, M.D. Perkins, G.S. Wilson, G. Gokulrangan, T. Williams, F. Afroz, C. Schoneich, T.S. Derric, C.R. Middaugh and S. Bogdanowich-Knipp, Anal. Chem., 71 (1999) 389R.
164 N. Sreerama and R.W. Woody, Anal. Biochem., 287 (2000) 252.
165 N.H. Andersen, Z. Liu and K.S. Prickett, FEBS Lett., 399 (1996) 47.
166 N. Sreerama, S.Y. Venyaminov and R.W. Woody, Anal. Biochem., 287 (2000) 243.
167 S.W. Provencher and J. Glöckner, Biochemistry, 20 (1981) 33.

168 N. Sreerama and R.W. Woody, Anal. Biochem., 209 (1993) 32.
169 W.C. Johnson Jr., Proteins Struct. Funct. Genet., 35 (1999) 307.
170 M.A. Andrade, P. Chacón, J.J. Merolo and F.Morán, Protein Eng., 6 (1993) 383.
171 B. Dalmas, G.J. Hynter and W.H. Bannister, Biochem. Mol. Biol. Int., 34 (1994) 17.
172 N. Sreerama and R.W. Woody, J. Mol. Biol., 242 (1994) 497.
173 L. Whitmore and B.A. Wallace, Nucleic Acid Res., 32 (2004) W668.
174 P. McPhie, Anal. Biochem., 293 (2001) 109.
175 V. Raussens, J.-M. Ruysschaert and E. Goormaghtigh, Anal. Biochem., 319 (2003) 114.
176 I.B. Grishina and R.W. Woody, Faraday Discuss., 99 (1994) 245.
177 J. Reed and T.A. Reed, Anal. Biochem., 254 (1997) 36.
178 C. Krittanai and W.C. Johnson Jr., Anal. Biochem., 253 (1997) 57.
179 N. Sreerama and R.W. Woody, Protein Sci., 12 (2003) 384.
180 P. Pancoska, V. Janota and T.A. Keiderling, Anal. Biochem., 267 (1999) 72.
181 B-Y. Lu and J-Y. Chang, Biochem. J., 364 (2002) 81.
182 H. Zhang, K. Kaneko, J.T. Nguyen, T.L. Livshits, M.A. Baldwin, F.E. Cohen, T.L. James and S.B. Prusiner, J. Mol. Biol., 250 (1995) 514.
183 I.V. Baskakov, G. Legname, S.B. Prusiner and F.E. Cohen, J. Biol. Chem., 376 (2001) 19687.
184 L.O. Tjernberg, A. Tjernberg, N. Bark, Y. Shi, B.P. Ruzsicska, Z. Bu, J. Thyberg and D.J.E. Callaway, Biochem. J., 366 (2002) 343.
185 C.M. Yip and J. McLaurin, Biophys. J., 80 (2001) 1359.
186 S.A. Kozin, S. Zirah, S. Rebuffat, G. Hui Bon Hoa and P. Debey, Biochem. Biophys. Res. Commun., 285 (2001) 959.
187 G. Li, P. Zhou, Z. Shao, X. Xie, X. Chen, H. Wang, L. Chunyu and T. Yu, Eur. J. Biochem., 268 (2001) 6600.
188 Z. Shao, F. Vollrath, Y. Yang and H.C. Thøgersen, Macromolecules, 36 (2003) 1157.
189 S.M. Kelly and N.C. Price, Biochim. Biophys. Acta, 1338 (1997) 161.
190 E. Chen, M.J. Wood, A.L. Fink and D.S. Kliger, Biochemistry, 37 (1998) 5589.
191 G.R. Jones and D.T. Clarke, Faraday Discuss., 126 (2004) 223.
192 Z.-J. Qin, D.-M. Hu, L. Shimada, T. Nakagawa, M. Arai, J.-M. Zhou and H. Kihara, FEBS Lett., 507 (2001) 299.
193 K. Kuwajima, G.V. Semisotnov, A.V. Finkehtein, S. Sugai and O.B. Ptitsyn, FEBS Lett., 334 (1993) 265.
194 F. Quaglia, G. De Rosa, E. Granata, F. Ungaro, E. Fattal and M. Immacolata La Rotonda, J. Control Release, 86 (2003) 267.
195 Y. Sadhale and J.C. Shah, Int. J. Pharm., 191 (1999) 51.
196 Y.M. Kwon, M. Baudys, K. Knutson and S.W. Kim, Pharm. Res., 18 (2001) 1754.
197 X.-C. Shen, H. Liang, J.-H. Guo, C. Song, X.-W. He and Y.-Z Yuan, J. Inorg. Biochem., 95 (2003) 124.
198 N. Fatemi and B. Sarkar, Environ. Health Perspect., 110 (2002) 695.
199 T. Kowalik-Jankowska, M. Ruta, K. Wiśniewska, L. Łankiewicz, J. Inorg. Biochem., 95 (2003) 270.
200 M.A. Schumacher, M.C. Miller and R.G. Brennan, EMBO J., 23 (2004) 1.
201 W. Huang, Z. Zhang, X. Han, J. Tang, Z. Peng, S. Dong and E. Wang, Biophys. Chem., 94 (2001) 165.
202 F. Zsila, Z. Bikádi and M. Simonyi, Biochem. Pharmacol., 64 (2002) 1651.
203 F. Zsila, Z. Bikádi and M. Simonyi, Biochem. Biophys. Res. Commun., 301 (2003) 776.
204 F. Zsila, Z. Bikádi and M. Simonyi, Biochem. Pharmacol., 65 (2003) 447
205 N. Sreerama and R.W. Woody, Protein Sci., 13 (2004) 100.
206 K. Park, A. Perczel, and G.D. Fasman, Protein Sci., 1 (1992) 1032.
207 B.A. Wallace, J.G. Lees, A.J.W. Orry, A. Lobley and R.W. Janes, Protein Sci., 12 (2003) 875.
208 D. Seebach, J.V. Schreiber, P.I. Arvidsson and J. Frackenpohl, Helv. Chim. Acta, 84 (2001) 271.
209 T.A. Martinek, G.K. Tóth, E. Vass, M. Hollósi, and F. Fülöp, Angew. Chem. Int. Ed., 41 (2002) 1718.
210 K.D. McReynolds and J. Gervay-Hague, Tetrahedron: Asymmetry, 11 (2000) 337.

211 A. Glättli, X. Daura, D. Seebach and W.F. van Gunsteren, J. Am. Chem. Soc., 124 (2002) 12972.
212 X. Daura, D. Bakowies, D. Seebach, J. Fleischhauer, W.F. van Gunsteren and P. Krüger, Eur. Biophys. J., 32 (2003) 661.
213 S.K. Mishra and P.C. Mishra, Spectrochim. Acta A, 57 (2001) 2433.
214 M.P. Fülscher, L. Serrano-Andrés and B.O. Roos, J. Am. Chem. Soc., 119 (1997) 6168.
215 A. Holmén, A. Broo, B. Albinsson and B. Nordén, J. Am. Chem. Soc., 119 (1997) 12240.
216 A. Broo and A. Holmén, J. Phys. Chem. A, 101 (1997) 3589.
217 R. Vianello and Z.B. Maksiè, Coll. Czech. Chem. Comm., 68 (2003) 2322.
218 J. Kypr and M. Vorliková, Biopolymers, 67 (2002) 275.
219 K.H. Johnson, D.M. Gray and J.C. Sutherland, Nucleic Acid Res., 19 (1991) 2275.
220 J.C. Maurizot, Chapter 25 in Ref. [7].
221 I. Kejnovská, J. Kypr and M. Vorlièková, Chirality, 15 (2003) 584.
222 F. Seela, C. Wei and A. Melenewski, Nucleic Acid Res., 24 (1996) 4940.
223 V. Dapiæ, V. Abdomeroviè, R. Marrington, J. Peberdy, A. Rodger, J.O. Trent and P.J. Bates, Nucleic Acid Res., 31 (2003) 2097.
224 L.M. Wilhelmsson, A. Holmén, P. Lincoln, P.E. Nielsen and B. Nordén, J. Am. Chem. Soc., 123 (2001) 2434.
225 L.M. Wilhelmsson, P. Sandrin, A. Holmén, B. Albinsson, P. Lincoln, P.E. Nielsen and B. Nordén, J. Phys. Chem., 107 (2003) 9094.
226 G.-S. Jiao and K. Burgess, Bioorg. Med. Chem. Lett., 13 (2003) 2785.
227 L. Venkatarangan, A. Sivaprasad, F. Johnson and K. Ashis, Nucleic Acid Res., 29 (2001) 1458.
228 R. Casale and L.W. McLaughlin, J. Am. Chem. Soc., 112 (1990) 5264.
229 I. Detmer, D. Summerer and A. Marx, Chem. Comm., (2002) 2314.
230 K.M. Ellemann Nielsen, M. Petersen, A.E. Håkansson, J. Wengel and J.P. Jacobsen, Chem. Eur. J., 8 (2002) 3001.
231 D. Renneberg and C.J. Leumann, J. Am. Chem. Soc., 124 (2002) 5993.
232 J. Jaumot, N. Escaja, R. Gargallo, E. Pedroso and R. Tauler, Nucleic Acid Res., 30 (2002) e92.
233 R. Gargallo, M. Vives, R. Tauler and R. Eritja, Biophys. J., 81 (2001) 2886.
234 M. Vives, R. Gargallo and R. Tauler, Anal. Biochem., 291 (2001) 1.
235 I.H.M. van Stokkum, H.J.W. Spoelder, M. Bloemendal, R. Grondelle and F.C.A. van Groen, Anal. Biochem., 191 (1990) 110.
236 V. Ivanov, K. Grzeskowiak and G. Zocchi, J. Phys. Chem., 107 (2003) 12847.
237 B.L. Varnado, C.J. Voci, L.M. Mayer and J.K. Coward, Bioorg. Chem., 28 (2000) 395.
238 T.D. McGregor, W. Bousfeld, Y. Qu and N. Farrel, J. Inorg. Biochem., 91 (2002) 212.
239 P.R. Hoyne, A.M. Gacy, C.T. McMurray and L.J. Maher III, Nucleic Acid Res., 28 (2000) 770.
240 S. Ivanov, A. Yakov, J.-R. Bertrand, C. Malvy and M.B. Gottikh, Nucleic Acid Res., 31 (2003) 4256.
241 M. Durand, E. Seche and J.C. Maurizot, Eur. Biophys. J., 30 (2002) 625.
242 J. Ren, X. Qu, J.O. Trent and J.B.Chaires, Nucleic Acid Res., 30 (2002) 2307.
243 P.E. Schipper, B. Nordén and F. Tjerneld, Chem. Phys. Lett., 70 (1980) 17.
244 R.F. Pasternack, Chirality, 15 (2003) 329.
245 M. Tabata, A.K. Sarker, E. Nyarko, J. Inorg. Biochem., 94 (2003) 50.
246 S. Allenmark, Chirality, 15 (2003) 409.
247 H.G. Brittain, Tech. Instrum. Anal. Chem., 14 (1994) 307.
248 M. Simonyi, Z. Bikadi, F. Zsila and J. Deli, Chirality, 15 (2003) 680; F. Zsila, J. Deli, Z. Bikadi and M. Simonyi, Chirality, 13 (2001) 739; F. Zsila, Z. Bikadi, J. Deli and M. Simonyi, Chirality, 13 (2001) 446.
249 J. Karolin, L.B.-A. Johansson, U. Ring and H. Langhals, Spectrochim. Acta A, 52 (1996) 747.
250 R. Aiyama, H. Nagai, S. Sawada, T. Yokokura, H. Itokawa and M. Nakanishi, Chem. Pharm. Bull., 40 (1992) 2810.
251 A.R.A. Palmans, J.A.J.M. Vekemans, E.E. Havinga and E.W. Meijer, Angew. Chem. Int. Ed., 36 (1997) 2648.

252 T. Giorgi, S. Lena, P. Mariani, M.A. Cremonini, S. Masiero, S. Pieraccini, J.P. Rabe, P. Samori, G.P. Spada and G. Gottarelli, J. Am. Chem. Soc., 125 (2003) 14741.
253 C. Nuckolls, T.J. Katz, G. Katz, P.J. Collings and L. Castellanos, J. Am. Chem. Soc., 121 (1999) 79.
254 M.S. Gin, T. Yokozawa, R.B. Prince and J.S. Moore, J. Am. Chem. Soc., 121 (1999) 2643.
255 K. Murata, A. Ayoki, T. Suzuki, T. Harada, H. Kuwabata, T. Komori, F. Ohseto, K. Ueda and S. Shinkai, J. Am. Chem. Soc., 116 (1994) 6664.
256 J.J. van Gorp, J.A.J.M. Vekemans and E.W. Meijer, J. Am. Chem. Soc., 124 (2002) 14759.
257 S. Kawano, S. Tamaru, N. Fujita and S. Shinkai, Chem. Eur. J., 10 (2004) 343.
258 H.-G. Kuball and T. Höfer, Chapter 5 in Ref. [7].
259 Y. Shindo and M. Nishio, Biopolymers, 30 (1990) 25.
260 M.M. Green, H. Ringsdorf, J. Wagner and R. Wüstefeld, Angew. Chem. Int. Ed., 29 (1990) 1478.
261 J.E. Forman, R.E. Barrans and D.A. Dougherty, J. Am. Chem. Soc., 117 (1995) 9213.
262 L. Somogyi, E. Samu, P. Huszty, A. Lazar, J.G. Angyan, P.R. Surjan and M. Hollosi, Chirality, 13 (2001) 109; V. Farkas, L. Szalay, E. Vass, M. Hollosi, G. Horvath and P. Huszty, Chirality, 15 (2003) S65.
263 Y.A. Zhdanov, Yu. E. Alekseev, E. Kompantseva and E. N. Vergeichik, Russ. Chem. Rev., 61 (1992) 1025.
264 K. Harata and H. Uedaira, Bull. Chem. Soc. Jpn., 48 (1975) 375.
265 M. Kamiya, S. Mitsuhashi, M. Makino and H. Yoshioka, J. Phys. Chem., 96 (1992) 95.
266 M. Kodaka, J. Am. Chem. Soc., 115 (1993) 3702.
267 M. Kodaka, J. Phys. Chem. A, 102 (1998) 8101.
268 K. Kano, M. Tatsumi and S. Hashimoto, J. Org. Chem., 56 (1991) 6579.
269 D.A. Lightner, J.K. Gawronski and K. Gawronska, J. Am. Chem. Soc., 107 (1985) 2456.
270 T. Sanji, A. Yoshiwara, H. Sakurai and M. Tanaka, Chem. Commun., (2003) 1506.
271 V. Buss and C. Reichardt, Chem. Commun., (1992) 1636.
272 T. Kuwabara, M. Takamura, A. Matsushita, H. Ikeda, A. Nakamura, A. Ueno and F. Toda, J. Org. Chem., 63 (1998) 8729.
273 T. Kuwabara, A. Nakamura, A. Ueno and F. Toda, J. Phys. Chem., 98 (1994) 6297.
274 A. Ueno, M. Fukushima and T. Osa, J. Chem. Soc. Perkin Trans., 2 (1990) 1067.
275 M. Narita, J. Itoh, T. Kikuchi and F. Hamada, J. Incl. Phen. Macrocyclic Chem., 42 (2002) 107.
276 K. Iwamoto, H. Shimizu, K. Araki and S. Shinkai, J. Am. Chem. Soc., 115 (1993) 3997.
277 T. Morozumi and S. Shinkai, Chem. Commun., (1994) 1219.
278 K. Kobayashi, Y. Asakawa, Y. Kikuchi, H. Toi and Y. Aoyama, J. Am. Chem. Soc., 115 (1993) 2648.
279 Y. Kikuchi, K. Kobayashi and Y. Aoyama, J. Am. Chem. Soc., 114 (1992) 1351.
280 Y. Tanaka, Y. Murakami and R. Kiko, Chem. Commun., (2003) 160.
281 G. Pescitelli, S. Gabriel, Y. Wang, J. Fleischhauer, R.W. Woody and N. Berova, J. Am. Chem. Soc., 125 (2003) 7613.
282 N. Yoshida, T. Ishizuka, A. Osuka, D.H. Jeong, H.S. Cho, D. Kim, Y. Matsuzaki, A. Nogami and K. Tanaka, Chem. Eur. J., 9 (2003) 58.
283 J.M. Ribo, J.M. Bofill, J. Crusatsand and R. Rubires, Chem. Eur. J., 7 (2001) 2733.
284 A. Harada, K. Shiotsuki, H. Fukushima, H. Yamaguchi and M. Kamachi, Inorg. Chem., 34 (1995) 1070.
285 T. Imada, H. Murakami and S. Shinkai, Chem. Commun., (1994) 1557.
286 M. Takeuchi, T. Imada and S. Shinkai, Angew. Chem. Int. Ed., 37 (1998) 2096.
287 X. Huang, B.H. Rickman, B. Borhan, N. Berova and K. Nakanishi, J. Am. Chem. Soc., 120 (1998) 6185; X. Huang, B. Borhan, B.H. Rickman, K. Nakanishi and N. Berova, Chem. Eur. J., 6 (2000) 216.
288 G. Proni, G. Pescitelli, X. Huang, N.Q. Quraishi, K. Nakanishi and N. Berova, Chem. Commun., (2002) 1590.
289 V.V. Borovkov, J.M. Lintuluoto and Y. Inoue, Org. Lett., 2 (2000) 1565; V.V. Borovkov, J.M. Lintuluoto and Y. Inoue, J. Am. Chem. Soc., 123 (2001) 2979.
290 J.M. Lintuluoto, V.V. Borovkov and Y. Inoue, J. Am. Chem. Soc., 124 (2002) 13676.

291 T. Hayashi, T. Aya, M. Nonoguchi, T. Mizutani, Y. Hisaeda, S. Kitagawa and H. Ogoshi, Tetrahedron, 58 (2002) 2803.
292 S.E. Boiadjiev and D.A. Lightner, J. Org. Chem., 68 (2003) 7591.
293 S.E. Boiadjiev and D.A. Lightner, Tetrahedron, 58 (2002) 7411.
294 D.A. Lightner, J.K. Gawronski and W.M.D. Wijekoon, J. Am. Chem. Soc., 109 (1987) 6354.
295 J. Gawronski, T. Połonski and D.A. Lightner, Tetrahedron, 46 (1990) 8053.
296 D.A. Lightner, J.K. Gawronski and K. Gawronska, J. Am. Chem. Soc., 107 (1985) 2456.
297 F. Zsila, FEBS Lett., 539 (2003) 85.
298 S.E. Boiadjiev, R.V. Person, G. Pusicha, C. Knobler, E. Maverick, K.N. Trueblood and D.A. Lightner, J. Am. Chem. Soc., 114 (1992) 10123.
299 D. Krois and H. Lehner, J. Chem. Soc. Perkin Trans., 2 (1995) 489.
300 E.E. Havinga, M.M. Bouman, E.W. Meijer, A. Pomp and M.M.J. Simenon, Synth. Met., 66 (1994) 93.
301 M.R. Majidi, L.A.P. Kane-Maguire and G.G. Wallace, Polymer, 35 (1994) 3113.
302 M. Bodner and M.P. Espe, Synth. Met., 135–136 (2003) 403.
303 G.-L. Yuan and N. Kuramoto, Polymer, 44 (2003) 5501.
304 S.-J. Su, M. Takeishi and N. Kuramoto, Macromolecules, 35 (2002) 5752.
305 E.V. Strounina, L.A.P. Kane-Maguire and G.G. Wallace, Synth. Met., 106 (1999) 129.
306 E. Yashima, T. Matsushima and Y. Okamoto, J. Am. Chem. Soc., 117 (1995) 11596; E. Yashima, T. Matsushima and Y. Okamoto, J. Am. Chem. Soc., 119 (1997) 6345.
307 E. Yashima, Y. Maeda and Y. Okamoto, Chem. Lett., (1996) 955; E. Yashima, Y. Maeda, T. Matsushima and Y. Okamoto, Chirality, 9 (1997) 593.
308 E. Yashima, T. Nimura, T. Matsushima and Y. Okamoto, J. Am. Chem. Soc., 118 (1996) 9800.
309 R. Nonokawa and E. Yashima, J. Am. Chem. Soc., 125 (2003) 1278; R. Nonokawa and E. Yashima, J. Polym Sci., A41 (2003) 1004.
310 N. Purdie, Tech. Instr. Anal. Chem., 14 (1994) 241.
311 H. G. Brittain, Chapter 29 in Ref. [7].
312 H. Hegedus, A. Gergely, T. Veress and P. Horvath, Analusis, 27 (1999) 458.
313 P. Gortazar, A. Roën and J. T. Vazquez, Chirality, 10 (1998) 507.
314 G. Palmer and M. Degli Esposti, Biochemistry, 33 (1994) 176.
315 M. Akasawa, H.M. Janssen, E.W. Meijer, D. Pasini and J.F. Stoddart, Eur. J. Org. Chem., (1998) 983.
316 J. Lah, N.M. Maier, W. Lindner and G. Vesnaver, J. Phys. Chem. B, 105 (2001) 1670.
317 G. Wulff, S. Krieger, B. Kühneweg and A. Steigel, J. Am. Chem. Soc., 116 (1994) 409.
318 P. Salvadori, C. Bertucci and C. Rosini, Chirality, 3 (1991) 376.
319 A. Gergely, Tech. Instrum. Anal. Chem., 14 (1994) 279.
320 D.R. Bobbit and S.W. Linder, Trends Anal. Chem., 20 (2001) 111.
321 P. Salvadori, L. DiBari and G. Pescitelli, Chapter 28 in Ref. [7].
322 K. Kudo, K. Ajima, M. Sakamoto, M. Saito, S. Morris and E. Castglioni, Chromatography, 20 (1990) 59.
323 G. Bringmann, K. Messer, M. Wohlfarth, J. Kraus, K. Dumbuya and M. Rückert, Anal. Chem., 71 (1999) 2678; G. Bringmann, K. Messer, W. Saeb, E.-M. Peters and K. Peters, Phytochemistry, 56 (2001) 387; G. Bringmann, M. Wohlfarth, H. Rischer, J. Schlauer and R. Brun, Phytochemistry, 61 (2002) 195.
324 S. Sato, K. Shimizu, J.-H. Kim, E. Ami, K. Akasaka and H. Ohrui, Chromatography, 20 (1999) 71; K. Shimizu, J.-H. Kim, K. Akasaka and H. Ohrui, Chirality, 11 (1999) 149; H. Meguro, J.-H. Kim, C. Bal, Y. Nishida and H. Ohrui, Chirality, 13 (2001) 441.
325 R. Kiesswetter, F. Brandl, N. Kastner-Pustet and A. Mannschreck, Chirality, 15 (2003) S40.
326 G. Brandl, F. Kastner, R. Fritsch, H. Zinner and A. Mannschreck, Monatsh. Chem., 123 (1992) 1059.
327 J.C. Slijkhuis, K.D. Hartog, C. van Alpen, L. Blok-Tip, P.M.J.M. Jongen and D. de Kaste, J. Pharm. Biomed. Anal., 32 (2003) 905.

328 D. Szegvari, P. Horvath, A. Gergely, S. Nemeth and S. Görög, Anal. Bioanal. Chem., 375 (2003) 713.
329 A. Gergely, F. Zsila, P. Horvath and G. Szasz, Chirality, 11 (1999) 741.
330 S. Caccamese, G. Principato, R. Jokela, A. Tolvanen and D.D. Belle, Chirality, 13 (2001) 691.
331 L. Thunberg, S. Andersson, S. Allenmark and J. Vessman, J. Pharm. Biomed. Anal., 27 (2002) 431.
332 H. Kanazawa, A. Tsubayashi, Y. Nagata, Y. Matsushima, C. Mori, J. Kizu and M. Higaki, J. Chromatography, A948 (2002) 303.
333 S. Caccamese, L. Manna and G. Scivoli, Chirality, 15 (2003) 661.
334 M.G. Finn, Chirality, 14 (2002) 534.
335 P. Horvath, A. Gergely and B. Noszal, Talanta, 44 (1997) 1479.
336 L. Chen, Y. Zhao, F. Gao and M. Garland, Appl. Spectrosc., 57 (2003) 797.
337 G. Beke, A. Gergely, G. Szasz, A. Szentesi, J. Nyitray, O. Barabas, U. Harmath and P. Matyus, Chirality, 14 (2002) 365.
338 M.T. Miller, Z. Ge and B. Mao, Chirality, 14 (2002) 659.
339 M.T. Reetz, K.M. Kühling, H. Hinrichs and A. Deege, Chirality, 12 (2000) 479.
340 K. Mikami, R. Angeland, K. Ding, A. Ishii, A. Tanaka, N. Sawada, K. Kudo and M. Senda, Chem. Eur. J., 7 (2001) 730.
341 K. Kudo, K. Ajima, M. Sakamoto and M. Saito, Chromatography, 19 (1998) 138.
342 R.C. Williams, J.F. Edwards, A.S. Joshi and A.-F. Aubry, J. Pharm. Biomed. Anal., 25 (2001) 501.
343 T. Hattori, Y. Minato, S. Yao, M.G. Finn and S. Miyano, Tetrahedron Lett., 42 (2001) 8015.
344 E. Bossu, V. Cotichini, G. Gostoli and A. Farina, J. Pharm. Biomed. Anal., 26 (2001) 837.
345 C. Bertucci, V. Andrisano, V. Cavrini and E. Castiglioni, Chirality, 12 (2000) 84.
346 K. Wu, L. Geng, M. Joseph and L.B. McGown, Anal. Chem., 65 (1993) 2339.
347 N. Chen, K. Ikemoto, T. Sugimoto, S. Murata, H. Ichinose and T. Nagatsu, Heterocycles, 56 (2002) 387.
348 J.-G. Dong, A. Wada, T. Takakuwa, K. Nakanishi and N. Berova, J. Am. Chem. Soc., 119 (1997) 12024; T. Nehira, C.A. Parish, S. Jockusch, N.J. Turro, K. Nakanishi and N. Berova, J. Am. Chem. Soc., 121 (1999) 8681.
349 L. Geng and L. B. McGown, Anal. Chem., 66 (1994) 3243.
350 J.G. Lees and B.A. Wallace, Spectroscopy, 16 (2002) 121; B.A. Wallace and R.W. Janes, Biochem. Soc. Trans., 31 (2003) 631.
351 D.S. Kliger and J.W. Lewis, Chapter 9 in Ref. [7].
352 R.A. Goldbeck, S.C. Bjorling and D.S. Kliger, Proc. SPIE – Int. Soc. Opt. Eng., 1432 (1991) 14; J.W. Lewis, R.A. Goldbeck, D.S. Kliger, X. Xie, R.C. Dunn and J.D. Simon, J. Phys. Chem., 96 (1992) 5243; R.A. Goldbeck and D.S. Kliger, Methods Enzymol., 226 (1993) 147; C.-F. Zhang, J.W. Lewis, R. Cerpa, I.D. Kuntz and D.S. Kliger, J. Phys. Chem., 97 (1993) 5499; E. Chen, R.A. Goldbeck and D.S. Kliger, Annu. Rev. Biophys. Biomol. Struct., 26 (1997) 327; R.A. Goldbeck and P.S. Kliger, Proc. SPIE – Int. Soc. Opt. Eng., 3256 (1998) 15.
353 X. Xie and J.D. Simon, Proc. SPIE – Int. Soc. Opt. Eng., 1204 (1990) 66; J.D. Simon, X. Xie and R.C. Dunn, Proc. SPIE – Int. Soc. Opt. Eng., 1432 (1991) 211.

© 2006 Elsevier B.V. All rights reserved.
Chiral Analysis
K.W. Busch and M.A. Busch, Eds.

CHAPTER 14

Determination of molecular stereochemistry using optical rotatory dispersion, vibrational circular dichroism and vibrational Raman optical activity

Prasad L. Polavarapu

Department of Chemistry, Vanderbilt University, Nashville, TN 37235, USA

14.1 OPTICAL ROTATORY DISPERSION

14.1.1 Introduction

Optically active or chiral substances rotate the plane of polarization of linearly polarized light passing through a sample of that substance. Such rotation is referred to as optical rotation. At a given concentration, wavelength and temperature, the amount of rotation depends on the molecular structure of that substance. Conversely, it should be possible to determine the molecular structure from the measured optical rotation for an optically active sample.

The phenomenon of rotation of plane-polarized light by chemical samples was discovered more than 200 years ago [1,2]. The "specific rotatory power" or "specific rotation," $[\alpha]$, of a liquid is defined as follows:

$$[\alpha] = \alpha / l\rho, \tag{1}$$

where α is the measured optical rotation in degrees, l is the path length of light traversing through the liquid sample, in decimeters (1 decimeter = 100 mm), and ρ is the density of liquid (in g/cc). For solutions, specific rotation is defined as,

$$[\alpha] = 100\,\alpha / lc, \tag{2}$$

where c is grams of chemical substance in 100 cc of solution. An alternate form of Eq. (2) is,

$$[\alpha] = \alpha / lc', \tag{3}$$

where c' is grams of chemical substance in 1 cc of solution. Sometimes, molar rotation $[\Phi]$, a modified version of specific rotation, defined as,

$$[\Phi] = [\alpha]M/100 \tag{4}$$

is reported. In Eq. (4), M is the molecular weight of the chemical substance.

Specific rotation depends on concentration and temperature of the solution, in addition to the wavelength of measurement. Thus the wavelength (λ) and temperature (t in°C) of the measurements are indicated by appending subscript and superscript, respectively to $[\alpha]$ as $[\alpha]_\lambda^t$. The corresponding notation for observed rotation is α_λ^t. The concentration of the solution, and the solvent used, are specified separately.

A plot of specific rotation as a function of wavelength, at a given temperature, is referred to as optical rotatory dispersion (ORD) curve.

Molecules belonging to the C_n, D_n, T, O and I point groups (referred to as chiral point groups) are optically active and can exhibit optical rotation. Such molecules are referred to as chiral [3], or optically active, molecules. The connection between specific rotation and molecular structure was addressed by Pasteur, van't Hoff and Le Bel [1,2]. But it was not possible at that time to determine the molecular stereochemistry (absolute configuration and conformations) from specific rotations. The quantum mechanical equation for specific rotation was developed by Rosenfeld [4–6], but its implementation proved to be difficult at that time. As a result, empirical hypotheses, such as Whiffen's empirical model [7] and Brewster's empirical rules [8], or models based on semi-classical theories such as, Kirkwood's polarizability theory [9], and Applequist's atom–dipole interaction model [10], were widely used in the twentieth century.

The lack of a reliable method to correlate the observed specific rotation with molecular structure restricted the use of specific rotation as a structural tool. Hence much of the work on specific rotations was limited to teaching, analytic and synthetic laboratories [11–13]. This status has changed dramatically since 1997, following the first quantum mechanical prediction of specific rotation [14]. Standard quantum mechanical programs [15–17] are now available to calculate the specific rotation at wavelengths of interest for a given structure. This development makes it possible now for any practising chemist to utilize ORD for determining molecular stereochemistry.

14.1.2 Experimental methods

The concepts for an instrument to measure the optical rotation of a sample are very simple, at least, in principle (see Fig. 14.1). First, a light component of appropriate wavelength must be generated. This can be achieved in one of the following way by: (a) passing the polychromatic light through an appropriate monochromator; (b) passing the polychromatic light through an appropriate narrowband wavelength filter; (c) using monochromatic light source such as a laser beam. Then the second step is to linearly polarize this chosen light component which can be done by using an appropriate linear polarizer, referred to as input polarizer. This linearly polarized light passes through the sample held in an appropriate sample cell, then through a linear polarization analyzer, referred to as output analyzer and finally to an appropriate light detector. When the

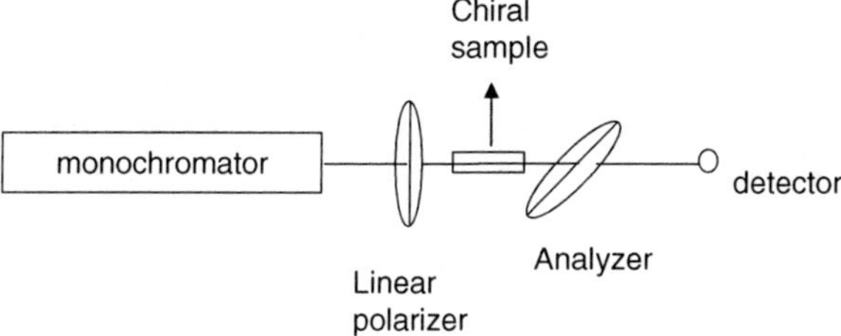

Fig. 14.1. Schematic of a polarimeter for measuring optical rotation.

sample is optically inactive or the sample cell is empty, and the output analyzer axis is perpendicular to that of the input polarizer, no or minimal light reaches the detector. This position is referred to as the null point. When an optically active sample is introduced into the sample cell, the linear polarization of the light is rotated by the sample, so the analyzer position no longer represents the null point. The analyzer axis must be rotated to a new position to get the null point. The angle by which the analyzer axis is rotated to find this null point is the measured optical rotation angle, α. The angle is positive if the rotation is clockwise and negative if the rotation is counter-clockwise when viewed into the incoming light [13]. The specific rotation is then deduced using Eq. (1), (2) or (3).

Different versions of instruments [18–21], referred to as polarimeters or spectro-polarimeters, based on these concepts are commercially available. Although the commercial instruments can be used to measure the optical rotation of liquid or liquid solution samples, special instruments are needed to do the same in the vapor phase for samples with low vapor pressures [22].

14.1.2.1 Calibration of polarimeters

The commercial polarimeters can be purchased with Quartz plates as calibration standards traceable to the National Institute of Standards and Technology (NIST), for verifying the instrumental accuracy. Alternately, high-purity sucrose and dextrose samples with rotation value specified by the NIST can also be purchased from the manufacturers. The instrument should be checked periodically with one of these calibration standards.

14.1.2.2 How much sample is needed for a measurement?

The answer to this question depends on several factors because the optical rotation of a sample depends on the molecular structure, concentration and wavelength of measurement. Typical amounts needed for optical rotation measurements are in the range of micrograms to milligrams.

14.1.2.3 Sample cells

Polarimeter cells can be equipped with inexpensive glass windows, which are sufficient for the measurements in the visible region. For wavelengths shorter than 400 nm, fused silica windows are appropriate. Cylindrical cells are generally used for optical rotation measurements because it is useful to have the light pass through a maximum possible amount of the sample. The cylindrical cells are commercially available in various lengths, starting from 25 to 200 mm. The longer the cell, the larger will be the observed rotation, and therefore better accuracy. However, colored samples will have larger absorption of visible light, so the amount of light reaching the detector will be small and the measured rotation may not be accurate. In such cases one may have to use the shorter length cell. The measured rotation is independent of the bore diameter of the cell. However, larger diameter bore allows the light to pass through the cell unobstructed thereby minimizing the noise, and smaller diameter may block the light thereby increasing the noise in measurement.

One of the vexing problems in measurements with a polarimeter using cylindrical cells is the formation of bubbles when the cell (especially smaller cells) is filled with a liquid solution. If a bubble is formed in the cell and that bubble stays in the light path, then the optical rotation reading will fluctuate significantly and becomes unstable. Thus care must be taken to avoid bubbles in filling the cell. Some manufacturers supply cells with a bubble trap to avoid such problems. Alternately, cells equipped with a central funnel can be used to avoid bubbles. However, funneled cells may not be appropriate for volatile solutions because concentration can keep changing during the measurement.

14.1.2.4 Solvent dependence of optical rotation

The magnitude of optical rotation can vary with the solvent, because of the differences in refractive indices of different solvents. Sometimes the sign of optical rotation can also reverse [23] in different solvents, which is due to different conformations of the sample molecules in different solvents. The measurement of optical rotation in different solvents is strongly recommended.

14.1.2.5 Temperature dependence of optical rotation

When sample molecules exist in different conformations, the populations of these conformations can change with temperature, and as a result the observed rotation changes with temperature. Even for a sample with single conformer molecules, the optical rotation can change with temperature, although only slightly, because optical rotation is inherently dependent on temperature [24]. Therefore, the temperature at which optical rotation is measured should be specified.

14.1.2.6 Concentration dependence of optical rotation

Optical rotation changes with concentration (higher concentration sample will have a larger optical rotation than dilute concentration samples). The dependence of optical rotation on concentration does not have to be linear. It can be a quadratic function or even a polynomial function of concentration [23,25]. Nevertheless, optical rotation should approach zero as concentration approaches zero. Note however that specific rotation does not have to become zero as the concentration approaches zero; instead specific rotation extrapolated to zero concentration is called intrinsic rotation and designated as $\{\alpha\}$. It is recommended that optical rotation be measured at four or more different concentrations to determine the functional dependence of optical rotation on concentration.

14.1.2.7 At what wavelengths are the measurements needed?
The optical rotation values are commonly reported at the sodium D line (589 nm). The optical rotation magnitude generally increases as the wavelength decreases (and approaches that of an electronic transition) and changes its sign at the center of an electronic absorption band. Most commercial polarimeters are equipped to make optical rotation measurements at several discrete wavelengths. The most commonly available wavelengths are: 633, 589, 546, 435, 405 and 365 nm. It is recommended to measure [26] the optical rotation at several wavelengths and obtain the ORD curve (*vide infra*). Some commercial instruments provide continuous variation in wavelength and such instruments can be used to measure a complete ORD spectrum.

14.1.3 Theoretical methods

14.1.3.1 Empirical and semi-classical models
The empirical hypotheses depend on van't Hoff's principle [27] of optical superposition, according to which the specific rotation of a chiral compound can be written as a sum of the contributions from individual chiral centers or segments in the molecule. Whiffen [7] adopted this concept for dihedral segments in a molecule, while Brewster [8] adopted this concept for both chiral centers and dihedral segments. The semi-classical models of Kirkwood [9] and Applequist [10], on the other hand, make use of electromagnetic theory. These methods were popular in the earlier decades, but they are not recommended at this time because more reliable quantum mechanical methods are now available for routine predictions.

14.1.3.2 Quantum mechanical methods
The electric and magnetic dipole moments, μ_α and m_α respectively, in the presence of static electric field F_α, magnetic field B_α and their time derivatives $\dot{F}_\alpha$ and $\dot{B}_\alpha$, are given as [28,29],

$$\mu_\alpha = \mu_\alpha^{\mathrm{o}} + \alpha_{\alpha\beta}F_\beta + \omega^{-1}G'_{\alpha\beta}\dot{B}_\beta + \cdots \tag{5}$$

$$m_\alpha = m_\alpha^{\mathrm{o}} - \omega^{-1}G'_{\alpha\beta}\dot{F}_\alpha + \cdots \tag{6}$$

In Eqs. (5) and (6) $\alpha_{\alpha\beta}$ and $G'_{\alpha\beta}$ represent, respectively, the electric dipole–electric dipole and electric dipole–magnetic dipole polarizability tensors; ω is the angular frequency $2\pi\nu$; μ_α^{o} and m_α^{o} represent, respectively, the electric and magnetic dipole moments in the absence of external fields. The electric dipole–magnetic dipole polarizability tensor $G'_{\alpha\beta}$is given as [28,29],

$$G'_{\alpha\beta} = \frac{-4\pi}{h}\sum_{\mathrm{n}\neq\mathrm{s}}\frac{\omega}{\omega_{\mathrm{ns}}^2 - \omega^2}\mathrm{Im}\{\langle\psi_{\mathrm{s}}^{\mathrm{o}}|\tilde{\mu}_\alpha|\psi_{\mathrm{n}}^{\mathrm{o}}\rangle\langle\psi_{\mathrm{n}}^{\mathrm{o}}|\tilde{m}_\beta|\psi_{\mathrm{s}}^{\mathrm{o}}\rangle\}, \tag{7}$$

where $\tilde{\mu}_\alpha$ is the electric dipole moment operator; $\tilde{m}_\beta$ is the magnetic dipole moment operator; $\psi_{\mathrm{s}}^{\mathrm{o}}$ and $\psi_{\mathrm{n}}^{\mathrm{o}}$ represent the ground and excited electronic state wave functions, respectively; $\omega_{\mathrm{ns}} = 2\pi(E_{\mathrm{n}}^{\mathrm{o}} - E_{\mathrm{s}}^{\mathrm{o}})/h$ with $E_{\mathrm{n}}^{\mathrm{o}}$ and $E_{\mathrm{s}}^{\mathrm{o}}$ representing the unperturbed energies of states n and s respectively. The sum in Eq. (7) extends over infinite number of excited electronic states. The optical rotatory parameter β is defined as,

$$\beta = -\frac{1}{3}\omega^{-1}(G'_{xx} + G'_{yy} + G'_{zz}) \tag{8}$$

This parameter is related to the optical rotation ϕ (in radians $\cdot$ cm^{-1}) as [6],

$$\phi = \frac{4\pi N\omega^2\beta}{c^2}, \tag{9}$$

where N represents the number of molecules per unit volume. The experimental optical rotations are converted and reported as specific rotation $[\alpha]$, in units of deg $\cdot$ cc/dm $\cdot$ g. The corresponding theoretical quantity is $[\alpha] = 3600\phi V_m/2\pi M$ where V_m and M are, respectively, the molar volume and molar mass. With these definitions, the calculated specific rotation is given as [14],

$$[\alpha] = 13.43 \times 10^{-5}\frac{\beta\bar{\nu}^2}{M} \tag{10}$$

with β in units of bohr^4 (1 bohr $= 5.29177\times10^{-11}$m), M in g/mol and $\bar{\nu}$ (the wave number where optical rotation is measured) in cm^{-1}.

A method to evaluate Eq. (7) in the static limit has been developed by Amos [30] as early as 1982. Jorgensen and coworkers [31] have developed a linear response or frequency-dependent (dynamic) method to evaluate Eq. (7) in 1994. However, the use of Eq. (7) for the calculation of specific rotation has not been reported until the first article on the subject appeared in 1997 [14].

The quantum mechanical calculations of specific rotation can now be carried out using different theoretical methods.

(a) In the static method [30], the approximation involved is that the wavelength of measurement is considered to be much larger than that of first electronic transition. Because of this approximation, the static method is not suitable for calculating ORD. This limitation was overcome in the dynamic method [31] where wavelength dependence is incorporated in evaluating Eq. (7) The dynamic method is recommended for routine use.

(b) The use of ordinary Gaussian functions (known as non-London orbitals and as field-independent atomic orbitals (FIAOs)) make the results gauge origin-dependent, while the use of London orbitals (also referred to as gauge including atomic orbitals (GIAOs)), in these calculations results in gauge origin-independent predictions. Although the initial calculations [14] used the static method and non-London orbitals, they were later verified using the dynamic method and London orbitals [32]. The use of London orbitals is recommended for routine use.

(c) Specific rotation can be calculated at different theoretical levels, such as Hartree–Fock (HF) method, complete active space self-consistent filed (CASSCF) method [33], density functional theory (DFT) [34], coupled cluster (CC) method [35] or time-dependent gauge invariant method [36]. Predictions of specific rotations using HF or DFT theory can be obtained routinely using currently available programs, while the remaining methods (CC, CASSCF) are meant for specialists. The HF theory does not include the effects of electron correlation, while DFT incorporates these effects using density functionals. Thus DFT is recommended for routine use. The specific rotation can also be derived from DFT-predicted electronic circular dichroism using either Kramers–Kronig transform or sum-over states methods [37].

A complete ORD curve can now be predicted using quantum mechanical methods. Although initial calculations of ORD [38] could not be carried out near the resonance frequencies (that is the frequency where an electronic transition occurs), DFT method using London orbitals has now been developed [39] to overcome this limitation. Thus, it is recommended that an ORD curve through the electronic absorption region, (instead of specific rotation at a single wavelength) be predicted using quantum mechanical methods.

The predicted specific rotations are normally obtained for isolated molecules, while the experimental measurements are done for liquid solutions. To account for solvent influence, a polarizable continuum (PCM) method [40] has been incorporated into quantum theoretical predictions of specific rotation, but much progress is required for accurately predicting solvent influence on specific rotations. An alternative to incorporating solvent influence in theoretical calculations is to determine the solvent influence experimentally. For this purpose, one would measure intrinsic rotation (specific rotation at infinite dilution) [23,25] in different solvents. Intrinsic rotation provides the specific rotation value of a solute molecule in solvent cage, so solute–solute interactions are absent in the experimentally measured intrinsic value. When intrinsic rotations are measured in different solvents, the differences among experimental intrinsic rotation values provide the magnitude of solvent cage influence. Such studies [23,25] are just beginning to appear.

14.1.4 Protocol for determining molecular stereochemistry

The following protocol is recommended, although it has not been strictly followed in the literature.

(1) Measure the optical rotation of a sample of interest at a given wavelength and several low concentrations. Convert the measured optical rotation to specific rotation value, using Eq. (1), (2) or (3). From these data determine the intrinsic rotation (specific rotation at infinite dilution). If the molecules of a sample of interest can (or suspected to) exist in different conformations, then it would also be useful to measure the optical rotation as a function of solvent. If the sign of optical rotation at a given wavelength changes in different solvents, that would be an indication that the predominant conformations change in different solvents. Then the sign and magnitude of specific rotation obtained in an inert solvent are to be preferred for comparison with predictions for isolated molecules. If the sign of optical rotation changes with concentration, then that would be an indication that the predominant conformation also changes with

concentration. In any case, the sign and magnitude of specific rotation obtained at very dilute concentration (or at infinite dilution) and in inert solvents are to be preferred for comparison with the theoretical predictions for isolated molecule. These measurements are to be repeated at several different discrete wavelengths, so that the experimental ORD spectrum can be obtained. Note that as the wavelength approaches that of an electronic absorption band, the magnitude of rotation increases and changes sign at the center of that electronic absorption band.

(2) Undertake quantum mechanical predictions. Before undertaking optical rotation calculations, it is necessary to find out if the electronic transition wavelengths calculated at a given theoretical level match with those observed in the experimental electronic spectrum. If they do not, which is often the case, then optical rotation should not be calculated at the same wavelengths as those used in the experimental measurement. Instead, optical rotation calculations should be done at wavelengths shifted [37] from those in the experimental measurement. The minimum theoretical level recommended for the prediction of specific rotation is the use of B3LYP density functional and 6-31G* basis set.

   (a) For single conformer molecules, the starting geometry can be obtained in a straightforward manner using a molecule building program. In doing so, one chooses arbitrarily either one of the two possible absolute configurations. This geometry is used as the starting point and is further optimized to obtain the minimum energy geometry. Using this optimized geometry, specific rotation is calculated at the wavelengths of interest. Compare the predicted ORD with the experimental ORD. If the predicted sign of ORD matches that of experimentally measured ORD sign, then the absolute configuration used in the calculations is assigned to the experimental sign of optical rotation. On the contrary, if the predicted sign of ORD is opposite to that of experimentally measured ORD sign, then the absolute configuration opposite to that used in the calculations is assigned to the experimental sign of optical rotation.

   (b) For multiple conformer molecules with a single chiral center, explore all feasible conformations of this molecule. Optimize the geometry for each such conformation, making sure that all conformations have the same absolute configuration. Compare the electronic energies of optimized conformations and estimate the populations using these electronic energies. If Gibbs energies are available, for example from vibrational frequency calculations (see Section 14.2), then Gibbs energies are to be used instead of electronic energies. The energies are normally reported in atomic units, and they can be converted to kcal/mol by multiplying with 627.45. The fractional population, P_i, of conformer "i" is obtained as $P_i = N_i/N_o = e^{-(E_i-E_o)/RT}$ (with the constraint that $\sum_i P_i = 1$), where E_i is the electronic energy of conformer "i" and E_o is that of lowest energy conformer, R is gas constant (1.987×10^{-3} kcal/mol·K) and T is the temperature in absolute scale. For all conformations with population greater than ~10%, calculate the specific rotation at the wavelengths of interest using the optimized geometry.

Using the predicted specific rotations and populations of all conformers, obtain the population-weighted sum of specific rotations at each of the wavelengths of interest. Compare the predicted ORD with the experimental ORD. If the sign of predicted ORD matches that of experimental ORD, then the absolute configuration used in the calculations is assigned to the experimental sign of optical rotation. On the contrary, if the sign of predicted ORD is opposite to that of experimental ORD, then the absolute configuration opposite to that used in the calculations is assigned to the experimental sign of optical rotation.

(c) If the molecule of interest has n chiral centers, then one has to consider 2^n diastereomers, of which one-half are mirror images of the other half. Then one has to repeat step (b) for each of the 2^{n-1} diastereomers. Since the experimentally measured sign of ORD here can agree with that of predicted ORD for more than one diastereomer, comparison should take into account the magnitudes of specific rotation as well.

If the magnitudes of experimental or predicted specific rotation are small, then additional calculations at a higher theoretical level should be undertaken to check for consistency in the calculations.

14.1.5 Applications

The specific rotations predicted for different molecules using quantum mechanical methods have been quite successful. As an example, the absolute configuration of $(+)_{589}$-bromochlorofluoromethane, where $(+)_{589}$ refers to positive specific rotation in cyclohexane solvent at 589 nm, has been established as (*S*) using the specific rotations calculated at very high levels of theory [41]. The experimental ORD curve of $(+)_{589}$-bromochlorofluoromethane and that calculated for (*S*)-bromochlorofluoromethane using B3LYP functional and very large aug-cc-pVTZ basis set, in the 350–600 nm region, are compared in Fig. 14.2. Calculations and measurements of specific rotation at several

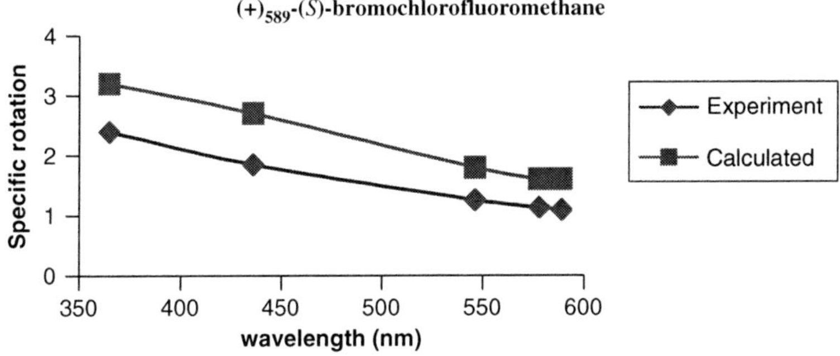

Fig. 14.2. Comparison of experimental specific rotation for $(+)_{589}$-bromochlorofluoromethane with B3LYP/aug-cc-pVTZ calculated specific rotation for (*S*)-bromochlorofluoromethane [41].

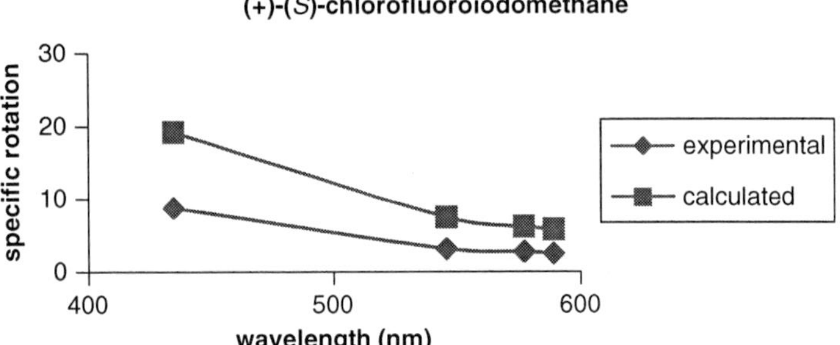

Fig. 14.3. Comparison of experimental specific rotation for (+)-(*S*)-chlorofluoroiodomethane with B3LYP/DGDZVP calculated specific rotation for (*S*)-chlorofluoroiodomethane [42].

wavelengths have been important to establish the reliability of absolute configuration assignment owing to the extremely small specific rotation magnitude for this molecule. Furthermore, calculations were also done using several large basis sets to confirm the reliability of predicted sign for specific rotation.

Another example is the determination of absolute configuration of chlorofluoroiodomethane [42]. The experimental ORD for the enantiomer with positive rotation in cyclohexane and calculated ORD for (*S*) enantiomer (with B3LYP functional and DGDZVP basis set) in the ~400–600 nm region are compared in Fig. 14.3. From this comparison, the absolute configuration of chlorofluoroiodomethane can be assigned as $(+)_{589}$-(*S*).

Although the number of molecules for which the quantum mechanical specific rotations have been compared to the corresponding experimental specific rotations, is quite large, most comparisons were done at a single wavelength (589 nm). There are sufficient number of examples in the literature where ORD, rather than a single wavelength measurement/calculation, has been successfully used to predict or confirm the absolute configuration and predominant conformations of organic molecules. Instead of discussing these examples, the reader is directed to a list of articles, summarized in (Appendix A), that dealt with the determination of molecular stereochemistry using specific rotation.

14.1.6 Commercial polarimeters and quantum mechanical software

***Commercial suppliers of Polarimeters in the USA (in alphabetical order)*:**

(1) Jasco Inc, 8649 Commerce Dr, Easton, MD 21602, USA
www.jascoinc.com

(2) Rudolph Instruments, Inc, 40 Pier Lane, Fairfield, NJ 07004-2113, USA
www.rudolphinst.com

(3) Rudolph Research Analytical, 354 Route 206, Flanders, NJ 07836, USA
www.rudolphresearch.com

***Quantum mechanical software*:**

Commercial: (a) *Gaussian 03*, Gaussian Inc., Wallingford, CT 06492, USA, www.gaussian.com
(b) *Turbomole*, www.turbomole.com
(c) *Amsterdam density functional program*, www.scm.com.

Freeware: (a) *Dalton program*, www.kjemi.uio.no/software/dalton/
(b) *PSI 3.2*, www.psicode.org

14.2 VIBRATIONAL CIRCULAR DICHROISM

14.2.1 Introduction

Differential absorption of left and right circularly polarized light is referred to as circular dichroism (CD), which is exhibited by all optically active or chiral molecules. When light in the visible or ultraviolet region is passed through a chemical sample of interest, the absorption of that light results in molecules going from the ground electronic state to an excited electronic state. Differential absorption of left and right circularly polarized visible/ultraviolet light is then referred to as electronic circular dichroism (ECD), because this differential absorption results from molecular electronic transitions. ECD is a well-established spectroscopic method [43] that has been in use for several decades.

Infrared (IR) region of electromagnetic radiation is divided into three sub-regions, namely near-infrared (~1–2.5 μm), mid-infrared (~2.5–25 μm) and far-infrared (~25–100 μm). The absorption of infrared light results in a vibrational transition (molecules going from the ground vibrational state to an excited vibrational state, both states belonging to the same ground electronic state). When left and right circularly polarized infrared light is passed through a sample, the resulting circular dichroism is referred to as infrared vibrational circular dichroism (VCD) [44]. Thus, VCD can be viewed as a combination of two separate techniques, namely infrared and circular dichroism spectroscopies. Although VCD method is not as old as either CD or infrared spectroscopy (first VCD measurement was reported in the early seventies), mid-infrared VCD has emerged as a method with distinct advantages for determining molecular stereochemistry.

The successful emergence of VCD in the mid-infrared as a viable technique for chiral structural analysis results from the advances made both in instrumentation and in quantum mechanical predictions of vibrational properties. These two developments together brought this area to the research forefront. The advantages resulting from shifting the region of measurement of CD from the visible to the mid-infrared region are as follows:

(a) Since every molecule has $3N-6$ vibrations, where N is the number of atoms in the molecule, that many vibrational transitions, in principle, are available for CD investigations. It may not be possible to access all of the $(3N-6)$ vibrational transitions in practice, but the number of accessible transitions is much larger than that of electronic transitions. In arriving at the final conclusion regarding the three-dimensional molecular structure, more stringent requirements result from the need to analyze and interpret a larger data set and hence the probability for that conclusion to go wrong is minimized.

(b) The experimental vibrational absorption and VCD spectra in the mid-infrared region can be successfully predicted using quantum mechanical calculations. In other words, for a given structure, the experimental and quantum mechanically predicted mid-infrared VCD spectra are in closer agreement to each other. The established reliability in the quantum mechanical predictions of mid-infrared VCD has been a crucial factor for successful chiral structural analysis using VCD in the mid-infrared. For a molecule with an unknown structure, if quantum mechanically predicted mid-infrared VCD spectra for different possible structures are compared to the corresponding experimental spectra then the predicted spectrum that matches best with the experimental spectrum will reveal the structure of that molecule.

The disadvantages associated with mid-infrared VCD measurements are as follows:

(a) Longer wavelengths associated with the mid-infrared region probe the molecular chirality less efficiently compared to shorter wavelengths associated with the visible region. As a result, circular dichroism signals in the mid-infrared region are much smaller (by about three orders of magnitude) than those in the visible region;

(b) The detectors used for mid-infrared spectral detection are inherently less sensitive than the corresponding detectors used for visible spectral detection. As a result, the signal-to-noise ratio associated with circular dichroism in the mid-infrared is lower than that in the visible region. These disadvantages translate into the requirements for longer data collection times and larger sample concentrations.

It is important to note that the advantages outweigh the disadvantages. As a result, VCD in the mid-infrared region has now acquired the status of a reliable and convenient tool for determining molecular stereochemistry.

14.2.2 Experimental methods

The principles of measurement of circular dichroism in the visible spectral region, probing electronic transitions, are well known. The same principles apply for measuring circular dichroism in the mid-infrared region, except that here the measurements are done in the mid-infrared spectral region where fundamental vibrational transitions are probed. In the early developmental stages [45], instruments for measuring VCD in the mid-infrared were built in individual laboratories. They were based on the concept of dispersive mid-infrared spectrometers, where polychromatic light from a mid-infrared source is dispersed using a grating in the monochromator. These dispersive instruments are no longer in wide use. One typical drawback of dispersive instruments is attributed to a significant light loss at the entrance and exit slits of the monochromator. This throughput disadvantage is avoided in Fourier transform infrared (FTIR)

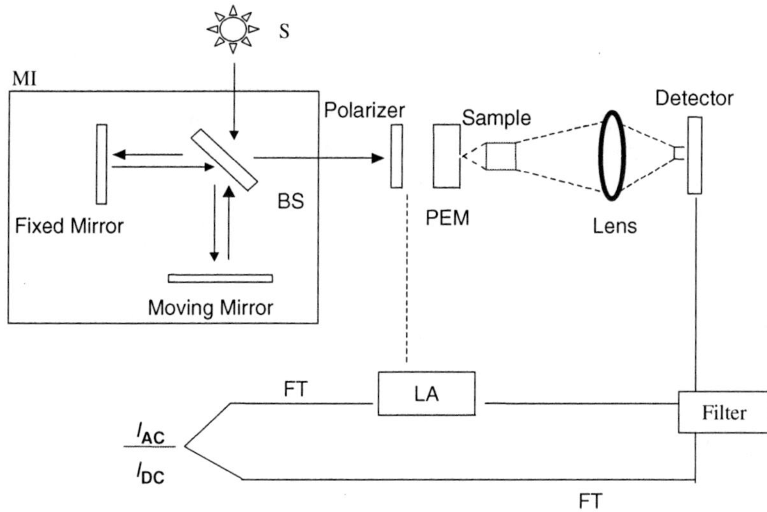

Fig. 14.4. A schematic of Fourier transform VCD spectrometer.

spectrometers. Furthermore in FTIR spectrometers, the nature of interference phenomenon permits the detection of light at all wavelengths simultaneously. These two advantages have lead to the adoption of FTIR spectrometers for VCD measurements in the mid-infrared [46]. During the initial period of these developments, VCD spectra for single enantiomers could not be obtained without measuring the same for opposite enantiomer or racemic mixture, due to artifacts. Minimization of artifacts in the VCD measurements has been the primary focus for several intervening years. With the development of approaches to minimize the artifacts [47], and the ability to record the mid-infrared VCD spectra for single enantiomers, commercial instruments for measuring VCD in the mid-infrared began to emerge (see Section 14.2.6). Most instruments in current use are based on the Fourier transform method [48]. A general schematic of such an instrument is shown in Fig. 14.4. Light from a polychromatic mid-infrared source (S) is passed through a Michelson interferometer (MI), which contains a beamsplitter (BS), fixed mirror and a movable mirror. The light intensity exiting the interferometer is modulated at different frequencies for different wavelengths. These frequencies, called Fourier frequencies, depend on the velocity of moving mirror in the interferometer. The light exiting the interferometer is circularly polarized using a linear polarizer and photoelastic modulator (PEM). The PEM converts the linearly polarized light into alternating left and right circularly polarized light components. The modulation between circularly polarized components takes place at a frequency (f_m) that is characteristic of the optical element in PEM (usually f_m ~37 kHz for ZnSe optical element). This alternating circularly polarized light is passed through the sample of interest and then focused on to a liquid nitrogen-cooled detector using a lens. The use of lenses, to manipulate the light after it was circularly polarized, has been a key factor in reducing the artifact signals [47]. The electronic signal at the detector contains a high-frequency (referred to as AC) component riding on top of a low-frequency (referred to as DC) component. The high-frequency component arises from differential absorption of left and right circularly polarized light components and the low-frequency component comes from intensity modulation in the interferometer. These two components are separated using an electronic filter. The high-frequency component is processed by a lock-in amplifier (LA), which is tuned to the frequency of PEM, and then Fourier transformed. The low-frequency component is Fourier transformed separately. Circular dichroism signal is extracted as the ratio of the Fourier transformed AC component (I_{AC}) to the Fourier transformed DC component (I_{DC}). It is worth noting that during the measurement of VCD, the corresponding absorption spectrum is also obtained. As a result both vibrational absorption and circular dichroism spectra are measured for each sample. The recent availability of commercial instruments (see Section 14.2.6) for measuring VCD, based on this Fourier transform method, has been a major factor for the increasing adaptation of this technique. Some basic considerations needed for VCD measurements are as follows:

14.2.2.1 Calibration of VCD spectrometers

The experimental magnitudes of VCD bands have to be calibrated using a multiple wave plate and analyzer [48,49]. The manufacturers of VCD instruments should provide the details on calibration procedure. The spectra obtained with multiple wave plate and analyzer are normally stored on the computer and can be retrieved for day to day calibration. A new calibration measurement is needed whenever optics or electronics in VCD instrument are modified or replaced. Otherwise, a new calibration measurement is recommended at periodic intervals.

14.2.2.2 How much sample is needed for a measurement?

The required amount of a sample is dictated by the strength of the sample absorption bands. As a rule of thumb one needs to keep the absorbance of a band of interest to be less than 1.0 and greater than 0.1, with optimum absorbance being in the range of 0.3–0.6. Typical sample concentrations required are ~4–100 mg/ml for the solution phase studies and ~1–5 mg/ml for the film studies. Typical sample volumes required are ~25 to several hundred microliters for the solution studies and ~100–500 μl for the film studies.

14.2.2.3 Sample cell

Different types of sample cells are available for infrared spectroscopy, depending on the state of the sample (gas, liquid or solid). Only a limited number of VCD measurements have been carried out for the gas phase samples, because most chiral samples do not have enough vapor pressure to undertake the VCD measurements. A majority of the reported VCD measurements to date have been done for liquid solutions. Solid samples (such as KBr pellets) are not normally used due to possible artifacts. However, VCD measurements have been successfully made for the film samples, where the films were prepared by evaporating the solution using either drop-cast method [50] or spray method [51] on an IR transparent plate. When the film samples are used, it is necessary to check for consistency in the VCD spectra when the film is rotated by 45 or 90° around the light beam axis. VCD measurements have also been reported for solid samples ground as oilmulls [52].

For film samples, a single IR transparent plate is all that is needed for holding the sample. For liquid samples three different types of cells are available: (a) fixed path length cells; (b) demountable cells; (c) variable path length cells. The latter two types of cells are generally used for VCD measurements. Since a proper combination of sample concentration and path length is needed to obtain the desired absorbance, the variable path length cells are convenient to vary the path length at a given concentration. Variable path length cells permit the use of large path lengths, but too large a path length may cause solvent absorption bands to dominate. However variable path length cells are expensive and care must be exercised in the maintenance of these cells. Demountable cells are economical, but assembling the cell with a proper spacer, and to ensure that the sample solution does not leak/evaporate, is a time-consuming process and needs some practice. Spacers with thickness from 6 to 500 μm are available for demountable cells. Note that for aqueous solutions, it is necessary to use 6 μm spacer (so that water absorption is minimized) and at the same time, high sample concentration should be used. For the gas phase VCD studies [53], a simple evacuable single-pass cell is to be preferred and multiple-pass cells with reflective mirrors are not recommended.

For any of the cells mentioned above, it is preferable to have the cell windows made of BaF_2 for the mid-infrared region. One may also use CaF_2 windows, but the transmission of CaF_2 does not permit measurements below $\sim 1100\,cm^{-1}$. Although KBr or KCl windows can also be used, their hygroscopic nature renders them inconvenient for routine use. Higher refractive index associated with ZnSe and KRS-5 windows may result in loss of throughput and in artifact signals.

14.2.2.4 Solvent considerations

Solvent should be the one that dissolves the sample, and does not have absorption interference in the region of measurement. If a solvent has rich mid-infrared spectrum in the region of interest, then that solvent should be avoided. Inert solvents such as CCl_4, CS_2 are ideal choices for the mid-infrared region. However, polar solvents such as CH_3OH, $CHCl_3$ and CH_2Cl_2 can also be used but, to avoid interference with sample absorption bands, their deuterated analogs are sometimes beneficial to use. It is recommended that the infrared absorption spectrum of solvent of interest be evaluated to determine the regions of minimal interference from solvent.

14.2.2.5 Temperature dependence of VCD

In the harmonic approximation, the fundamental vibrational spectral intensities of a fixed conformer molecule are temperature independent [54]. Thus, any temperature-dependent variations in the VCD spectrum may be attributed to the presence of multiple conformers and to the shift in conformer equilibrium as a function of temperature. The temperature-dependent VCD would be useful to study the change in population of conformers. For increasing the temperature, cells embedded with heating cartridges can be purchased. For lowering the temperature, jacketed cells (for circulating a coolant) are available, but to avoid water condensation on the windows an evacuable outer jacket is needed.

14.2.2.6 Concentration dependence of VCD

Concentrations that lead to sample aggregation or intermolecular hydrogen bonding should not be used, to avoid the complexities in interpreting the spectra resulting from such effects. Thus, in general, dilute solutions are to be preferred. However, a balance needs to be found between concentration of the sample and path length of the cell in order to maintain the desired absorbance of vibrational bands and to avoid the interference from solvent bands. These demands may restrict the VCD studies for certain classes of molecules. For example, to eliminate intermolecular hydrogen bonding in carboxylic acids one has to use extremely dilute solutions, but such low concentrations require larger path lengths where solvent absorption interference precludes VCD measurements. One approach to overcome this limitation is to chemically convert the carboxylic acid to corresponding methyl ester [55] and perform the VCD measurements and calculations for this ester.

14.2.2.7 In which regions are the measurements to be made?

In the near-infrared region, vibrational transitions are mostly overtones and combinations. The interpretation of VCD associated with overtones and combination bands is not straightforward [56]. On the other hand, the VCD measurements in the far-infrared region [57] are not yet practical. VCD in the mid-infrared region arises from the fundamental vibrational transitions and can be predicted reliably using quantum mechanical programs. Thus, VCD measurements in the mid-infrared region represent the best choice for determining molecular stereochemistry.

14.2.3 Theoretical methods

14.2.3.1 Empirical and semi-classical models

Numerous empirical or semi-classical models have been developed over the years for interpreting VCD spectra. But these methods do not have sufficient accuracy in obtaining reliable predictions. For this reason, these methods will not be emphasized or described here.

14.2.3.2 Quantum mechanical methods

The theoretical quantities appropriate for vibrational absorption and VCD intensities of fundamental transitions are dipole strength and rotational strength, respectively. They are given as:

Dipole strength,

$$D_{01} = \left| < \psi_0 | \tilde{\mu}_\alpha | \psi_1 > \right|^2 \tag{11}$$

Rotational strength,

$$R_{01} = \mathrm{Im}[< \psi_0 | \tilde{m}_\alpha | \psi_1 > < \psi_1 | \tilde{\mu}_\alpha | \psi_0 >], \tag{12}$$

where ψ_0 and ψ_1 are, respectively the wavefunctions of ground and excited vibrational states; $\tilde{\mu}_\alpha$ and $\tilde{m}_\alpha$ are, respectively the molecular electric dipole and magnetic dipole moment operators. In Eq. (12), "Im" stands for "imaginary part of." The advances made in quantum mechanics, in the last two decades [58a–c], paved the way for successful predictions of vibrational absorption and circular dichroism spectra in the mid-infrared region. Both commercial and freeware quantum mechanical programs (see Section 14.2.6) are available for predicting vibrational absorption and circular dichroism spectra. These programs are designed to be simple enough for non-specialists to use.

If the goal is to determine the absolute configuration, then it is sufficient to compare the predicted and experimental mid-infrared VCD signs for all of the vibrational bands measured. A quantitative comparison of VCD magnitudes is not needed. However, if the goal is to establish the populations of individual conformers, besides absolute configuration, then one has to make a quantitative comparison between predicted and experimental VCD band intensities. It may happen sometimes that the relative populations of conformers for isolated molecule may be different from those in the solution phase. In such cases, the relative intensities of bands in the predicted and experimental spectra would not agree, although VCD band signs may still match. Another level of analysis can be undertaken here by using the predicted vibrational band intensities for different conformers and experimental band intensities, in conjunction with regression methods, for quantitative determination of conformer populations [25]. This is an area that is beginning to be explored.

14.2.4 Protocol for determining molecular stereochemistry

(1) Measure the experimental absorption and VCD spectra. The experimental absorption spectra are obtained as absorbance (labeled A) and VCD spectra as differential absorbance (labeled ΔA) respectively. Convert the units to molar absorptivity (ε and $\Delta\varepsilon$, respectively) in units of $L{\cdot}mol^{-1}{\cdot}cm^{-1}$ by dividing the spectra with concentration (in $mol{\cdot}L^{-1}$) and path length (in cm).

(2) Predict the vibrational absorption and VCD spectra using quantum mechanical methods. To predict the vibrational absorption and circular dichroism spectra, one starts by choosing: (a) the theoretical level at which the predictions are to be made; (b) a plausible geometry for the molecule. A reliable theoretical level has been the density functional method [58d], with the B3LYP functional and 6-31G* basis set being widely used. In some cases, it might be necessary to increase the basis set size from 6-31G* to a larger basis set, but in most instances a satisfactory prediction can be obtained at the B3LYP/6-31G* level. For single conformer molecules, the starting geometry can be obtained in a straightforward manner using a molecule building program. In doing so, one chooses an arbitrary absolute configurations. This geometry is used as the starting point and is further optimized to obtain the minimum energy geometry. At this optimized geometry, for the chosen absolute configuration, vibrational absorption and circular dichroism intensities for all vibrational bands are calculated. The calculated vibrational frequencies are usually larger than the experimental frequencies. For this reason, calculated vibrational frequencies are multiplied by a constant

(whose magnitude depends on the basis set used in the calculation). The calculated absorption intensities are normally listed as IR intensities (in units of km/mol) or as dipole strength (in units of 10^{-40} esu^{2}cm^2), while VCD intensities are listed as rotational strength (in units of 10^{-44} esu^{2}cm^2). Convert these to peak intensities, ε_o (in units of L·mol^{-1}·cm^{-1}) and $\Delta\varepsilon_o$ (in units of 10^{-4}L·mol^{-1}·cm^{-1}) using the following equations:

$$\varepsilon_o = \frac{100}{2.303\,\pi\Delta} \times I, \tag{13}$$

where I is absorption intensity in km/mol; Δ is the bandwidth.

$$\Delta\varepsilon_o = \frac{\nu_o}{23.03\Delta\pi} \times R, \tag{14}$$

where ν_o is the calculated vibrational frequency in cm^{-1} and R is the rotational strength in 10^{-44} esu^{2}cm^2. Using these converted intensities, and calculated band positions, simulate a band profile using Lorentzian band shapes and appropriate band width (normally 5–10 cm^{-1}) as follows:

$$\varepsilon(\nu) = \frac{\Delta^2}{\Delta^2 + (\nu - \nu_o)^2}\varepsilon_o, \tag{15}$$

where $\varepsilon(\nu)$ is the absorption intensity at a given frequency ν.

$$\Delta\varepsilon(\nu) = \frac{\Delta^2}{\Delta^2 + (\nu - \nu_o)^2}\Delta\varepsilon_o \tag{16}$$

where $\Delta\varepsilon(\nu)$ is the VCD intensity at a given frequency ν. Apply Eqs. (13)–(16), for all calculated vibrations in a given region of interest and add all band profiles to generate the calculated absorption and VCD spectra. Then compare the calculated and observed spectra. If the predicted VCD band signs match the corresponding experimentally observed VCD band signs, then the absolute configuration of the experimentally investigated molecules is assigned as that used in the calculations. If the predicted signs and experimental signs are opposite to each other, then the opposite absolute configuration is assigned to the experimentally investigated molecules.

For multiple conformer molecules, one has to consider all possible conformers. Initial geometries for these conformers can be obtained using molecule building programs. These starting geometries are further optimized using geometry optimization methods. From the energies at the optimized geometries one can estimate their approximate relative populations. For the conformers which have dominant populations, a second calculation that predicts band positions, vibrational absorption and VCD intensities is carried out. This calculation also provides Gibbs free energies, which must be used for correct relative populations. The energies are normally reported in atomic units, and they can be converted to kcal/mol by multiplying with 627.45. Then population weighted spectra are generated from these individual conformer calculations and compared to the experimentally observed VCD spectrum to establish the absolute configuration, as for single conformer molecules.

It is necessary to note that if there are n chiral centers in the molecule, then one has to consider 2^n diastereomers, of which one-half are mirror images of the other half. Furthermore if there are m conformers for each diastereomer, then one has to undertake $m \times 2^{n-1}$ calculations before comparing the predicted VCD spectrum with the experimentally observed VCD spectrum. Thus, the level of complexity in analysis increases with the increasing number of conformers and/or chiral centers in a given molecule. However, VCD in the mid-infrared region is not limited to small molecules. Several large size chiral pharmaceuticals and agrochemicals have been investigated successfully, with the limitation arising from the available computing power. In most cases reported in the literature, the quantum mechanical calculations of vibrational absorption and VCD spectra have been carried out on desktop computers.

The theoretical calculations are usually carried out for isolated molecules, while the experimental spectra are obtained in the solution phase. As long as non-polar solvents are used and intermolecular hydrogen bonds or aggregates do not form at the concentrations used, the comparison of predicted mid-infrared VCD spectrum for isolated molecule with the corresponding experimentally observed spectrum is sufficient for determining the absolute configuration. If intermolecular hydrogen bonding is unavoidable, as for example in carboxylic acids, then one should undertake calculations for dimers. As an alternative, the corresponding methyl esters can be investigated in monomeric form.

14.2.5 Applications

The vibrational absorption and VCD spectra predicted for different molecules using quantum mechanical methods at the DFT (using B3LYP functional) level have been quite successful. As an example, the experimental and predicted spectra [55] for 2-(4-chloro-2-methylphenoxy) propanoic acid methyl ester are shown in Fig. 14.5. There are 81 possible conformations for this molecule. The predicted spectra were obtained as the population-weighted spectra of the four lowest energy conformations. Excellent agreement can be seen between VCD spectrum observed for (+)-enantiomer and that calculated for (*R*)-configuration. Thus, the absolute configuration of 2-(4-chloro-2-methylphenoxy) propanoic acid methyl ester can be confidently assigned [55] as (+)-(*R*).

There are large number of examples in the literature for using VCD to determine the three-dimensional molecular structures in the solution phase. Instead of discussing these examples, the reader is directed to a list of articles, summarized in (Appendix B), that dealt with the determination of absolute configuration and/or predominant conformations using VCD. It should be noted that VCD applications with quantum mechanical predictions have been extended to the determination of secondary structures in peptides and proteins by Keiderling and coworkers [59a] and in nucleic acids by Wieser and coworkers [59b]. However, such applications are beyond the scope of this chapter and are not included in the list of articles provided in Appendix B.

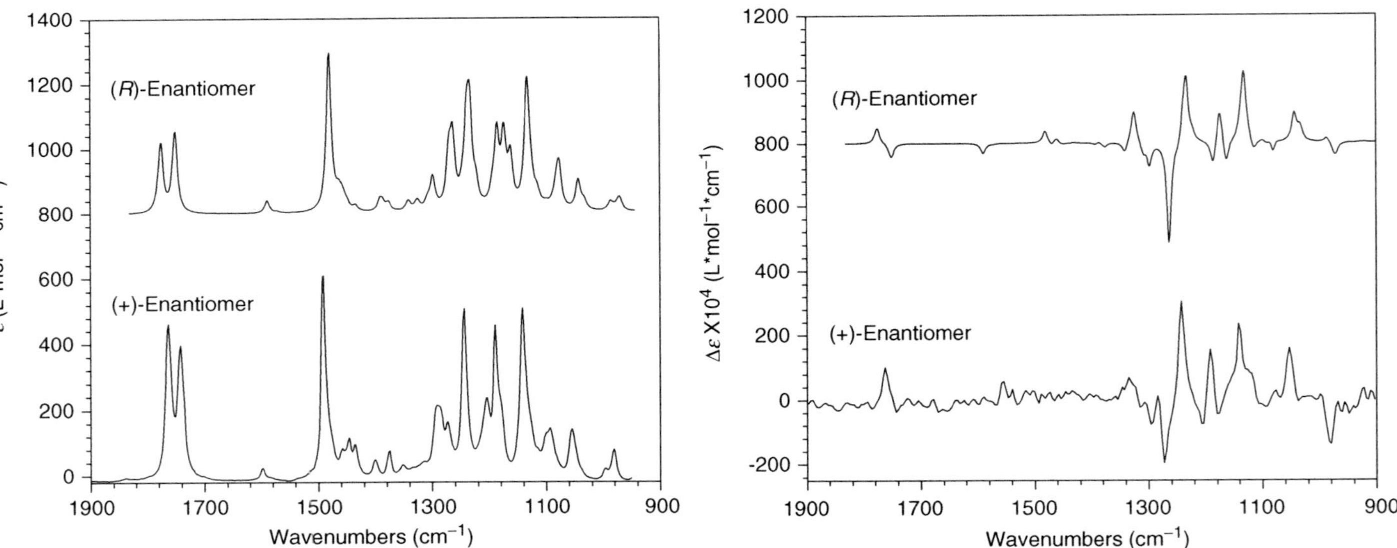

Fig. 14.5. Absorption (left panel) and VCD (right panel) spectra for 2-(4-chloro-2-methylphenoxy) propanoic acid methyl ester. The experimental spectra were obtained for (+)-enantiomer and calculated spectra were obtained for (*R*)-configuration at B3LYP/6-31G* level [55].

14.2.6 Commercial VCD spectrometers and quantum mechanical software

Commercial suppliers of VCD spectrometers (in alphabetical order):

(1) BioTools Inc., 950 N Rand Rd Unit 123, Wauconda, IL 60084, USA
www.btools.com
(2) Bruker Optics Inc., 19 Fortune Drive, Manning Park, Billerica, MA 01821 – 3991, USA
www.brukeroptics.com
(3) Jasco Inc, 8649 Commerce Dr, Easton, MD 21602, USA
www.jascoinc.com

Quantum mechanical software:

Commercial: *Gaussian 03*, Gaussian Inc., Wallingford, CT 06492, USA
www.gaussian.com
Freeware: *Dalton, a molecular electronic structure program*, University of Oslo
www.kjemi.uio.no/software/dalton/

14.2.7 Books on VCD

(1) P.L. Polavarapu, Vibrational Spectra: Principles and Applications with Emphasis on Optical Activity, Elsevier, Amsterdam, 1998.
(2) M. Diem, Introduction to Modern Vibrational Spectroscopy, John Wiley & Sons, New York, 1993.
(3) L.D. Barron, Molecular Light Scattering and Optical Activity, 2nd Ed., Cambridge University Press, Cambridge, 2004.

14.3 VIBRATIONAL RAMAN OPTICAL ACTIVITY

14.3.1 Introduction

Unlike in ECD and VCD phenomena, where differential absorption contains stereochemical information, in vibrational Raman optical activity (VROA) phenomenon stereochemical information is contained in the differential Raman scattering by chiral molecules. The vibrational Raman scattering involves vibrational transitions, just as in vibrational absorption, but the molecular parameters responsible for these two phenomena are different. The first theoretical formalism predicting the existence of VROA phenomenon was developed [60] by Barron and Buckingham in 1971. Subsequently, experimental efforts were made to observe this phenomenon by different research groups. Although the initial experimental attempts proved to be unreliable, Barron, Bogaard and Buckingham were able to measure [61] genuine and reproducible VROA signals in 1973. These observations were later confirmed in 1975 by Hug et al. [62]. This period marked the beginning of VROA spectroscopy. The experimental

VROA measurements in this period of development were both laborious and hard (it used to take 24 h to record just 300 cm^{-1} wide VROA spectrum), because: (a) VROA signals are very small (10^{-4}–10^{-5} of Raman intensity); (b) technology of that time was not sufficiently advanced; (c) a clear understanding of the factors responsible for artifacts was lacking. Despite the hardships involved in VROA measurements, a few research groups, in particular that of Barron, continued to advance this field by undertaking additional experimental VROA measurements. Since these humble beginnings, stunning progress has been made both in the experimental and theoretical aspects, to the point of turning VROA into a practical spectroscopic technique, as described below.

A significant advance in the experimental VROA measurements during the intervening years was the utilization of multi-channel detectors in combination with a spectrograph. The use of OMA vidicon tube as a multi-channel detector by Moskovits and coworkers [63] lead to a tenfold increase in speed over their earlier scanning instrument. The self-scanning silicon photodiode array utilized by Hug and Surbeck [64], demonstrated further increase in the speed for VROA measurements. The one developed [65] in Barron's laboratory provided state-of-the-art quality VROA spectra and generated a large collection of VROA spectra. The increase in speed that resulted from the use of multi-channel detectors also facilitated the analysis of the origin of artifacts [66] and investigation of new experimental designs [67–71]. Most of the VROA measurements until 1989 were made in the 90° scattering geometry with the incident light modulated between left and right circularly polarized light and collecting the scattered polarization parallel to the scattering plane (this is referred to as depolarized VROA), as this arrangement is less susceptible to artifacts. In 90° scattering geometry, when the incident light is circularly polarized and scattered polarization perpendicular to the scattering plane is collected, the corresponding measurement is referred to as polarized VROA, which was hard to measure because of the large artifacts associated with the polarized Raman bands. With the increased understanding of artifacts and their control, it became possible to measure [64,67] polarized VROA. Hecht and Barron were also able to measure another form of VROA, known as magic angle VROA [68], which depended only on the electric dipole–magnetic dipole polarizability contribution and not on the electric dipole–electric quadrupole polarizability contribution. Realizing that in backscattering geometry VROA intensity is 4 times that of polarized 90° VROA, conventional Raman intensity increases twofold and artifacts are reduced over those in 90° scattering geometry, Barron et al. initiated [69] backscattering VROA measurements for achieving higher signal-to-noise ratio.

The VROA measurements discussed so far utilized the circularly polarized incident light and are now referred to as incident circular polarization (ICP) measurements: for 90° geometry they are ICP-depolarized VROA (where scattered light with polarization parallel to the scattering plane is collected), and ICP-polarized VROA (where scattered light with polarization perpendicular to the scattering plane is collected); for backscattering they are referred to as ICP-backscattering VROA.

Another significant development is the realization by Nafie and coworkers that, instead of circularly polarizing the incident light, one can analyze the Raman light scattered by chiral molecules for circular polarizations using *linearly* polarized incident light. These measurements are referred to as scattered circular polarization (SCP) measurements [70]; for 90° scattering geometry they are called SCP-VROA in 90° geometry and for backscattering geometry they are called SCP-VROA in backscattering geometry. Another variation is to use circularly polarized incident light and analyze the Raman light scattered by chiral molecules for circular polarizations. These measurements are referred to as dual circular polarization (DCP)-VROA [71]. There are two versions of DCP measurements: one where incident and scattered circular polarizations are in-phase (DCP_I) and another where they are out-of-phase (DCP_{II}).

The next important advances in experimental VROA measurements came from the availability of charge-coupled devices (CCDs) as Raman detectors. As CCDs have high quantum efficiency, low read-out noise and low dark current at cryogenic temperatures, much higher signal-to-noise ratio was achieved [72]. Incorporating the CCD detectors into an ICP-backscattering instrument, Barron et al. were able to measure VROA in biological molecules [72]. Demonstration of the feasibility of application of VROA to biological molecules has changed the course of VROA spectroscopy. Since then CCD detectors were incorporated into SCP and DCP-VROA [73–75], ICP-depolarized VROA [76] instruments.

All these past and some new developments were incorporated recently by Hug et al. [77] into a unique new design with high throughput for SCP-VROA measurements. This latest development in instrumentation represents state-of-the-art in VROA measurements which is described in Section 14.3.2.

14.3.2 Experimental methods

The latest VROA instrument developed by Hug et al. [77] provided the basis for a commercial VROA instrument (see Section 14.3.6) that was introduced recently. This instrument uses backscattering geometry, with linearly polarized incident light, and digitizes the two circularly polarized Raman-scattered components simultaneously, as shown in Fig. 14.6. A concise description of the instrument is as follows:

The monochromatic laser light is linearly polarized by a polarizer (P) and directed to the sample using a small right-angle prism. The backscattered light from the sample is collected and passed through a circular analyzer, which is composed of a quarter wave plate (QW) and linear polarizer. The linear polarizer here is a polarizing cube beam splitter (PS), which directs the orthogonal linear polarizations in two different directions that are at 90° to each other. These two orthogonal directions are referred to as two channels. Light in each channel is collected through a fiber optic cable (FC), and these two fiber optic cables are brought to the entrance of a spectrograph and juxtaposed. The Raman light dispersed by this spectrograph (SG) is then detected by a CCD detector. The CCD detector pixels are binned so that in each column of CCD detector pixels, the top half of the pixels are grouped to give one signal and the bottom half of the pixels are grouped separately to give another signal. The image of the light is processed such that the light coming from one channel is detected by the top half of the pixels and that from the second channel is detected by the bottom half of the pixels. Thus, the two circular polarization components scattered by a chiral sample are digitized simultaneously in

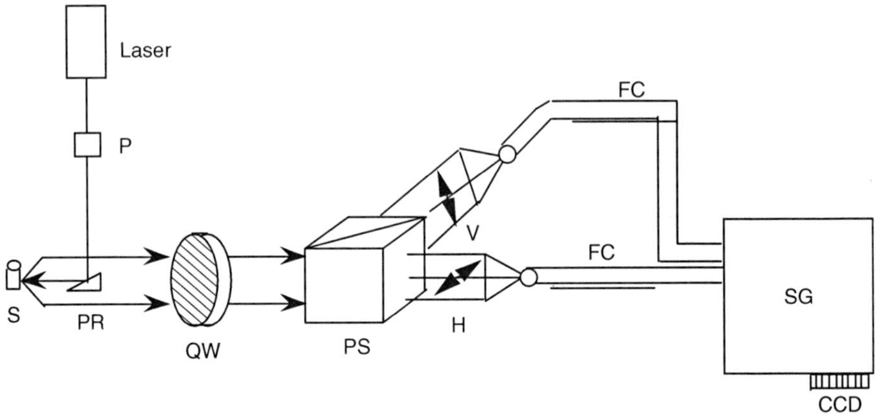

Fig. 14.6. Optical layout of the VROA instrument [77]. P: polarizer; PR: right-angle prism; S: sample; QW: quarter wave plate; PS: polarizing cube beam splitter; H: horizontal polarization; V: vertical polarization; FC: fiber optic cable: SG: spectrograph; CCD: charge coupled detector.

two different channels by the CCD detector. The difference between, and the sum of, these two channels can be generated using appropriate software. This dual channel design has certain distinct advantages:

(1) Any laser intensity fluctuations will affect the two channels similarly, so the effect of laser flicker noise on VROA is eliminated. The laser intensity fluctuations used to be a major source of problems in all previous designs due to long data collection times involved.
(2) Scattering artifacts arising from particulate matter (such as dust etc.) are also eliminated in the simultaneous detection of right and left circularly polarized scattered light.
(3) Thermal Schlieren affects are also eliminated.

These advantages were demonstrated recently by constructing [77] a SCP-backscattering VROA instrument of this design and performing the VROA measurements on (−)-(M)-σ-[4]Helicene [78]. Typical data collection times and laser power used were 40 min and 115 MW, respectively.

14.3.2.1 Sample cell

All of the reported VROA measurements were done for either neat liquids or liquid solutions. Solid samples are not normally investigated. Gas samples do not have enough density of molecules for VROA measurements. A quartz cuvette or capillary is normally used for holding the samples in VROA measurements.

14.3.2.2 How much sample is needed for a measurement?

In principle, one would require only a small amount of sample whose volume matches that of laser beam diameter incident on the sample. Milligram amount of samples are sufficient with sample volume in the microliter range.

14.3.2.3 Solvent considerations

In most cases, neat liquid samples were used. In the case of biological samples, measurements were done for aqueous solutions. Water is an excellent solvent for Raman spectroscopy because water has a weak Raman spectrum. Organic solvents can certainly be used, but here one should be aware of interfering Raman bands of the solvent. Special precautions are needed to avoid/suppress interfering fluorescence from the impurities.

14.3.2.4 Which laser wavelength is to be used?

Since scattering intensity is inversely proportional to the fourth power of incident wavelength, shorter incident wavelength is to be preferred. But when the incident wavelength is close to that of an electronic transition, resonance effect can lead to sample "burning," which should be avoided. A laser wavelength in the region of ~400–500 nm is commonly used.

14.3.3 Theoretical methods

14.3.3.1. Empirical and semi-classical models

Several empirical or semi-classical models have been developed over the years for interpreting VROA spectra. But these methods do not have sufficient accuracy in obtaining reliable predictions. For this reason, these methods will not be emphasized or described here.

14.3.3.2 Quantum mechanical methods

The predictions of VROA spectra require three different polarizability derivatives, namely $(\partial\alpha_{\alpha\beta}/\partial Q_i)$, $(\partial G'_{\alpha\beta}/\partial Q_i)$ and $(\partial A_{\alpha\beta\gamma}/\partial Q_i)$, where $\alpha_{\alpha\beta}$, $G'_{\alpha\beta}$ and $A_{\alpha\beta\gamma}$ are, respectively the electric dipole–electric dipole, electric dipole–magnetic dipole and electric dipole–electric quadrupole polarizability tensors given below; Q_i is the normal coordinate for ith vibration. The prediction of Raman intensities requires only the normal coordinate derivatives of electric dipole–electric dipole polarizability, while the prediction of VROA requires all three polarizability derivatives. The quantum mechanical expression for $G'_{\alpha\beta}$ is given by Eq. (7) while those for $\alpha_{\alpha\beta}$ and $A_{\alpha\beta\gamma}$ are as follows:

$$\alpha_{\alpha\beta} = \frac{4\pi}{h} \sum_{n \neq s} \frac{\omega_{ns}}{\omega_{ns}^2 - \omega^2} \operatorname{Re}\{\langle \Psi_s^o | \tilde{\mu}_\alpha | \Psi_n^o \rangle \langle \Psi_n^o | \tilde{\mu}_\beta | \Psi_s^o \rangle \tag{17}$$

$$A_{\alpha\beta\gamma} = \frac{4\pi}{h} \sum_{n \neq s} \frac{\omega_{ns}}{\omega_{ns}^2 - \omega^2} \operatorname{Re}\{\langle \Psi_s^o | \tilde{\mu}_\alpha | \Psi_n^o \rangle \langle \Psi_n^o | \tilde{\Theta}_{\beta\gamma} | \Psi_s^o \rangle \tag{18}$$

In Eqs. (17) and (18), $\tilde{\Theta}$ is the electric quadrupole moment operator and the remaining terms have the same meaning as in Eq. (7). In practice, one determines the Cartesian polarizability derivatives, $(\partial\alpha_{\alpha\beta}/\partial X_A)$, $(\partial G'_{\alpha\beta}/\partial X_A)$ and $(\partial A_{\alpha\beta\gamma}/\partial X_A)$, where X_A is a Cartesian displacement of atom A, and converts them to $(\partial\alpha_{\alpha\beta}/\partial Q_i)$, $(\partial G'_{\alpha\beta}/\partial Q_i)$ and $(\partial A_{\alpha\beta\gamma}/\partial Q_i)$ using normal coordinate vectors called **S** vectors [54]. Analytic quantum mechanical procedures to evaluate the derivatives $(\partial\alpha_{\alpha\beta}/\partial X_A)$ and

$(\partial A_{\alpha\beta\gamma}/\partial X_A)$ have been developed [79], but that for $(\partial G'_{\alpha\beta}/\partial X_A)$ has not yet been developed. For this reason, all three derivative tensors are evaluated numerically by calculating the tensors $\alpha_{\alpha\beta}$, $G'_{\alpha\beta}$ and $A_{\alpha\beta\gamma}$ at equilibrium and displaced nuclear geometries. Thus VROA calculation can be carried out in two parts: one part involving the calculation of **S** vectors and another part involving the calculation of cartesian polarizability derivative tensors $(\partial\alpha_{\alpha\beta}/\partial X_A)$, $(\partial G'_{\alpha\beta}/\partial X_A)$ and $(\partial A_{\alpha\beta\gamma}/\partial X_A)$. These two parts can be carried out at two different theoretical levels, if desired. The first part involving the calculation of **S** vectors is done at a theoretical level where molecular geometry is also optimized. For the second part, a different theoretical level may be used, but the same molecular geometry as that used in the first part must be used.

The first *ab initio* method for VROA predictions was identified and implemented [80] in 1989. This method was based on the quantum mechanical advances made earlier by Amos [30] for *ab initio* calculation of electric dipole–magnetic dipole polarizability tensor. It is important to note that the *ab initio* VROA calculations were used in determining the absolute configuration of bromochlorofluoromethane for the first time [26].

The electric dipole–magnetic dipole polarizability tensor derivatives required for the VROA calculations were obtained in the initial stages using the static method of Amos [30] and non-London orbitals. The gauge origin independence and wavelength dependence for this tensor derivative was later achieved by Helgaker et al. [31] by using the London orbitals and the dynamic method. Since then several VROA calculations have been performed using this method at the Hartree–Fock level, although some times normal coordinates obtained at the density functional theoretical level were used (see Section 14.3.6). However, recent implementation [81] of DFT methods for calculating the needed tensors made it possible to predict VROA spectra [41,81] with improved accuracy.

14.3.4 Protocol for determining molecular stereochemistry

The protocol for determining molecular stereochemistry using VROA is similar to that described for VCD.

(1) Measure the experimental vibrational Raman and VROA spectra. The intensities in these spectra are normally in photoelectron counts and represent relative intensities (not absolute) because scattering intensities depend on laser power, wavelength and scattering collection angle.

(2) Predict the vibrational Raman and VROA spectra using quantum mechanical methods. To predict these spectra, one starts by choosing appropriate theoretical level for calculating the **S** vectors. For single conformer molecules, for a chosen absolute configuration the geometry is optimized at this theoretical level. At this optimized geometry, vibrational frequencies are calculated. The calculated vibrational frequencies are usually larger than the experimental frequencies. For this reason, calculated vibrational frequencies are multiplied by a constant

(whose magnitude depends on the basis set used in the calculation). This vibrational frequency calculation provides the needed **S** vectors. Using the same molecular geometry as that used in the calculation of **S** vectors, Cartesian polarizability derivative tensors $(\partial\alpha_{\alpha\beta}/\partial X_A)$, $(\partial G'_{\alpha\beta}/\partial X_A)$ and $(\partial A_{\alpha\beta\gamma}/\partial X_A)$ are calculated and converted to normal coordinate derivatives, $(\partial\alpha_{\alpha\beta}/\partial Q_i)$, $(\partial G'_{\alpha\beta}/\partial Q_i)$ and $(\partial A_{\alpha\beta\gamma}/\partial Q_i)$. The invariants of these normal coordinate derivative tensors, $\bar{\alpha}_i^2$, β_i^2, $\bar{\alpha}_i\bar{G}'_i$, γ_i^2 and δ_i^2 are then extracted as follows:

$$\bar{\alpha}_i^2 = \frac{1}{9}\left(\frac{\partial\alpha_{\alpha\alpha}}{\partial Q_i}\right)\left(\frac{\partial\alpha_{\beta\beta}}{\partial Q_i}\right) \tag{19}$$

$$\gamma_i^2 = \frac{1}{2}\left[3\left(\frac{\partial\alpha_{\alpha\beta}}{\partial Q_i}\right)\left(\frac{\partial G'_{\alpha\beta}}{\partial Q_i}\right) - \left(\frac{\partial\alpha_{\alpha\alpha}}{\partial Q_i}\right)\left(\frac{\partial G'_{\beta\beta}}{\partial Q_i}\right)\right] \tag{20}$$

$$\bar{\alpha}_i\bar{G}'_i = \frac{1}{9}\left(\frac{\partial\alpha_{\alpha\alpha}}{\partial Q_i}\right)\left(\frac{\partial G'_{\beta\beta}}{\partial Q_i}\right) \tag{21}$$

$$\beta_i^2 = \frac{1}{2}\left[3\left(\frac{\partial\alpha_{\alpha\beta}}{\partial Q_i}\right)\left(\frac{\partial\alpha_{\alpha\beta}}{\partial Q_i}\right) - \left(\frac{\partial\alpha_{\alpha\alpha}}{\partial Q_i}\right)\left(\frac{\partial\alpha_{\beta\beta}}{\partial Q_i}\right)\right] \tag{22}$$

$$\delta_i^2 = \frac{\omega}{2}\left(\frac{\partial\alpha_{\alpha\beta}}{\partial Q_i}\right)\left(\frac{\partial A_{\gamma\delta\beta}}{\partial Q_i}\right) \tag{23}$$

In Eqs. (19)–(23), Cartesian summation convention is used. The predicted vibrational Raman and ROA intensities for *i*th band are determined by Raman scattering activity S_i, and ROA activity P_i,

$$S_i = \left(a_1\,\bar{\alpha}_i^2 + a_2\beta_i^2\right) \tag{24}$$

$$P_i = \left(a_3\bar{\alpha}_i\bar{G}'_i + a_4\gamma_i^2 + a_5\delta_i^2\right), \tag{25}$$

where coefficients a_1 through a_5 are determined by the geometry and polarizations used for the experimental measurements. It should be noted that the observed spectral intensities represent scattering cross sections, which depend on the incident laser wavelength and vibrational frequency. For this reason, the above-mentioned vibrational Raman and ROA activities are to be multiplied by $(\nu_o - \nu_i)^4/1 - e^{hc\nu_i/kT}\ h/8\pi^2 c\nu_i$, where ν_o and ν_i respectively, are the laser and vibrational frequencies (in cm^{-1}) and h, c, and k are universal constants. For comparing the relative intensities in a small vibrational frequency region this multiplicative factor does not make much difference, but when a wide spectral region is compared this multiplicative factor should be taken into account. For simulating the predicted vibrational Raman and ROA spectra one would assign a band profile for each vibrational band (as in the case of VCD) and sum over all band profiles.

For determining the molecular stereochemistry, the experimental and predicted VROA spectra are compared. For molecules with multiple conformers or chiral centers, the same process as that used for VCD calculations (that is, spectra for different diastereomers and population-weighted spectra for different conformers) is needed. If the predicted VROA band signs match the corresponding experimentally observed VROA band signs, then the absolute configuration of the experimentally investigated molecules is assigned as that used in the calculations. If the predicted signs and experimental signs are opposite to each other, then the opposite absolute configuration is assigned to the experimentally investigated molecules.

14.3.5 Applications

The vibrational Raman and VROA spectra were predicted for a limited number of chiral molecules using quantum mechanical methods. In these limited cases, satisfactory agreement was found between the predicted and experimental spectra. While most of these predictions were obtained at the Hartree–Fock level of theory, a few predictions were obtained at the level of DFT (see Appendix C). A list of molecules for which the experimental and corresponding quantum mechanical VROA spectra were obtained is given in the Table 14.1. Among these molecules, absolute configuration has been determined for the first time for bromochlorofluoromethane using VROA and specific rotation. For the remaining molecules their known absolute configurations have been confirmed using VROA.

As an example, the VROA spectra for (−)-(*R*)-isoflurane are shown in Fig. 14.7. The experimental ICP-depolarized VROA spectrum was reported by Barron and coworkers [82] for (−)-isoflurane. The predicted vibrational Raman and ROA spectra were obtained in our laboratory at the HF level of theory using 6-31G* basis set for two conformers, labeled as conformer 1 and 2 [83], of isoflurane. Assuming equal populations for these two conformers, the population-weighted VROA spectrum for (*R*)-enantiomer is compared to the experimental VROA spectrum for (−)-enantiomer. The signs of VROA bands in the predicted and experimental spectra are in agreement for most of the bands

Table 14.1 Molecules investigated for stereochemistry using quantum mechanical and experimental vibrational Raman optical activity

1-phenylethanol	desflurane
2,3-*trans*-dimethyloxirane	galaxolide
2,3-*trans*-dimethylthiirane	isoflurane
3,6-dimethyl-1,4-dioxane-2,5-dione	isotopomers of alanine
3-methylcyclohexanone	methyloxirane
3-methylcyclopentanone	methylthiirane
4-methylisochromane	*N*-acetyl-*N*′-methyl-L-alaninamide
alanine	sigma-[4]Helicene
alanyl-alanine	spiro-[2,2]pentane-1,4-diene
alpha-pinene	substituted oxiranes
bromochlorofluoromethane	tartaric acid
cyclo(L-Pro-L-Pro)	*trans*-pinane

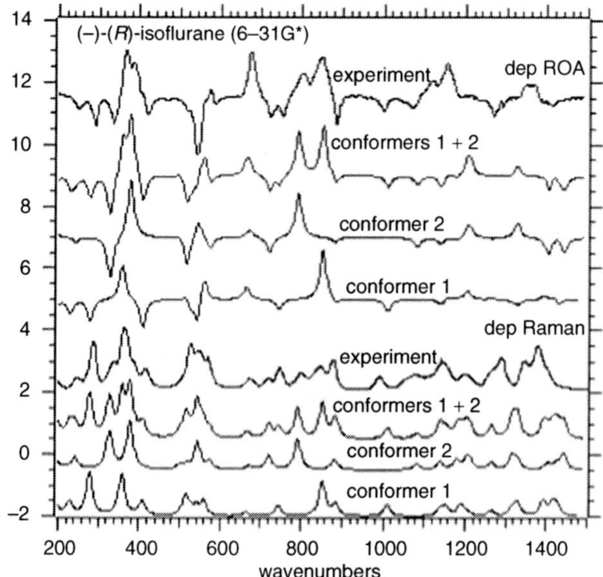

Fig. 14.7. Predicted depolarized vibrational Raman and ROA spectra for two individual conformers and their sum for (*R*)-isoflurane and the corresponding experimental spectra for (−)-isoflurane [82]. Conformers 1 and 2 are defined in Ref. [83]. The intensities (*y*-axis) are in arbitrary units, so only the relative intensities are meaningful.

confirming the assignment of absolute configuration as (−)-(*R*). To accurately reproduce the relative magnitudes of band intensities, higher level calculations would be needed.

Recent reviews [84] emphasizing the developments in VROA should be consulted for additional details on the chiroptical applications of VROA spectrosocpy.

14.3.6 Commercial VROA spectrometers and quantum mechanical software

***Commercial suppliers of VROA spectrometers*:**

(1) BioTools Inc., 950 N Rand Rd Unit 123, Wauconda, IL 60084, USA
www.btools.com

***Quantum mechanical software*:**

Commercial: *Gaussian 03*, Gaussian Inc., Wallingford, CT 06492, USA.
www.gaussian.com

Freeware: *Dalton, a molecular electronic structure program*, University of Oslo
www.kjemi.uio.no/software/dalton/

14.3.7 Books on VROA

(1) L.D. Barron, Molecular Light Scattering and Optical Activity, 2nd Ed., Cambridge University Press, Cambridge, 2004.

(2) P.L. Polavarapu, Vibrational Spectra: Principles and Applications with Emphasis on Optical Activity, Elsevier, Amsterdam, 1998.
(3) M. Diem, Introduction to Modern Vibrational Spectroscopy, John Wiley & Sons, New York, 1993.

14.4 SUMMARY

In this chapter, three different chiroptical methods for determining molecular stereochemistry (absolute configuration and predominant conformations) have been summarized. All three methods represent useful and powerful stereochemical approaches. However, it is not possible for a given researcher to have access to all three methods and may have to choose or depend on one approach. For newcomers this choice might be difficult without additional information. Based on the author's personal experience in these three research areas the relative merits of these approaches can be stated as follows:

(a) Instrumentation is inexpensive for ORD measurements. Also, ORD is easy to measure and also easy to calculate using available quantum mechanical programs. However, to obtain reliable predictions it may be necessary to obtain predictions at the higher levels of theory (such as B3LYP functional and aug-cc-pVDZ basis set). Even though this has been accomplished for several molecules using moderate levels of theory (such as B3LYP functional and 6-31G* basis set), in the case of molecules with unknown absolute configuration one would not know *a priori* which level of theory might be appropriate.
(b) Vibrational circular dichroism also became easy to measure with the availability of commercial instrumentation, but the instrumentation is more expensive than that of ORD. More expertise than that needed for ORD would be needed for experimental measurements. Available quantum mechanical programs can be used for routine VCD predictions but here also some background in spectroscopy would be appropriate. Nevertheless, satisfactory VCD predictions can be obtained at a moderate level of theory (such as B3LYP functional and 6-31G* basis set).
(c) For VROA, introduction of commercial instrumentation made the measurements easier but still measurements are a bit more involved than those for ORD or VCD. VROA instrumentation is more expensive than that of either ORD or VCD. Quantum mechanical programs are also available for VROA predictions but VROA calculations are more involved than those of either ORD or VCD. More expertise than that needed for ORD or VCD would be needed for both experimental measurements and theoretical calculations. These comments hopefully will help the newcomers to choose one or more of the three methods for their research in molecular stereochemistry.

14.5 ACKNOWLEDGMENTS

I thank Professor L. D. Barron for the VROA spectrum of (−)-isoflurane. This work was supported by grants from NSF (CHE0092922) and Vanderbilt University.

REFERENCES

1 H. Landolt, The Optical Rotating Power of Organic Substances and its Applications, The Chemical Publishing Co, Easton, PA, 1902.

2 T.M. Lowry, Optical Rotatory Power, Dover Publications Inc., New York, 1964; A. Lakhtakia, Selected papers on natural optical activity, Bellingham, SPIE optical engineering press, 1990.

3 The word chiral (Greek word meaning hand) was introduced by Lord Kelvin. See J.M. Hicks, Chirality: Physical Chemistry, ACS Symposium Series 810. Americal Chemical Society, Washington DC, 2002.

4 L. Rosenfeld, Quantenmechanische theorie der naturlichen optischen aktivitat von flussigkeiten und gasen. Z. Physik, 52 (1929) 161–174.

5 D.J. Caldwell and H. Eyring, The Theory of Optical Activity, Wiley, New York, 1971.

6 H. Eyring, J. Walter and G.E. Kimball, Quantum Chemistry, John Wiley & Sons, New York, 1944.

7 D.H. Whiffen, Optical rotation and geometrical structure. Chem. Ind., (1956) 964–968.

8 J.H. Brewster, A useful model of optical activity. I. Open chain compounds. J. Am. Chem. Soc., 81 (1959) 5475–5483.

9 J.G. Kirkwood, On the theory of optical rotatory power. J. Chem. Phys., 5 (1937) 479–491.

10 J. Applequist, On the polarizability theory of optical rotation. J. Chem. Phys., 58 (1973) 4251–4259.

11 D.P. Shoemaker, C.W. Garland and J.W. Nibler, Experimental Physical Chemistry. McGraw Hill, New York, 1989.

12 C. Zhao and P.L. Polavarapu, Comparative evaluation of vibrational circular dichroism and optical rotation for determination of enantiomeric purity. Appl. Spectrosc., 55 (2001) 913–918.

13 E.L. Eliel, S.H. Wilen and M.P. Doyle, Basic Organic Stereochemistry, Wiley, New York, 2001.

14 P.L. Polavarapu, Ab initio molecular optical rotations and absolute configurations. Mol. Phys., 91 (1997) 551–554.

15 R.D. Amos and J.E. Rice, CADPAC: The Cambridge analytic derivative package, Issue 4.0.

16 Dalton, a molecular electronic structure program, University of Oslo, Oslo, 2001, www.kjemi.uio.no/software/dalton/.

17 Gaussian 03, Gaussian, Inc.; www.gaussian.com.

18 C. Djerassi, Optical Rotatory Dispersion, McGraw Hill, New York, 1960.

19 P. Crabbe, Optical Rotatory Dispersion and Circular Dichroism in Organic Chemistry, Holden-Day, San Francisco, 1965.

20 For digital polarimeters, see Section 1.7

21 For commercial laser polarimeters, see www.pdr-chiral.com.

22 T. Miller, K.B. Wiberg and P.H. Vaccaro, Cavity ring-down polarimetry: a new scheme for probing circular birefringence and circular dichroism in the gas phase. J. Phys. Chem. A, 104 (2000) 5959–5968.

23 P.L. Polavarapu, A. Petrovic and F. Wang, Intrinsic rotation and molecular structure. Chirality, 15 (2003) S143–S149; 15 (2003) 801.

24 K.B. Wiberg, P.H. Vaccaro and J.R. Cheeseman, Conformational effects on optical rotation. 3-Substituted 1-butenes. J. Am. Chem. Soc., 125 (2003) 1888–1896.

25 J. He, A. Petrovic and P.L. Polavarapu, Quantitative determination of conformer populations: assessment of specific rotation, vibrational absorption and vibrational circular dichroism in substituted butynes. J. Phys. Chem., 108 (2004) 1671–1680.

26 J. Costante, L. Hecht, P.L. Polavarapu, A. Collet and L.D. Barron, Absolute configuration of bromochlorofluoromethane from experimental and ab initio theoretical vibrational Raman optical activity. Angew. Chem., 36 (1997) 885–887.

27 J.H. Van't Hoff, Die Lagerung der atome in Raume. Vieweg. Braunschweig, 8 (1908) 95–97.

28 L.D. Barron, Molecular Light Scattering and Optical Activity, Cambridge University press, Cambridge, 1982.

29 A.D. Buckingham and G.C. Longuet-Higgins, The quadrupole moments of dipolar molecules. Mol. Phys., 14 (1968) 63–72.

30 R.D. Amos, Electric and magnetic properties of CO, HF, HCl and CH_3F. Chem. Phys. Lett., 87 (1982) 23–26.

31 T. Helgaker, K. Ruud, K.L. Bak and P. Jorgensen, Vibrational Raman optical activity calculations using London atomic orbitals. Farad. Discuss., 99 (1994) 165–180.

32 P.L. Polavarapu and C. Zhao, A comparison of ab initio optical rotations obtained with static and dynamic methods. Chem. Phys. Lett., 296 (1998) 105–110.

33 P.L. Polavarapu, D.K. Chakraborty and K. Ruud, Molecular optical rotations: an evaluation of the semiempirical models. Chem. Phys. Lett., 319 (2000) 595–600.

34 K. Yabana and G.F. Bertsch, Application of time-dependent density functional theory to optical activity. Phys. Rev. A, 60 (1999) 1271–1279; J.R. Cheeseman, M.J. Frisch, F.J. Devlin and P.J. Stephens, Hartree-Fock and density functional theory ab initio calculation of optical rotation using GIAOs: basis set dependence. J. Phys. Chem. A, 104 (2000) 1039–1046; S. Grimme, Calculation of frequency dependent optical rotation using density functional response theory. Chem. Phys. Lett., 339 (2001) 380–388; J. Autschbach, S. Patchkovskii, T. Ziegler, S.J.A. van Gisbergen and E.J. Baerends, Chiroptical properties from time-dependent density functional theory. II. Optical rotations of small to medium size organic molecules. J. Chem. Phys., 117 (2002) 581–592.

35 K. Ruud and T. Helgaker, Optical rotation studied by density-functional and coupled-cluster methods. Chem. Phys. Lett., 352 (2002) 533–539.

36 M.P. Cayere, M. Rerat and A. Dargelos, Theoretical treatment of the electronic circular dichroism spectrum and the optical rotatory power of H_2S_2. Chem. Phys., 226 (1998) 297–306.

37 P.L. Polavarapu, Kramers-Kronig transformation for optical rotatory dispersion studies. J. Phys. Chem. A, 109 (2005) 7013–7023.

38 P.L. Polavarapu and C. Zhao, Ab initio predictions of anomalous optical rotatory dispersion. J. Am. Chem. Soc., 121 (1999) 246–247.

39 P. Norman, K. Ruud and T. Helgaker, Density-functional theory calculations of optical rotatory dispersion in the nonresonant and resonant frequency regions. J. Chem. Phys., 120 (2004) 5027–5035.

40 B. Mennucci, J. Tomasi, R. Cammi, J.R. Cheeseman, M.J. Frisch, F.J. Devlin, S. Gabriel and P.J. Stephens, Polarizable continuum model (PCM) calculations of solvent effects on optical rotations of chiral molecules. J. Phys. Chem., 106 (2002) 6102–6113.

41 P.L. Polavarapu, Absolute configuration of bromochlorofluoromethane. Angew. Chem., 41 (2002) 4544–4546.

42 J. Crassous, Z. Jiang, V. Schurig and P.L. Polavarapu, Preparation of (+)-chlorofluoroiodomethane, determination of its enantiomeric excess and of its absolute configuration. Tetrahed. Asymmetry, 15 (2004) 1995–2001.

43 N. Berova, K. Nakanishi and R.W. Woody, Circular Dichroism: Principles and Applications, John Wiley & Sons, New York, 2000.

44 G. Holzwarth, E.C. Hsu, H.S. Mosher, T.R. Faulkner and A. Moscowitz, Infrared circular dichroism of carbon-hydrogen and carbon-deuterium stretching modes. Observations. J. Am. Chem. Soc., 96 (1974) 251–252.

45 L.A. Nafie, T.A. Keiderling and P.J. Stephens, Vibrational circular dichroism. J. Am. Chem. Soc., 98 (1976) 2715–2723.

46 L.A. Nafie, M. Diem and D.W. Vidrine, Fourier transform infrared vibrational circular dichroism. J. Am. Chem. Soc., 101 (1979) 496–498.

47 P. Malon and T.A. Keiderling, A solution to the artifact problem in Fourier-transform vibrational circular dichroism. Appl. Spectrosc., 42 (1988) 32–38; G.C. Chen, P.L. Polavarapu and S. Weibel, New design for Fourier-transform infrared vibrational circular dichroism spectrometers. Appl. Spectrosc., 48 (1994) 1218; D. Tsankov, T. Eggiman and H. Weiser, Alternative dosing for improved FTIR/VCD capabilities. Appl. Spectrosc., 49 (1995) 132–138.

48 L.A. Nafie and D.W. Vidrine, in: J.R. Ferraro and L.J. Basile (Eds.), Fourier Transform Infrared Spectroscopy, Academic Press, New York, 1982; P.L. Polavarapu, in: J.R. Ferraro and L.J. Basile (Eds.), Fourier Transform Infrared Spectroscopy, Academic Press, New York, 1985; T.A. Keiderling, in: J.R. Ferraro and K. Krishnan (Eds.), Fourier Transform Infrared Spectroscopy, Academic Press, New York, 1990.

49 J.C. Cheng, L.A. Nafie and P.J. Stephens, Polarization scrambling using a photoelastic modulator: application to circular dichroism measurement. J. Opt. Soc. Am., 65 (1975) 1031–1035.

50 G. Shanmugam and P.L. Polavarapu, Vibrational circular dichroism of protein films. J. Am. Chem. Soc., 126 (2004) 10292–10295; G. Shanmugam and P.L. Polavarapu, Vibrational circular dichroism spectra of protein films: thermal denaturation of bovine serum albumin. Biophys. Chem., 111 (2004) 73–77; G. Shanmugam and P.L. Polavarapu, Structure of Ab(25-35) peptide in different environments. Biophys. J., 87 (2004) 622–630.

51 R.A. Lombardi, X. Cao and L.A. Nafie, Detection of chirality in solid-state samples using Fourier transform vibrational circular dichroism. 16th International Symposium on Chirality, New York, 2004.

52 L.A. Nafie, Advanced applications of mid-IR and near-IR vibrational circular dichroism: new spectral regions, reaction monitoring and quality control. 16th International Symposium on Chirality, New York, 2004.

53 P.L. Polavarapu and D.F. Michalska, Vibrational circular dichroism in (S)-(-)-epoxypropane. Measurement in vapor phase and verification of the perturbed degenerate mode theory. J. Am. Chem. Soc., 105 (1983) 6190–6191; P.L. Polavarapu, Rotational-vibrational circular dichroism. Chem. Phys. Lett., 161 (1989) 485–490.

54 E.B. Wilson, J.C. Decius and P.C. Cross, Molecular Vibrations, Dover Publications, New York, 1955.

55 (a) J. He, F. Wang and P.L. Polavarapu, Absolute configurations of chiral herbicides determined from vibrational circular dichroism. Chirality, 17 (2005) S1–S8; (b) J. He and P.L. Polavarapu, Determination of the absolute configuration of chiral alpha-aryloxypropanoic acids using vibrational circular dichroism studies: 2-(2-chlorophenoxy) propanoic acid and 2-(3-chlorophenoxy) propanoic acid. Spectrochim. Acta. A, 61 (2005) 1327–1334; (c). J. He and P.L. Polavarapu, Determination of inter-molecular hydrogen bonded conformers of a-aryloxypropanoic acids using density functional theory predictions of vibrational absorption and vibrational circular dichroism spectra. J. Chem. Theory and Computation, 1 (2005) 506–514.

56 P.L. Polavarapu, Vibrational optical activity of anharmonic oscillator. Mol. Phys., 89 (1996) 1503–1510; S. Abbate, G. Longhi and C. Santina, Theoretical and experimental studies for the interpretation of vibrational circular dichroism spectra in the CH-stretching overtone region. Chirality, 12 (2000) 180–190.

57 P.L. Polavarapu and Z. Deng, Measurement of vibrational circular dichroism below 600 cm^{-1}: Progress towards meeting the challenge. Appl. Spectrosc., 50 (1996) 686–692.

58 (a) P.A. Galwas, On the Distribution of Optical Polarization in Molecules, Chapter 6. Ph.D. Thesis, Cambridge University, Cambridge, 1983; (b) A.D. Buckingham, P.W. Fowler and P.A. Galwas, Velocity-dependent property surfaces and the theory of vibrational circular dichroism. Chem. Phys., 112 (1987) 1–14; (c) P.J. Stephens, Theory of vibrational circular dichroism. J. Phys. Chem., 89 (1985) 748–752; (d) J.R. Cheeseman, M.J. Frisch, F.J. Devlin and P.J. Stephens, Ab initio calculation of atomic axial tensors and vibrational rotational strengths using density functional theory. Chem. Phys. Lett., 252 (1996) 211–220.

59 (a) J. Kubelka, R. Huang and T.A. Keiderling, Solvent effects on IR and VCD spectra of helical peptides: DFT-based static spectral simulations with explicit water. J. Phys. Chem. B, 109 (2005) 8231–8243; (b) V. Andrushchenko, H. Wieser and P. Bour, RNA structural forms studied by vibrational circular dichroism: ab initio interpretation of the spectra. J. Phys. Chem. B, 108 (2004) 3899–3911.

60 L.D. Barron and A.D. Buckingham, Rayleigh and Raman scattering from optically active molecules. Mol. Phys., 20 (1971) 1111–1119.

61 L.D. Barron, M.P. Bogaard and A.D. Buckingham, Raman scattering of circularly polarized light by optically active molecules. J. Am. Chem. Soc., 95 (1973) 603–605; L.D. Barron, M.P. Bogaard and A.D. Buckingham, Differential scattering of right and left circularly polarized light by asymmetric molecules. Nature, 241 (1973) 113–114.

62 W. Hug, S. Kint, G.F. Bailey and J.R. Scherer, Raman circular intensity differential spectroscopy. The spectra of (−)-a-pinene and (+)-a-phenylethylamine. J. Am. Chem. Soc., 97 (1975) 5589–5590.

63 T. Brocki, M. Moskovits and B. Bosnich, Vibrational optical activity. Circular differential Raman scattering from a series of chiral terpenes. J. Am. Chem. Soc., 102 (1980) 495–500.

64 W. Hug and H. Surbeck, Vibrational Raman optical activity spectra recorded in perpendicular polarization. Chem. Phys. Lett., 60 (1979) 186–192.

65 L.D. Barron, J.F. Torrance and D.J. Cutler, A new multichannel Raman optical activity instrument. J. Raman Spectrosc., 18 (1987) 281–287.

66 W. Hug, Optical artefacts and their control in Raman circular difference scattering. Appl. Spectrosc., 35 (1981) 115–124; L.D. Barron and J. Vrbancich, On the theory of the dominant artefacts in natural and magnetic Raman optical activity. J. Raman Spectrosc., 15 (1984) 47–50; J.R. Escribano, The influence of a finite collection angle on Rayleigh and Raman optical activity. Chem. Phys. Lett., 121 (1985) 191–193; L. Hecht, B. Jordanov and B. Schrader, Appl. Spectrosc., 41 (1987) 295–307; L. Hecht and L.D. Barron, An analysis of modulation experiments for Raman optical activity. Appl. Spectrosc., 44 (1990) 483–491; D. Che and L.A. Nafie, Theory and reduction of artefacts in incident, scattered and dual circular polarization forms of Raman optical activity. Appl. Spectrosc., 47 (1993) 544–555.

67 L.D. Barron, L. Hecht and S.M. Blyth, Polarized Raman optical activity of menthol and related molecules. Spectrochim. Acta, 45A (1989) 375–379.

68 L. Hecht and L.D. Barron, Magic angle Raman optical activity: beta-pinene and nopinone. Spectrochim. Acta, 45A (1989) 671–674.

69 W. Hug, Instrumental and theoretical advances in Raman optical activity in: J. Lascombe and P.V. Huong (Eds.), Raman Spectroscopy, Wiley, Chichester, 1982; L. Hecht, L.D. Barron and W. Hug, Vibrational Raman optical activity in backscattering. Chem. Phys. Lett., 158 (1989) 341–344.

70 K.M. Spencer, T.B. Freedman and L.A. Nafie, Scattered circular polarization Raman optical activity. Chem. Phys. Lett., 149 (1988) 367–374.

71 L.A. Nafie and T.B. Freedman, Dual circular polarization Raman optical activity. Chem. Phys. Lett., 154 (1989) 260–266.

72 L. Hecht, L.D. Barron, A.R. Gargaro, Z.Q. Wen and W. Hug, Raman optical activity instrument for biochemical studies. J. Raman Spectrosc., 23 (1992) 401–411.

73 L. Hecht, D. Che and L.A. Nafie, A new scattered circular polarization Raman optical activity instrument equipped with a charge-coupled-device detector. Appl. Spectrosc., 45 (1991) 18–25.

74 D. Che, L. Hecht and L.A. Nafie, Dual and incident circular polarization Raman optical activity backscattering of (−)-trans-pinane. Chem. Phys. Lett., 180 (1991) 182–189.

75 M. Vargek, T.B. Freedman and L.A. Nafie, Improved backscattering dual circular polarization Raman optical activity spectrometer with enhanced performance for biomolecular applications. J. Raman Spectrosc., 28 (1997) 627–633.

76 L. Hecht and L.D. Barron, Instrument for natural and magnetic Raman optical activity studies in right-angle scattering. J. Raman Spectrosc., 25 (1994) 443–451.

77 W. Hug and G. Hangartner, A novel high-throughput Raman spectrometer for polarization difference measurements. J. Raman Spectrosc., 30 (1999) 841–852.

78 W. Hug, G. Zuber, A.D.Meijere, A.F. Khlebnikov and H.-J.Hansen, Raman optical activity of a purely sigma-bonded helical chromophore (-)-(M)-sigma-[4]Helicene. Helvetica Chim. Acta, 84 (2001) 1–21.

79 (a) M.J. Frisch, Y. Yamaguchi, J.F. Gaw, H.F. Schaeffer II and J.S. Binkley, Analytic Raman intensities from molecular electronic wavefunctions. J. Chem. Phys., 84 (1986) 531–532; (b) R.D. Amos, Calculation of polarizability derivatives using analytic gradient methods. Chem. Phys. Lett., 124 (1986) 376–381; (c) O. Quinet and B.J. Champagne, Time-dependent Hartree-Fock schemes for analytical evaluation of the Raman intensities. Chem. Phys., 115 (2001) 6293–6299; (d) O. Quinet, V. Liegeois and B.J. Champagne, TDHF evaluation of the dipole-quadrupole polarizability and its geometrical derivatives. J. Chem. Theory Comput., 1 (2005) 444–452.

80 (a) P.K. Bose, L.D. Barron and P.L. Polavarapu, Ab initio and experimental vibrational Raman optical activity in (+)-(R)-methylthiirane. Chem. Phys. Lett., 155 (1989) 423–429; (b) P.L. Polavarapu, Ab initio vibrational Raman and Raman optical activity spectra. J. Phys. Chem., 94 (1990) 8106–8112.

81 K. Ruud, T. Helgaker and P. Bour, Gauge-origin independent density-functional theory calculations of vibrational Raman optical activity. J. Phys. Chem. A, 106 (2002) 7448–7455.
82 L. Hecht and L.D. Barron, Recent developments in Raman optical activity instrumentation. Farad. Discuss., 99 (1994) 35–47.
83 P.L. Polavarapu, A. Cholli and G. Vernice, Absolute configuration of isoflurane. J. Am. Chem. Soc., 114 (1992) 10953–10954.
84 (a) L.D. Barron, L. Hecht, I.H. McColl and E.W. Blanch, Raman optical activity comes of age. Mol. Phys., 102 (2004) 731–744; (b) M. Pecul and K. Ruud, Ab initio calculation of vibrational Raman optical activity. Int. J. Quantum Chem., 104 (2005) 816–829.

APPENDIX

A. Literature articles on quantum mechanically predicted specific rotations for evaluating molecular stereochemistry

J. Autschbach, S. Patchkovskii, T. Ziegler, S.J.A. van Gisbergen and E.J. Baerends, Chiroptical properties from time-dependent density functional theory. II. Optical rotations of small to medium size organic molecules. J. Chem. Phys., 117 (2002) 581–592.

J.R. Cheeseman, M.J. Frisch, F.J. Devlin and P.J. Stephens, Hartree-Fock and density functional theory ab initio calculation of optical rotation using GIAOs: Basis set dependence. J. Phys. Chem. A, 104 (2000) 1039–1046.

S. Coriani, M. Pecul, A. Rizzo, P. Jorgensen and M. Jaszunski, Ab initio study of magnetochiral birefringence. J. Chem. Phys., 117 (2002) 6417–6428.

J. Crassous, Z. Jiang, V. Schurig and P.L. Polavarapu, Preparation of (+)-chlorofluoroiodomethane, determination of its enantiomeric excess and of its absolute configuration. Tetrahed. Asymmetry, 15 (2004) 1995–2001.

T.D. Crawford, L.S. Owens, M.C. Tam, P.R. Schreiner and H. Koch, Ab initio calculation of optical rotation in (P)-(+)-[4]triangulane. J. Am. Chem. Soc., 127 (2005) 1368.

E. Giorgio, C. Minichino, R.G. Viglione, R. Zanasi and C. Rosini, Assignment of the molecular absolute configuration through the ab initio Hartree-Fock calculation of the optical rotation: Can the circular dichroism data help in reducing basis set requirements? J. Org. Chem., 68 (2003) 5186–5192.

E. Giorgio, C. Rosini, R.G. Viglione and R. Zanasi, Calculation of gas phase specific rotation of (S)-propylene oxide at 355 nm. Chem. Phys. Lett., 376 (2003) 452–456.

E. Giorgio, N. Parrinello, A. Caccamese and C. Rosini, Non-empirical assignment of the absolute configuration of (−)-naringenin, by coupling the exciton analysis of the circular dichroism spectrum and the ab initio calculation of the optical rotatory power. Org. Biomol. Chem., 2 (2004) 3602–3607.

E. Giorgio, R.G. Viglione and C. Rosini, Assignment of absolute configuration of large molecules by ab initio calculation of rotatory power within small basis set scheme. The case of some biologically active natural products. Tetrahed: Asymmetry, 15 (2004) 1979–1986.

E. Giorgio, R.G. Viglione, R. Zanasi and C. Rosini, Ab initio calculation of optical rotatory dispersion (ORD) curves: A simple and reliable approach to the assignment of the molecular absolute configuration. J. Am. Chem. Soc., 126 (2004) 12968–12976.

E. Giorgio, L. Maddau, E. Spanu, A. Evidente and C. Rosini, Assignment of the absolute configuration of (+)-diplopyrone, the main phytotoxin produced by Diplodia mutila, the pathogen of the cork oak decline, by a nonempirical analysis of its chiroptical properties. J. Org. Chem., 70 (2005) 7–13.

E. Giorgio, M. Roje, K. Tanaka, Z. Hamersak, V. Sunjic, K. Nakanishi, C. Rosini and N. Berova, Determination of the absolute configuration of flexible molecules by ab initio O.R.D calculations: A case study with cytoxazones and isocytoxazones. J. Org. Chem., 70 (2005) 6557–6563.

M. Goldsmith, N. Jayasuriya, D.N. Beratan and P. Wipf, Optical rotation of noncovalent aggregates. J. Am. Chem. Soc., 125 (2003) 15696–15697.

S. Grimme, Calculation of frequency dependent optical rotation using density functional response theory. Chem. Phys. Lett., 339 (2001) 380–388.

S. Grimme, A. Bahlmann and G. Haufe, Ab initio calculations for optical rotations of conformationally flexible molecules: A case study on six-, seven-, and eight-membered fluorinated cycloalkanol esters. Chirality, 14 (2002) 793–797.

S. Grimme, F. Furche and R. Ahlrichs, An improved method for density functional calculations of the frequency-dependent optical rotation. Chem. Phys. Lett., 361 (2002) 321–328.

J. He, A. Petrovic and P.L. Polavarapu, Quantitative determination of conformer populations: Assessment of specific rotation, vibrational absorption and vibrational circular dichroism in substituted butynes. J. Phys. Chem., 108 (2004) 1671–1680.

J. He, A.G. Petrovic and P.L. Polavarapu, Determining the conformer populations of (R)-(+)-3-methylcyclopentanone using vibrational absorption, vibrational circular dichroism, and specific rotation. J. Phys. Chem. B, 108 (2004) 20451–20457.

L. Hecht, J. Costante, P.L. Polavarapu, A. Collet and L.D. Barron, Absolute configuration of bromochlorofluoromethane from experimental and ab initio theoretical vibrational Raman optical activity. Angew. Chem., 36 (1997) 885–887.

R.K. Kondru, P. Wipf and D.N. Beratan, Atomic contributions to the optical rotation angle as a quantitative probe of molecular chirality. Science, 282 (1998) 2247–2250.

R.K. Kondru, P. Wipf and D.N. Beratan, Theory assisted determination of absolute stereochemistry for complex natural products via computation of molar rotation angles. J. Am. Chem. Soc., 120 (1998) 2204–2205.

R.K. Kondru, C.H.T. Chen, D.P. Curan, D.N. Beratan and P. Wipf, Determination of the absolute configuration of 1,3,5,7-tetramethyl-1,3-dihydroindol-2-one by optical rotation computation. Tetrahed: Asymmetry, 10 (1999) 4143–4150.

R.K. Kondru, P. Wipf and D.N. Beratan, Structural and conformational dependence of optical rotation angles. J. Phys. Chem. A, 103 (1999) 6603–6611.

R.K. Kondru, D.N. Beratan, G.K. Friestad, A.B. Smith III and P. Wipf, Chiral action at a distance: Remote substituent effects on the optical activity of calyculins A and B. Org. Lett., 2 (2000) 1509–1512.

J. Kongsted, T.B. Pedersen, M. Strange, A. Osted, A.E. Hansen, K.V. Mikkelsen, F. Pawlowski, P. Jorgensen and C. Hattig, Coupled cluster calculations of optical rotation of S-propylene oxide in gas phase and solution. Chem. Phys. Lett., 401 (2005) 385–392.

T. Kuwada, M. Fukui, T. Hata, T. Choshi, J. Nobuhiro, Y. Ono and S. Hibino, The absolute configuration of (+)-oxopropaline D by theoretical calculation of specific rotation and asymmetric synthesis. Chem. Pharm. Bull., 51 (2003) 20–23.

A. Ligabue, P. Lazzeretti, M.P. Beccar Varela and M.B. Ferraro, On the resolution of the optical rotatory power of chiral molecules into atomic terms. A study of hydrogen peroxide. J. Chem. Phys., 116 (2002) 6427–6434.

D. Marchesan, S. Coriani, C. Forzato, P. Nitti, G. Pitacco and K. Ruud, Optical rotation calculation of a highly flexible molecule: The case of paraconic acid. J. Phys. Chem. A, 109 (2005) 1449–1453.

D.M. McCann, P.J. Stephens and J.R. Cheeseman, Determination of absolute configuration using density functional theory calculation of optical rotation: Chiral alkanes. J. Org. Chem., 69 (2004) 8709–8717.

B. Mennucci, J. Tomasi, R. Cammi, J.R. Cheeseman, M.J. Frisch, F.J. Devlin, S. Gabriel and P.J. Stephens, Polarizable continuum model (P.C.M) calculations of solvent effects on optical rotations of chiral molecules. J. Phys. Chem., 106 (2002) 6102–6113.

T. Miller, K.B. Wiberg and P.H. Vaccaro, Cavity ring-down polarimetry: A new scheme for probing circular birefringence and circular dichroism in the gas phase. J. Phys. Chem. A, 104 (2000) 5959–5968.

T. Miller, K.B. Wiberg, P.H. Vaccaro, J.R. Cheeseman and M.J. Frisch, Cavity ring-down polarimetry (CRDP): Theoretical and experimental characterization. J. Opt. Soc. Am. B, 19 (2002) 125–141.

P. Norman, K. Ruud and T. Helgaker, Density-functional theory calculations of optical rotatory dispersion in the nonresonant and resonant frequency regions. J. Chem. Phys., 120 (2004) 5027–5035.

M. Pecul, K. Ruud and A. Rizzo, Conformational effects on the optical rotation of alanine and proline. J. Phys. Chem. A, 108 (2004) 4269–4276.

T.B. Pedersen, A.M.J. Sanchez de Meras and H. Koch, Polarizability and optical rotation calculated from the approximate coupled cluster singles and doubles C.C.2 linear response theory using Cholesky decomposition. J. Chem. Phys., 120 (2004) 8887–8897.

T.B. Pedersen, H. Koch, L. Boman and A.M.J. Sanchez de Meras, Origin invariant calculation of optical rotation without recourse to London orbitals. Chem. Phys. Lett., 393 (2004) 319–326.

A.G. Petrovic, P.L. Polavarapu, J. Drabowicz, Y. Zhang, O.J. McConnell and H. Duddeck, Absolute configuration of C2 symmetric spiroselenurane: 3,3,3′,3′-tetramethyl-1,1′-spirobi[3H,2,1]-benzoxaselenole. Chem. Eur., 11 (2005) 4257–4262.

A.G. Petrovic, J. He, P.L. Polavarapu, L.S. Xiao and D.W. Armstrong, Absolute configuration and predominant conformations of 1,1-dimethyl-2-phenylethyl phenyl sulfoxide. Org. Biomol. Chem., 3 (2005) 1977–1981.

P.L. Polavarapu, Molecular optical rotations and structures. Tetrahed. Asymmetry, 8 (1997) 3397–3401.

P.L. Polavarapu, Ab initio molecular optical rotations and absolute configurations. Mol Phys., 91 (1997) 551–554.

P.L. Polavarapu, Absolute configuration of bromochlorofluoromethane. Angewandte Chem., 41 (2002) 4544–4546.

P.L. Polavarapu, Optical rotation: Recent advances in determining the absolute configuration. Chirality, 14 (2002) 768–781, 15 (2003) 284–285.

P.L. Polavarapu, Kramers-Kronig transformation for optical rotatory dispersion studies. J. Phys. Chem. A, 109 (2005) 7013–7023.

P.L. Polavarapu and D.K. Chakraborty, Absolute stereochemistry of chiral molecules from ab initio theoretical and experimental molecular optical rotations. J. Am. Chem. Soc., 120 (1998) 6160–6164.

P.L. Polavarapu and D.K. Chakraborty, Ab initio theoretical optical rotations of small molecules. Chem. Phys., 240 (1999) 1–8.

P.L. Polavarapu and C. Zhao, A comparison of ab initio optical rotations obtained with static and dynamic methods. Chem. Phys. Lett., 296 (1998) 105–110.

P.L. Polavarapu and C. Zhao, Ab initio predictions of anomalous optical rotatory dispersion. J. Am. Chem. Soc., 121 (1999) 246–247.

P.L. Polavarapu, D.K. Chakraborty and K. Ruud, Molecular optical rotations: An evaluation of the semiempirical models. Chem. Phys. Lett., 319 (2000) 595–600.

P.L. Polavarapu, A. Petrovic and F. Wang, Intrinsic rotation and molecular structure. Chirality, 15 (2003) S143–S149; Chirality, 15 (2003) 801.

P.L. Polavarapu, F. Wang and J. Drabowicz, Determination of absolute configuration of chiral phosphorous compounds from optical rotations (unpublished results).

P.L. Polavarapu, J. He, J. Crassous and K. Ruud, Absolute configuration of C_{76} from optical rtatory dspersion. Chem. Phys. Chem., 6(12) (2005) 25.

S. Ribe, R.K. Kondru, D.N. Beratan and P. Wipf, Optical rotation computation, total synthesis and stereochemistry assignment of the marine natural product pitimide A. J. Am. Chem. Soc., 122 (2000) 4608–4617.

B.C. Rinderspacher and P.R. Schreiner, Structure-property relationships of prototypical chiral compounds: Case studies. J. Phys. Chem. A, 108 (2004) 2867–2870.

K. Ruud and T. Helgaker, Optical rotation studied by density-functional and coupled-cluster methods. Chem. Phys. Lett., 352 (2002) 533–539.

K. Ruud and R. Zanasi, The importance of molecular vibrations: The sign change of the optical rotation of methyloxirane. Angewandte Chem. Int. Ed., 44 (2005) 3594–3596.

K. Ruud, P.R. Taylor and P. Astrand, Zero-point vibrational effects on optical rotation. Chem. Phys. Lett., 337 (2001) 217–223.

K. Ruud, P.J. Stephens, F.J. Devlin, P.R. Taylor, J.R. Cheeseman and M.J. Frisch, Coupled cluster calculations of optical rotation. Chem. Phys. Lett., 373 (2003) 606–614.

P.J. Stephens, F.J. Devlin, J.R. Cheeseman, M.J. Frisch, B. Mennucci and J. Tomasi, Prediction of optical rotation using density functional theory: 6,8-dioxabicyclo[3.2.1]octanes. Tetrahed. Asymmetry, 11 (2000) 2443–2448.

P.J. Stephens, F.J. Devlin, J.R. Cheeseman and M.J. Frisch, Calculation of optical rotation using density functional theory. J. Phys. Chem. A, 105 (2001) 5356–5371.

P.J. Stephens, F.J. Devlin, J.R. Cheeseman and M.J. Frisch, Ab initio prediction of optical rotation: a comparison of density functional theory and Hartree-Fock methods for three 2,7,8-trioxabicyclo[3.2.1]octanes. Chirality, 14 (2002) 288–296.

P.J. Stephens, F.J. Devlin, J.R. Cheeseman, M.J. Frisch and C. Rosini, Determination of absolute configuration using optical rotation calculated using density functional theory. Org. Lett., 4(26) (2002) 4595–4598.

P.J. Stephens, F.J. Devlin, J.R. Cheeseman, M.J. Frisch, O. Bortolini and P. Beese, Determination of absolute configuration using ab initio calculation of optical rotation. Chirality, 15 (2003) S57–S64.

P.J. Stephens, D.M. McCann, E. Butkus, S. Stoncius, J.R. Cheeseman and M.J. Frisch, Determination of absolute configuration using concerted ab initio DFT calculations of electronic circular dichroism and optical rotation: Bicyclo[3.3.1]nonane diones J. Org. Chem., 69(6) (2004) 1948–1958.

P.J. Stephens, D.M. McCann, F.J. Devlin, J.R. Cheeseman and M.J. Frisch, Determination of the absolute configuration of [3(2)](1,4) barrelenophanedicarbonitrile using concerted time-dependent density functional theory calculations of optical rotation and electronic circular dichroism. J. Am. Chem. Soc., 126 (2004) 7514–7521.

P.J. Stephens, D.M. McCann, J.R. Cheeseman and M.J. Frisch, Determination of absolute configurations of chiral molecules using ab initio time-dependent density functional theory calculations of optical rotation: How reliable are absolute configurations obtained for molecules with small rotations? Chirality, 17 (2005) S52–S64.

M.C. Tam, N.J. Russ and T.D. Crawford, Coupled cluster calculation of optical rotatory dispersion of (S)-methyloxirane. J. Chem. Phys., 121 (2004) 3550–3557.

S.B. Vogensen, J.R. Greenwood, A.R. Varming, L. Brehm, D.S. Pickering, B. Nielsen, T. Liljefors, R.P. Clausen, T.N. Johansen and P. Krogsgaard-Larsen, A stereochemical anomaly: The cyclised (R)-AMPA analogue (R)-3-hydroxy-4,5,6,7-tetrahydroisoxazolo[5,4-c]pyridine-5-carboxylic acid [(R)-5-HPCA] resembles (S)-AMPA at glutamate receptors. Org. Biomol. Chem., 2 (2004) 206–213.

F. Wang, Y. Wang, P.L. Polavarapu, T. Li, J. Drabowicz, K.M. Pietrusiewicz and K. Zygo, Absolute configuration of tert-butyl-1-(2-methylnaphthyl)phosphineoxide. J. Org. Chem., 67 (2002) 6539–6541.

F. Wang, P.L. Polavarapu, J. Drabowicz, P. Kielbasinski, M.J. Potrzebowski, M. Mikolajczyk, M.W. Wieczorek, W. Majzner and I. Lazewska, Solution and crystal structures of chiral molecules can be significantly different: Tert-butylphenylphosphinoselenoic acid. J. Phys. Chem. A, 108 (2004) 2072–2079.

K.B. Wiberg, P.H. Vaccaro and J.R. Cheeseman, Conformational effects on optical rotation. 3-Substituted 1-butenes. J. Am. Chem. Soc., 125 (2003) 1888–1896.

K.B. Wiberg, Y. Wang, M.J. Murphy and P.K. Vaccaro, Temperature dependence of optical rotation: a-pinene, b-pinene, pinane, camphene, camphor and fenchone. J. Phys. Chem. A, 108 (2004) 5559–5563.

K.B. Wiberg, Y. Wang, P.H. Vaccaro, J.R. Cheeseman, G. Trucks and M.J. Frisch, Optical activity of 1-butene, butane and related hydrocarbons. J. Phys. Chem., 108 (2004) 32–38.

G. Zuber, M. Goldsmith, D.N. Beratan and P. Wipf, Assignment of the absolute configuartion of [n]-ladderanes by TD-DFT optical rotation calculations. Chirality, 17 (2005) 507–510.

B. Literature articles (1999–2005) on quantum mechanically predicted VCD spectra for evaluating molecular stereochemistry

A. Aamouche, F.J. Devlin and P.J. Stephens, Conformations of chiral molecules in solution: ab initio vibrational absorption and circular dichroism studies of 4,4a,5,6,7,8-hexahydro-4a-methyl-2(3H)-naphthalenone and 3,4,8,8a-tetrahydro-8a-methyl-1,6(2H,7H)-naphthalenedione. J. Am. Chem. Soc., 122 (2000) 7358–7367.

A. Aamouche, F.J. Devlin and P.J. Stephens, Structure, vibrational absorption and circular dichroism spectra, and absolute configuration of Troger's base. J. Am. Chem. Soc., (2000) 2346–2354.

A. Aamouche, F.J. Devlin, P.J. Stephens, J. Drabowicz, B. Bujnicki and M. Mikolajczyk, Vibrational circular dichroism and absolute configuration of chiral sulfoxides: tert-butyl methyl sulfoxide. Chem. Eur. J., 6 (2000) 4479–4486.

V. Andrushchenko, J.L. McCann, J.H. van de Sande and H. Wieser, Determining structures of polymeric molecules by vibrational circular dichroism (VCD) spectroscopy. Vib. Spectrosc. 22 (2000) 101–109.

C.S. Ashvar, F.J. Devlin and P.J. Stephens, Molecular structure in solution: An ab initio vibrational spectroscopy study of phenyloxirane. J. Am. Chem. Soc., 121 (1999) 2836–2849.

D. Bas, T. Burgi, J. Lacour, J. Vachon and J. Weber, Vibrational and electronic circular dichroism of Δ-TRISPHAT [Tris(tetrachlorobenzenediolato)-phosphate(V)] anion. Chirality, 17 (2005) S143–S148.

L.A. Bodack, T.B. Freedman, B.Z. Chowdhry and L.A. Nafie, Solution conformations of cyclosporins and magnesium-cyclosporin complexes determined by vibrational circular dichroism. Biopolymers, 73 (2004) 163–177.

P. Bour, K. Zaruba, M. Urbanova, V. Setnicka, P. Matejka, Z. Fiedler, V. Kral and K. Volka, Vibrational circular dichroism of tetraphenylporphyrin in peptide complexes? A computational study. Chirality, 12 (2000) 191–198.

P. Bour, H. Navratilova, V. Stnicka, M. Urbanova and K. Volka, (3R,4S)-4-(4-fluorophenyl)-3-hydroxymethyl-1-methylpiperidine: Conformation and structure monitoring by vibrational circular dichroism. J. Org. Chem., 67 (2002) 161–168.

T. Buffeteau, L. Ducasse, A. Brizard, I. Huc and R. Oda, Density functional theory calculation of vibrational absorption and circular dichroism spectra of dimethyl-L-tartrate. J. Phys. Chem. A, 108 (2004) 4080–4086.

E. Buktus, A. Zilinskas, S. Stoncius, R. Rozenbergas, M. Urbanova, V. Setnicka, P. Bour and K. Volka, Synthesis and chiroptical properties of enantiopure tricyclo[4.4.0.03,8]nonane-4,5-dione (twistbrendanedione). Tetrahed. Asymmetry, 13 (2002) 633–638.

T. Burgi, A. Vargaa and A. Baker, VCD spectroscopy of chiral cinchona modifiers used in heterogeneous enantioselective hydrogenation: Conformation and binding of non-chiral acids. J. Chem. Soc. Perkin Trans., 2 (2002) 1596–1601.

C. Cappelli, S. Monti and A. Rizzo, Effect of the environment on vibrational infrared and circular dichroism spectra of (S)-proline. Int. J. Quantum Chem., 104 (2005) 744–745.

R.H. Cichewicz, L.J. Clifford, P.R. Lassen, X. Cao, T.B. Freedman, L.A. Nafie, J.D. Deschamps, V.A. Kenyon, J.R. Flanary, T.R. Holman and P. Crews, Stereochemical determination and bioactivity assessment of (S)-(+)-curcuphenol dimers isolated from the marine sponge Didiscus aceratus and synthesized through laccase biocatalysis. Bioorg. Med. Chem., 13 (2005) 5600–5612.

F.J. Devlin and P.J. Stephens, Conformational analysis using ab initio vibrational spectroscopy: 3-methylcyclohexanone. J. Am. Chem. Soc., 121 (1999) 7413–7414.

F.J. Devlin, P.J. Stephens, P. Scafato, S. Superchi and C. Rosini, Determination of absolute configuration using vibrational circular dichroism spectroscopy: The chiral sulfoxide 1-thiochroman S-oxide. Tetrahedr. Asymmetry, 12 (2001) 1551–1558.

F.J. Devlin, P.J. Stephens, C. Osterle, K.B. Wiberg, J.R. Chesseman and M.J. Frisch, Configurational and conformational analysis of chiral molecules using IR and VCD spectroscopies: spiropentylcarboxylic acid methyl ester and spiropentyl acetate. J. Org. Chem., 67 (2002) 8090–8096.

F.J. Devlin, P.J. Stephens, P. Scafato, S. Superchi and C. Rosini, Conformational analysis using infrared vibrational circular dichroism spectroscopies: the chiral cyclic sulfoxides 1-thiochroman-4-one S-oxide, 1-thiaindan S-oxide and 1-thiochroman S-oxide. J. Phys. Chem. A, 106 (2002) 10510–10524.

F.J. Devlin, P.J. Stephens, P. Scafato, S. Superchi and C. Rosini, Determination of absolute configuration using vibrational circular dichroism spectroscopy: the chiral sulfoxide 1-thiochromanone S-oxide. Chirality, 14 (2002) 400–406.

F.J. Devlin, P.J. Stephens and P. Besse, Conformational rigidification via derivatization facilitates the determination of absolute configuration using chiroptical spectroscopy: a case study of the chiral alcohol endo-borneol. J. Org. Chem., 70 (2005) 2980–2993.

F.J. Devlin, P.J. Stephens and P. Besse, Are the absolute configurations of 2-(1-hydroxyethyl)-chromen-4-one and its 6-bromo derivative determined by X-ray crystallography correct? A vibrational circular dichroism study of their acetate derivatives. Tetrahedr. Asymmetry, 16 (2005) 1557–1566.

J. Dobler, N. Peters, C. Larsson, A. Bergman, E. Geidel and H. Huhnerfuss, The absolute structures of separated PCB-methylsulfone enantiomers determined by vibrational circular dichroism and quantum mechanical calculations, J. Mol. Struct. (Theochem), 586 (2002) 159–166.

J. Drabowicz, B. Dudzinski, M. Mikolajczyk, F. Wang, A. Dehlavi, J. Goring, M. Park, C. Rizzo, P.L. Polavarapu, P. Biscarini, M.W. Wieczorek and W.R. Majzner, Absolute configuration, predominant conformations and vibrational circular dichroism spectra of enantiomers of n-butyl-t-butyl sulfoxide. J. Org. Chem., 66 (2001) 1122–1129.

D. Dunmire, T.B. Freedman, L.A. Nafie, C. Aeschlimann, J.G. Gerber and J. Gal, Determination of absolute configuration of the antifungal agents ketoconazole, itraconazole, and micoconazole with vibrational circular dichroism. Chirality, 17 (2005) S101–S108.

A.B. Dyatkin, T.B. Freedman, X. Cao, R.K. Dukor, B.E. Maryanoff, C.A. Maryanoff, J.M. Matthews, R.D. Shaw and L.A. Nafie, Determination of the absolute configuration of a key tricyclic component of a novel vasopressin receptor antagonist by use of vibrational circular dichroism. Chirality, 14 (2002) 215–219.

T.B. Freedman, X. Cao, D.A. Young and L.A. Nafie, Density functional theory calculations of vibrational circular dichroism in transition metal complexes: Identification of solution conformations and mode of chloride ion association for (+)-tris-(ethylenediaminato)cobalt(III). J. Phys. Chem. A, 106 (2002) 3560–3565.

T.B. Freedman, R.K. Dukor, J.C.M. van Hoof, E.R. Kellenbach and L.A. Nafie, Determination of the absolute configuration of (−)-mirtazapine by vibrational circular dichroism. Helv. Chim. Acta, 85 (2002) 1160–1165.

T.B. Freedman, X. Cao, R.K. Dukor and L.A. Nafie, Absolute configuration determination of chiral molecules in the solution state using vibrational circular dichroism. Chirality, 15 (2003) 743–758.

T.B. Freedman, X. Cao, L.A. Nafie, M. Kalbermatter, A. Linden and A.J. Rippert, An unexpected atropisomerically stable 1,1-biphenyl at ambient temperature in solution elucidated by vibrational circular dichroism (V.C.D). Helv. Chim. Acta, 86 (2003) 3141–3155.

T.B. Freedman, X. Cao, R. Oliveira, Q.B. Cass and L.A. Nafie, Determination of absolute configuration and solution conformation of gossypol by vibrational circular dichroism. Chirality, 15 (2003) 196–200.

T.B. Freedman, X. Cao, A. Rajca, H. Wang and L.A. Nafie, Determination of absolute configuration in molecules with chiral axes by vibrational circular dichroism: a C2 symmetric annelated heptathiophene and a D2-symmetric dimer of 1,1′-binaphthyl. J. Phys. Chem., 107 (2003) 7692–7696.

T.B. Freedman, X. Cao, L.A. Nafie, A. Solladie-Cavallo, L. Jierry and L. Bouerat, VCD configuration of enantiopure/-enriched tetrasubstituted a-fluoro cyclohexanones and their use for epoxidation of trans-olefins. Chirality, 16 (2004) 467–474.

T.B. Freedman, X. Cao, L.A. Nafie, M. Kalbermatter, A. Linden and A.J. Rippert, Determination of the atropisomeric stability and solution conformation of asymmetrically substituted biphenyls by means of vibrational circular dichroism (VCD), Helvetica Chim. Acta, 88 (2005) 2302–2314.

T. Furo, T. Mori, T. Wada and Y. Inoue, Absolute configuration of chiral [2.2]paracylophanes with intramolecular charge-transfer interaction. Failure of the exciton chirality method and use of the sector rule applied to the cotton effect of CT transition. J. Am. Chem. Soc., 127 (2005) 8242–8243.

J. He and P.L. Polavarapu, Determination of inter-molecular hydrogen bonded conformers of a-aryloxypropanoic acids using density functional theory predictions of vibrational absorption and vibrational circular dichroism spectra. J. Chem. Theory Comput., 1 (2005) 506–514.

J. He and P.L. Polavarapu, Determination of the absolute configuration of chiral a-aryloxypropanoic acids using vibrational circular dichroism studies: 2-(2-chlorophenoxy) propanoic acid and 2-(3-chlorophenoxy) propanoic acid. Spectrochim. Acta A, 61 (2005) 1327–1334.

J. He, A. Petrovic and P.L. Polavarapu, Quantitative determination of conformer populations: Assessment of specific rotation, vibrational absorption and vibrational circular dichroism. J. Phys. Chem., 108 (2004) 1671–1680.

J. He, F. Wang and P.L. Polavarapu, Absolute configurations of chiral herbicides determined from vibrational circular dichroism. Chirality, 17 (2005) S1–S8.

H. Izumi, S. Futamura, L.A. Nafie and R.K. Dukor, Determination of molecular stereochemistry using vibrational circular dichroism spectroscopy: Absolute configuration and solution conformation of 5-formyl-cis-cis-1,3,5-trimethyl-3-hydroxymethylcyclohexane-1-carboxylic acid lactone. The Chemical Record, 3 (2003) 112–119.

T. Kuppens, W. Langenaeker, J.P. Tollenaere and P. Bultinck, Determination of the stereochemistry of 3-hydroxymethyl-2,3-dihydro-[1,4]dioxino[2,3-b]-pyridine by vibrational circular dichroism and the effect of DFT integration grids. J. Phys. Chem. A, 107 (2003) 542–553.

K. Monde, T. Taniguchi, N. Miura, S. Njishimura, N. Harada, R.K. Dukor and L.A. Nafie, Preparation of cruciferous phytoalexin related metabolites, (−)-dioxibrassinin and (−)-3-cyanomethyl-3-hydroxyoxindole, and determination of their absolute configurations by vibrational circular dichroism (VCD). Tetrahedron Lett., 44 (2003) 6017–6020.

T. Mori, H. Izumi and Y. Inoue, Chiroptical properties of organic radical cations. The electronic and vibrational circular dichroism spectra of a-tocopherol derivatives and sterically hindered chiral hydroquinone ethers. J. Phys. Chem. A, 108 (2004) 9540–9549.

A. Petrovic, P.L. Polavarapu, J. Drabowicz, Y. Zhang, O.J. McConnell and H. Duddeck, Absolute configuration of C2 symmetric spiroselenurane: 3,3,3′,3′-tetramethyl-1,1′-spirobi[3H,2,1]-benzoxaselenole. Chem. Eur., 11 (2005) 4257–4262.

A. Petrovic, J. He, P.L. Polavarapu, L.S. Xiao and D.W. Armstrong, Absolute configuration and predominant conformations of 1,1-dimethyl-2-phenylethyl phenyl sulphoxide. Org. Biomol. Chem., 3 (2005) 1977–1981.

P.L. Polavarapu and J. He, Chiral analysis using circular dichroism spectroscopy in the mid-infrared. Analytical Chem., 76 (2004) 61A–67A.

P.L. Polavarapu, C. Zhao and K. Ramig, Vibrational circular dichroism, absolute configuration and predominant conformations of volatile anesthetics: 1,2,2,2-tetrafluoroethyl methyl ether. Tetrahedr. Asymmetry, 10 (1999) 1099–1106.

P.L. Polavarapu, C. Zhao, A. Cholli and G. Vernice, Vibrational circular dichroism, absolute configuration and predominant conformations of volatile anesthetics: Desflurane. J. Phys. Chem. B, 103 (1999) 6127–6132.

G. Roda, P. Conti, M. De Amici, J. He, P.L. Polavarapu and C. De Micheli, Enantiopure stereoisomeric homologues of glutamic acid: Chemoenzymatic synthesis and assignment of their absolute configurations. Tetrahedr. Asymmetry, 15 (2004) 3079–3090.

V. Setnicka, M. Urbanova, P. Bour, V. Kral and K. Volka, Vibrational circular dichroism of 1,1′-binaphthyl derivatives: Experimental and theoretical study. J. Phys. Chem. A, 105 (2001) 8931–8938.

A. Solladie-Cavallo, C. Marsol, G. Pescitelli, L. Di Bari, P. Salvadori, X. Huang, N. Fujioka, N. Berova, X. Cao, T.B. Freedman and L.A. Nafie, (R)-(+)- and (S)-(−)-1-(9-phenanthryl)ethylamine: Assignment of absolute configuration by CD Tweezer and VCD methods, and difficulties encountered with the CD exciton chirality method. Eur. J. Org. Chem., 11 (2002) 1788–1796.

A. Solladie-Cavallo, C. Marsol, M. Yaakoub, K. Azyat, A. Klein, M. Roje, C. Suteu, T.B. Freedman, X. Cao and L.A. Nafie, Erythro-1-naphthyl-1-(2-piperidyl)methanol: Synthesis resolution, NMR relative configuration and VCD absolute configuration. J. Org. Chem., 68 (2003) 7308–7315.

P.J. Stephens and F.J. Devlin, Determination of the structure of chiral molecules using ab initio vibrational circular dichroism spectroscopy. Chirality, 12 (2000) 172–179.

P.J. Stephens, A. Aamouche, F.J. Devlin, S. Superchi, M.I. Donnoli and C. Rosini, Determination of absolute configuration using vibrational circular dichroism spectroscopy: The chiral sulfoxide 1-(2-methylnaphthyl)methyl sulfoxide). J. Org. Chem., 66 (2001) 3671–3677.

P.J. Stephens, D.M. McCann, F.J. Devlin, T.C. Flood, E. Buktus, S. Stoncius, J.R. Cheeseman, Determination of molecular structure using vibrational ciorcular dichroism spectroscopy: The keto-lactone product of Baeyer-Villiger oxidation of (+)-(1R,5S)-bicyclo[3.31]nonane-2,7-dione. J. Org. Chem., 70 (2005) 3903–3913.

H.Z. Tang, B.M. Novak, J. He and P.L. Polavarapu, A thermal and solvocontrollable cylindrical nanoshutter based on a single screw-sense helical polyguanidine. Angew Chem. Int. Ed. 44 (2005) 7298–7301.

M. Urbanova, V. Setnicka, P. Bour, H. Navratilova and K. Volka, Vibrational circular dichroism spectroscopy study of paroxetine and femoxetine precursors. Biopolymers (Biospectroscopy), 67 (2002) 298–301.

M. Urbanova, V. Setnicka, F. Devlin and P.J. Stephens, Determination of molecular structure in solution using vibrational circular dichroism spectroscopy: The supramolecular teramer of S-2,2′-dimethyl-biphenyl-6,6′-dicarboxylic acid. J. Am. Chem. Soc., 127 (2005) 6700–6711.

C. Vanda, F. Peri, S. Pollicino, A. Ricci, F. Devlin, P.J. Stephens, F. Gasparrini, R. Rompietti and C. Villani, Synthesis, chromatographic separation, vibrational circular dichroism spectroscopy, and ab initio DFT studies of chiral thiepane tetraol derivatives. J. Org. Chem., 70 (2005) 664–669.

F. Wang and P.L. Polavarapu, Vibrational circular dichroism, predominant conformations and hydrogen bonding in (S)-(−)-3-butyn-2-ol. J. Phys. Chem. A, 104 (2000) 1822–1826.

F. Wang and P.L. Polavarapu, Conformational stability of (+)-epichlorohydrin. J. Phys. Chem. A, 104 (2000) 6189–6196.

F. Wang and P.L. Polavarapu, Vibrational circular dichroism, predominant conformations and intermolecular interactions in (R)-(−)-2-butanol. J. Phys. Chem. A, 104 (2000) 10683–10687.

F. Wang and P.L. Polavarapu, Predominant conformations of (2R,3R)-(−)-2,3-butanediol. J. Phys. Chem. A, 105 (2001) 6991–6997.

F. Wang, P.L. Polavarapu, J. Drabowicz and M. Mikolajczyk Absolute configurations. predominant conformtaions and tautomeric structures of eantiomeric t-butylphenylphosphine oxides. J. Org. Chem., 65 (2000) 7561–7565.

F. Wang, H. Wang, P.L. Polavarapu and C.J. Rizzo, Absolute configuration and conformational stability of (+)-2,5-dimethylthiolane and (−)-2,5-dimethylsulfolane. J. Org. Chem., 66 (2001) 3507–3512.

F. Wang, P.L. Polavarapu, J. Drabowicz, M. Mikolazczyk and P. Lyzwa, Absolute configurations, predominant conformations and tautomeric structures of enantiomeric tert-butylphenylphosphinothioic acid. J. Org. Chem., 66 (2001) 9015–9019.

F. Wang, P.L. Polavarapu, F. Lebon, G. Longhi, S. Abbate and M. Catellani, Absolute configuration and conformational stability of (S)-(+)-3-(2-methyl butyl)thiophene and (+)-3,4-di[(S)-2-methylbutyl)]thiophene and their polymers. J. Phys. Chem. A, 106 (2002) 5918–5923.

F. Wang, P.L. Polavarapu, F. Lebon, G. Longhi, S. Abbate and M. Catellani, Conformational analysis of (S)-(+)-1-bromo-2-methylbutane and the influence of bromine on conformational stability. J. Phys. Chem., 105 (2002) 12365–12369.

F. Wang, P.L. Polavarapu, V. Schurig and R. Schmidt, Absolute configuration and conformational stability of a degradation product of inhalation anesthetic sevoflurane: A vibrational circular dichroism study. Chirality, 14 (2002) 618–624.

F. Wang, Y. Wang, P.L. Polavarapu, T. Li, J. Drabowicz, K.M. Pietrusiewicz and K. Zygo, Absolute configuration of tert-butyl-1-(2-methylnaphthyl)phosphineoxide. J. Org. Chem., 67 (2002) 6539–6541.

F. Wang, P.L. Polavarapu, J. Drabowicz, P. Kielbasinski, M.J. Potrzebowski, M. Mikolajczyk, M.W. Wieczorek, W.W. Majzner and I. Lazewska, Solution and crystal structures of chiral molecules can be significantly different: tert-butylphenylphosphinoselenoic acid. J. Phys. Chem., 108 (2004) 2072–2079.

C. Zhao, P.L. Polavarapu, H. Grosenick and V. Schurig, Vibrational circular dichroism, absolute configuration and predominant conformations of volatile anesthetics: Enflurane. J. Mol. Struct., 550 (2000) 105–115.

C. Literature articles on quantum mechanically predicted VROA spectra for evaluating molecular stereochemistry

L.D. Barron, A.R. Gargaro, L. Hecht and P.L. Polavarapu, Experimental and ab initio theoretical vibrational Raman optical activity of alanine. Spectrochim. Acta, 47A (1991) 1001–1016.

L.D. Barron, A.R. Gargaro, L. Hecht, P.L. Polavarapu and H. Sugeta, Experimental and ab initio theoretical vibrational Raman optical activity of tartaric acid. Spectrochim. Acta, 48A (1992) 1051–1066.

T.M. Black, P.K. Bose, P.L. Polavarapu, L.D. Barron and L. Hecht, Vibrational optical activity in trans-2,3-dimethyloxirane. J. Am. Chem. Soc., 112 (1990) 1479–1489.

P.K. Bose, L.D. Barron and P.L. Polavarapu, Ab initio and experimental vibrational Raman optical activity in (+)-(R)-methylthiirane. Chem. Phys. Lett., 155 (1989) 423–429.

P.K. Bose, P.L. Polavarapu, L.D. Barron and L. Hecht, Ab initio and experimental Raman optical activity in (+)-(R)-methyloxirane. J. Phys. Chem., 94 (1990) 1734–1740.

P. Bour, Calculation of the Raman optical activity via the sum-over-states expansion. Chem. Phys. Lett., 288 (1998) 363–370.

P. Bour, Computations of the Raman optical activity via the sum-over-states expansions. J. Comput. Chem., 22 (2001) 426–435.

P. Bour, V. Baumruk and J. Hanzlikova, Measurement and calculation of the Raman optical activity of alpha-pinene and trans-pinane. Collect. Czech. Chem. Commun., 62 (1997) 1384–1395.

P. Bour, J. Kapitan and V. Baumruk, Simulation of the Raman optical activity of L-alanyl-L-alanine, J. Phys. Chem. A, 105 (2001) 6362–6368.

P. Bour, V. Sychrovsky, P. Malon, J. Hanzlikova, V. Baumruk, J. Pospisek and M. Budesinsky, Conformation of the dipeptide cyclo(L-pro-L-pro) monitored by the nucelar magnetic resonance and Raman optical activity sepctra. Experimental and ab initio computational study. J. Phys. Chem. A, 106 (2002) 7321–7327.

Z. Deng, P.L. Polavarapu, S.J. Ford, L. Hecht, L.D. Barron, C.S. Ewig and K. Jalkanen, The solution phase conformations of N-acetyl-N′-methyl-L-alaninamide from vibrational Raman optical activity. J. Phys. Chem., 100 (1996) 2025–2034.

L. Hecht, J. Costante, P.L. Polavarapu, A. Collet and L.D. Barron, Absolute configuration of bromochlorofluoromethane from experimental and ab initio theoretical vibrational Raman optical activity. Angew. Chem., 36 (1997) 885–887.

W. Hug, G. Zuber, A.D. Meijere, A.F. Khlebnikov and H.J. Hansen, Raman optical activity of a purely sigma-bonded helical chromophore (−)-(M)- sigma -[4]Helicene. Helvetica Chim. Acta, 84 (2001) 1–21.

K.J. Jalkanen, R.M. Nieminen, K. Frimand, J. Bohr, H. Bohr, R.C. Wade, E. Tajkhorshid and S. Suhai, A comparison of aqueous solvent models used in the calculation of Raman and ROA spectra of L-alanine. Chem. Phys., 265 (2001) 125–151.

K.J. Jalkanen, R.M. Nieminen, M. Kanpp-Mohammady and S. Suhai, Vibrational analysis of various isotopomers of L-alanyl-L-alanine in aqueous solution: Vibrational absorption, vibrational circular dichroism, Raman and Raman optical activity spectra. Int. J. Quantum Chem., 92 (2003) 239–259.

N.A. Macleod, P. Butz, J.P. Simons, G.H. Grant, C.M. Baker and G.E. Tranter, Structure, electronic circular dichroism and Raman optical acitivity in the gas phase and in solution: A computational and experimental investigation. Phys. Chem. Chem. Phys. 7 (2005) 1432–1440.

M. Pecul, A. Rizzo and J. Leszczynski, Vibrational Raman and Raman optical activity spectra of D-lactic acid, D-lactate and D-glyceraldehyde. J. Phys. Chem. A, 106 (2002) 11008–11016.

P.L. Polavarapu, Ab initio Raman and Raman optical activity spectra. J. Phys. Chem., 94 (1990) 8106–8112.

P.L. Polavarapu, Absolute configuration of bromochlorofluoromethane. Angewandte Chem. Int. Ed., 41 (2002) 4544–4546.

P.L. Polavarapu and Z. Deng, Structural determinations from vibrational Raman optical activity: from a single peptide group to b-turns. Faraday Disc., 99 (1994) 151–163.

P.L. Polavarapu, T.M. Black, L.D. Barron and L. Hecht, Vibrational Raman optical activity in (R)-(+)-3-methylcyclohexanone: Experimental and ab initio theoretical studies and the origin of unusual couplets. J. Am. Chem. Soc., 115 (1993) 7736–7742.

P.L. Polavarapu, P.K. Bose, L. Hecht and L.D. Barron, Vibrational Raman optical activity in (R)-(+)-3-methylcyclopentanone: Experimental and ab initio theoretical studies. J. Phys. Chem., 97 (1993) 11211–11215.

P.L. Polavarapu, L. Hecht and L.D. Barron, Vibrational Raman optical activity in substituted oxiranes. J. Phys. Chem., 97 (1993) 1793–1799.

P.L. Polavarapu, S.T. Pickard, H.E. Smith, T.M. Black, L.D. Barron and L. Hecht, Determination of absolute configurations from vibrational Raman optical activity: Trans-2,3-dimethylthiirane. Talanta, 40 (1993) 545–549.

M. Reiher, V. Liegeois and K. Ruud, Basis set and density functional dependence of vibrational Raman optical activity calculations. J. Phys. Chem. A, 109 (2005) 7567–7574.

K. Ruud, T. Helgaker and P. Bour, Gauge-origin independent density-functional theory calculations of vibrational Raman optical activity. J. Phys. Chem. A, 106 (2002) 7448–7455.

C.N. Tam, P. Bour and T.A. Keiderling, Vibrational optical activity of (3S,6S)-3,6-dimethyl-1,4-dioxane-2,5-dione. J. Am. Chem. Soc., 118 (1996) 10285–10293.

G.S. Yu, T.B. Freedman, L.A. Nafie, Z. Deng and P.L. Polavarapu, Experimental measurement and ab initio calculation of Raman optical activity of L-alanine and its deuterated isotopomers. J. Phys. Chem., 99 (1995) 835–843.

G. Zuber and W. Hug, Rarefied basis sets for the calculation of optical tensors. 1. The importance of gradients on hydrogen atoms for the Raman scattering tensor. J. Phys. Chem. A, 108 (2004) 2108–2118.

G. Zuber and W. Hug, Computational interpretation of vibrational optical activity: The ROA spectra of (4S)-4-methylisochromane and the (4S)-isomers of galaxolide. Helv. Chim. Acta, 87 (2004) 2208–2234.

G. Zuber, M. Goldsmith, D.N. Beratan and P. Wipf, Towards Raman optical activity calculations of large molecules. Chem. Phys. Chem., 6 (2005) 595–597.

© 2006 Elsevier B.V. All rights reserved.
Chiral Analysis
K.W. Busch and M.A. Busch, Eds.

CHAPTER 15

Vibrational optical activity in chiral analysis

Laurence A. Nafie[1,2] and Rina K. Dukor[2]

[1]*Department of Chemistry, 1-014 CST, Syracuse University, Syracuse, New York, NY 13244, USA*
[2]*BioTools, Inc. 950 N. Rand Road, Unit 123, Wauconda, Illinois, IL 60084, USA*

15.1 INTRODUCTION

Chiral analysis encompasses all analytical techniques focused on quantitative measures of the chiral properties of molecules. The properties unique to chiral molecules are absolute configuration (AC) and enantiomeric excess (EE). As will be discussed below, vibrational optical activity (VOA) can provide information on samples of chiral molecules by direct spectroscopic investigation without modifying the molecules by reacting them, sometimes irreversibly, with other chiral molecules or substrates. From a molecular point of view, VOA chiral analysis may be regarded as non-invasive in that the molecule is not changed or altered from its natural solution or solid-state condition in order to carry out the analysis. An important property of VOA is its sensitivity to the conformational states of chiral molecules, both in solution and in the solid state. Beyond the determination of the absolute chirality of a molecule, VOA can be used to specify these states, in conjunction with quantum mechanical calculations, and determine populations of solution-state conformers that are interconverting faster than the nuclear magnetic resonance (NMR) time scale. VOA can also be used to probe the solid-state conformations of chiral molecules as well as their polymorphic crystal forms. While this information is not strictly chiral in nature, VOA has extraordinary sensitivity to both the absolute structure and the conformation of chiral molecules, and this aspect of chiral analysis by VOA should not be overlooked.

The most important physical characteristic of any chiral molecule of a known primary structure is its absolute configuration. While this can often be achieved with optical rotation (OR) or circular dichroism from electronic transitions (ECD) by classical methods [1,2], the results are sometimes not definitive and may depend on rules or trends for which there are exceptions. When exceptions occur for the first time, a mistake in prediction usually occurs. Instead, standard analytical methodology has turned almost exclusively to anomalous single-crystal X-ray diffraction for the *a priori* unambiguous determination of the AC of a chiral molecule [3]. In some cases, though, it is difficult to obtain single crystals with a heavy atom or of sufficient quality for X-ray analysis, including molecules that are liquids or oils in their pure states and can therefore never crystallize. Vibrational circular dichroism (VCD), together with Raman optical activity (ROA), offers an alternative route to the *a priori* unambiguous determination of AC. The method, as explained in more detail below, involves the comparison of measured and calculated VCD spectra across a range of frequencies for the molecule in question. This method has proven to be highly successful, and the ACs of literally hundreds of molecules have been solved over the past several years [4–7]. Most major pharmaceutical companies today use, or are exploring the use of, VCD to determine the AC of new chiral drug molecules or their precursors [8–12].

The second important characteristic of chiral molecules is the EE [13], of the constituent molecules. When new chiral molecules are brought to market as active pharmaceutical ingredients (APIs), the Food and Drug Administration (FDA) requires proof of the absence, usually at the level of a few tenths of a percent, that there are no significant amounts of the other enantiomer of the API present in the product. The other enantiomer can in some cases cause serious undesirable side effects. The standard method today for the determination of EE is chiral chromatography [13]. In this method, a separation of enantiomers is achieved with the use of a chiral stationary phase. Once the separation of the enantiomers is achieved for a racemic sample of the chiral molecule, the sensitivity of the method to the presence of the undesired enantiomer can be established. Sensitivities on the order of one-tenth of a percent are typical for chiral HPLC.

Both VCD and ROA, offer interesting alternatives to chiral HPLC for the determination of EE of chiral molecules. Since the intensity of the VCD spectrum relative to its parent infrared (IR) absorption spectrum varies linearly with EE, from zero for a racemic mixture to the maximum size for the pure enantiomer sample, once calibrated, VCD or ROA can be used to determine EE without the step of separating enantiomers. Furthermore, because the IR or Raman always accompanies the VOA, and because the VOA spectrum, obtained with a Fourier transform or a multi-channel spectrometer, typically embraces hundreds of wavenumber frequencies and dozens of vibrational bands, structure determination is obtained simultaneously with EE determination. By contrast, bands eluting from an HPLC typically must be analyzed later by some other technique to verify the structural identity of the compound of each eluted band. As we shall demonstrate below, this simultaneous sensitivity to both structure and chirality across wide spectral ranges permits the use of VCD to monitor both the mole fraction and %EE of more than one chiral molecule undergoing reaction as a function of time [14]. This kind of information is currently impossible to obtain in real-time by any other technique and permits access to a deeper understanding of the course of the reactions of chiral molecules both for studying reaction dynamics and for optimizing the production of large batches of chiral molecules.

VCD and ROA can also be used to study the conformations of large biological molecules, such as peptides, proteins, carbohydrates, glycoproteins, and nucleic acids, and are starting to be used to study the conformational structures of protein pharmaceuticals in various stages of development and formulation [15–19]. The current spectroscopic range of VCD is being extended into the near-IR to better be able to routinely monitor the protein secondary structure and ensure the pharmaceutical viability of these new classes of drugs [20]. ROA can now be used to classify the folding family of newly isolated or engineered proteins using factor analysis routines, and this methodology has been extended to the classification of the coat protein structure of viruses [21,22].

VCD has recently been extended from solution-state studies to various forms of solids, including KBr pellets [23], mulls, films [24–27], and spray-dried powders [28]. This greatly extends the utility of VCD to monitor broader classes of pharmaceutical products and opens the way for using VCD to study APIs in the presence of both achiral and chiral excipients in the solid phase, and even in final formulated pills and tablets.

In all areas of application, except the studies of conformations of biological molecules, ROA currently lags VCD in development and use. But as demonstrated below, instrumentation for measurement [29–31] and software for carrying out *ab initio* calculations are now available commercially for both VCD and ROA [32]. In principle, there is no reason why ROA will not catch up to VCD as a reliable method for chiral analysis, in the same way that neither IR nor Raman are regarded as more important than the other as achiral probes of the structure or physical state of molecular samples. Each has relative advantages and disadvantages, and strengths and weaknesses. What is clear is that VCD and ROA now provide very powerful supplements to IR and Raman, as well as the entire range of other analytical probes for monitoring and characterizing chiral molecules. It is important to keep in mind the power of vibrational spectroscopy in general, and VOA in particular, to characterize the structure and stereospecific properties of chiral molecules.

15.2 DEFINITIONS OF VIBRATIONAL OPTICAL ACTIVITY

Vibrational optical activity (VOA) consists of two complementary spectroscopic methods, vibrational circular dichroism (VCD) and vibrational Raman optical activity (ROA). For both VCD and ROA, one measures the differential response of a chiral molecule to left *versus* right circularly polarized radiation associated with a vibrational transition in a molecule.

VCD consists of a single form derived from IR absorption. One could imagine the fluorescence forms of VCD, such as the fluorescence detected VCD or circularly polarized vibrational fluorescence, but fluorescence intensity in the IR is very weak and this has not been observed. In addition, one could measure vibrational optical rotatory dispersion (VORD) in the IR, but again this has not been carried out for solutions of chiral molecules in part because ORD spectra are difficult to interpret. On the other hand, there are four forms of circularly polarized (CP) ROA and four forms of linearly polarized (LP) ROA derived from their associated ordinary Raman scattering. Further, ROA and Raman can be observed using a variety of scattering geometries and polarization modulation schemes. Other forms of VOA may well evolve in the future, but for the present VCD and ROA are the two overall methods. The general definition of VOA is the differential interaction of a chiral molecule with left *versus* right circularly polarized radiation during vibrational excitation. In particular, there is one form of VCD and four forms of CP-ROA [33,34]. The additional forms of CP-ROA arise due to the flexibility of modulating between left and right circular polarization states of the incident laser beam or the scattered Raman radiation, or both, either in-phase (right–right minus left–left) or out-of-phase (right–left minus left–right). These respectively are called incident circular polarization Raman optical activity (ICP-ROA), scattered circularly polarization ROA (SCP-ROA), and two forms of dual circular polarization ROA (DCP_I ROA and DCP_{II} ROA). Definitions for these five forms of VOA are given below. Note that the convention for VCD is left minus right, whereas for ROA it is right minus left, a curiosity in the history of the development of these two forms of VOA [35,36].

$$\begin{aligned}
&\text{VCD} && \Delta A(\nu) = A_L(\nu) - A_R(\nu) \\
&\text{ICP-ROA} && \Delta I_\alpha(\nu) = I_\alpha^R(\nu) - I_\alpha^L(\nu) \\
&\text{SCP-ROA} && \Delta I^\alpha(\nu) = I_R^\alpha(\nu) - I_L^\alpha(\nu) \\
&\text{DCP}_\text{I}\text{-ROA} && \Delta I_\text{I}(\nu) = I_R^R(\nu) - I_L^L(\nu) \\
&\text{DCP}_\text{II}\text{-ROA} && \Delta I_\text{II}(\nu) = I_L^R(\nu) - I_R^L(\nu).
\end{aligned}$$

Close analogies exist between the ICP and SCP forms of ROA and two forms of electronic optical activity in fluorescence or luminescence. The background fluorescence associated with ICP-ROA corresponds to the fluorescence detected circular dichroism (FDCD) [37] since in this measurement the total fluorescence is measured while modulating the incident radiation between left and right circular polarization states. Alternatively, the fluorescence background associated with SCP-ROA is usually called circularly polarized emission (CPE) or circularly polarized luminescence (CPL) [38,39].

While only one form of VCD has ever been measured, all four forms of ROA have been measured and published. VCD was first measured in 1974 [40], and confirmed one year later [41,42], using dispersive grating instrumentation. This was followed by the first demonstrations of Fourier transform VCD measurements from our Syracuse laboratory starting in 1979 [43,44].

ICP-ROA was the first form of ROA to be discovered and developed in the early to mid-1970s [45]. SCP-ROA was predicted to exist theoretically as a quantity originally called the degree of circularity (of the scattered radiation) [46], but it was not until 1988 that a practical method for the measurement of this form of ROA, now called SCP-ROA, was first devised and demonstrated in our laboratory [47]. Shortly thereafter, we predicted two new forms of ROA, the two forms of DCP-ROA, to exist as distinct from the ICP and SCP forms of ROA [48]. DCP_I-ROA was first measured in backscattering in our laboratory in 1991 [49] and we succeeded to isolate the DCP_{II} form of ROA, which only becomes non-zero as pre-resonance enhancement of Raman intensities becomes significant, a few years later [50]. LP-ROA has been predicted theoretically and can only exist close to resonance Raman conditions, and then only for right-angle scattering [51,52]. While ICP-ROA is the old known form of ROA, the first, and to date only, commercial ROA spectrometer, is designed to measure SCP-ROA [53]. In the following section, we explain the fundamentals of the measurement and calculation of VOA spectra before going on to particular applications of chiral analysis.

15.3 MEASUREMENT OF VIBRATIONAL OPTICAL ACTIVITY

Because VCD and ROA intensities represent very small differences in the parent IR absorption or Raman scattering, respectively, neither can be measured without specialized instrumentation, which is why neither was discovered until the 1970s. VCD intensities are typically four to five orders of magnitude smaller than the parent IR spectra, and similarly ROA intensities are three to four orders of magnitude smaller than the parent Raman-scattering spectra. The major obstacle to overcome in the measurement of VOA is to make sure that the spectra are free of offsets or distortions arising from optical components in the instrumentation that interfere with the true VOA spectrum. Various approaches to eliminate or reduce these spectral artifacts significantly have been devised and implemented for the commercially available instruments that measure VCD [54–56] and ROA [57–59].

While the measurement of VCD can be carried out on either dispersive grating or Fourier transform infrared (FT-IR) instruments, all commercially available VCD spectrometers are based on the Fourier transform design pioneered at Syracuse University [43,44]. The first and the only company to offer a dedicated, factory-aligned FT-IR VCD spectrometer, the Chiral*IR*, was BioTools, Inc., currently of Wauconda, Illinois, in 1997. The spectrometer is manufactured for BioTools by ABB Bomem (now ABB Process Analytics), Inc. of Quebec, Canada. Since then, several other major manufacturers of FT-IR spectrometers, such as Varian/Digilab, Thermo/Nicolet, and Bruker, have offered, or do offer, a VCD accessory bench which requires on-site alignment after installation. The measurement procedure for VCD involves preparing a sample with an average transmission in the regions of interest roughly between 10 and 90%, with 40% optimum for signal-to-noise ratio, or an absorbance of approximately 0.4. This is achieved by adjusting the path length or concentration until the desired absorbance is obtained resulting in an IR spectrum of high quality. Using standard sampling accessories, a good VCD spectrum can typically be obtained from a few milligrams of sample.

More recently, BioTools, Inc. has offered an upgrade for the Chiral*IR* that consists of a dual polarization modulation accessory bench that automatically corrects VCD baseline offsets in real-time [55]. The wavenumber frequency range of the standard mid-IR VCD spectrometer from 800 to 2000 cm^{-1} has been extended in optimized stages to 4000 cm^{-1} covering the hydrogen-stretching region where VCD was first discovered, to 7000 cm^{-1} covering the first combination and overtone region [60], and then on to 10,000 and even 14,000 cm^{-1} that actually connects to the long-wavelength end of the visible region of the spectrum where electronic transitions and ECD is measured [20]. Examples of VCD in all of these regions will be presented by example or application below.

ROA spectra can now also be measured using a commercially available instrument [53]. BioTools, Inc. announced the availability of the Chiral*RAMAN* SCP-ROA spectrometer in 2003 and has placed over ten such instruments throughout the world over the last three years. This instrument, based on the recent design of Werner Hug, features active real-time artifact suppression optics, fiber optic image transformation and simultaneous collection of left and right circularly polarized scattered Raman radiation on a back-thinned charge coupled device (CCD) detector [57–59]. The resulting Raman spectrometer is nearly an order of magnitude faster than all previous ROA spectrometers. As a side benefit, the ordinary Raman performance is so exceptional that Raman spectra of proteins in water at moderate concentrations have been obtained with signal-to-noise ratios in the range of 100 in less than 150 ms, the current minimum exposure time of the CCD camera electronics. Because of the ease of sampling of all classes of biological molecules, ROA holds great promise as a sensitive probe of the conformational state of these molecules. A number of applications of ROA to problems of biological interest, including unfolded proteins and viruses, have appeared in recent years [18,19,21,22,53,61–73]. Biological applications featuring VCD are also prominent in the recent literature, typically for smaller molecules, such as peptides and polypeptides [74–84].

15.4 THEORETICAL BASIS OF VIBRATIONAL OPTICAL ACTIVITY

An appreciation of the theoretical aspects of VOA is an important prerequisite to its full utilization for chiral analysis. This is important not only for the design and execution of experiments for particular applications, but also for the understanding of how to bring together measured and calculated VOA spectra for the elucidation of molecular properties such as absolute configuration and solution-state conformation.

The development of theoretical methodology for VCD and ROA has followed different paths over the past twenty-five years. The formal theory of ICP-ROA was published prior to the first experimental observation [85], whereas a complete formal theory of VCD was not achieved until nearly a decade after the first experimental observation, due to the need to include non-Born–Oppenheimer coupling effects into the expression for the magnetic dipole transition moment [86,87]. On the other hand, VCD was first calculated using *ab initio* quantum mechanical methods [88,89] before this was achieved for ROA [90]. Because VCD passed the key experimental and theoretical hurdles, such as commercial availability of instrumentation and software for calculations, at an earlier stage than for ROA, and because it is inherently simpler both to measure and to calculate, there is a more extensive literature of VCD, compared to that of ROA. This literature for both VCD and ROA covers measurement techniques, theoretical calculations and applications to molecules of biological and pharmaceutical interest. Today, both VCD and ROA possess the advantage of a fully dedicated, commercially available instrument [29–31] as well as commercially available software for the calculation of VCD and ROA intensities from the first principles [32].

15.4.1 Theory of vibrational circular dichroism

VCD is an extension of ECD from electronic to vibrational transitions [33,34,91]. The differential absorbance of left and right circularly polarized IR radiation by a chiral molecule during vibrational excitation is measured as $\Delta A = A_L - A_R$ for absorbance or $\Delta\varepsilon = \varepsilon_L - \varepsilon_R$ for molar absorptivity, where anisotropy ratios, $g = \Delta A/A = \Delta\varepsilon/\varepsilon$, are typically in the range from 10^{-3} to 10^{-6}, a factor of 10–100 smaller than for electronic CD. The integrated IR absorption and VCD intensities are proportional to the dipole strength (D), and rotational strength (R), respectively, with $g = 4R/D$.

IR intensities depend on the absolute square of the electric-dipole transition moment of the molecule given by,

$$D_{\mathrm{r}}^{\mathrm{a}}(g0 \rightarrow g1) = \left|\left\langle \tilde{\Psi}_{\mathrm{g1}}^{\mathrm{a}} | \hat{\mu}_{\mathrm{r}} | \tilde{\Psi}_{\mathrm{g0}}^{\mathrm{a}} \right\rangle\right|^2 \tag{1}$$

and VCD intensity arises from the imaginary part of the scalar product of the electric- and magnetic-dipole transition moments of the molecule given by,

$$R_{\mathrm{r}}^{\mathrm{a}}(g0 \rightarrow g1) = \mathrm{Im}\left(\left\langle \tilde{\Psi}_{\mathrm{g0}}^{\mathrm{a}} | \hat{\mu}_{\mathrm{r}} | \tilde{\Psi}_{\mathrm{g1}}^{\mathrm{a}} \right\rangle \bullet \left\langle \tilde{\Psi}_{\mathrm{g1}}^{\mathrm{a}} | \hat{m} | \tilde{\Psi}_{\mathrm{g0}}^{\mathrm{a}} \right\rangle\right) \tag{2}$$

for a fundamental vibrational transition between the ground and the first excited vibrational states, $\tilde{\Psi}_{\mathrm{g0}}^{\mathrm{a}}$ and $\tilde{\Psi}_{\mathrm{g1}}^{\mathrm{a}}$, of normal mode "a" in the ground electronic state "g". The position-form electric dipole moment operator ($\hat{\mu}_{\mathrm{r}}$) and the magnetic dipole moment operator ($\hat{m}$) consist of electronic and nuclear contributions for electrons j with position $\boldsymbol{r}_j$, velocity $\dot{\boldsymbol{r}}_j$, mass m and charge $-e$, and nuclei J with position $\boldsymbol{R}_J$, velocity $\dot{\boldsymbol{R}}_j$, mass M_J, and charge $Z_J e$,

$$\hat{\mu}_{\mathrm{r}} = \hat{\mu}_{\mathrm{r}}^{\mathrm{E}} + \hat{\mu}_{\mathrm{r}}^{\mathrm{N}} = -\sum_j e\boldsymbol{r}_j + \sum_J Z_J e \boldsymbol{R}_J \tag{3}$$

$$\begin{aligned} \hat{m} = \hat{m}^{\mathrm{E}} + \hat{m}^{\mathrm{N}} &= -\sum_j \frac{e}{2c}\boldsymbol{r}_j \times \dot{\boldsymbol{r}}_j + \sum_J \frac{Z_J e}{2c}\boldsymbol{R}_J \times \dot{\boldsymbol{R}}_J \\ &= \sum_j \frac{ie\hbar}{2mc}\boldsymbol{r}_j \times \frac{\partial}{\partial \boldsymbol{r}_j} - \sum_J \frac{ie\hbar Z_j}{2M_J c}\boldsymbol{R}_J \times \frac{\partial}{\boldsymbol{R}_J} \end{aligned} \tag{4}$$

The theory of IR absorption for a vibrational transition within a given electronic state, usually the ground electronic state of the molecule, is straightforward. One invokes a separation of the electronic and vibrational parts of the wavefunctions $\tilde{\Psi}_{\mathrm{g0}}^{\mathrm{a}}$ and $\tilde{\Psi}_{\mathrm{g1}}^{\mathrm{a}}$ by implementing the Born–Oppenheimer (BO) approximation. At this level, one obtains the correlation between the positions of the nuclei and the electron probability density of the molecule. While this is sufficient for the position formulation of the dipole strength with the electric dipole moment operator given in Eq. (3), the magnetic dipole transition moment in Eq. (4), necessary for VCD, has nuclear and electronic velocity operators, and the electronic contribution to the vibrational magnetic dipole transition moment vanishes within the BO approximation. To solve this unrealistic description, the lowest order correction to the BO approximation is necessary [86]. Further, it has been shown that this lowest order non-BO contribution to the magnetic dipole transition moment, and also the velocity formulation of the electric dipole transition moment, carries the exact correlation needed between nuclear velocities and vibrationally generated current density in molecules [87]. With these non-BO contributions in place, a complete vibronic coupling theory was available for implementation using quantum chemistry programs. The implementation of these basic theoretical expressions is a subject unto itself, and descriptions at various levels can be found in articles and reviews on the theoretical formulation and calculation of VCD [34,88,89,92,93]. Most recently, the vibronic theory of VCD was extended to the case of VCD intensities in molecules with low-lying electronic states, but this theory has not yet been implemented for theoretical calculations [94].

15.4.2 Theory of Raman optical activity

The theory of ROA is intrinsically more complex than that of VCD [33,95–97]. The complexity arises from the fact that two photons instead of one are required to complete a Raman scattering event. The theory changes with variations in the relative geometry of the incident and scattered beams and the polarization modulation schemes of the two photons. As discussed in the introduction, there are four basic forms of circular polarization—ROA, ICP, SCP, and DCP_I and DCP_{II} ROA. In addition, for each of these forms there are three basic geometrical setups, right-angle scattering, backscattering and forward scattering. Finally, as with ordinary Raman scattering, there are three basic cases: the general theory, appropriate for all circumstances, the far-from-resonance (FFR) limit, for transparent samples when using visible excitation, and the single-electronic-state (SES) resonance case when there is direct resonance with a single electronic state [98]. In addition to these forms of ROA, there are four forms of linear polarization ROA; these are more specialized forms of ROA that require near or direct resonance conditions to be observed [52]. In this section, we summarize and discuss the expression for the most common forms and limiting cases of ROA. For the sake of conciseness, we consider explicitly only expressions for unpolarized backscattering ICP_u ROA and backscattering DCP_I ROA. These account for nearly all the ROA spectra measured in recent years. Other forms have been measured on occasion, but mainly for theoretical interest.

The major expressions for all cases of circular and linear polarization ROA and associated Raman intensity have been published previously [96]. Here, we describe qualitatively only the general form of the theory, and the interested reader may seek the mathematical expressions elsewhere if desired. In the general case, which includes the possibility of direct resonance interactions, the Raman tensor is not symmetric. This results in three Raman tensor invariants, the isotropic Raman invariant (α^2), the symmetric anisotropic Raman invariant $(\beta_s(\tilde{\alpha})^2)$; and the antisymmetric Raman invariant $(\beta_a(\tilde{\alpha})^2)$, where the tilde above a symbol indicates a complex quantity. The last of these invariants is only active under resonance conditions and is responsible for the phenomena of inverse polarized Raman scattering. The three Raman invariants are appropriate linear combinations of products of Raman tensor elements. For ROA in the general case there are ten ROA invariants, five in Roman font, conceptually associated with ICP-ROA, and five in script font, conceptually associated with SCP-ROA, where the difference between these two is the order of the operators in contributing tensors. The five types of invariants are the isotropic ROA invariants, αG and $\alpha\mathrm{G}$, the symmetric and antisymmetric magnetic-dipole ROA invariants, $\beta_{\mathrm{s}}(\tilde{G})^2$, $\beta_{\mathrm{a}}(\tilde{G})^2$, $\beta_{\mathrm{s}}(\tilde{\mathrm{G}})^2$, and $\beta_{\mathrm{a}}(\tilde{\mathrm{G}})^2$, and the symmetric and antisymmetric electric-quadrupole ROA invariants, $\beta_{\mathrm{s}}(\tilde{A})^2$, $\beta_{\mathrm{a}}(\tilde{A})^2$, $\beta_{\mathrm{s}}(\tilde{\mathrm{A}})^2$, and $\beta_{\mathrm{a}}(\tilde{\mathrm{A}})^2$. Analogously, the ROA invariants are appropriate linear combinations of products of the ordinary Raman tensor elements with ROA tensor elements, where in the latter, one of the two electric-dipole moment operators in the Raman tensor is substituted with a magnetic-dipole moment operator or and electric-quadrupole-moment operator. This is analogous to the rotational strength invariant of VCD, which differs from the dipole strength by substitution of a magnetic-dipole moment operator for one of the two electric-dipole moment operators in the absolute square of the electric dipole transition moment. Ordinary Raman intensities are appropriate linear combinations of Raman tensor invariants and similarly ROA intensities are linear combinations of ROA invariants.

The original theory of ROA was couched in theoretical expressions of the FFR approximation. In this limit, the Raman tensor becomes symmetric and the number of Raman invariants reduces from three to two, α^2 and $\beta(\alpha)^2$, where the antisymmetric anisotropic Raman invariant vanishes and the remaining anisotropic invariant is symmetric and needs no subscript. More dramatically, in the FFR limit the number of ROA invariants reduces from ten to three. Half of these vanish when the difference between the script and bold ROA tensors vanishes owing to the FFR equivalence of the incident and scattered radiation interactions. A further reduction occurs due to the vanishing of the remaining antisymmetric invariants, leaving only $\alpha G'$, $\beta(G')^2$, and $\beta(A)^2$. The prime on the magnetic dipole optical activity tensor symbol indicates imaginary part, which produces a real observable quantity as in the expression for the rotational strength in Eq. (2).

Theoretical expressions for these two Raman and three ROA invariants in the FFR approximation are given by

$$\alpha^2 = \frac{1}{9}\alpha_{\alpha\alpha}\alpha_{\beta\beta} \tag{5}$$

$$\beta(\alpha)^2 = \frac{1}{2}(3\alpha_{\alpha\beta}\alpha_{\alpha\beta} - \alpha_{\alpha\alpha}\alpha_{\beta\beta}) \tag{6}$$

$$\alpha G' = \frac{1}{9}\alpha_{\alpha\alpha}G'_{\beta\beta} \tag{7}$$

$$\beta(G')^2 = \frac{1}{2}(3\alpha_{\alpha\beta}G'_{\alpha\beta} - \alpha_{\alpha\alpha}G'_{\beta\beta}) \tag{8}$$

$$\beta(A)^2 = \frac{1}{2}\omega_0\alpha_{\alpha\beta}\varepsilon_{\alpha\gamma\delta}A_{\gamma,\delta\beta}, \tag{9}$$

where the polarizability and optical activity tensors are given by the following quantum mechanical expressions that, unlike IR and VCD, involve no contributions from the nuclear motion,

$$\alpha_{\alpha\beta} = \frac{2}{\hbar}\sum_{j\neq n}\frac{\omega_{jn}}{\omega_{jn}^2 - \omega_0^2}\mathrm{Re}[\langle n|\hat{\mu}_\alpha|j\rangle\langle j|\hat{\mu}_\beta|n\rangle] \tag{10}$$

$$G'_{\alpha\beta} = \frac{-2}{\hbar}\sum_{j\neq n}\frac{\omega_0}{\omega_{jn}^2 - \omega_0^2}\mathrm{Im}[\langle n|\hat{\mu}_\alpha|j\rangle\langle j|\hat{m}_\beta|n\rangle] \tag{11}$$

$$A_{\alpha,\beta\gamma} = \frac{2}{\hbar}\sum_{j\neq n}\frac{\omega_0}{\omega_{jn}^2 - \omega_0^2}\mathrm{Re}\left[\langle n|\hat{\mu}_\alpha|j\rangle\left\langle j\middle|\hat{\Theta}_{\beta\gamma}\middle|n\right\rangle\right] \tag{12}$$

Here, repeated Greek subscripts are summed over all three Cartesian coordinates and the symbol $\hat{\Theta}_{\beta\gamma}$ represents the electric-quadrupole operator.

With these Raman and ROA invariants, we can write the following expressions for the Raman and ROA intensities associated with unpolarized ICP (ICP_u) scattering and DCP_I scattering [49],

$$I_u^R(180°) - I_u^L(180°) = \frac{8K}{c}[12\beta(G')^2 + 4\beta(A)^2] \tag{13}$$

$$I_u^R(180°) + I_u^L(180°) = 4K[45\alpha^2 + 7\beta(\alpha)^2] \tag{14}$$

and

$$I_R^R(180°) - I_L^L(180°) = \frac{8K}{c}[12\beta(G')^2 + 4\beta(A)^2] \tag{15}$$

$$I_R^R(180°) + I_L^L(180°) = 4K[6\beta(\alpha)^2], \tag{16}$$

where K is a constant associated with the scattering of radiation and c is the speed of light in vacuum. From these expressions, it can be seen that the ROA intensities for these two backscattering setups are the same and involve only the two anisotropic ROA invariants. These intensities are significantly larger than the depolarized ROA intensities associated with right-angle scattering, and hence the increased use of backscattering ROA in recent years. It is also clear that even though the ROA intensities of ICP_u and DCP_I are identical, their Raman intensities differ. The reason for this is that DCP_I scattering includes an additional polarization discrimination of the scattering radiation before detection whereas in ICP_u scattering, all the scattered light coming from the sample is measured. The additional Raman scattered radiation in ICP_u is actually polarized DCP_{II}-Raman backscattering which, in the FFR approximation carries zero ROA intensity. Hence no additional ROA intensity is present, but strongly polarized DCP_{II}-Raman intensity is measured in ICP_u ROA that is not measured for the same ROA spectrum using the DCP_I setup. The additional Raman intensity for ICP_u necessarily increases the noise in the measured ROA spectrum of a given magnitude. It is interesting to note that DCP_I-Raman intensity corresponds to the classical depolarized Raman intensity, whereas ICP_u-Raman intensity is the same as the corresponding polarized Raman intensity, and the ratio between these two intensities is the classical depolarization ratio for unpolarized incident radiation, yielding a value of 6/7 for depolarized bands. Aside from the theoretical noise advantage for depolarized bands, there is a significant advantage for DCP_I ROA in the case of strongly polarized bands which can have substantially higher noise, and the potential for interfering artifacts, in the ICP_u ROA measurement.

15.5 CALCULATION OF VIBRATIONAL OPTICAL ACTIVITY

For more than a decade following the discovery of ROA and VCD, the interpretation of VOA spectra was carried out on an empirical basis by correlation of spectra to structure or by means of model calculations which in turn were based on empirical concepts and parameters. With the appearance of *ab initio* quantum mechanical routines for the accurate calculation of VOA spectra, a new era emerged in the interpretation, analysis, and practical application of these spectra.

The *ab initio* calculation of VCD was pioneered by Stephens, Rauk, and Nafie and Freedman [86,88,92,99]. A significant advance in the formulation of VCD occurred with the introduction of gauge-invariant atom orbitals (GIAOs) [93,100] also known as London atomic orbitals (LAOs) [101]. Commercial software for carrying out *ab initio* quantum mechanical calculations of VCD became available about the same time as VCD instrumentation with the release of Gaussian 98 by Gaussian Inc. currently located in New Haven, CT. The release of Gaussian 03 in the summer of 2003 added many more features for VCD calculations [32] including full simulation of the spectrum using GaussView 3 for the ease of comparison of the calculated VCD and IR spectra to the measured ones. Extensive testing has shown that the most efficient level of calculation for the simulation of VCD spectra of typical organic molecules is density functional theory (DFT) using GIAOs with hybrid functionals such as B3LYP and a basis set of 6–31G(d) or higher [4,5,7].

The *ab initio* calculation of ROA has been pioneered primarily by Polarvarapu [90], Helgekar and coworkers [102], and more recently by Hug [103,104]. The latest release of Gaussian 03 now includes subroutines for running ROA in the far-from-resonance approximation. With this program, it is possible to calculate ROA spectra for typical experimental setups as well as all Raman and ROA tensor quantities from which any form of ROA and any experimental measurement can be calculated. At this time, the ROA tensor routines are not yet programmed with analytical field derivatives and hence must be determined by finite difference calculations. In addition, it has been found that higher level basis sets, including diffuse functions, are important for the accurate calculation of ROA intensities [105–110]. As a result, calculated ROA intensities require a greater time investment compared to the corresponding VCD calculations for the same molecule, and consequently the upper size limit of a molecule for which ROA can be calculated is smaller than that of VCD. This is an intrinsic advantage of VCD calculations over those for the corresponding ROA case, and hence it is likely that VCD will always hold an advantage over ROA in terms of speed

References pp. 536–544

or accuracy for the VOA calculation of a particular molecule with a similar degree of accuracy, despite increases in the computational speed and efficiency in the future.

With the advent of powerful software packages for the application of quantum mechanical calculations of molecular properties and the rapidly increasing performance of modern computer hardware, it is now possible to simulate the spectra of molecules from the first principles and compare these theoretical spectra with the traditional experimental ones. While this can be carried out for electronic CD (and OR), as well as for VCD and ROA spectra, there are a number of advantages inherent in vibrational spectra that make this comparison of theory to experiment more informative and accurate than the corresponding comparisons for electronic spectra.

First, vibrational spectra typically contain many more transitions through which the molecular stereochemistry can be probed. Each molecule of N atoms contains $3N-6$ degrees of vibrational freedom and most of these vibrational transitions can be observed with IR or Raman instrumentation. Thus, VCD and ROA spectra contain many vibrational bands representing all parts of the molecule.

Second, VCD and non-resonant ROA sample the ground electronic state properties of molecules. The molecular transition involves only the modulation of the equilibrium ground state and hence it is easier to calculate from a quantum mechanical standpoint. To the extent, the excited electronic states enter the theoretical description of VCD and ROA intensities, they do so as sums over all allowed electronic states and the detailed description of individual excited electronic states is not a critical part of the calculation. On the other hand, accurate electronic CD calculations require an accurate description of both the ground electronic state of the molecule, the starting point, but also of each excited electronic state for which a CD band is to be calculated. Further, an accurate simulation of an ECD spectrum requires some description of the band shapes and the underlying vibronic substructure of the electronic state, which is currently beyond the available theoretical technology.

Third, VCD and ROA spectra are not restricted to the presence in the molecule of a chromophore that will report on the molecular stereochemistry from a particular and sometimes remote vantage point. Virtually all chiral molecules exhibit VCD and ROA spectra that have bands in an expected intensity range. Molecules do not need to be modified in order to exhibit VCD or ROA spectra that are useful for analysis or comparison to theoretical calculations.

Despite the many advantages of VCD and ROA, they, like all the methods in molecular spectroscopy, play a complementary role relative to other techniques such as electronic CD, NMR, X-ray crystallography, and fluorescence, in gaining a deeper understanding of the structure and conformation of molecules. The principal contribution of VCD and ROA is that they are structure-rich probes of the stereochemistry of molecules in solution and disordered states. They can resolve multiple conformational states that are interchanging on time scales in the sub-picosecond region, and they show great promise for the accurate comparison of *ab initio* simulations to experimental measurements. It is clear now that, with further improvements and availability of instrumentation and software, VCD and ROA will emerge in the coming years as incisive, important tools for the precise determination of molecular stereochemistry in the solution and solid states.

15.6 DETERMINATION OF ABSOLUTE CONFIGURATION

The determination of the AC of chiral molecules has a long history in the field of molecular stereochemistry. X-ray crystallography is the method used to obtain the most definitive information about the AC of a chiral molecule. Additionally, AC can often be deduced from the knowledge of reaction mechanisms in organic chemistry applied to well-characterized transformations from starting materials of known absolute configuration to the final products.

OR and ECD can also be used to predict absolute configuration, usually based on rules for the sign of the rotation angle or ECD bands. Nuclear magnetic resonance (NMR) is blind to chirality, but ancillary methods for deducing absolute configuration using chiral shift reagents or chemical derivatives have been devised. However, whenever methods such as organic reaction mechanisms, OR, ECD, or NMR are relied upon fully over time, exceptions inevitably arise that result in erroneous predictions.

Conversely, the Bijovet method of X-ray crystallography, with many technical improvements in X-ray technology in recent years, has become the recognized standard for the *a priori* determination of the AC of chiral molecules [3]. There is, however, one major drawback with X-ray crystallography. A pure single crystal of the sample is required. In the case of liquids or oils, X-ray crystallography is precluded and other less definitive methods must be used if they possess the capability. In some other cases, crystals cannot be grown within reasonable periods of time, thereby either precluding this method or waiting for months to years for crystals. Another recommended requirement of the Bijovet method is the inclusion in the crystal of a heavy atom. When this is not the case, the determination of the phase required for the X-ray analysis is more difficult and sometimes mistakes are made, as was recently discovered by comparing the X-ray prediction to the results of a VCD determination of the AC [111].

Within the past decade, VCD has emerged as a powerful new method for the determination of the AC of chiral molecules. All molecules absorb radiation in the IR region, where their absorption pattern across the spectrum serves as a rich fingerprint of molecular structure and shape. In addition, all chiral molecules have a VCD spectrum that consists of an even more powerful fingerprint spectrum of the structure and shape of the molecule. The additional power afforded to VCD is due to its stereospecific sensitivity. Molecules with opposite absolute configuration, pairs of enantiomers, have the same IR spectrum, but opposite VCD spectra. This point is illustrated in Fig. 15.1, where the IR and VCD spectra of the (+)- and (−)-enantiomers of camphor in the mid-IR region of the spectrum are presented.

Although the magnitude of the VCD spectrum is approximately 10,000 times smaller than that of the parent IR spectrum, with modern FT-VCD instrumentation, the measurement of a VCD spectrum is routine. Each absorbance band in the IR spectrum has a corresponding VCD band. There is no correlation between strong IR and strong VCD bands, but each band in the VCD spectrum reports on the structure of the molecule as well as its AC. Because VCD measurements can be carried out in the solution phase, and because the results are direct with rapid turnaround,

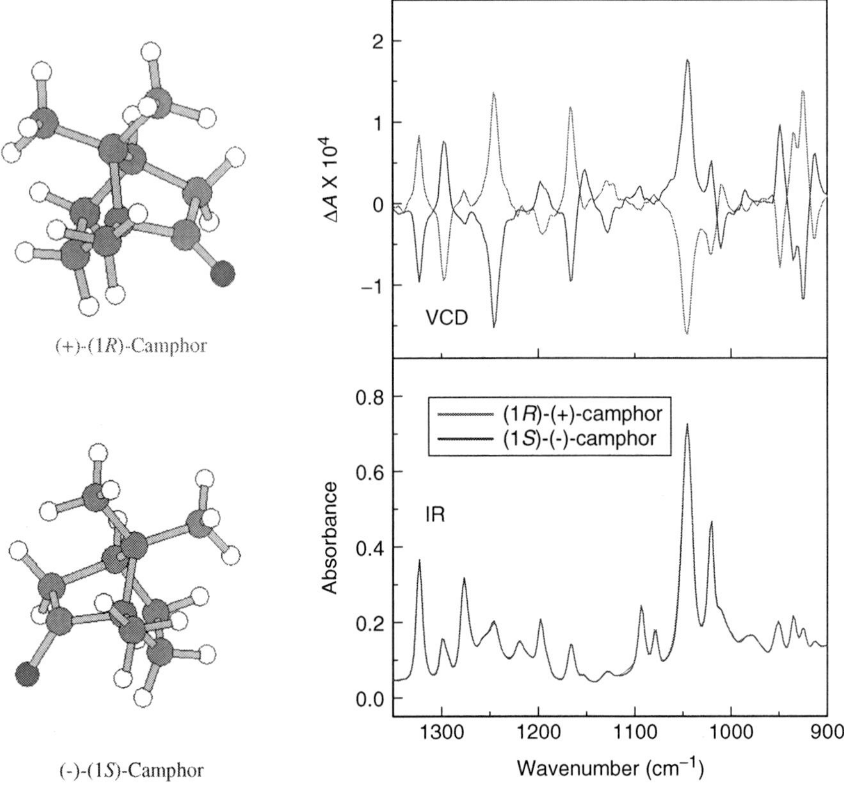

Fig. 15.1. The VCD and IR spectra of (+)-*R*-camphor (upper structure) and (−)-*S*-camphor (lower structure).

the VCD method of AC determination stands as an alternative to X-ray crystallography for the routine determination of AC in chiral molecules.

The AC of a chiral molecule is determined from VCD by comparing the results of an experimental measurement of the IR and VCD spectra of the chiral molecule with the quantum mechanical calculation of the same IR and VCD spectra for a particular choice of the AC of that molecule [4–7]. The sample is measured as a neat liquid or as a solution in a suitable solvent. Crystallization is not required, and the VCD measurement takes anywhere from 5 min to several hours depending on the quality of the spectrum desired and intensity of the VCD spectrum relative to the noise level. The calculation for a particular conformation typically takes on the order of a day of computer time depending on the size and complexity of the molecule and the speed and memory capacity of the computer. The ACs of molecules with up to 100 heavy atoms, where a heavy atom is any element beyond hydrogen, have now been determined by VCD. If the measured and calculated VCD spectra agree sign-for-sign for the major bands across the entire spectrum, and the relative magnitudes of those bands in both the VCD and IR spectra agree, then the AC is deduced without ambiguity or reference to any prior structure or calculation. If the bands across the VCD spectrum are opposite in sign, then the wrong enantiomer was chosen for the calculation. Calculation of the mirror-image structure then produces the desired agreement and again the AC is determined unambiguously.

Over the past four years, nearly every major pharmaceutical company in the world has either purchased a VCD spectrometer for AC determination or outsourced the determination of the AC of selected molecules by VCD through collaboration, feasibility study, or contract [8]. Some companies from which researchers have reported the results of VCD measurements at scientific meetings include Organon, GlaxoSmithKline, AstraZeneca, Johnson & Johnson Pharmaceutical Research & Development, Bristol Myers Squibb, Pfizer, Abbott and Wyeth Laboratories. Through this activity, the absolute configurations of hundreds of chiral molecules have been determined over the past four years. Currently, confidentiality restricts much of this work from public disclosure, but a number of articles have been published after the results were approved for dissemination [9–12].

The simplest application of VCD to the determination of AC is for structures that are rigid to the degree that only one conformation is dominant in solution. In Fig. 15.2, we provide such an example where the AC of a newly synthesized chiral iminolactone (1,4-oxazin-2-one), was determined by VCD spectroscopy [112].

The stereo-specific line structures as well as its space filling optimized geometry from the DFT calculation are shown. The comparison of the measured (+)-enantiomer with the calculated *R*-enantiomer is also shown. The close agreement of all the measured features, together with the agreement of the intensity pattern of the IR spectrum, make available the unambiguous determination of the AC of this molecule.

A more challenging example is the assignment of AC of the aryl axial chirality of the two atropisomers named dicurcuphenol B and C, isolated from the marine sponge *Didiscus aceratus*, the structures of which are shown in Fig. 15.3 [113].

The VCD and IR of these two isolates are also shown in this figure. The dicurcuphenols B and C are diasteriomers in that they possess a common chiral tail that breaks the mirror symmetry of the VCD of these two molecules. In addition, this tail contains a high degree of conformational freedom which is undesirable from a computational point of view. The strategy adopted was to subtract the VCD of the two dicurcuphenols, thereby isolating the contributions of opposite VCD originating from the axial chirality. In Fig. 15.4, the comparison of the subtracted VCD to that calculated for the rigid chiral model compound (*M*)-10, which is the structure of dicurcuphenol minus its chiral tail, is shown.

The excellent agreement allows the axial chirality of dicurcuphenol B to be assigned as *M*, therefore dicurcuphenol C has *P* axial chirality.

In Fig. 15.5, results are presented for the determination of the AC of the molecule McN 5652-X, a high-affinity ligand for transport of seratonin in the brain, for which more than one conformer contributes to the observed VCD spectrum [12].

The theoretical analysis of this compound yielded two low-energy conformers, SRa and SRb, shown in the figure. Separate IR and VCD spectra were calculated for each conformer. Comparison of the calculated VCD for the two conformers with the experimental VCD spectra identifies bands in the experimental spectrum associated

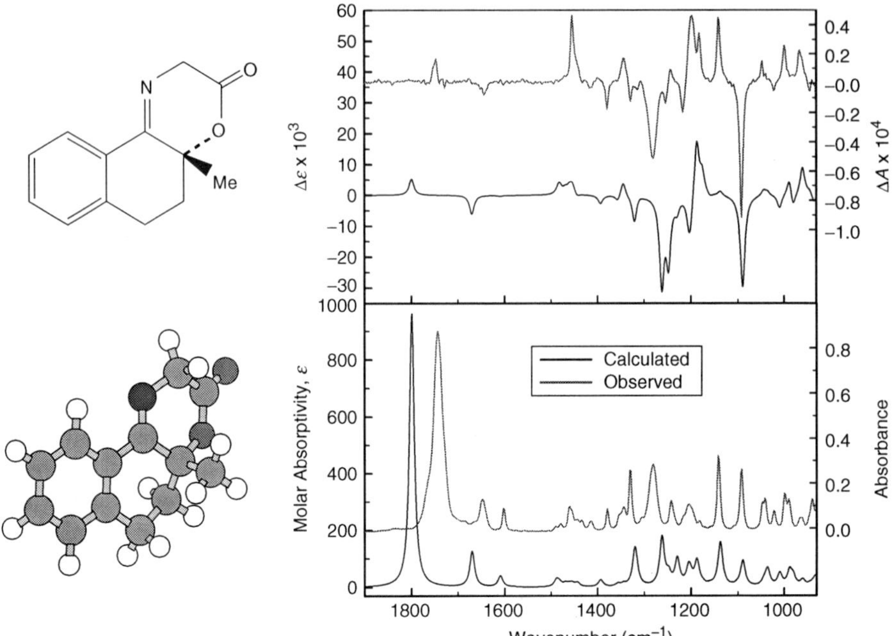

Fig. 15.2. Line structure and optimized geometry of a new chiral iminolactone with comparison of observed (+)-enantiomer VCD and IR to the calculated *R*-enantiomer VCD and IR as a proof of absolute configuration.

with each of the conformers. Many VCD bands for the two conformers are the same and are also seen in the experimental spectrum. From the agreement in sign of all the major VCD features and the relative intensities in the experimental and calculated spectra for both IR and VCD, the absolute configuration of (+)-McN 5652-X can be assigned to be 6*S*,10*R*, as shown in the structure in Fig. 15.5. This example illustrates that from VCD analysis it is possible to determine the solution-state conformation of a molecule as well as its absolute configuration. In fact, agreement between the experimental and calculated IR and VCD spectra cannot be achieved unless the correct conformational states of the molecule are found and used as the basis for the calculations. Hence, VCD analysis provides not only the absolute configuration, but in addition, the solution-state conformation or conformational population. Within the past several years, a number of examples for the determination of AC using VCD have appeared in the literature [11,111,114–136]. The extent and variety of these examples are testaments to the growing importance of the use of VCD for the determination of AC in chiral molecules.

The methodology for the determination of absolute configuration using ROA is identical to that for VCD. Presently, the measurement of the ROA of chiral molecules in non-aqueous solution is problematic due to interference from strong solvent Raman bands that can saturate the CCD detector in the case of dilute solutions. Much better

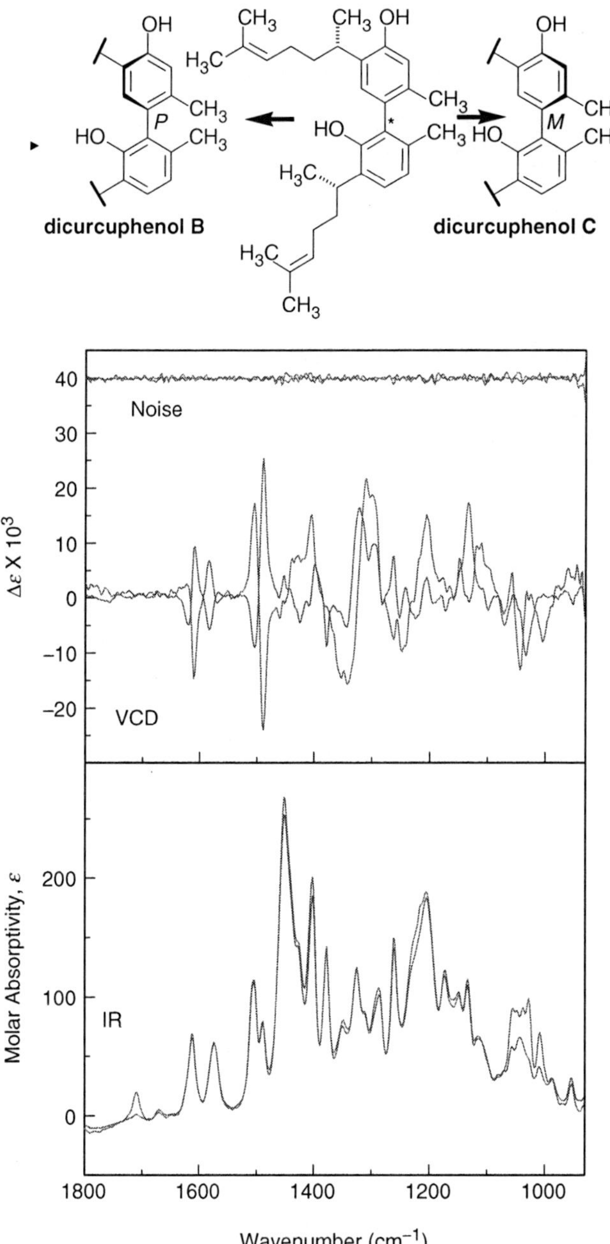

Fig. 15.3. Sturctures, IR and VCD of the two atropisomers dicurcuphenol B and C isolated from the marine sponge *Didiscus aceratus*.

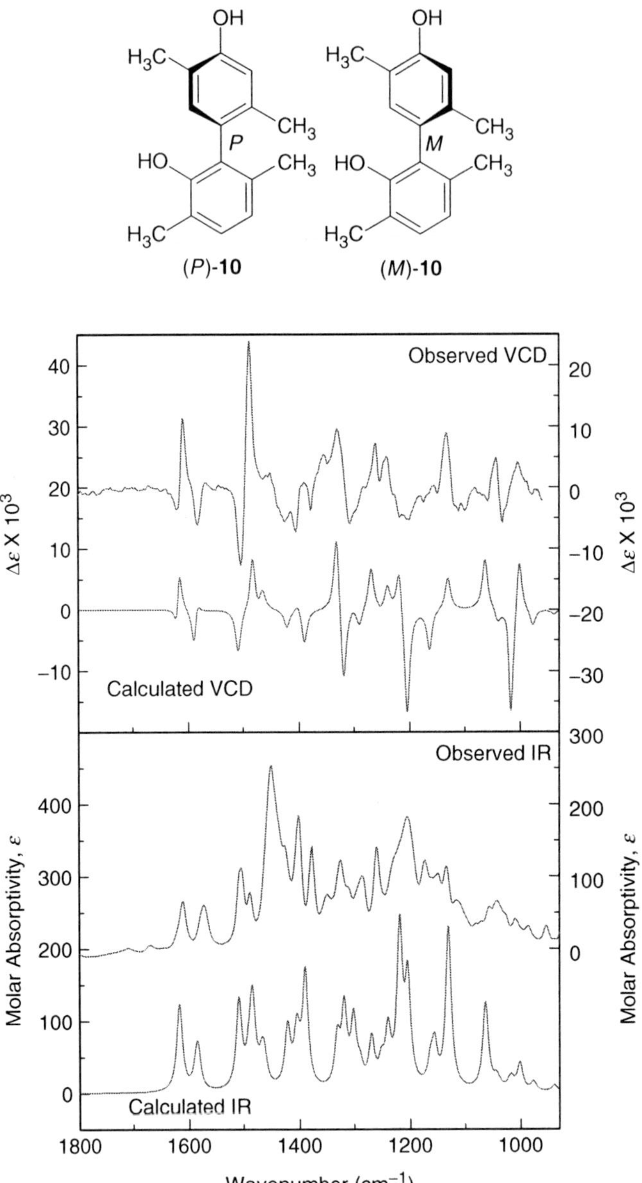

Fig. 15.4. Structures of the model phenol atropisomers (*P*)-**10** and (*M*)-**10** used for theoretical VCD calculations and comparison of the calculated IR and VCD spectra of (*P*)-**10** with the difference in VCD spectrum of the atropisomers, curcuphenol B minus curcuphenol C shown divided by two, thereby establishing the absolute aryl axial chirality of the latter isolated natural products.

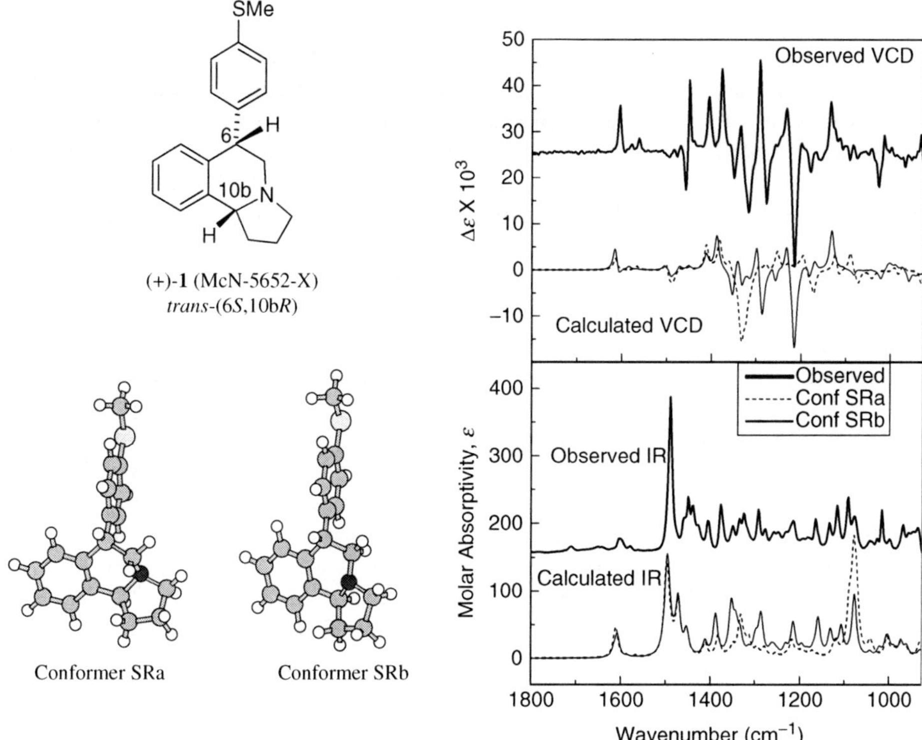

Fig. 15.5. Comparison of the measured and calculated VCD and IR spectra of (+)-McN 5652-X. The calculated spectra are for two conformers, SRa and SRb.

for ROA analysis are compounds whose natural solvent is water, such as amino acids, sugars, and pharmaceutical molecules soluble in H_2O at any pH. Water has a relatively weak Raman scattering spectrum and hence poses no serious problem to the measurement of the ROA spectrum. The procedure to determine the absolute configuration is essentially the same as for VCD, namely to start with an *ab initio* computational analysis of the Raman and ROA spectra of the preferred conformer or conformers of the molecule. Comparison of the experimental Raman and ROA spectra to the corresponding calculated spectra yields the AC. A few examples of the comparison of measured and *ab initio* calculated ROA spectra have appeared in the literature [103–105,110,137,138]. Results to date have demonstrated that a more extensive basis set using diffuse basis sets is necessary to capture an accurate Raman and ROA spectral response to the vibrational modes of the molecule. This, together with the limitations in the implementation of ROA subroutines in Gaussian 03 limit, the size of the molecules that can currently be considered, and hence ROA is currently more limited in its range of application to absolute configuration than is VCD. This discrepancy is likely to continue in the future since it is intrinsic to the difference in complexity and computation demands of ROA compared to those of VCD.

Although VCD and ROA are the most powerful and effective methods for the determination of absolute configuration in chiral pharmaceuticals using optical activity, at least two other methods are currently under development and exploration. The first of these is the determination of absolute configuration by measured and *ab initio* calculated OR [139–146]. While the experimental data are more widely available than experimental VCD spectra, the calculations require a higher level of quality for good reliability and only one datum is produced rather than an entire spectrum, so there is no way to be sure that good agreement between the experiment and calculation has been achieved. VOA, on the other hand, requires not only sign agreement, but also spectral confirmation through corresponding relative intensities of the IR/Raman and VCD/ROA bands for which the correct conformation, or distribution of conformers, has been determined and which forms the basis of the comparison between the measured and calculated results.

The second approach is the corresponding analysis for electronic circular dichroism (ECD) [147,148]. For the comparison of experiment to theory, typically only a few ECD bands are available. Again, higher level calculations than those needed for VCD analysis are required for good reliability, but here, predicting correctly the relative signs for two or more transitions gives some confidence of having a valid analysis. The software for calculating OR and ECD is available in Gaussian 03 [32]. Here again, calculation of the UV-visible absorption and ECD spectra for one-to-several excited electronic state transitions offers no definitive proof that the correct solution-state conformation, or distribution of conformations, is being used for the calculated ECD spectrum. Nevertheless, there is now a growing awareness of the value of using any two or all three of these *ab initio* methods, VCD, OR, and ECD, for the determination of AC from the experimentally determined data. In particular, there is value to including VCD or ROA in the determination to ensure that the conformational state or states of the molecule have been correctly identified. If the same prediction of absolute configuration, using the correct conformations, emerges from such a multiple analysis, an even higher level of confidence in prediction is achieved.

As technology advances in the future to make available even more efficient instrumentation for the measurement and *ab initio* computation of VCD, ROA, OR, and ECD spectra, the determination of absolute configuration of chiral pharmaceuticals will be increasingly carried out by these methods because of their ease of sample preparation and time to final results.

15.7 DETERMINATION OF ENANTIOMERIC EXCESS OF MULTIPLE CHIRAL SPECIES

Another important property of a sample of optically active molecules is the EE of each chiral species present. Using conventional techniques, the fractional composition and the EEs of each species cannot be determined in the intact sample for reasons explained below. The traditional method for the determination of EE is chiral chromatography [13]. This is routinely accomplished within an accuracy of 0.1%, approximately an order of magnitude better than that demonstrated by VCD or ROA to date. On the other hand, VOA has demonstrated a sensitivity in the range of one-to-several %EE. As such, VOA should not be considered as a replacement of chiral chromatography for the definitive measurement of the %EE of any sample. Nevertheless, VOA possesses several very important unique advantages over chiral chromatography as it is currently practised. The most unique of these is the potential of VOA to be used as a real-time *in situ* monitor of the EE of multiple chiral species in a reacting medium [14]. The second advantage of VOA is the determination of EE of chiral-active ingredients in various states or environments such as liquids, solutions, solids, and a variety of formulated forms in the presence of excipients [149] as tablets, powders, or liquids. All other methods of EE determination, currently devised or in use, require pure dissolved samples, and in the case of chiral chromatography require separation of enantiomers.

The power of VOA to enable these *in situ* measurements relies on two important facts. The first is that the measurement of the VOA is always accompanied by a simultaneous measure of the parent IR or Raman spectrum. This provides an independent spectral measurement that contains information about the structural identity, composition, and state of the sample being measured. Second, and perhaps more important, the magnitude of the VCD or ROA is measured relative to the corresponding IR or Raman spectrum. Any changes in concentration, path length, density, or temperature that may affect the magnitude of the measured VOA spectrum can be automatically normalized out by corresponding changes in the parent vibrational spectrum. Absolute values of the EE can be obtained in this way for any species under any conditions following a single calibration measurement of the VOA relative to the parent IR or Raman for a known value of EE.

Recently, a flow-cell apparatus was constructed for the continuous measurement of VCD for a sample changing with time whereby the sample cell is never removed from the instrument [14]. This apparatus has been used in various ways to demonstrate the potential for VCD to monitor a reaction mixture that may be changing in time in both the composition and EE of each chiral species present. Initial studies focused just on following changes in EE of a single species, such as α-pinene or camphor in CCl_4 solution. Here, a starting solution of EE was first measured and then in successive steps measured amounts of a solution of the opposite enantiomer at the same concentration of the opposite enantiomer were first mixed with the original solution, circulated through the flow cell, and then measured for the new EE. The resulting sets of VCD spectra, varying from 100% EE in ten steps or so to a few %EE, could be analyzed using chemometric partial least squares (PLS) analysis to an accuracy of 1–2% EE for measurement times of 10 or 20 min and a spectral resolution of $4\,cm^{-1}$. Subsequently, we found that similar accuracy could be achieved for measurement times as short as 1 min if the resolution was lowered to 8 or $16\,cm^{-1}$.

A more challenging experiment is to use the FT-VCD-PLS method to follow changes in a two-component mixture, where both the mole fraction and the %EE of each species is changing as a function of time. The two chiral molecules chosen were (1*S*)-(–)-camphor and {(1*S*)-endo}-(–)-borneol, where the structures and absolute configuration of these molecules were [14],

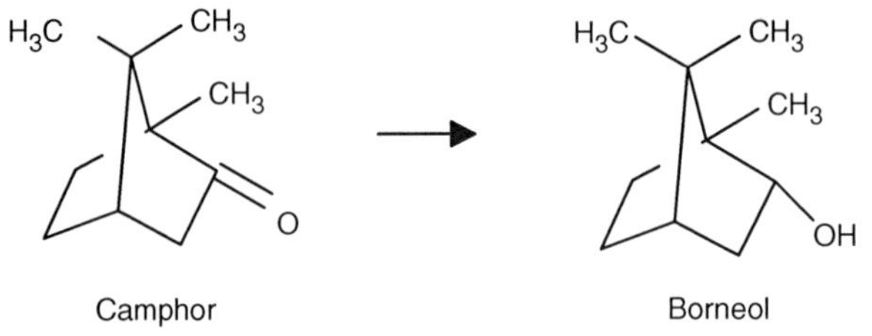

Camphor → Borneol

It is clear that these molecules differ only in one functional group that is a carbonyl group in camphor and a hydroxyl group in borneol. We used the same mixing apparatus described above for these studies.

Before carrying out the desired mixing experiment, a series of carefully designed calibration measurements were carried out to establish a training set with which the mixing experiments were to be analyzed using PLS. Mixtures of camphor and borneol, each at 100% EE were prepared across the full range of expected concentrations for the mixing experiments. This included the IR and VCD spectra of pure camphor and borneol, as shown in Fig. 15.6. From this figure, it can be seen that while the IR and VCD of these two molecules are distinct from one another, they do possess a high degree of similarity. This not only adds a challenge to the experiment as a whole, but also a degree of realism since any pair of molecules that are related as reactant and product are likely to have similar IR spectra, and quite possibly similar VCD spectra as well. The resulting set of training spectra were subjected to the cross-correlation regression analysis using PLS.

Following the calibration measurements, a mixing experiment involving each species at 100% EE was set up in two parts, one starting from camphor and adding borneol and the other with the solutions reversed. By combining these two studies, a simulation of a complete reaction starting with pure camphor and ending with pure borneol was achieved. Since the solutions were all at 100% EE, the sets of both the IR and the VCD can be used independently to predict the mole fraction of the mixing experiments, where, because of higher signal quality, the errors in prediction for the IR were approximately an order of magnitude better than those for VCD. Taking the ratio of these two determinations for each mixing point allowed prediction of the %EE of each species, which was known by design to be 100% EE. Errors of prediction, based on the ratio of the IR and VCD determinations were between 2 and 3% EE for each species.

With these calibrations and preliminary measurements completed, a more general mixing experiment was undertaken. The original solution was 9 ml 0.600 M at 100.00% EE [(1*S*)-endo]-(–)-borneol and 0.400 M at 100.00% EE (1*S*)-(–)-camphor mixture in CCl_4 solution. The add-ins were 0.400 M at 0.00% EE [(1*S*)-endo]-(–)-borneol and 0.600 M at 0.00% EE (1*S*)-(–)-camphor in CCl_4 solution. The volume of add-in was 1 ml for each injection. In total, ten samples were prepared and measured by this flow-cell sampling method and the IR and VCD spectra generated by these experiments are displayed in Fig. 15.7.

The final solution ended at mole fractions of camphor and borneol each at 0.500 and the %EE for camphor 40% and that of borneol 60%. For this system, to obtain

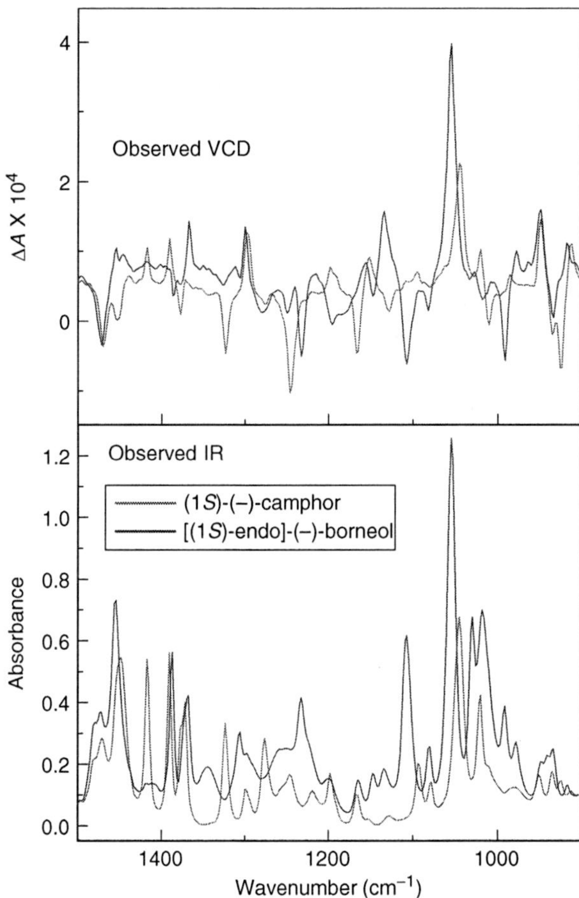

Fig. 15.6 Comparison of the IR and VCD spectra of (1*S*)-(−)-camphor and [(1*S*)-endo]-(−)-borneol, where both are 1.000 M solutions in CCl_4.

the %EE of camphor and borneol at each step, both the IR and VCD spectra are required, since during the experiment, both the mole fractions and the EEs of the two components change. After these spectra were analyzed using PLS chemometric software, it was found that the EEs of both camphor and borneol could be determined at each mixing point with an accuracy of approximately 2–3%.

The general method for analyzing these spectra to determine mole fractions and %EE at each mixing stage is as follows: in principle, multiple species can be present, each with changing mole fraction and changing %EE. The IR spectra of each species at known concentration can be used to determine the mole fraction of each species present in the mixture. The VCD spectra of each species at known %EE, say 100%, can be used to determine the *apparent* mole fractions of each of the chiral species present. These apparent mole fractions are less than the actual mole fractions, depressed by %EE values less than 100%. By normalizing the *apparent*

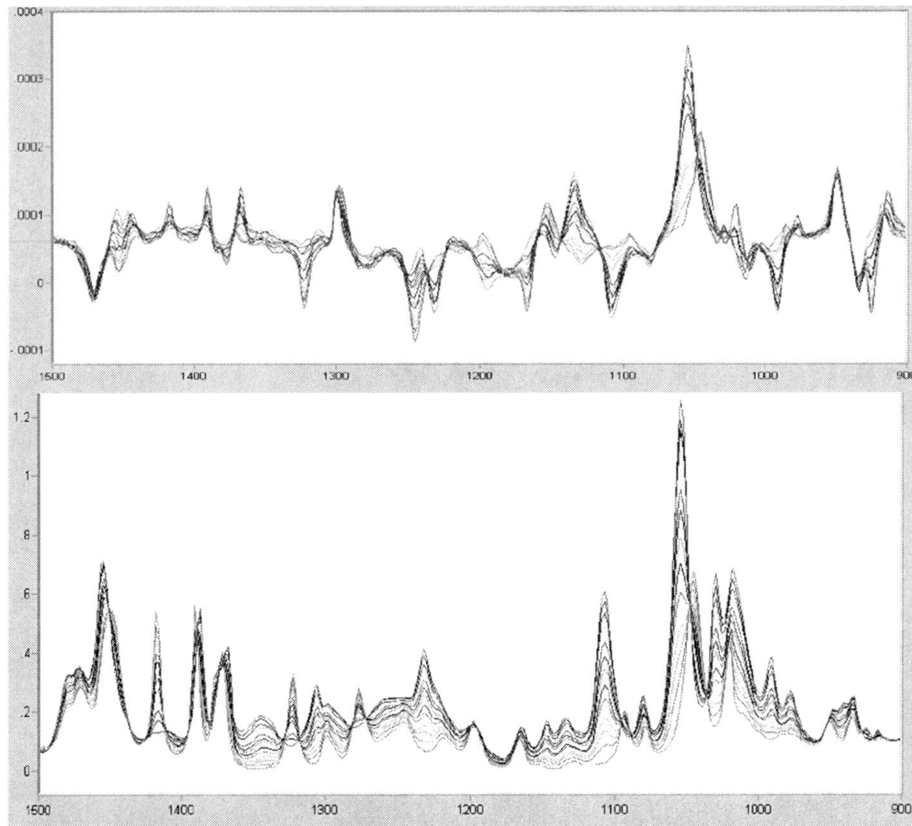

Fig. 15.7. Ten correlated sets of VCD and IR spectra for flow-cell measurements of mixtures of (1*S*)-(−)-camphor and [(1*S*)-endo]-(−)-borneol in CCl_4, where the IR spectral changes reflect only the changing composition between camphor and borneol whereas the VCD spectra change as the result of both changing composition and change EE of both species.

mole fractions by those obtained from the IR for each species, one is left with population coefficients between 0 and 1 that represent the EE for that species, and after multiplying by 100, the %EE of each species is determined.

The results presented here demonstrate that FT-VCD can be used to monitor simultaneously the %EE of multiple chiral molecules as a function of time. Although the intrinsic signal-to-noise ratio of VCD is not as high as other traditional monitors of the kinetics of chiral molecules, there are a number of advantages of FT-VCD that arise from its information content and from the multiplex nature of its measurement process. FT-VCD spectra typically contain many bands representing different vibrational modes from all portions of the molecule. For VCD, there is no concern about the concept of chromophore since all molecules have “vibrational chromophores” representing all structural locations in the molecule. Since different molecules have different vibrational bands and different vibrational frequencies for these bands, it is often possible to identify individual peaks in a mixture that belong to particular molecules. In the case of monitoring the %EE of a single species, the multiplex advantage of FT-VCD means that each point in the spectrum, of which there may be hundreds or thousands across the spectrum, represents an independent measure of the %EE. Combining these data points using PLS, with higher weighting accorded to regions of higher signal-to-noise ratio, the disadvantage of relative low signal-to-noise ratio at individual spectral locations is in large measure overcome.

These studies have been extended in two ways. One is a spectral extension of FT-VCD from the mid-IR into the near-IR region above 4000 cm^{-1}. As an example, the mid-IR FT-VCD spectra are compared to various regions of the near-IR FT-VCD for a 1.0 M solution of (1*R*)-(−)-camphorquinone in $CDCl_3$ solution [20].

The first spectral segment covers the typical region of the mid-IR from 800 to 1600 cm^{-1} using a liquid-nitrogen cooled HgCdTe at a path length of 100 μm. For the region of CH-stretching modes, a thermoelectrically cooled HgCdTe photovoltaic detector with a cut-off near 2000 cm^{-1} was employed and a sample path length of 50 μm. Beyond 3800 cm^{-1} much longer path lengths are needed and room-temperature detectors are now sufficient for optimum signal quality. Here, relatively weak combination and overtone bands from predominantly hydrogen-stretching modes occur. The first region from 3800 to 6200 cm^{-1}, using an InGaAs detector and a path length of 2 mm, contains combination bands that represent one CH stretching and one CH bending mode below 5000 cm^{-1} and overtones of the CH stretching modes between 5500 and 6000 cm^{-1}. In the next region between 6000 and 10,000 cm^{-1}, a path length of 20 mm and a Ge detector is used. Here, the bands represent the second combination band region (2 CH stretches and one CH bend) and the second CH-stretching overtone region (3 CH stretches). We have noted certain similarities among some of these regions in terms of the relative intensities and sign patterns of the VCD.

As can be seen in Fig. 15.8, the near-IR has the intrinsic advantage of longer sampling path lengths and, as a result, greater ease of sampling. The experiments described above for %EE determination of one- and two-component samples with a flow cell have been repeated for the near-IR [150]. In this region, it has been found that approximately nearly the same level of accuracy in the prediction of %EE can be achieved. Because detectors and VCD baselines are more sensitive to environmental perturbations in the near-IR, a high level of chemometric analysis and more extensive calibration measurements are needed to achieve similar results, where we find errors in the 2–3% range, rather than the 1–2% range found in the mid-IR. Since the intrinsic bandwidths are broader in the near-IR, a resolution of 16 cm^{-1} was used instead of the 4 cm^{-1} used in the mid-IR region. Again, lowering the resolution further to 32 or 64 cm^{-1} provides the opportunity to achieve the same accuracy at shorter collection times approaching 1 min per measurement.

Following the extension of our FT-VCD instrumentation into the near-IR region, flow-cell mixing measurements have been extended to the reaction of real molecules in a reaction vessel. Here, the reacting solution is circulated through a flow cell without the need to add in a secondary solution to achieve a change in mole fraction or %EE or both. One such experiment is the epimerization of 2,2-dimethyl-1,3-dioxolane-4-methanol (DDM) carried out in the near-IR using an InGaAs detector with a spectral

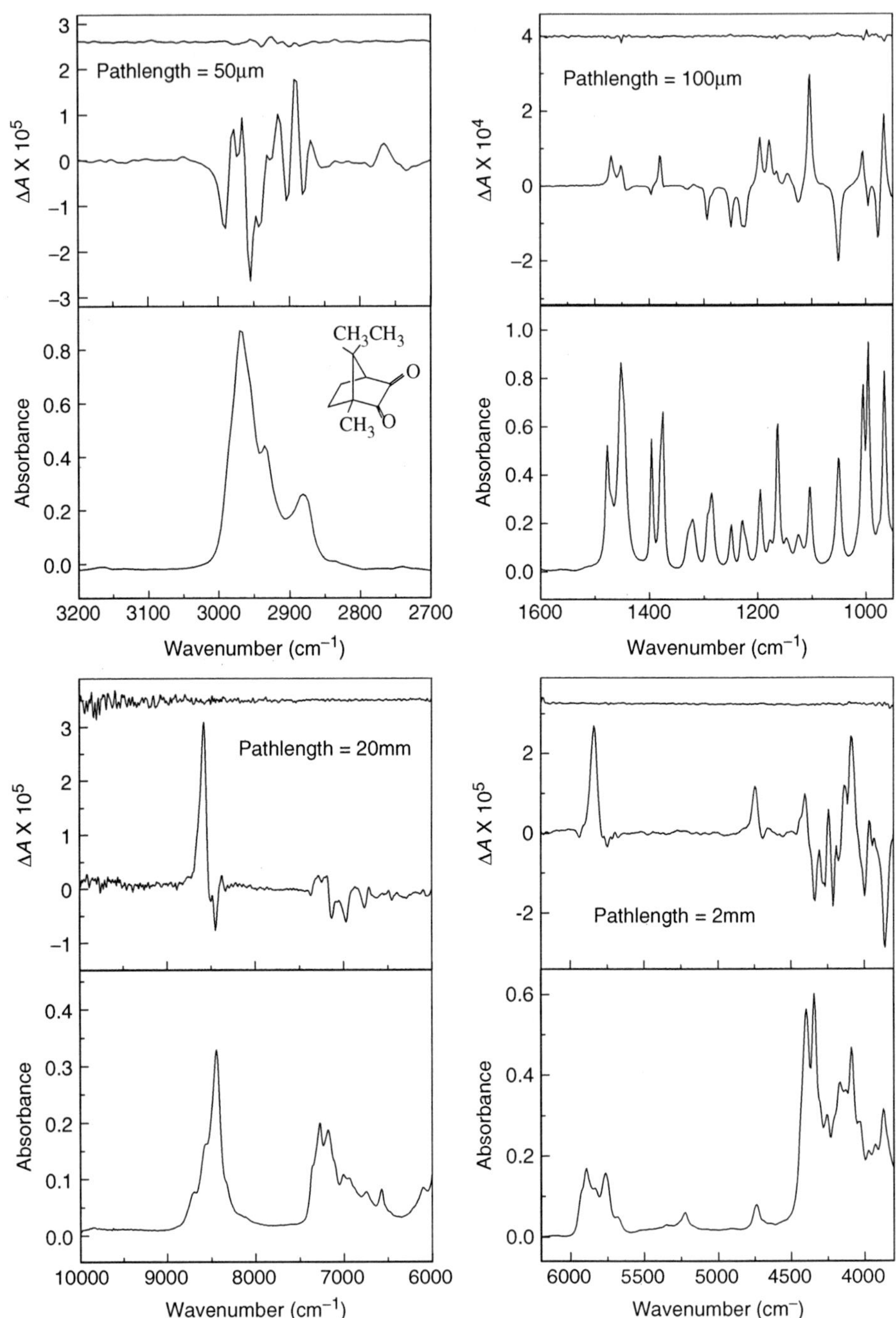

Fig. 15.8. VCD and IR absorption spectra of 1.0 M (1*R*)-(−)-camphorquinone in $CDCl_3$ solution.

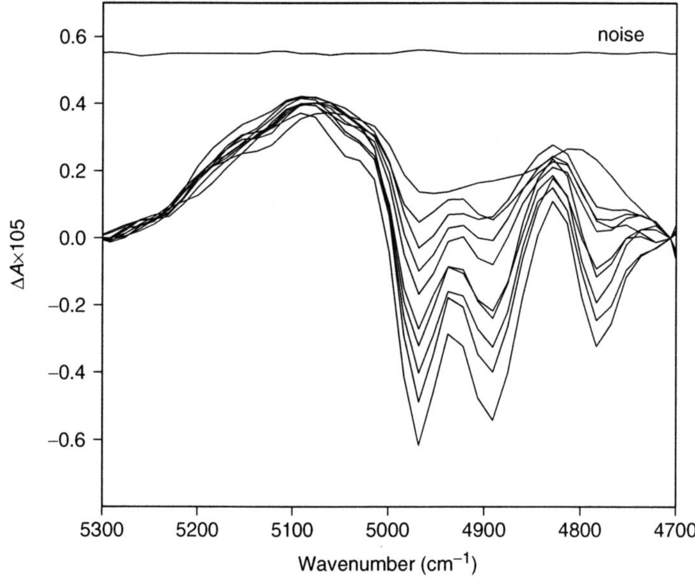

Fig. 15.9. NIR-VCD spectra for the calibration samples in toluene displayed in the range of 5300–4700 cm^{-1} at a resolution of 32 cm^{-1}.

resolution of 32 cm^{-1}. The structure and epimerization pathway of this molecule, together with a set of near-IR VCD spectra are shown in Fig. 15.9. [151].

The reaction intermediate is a protonated species which has longer lifetimes in solvents that give greater stability to this intermediate. Once the intermediate forms, it has two equally probable pathways back to either the original structure or its enantiomer. In Fig. 15.10, the reaction kinetics of %EE determined by the FT-VCD-PLS method is shown for three different solvents.

Of these solvents, methylcyclohexane best stabilizes the intermediate and has the fastest loss of %EE. Toluene and carbon tetrachloride show nearly the same half-life, but have distinctly different kinetic detail that is not understood at present.

In addition to this study, the reaction of 2-butanol to 2-chlorobutane by the addition of thionyl chloride has been monitored using the near-IR FT-VCD-PLS method described here [151]. It was possible to confirm that the reaction proceeded from reactant to product under the conditions tested, without loss of chirality at the 3%EE confidence level, while simultaneously monitoring the change in mole fraction from pure butanol to nearly pure chlorobutane.

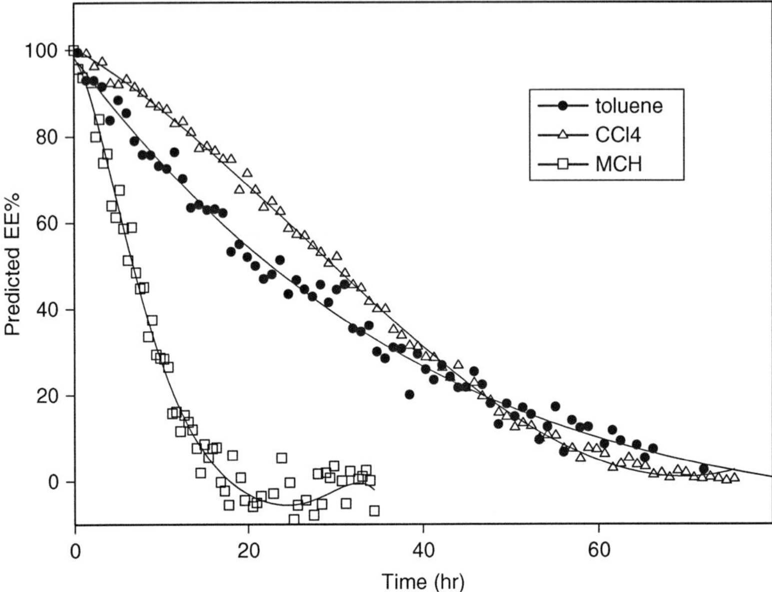

Fig. 15.10. %EE of (*S*)-DDM as a function of reaction time under the treatment of trifluoroacetic acid in toluene, carbon tetrachloride, and methylcyclohexane.

15.8 VIBRATIONAL OPTICAL ACTIVITY OF SOLIDS AND FORMULATED PRODUCTS

Most VCD studies to date have focused on samples that are neat liquids or solutions. Several early investigations of solid-phase mulls and films have been published but without much success [152–155]. The main reason for not continuing these studies at the time was that, due to poor VCD artifact control, spectra proved difficult to reproduce because of small differences in birefringence between samples. More recently, with the improvements in artifact level and control made possible by the use of the dual photoelastic modulator (PEM) method [55], and greater care in sampling, solid-phase VCD studies have resumed. Recently, we published a study of a spin-coated conducting organic polymer doped with a chiral camphor-based molecule [156]. The spin-coating process led to a sample film that had virtually no orientation dependence and as a result no linear birefringence artifacts. Several additional VCD studies of films have been published recently using a dual PEM setup that report success for a variety of samples including proteins and carbohydrates [24–27]. In most cases, the film VCD was very close to the VCD measured in solution. The main advantage of using films is that aqueous solvent absorption is avoided and roughly two orders of magnitude less sample is needed to obtain a high-quality VCD spectrum.

Mulls and KBr pellets are alternatives to films for VCD studies. Mulls preserve the original crystal form of the solid, but as with IR absorption, scattering from the small crystal particles must be avoided. This is achieved for mulls by grinding the particles to a distribution with an average crystal size less than the wavelength of the light being used and then adding an index-matching fluid to decrease the contrast between the solid particles and their immediate optical environment. The tell-tale signature of the presence of a scattering contribution is the shape of the IR and VCD bands. Called the Christiansen effect [152], scattering produces IR bands that have a sharp rise on the high-frequency side of a peak, where intensity has been reduced and an extended tail to low frequency where intensity has been added. The corresponding effect on positive VCD bands is a scattering contribution that produces a negative VCD contribution to high frequency and a positive one to low frequency, superimposed on the normal VCD spectrum. The contributions are reversed in sign for negative VCD bands. As a result, VCD spectra with high scattering contributions can exhibit bisignate couplets where, in the absence of scattering only a monosignate band is observed.

An advantage of solid-phase sampling in mulls is that it permits different crystal forms of a molecule to be compared. As an example, in Fig. 15.11, the IR and VCD spectra of propranolol in two different solid forms are presented.

The band shapes for both the IR spectra are symmetric indicating the absence of observable levels of scattering. The IR and VCD spectra show a high degree of similarity and at the same time some significant differences can be observed. For example, a band in the hydrochoride salt appears at 1320 cm^{-1} with significant VCD that is absent in the free base, and there is a very strong VCD couplet in the free base centered just above 1100 cm^{-1}, which is only a small couplet in the hydrochloride salt even though the corresponding IR bands for these two couplets are almost the same. Detailed interpretation of these spectra must await the development of accurate *ab initio* descriptions of the unit cell of these solids with periodic boundary conditions describing the interactions of the cell with the neighboring unit cells.

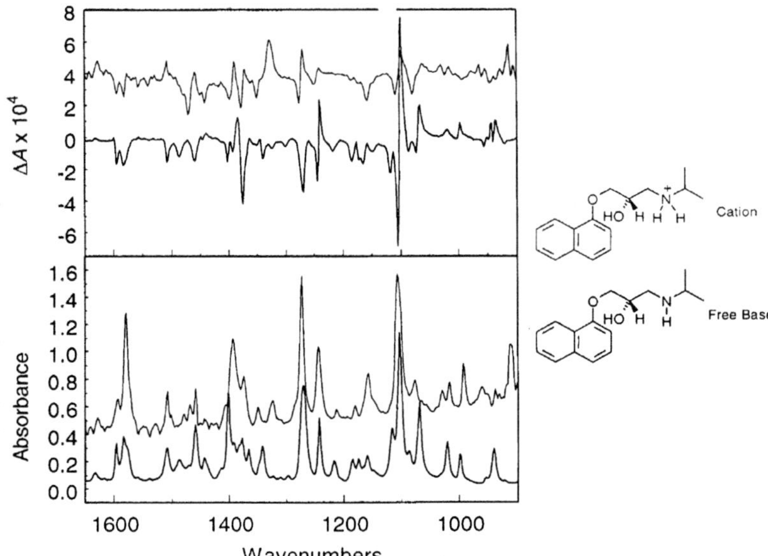

Fig. 15.11. Solid-phase mull IR and VCD of propranolol as the hydrochloride chloride salt (grey, upper) and as the free base (black, lower).

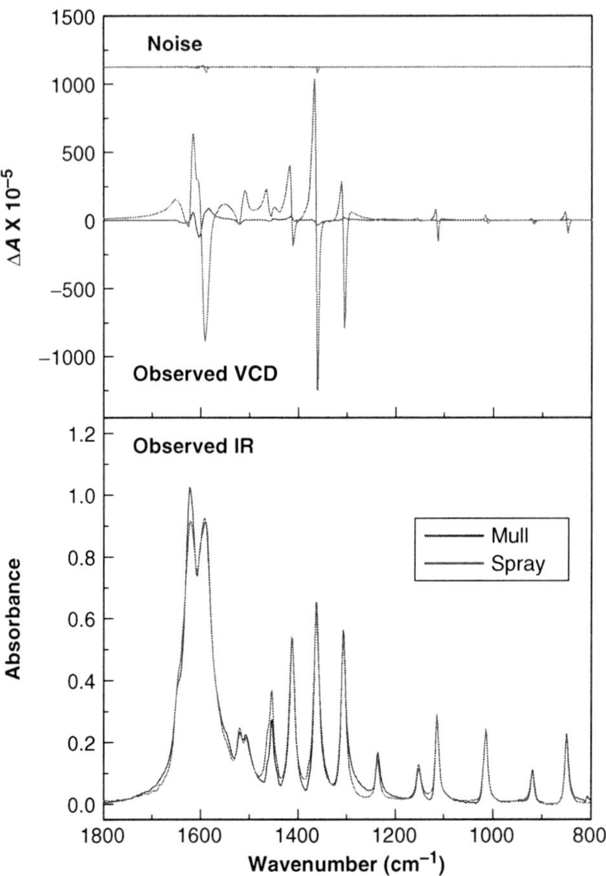

Fig. 15.12. Comparison of the VCD and IR of L-alanine as a mull and as a spray-dried film.

Recently, it has been discovered that spray-dried films produce very large VCD intensities compared with mulls and KBr pellets for small molecules of biological significance such as amino acids, small peptides, and active pharmaceuticals [28]. The films are produced from aqueous solutions that are sprayed as a fine mist toward an IR window, such as BaF_2, which has been heated to approximately 80°C. The mist vaporizes rapidly to a crystalline film upon contact with the window. As an example in Fig. 15.12, we show the spray-dried film of L-alanine compared with the corresponding mull.

The spray-dried sample has VCD intensity that is nearly two orders of magnitude larger than that of the mull for nearly identical IR absorbance spectra. It appears that a different crystal form of L-alanine forms on the window under these conditions of rapid drying that is different from commercially available crystalline powders and used for the mull. The most intense VCD band has a ratio of the VCD to IR intensity of approximately 0.02 that requires an extent of chiral order in the sample on the order of tens of alanine molecules. This illustrates how VCD can be used to probe different crystal structures of the same molecule, and in this case the VCD spectrum reveals that spray-dried crystals of L-alanine, and the other amino acids, have a long range chiral order that extends well beyond that of the ordinary L-alanine crystals produced from precipitating solutions.

With the ability to observe VCD in solids comes the opportunity to observe VCD in mixtures of active pharmaceutical ingredients (APIs) and excipients. Since some excipients are chiral, such as dextrose, mannitol, and cyclodextrins, interactions between APIs and excipients that change the state, absolute configuration, or EE of the API cannot automatically be ruled out. Thus, VCD offers the opportunity to investigate the crystal state and all chiral properties of APIs *in situ* as the final formulated products. Additionally, preliminary results show the potential of VCD to monitor not only the state of the API, but also any chiral precipitates that may be present [157]. Research toward the goal of the complete chiral determination of solid-phase pharmaceutical products is currently in progress, and the prognosis for success is excellent as no serious obstacles are currently seen that prevent progress in this direction.

The discussion of solid VOA to this point has focused on VCD. ROA spectra of solids have yet to be reported, however nothing *a priori* precludes this type of measurement since it is well known that Raman scattering can be successfully measured for solids. Scattering in the visible region is much more of a problem that it is in the IR region. On the other hand, scattering is virtually all Rayleigh scattering, and with proper optical filtering and avoiding fluorescent solids, ROA measurements in the solid phase should be reported in the future. Such a capability would add yet another probe of chirality to pharmaceutical products with chiral APIs and chiral excipients, if used.

15.9 SUMMARY AND CONCLUSIONS

VOA spectra are rich in structural and stereo-specific information coming from vibrational chromophores extending over the entire molecular frame. The theory of VCD and ROA has advanced sufficiently in recent years to enable the accurate *ab initio* calculation of observed spectra for small- to medium-sized molecules. This opens the way for the use of VOA to solve the stereochemical problems involving absolute configuration and solution conformation in a direct way without the need to modify the molecule, to calculate excited electronic state properties or to crystallize the sample. The development of flow-cell techniques makes possible the monitoring of the reaction kinetics of chiral molecules in solution. Because both FT-VCD and CCD-ROA are multiplex techniques, it is possible to follow in real time the evolution of the mole fraction and EE of multiple species as a reaction proceeds. No other forms of optical activity currently possess this capability. Extension of VCD into the near-IR region further enhances the potential for VCD to be used routinely to monitor processes involving the synthesis, manufacture, and processing of pharmaceutical products containing chiral APIs. The commercialization of hardware and software for the measurement and calculation of VCD and ROA opens the way for these two forms of

VOA to be broadly available for all those interested in solving problems involving molecular chirality and chiral analysis [8,61].

REFERENCES

1 P. Salvadori and F. Ciardelli (Eds.), Optical Rotatory Dispersion and Circular Dichroism; Heyden and Sons, Ltd., London, 1973, pp. 3–24.

2 L.D. Barron, and L. Hecht, Vibrational Raman optical activity: from fundamentals to biological applications, in: N. Berova, K. Nakanishi and R.W. Woody (eds.), Circular Dichroism: Principles and Applications, 2nd Ed. Wiley-VCH, New York, 2000, pp. 667–701.

3 E.F. Paulus and A. Gieren, Structure analysis by diffraction, in: H. Gunzler and A. Williams (Eds.), Handbook of Analytical Techniques Vols. 1, 2, Wiley-VCH, Weinheim, 2001, pp. 400–403.

4 P.J. Stephens and F.J. Devlin, Determination of the structure of chiral molecules using ab initio vibrational circular dichroism spectroscopy. Chirality, 12 (2000) 172–179.

5 T.B. Freedman, X. Cao, R.K. Dukor and L.A. Nafie, Absolute configuration determination in the solution state using vibrational circular dichroism. Chirality, 15 (2003) 734–758.

6 P.L. Polavarapu and J. He, Chiral analysis using mid-IR vibrational CD spectroscopy. Anal. Chem., 76 (2004) 61A–67A.

7 P.J. Stephens, Vibrational circular dichroism spectroscopy: a new tool for the stereochemical characterization of chiral molecules. Comp. Med. Chem. for Drug Discovery, (2004) 699–725.

8 A.M. Rouhi, A dream realized: spectroscopic tool championed by two chemists makes it easier to determine absolute configurations. Chem. Eng. News, 83 (2005) 32–33.

9 T.B. Freedman, R.K. Dukor, P.J.C.M. van Hoof, E.R. Kellenbach and L.A. Nafie, Determination of the absolute configuration of (−)-mirtazapine by vibrational circular dichroism. Helv. Chim. Acta, 85 (2002) 1160–1165.

10 A.B. Dyatkin, T.B. Freedman, X. Cao, R.K. Dukor, B.E. Maryanoff, C.A. Maryanoff, J.M. Matthews, R.D. Shah and L.A. Nafie, Determination of the absolute configuration of a key tricyclic component of a novel vasopressin receptor antagonist by use of vibrational circular dichroism. Chirality, 14 (2002) 215–219.

11 A. Holmen, J. Oxelbark and S. Allenmark, Direct determination of the absolute configuration of a cyclic thiolsulfinate by VCD spectroscopy. Tetrahedron: Asymmetry, 14 (2003) 2267–2269.

12 B.E. Maryanoff, D.F. McComsey, R.K. Dukor, L.A. Nafie, T.B. Freedman, X. Cao and V.W. Day, Structural studies on McN-5652-X, a high-affinity ligand for the serotonin transporter in mammalian brain. Bioorga. Med. Chem., 11 (2003) 2463–2470.

13 R.D. Shah and L.A. Nafie, Spectroscopic methods for determining enantiomeric purity and absolute configuration in chiral pharmaceutical molecules. Curr. Opin. Drug Discovery Developing, 4 (2001) 764–775.

14 C. Guo, R.D. Shah, R.K. Dukor, X. Cao, T.B. Freedman and L.A. Nafie, Determination of enantiomeric excess in samples of chiral molecules using fourier transform vibrational circular dichroism spectroscopy: simulation of real-time reaction monitoring. Anal. Chem., 76 (2004) 6956–6966.

15 R.K. Dukor and L.A. Nafie, Vibrational optical activity of pharmaceuticals and biomolecules, in: R.A. Meyers (Ed.), Encyclopedia of Analytical Chemistry, John Wiley and Sons, Chichester, UK, 2000, pp. 662–676.

16 L.A. Nafie and T.B. Freedman, Biological and pharmaceutical applications of vibrational optical activity, in: H.-U. Gremlich and B. Yan (Eds.), Infrared and Raman Spectroscopy of Biological Materials, Marcel Dekker, Inc., New York, 2001, pp. 15–54.

17 T.A. Keiderling, Vibrational circular dichroism of peptides and proteins. Survey of techniques, qualitative and quantitative analyses, and applications. in: H.-U. Gremlich and B. Yan (Eds.), Infrared and Raman spectroscopy of biological materials, Marcel Dekker, Inc., New York, 2001, pp. 55–100.

18 L.D. Barron, E.W. Blanch, I.H. McColl, C.D. Syme, L. Hecht and K. Nielsen, Structure and behaviour of proteins, nucleic acids and viruses from vibrational Raman optical activity. Spectroscopy, 17 (2003) 101–126.
19 L.D. Barron, L. Hecht, I.H. McColl and E.W. Blanch, Raman optical activity comes of age. Mol. Phys., 102 (2004) 731–744.
20 X. Cao, R.D. Shah, R.K. Dukor, C. Guo, T.B. Freedman and L.A. Nafie, Extension of Fourier transform vibrational circular dichroism into the near-infrared region: continuous spectral coverage from 800 to 10 000 cm^{-1}. Appl. Spectrosc., 58 (2004) 1057–1064.
21 E.W. Blanch, L. Hecht and L.D. Barron, Vibrational Raman optical activity of proteins, nucleic acids, and viruses. Methods, 29 (2003) 196–209.
22 E.W. Blanch, I.H. McColl, L. Hecht, K. Nielsen and L.D. Barron, Structural characterization of proteins and viruses using Raman optical activity. Vib. Spectrosc., 35 (2004) 87–92.
23 X. Cao, R.K. Dukor, J. Carptenter and L.A. Nafie, Secondary structural states of proteins and protein-excipient mixtures probed by vibrational circular dichroism in KBr pellets. J. Pharm. Sci., (2005) In preparation.
24 G. Shanmugam and P.L. Polavarapu, Vibrational circular dichroism of protein films. J. Amer. Chem. Soc., 126 (2004) 10292–10295.
25 G. Shanmugam and P.L. Polavarapu, Vibrational circular dichroism spectra of protein films: thermal denaturation of bovine serum albumin. Biophys. Chem., 111 (2004) 73–77.
26 G. Shanumgam and P.L. Polavarapu, Structure of a beta(25–35) peptide in different environments. Biophysical, 87 (2004) 622–630.
27 A.G. Petrovich, P.K. Bose and P.L. Polavarapu, Vibrational circular dichroism of carbohydrate films formed from aqueous solutions. Carbohyd. Res., 339 (2004) 2713–2720.
28 R.A. Lombardi, X. Cao, S.S. Kim, R.K. Dukor and L.A. Nafie, Detection of chirality in solid-state amino acids using Fourier transform vibrational circuluar dichroism, Abstract: BIOL-195, 226th ACS National meeting, New York, 2003.
29 H. Buijs and L.A. Nafie, (1997) Determination of Enantiomeric Purity of Chiral Pharmaceutical Molecules using Fourier Transform Vibrational Circular Dichroism Spectroscopy. Book of Abstracts for Pittcon '97, Abstract #1206.
30 L.A. Nafie, F. Long, T.B. Freedman, H. Buijs, A. Rilling, J.-R. Roy and R.K. Dukor, The determination of enantiomeric purity and absolute configuration by vibrational circular dichroism spectroscopy, in: J.A.d. Haseth (Ed.), Fourier transform spectroscopy: 11th International conference, Vol. 430, Amer. Inst. of Phys., Woodbury, NY, 1998, pp. 432–434.
31 Anon, First commercial spectrometer for the measurement of Raman optical activity. Appl. Spectrosc., 58 (2004) 160A.
32 M.J. Frisch, G.W. Trucks, H.B. Schlegel, G.E. Scuseria, M.A. Robb, J.R.J.A. Cheeseman, J. Montgomery, T. Vreven, K.N. Kudin, J.C. Burant, J.M. Millam, S.S. Iyengar, J. Tomasi, V. Barone, B. Mennucci, M. Cossi, G. Scalmani, N. Rega, G.A. Petersson, H. Nakatsuji, M. Hada, M. Ehara, K. Toyota, R. Fukuda, J. Hasegawa, M. Ishida, T. Nakajima, Y. Honda, O. Kitao, H. Nakai, M. Klene, X. Li, J.E. Knox, H.P. Hratchian, J.B. Cross, C. Adamo, J. Jaramillo, R. Gomperts, R.E. Stratmann, O. Yazyev, A.J. Austin, R. Cammi, C. Pomelli, J.W. Ochterski, P.Y. Ayala, K. Morokuma, G.A. Voth, P. Salvador, J.J. Dannenberg, V.G. Zakrzewski, S. Dapprich, A.D. Daniels, M.C. Strain, O. Farkas, D.K. Malick, A.D. Rabuck, K. Raghavachari, B. Foresman, J.V. Ortiz, Q. Cui, A.G. Baboul, S. Clifford, J. Cioslowski, B.B. Stefanov, G. Liu, A. Liashenko, P. Piskorz, I. Komaromi, R.L. Martin, D.J. Fox, T. Keith, Al-M.A. Laham, C.Y. Peng, A. Nanayakkara, M. Challacombe, P.M.W. Gill, B. Johnson, W. Chen, M.W. Wong, C. Gonzalez and J.A. Pople, Gaussian 03 Revision B.03 edit. Gaussian, Inc., Pittsburgh, PA 2003.
33 L.A. Nafie, Infrared and Raman vibrational optical activity: Theoretical and experimental aspects. Ann. Rev. Phys. Chem., 48 (1997) 357–386.
34 L.A. Nafie and T.B. Freedman, Theory of vibrational optical activity. in: K. Nakanishi, N. Berova and R. Woody, (Eds.), Circular Dichroism: Principles and Applications, 2nd Ed., Wiley-VCH, New York, 2000, pp. 97–131.

35 L.D. Barron and J. Vrbancich, On the sign convention for Raman optical activity. Chem. Phys. Lett., 102(2,3) (1983) 285–286.
36 L.A. Nafie, An alternative view on the sign convention of Raman optical activity. Chem. Phys. Lett., 102 (1983) 287–288.
37 D.H.I. Turner, J. Tinoco and M. Maestre, Fluorescence detected circular dichroism. J. Am. Chem. Soc., 96 (1974) 4340–4342.
38 F.S. Richardson, Circular polarization differentials in the luminescence of chiral systems, in: S.F. Mason (Ed.), Optical Acitivity and Chiral Discrimination, D. Reidel Pubishing Co., Dordrecht, 1979, pp. 189–217.
39 J.P. Riehl, Chrioptical studies of molecules in electronically excited states. in: N. Purdie and H.G. Brittain, (Eds.), Analytical Applications of Circular Dichroism, Elsevier, Amsterdam, (1994) pp. 207–240.
40 G. Holzwarth, E.C. Hsu, H.S. Mosher, T.R. Faulkner and A. Moscowitz, Infrared circular dichroism of carbon-hydrogen and carbon-deuterium stretching modes. Observations, J. Am. Chem. Soc., 96 (1974) 251–252.
41 L.A. Nafie, J.C. Cheng and P.J. Stephens, Vibrational circular dichroism of 2,2,2-Trifluoro-1-phenylethanol. J. Am. Chem. Soc., 97 (1975) 3842.
42 L.A. Nafie, T.A. Keiderling and P.J. Stephens, Vibrational circular dichroism. J. Am. Chem. Soc., 98 (1976) 2715–2723.
43 L.A. Nafie, M. Diem and D.W. Vidrine, Fourier transform infrared vibrational circular dichroism. J. Am. Chem. Soc., 101 (1979) 496–498.
44 L.A. Nafie and M. Diem, Theory of high frequency differential interferometry: Application to Infrared circular and linear dichroism via Fourier transform spectroscopy. Appl. Spectrosc., 33 (1979) 130–135.
45 L.D. Barron, M.P. Bogaard and A.D. Buckingham, Raman scattering of circularly polarized light by optically active molecules. J. Am. Chem. Soc., 95 (1973) 603–605.
46 L.D. Barron, Molecular light scattering and optical activity, second edition, Cambridge University Press, Cambridge, (2004).
47 K.M. Spencer, T.B. Freedman and L.A. Nafie, Scattered circular polarization Raman optical activity. Chem. Phys. Lett., 149 (1988) 367–374.
48 L.A. Nafie and T.B. Freedman, Dual circular polarization raman optical activity. Chem. Phys. Lett., 154 (1989) 260–266.
49 D. Che, L. Hecht and L.A. Nafie, Dual and incident circular polarization Raman optical activity backscattering of (−)-trans-pinane. Chem. Phys. Lett., 180 (1991) 182–190.
50 G.-S. Yu and L.A. Nafie, Isolation of preresonance and out-of-phase dual circular polarization Raman optical activity. Chem. Phys. Lett., 222 (1994) 403–410.
51 L. Hecht and L.A. Nafie, Linear Polarization Raman optical activity: A new form of natural optical activity. Chem. Phys. Lett., 174(6) (1990) 575–582.
52 Hecht, L. and L.D. Barron, Linear polarization Raman optical activity: The importance of the non-resonant term in the Kramers-Heisenberg-Dirac dispersion formula under resonance conditions. Chem. Phys. Lett., 225 (1994) 519–524.
53 L. Nafie, R.K. Dukor, L. Barron, L. Hecht, G. Hangartner and W. Hug, 31st FACSS Conference, Portland, Oregon, (2004).
54 L.A. Nafie, Vibrational circular dichroism: spectrometer, in: J.C. Lindon, G.E. Tranter and J.L. Holmes (Eds.), Encyclopedia of Spectroscopy and Spectrometery, Academic Press, Ltd, London, (1999) pp. 2391–2402.
55 L.A. Nafie, Dual polarization modulation: Real-time, spectral multiplex separation of circular dichroism from linear birefringence spectral intensities. Appl. Spectrosc., 54 (2000) 1643–1645.
56 L.A. Nafie, H. Buijs, A. Rilling, X. Cao and R.K. Dukor, Dual source Fourier transformation polarization modulation spectroscopy: an improved method for the measurement of circular and linear dichroism. Appl. Spectrosc., 58 (2004) 647–654.

57 W. Hug, Raman optical activity: Spectrometers, in: J. Lindon, G. Tranter and J. Holmes (Eds.), Encylcopedia of Spectroscopy and Spectrometry, Academic Press, Ltd., London, (1999) pp. 1966–1976.

58 W. Hug and G. Hangartner, A very high throughput Raman and Raman optical activity spectrometer. J. Raman Spectrosc., 30 (1999) 841–852.

59 W. Hug, Virtual enantiomers as the solution of optical activitiy's deterministic offset problem. Appl. Spectrosc., 57 (2003) 1–13.

60 L.A. Nafie, R.K. Dukor, Roy, J.-R., Rilling, A., Cao, X. and H. Buijs, Observation of Fourier transform near-infrared vibrational circular dichroism to 6150 cm-1. Appl. Spectrosc., 57 (2003) 1245–1249.

61 M. Freemantle, An incisive probe for biomolecules: Raman optical activity spectroscopy yields information on biomolecular structure. Chem. Eng. News., 81 (2003) 36–39.

62 L.D. Barron, E.W. Blanch and L. Hecht, Unfolded proteins studied by Raman optical activity. Adv. Prot. Sci., 62 (2002) 51–90.

63 I.H. McColl, E.W. Blanch, L. Hecht and L.D. Barron, A study of α-helix hydration in polypeptides, proteins, and viruses using vibrational Raman optical activity. J. Amer. Chem. Soc., 126 (2004) 8181–8188.

64 I.H. McColl, E.W. Blanch, L. Hecht, N.R. Kallenbach and L.D. Barron, Vibrational Raman optical activity characterization of poly(L-proline) II helix in alanine oligopeptides. J. Amer. Chem. Soc., 126 (2004) 5076–5077.

65 F. Zhu, N.W. Isaacs, L. Hecht and L.D. Barron. Polypeptide and carbohydrate structure of an intact glycoprotein from Raman optical activity. J. Amer. Chem. Soc., 127 (2005) 6142–6143.

66 E.W. Blanch, A.C. Gill, A.G.O. Rhie, J. Hope, L. Hecht, K. Nielsen and L.D. Barron, Raman Optical Activity Demonstrates Poly(l-proline) II Helix in the n-terminal Region of the Ovine Prion Protein: Implications for Function and Misfunction. J. Mol. Biol., 343 (2004) 467–476.

67 C. Toniolo, F. Formaggio, S. Tognon, Q.B. Broxterman, B. Kaptein, R. Huang, V. Setnicka, T.A. Keiderling, I.H. McColl, L. Hecht and L.D. Barron, The complete chirospectroscopic signature of the peptide 310-helix in aqueous solution. Biopolymers, 75 (2004) 32–45.

68 I.H. McColl, E.W. Blanch, A.C. Gill, A.G.O. Rhie, M.A. Ritchie, L. Hecht, K. Nielsen and L.D. Barron, A new perspective on b-sheet structures using vibrational Raman optical activity: from poly(L-lysine) to the prion protein. J. Amer. Chem. Soc., 125 (2003) 10019–10026.

69 E.W. Blanch, D.J. Robinson, L. Hecht and L.D. Barron, A comparison of the solution structures of tobacco rattle and tobacco mosaic viruses from Raman optical activity. J. Gen. Virolo. 82 (2001) 1499–1502.

70 E. Smyth, C.D. Syme, E.W. Blanch, L. Hecht, M. Vasak and L.D. Barron, Solution structure of native proteins with irregular folds from Raman optical activity. Biopolymers, 58 (2001) 138–151.

71 E.W. Blanch, D.J. Robinson, L. Hecht, C.D. Syme, K. Nielsen and L.D. Barron, Solution structures of potato virus X and narcissus mosaic virus from Raman optical activity. J. Gen. Virol., 83 (2002) 241–246.

72 C.D. Syme, E.W. Blanch, C. Holt, R. Jakes, M. Goedert, L. Hecht and L.D. Barron, A Raman optical activity study of rheomorphism in caseins, synucleins and tau. New insight into the structure and behaviour of natively unfolded proteins. Eur. J. Biochem. 269 (2002) 148–156.

73 J. Kapitan, V. Baumruk, V. Gut, J. Hlavacek, H. Dlouha, M. Urbanova, E. Wuensch and P. Malon, Raman optical activity of the central part of hinge peptide. Collect. Czech. Chem. Commun., 70 (2005) 403–409.

74 P. Bour and T.A. Keiderling, Structure, spectra and the effects of twisting of b-sheet peptides. A density functional theory study. Theochem, 675 (2004) 95–105.

75 R. Huang, J. Kubelka, W. Barber-Armstrong, R.A.G.D. Silva, M. Decatur Sean and A. Keiderling Timothy, Nature of vibrational coupling in helical peptides: an isotopic labeling study. J. Amer. Chem. Soc., 126 (2004) 2346–54.

76 T.A. Keiderling and Q. Xu, Optical spectra of unfolded proteins: a partially ordered polymer problem. Macromol. Symp., 220 (2005) 17–31.

77 J. Kubelka, R. Huang and T.A. Keiderling, Solvent effects on IR and VCD spectra of helical peptides: DFT-based static spectral simulations with explicit water. J. Phys. Chem. B, 109 (2005) 8231–8243.

78 R. Schweitzer-Stenner, Secondary structure analysis of polypeptides based on an excitonic coupling model to describe the band profile of amide I′ of IR, Raman and vibrational circular dichroism spectra. J. Phys. Chem. B, 108 (2004) 16965–16975.

79 F. Eker, X. Cao, L. Nafie, Q. Huang and R. Schweitzer-Stenner, The structure of alanine based tripeptides in water and dimethyl sulfoxide probed by vibrational spectroscopy. J. Phys. Chem. B, 107 (2003) 358–365.

80 F. Eker, K. Griebenow, X. Cao, A. Nafie Laurence and R. Schweitzer-Stenner, Tripeptides with ionizable side chains adopt a perturbed polyproline II structure in water. Biochemistry, 43 (2004) 613–621.

81 R. Schweitzer-Stenner, F. Eker, K. Griebenow, X. Cao and A. Nafie Laurence, The conformation of tetraalanine in water determined by polarized Raman, FT-IR, and VCD spectroscopy. J. Amer. Chem. Soc., 126 (2004) 2768–2776.

82 R. Schweitzer-stenner, F. Eker, A. Perez, K. Griebenow, X. Cao and L.A. Nafie, The structure of tri-proline in water probed by polarized Raman, Fourier transform infrared, vibrational circular dichroism, and electric ultraviolet circular dichroism spectroscopy. Biopolymers, 71 (2003) 558–568.

83 F. Eker, X. Cao, L. Nafie and R. Schweitzer-Stenner, Tripeptides adopt stable structures in water. A combined polarized visible Raman, FTIR and VCD spectroscopy study. J. Amer. Chem. Soc., 124 (2002) 14330–14341.

84 F. Eker, K. Griebenow, L.A. Nafie, X. Cao and R. Schweitzer-Stenner, Preferred peptide backbone conformations in the unfolded state revealed by the structural analysis of alanine-based AXA tripeptides in the solution state. Proc. Natl. Acad. Sci. USA, 101 (2004) 10054–10059.

85 L.D. Barron and A.D. Buckingham, Rayleigh and Raman Scattering from optically active molecules. Mol. Phys., 20(6) (1971) 1111–1119.

86 L.A. Nafie and T.B. Freedman, Vibronic coupling theory of infrared vibrational intensities. J. Chem. Phys., 78 (1983) 7108–7116.

87 L.A. Nafie, Adiabatic behavior beyond the Born-Oppenheimer approximation. Complete adiabatic wavefunctions and vibrationally induced electronic current density. J. Chem. Phys., 79 (1983) 4950–4957.

88 P.J. Stephens, Theory of vibrational circular dichroism. J. Phys. Chem., 89 (1985) 748–752.

89 P.J. Stephens and M.A. Lowe, Vibrational circular dichroism. Ann. Rev. Phys. Chem., 36 (1985) 213–241.

90 P.L. Polavarapu, Ab initio Raman and Raman optical activity spectra. J. Phys. Chem., 94 (1990) 8106–8112.

91 L. Nafie, Vibrational optical activity. Appl. Spectrosc., 50(5) (1996) A14–A26.

92 A. Rauk, Vibrational circular dichroism intensitites: ab initio calculations, in: P.G. Mezey (Ed.), New Developments in Molecular Chirality, Kluwer Academic Publishers, Netherlands, 1991, pp. 57–92.

93 L.A. Nafie, Velocity-gauge formalism in the theory of vibrational circular dichroism and infrared absorption. J. Chem. Phys., 96 (1992) 5687–5702.

94 L.A. Nafie, Theory of vibrational circular dichroism and infrared absorption: Extension to molecules with low-lying excited electronic states. J. Phys. Chem. A, 108 (2004) 7222–7231.

95 L. Hecht and L.A. Nafie, Theory of natural Raman optical activity I. Complete circular polarization formalism. Mol. Phys., 72 (1991) 441–469.

96 L.A. Nafie and D. Che, Theory and measurement of Raman optical activity, in: M. Evans and S. Kielich (Eds.), Modern Nonlinear Optics, Part 3, Vol. 85, John Wiley & Sons, New York, 1994, pp. 105–149.

97 L.A. Nafie, G.-S. Yu, X. Qu and T.B. Freedman, Comparison of IR and Raman forms of vibrational optical activity. Faraday Discuss. 99 (1994) 13–34.

98 L.A. Nafie, Theory of resonance Raman optical activity: The single-electronic state limit. Chem. Phys., 205 (1996) 309–322.

99 T.B. Freedman and L.A. Nafie, Theoretical formalism and models for vibrational circular dichroism intensity, in: M. Evans and S. Kielich (Eds.), Modern Nonlinear Optics, Part 3, Vol. 85, John Wiley & Sons, New York, 1994, pp. 207–263.

100 K.L. Bak, F.J. Devlin, C.S. Ashvar, P.R. Taylor, M.J. Frisch and P.J. Stephens. Ab-initio calculation of vibrational circular-dichroism spectra using gauge-invariant atomic orbitals. J. Phys. Chem., 99 (1995) 14918–14922.

101 K.L. Bak, P. Jorgensen, T. Helgaker and K. Ruud, Basis-set convergence and correlation-effects in vibrational circular-dichroism calculations using London atomic orbitals. Faraday Discuss, (1994) 121–129.

102 T. Helgaker, K. Ruud, K.L. Bak, P. Jorgensen and J. Olsen, (1994). Vibrational Raman optical-activity calculations using London atomic orbitals. Faraday Discuss, 165–180.

103 W. Hug, Visualizing Raman and Raman optical activity generation in polyatomic molecules. Chem. Phys., 264 (2001) 53–69.

104 W. Hug, G. Zuber, A. De Meijere, A.F. Khlebnikov and H.-J. Hansen, Raman optical activity of a purely σ-bonded helical chromophore: (-)-(M)-σ-[4]helicene. Helv. Chim. Acta, 84 (2001) 1–21.

105 M. Pecul and A. Rizzo, Raman optical activity spectra: basis set and electron correlation effects. Mol. Phys., 101 (2003) 2073–2081.

106 M. Pecul and K. Ruud, Ab initio calculation of vibrational Raman optical activity. Int. J. Quantum. Chem., 104 (2005) 816–829.

107 M. Reiher, V. Liegeois and K. Ruud, Basis set and density functional dependence of vibrational Raman optical activity calculations. J. Phys. Chem. A, 109 (2005) 7567–7574.

108 K. Ruud, T. Helgaker and P. Bour, Gauge-origin independent density-functional theory calculations of vibrational Raman optical activity. J. Phys. Chem. A, 106 (2002) 7448–7455.

109 G. Zuber, M.-R. Goldsmith, D.N. Beratan and P. Wipf, Towards Raman optical activity calculations of large molecules. Chem. Phys. Chem., 6 (2005) 595–597.

110 G. Zuber and W. Hug, Computational interpretation of vibrational optical activity: The ROA spectra of (4S)-4-methylisochromane and the (4S)-isomers of Galaxolide. Helv. Chim. Acta, 87 (2004) 2208–2234.

111 F.J. Devlin, P.J. Stephens and P. Besse, Are the absolute configurations of 2-(1-hydroxyethyl)-chromen-4-one and its 6-bromo derivative determined by X-ray crystallography correct? A vibrational circular dichroism study of their acetate derivatives. Tetrahedron: Asymmetry, 16 (2005) 1557–1566.

112 A. Solladie-Cavallo, O. Sedy, M. Salisova, M. Biba, C.J. Welch, L. Nafie and T. Freedman, A chiral 1,4-oxazin-2-one: asymmetric synthesis versus resolution, structure, conformation and VCD absolute configuration. Tetrahedron: Asymmetry, 12 (2001) 2703–2707.

113 H. Cichewicz Robert, J. Clifford Laura, R. Lassen Peter, X. Cao, B. Freedman Teresa, A. Nafie Laurence, D. Deschamps Joshua, A. Kenyon Victor, R. Flanary Jocelyn, R. Holman Theodore and P. Crews, Stereochemical determination and bioactivity assessment of (S)-(+)-curcuphenol dimers isolated from the marine sponge Didiscus aceratus and synthesized through laccase biocatalysis. Bioorg. Med. Chem., 13 (2005) 5600–5612.

114 P.J. Stephens, D.M. McCann, F.J. Devlin, T.C. Flood, E. Butkus, S. Stoncius and J.R. Cheeseman, Determination of molecular structure using vibrational circular dichroism spectroscopy: the keto-lactone product of Baeyer-Villiger oxidation of (+)-(1R,5S)-bicyclo[3.3.1]nonane-2,7-dione. J. Org. Chem., 70 (2005) 3903–3913.

115 H. Pham-Tuan, C. Larsson, F. Hoffmann, A. Bergman, M. Froba and H. Huhnerfuss,. Enantioselective semipreparative HPLC separation of PCB metabolites and their absolute structure elucidation using electronic and vibrational circular dichroism. Chirality, 17 (2005) 266–280.

116 G. Petrovic Ana, L. Polavarapu Prasad, J. Drabowicz, Y. Zhang, J. McConnell Oliver, and H. Duddeck, Absolute configuration of C2-symmetric spiroselenurane: 3,3,3′,3′-tetramethyl-1,1′-spirobi[3 H,2,1]benzoxaselenole. Chemistry (Weinheim an der Bergstrasse, Germany), 11 (2005) 4257–4262.

117 G. Petrovic Ana, J. He, L. Polavarapu Prasad, S. Xiao Ling and W. Armstrong Daniel, Absolute configuration and predominant conformations of 1,1-dimethyl-2-phenylethyl phenyl sulfoxide. Org. Biomol. Chem. 3 (2005) 1977–1981.

118 J. He, F. Wang and L. Polavarapu Prasad, Absolute configurations of chiral herbicides determined from vibrational circular dichroism. Chirality, 17 (2005) Suppl, S1–S8.

119 J. He and L. Polavarapu Prasad, Determination of the absolute configuration of chiral alpha-aryloxypropanoic acids using vibrational circular dichroism studies: 2-(2-chlorophenoxy) propanoic acid and 2-(3-chlorophenoxy) propanoic acid. Spectrochim. Acta A, Mol. Biomol. Spectrosc., 61 (2005) 1327–1334.

120 J. He and P.L. Polavarapu, Determination of Intermolecular Hydrogen Bonded Conformers of a-Aryloxypropanoic Acids Using Density Functional Theory Predictions of Vibrational Absorption and Vibrational Circular Dichroism Spectra. J. Chem. Theory and Comput., 1 (2005) 506–514.

121 D. Dunmire, B. Freedman Teresa, A. Nafie Laurence, C. Aeschlimann, G. Gerber John and J. Gal, Determination of the absolute configuration and solution conformation of the antifungal agents ketoconazole, itraconazole, and miconazole with vibrational circular dichroism. Chirality, 17 (2005) Suppl S101–S108.

122 F.J. Devlin, P.J. Stephens and P. Besse, Conformational rigidification via derivatization facilitates the determination of absolute configuration using chiroptical spectroscopy: a case study of the chiral alcohol endo-borneol. J. Org. Chem., 70 (2005) 2980–2993.

123 V. Cere, F. Peri, S. Pollicino, A. Ricci, J. Devlin Frank, J. Stephens Philip, F. Gasparrini, R. Rompietti and C. Villani, Synthesis, chromatographic separation, vibrational circular dichroism spectroscopy, and ab initio DFT studies of chiral thiepane tetraol derivatives. J. Org. Chem., 70 (2005) 664–669.

124 D. Bas, T. Burgi, J. Lacour, J. Vachon and J. Weber, Vibrational and electronic circular dichroism of delta-TRISPHAT [tris(tetrachlorobenzenediolato)phosphate(V)] anion. Chirality, 17 (2005) Suppl, S143–S148.

125 T. Taniguchi, K. Monde, N. Miura and S.-I. Nishimura, A characteristic CH band in VCD of methyl glycosidic carbohydrates. Tetrahedron Lett., 45 (2004) 8451–8453.

126 T. Kuppens, P. Bultinck and W. Langenaeker, Determination of absolute configuration via vibrational circular dichroism. Drug Discovery Today: Technologies, 1 (2004) 269–275.

127 J. He, A. Petrovich and P.L. Polavarapu, Quantitative determination of conformer populations: assessment of specific rotation, vibrational absorption, and vibrational circular dichroism in substituted butynes. J. Phys. Chem. A, 108 (2004) 1671–1680.

128 T.B. Freedman, X. Cao, L.A. Nafie, A. Solladie-Cavallo, L. Jierry and L. Bouerat, VCD configuration of enantiopure/-enriched tetrasubstituted a-fluoro cyclohexanones and their use for epoxidation of trans-olefins. Chirality, 16 (2004) 467–474.

129 T. Buergi, A. Urakawa, B. Behzadi, K.-H. Ernst and A. Baiker, The absolute configuration of heptahelicene: a VCD spectroscopy study. New J. Chem., 28 (2004) 332–334.

130 A. Solladie-Cavallo, C. Marsol, M. Yaakoub, K. Azyat, A. Klein, M. Roje, C. Suteu, T.B. Freedman, X. Cao and L.A. Nafie, Erythro-1-naphthyl-1-(2-piperidyl)methanol: Synthesis, resolution, NMR relative configuration, and VCD absolute configuration. J. Org. Chem., 68 (2003) 7308–7315.

131 K. Monde, T. Taniguchi, N. Miura, S.-I. Nishimura, N. Harada, R.K. Dukor and L.A. Nafie, Preparation of cruciferous phytoalexin related metabolites, (−)-dioxibrassinin and (−)-3-cyanomethyl-3-hydroxyoxindole, and determination of their absolute configurations by vibrational circular dichroism (VCD). Tetrahedron Lett., 44 (2003) 6017–6020.

132 H. Izumi, S. Futamura, L.A. Nafie and R.K. Dukor, Determination of molecular stereochemistry using vibrational circular dichroism spectroscopy: absolute configuration and solution

conformation of 5-formyl-cis,cis-1,3,5-trimethyl-3-hydroxymethylcyclohexane-1-carboxylic acid lactone. Chem. Rec., 3 (2003) 112–119.

133 B. Freedman Teresa, X. Cao, V. Oliveira Regina, B. Cass Quezia and A. Nafie Laurence, Determination of the absolute configuration and solution conformation of gossypol by vibrational circular dichroism. Chirality, 15 (2003) 196–200.

134 T.B. Freedman, X. Cao, A. Rajca, H. Wang and L.A. Nafie, Determination of absolute configuration in molecules with chiral axes by vibrational circular dichroism: a C2-symmetric annelated heptathiophene and a D2-symmetric dimer of 1,1′-binaphthyl. J. Phys. Chem. A, 107 (2003) 7692–7696.

135 T.B. Freedman, X. Cao, L.A. Nafie, M. Kalbermatter, A. Linden and A.J. Rippert, An unexpected atropisomerically stable 1,1-biphenyl at ambient temperature in solution, elucidated by vibrational circular dichroism (VCD). Helv. Chim. Acta., 86 (2003) 3141–3155.

136 I. Donnoli Maria, E. Giorgio, S. Superchi and C. Rosini, Circular dichroism spectra and absolute configuration of some aryl methyl sulfoxides. Org. Biomol. Chem., 1 (2003) 3444–3449.

137 K.J. Jalkanen, R.M. Nieminen, M. Knapp-Mohammady and S. Suhai, Vibrational analysis of various isotopomers of l-analyl-l-alanine in aqueous solution: vibrational absorption, vibrational circular dichroism, Raman, and Raman optical acticity spectra. Int. J. Quantum. Chem., 92 (2003) 239–259.

138 G. Zuber and W. Hug, Rarefied basis sets for the calculations of optical tensors: 1. The importance of gradients on hydrogen atoms for the Raman scattering tensor. J. Phys. Chem. A., 108 (2003) 2108–2118.

139 P.L. Polavarapu and D.K. Chakraborty, Ab initio theoretical optical rotations of small molecules. Chem. Phys., 240 (1999) 1–8.

140 P.L. Polavarapu and C. Zhao, Ab initio predictions of anomalous optical rotatory dispersion. J. Am. Chem. Soc., 121 (1999) 246–247.

141 J.R. Cheeseman, M.J. Frisch, F.J. Devlin and P.J. Stephens, Hartree-Fock and density functional theory ab initio calculation of optical rotation using GIAOs: Basis set dependence. J. Phys. Chem. A, 104 (2000) 1039–1046.

142 P.J. Stephens, F.J. Devlin, J.R. Cheeseman, M.J. Frisch, B. Mennucci and J. Tomasi, Prediction of optical rotation using density functional theory: 6,8-dioxabicyclo[3.2.1]octanes. Tetrahedron: Asymmetry, 11 (2000) 2443–2448.

143 P.J. Stephens, F.J. Devlin, J.R. Cheeseman and M.J. Frisch, The calculation of optical rotation using density functional theory. J. Phys. Chem. A, 105 (2001) 5356–5371.

144 P.L. Polavarapu, Optical Rotation: Recent advances in determining the absolute configuration. Chirality, 14 (2002) 768–781.

145 P.J. Stephens, F.J. Devlin, J.R. Cheeseman and M.J. Frisch, Ab initio prediction of optical rotation: comparison of density functional theory and Hartree-Fock methods for three 2,7,8-trioxabicyclo[3.2.1]octanes. Chirality, 14 (2002) 288–296.

146 P.J. Stephens, F.J. Devlin, J.R. Cheeseman, M.J. Frisch and C. Rosini, Determination of absolute configuration using optical rotation calculated using density functional theory. Org. Lett., 4 (2002) 4595–4598.

147 A.E. Hansen and K.L. Bak, Ab-initio calculations of electronic circular dichroism. Enantiomer 4 (1999) 455–476.

148 D.M. Rogers and J.D. Hirst, First-principle calculations of electronic circular dichroism in the near ultraviolet region. Biochemistry, 43 (2004) 11092–11102.

149 R.K. Dukor, X. Cao, J. Carpenter, R. LoBrutto, J. Pepper, T. Li and L.A. Nafie, Spectroscopic studies of protein formulations: avoidance of excipient interference using vibrational circular dichroism, (2005) (in preparation).

150 C. Guo, R.D. Shah, R.K. Dukor, X. Cao, T.B. Freedman and L.A. Nafie, Enantiomeric excess determination by fourier transform near-infrared vibrational circular dichroism spectroscopy: simulation of real-time process monitoring. Appl. Spectrosc., 59 (2005) 1114–1124.

151 C. Guo, Enantiomeric excess determination and reaction monitoring of chiral molecules using near-infrared and mid-infrared vibrational circular dichroism, Syracuse University (2004).

152 M. Diem, E. Photos, H. Khouri and L.A. Nafie, Vibrational circular dichroism in amino acids and peptides. 3. Solution and solid phase spectra of serine and alanine. J. Am. Chem. Soc., 101 (1979) 6829–6837.

153 A.C. Sen and T.A. Keiderling, Vibrational circular dichroism of polypeptides. III. Film studies of several.alpha.-helical and.beta.-sheet polypeptides. Biopolymers, 23 (1984) 1533–1545.

154 U. Narayanan, T.A. Keiderling, G.M. Bonora and C. Toniolo, Vibrational circular dichroism of polypeptides. IV. Film studies of L-alanine homo-oligopeptides. Biopolymers, 24 (1985) 1257–1263.

155 U. Narayanan, T.A. Keiderling, G.M. Bonora and C. Toniolo, Vibrational circular dichroism of polypeptides. 7. Film and solution studies of.beta.-sheet-forming homooligopeptides. J. Am. Chem. Soc., 108 (1986) 2431–2437.

156 D.M. Tigelaar, W. Lee, K.A. Bates, A. Saprigin, V.N. Prigodin, X. Cao, L.A. Nafie, M.S. Platz and A.J. Epstein, Role of solvent and secondary doping in polyaniline films doped with chiral camphorsulfonic acid: Preparation of a chiral metal. Chem. Mater. 14 (2002) 1430–1438.

157 R.K. Dukor, X. Cao, R. LoBrutto, J. Pepper and L.A. Nafie, Variations in excipient VCD as a function of forumulated active pharmaceutical ingredient (2005) (in preparation).

© 2006 Elsevier B.V. All rights reserved.
Chiral Analysis
K.W. Busch and M.A. Busch, Eds.

CHAPTER 16

Raman optical activity

Ewan W. Blanch[1,*], Lutz Hecht[2] and Laurence D. Barron[2]

[1]*Manchester Interdisciplinary Biocentre, The University of Manchester, 131 Princess Street, Machester M1 7ND, UK*
[2]*Department of Chemistry, University of Glasgow, Glasgow G12 8QQ, UK*

16.1 INTRODUCTION

Chiroptical spectroscopic techniques are particularly useful in chemistry and biology. The most widely known of these, optical rotation and ultraviolet circular dichroism (UVCD), measure electronic optical activity and are routinely applicable to most chiral molecules under a wide range of sample conditions. More recently, advances in instrument design and technology have led to the extension of optical activity measurements into the vibrational spectrum using both Raman and infrared techniques [1–9].

Conventional vibrational spectroscopic techniques are widely used as they are sensitive to a wealth of structural information as each contains contributions from all $3N-6$ normal modes of vibration in the case of nonlinear molecules. The combination of vibrational spectroscopy with stereochemical sensitivity greatly enhances the incisiveness of these techniques relative to conventional spectroscopies. As vibrational optical activity probes the stereochemistry of the molecular framework directly, whereas electronic optical activity probes stereochemistry indirectly through the interaction of a photon with a chromophore, far more stereochemical information is readily available using vibrational optical activity. This applies to both biological molecules, which are usually chiral, and smaller organic and even some inorganic molecules. The increase in stereochemical information provided by vibrational optical activity compared with conventional electronic optical activity is demonstrated by considering the example of *R*-(+)-3-methylcyclohexanone, a simple organic chiral molecule. The conventional electronic circular dichroism (ECD) spectrum of *R*-(+)-3-methylcyclohexanone contains a single band envelope in the near ultraviolet region originating in the chiral perturbation of the $\pi^* \leftarrow n$ transition of the carbonyl chromophore by the methyl group. Analysis of the spectrum to yield stereochemical information is achieved using the 'octant rule,' which relates the sign of the ECD band to the position of the perturbing

* Author for correspondence. E-mail: E.Blanch@umist.ac.uk

group around the carbonyl group. In comparison to this single band, a vibrational optical activity spectrum of *R*-(+)-3-methylcyclohexanone can contain up to 54 fundamental bands, each associated with one of the $3N-6$ normal modes of vibration. Each of these bands will contain information about the absolute configuration of that part of the molecule embraced by the corresponding vibrational mode.

Vibrational optical activity was first observed in chiral molecules in the liquid phase in 1973 by Barron and co-workers [10] using a Raman optical activity (ROA) technique which measures a small difference in the intensity of Raman scattering using right- and left-circularly polarized incident light [11,12]. These observations were confirmed as genuine by Hug and co-workers in 1975 [13]. A lack of sensitivity originally restricted ROA studies to favorable samples, such as small organic molecules in concentrated solutions or neat liquids, and the complementary technique of vibrational circular dichroism (VCD) found more widespread application. Although VCD is also a measure of vibrational optical activity, it differs from ROA as it is a measure of the differential absorption of left- and right-circularly polarized infrared radiation [4,14], whereas ROA is a measure of the differential inelastic scattering of right- and left-circularly polarized visible radiation. One form of ROA measurement is depicted in Fig. 16.1 and shows that for a vibrational transition of angular frequency ω_ν the ROA intensity is I^R-I^L, where I^R and I^L are the Stokes–Raman scattering intensities at the angular frequency $(\omega - \omega_\nu)$ in right- and left-circularly polarized incident light, respectively, of angular frequency ω. Measurement of the circularly polarized component in the scattered light using linearly polarized, or un-polarized incident radiation is an alternative strategy that generates similar information and will be described in greater detail in Section 16.2.5. The generation of ROA involves interactions between the chiral molecule and the incident radiation that are of lower order than those that generate VCD. As a result of this difference in mechanism, the ROA and VCD bands associated with a particular normal mode of vibration usually show completely different vibrational optical activity. Therefore, ROA and VCD are complementary chirally sensitive techniques, just as Raman scattering and infrared absorption are complementary conventional vibrational spectroscopic techniques.

The aim of this contribution is to outline the theory and practice of ROA spectroscopy through its application to a diverse range of molecules. Although these studies have historically centered on a small number of laboratories, recent advances in

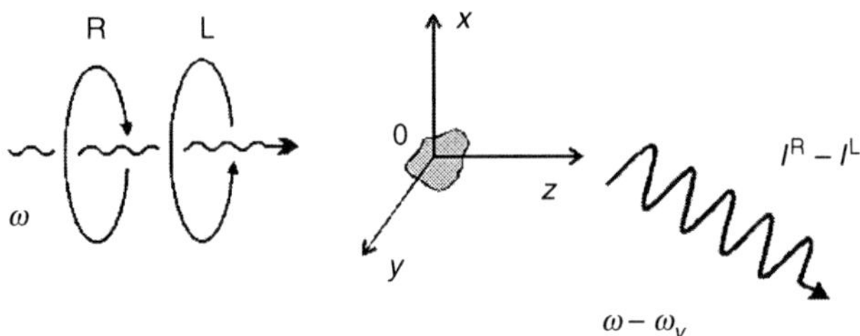

Fig. 16.1. The basic ROA experiment measures a small difference in the intensity of Raman scattering from chiral molecules in right- and left-circularly polarized light. Incident photons have an angular frequency ω and Raman scattered photons have an angular frequency $\omega - \omega_\nu$.

instrumentation and analysis have greatly promoted the applicability of ROA to a wide range of timely problems in chemistry and biology. The size of the ROA community has increased concomitantly over the last two years and this chapter indicates several possible directions for future research using this powerful technique.

16.2 FUNDAMENTAL THEORY OF ROA

16.2.1 Distinction between ROA and optical rotation

Optical rotation and circular dichroism are produced by changes in the polarization state of light transmitted through an optically active medium. As such, they are associated with refraction, a consequence of light scattering by the molecules in the medium. This is accompanied by Rayleigh and Raman scattering from the molecule in all directions. The phenomena of ROA and optical rotation (or CD) arise from fundamentally different mechanisms of interaction between chiral molecules and circularly polarized radiation.

Optical rotation is a measurement of birefringence in the medium and involves interference between unscattered waves and waves that are scattered in the forward direction by molecules in the medium. A simple two-group model can be used to understand the fundamental mechanism responsible for this process [9,15]. Consider the case of two achiral axially symmetric groups being held together in a twisted chiral arrangement, as shown in Fig. 16.2. An unscattered photon interferes at the detector with another photon that has sampled the chirality of the molecule by being deflected

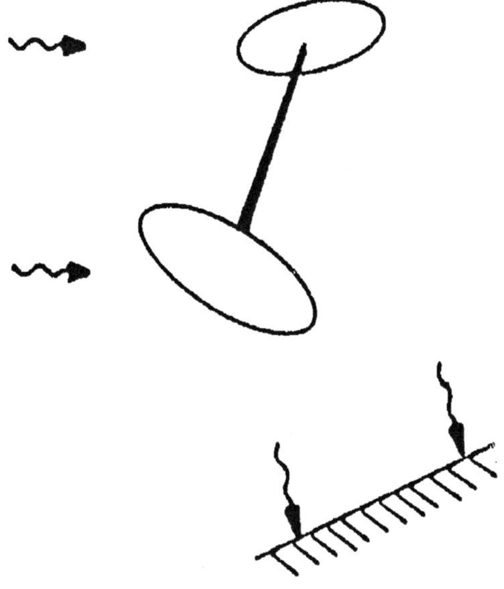

Fig. 16.2. Simple two-group model of scattering from two achiral subunits held in a twisted chiral arrangement.

from one group to the other before being scattered in the forward direction. However, for light scattering in directions other than the forward direction there are no photons transmitted to the detector. Therefore, in the two-group model of Rayleigh and Raman optical activity (ROA) the interference between photons scattered independently from the two groups provides information on chirality without the dynamic coupling between the two groups that is essential for conventional optical rotation. However, such coupling may make higher order contributions.

16.2.2 Electric and magnetic properties

Interaction of a light wave with a molecule sets bound charges into oscillation which results in secondary waves being scattered in all directions. We can consider this model of light scattering in terms of the characteristic radiation fields generated by the oscillating electric and magnetic multipole moments induced in the molecule by the incident light wave. The electric dipole, magnetic dipole and traceless electric quadrupole moments of a molecule, in Cartesian tensor notation and SI units, are defined, respectively, as [9,16]

$$\mu_\alpha = \sum\nolimits_i e_i r_{i\alpha}, \tag{1a}$$

$$m_\alpha = \sum\nolimits_i (e_i/2m_i)\varepsilon_{\alpha\beta\gamma} r_{i\beta} p_{i\gamma}, \tag{1b}$$

$$\Theta_{\alpha\beta} = (1/2)\sum\nolimits_i e_i(3r_{i\alpha}r_{i\beta} - r_i^2\delta_{\alpha\beta}), \tag{1c}$$

where particle i with position vector r_i has charge e_i, mass m_i and linear momentum p_i. The Greek subscripts denote vector or tensor components and can be equal to x, y or z; a repeated Greek suffix denotes Einstein summation over the three tensor components; $\delta_{\alpha\beta}$ is the unit second-rank symmetric tensor and $\varepsilon_{\alpha\beta\gamma}$ is the unit third-rank antisymmetric tensor.

The electric vector of a plane-polarized light wave with wavelength λ and angular frequency $\omega = 2\pi c/\lambda$ propagating in the direction of unit vector $\mathbf{n}$ with velocity c is given, in complex notation, by

$$\mathbf{E} = \mathbf{E}^{(0)} \exp[-i\omega(t - \mathbf{n}\cdot\mathbf{r}/c)], \tag{2}$$

where $\mathbf{E}^{(0)}$ is the field amplitude. In the far-from-resonance approximation, the real part of the electric vector together with the magnetic vector $\mathbf{B}$ of the light wave and the electric field gradient tensor $\nabla_\alpha \mathrm{E}_\beta$ induce the oscillating electric dipole, magnetic dipole and electric quadrupole moments, respectively, which are shown below, [9,16]

$$\mu_\alpha = \alpha_{\alpha\beta} + (1/\omega)G'_{\alpha\beta}\dot{B}_\beta + (1/3)A_{\alpha\beta\gamma}\nabla_\beta E_\gamma + \cdots, \tag{3a}$$

$$m_\alpha = -(1/\omega)G'_{\alpha\beta}\dot{E}_\beta + \cdots, \tag{3b}$$

$$\Theta_{\alpha\beta} = A_{\alpha\beta\gamma}E_\gamma + \cdots, \tag{3c}$$

where the fields and field gradients are evaluated at the same molecular origin as used to define the multipole moments. Quantum mechanical expressions for the dynamic molecular property tensors are obtained from time-dependent perturbation theory and are given by

$$\alpha_{\alpha\beta} = (2/\hbar)\sum\nolimits_{j\neq n}\left(\omega_{jn}\Big/[\omega_{jn}^2-\omega^2]\right)\mathrm{Re}\left(\langle n|\mu_\alpha|j\rangle\langle j|\mu_\beta|n\rangle\right), \tag{4a}$$

$$G'_{\alpha\beta} = -(2/\hbar)\sum\nolimits_{j\neq n}\left(\omega\Big/\left[\omega_{jn}^2-\omega^2\right]\right)\mathrm{Im}\left(\langle n|\mu_\alpha|j\rangle\langle j|m_\beta|n\rangle\right), \tag{4b}$$

$$A_{\alpha\beta\gamma} = -(2/\hbar)\sum\nolimits_{j\neq n}\left(\omega_{jn}\Big/[\omega_{jn}^2-\omega^2]\right)\mathrm{Re}\left(\langle n|\mu_\alpha|j\rangle\langle j|\Theta_{\beta\gamma}|n\rangle\right). \tag{4c}$$

The initial, as well as final, and the intermediate states of the molecule are represented by n and j, respectively, and $\omega_{jn} \equiv \omega_j - \omega_n$ is their angular frequency separation. The electric dipole–electric dipole tensor $\alpha_{\alpha\beta}$ is the polarizability and is responsible for a wide range of phenomena including van der Waals forces, refraction and both light and electron scattering; $G'_{\alpha\beta}$ is the electric dipole–magnetic dipole optical activity tensor whose isotropic part is responsible for optical rotation in fluids; and $A_{\alpha\beta\gamma}$ is the electric dipole–electric quadrupole tensor which makes additional contributions to optical rotation in oriented samples.

We can also consider the explicit expressions for conventional optical rotation observables in terms of the optical activity tensors defined above [9]. The Rosenfeld equation [17] for the optical rotation angle developed over a pathlength l in a transparent isotropic sample containing N chiral molecules per unit volume is expressed, in SI units, as

$$\Delta\vartheta = -\frac{1}{9}\omega\mu_0 lN\left(G'_{xx}+G'_{yy}+G'_{zz}\right). \tag{5}$$

The Buckingham–Dunn equation [18] gives the corresponding optical rotation angle for a sample such as a crystal in which the chiral molecules are oriented. For a light beam propagating along the positive z direction, we have

$$\Delta\theta = -\frac{1}{2}\omega\mu_0 lN\left[\frac{1}{3}\omega\left(A_{xyz}-A_{yxz}\right)+G'_{xx}+G'_{yy}\right]. \tag{6}$$

Although contributions to optical rotation from $A_{\alpha\beta\gamma}$ average to zero in isotropic samples, they contribute to ROA in isotropic samples to the same order as $G'_{\alpha\beta}$. This electric dipole–electric quadrupole contribution is a source of additional richness, and hence information, in ROA spectra that has no counterpart in either ECD or VCD spectra. Electric dipole–electric quadrupole optical activity, which was first observed in 1971 in the form of circular dichroism [19], has now come to prominence through natural X-ray circular dichroism, with the first observations being reported in 1998 [20–23]. Natural X-ray circular dichroism is largest in crystals because, in the X-ray region, the electric dipole–electric quadrupole contribution is large but the electric dipole–magnetic dipole contribution is small.

16.2.3 Discovery of ROA

The fundamental scattering mechanism responsible for ROA was discovered in 1969 by Atkins and Barron [24]. They showed that waves scattered *via* the polarizability and optical activity tensors of a molecule give rise to a dependence of the intensity of scattering on the degree of circular polarization of the incident light and a circular component in the scattered light. In 1971 Barron and Buckingham developed a more definitive version of this theory incorporating a classical electrodynamics expression for the electric field vector radiated by the oscillating electric dipole, magnetic dipole and electric quadrupole moments [25]. The dominant contributions to the scattering intensity were found to be dependent on α^2 and gave rise to conventional Rayleigh and Raman scattering. Additional contributions, however, originate in $\alpha G'$ and αA which have opposite signs in right- and left-circularly polarized light. These give rise to the fundamental pseudoscalar observable of a difference in scattering intensity between right- and left-circularly polarized light, $I^R - I^L$.

16.2.4 ROA observables

A more useful experimental observable than the raw ROA intensity, $I^R - I^L$, is the dimensionless circular intensity difference (CID or Δ) introduced by Barron and Buckingham and defined as [25]

$$\Delta = \frac{(I^R - I^L)}{(I^R + I^L)}. \tag{7}$$

The denominator is effectively the conventional Raman scattering intensity and normalizes the ROA measurement with respect to the absolute Raman intensities measured using different instruments, which can vary considerably.

Different experimental configurations may be used to perform ROA measurements. The scattering angle can be varied, with the most important scattering geometries being forward (0°), right-angle (90°) and backward (180°). There are two distinct forms of right-angle scattering in which a linear polarizer is placed in the scattered beam with its transmission axis aligned either perpendicular (x) or parallel (z) to the scattering plane (yz). These two measurements generate polarized and depolarized ROA, respectively, and correspond to the polarized and depolarized Raman scattering measurements used to determine the depolarization ratio in conventional Raman spectroscopy. In terms of the molecular polarizability, $\alpha_{\alpha\beta}$, and the electric dipole–magnetic dipole and electric dipole–electric quadrupole optical activity tensors $G'_{\alpha\beta}$ and $A_{\alpha\beta\gamma}$, the CID expressions for forward, backward, polarized and depolarized right-angle scattering from an isotropic sample for incident wavelengths much greater than the molecular dimensions are [7,9,11,26]

$$\Delta(0°) = \frac{8[45\alpha G' + \beta(G')^2 - \beta(A)^2]}{2c[45\alpha^2 + 7\beta(\alpha)^2]}, \tag{8a}$$

$$\Delta(180°) = \frac{48[\beta(G')^2 + \frac{1}{3}\beta(A)^2]}{2c[45\alpha^2 + 7\beta(\alpha)^2]}, \tag{8b}$$

$$\Delta_x(90°) = \frac{12[45\alpha G' + 7\beta(G')^2 + \beta(A)^2]}{c[45\alpha^2 + 7\beta(\alpha)^2]}, \tag{8c}$$

$$\Delta_z(90°) = \frac{12[\beta(G')^2 - \frac{1}{3}\beta(A)^2]}{6c\beta(\alpha)^2}. \tag{8d}$$

As these tensor component products are averaged over all orientations of the scattering molecule they are invariant to axis rotations. Specifically, the isotropic invariants of the polarizability tensor and the electric dipole–magnetic dipole optical activity tensor are

$$\alpha = \frac{1}{3}\alpha_{\alpha\alpha} = \frac{1}{3}(\alpha_{xx} + \alpha_{yy} + \alpha_{zz}), \tag{9a}$$

$$G' = \frac{1}{3}G'_{\alpha\alpha} = \frac{1}{3}\left(G'_{xx} + G'_{yy} + G'_{zz}\right), \tag{9b}$$

and

$$\beta(\alpha)^2 = \frac{1}{2}(3\alpha_{\alpha\beta}\alpha_{\alpha\beta} - \alpha_{\alpha\alpha}\alpha_{\beta\beta}), \tag{9c}$$

$$\beta(G')^2 = \frac{1}{2}\left(3\alpha_{\alpha\beta}G'_{\alpha\beta} - \alpha_{\alpha\alpha}G'_{\beta\beta}\right), \tag{9d}$$

$$\beta(A)^2 = \frac{1}{2}\omega\alpha_{\alpha\beta}\varepsilon_{\alpha\gamma\delta}A_{\gamma\delta\beta}, \tag{9e}$$

are the anisotropic invariants of the polarizability–polarizability and polarizability–optical activity tensor component products in which the tensor components are referred to molecule-fixed axes. Common factors in the numerators and denominators of the CIDs [9] have not been cancelled so that the relative sum and difference intensities may be directly compared.

One additional experimental configuration for ROA CID measurements should be discussed. By setting the transmission axis of the linear polaroid analyzer in the right-angle scattered beam at the 'magic angle' of $\pm\sin^{-1}(2/3)^{1/2} \approx \pm 54.74°$ to the scattering plane (yz), the contribution from the electric dipole–electric quadrupole ROA mechanism vanishes so that pure electric dipole–magnetic dipole ROA spectra may be measured [26,27]. The associated magic angle CID is

$$\Delta^*(90°) = \frac{(20/3)[9\alpha G' + 2\beta(G')^2]}{(10/3)c[9\alpha^2 + 2\beta(\alpha)^2]}. \tag{10}$$

Hug [8,28] found that the invariant combinations of tensor products appearing in the numerator and denominator of Eq. (10) are all that is measured in scattering cross sections integrated over all directions, and pointed out that this is reminiscent of the situation in natural optical rotation and circular dichroism of isotropic samples where the electric dipole–electric quadrupole contributions also average to zero.

16.2.5 The scattered circular polarization strategy

As well as the circular intensity difference, ROA is also manifest as a small circularly polarized component in the scattered beam [8,9,24,27,29]. Within the far-from-resonance approximation, measurement of this circular component in the scattered beam, which is called scattered circular polarization (SCP) ROA, provides equivalent information to the CID measurement, which is now specifically referred to as incident circular polarization (ICP) ROA. Furthermore, the simultaneous measurement of both ICP and SCP ROA, which is called dual circular polarization (DCP) ROA, contains several advantages [11,30,31]. DCP ROA measurements may be performed either as in-phase (DCP_I) or out-of-phase (DCP_{II}) combinations of the constituent ICP and SCP ROA measurements. The dimensionless ROA observable for the DCP_I experiment in backscattering, which is the most important experimental configuration since it eliminates all isotropic contributions, is [11,27,30,31]

$$\Delta^{DCP_I}(180^\circ) = \frac{48\left[\beta(G')^2 + \frac{1}{3}\beta(A)^2\right]}{12c\beta(\alpha)^2}. \tag{11}$$

These results apply specifically to elastic Rayleigh scattering. For inelastic Raman scattering the same basic CID expressions apply but with the molecular property tensors replaced by corresponding vibrational Raman transition tensors between the initial and final vibrational states n_v and m_v, respectively. Consequently, the properties $\alpha_{\alpha\beta}$, etc. are replaced by $\langle m_v|\alpha_{\alpha\beta}(Q)n_v\rangle$ etc., where $\alpha_{\alpha\beta}(Q)$, etc. are the effective polarizability and optical activity operators that depend parametrically on the normal vibrational coordinates Q so that, within the Placzek polarizability theory of the Raman effect [32], the ROA intensity depends on products such as $(\partial\alpha_{\alpha\beta}/\partial Q)_0(\partial G'_{\alpha\beta}/\partial Q)_0$ and $(\partial\alpha_{\alpha\beta}/\partial Q)_0\varepsilon_{\alpha\gamma\delta}(\partial A_{\gamma\delta\beta}/\partial Q)_0$.

The basic symmetry selection rules for natural Rayleigh and Raman optical activity directly follow from these results. For natural Rayleigh optical activity, the same components of $\alpha_{\alpha\beta}$ and $G'_{\alpha\beta}$ must span the totally symmetric irreducible representation; while for natural vibrational ROA the same components of $\alpha_{\alpha\beta}$ and $G'_{\alpha\beta}$ must span the irreducible representation of the normal vibrational coordinate under consideration. This is only possible in the chiral point groups C_n, D_n, O, T and I; they lack improper symmetry operations, so that polar tensors such as $\alpha_{\alpha\beta}$ and axial tensors such as $G'_{\alpha\beta}$ have the same transformation properties. Although $A_{\alpha\beta\gamma}$ does not transform in the same manner as $G'_{\alpha\beta}$, the second-rank axial tensor $\varepsilon_{\alpha\gamma\delta}A_{\gamma\delta\beta}$, which combines with $\alpha_{\alpha\beta}$ in the expressions for optically active scattering, transforms the same as $G'_{\alpha\beta}$.

For a molecule composed entirely of idealized axially-symmetric bonds, in which $\beta(G')^2 = \beta(A)^2$ and $\alpha G' = 0$, a simple bond polarizability theory [9,15,28] shows that

ROA is generated exclusively by anisotropic scattering and the forward and backward CID expressions in Eqs. (8a) and (8b) reduce to [9]

$$\Delta(0°) = 0, \tag{12a}$$

$$\Delta(180°) = \frac{32\beta(G')^2}{c[45\alpha^2 + 7\beta(\alpha)^2]}. \tag{12b}$$

Therefore, although conventional Raman scattering intensities are the same in the forward and backward directions, ROA intensity is maximized in backscattering and zero in forward scattering. These considerations lead to the important conclusion that backscattering boosts the ROA signal relative to the background Raman intensity and is therefore the best experimental strategy for most ROA studies, and is particularly valuable for ROA studies of biomolecules in aqueous solution due to the weak signals and high Raman backgrounds typical of these samples [26,33].

16.3 RESONANCE ROA (RROA)

The expressions for ROA observables presented so far are valid only for scattering at transparent frequencies. In the case of resonance scattering at absorbing frequencies, additional contributions to the CIDs can arise with complications from an interesting Stokes–antiStokes asymmetry [34–38]. Experimental resonance Raman (RR) intensities [38] may be quantitatively compared to theoretical values through construction of RR excitation profiles, which correspond to the spectral dependence of RR band intensities on the laser excitation frequency as it is tuned through a molecule's electronic absorption band. Particularly successful have been results obtained in the limit of resonance with a single electron state (SES) in which the ground and excited vibrational normal modes are assumed to be identical other than for a shift along the mode from the equilibrium nuclear position [37]. The precise equalities between Stokes ICP and antiStokes SCP observables have been deduced using time reversal symmetry [35,36] and RROA expressions can be determined for all of the various scattering conditions in the SES limit. For example, the RROA and RR invariant spectra for $\mathrm{DCP_I}$ backscattering, $\mathrm{DCP_I}$ ROA, are

$$I_{\mathrm{R}}^{\mathrm{R}}(180°) - I_{\mathrm{L}}^{\mathrm{L}}(180°) = \frac{96K}{c}\mathrm{Im}\left[(\tilde{\alpha}_{zz})_{g1,g0}(\tilde{G}_{zz})^*_{g1,g0}\right], \tag{13a}$$

$$I_{\mathrm{R}}^{\mathrm{R}}(180°) + I_{\mathrm{L}}^{\mathrm{L}}(180°) = 24K\left|(\tilde{\alpha}_{zz})_{g1,g0}\right|^2, \tag{13b}$$

where the subscripts on the intensity symbols refer to the polarization states of the scattered radiation. The corresponding invariants are

$$\left|(\alpha_{zz})_{g1,g0}\right|^2 = \frac{1}{\hbar}\left|(\mu)^0_{eg}\right|^4 U(\omega_0, \Omega_1), \tag{13c}$$

$$\mathrm{Im}\left[(\alpha_{zz})_{g1,g0}(G_{zz})^*_{g1,g0}\right] = \frac{1}{\hbar}\left\{|\mu|^2_{eg}\mathrm{Im}\left[(\vec{\mu})^0_{ge}(m)^0_{eg}\right]\right\}U(\omega_0, \Omega_1), \tag{13d}$$

where $(\mu)^0_{eg}$ and $(m)^0_{eg}$ are the amplitudes of the electric dipole transition moment and magnetic dipole transition moment between the ground state (g) and the resonantly excited electronic state (e), $U(\omega_0, \Omega_1)$ is a resonant lineshape function. The corresponding CID ratio is, therefore,

$$\Delta^{\mathrm{DCP_I}}(180°) = \left[\frac{I^{\mathrm{R}}_{\mathrm{R}}(180°) - I^{\mathrm{L}}_{\mathrm{L}}(180°)}{I^{\mathrm{R}}_{\mathrm{R}}(180°) + I^{\mathrm{L}}_{\mathrm{L}}(180°)}\right]$$

$$= -\left(\frac{4}{c}\right)\left[\frac{\mathrm{Im}(\mu)^0_{ge}(m)^0_{eg}}{\left|(\mu)^0_{eg}\right|^2}\right] \qquad (13e)$$

$$= -g_{eg},$$

and the normalized CID ratio has the same magnitude, but opposite sign, as the electronic CD anisotropy ratio for the resonant electronic state, g_{eg} [6]. Therefore, $\mathrm{DCP_I}$ ROA in the limit of strong resonance with a single electronic state is characterized by strong monosignate peaks, with the ratio of RROA intensity to the parent RR spectrum being the same magnitude, but opposite sign, to that of the ratio of the electronic CD of the resonant state to its parent absorption spectrum. In contrast, for the out-of-phase backscattering configuration, $\mathrm{DCP_{II}}$, the RR intensity is the same as for that shown in Eq. (13b), but there is no RROA intensity as the corresponding tensor invariants vanish in the ROA-SES limit. Therefore,

$$\Delta^{\mathrm{DCP_{II}}}(180°) = 0. \qquad (14)$$

For backscattering unpolarized ICP ROA the RROA intensity in the SES limit is the same as that already shown for $\mathrm{DCP_I}$ ROA in Eq. (13a), but the RR intensity is twice as large. Therefore, the CID expression is given by

$$\Delta^{\mathrm{u}}(180°) = -\frac{1}{2}g_{eg}. \qquad (15)$$

Similarly, the corresponding expressions for the resonance case in the SES limit for right-angle depolarized ROA are [37]

$$I^{\mathrm{R}}_z(90°) - I^{\mathrm{L}}_z(90°) = \frac{24K}{c}\mathrm{Im}\left[(\tilde{\alpha}_{zz})_{g1,g0}\left(\tilde{G}_{zz}\right)^*_{g1,g0}\right], \qquad (16a)$$

$$I^{\mathrm{R}}_z(90°) + I^{\mathrm{L}}_z(90°) = 12K\left|(\tilde{\alpha}_{zz})_{g1,g0}\right|^2, \qquad (16b)$$

$$\Delta^z(90°) = -\frac{1}{2}g_{eg}, \qquad (16c)$$

and those for the right-angle polarized ROA configuration are

$$I^{\mathrm{R}}_x(90°) - I^{\mathrm{L}}_x(90°) = \frac{48K}{c}\mathrm{Im}\left[(\tilde{\alpha}_{zz})_{g1,g0}\left(\tilde{G}_{zz}\right)^*_{g1,g0}\right] \qquad (17a)$$

$$I_x^{\mathrm{R}}(90^\circ) + I_x^{\mathrm{L}}(90^\circ) = 24K\left|(\tilde{\alpha}_{zz})_{g1,g0}\right|^2, \tag{17b}$$

$$\Delta^x(90^\circ) = -\frac{1}{2}g_{eg}. \tag{17c}$$

All other scattering configurations can be considered in this way, but a general result is that all forms of single modulation RROA (both ICP and SCP) yield a CID value of $\Delta = -(1/2)g_{eg}$, while for $\mathrm{DCP_I}$ RROA, $\Delta = -g_{eg}$, and for $\mathrm{DCP_{II}}$ RROA, $\Delta = 0$ [37].

A new optical activity phenomenon called linear polarization ROA can also arise under resonance conditions in right-angle scattering [35,36,39] but has not yet been observed. This measures intensity differences in the Raman scattered light associated with orthogonal linear polarization states at ±45° to the scattering plane in the incident or scattered radiation, or both simultaneously. The Stokes–Meuller formalism has been used to provide a unique description of these distinct ROA phenomena and to summarize all the different associated ROA measurement strategies [11,26]. Resonance ROA is still at an early stage of development compared with transparent ROA at visible excitation wavelengths. Nonetheless, as already shown, a useful theory of natural resonance ROA has been developed and the first experimental observations have been reported for the stereoisomers of both naproxen and its deuterated methyl ester [38].

Extension of RROA into the deep UV region below 300 nm, though this has not been reported as yet, has great potential for structural biology. Peptide bonds absorb strongly in the region of ~190–240 nm, known as the 'far UV,' while the aromatic side chains of proteins absorb strongly in the range of ~250–280 nm, known as the 'near UV,' and even down to 220 nm. UVCD spectroscopy, of course, operates at these incident wavelengths and ultraviolet RROA (UVRROA) provides a possible means to significantly decrease concentrations of the samples under study by a few orders of magnitude. Excitation of UVRROA at these wavelengths would also avoid fluorescence from the sample, a problem in studies of many proteins. Conventional non-resonant ROA of biological molecules is discussed in Section 16.9.

16.4 SECOND HARMONIC SCATTERING FROM CHIRAL INTERFACES

Analogs of ROA in scattering from chiral species adsorbed onto appropriate achiral substrates are especially interesting since genuine chiroptical observables may be generated *via* pure electric dipole–electric dipole scattering mechanisms [40], which can boost the observables by three-orders of magnitude compared with ROA from the bulk samples which, as already outlined, require electric dipole–magnetic dipole and electric dipole–electric quadrupole scattering mechanisms. This may be especially valuable in second harmonic scattering from chiral interfaces, where the linear polarization chiroptical measurements mentioned above are the favored experimental strategy since large effects may consequently be observed at transparent incident or scattered wavelengths [41], unlike with the more obvious circular polarization chiroptical measurements which require rigorous resonant or pre-resonant scattering conditions at or near electronic absorption frequencies [42].

16.5 MAGNETIC ROA

Magneto-optical rotation, also known as the Faraday effect, and magnetic field-induced CD are well characterized phenomena displayed by all molecules, both chiral and achiral, in a magnetic field aligned parallel to an incident light beam [43]. These phenomena are instances of magnetic optical activity in transmitted light. Barron and Buckingham predicted that all molecules, both chiral and achiral, should show an analogous phenomenon in scattered light, or more precisely, an ROA effect in right-angle scattering upon application of a static magnetic field parallel to the incident radiation [44–46]. This was first observed in the resonance Raman spectrum of ferrocytochrome c [47,48] and was thought to originate in Zeeman splitting of the orbitally degenerate excited resonant state in conjunction with a vibronic development of the Raman transition tensors [49]. Since then, there have been periodic investigations of the phenomenon of magnetic ROA, sometimes referred to as Raman electron paramagnetic resonance (Raman EPR) in the case of paramagnetic systems with spin or orbital degeneracy. Subsequent studies on the metal halides $IrCl_6^{2-}$, $FeBr_4^-$, $CuBr_4^-$ [50] and $IrBr_6^{2-}$ [51] indicated that magnetic ROA may also originate in 'off-diagonal' Raman transitions between the Zeeman-split components of the Kramers-degenerate initial and final Raman levels which arise in odd electron systems [52]. A more detailed analysis was performed in a study of the magnetic ROA of $OsBr_6^{2-}$, $IrCl_6^{2-}$, and $U(C_8H_8)_2$ (uranocene) which indicated the sensitivity of the data to the magnetic structure of each molecule [53].

Little further research was conducted in this area for some time. During the 1990s, Hoffman and colleagues investigated magnetic ROA from magnons and phonons in uniaxial crystals of rutile-structure transition metal fluorides [54,55]. Several of these authors have recently revisited this subject, with emphasis being placed on the effects of temperature and applied magnetic field on the magnetic ROA from antiferromagnetic FeF_2 [56]. The authors reported that the CIDs observed for phonons originated in birefringence and could be nulled by careful alignment of the crystal, however a small zero-field splitting of the two acoustic magnon branches was found to be reproducible and inversely related to temperature. This was rationalized in terms of an effective magnetic field from magnetic dipole–dipole interactions in the FeF_2 [57]. It will be interesting to see if this clear demonstration of the enhanced sensitivity of magnetic ROA over the standard polarized Raman technique for studying magnetic excitations leads to further research in this field. More generally, surprisingly few studies have been conducted using magnetic ROA despite the information contained on the magnetic structure of ground- and low-level excited electronic states, including the sign of the g-factor, and how the magnetic structure changes in an excited non-totally symmetric vibrational state. An obvious parallel to the recent development of 'natural' ROA for studying biological molecules is the potential of magnetic ROA as a resonance

Raman spin probe of sites of biological function in paramagnetic biological molecules under near-physiological conditions [45].

16.6 COMPUTATION OF ROA OBSERVABLES

The computation of ROA spectra is a challenging problem for several reasons. The tensors underlying ROA generation, already discussed, generally require the use of computationally expensive methods and as the property vanishes for a static electromagnetic field, either the static-limit approximation [58] or linear response methods must be used. Furthermore, the contribution of the electric dipole–magnetic dipole optical activity tensor, $G'_{\alpha\beta}$, makes approximate theoretical calculations gauge-origin dependent while the limitations of Hartree–Fock (HF) harmonic force fields can affect computed intensities. However, considerable progress has been made in recent years. In particular, the development of a framework for accurate *ab initio* calculations of VCD constitutes a significant achievement in quantum chemistry. Recent general surveys of the theory of vibrational optical activity may be found elsewhere [6,7,9,59,60].

The calculation of ROA observables has been performed using several different approaches. Models of ROA exist, such as the atom dipole interaction model [61] and the bond polarizability model [9,62], that break the molecule down into either its constituent atoms or bonds, respectively. In the bond polarizability model, for example, the molecule is decomposed into bonds or groups supporting local internal vibrational coordinates. In principle, given a normal coordinate analysis and a set of bond polarizability parameters, the ROA associated with every normal mode of vibration of a chiral molecule may be calculated [62]. However, due to the approximations inherent in these models, such calculations do not reproduce well the experimental data at all. *Ab initio* computations of ROA spectra are far superior and much easier to implement. Models can, nonetheless, provide valuable physical insight into the generation of infrared and Raman vibrational optical activity, which is often not transparent in the computationally far superior *ab initio* methods. However, in a new approach to obtain such physical insight from *ab initio* computations, the numerical information is converted into a pictorial representation that enables ROA generation to be visualized in terms of atomic contribution patterns [28,63].

Since the property tensors $G'_{\alpha\beta}$ and $A_{\alpha\beta\gamma}$ responsible for ROA are time-even, there is no fundamental problem in ROA theory analogous to that arising in VCD theory due to the time-odd nature of the magnetic dipole moment operator [64]. Consequently, *ab initio* calculations of ROA appeared several years before the first such VCD calculations. An *ab initio* method, based on calculations of $\alpha_{\alpha\beta}$, $G'_{\alpha\beta}$ and $A_{\alpha\beta\gamma}$ in a static approximation by Amos [58], and how these property tensors vary with the normal vibrational coordinates, was developed by Polavarapu in the late 1980s [65,66]. Although these early *ab initio* calculations of ROA spectra did not reach the high levels of accuracy that are now routinely achievable for calculations of VCD spectra, they nonetheless proved valuable. For example, the absolute configuration of the archetypal small chiral molecule CHFClBr, which had resisted assignment for over 100 years, was reliably determined from a comparison of the experimental and *ab initio* theoretical Raman optical activity spectra [7,67]. Analogous VCD studies of this molecule are unfavorable because the frequencies of most of the fundamental normal modes of vibration are too low to be experimentally accessible, whereas all are accessible in the ROA spectrum.

There have been continued developments of ROA computations for small molecules based on self-consistent-field (SCF) linear response theory [68] and on the sum over states formalism [69–71]. Pecul et al. [72] have recently reported the first calculations of ROA spectra of carbohydrates, also at the SCF level, in order to gain insight into the influence of conformation on the characteristic band patterns. Although these *ab initio* methods are now able to provide reasonable agreement with experimental ROA data [73], and they have been shown to be able to discriminate between possible conformers of a cyclic dipeptide [74] as well as provide some insight into the effects of conformation and local hydrogen-bonding on CIDs [72], they are not yet sufficiently accurate to provide detailed information on the generation of ROA in biological molecules. Similar limitations and the reliance on a continuum model of the aqueous environment also led to significant discrepancies between the ROA spectra computed by Jalkannen et al. [75] for dialanine using a combination of molecular mechanics and density-functional theory (DFT), and experimental results, particularly in the important extended amide III region.

However, by including rarified basis sets containing moderately diffuse p-type orbitals on hydrogen atoms, Zuber and Hug have demonstrated that *ab initio* ROA calculations of a similar high level of quality to those of VCD may now be achieved [76]. These basis sets allowed the accurate description of the gradients of the electronic tensors of these atoms. It has already been shown that this is the single most important requirement for accurate computations of Raman and ROA spectra of organic molecules [8,28]. This is because hydrogen nuclei are the most mobile nuclei in both vibrations associated with hydrogen atoms and those of skeletal modes. It is probably not a coincidence that in many systems, ranging from small chiral organic molecules to proteins, those vibrations with large contributions from C–H deformations often generate large ROA signals. Interestingly, Zuber and Hug note that for molecules of practical interest, the electronic tensor calculations are now sufficiently optimized for most purposes and further improvements in the accuracy of simulations of ROA spectra will probably focus on increasing the precision of the vibrational contributions [76].

16.6.1 ROA of archetypal helical molecules

Consideration has been recently given to the generation of ROA in inherently chiral chromophores. Hug et al. [63] performed an extensive combined experimental and theoretical ROA study of (*M*)-trispiro[2.1.1.2]nonane, also known as (−)-(*M*)-σ-[4] helicene, which consists of a chiral arrangement of achiral cyclopropane subunits. Very large Δ values of up to 5×10^{-3} were observed in SCP backscattering for this helicene indicating it may represent the archetype helical chromophore for ROA and VCD, just as [6]helicene does for electronic optical activity. (−)-(*M*)-σ-[4]Helicene may also serve as the prototype for future ROA studies on other helical molecules with increasing numbers of spirocyclopropane groups. An excellent level of agreement was found between the experimental Raman and ROA spectra of (−)-(*M*)-σ-[4]helicene and calculations at a combined DFT and HF levels. Despite the difficulties in computing accurate ROA spectra, particularly in regions where extensive cancellation of large oppositely signed contributions makes the calculations highly sensitive to small variations in band shapes, the predicted results were the most accurate so far reported for a medium-to-large sized molecule. This mode-type analysis appears to be a particularly important advancement in this field and a more useful route to a qualitative understanding of both Raman and ROA spectra of inherently chiral molecules than standard vibrational decomposition methods.

16.7 INSTRUMENTATION

Since ROA is maximized in backscattering, a backscattering geometry has proven to be essential for the routine measurement of ROA spectra of biomolecules in aqueous solution. The optical layout of the backscattering ROA instrument, based on the ICP measurement strategy, used in the Glasgow laboratory for the past decade is depicted schematically in Fig. 16.3 [77,78]. A visible argon ion laser beam is gently focused into the liquid sample which is contained in a small rectangular fused quartz cell. The cone of backscattered light is reflected off a 45° mirror, which has a small central hole drilled to allow passage of the incident laser beam, through an edge filter to remove the Rayleigh line and into the collection optics of a single grating spectrograph. This is a customized version of a fast imaging spectrograph containing a highly efficient volume holographic transmission grating. The detector is a cooled back-thinned charge coupled device (CCD) camera with a quantum efficiency of ~80% over the spectral range in our studies. Operation of the CCD camera in multichannel mode allows the full spectral range to be

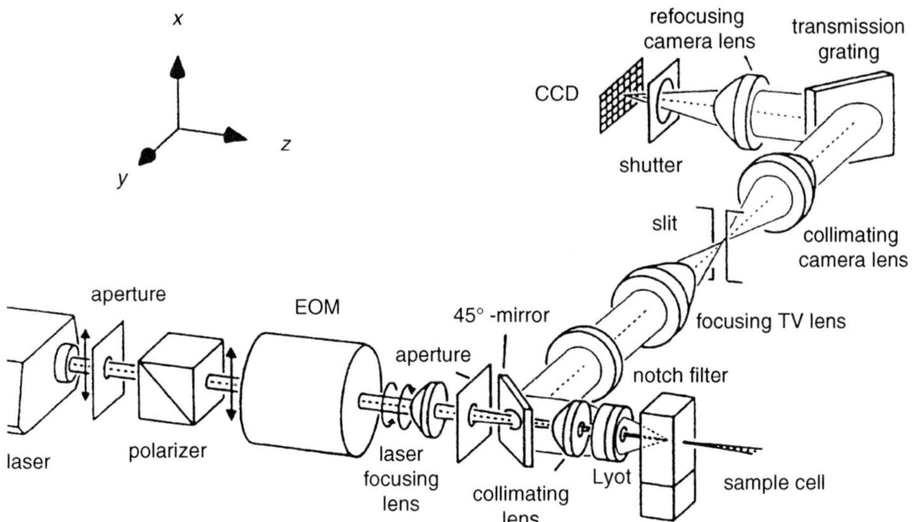

Fig. 16.3. Optical layout of the current backscattering ICP instrument used in Glasgow.

measured in a single acquisition. To measure the small ROA signals, the spectral acquisition is synchronized with an electro-optic modulator (EOM) used to switch the state of polarization of the incident laser beam between right- and left-circular states at a suitable rate.

Spectra are displayed in analog-to-digital converter (ADC) units as a function of the Stokes–Raman wavenumber shift with respect to the exciting laser line. The green 514.5 nm line of an argon-ion laser was used as this provides a compromise between reduced fluorescence from sample impurities with increasing wavelength and the increased scattering efficiency with decreasing wavelength due to the Rayleigh's λ^{-4} law. Typical laser power at the sample is ~700 mW and sample concentrations of proteins, polypeptides and nucleic acids are ~30–100 mg/ml while those of intact viruses are ~5–30 mg/ml. Under these conditions ROA spectra such as those presented here may be obtained in ~5–24 h for proteins and nucleic acids and ~1–4 days for intact viruses. Measurements can also be performed over the temperature range of ~0–60°C by directing dry air downwards over the sample cell from a device employed to cool protein crystals in X-ray diffraction experiments, in order to study dynamic behavior. This ICP ROA instrument has recently been upgraded with a solid-state Nd/YAG laser operating at 532 nm together with the insertion of counter-rotating half-wave plates into the incident laser beam after the EOM to average the azimuths of any residual linear contaminants in the right- and left-circular polarization states of the laser beam to help suppress artefacts arising from linear birefringence.

Although the design of the ICP ROA instrument described above will continue to be useful, a completely new design of ROA instrument with significant advantages has recently been developed [8,79,80]. This new instrument is based on measuring the circularly polarized component in the Raman scattered light (SCP strategy) instead of the Raman intensity difference in right- and left-circularly polarized incident light (ICP strategy) employed in the Glasgow instrument. This enables 'flicker noise' arising from dust particles, density fluctuations, laser power fluctuations, etc. to be eliminated since the intensity difference measurements required to extract the circularly polarized components of the scattered beam are taken between two orthogonal components of the scattered light measured during the same acquisition period. The flicker noise consequently cancels out, resulting in superior signal-to-noise characteristics. A commercial instrument based on this new design that also incorporates a sophisticated artefact suppression protocol based on a 'virtual enantiomers' approach to eliminate deterministic offsets and which greatly facilitates the routine acquisition of reliable ROA spectra [81] has recently become available from BioTools, Inc. This instrument has been shown to be able to obtain ROA spectra of unsurpassed quality on as little as microgram quantities of chemical samples in disposable capillary tubes, emphasizing the ease and reliability of such measurements [81].

In fact, an almost arbitrarily high level of precision and signal-to-noise ratio may be reached with this new instrument. The opportunity this presents for simplifying the collection of high quality ROA data will have considerable benefits for future research. Studies can now be performed on as little as one milligram of sample or less and the expensive sample cells frequently used in the past may not be required [79]. Although specialized sample cells and several milligrams of sample may still be required for at least some biological molecules, this is still a considerable improvement on the previous situation. This further increases the potential of ROA as a widely applicable probe of biological molecules, where it can be difficult to obtain in milligram quantities. Because of the great improvements in signal-to-noise ratio and artefact control of the new instrument, there is also the potential to extend practical ROA spectroscopy to longer wavelengths and so circumvent the fluorescence problems common to optical spectroscopic studies of biological molecules in the visible region [82,83].

16.8 STUDIES ON SMALL MOLECULES

A large number of small chiral organic molecules were studied in the early days of ROA, either as neat liquids or concentrated solutions. Stereochemical information was deduced by comparing ROA band patterns with those of related molecules of known structure [84], or by the application of simple model theories such as the two-group model [9,15] and the inertial model [9,85]. Because ROA spectra may be acquired down to $\sim 100\,\mathrm{cm}^{-1}$ and below, interesting low-wavenumber modes of vibration such as methyl torsions [9,85] may be studied as well as low-wavenumber skeletal modes that influence conformational transitions [86]. Another valuable application of ROA is in the measurement of enantiomeric excess in samples containing unequal amounts of mirror-image enantiomers and which, in favorable circumstances, may be determined to an accuracy of $\sim 0.1\%$ [87–89].

Although many ROA studies have been performed on small chiral organic molecules [9,12,90–93], particularly during the 1970s and 1980s, these will not be revisited here. Instead, we shall outline one example, namely a comparison of backward- and forward-scattered ROA spectra of β-pinene [94], since it can be rationalized using simple symmetry arguments. As displayed in Fig. 16.4, there is a large couplet, negative at low-wavenumber and positive at high-wavenumber, in the ROA spectrum of the (1*S*,5*S*)-enantiomer collected in the forward direction but not in the backward direction, that is associated with bands observed at ~ 716 and $765\,\mathrm{cm}^{-1}$ in the parent Raman spectrum. This couplet also appears in the polarized but not the depolarized ROA spectrum measured in scattering at 90°. It follows from Eqs. (8a–d) that these observations may only be reconciled if this couplet originates in pure isotropic scattering. It is possible to understand qualitatively how this isotropic ROA may be generated by considering the symmetry aspects of the optical activity tensors intrinsic to the exocyclic olefinic group $C{=}CH_2$. The olefinic methylene twist makes a significant contribution to the Raman band at $\sim 716\,\mathrm{cm}^{-1}$, whereas the Raman band at $\sim 765\,\mathrm{cm}^{-1}$ originates in a pinane-type skeletal vibration, so this large ROA couplet appears to originate in coupling between these two modes. The methylene twist transforms as A_u in the D_{2h} point group of ethene itself, and as A_2 in a structure of C_{2v} symmetry, both irreducible representations of which are spanned by the tensor components G'_{XX}, G'_{YY} and G'_{ZZ}. Therefore, a fundamental vibrational Raman scattering transition associated with the methylene twist is allowed through G', the isotropic part of the axial electric dipole–magnetic dipole–optical activity tensor, even in the parent structure of highest symmetry (D_{2h}). This transition may therefore be expected to show significant isotropic ROA if the effective symmetry of the

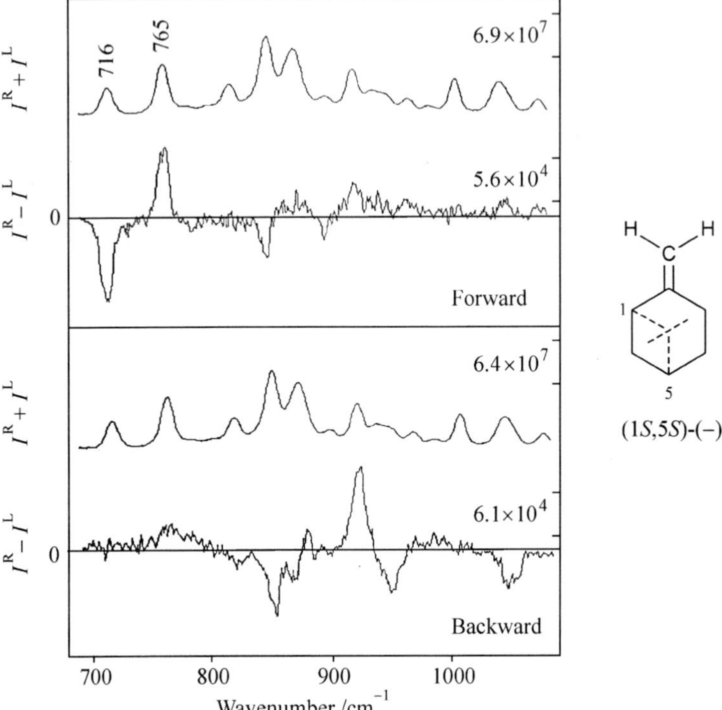

Fig. 16.4. The forward scattered (top pair) and backscattered (bottom pair) ICP Raman ($I^R + I^L$) and ROA ($I^R - I^L$) spectra of a neat liquid sample of (1*S*,5*S*)-(−)-β-pinene adapted from ref. [94].

olefinic group is sufficiently reduced to that of a chiral point group as in β-pinene, which has C_1 symmetry, in which α, the isotropic part of the polar polarizability tensor, can also contribute to Raman scattering in the methylene twist. Alternatively, the major contribution from α may arise through the pinane-type skeletal mode with which the methylene twist is coupled.

16.9 ENHANCED SENSITIVITY OF ROA TO BIOMOLECULAR STRUCTURE AND DYNAMICS

The normal vibrational modes of biopolymers, such as proteins and nucleic acids, are usually highly complex as they contain contributions from local vibrational coordinates in both the backbone and the side chains. ROA is able to cut through this complexity as the largest signals in an ROA spectrum are those that originate from vibrational coordinates which sample the most rigid and chiral parts of the molecule [78]. Typically, these reside in the polypeptide backbone and so give rise to band patterns that are characteristic of the backbone conformation, a schematic representation of which is shown in Fig. 16.5. For proteins, these band patterns will be characteristic of secondary loop and turn structures. The parent conventional Raman spectra of proteins,

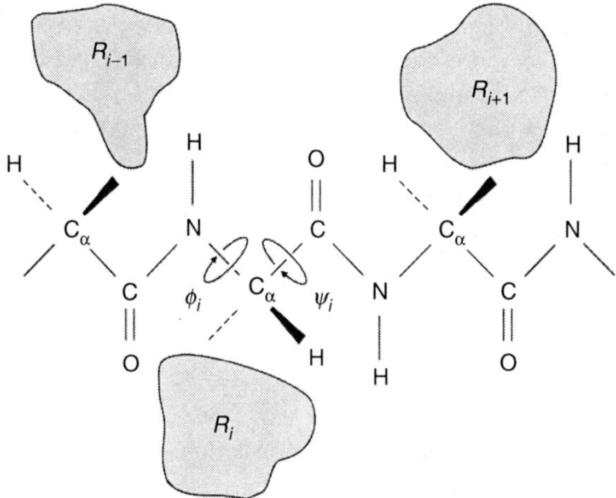

Fig. 16.5. A schematic representation of the polypeptide backbone of a protein, illustrating the Ramachandran ϕ, ψ angles and the amino acid side chains R_{i-1}, R_i and R_{i+1}.

by comparison, are dominated by bands arising from the side chains of the constituent amino acids, and these often obscure the Raman bands from the polypeptide backbone.

The timescale of the Raman scattering event is far shorter than that of the fastest fluctuations in the conformations of biological molecules ($\sim 3.3 \times 10^{-14}$ s for a vibration with Stokes wavenumber shift of 1000 cm^{-1} excited in the visible region). Consequently, the ROA spectra of biological molecules are a superposition of 'snapshot' spectra from all distinct conformations of the molecules at equilibrium [78]. As ROA intensity is dependent on absolute chirality, the contributions from enantiomeric structures that will arise as a mobile structure explores the range of accessible conformations will tend to cancel each other. Therefore, ROA exhibits an enhanced sensitivity to dynamic aspects of the behavior of biological molecules. Conventional Raman band intensities, and other non-pseudoscalar observables that are 'blind' to chirality, are generally additive and so are less sensitive to conformational mobility. Electronic optical activity manifested in the form of UVCD also demonstrates an enhanced sensitivity to the behavior of chiral molecules, but to a much lesser extent than ROA due to its dependence on electronic transitions.

16.9.1 ROA spectra of polypeptides and proteins

Normal vibrational modes of the backbone in polypeptides and proteins are typically associated with three main regions of the Raman spectrum [95,96]. These are the backbone skeletal stretch region ~870–1150 cm^{-1}, where bands originate principally in C_α–C, C_α–C_β and C_α–N stretch coordinates; the amide III region ~1230–1310 cm^{-1}, where bands are thought to mainly involve the in-phase combination of largely the N–H in-plane deformation with the C_α–N stretch; and the amide I region ~1630–1700 cm^{-1}, where bands originate mostly in the C=O stretch. Diem has shown, however, that the amide III region, in small peptides at least, involves far more mixing between the N–H and C_α–H deformations than originally thought, and so should be extended to at least 1340 cm^{-1} [1]. This extended amide III region is particularly important in ROA studies of the polypeptide backbone because the coupling between N–H and C_α–H deformations is highly sensitive to geometry and generates a rich and informative band structure. Excellent accounts of the extended amide III, and other, vibrational modes of short peptides have been written by Schweitzer–Stenner et al. [97,98]. As described in Section 16.9.3.1, a number of useful ROA bands do nonetheless arise from vibrations in the side chains of amino acids, many of which have been identified in conventional Raman spectra [95,96,99].

16.9.2 Poly(L-lysine)

The sensitivity of ROA bands to polypeptide backbone conformation is best illustrated through consideration of the spectra of homopolypeptide structural models. Poly(L-lysine) is an excellent example as it adopts well-defined conformations under specific conditions of temperature and pH and has been widely used as a model for the spectroscopic identification of secondary structure sequences in proteins [78,100–104]. At alkaline pH, poly(L-lysine) has neutral side chains and, at low temperature, supports a stable α-helical conformation stabilized by internal hydrogen bonds. Under the same alkaline conditions but at higher temperatures, poly(L-lysine) supports β-sheet structures. At neutral and acidic pH, the side chains in poly(L-lysine) are charged and so repel each other, leading to adoption of a disordered structure. The Raman (top of each pair) and backscattered ROA (bottom of each pair) spectra for these three conformations of poly(L-lysine) are presented in Fig. 16.6. Although accounts of the assignment of polypeptide and protein ROA bands for various structural elements have been given elsewhere [78,105–107], these will be briefly summarised here. As many more ROA spectra have been recorded in recent years, some early band assignments have been revised [78] and are still being further refined as more data are accumulated.

16.9.2.1 α-Helical poly(L-lysine)

The α-helix is one of the most common secondary structure motifs found in proteins and polypeptides and comprises a single strand of the polypeptide chain in a helical form with a right handed twist and is stabilized mainly by hydrogen bonds between C=O_i and N–H_{i+4} groups within the same chain. The extended amide III region of the ROA spectrum of α-helical poly(L-lysine) is dominated by the two positive bands that occur at ~1297 and 1341 cm^{-1}. The band at higher wavenumber is assigned to a hydrated form of α-helix, while the smaller band at lower wavenumber appears to be associated with α-helix in a more hydrophobic environment. Conventional Raman bands at similar wavelengths have been reported in studies of α-helix in polypeptides and filamentous bacteriophages. A Raman study of α-helical poly(L-alanine) by Lee and Krimm [108] provided definitive assignments of several normal modes of vibration in the extended amide III region. These normal modes were shown to transform variously as the A, E_1 and E_2 symmetry species of the point group of a model infinite regular helix. It appears that the ROA bands assigned to α-helix in this region may be related to several of these normal modes identified by Lee and Krimm [108], with the ROA band positions and

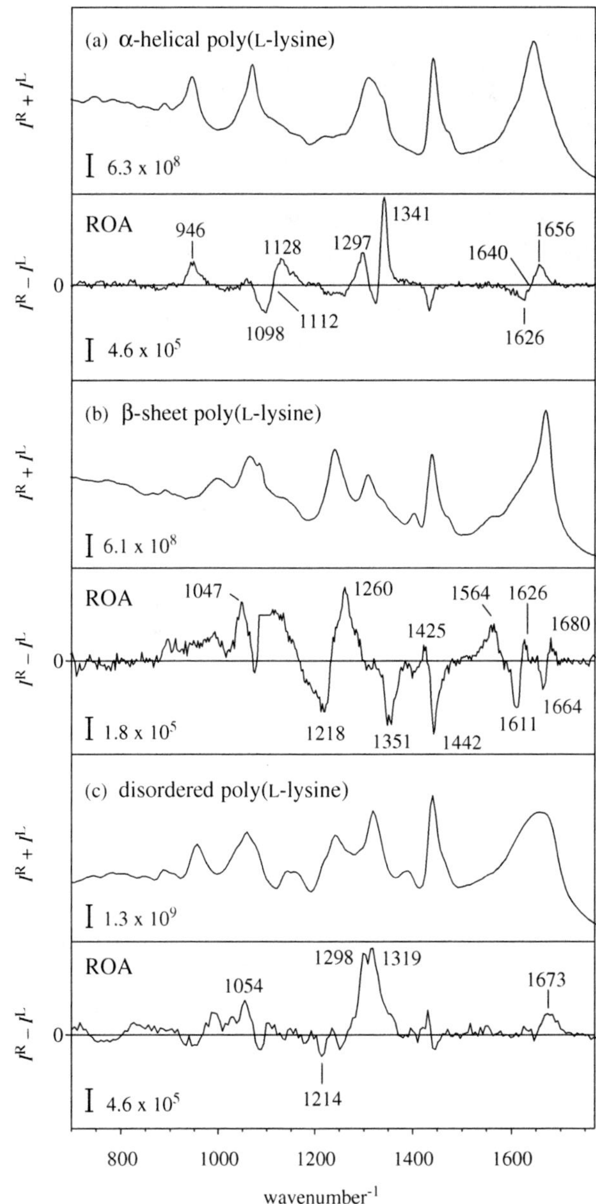

Fig. 16.6. Backscattered ICP Raman ($I^R + I^L$) and ROA ($I^R - I^L$) spectra of (a) α-helical, (b) β-sheet, and (c) disordered poly(L-lysine).

intensities being a function of the perturbations (which can be geometric, due to various types of hydration or both) to which the helical sequences are subjected.

The relative intensities of these two ROA bands appear to correlate with the exposure of the polypeptide backbone to the solvent within the elements of α-helix. For example, the positive ~1340 cm^{-1} α-helix ROA band assigned to hydrated α-helix completely disappears when the polypeptide or protein is dissolved in D_2O [78]. This indicates that N–H deformations of the peptide backbone make a significant contribution to the generation of the ~1340 cm^{-1} ROA band because the corresponding N–D deformations contribute to normal modes in a spectral region several hundred wavenumbers lower. Furthermore the corresponding sequences in proteins are exposed to solvent, rather than being buried in hydrophobic regions where amide protons can take weeks or longer to exchange. In comparison, although the positive ~1300 cm^{-1} α-helix ROA band also changes in D_2O, again suggesting some involvement of N–H deformations, in systems containing a significant amount of α-helix in a protected hydrophobic environment quite a lot of intensity is still retained. Further insight into the nature of these hydrated and hydrophobic variants of α-helix is provided by electron spin resonance studies of double spin-labelled alanine rich peptides [109,110] which identified a new and more open conformation of the α-helix. Computer modelling studies indicated that this more open geometry retains the internal hydrogen bonding scheme but changes the C=O···N angles which results in a splaying of the backbone amide carbonyls away from the helix axis and into the surrounding solvent. This more open conformation may, therefore, be the preferred conformation in aqueous solution as it would allow external hydrogen bonding to solvent water molecules. Consequently, an equilibrium would exist between the canonical form of α-helix, which may be responsible for the lower wavelength band ~1300 cm^{-1}, and this more open form, which may be responsible for the band ~1340 cm^{-1}. The former would be favored in a hydrophobic environment and the latter would be favored in a hydrophilic, or other hydrogen bonding environment [111].

The small amide I couplet, negative at ~1626 cm^{-1} and positive at ~1656 cm^{-1} is at a position characteristic of α-helix, though this couplet can be considerably more intense in α-helical proteins. This signature accords with the wavenumber range ~1645–1655 cm^{-1} for α-helix bands in conventional Raman spectra [96]. Positive ROA intensity in the range ~870–950 cm^{-1} also appears to be a signature of α-helix. The exact position and bandshapes of these ROA signatures vary in different α-helical polypeptides and proteins, possibly indicating heterogeneity within the α-helical sequences or the influence of side chains on the normal vibrational modes responsible. It is well known that the vibrations of nonaromatic side chains can couple with the backbone stretch vibrations in the backbone skeletal stretch region [99].

16.9.2.2 β-Sheet poly(L-lysine)

The β-sheet conformation is built up from two or more different adjacent strands of the polypeptide and is principally stabilized by cross-strand C=O to N–H hydrogen bonds. The constituent β-strands may be from either the same polypeptide chain or from different polypeptide chains. Distinct hydrogen-bonding patterns are associated with parallel or anti-parallel alignments of the individual strands and these may also be separated by various loops and turns. Consequently, there is far more structural diversity found within β-sheet structures than α-helical structures, including a range of curved and twisted surfaces depending on the composition and environment of the β-sheet. There are band assignments for β-sheet structures in the literature for most spectroscopic techniques, including infrared absorption (IR) [112], conventional transparent non-resonant [113] and ultraviolet [114] resonant Raman scattering, UVCD [115] and VCD [116], although these techniques have difficulty in discriminating between the different types of β-sheet structure, partly because of overlap of the characteristic β-sheet bands with those from other secondary structures in protein spectra. However, recent studies indicate that IR and UVCD can discriminate between at least some types of β-sheet in model polypeptides [117,118].

Until recently, the characterization of β-sheet structures in model polypeptides using ROA was complicated by the linear birefringence manifested by the gel-like samples formed by polypeptides at the high concentrations required for these experiments. Improvements in instrumentation and the careful control of experimental conditions has now overcome these difficulties and the Raman and ROA spectra of β-sheet poly(L-lysine) are shown in Fig. 16.6(b). The extended amide III region is dominated by a couplet, negative at ~1218 cm^{-1} and positive at ~1260 cm^{-1}. Negative ROA bands in the region ~1218–1247 cm^{-1} have been reported for β-structure [78] and these correlate with the identification of β-sheet amide III bands in the region ~1230–1245 cm^{-1} from conventional Raman spectroscopy [97,119]. This couplet is similar to those observed in proteins containing β-sheet, as will be discussed in Section 16.9.3, but is shifted by ~30 cm^{-1} to lower wavenumber. Recent deuteration studies [107] confirm that a significant contribution is made by N–H deformations to the vibrational mode responsible for this couplet. The large negative band at ~1351 cm^{-1} is assigned to β-turn structures [107].

A negative-positive-negative-positive band pattern is observed in the wavenumber region ~1600–1690 cm^{-1} which is very different to the band patterns typically observed in this region for β-sheet proteins. These take the form of a couplet, negative at low wavenumber and positive at high wavenumber, shifted by ~10 cm^{-1} to higher wavenumber than the amide I couplet that distinguishes α-helix, as shown in Fig. 16.6(b). The suppressed intensity of the amide I ROA couplet centred at ~1672 cm^{-1} for poly(L-lysine) compared to those typically observed for β-sheets in proteins, which occur roughly at the same position, parallels the weak amide I VCD observed for such model β-sheet structures compared to the more intense signals observed by VCD in β-sheet proteins [117]. It is thought that the 'planar' nature of the component strands within the flat multistranded β-sheet supported by poly(L-lysine) under these conditions gives rise to a smaller intrinsic skeletal chirality than that occuring in the twisted strands characteristic of the more irregular β-sheet structures found in proteins [117]. This would result in the weaker ROA and VCD band intensities observed. The more intense ROA couplet, negative at ~1611 cm^{-1} and positive at ~1626 cm^{-1}, does not match any clear parent Raman bands in the corresponding Raman spectrum, unlike the other ROA bands already described. Intense IR bands in the region ~1610–1620 cm^{-1} have been reported for β-sheet polypeptides and some denatured proteins [118], however, this would suggest that corresponding modes cannot be observed in conventional Raman spectroscopy but may be detected in ROA.

Another feature in the β-sheet poly(L-lysine) ROA spectrum that is not usually found for β-sheet proteins is the positive band ~1564 cm^{-1}. This is assigned to an amide II vibration, and these occur in the range ~1510–1570 cm^{-1} and originate in the out-of-phase combination of the N–H in-plane deformation with the C–N stretch.

Amide II vibrations are usually very weak in the conventional Raman spectra of proteins [113] but can be more intense in ultraviolet resonance Raman (UVRR) spectra [114]. Weak bands with positive ROA intensity have also been observed in this region for two β-sheet proteins, namely concanavalin A and the coat protein of the MS2 bacteriophage [105,120]. The positive ROA band ~1047 cm^{-1} is also assigned to β-structure and similar bands are observed in proteins containing β-sheet.

16.9.2.3 Disordered poly(L-lysine)

The details of the ROA spectrum of disordered poly(L-lysine), shown in Fig. 16.6c, are clearly different from those of the α-helical and β-sheet forms. The weak positive band at ~1673 cm^{-1} is typical of the bands observed in the amide I region for disordered polypeptides and proteins [105,106,121–123]. In the extended amide III region, the ROA spectrum is dominated by the positive band at ~1319 cm^{-1}. Largely on the basis of UVCD and VCD evidence, disordered poly(L-lysine) is thought to contain substantial amounts of the polyproline II (PPII) helical conformation, possibly in the form of short segments interspersed with residues having other conformations [103,116,124,125]. Therefore, we consider these bands, especially the positive band at ~1319 cm^{-1}, as being characteristic of PPII structure. This assignment has been fully established by a recent study on short alanine peptides [126]. PPII appears to be a particularly important structural motif in many unfolded polypeptides and proteins and to perform a special role in the function of a wide range of proteins [127–133]. To date there have been no reliable *ab initio* computations of the ROA spectrum of PPII helix, so our assignment of strong positive ROA at ~1319 cm^{-1} to PPII structure relies mainly on the evidence outlined above.

16.9.3 ROA spectra of proteins

A number of studies on model structures [107,111,126,134], similar to those on poly(L-lysine) outlined above, have provided a basis for more detailed investigations on proteins. We present here a brief summary of the current situation and describe in greater detail several representative protein ROA spectra as an illustration of the great potential of this technique for probing protein structure and dynamics.

Fig. 16.7(a) shows the backscattered Raman and ROA spectra of the α-helical protein human serum albumin (HSA) [78]. The main features of the ROA spectrum are in good agreement with a SCOP analysis of the Protein Data Bank (PDB) X-ray crystal structure 1ao6, which reports 69.2% α-helix, 1.7% 3_{10}-helix and the remainder being made up of loops and turns. The strong sharp positive band at ~1340 cm^{-1} and the weaker positive band at ~1300 cm^{-1} are assigned to α-helix in hydrated and hydrophobic environments, respectively, as described in Section 16.9.2.1. The amide I ROA couplet, negative at ~1640 cm^{-1} and positive at ~1665 cm^{-1}, and the positive ROA intensity in the range ~870–950 cm^{-1} are also characteristic signatures of α-helix [78,105,106].

Fig. 16.7(b) depicts the Raman and ROA spectra of the β-sheet protein jack bean concanavalin A [105] which contains, according to the PDB X-ray crystal structure 2cna, 43.5% β-strand, 1.7% α-helix and 1.3% 3_{10}-helix in a jelly roll β-barrel with the

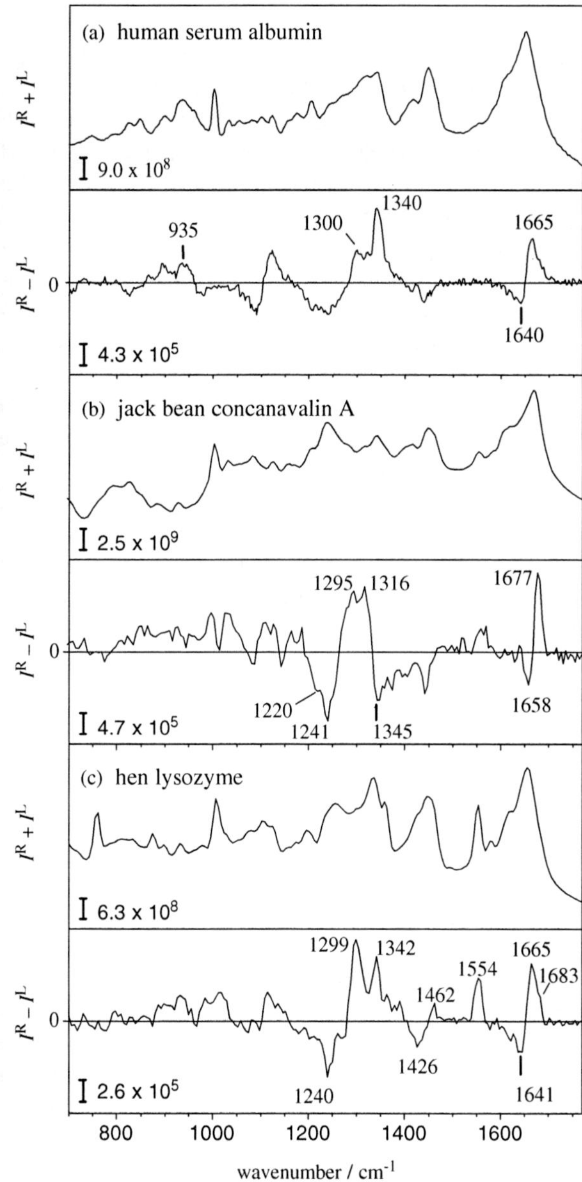

Fig. 16.7. Backscattered ICP Raman ($I^R + I^L$) and ROA ($I^R - I^L$) spectra of (a) human serum albumin, (b) jack bean concanavalin A, and (c) hen lysozyme at ~pH 5.4 and 20°C.

remainder being hairpin bends and long loops. The sharp negative band at ~1241 cm^{-1} originates in the β-structure and as the true signature of β-sheet in the amide III region appears to be a couplet, negative at low-wavenumber and positive at high-wavenumber, we also assign the positive band ~1295 cm^{-1} in the spectrum as being the high-wavenumber signal of the β-sheet couplet for this protein. The negative peak of the couplet appears to be constrained either to the region ~1220 cm^{-1} or the region ~1240 cm^{-1} in proteins, while the positive peak of this couplet appears to be more variable and appears in the range ~1260–1295 cm^{-1}. Side chain interactions may influence the position of this positive band, and bands from loops and turns may also contribute to this region. The amide I couplet, negative at ~1658 cm^{-1} and positive at ~1677 cm^{-1}, and bands in the backbone skeletal stretch region are also consistent with β-sheet structure [78,105,106,126]. Negative ROA bands in the range ~1340–1380 cm^{-1} originate in β-hairpin bends; an example appears at ~1345 cm^{-1} in the spectrum of concanavalin A. This signature allows ROA to differentiate between parallel and antiparallel types of β-sheet as only the latter usually contain hairpin bends while the ends of strands in the former are usually connected by α-helical sequences. Many β-sheet proteins also show a strong positive ROA band at ~1314–1325 cm^{-1} assigned to the PPII helix conformation. These correspond to the PPII helical elements known to occur in some of the longer loops between elements of secondary structure [135,136]. An example of such a PPII signal is observed at ~1316 cm^{-1} in the ROA spectrum of concanavalin A.

Fig. 16.7(c) shows the Raman and ROA spectra of the α+β protein hen lysozyme [78,105], which has a very different fold to those of HSA and concanavalin A. This is reflected in the large differences between their ROA spectra. Hen lysozyme contains 28.7% α-helix, 10.9% 3_{10}-helix and 6.2% β-sheet according to the PDB X-ray crystal structure 1lse, which is consistent with the presence of the positive ROA bands assigned to hydrophobic and hydrated α-helix observed at ~1299 and 1342 cm^{-1}, respectively, as well as the sharp negative band ~1240 cm^{-1} being indicative of β-structure. Proteins with a lysozyme-type fold often contain an unusually high 3_{10}-helix component. It is possible that the positive band ~1299 cm^{-1} contains a contribution from a 3_{10}-helix band, with a positive shoulder occurring at ~1295 cm^{-1}. There may also be bands contributing in this region from turns. A definitive analysis of all ROA bands in the extended amide III region has not been possible to date due to the difficulty of deconvoluting the complex, bisignate spectra.

The amide I couplet, negative at ~1641 and positive at 1665 cm^{-1} with a small shoulder at ~1683 cm^{-1}, also indicates the presence of both α-helix and a lesser amount of β-sheet. Since ROA spectra contain bands characteristic of loops and turns as well as bands indicative of secondary structure, they provide information about the overall three-dimensional (3D) solution structure of a protein [78,105]. Therefore, it is possible to deduce motif types of supersecondary elements and even tertiary fold information from ROA spectra. Protein fold recognition and structure elucidation promise to be particularly exciting areas of future research for chiroptical spectroscopic techniques in general, and ROA in particular.

To date, all ROA spectra of proteins reported in the literature have been measured at Glasgow University but at the time of writing several other research groups were beginning to perform similar studies. As an example, Fig. 16.8 presents the Raman and ROA spectra of the F10H mutant of the Arc repressor protein, measured by Professor V. Baumruk and Mr. J. Kapitan at Charles University, Prague. Arc is a small homodimeric repressor of the ribbon-helix-helix family of transcription factors [137]. No atomic resolution structure has been reported for this mutant but the details of the ROA spectrum of Arc are clearly very similar to those for the spectra of

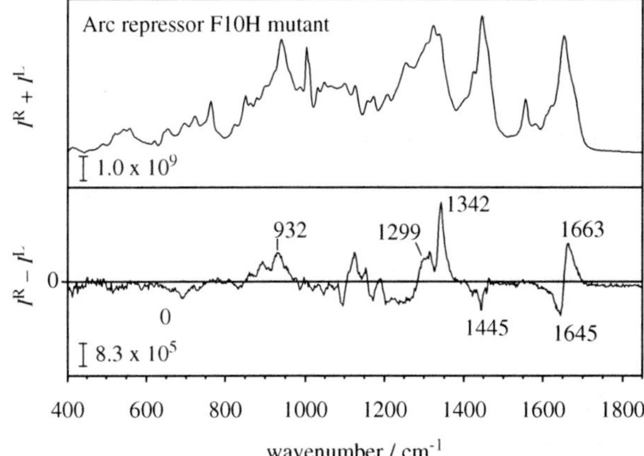

Fig. 16.8. Backscattered ICP Raman (I^R+I^L) and ROA (I^R-I^L) spectra of the F10H mutant of Arc repressor protein at ~pH 7.5 and 20°C.

α-helical polylysine, presented in Fig. 16.6(a), and the α-helical protein HSA, shown in Fig. 16.7(a), suggesting a large amount of α-helix with little β-sheet and no hairpin bends.

16.9.3.1 Side chain ROA bands

Bands from side chains are usually not as prominent in protein ROA spectra as they are in conventional protein Raman spectra. Differences between residue compositions, with attendant differences in side chains, may be responsible for some of the small variations observed in the characteristic ROA bands observed for α-helix, β-sheet and other structural motifs. There are, however, two distinct regions where side chain vibrations give rise to clear ROA signatures.

There are often weak ROA bands in the range ~1400–1480 cm^{-1}, associated with strong parent Raman bands [78,105,106], which originate mainly in CH_2 and CH_3 side chain deformations and examples of these may be observed in the ROA spectra of all three proteins displayed in Fig. 16.7(a–c). On the low-wavenumber side of this couplet there is often another relatively weak couplet, positive at low wavenumber and negative at high wavenumber, which originates in tryptophan vibrations. ROA intensities in this region are usually greatly enhanced in D_2O solution due to mixing with amide II vibrations [78]. These enhanced signals in D_2O could have significant diagnostic potential as they reflect both backbone and side chain conformations.

The region from ~1530–1580 cm^{-1} may contain large ROA bands originating in tryptophan vibrations. An example can be observed in the ROA spectra of hen lysozyme, in Fig. 16.7(c), which is assigned to the W3-type vibrational mode of the indole ring of tryptophan residues. Miura et al. [138] found that the magnitude of the torsion angle $\chi^{2,1}$ of the tryptophan side chain, which describes the orientation of the indole ring with respect to the local peptide backbone, can be deduced from the position of the corresponding conventional Raman band. Similarly, the magnitude of the torsion angle can be determined from the position of the W3-type vibrational mode in the

ROA spectrum but with the added advantage of a possible determination of its sign from the measured sign of the ROA band and hence the deduction of the absolute stereochemistry of the tryptophan side chain [139]. This information is not normally available other than from a structure determined at atomic resolution. Hen lysozyme contains six tryptophan residues, four with positive $\chi^{2,1}$ values and two with negative $\chi^{2,1}$ values according to the PDB structure 1lse. Partial cancellation of the resulting ROA signals yields the positive band observed in Fig. 16.7(c) at ~1554 cm^{-1} [140]. The W3 ROA band may consequently be also used as a probe of conformational heterogeneity among tryptophan residues in disordered protein sequences as the cancellation of signals with opposte signs results in a loss of ROA intensity. This is similar to the disappearance of near UVCD bands from aromatic residues upon the loss of tertiary structure. Therefore, these two techniques provide complementary views of order/disorder transitions as ROA probes the intrinsic skeletal chirality of the tryptophan side chain while UVCD probes the chirality in the immediate vicinity of the aromatic chromophore.

16.9.3.2 Studies on unfolded proteins

As yet it has not been possible to obtain useful ROA spectra of fully unfolded denatured states of proteins which have well-defined tertiary folds in the native state. This is because the intense Raman bands from chemical denaturants typically used at high concentration preclude ROA measurements, while thermally unfolded proteins often give rise to intense Rayleigh light scattering due to aggregation. There have been, however, several studies on partially unfolded denatured protein states associated with molten globules and reduced proteins [78,122,140,141], and also on proteins which are unfolded in their native biologically active states [121–123,134].

Molten globules are partially unfolded denatured protein states, stable at equilibrium, with well-defined secondary structure but lacking the specific tertiary interactions characteristic of the native state [142,143]. A widely studied molten globule is that supported by bovine α-lactalbumin at low pH and is called the A-state. Native bovine α-lactalbumin has the same fold and very similar X-ray crystal structure to that of hen lysozyme and this is apparent from its ROA spectrum [140] presented in Fig. 16.9(a), which is similar to that of hen lysozyme presented in Fig. 16.7(c). However, differences in detail between the spectra are clear, highlighting the sensitivity of ROA to small differences in structure between proteins with the same fold. The Raman and ROA spectra of A-state bovine α-lactalbumin are displayed in Fig. 16.9(b). The sensitivity of ROA to the complexity of order in molten globule states is indicated by the large differences observed here between the ROA spectra of the native and A-states of bovine α-lactalbumin as opposed to the small differences between the corresponding parent Raman spectra. Much of the ROA band structure in the extended amide III region of the A-state spectrum, such as the ~1340 cm^{-1} band assigned to hydrated α-helix, has disappeared and is replaced by a large couplet consisting of a single negative band observed at ~1236 cm^{-1}, probably originating in β-structure, and two distinct positive bands at ~1297 and 1312 cm^{-1}. In the amide I region the positive signal of the couplet has shifted by ~10 cm^{-1} to higher wavenumber compared to that for the native state. The signal originating from the W3 vibrational mode of tryptophan residues at ~1551 cm^{-1} has almost completely disappeared in the ROA spectrum of the A-state and been replaced by

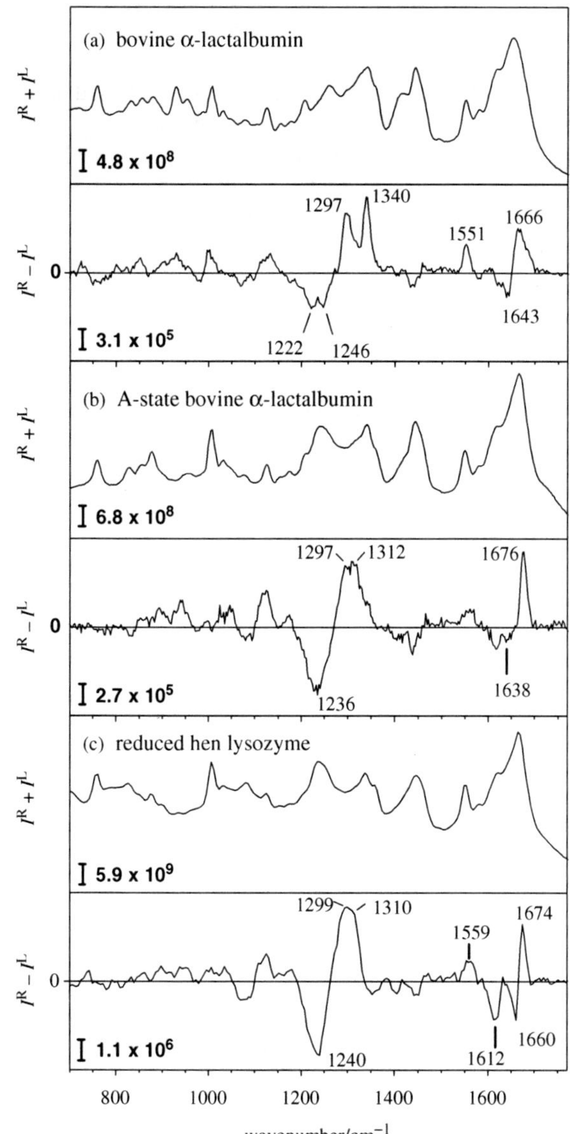

Fig. 16.9. Backscattered ICP Raman (I^R+I^L) and ROA (I^R-I^L) spectra of (a) native bovine α-lactalbumin at ~pH 4.6 and 20°C, (b) A-state bovine α-lactalbumin at ~pH 1.9 and 2°C, and (c) reduced hen lysozyme at ~pH 2.4 and 20°C.

a weak positive band at ~1560 cm^{-1}, indicating conformational heterogeneity amongst these side chains associated with a loss of the characteristic tertiary interactions found in the native state.

The Raman and ROA spectra of reduced hen lysozyme, in which the disulphide bonds are reduced, are shown in Fig. 16.9(c) [122]. There are significant changes in the ROA spectrum of reduced lysozyme compared with that of the native state shown in Fig. 16.7(c) indicating the loss of much of the native structure. Disappearance of the positive band at ~1340 cm^{-1} implies that none of the hydrated α-helix found in the native state persists in the reduced form. The positive signal of the amide I couplet has also shifted by ~9 cm^{-1} to higher wavenumber and becomes quite sharp indicating the presence of a significant amount of β-structure, perhaps due to aggregation. The couplet from ~1426 to 1462 cm^{-1} in the spectrum of the native state originating in aliphatic side chains and the band at ~1554 cm^{-1} assigned to tryptophan side chains are all greatly diminished, presumably due to conformational heterogeneity. Although of similar appearance, the ROA spectra of different denatured proteins display differences in detail and these may reflect the different residue compositions and their different Φ,ψ propensities.

Human lysozyme has a similar stability to hen lysozyme with regard to temperature and low pH but its thermal denaturation behavior is subtly different. Below ~pH 3.0 the thermal denaturation of human lysozyme is not a two-state process. Unlike hen lysozyme, it supports a partially folded molten globule-like state at elevated temperatures [144]. Incubation at 57°C and pH 2.0, under which conditions the partially folded state is the most highly populated, induces the formation of amyloid fibrils. Incubation at 70°C and pH 2.0, under which conditions the fully denatured state is the most highly populated, leads to the formation of amorphous aggregates [145].

The Raman and ROA spectra of the native and partially folded prefibrillar-amyloidogenic intermediate states of human lysozyme are shown in Figs. 16.10(a) and (b), respectively [140]. The structure and ROA spectrum of the native state of human lysozyme are similar to those of hen lysozyme, shown in Fig. 16.7(c). However, large changes are apparent in the ROA spectrum of the prefibrillar intermediate. Principally, loss of the positive ~1345 cm^{-1} band assigned to hydrated α-helix and the appearance of a new positive band at ~1325 cm^{-1} assigned to PPII helix suggests that hydrated α-helix has undergone a conformational change to PPII structure. The disappearance of the positive ~1550 cm^{-1} tryptophan band again indicates that major conformational changes have occurred among the five tryptophan residues, four of which lie within the α-helical domain. Thus, the ROA spectra suggest that the α-domain destabilizes and undergoes a conformational change in the prefibrillar intermediate and that PPII helix may be a critical conformational element involved in amyloid fibril formation [140]. There is no sign of an increase in β-sheet content in the intermediate from the ROA spectra. Hen lysozyme, which has a much lower propensity to form amyloid fibrils than the human variant [146], is virtually native-like under the same conditions of low pH and elevated temperature, which was verified by the corresponding ROA spectra [140].

ROA studies have also been performed on natively unfolded proteins including the bovine milk caseins [121], several wheat prolamins [123] and the human recombinant synuclein and tau brain proteins, some of which are associated with neurodegenerative disease [121]. In each case, the ROA spectra are all dominated by a strong positive band at ~1316–1322 cm^{-1} assigned to PPII helix. Absence of a well-defined amide I ROA couplet in these spectra indicates lack of secondary structure. Although their ROA spectra are similar, variations of detail reflect differences in residue composition and minor differences in structural elements. Our studies on the caseins, synucleins and tau

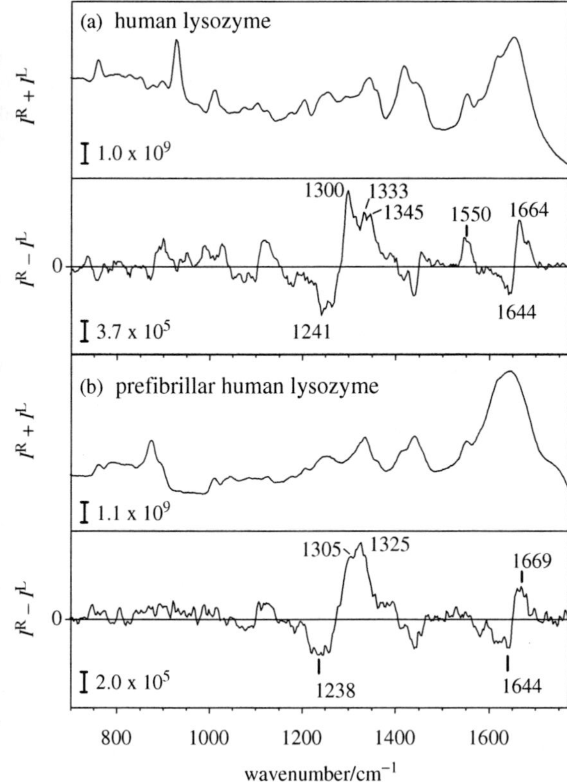

Fig. 16.10. Backscattered ICP Raman ($I^R + I^L$) and ROA ($I^R - I^L$) spectra of (a) native human lysozyme at ~pH 5.4 and 20°C, and (b) the prefibrillar intermediate of human lysozyme at ~pH 2.0 and 57°C.

suggest that these proteins may be more realistically classified as being 'rheomorphic,' meaning flowing shape [147], rather than 'random coil.' The rheomorphic state is distinct from the molten globule state which is a more compact entity containing a hydrophobic core and a significant amount of secondary structure [142,143].

The conformational plasticity supported by the mobile regions within native proteins, partially denatured protein states, and natively unfolded proteins, underlies many of the conformational (protein misfolding) diseases [148,149], many of which involve amyloid fibril formation. As PPII helical structure is extended, flexible and hydrated, it imparts a plastic open character to the structure of a protein and may be implicated in the formation of regular fibrils in the amyloid diseases [121,140,150,151]. PPII helix is not readily amenable to traditional methods of structure determination due to its inherent flexibility and the lack of intrachain hydrogen bonding, which has hindered its recognition in globular proteins until recently [127–133]. The PPII conformation is indistinguishable from an irregular backbone structure for free peptides in solution by ^{1}H NMR spectroscopy [152]. PPII helix may be recognized in polypeptides by UVCD [103,124, 152] and VCD [116,128,153] and in proteins from the deconvolution of UVCD spectra [154]. In contrast, ROA provides a clear and characteristic signature of PPII helix in both proteins and polypeptides. However, of the milk and brain proteins studied by ROA displaying a high PPII helix content, only α-synuclein and tau are fibrillogenic and associated with disease. Other properties of the constituent residues, such as mean hydrophobicity and charge [121], probably also play an important role in fibril formation.

16.9.4 Studies on nucleic acids

The study of the structure and function of nucleic acids remains a central problem in molecular biology. Although far fewer ROA studies have been performed on nucleic acids than proteins, the results obtained indicate that ROA appears to be sensitive to three different sources of chirality in nucleic acids: the chiral base-stacking arrangement of intrinsically achiral base rings, the chiral disposition of the base and sugar rings with respect to the C–N glycosidic link, and the inherent chirality associated with the asymmetric centres of the sugar rings. Early studies on pyrimidine nucleosides [155] and polyribonucleotides [156–158] provide a basis for the interpretation of ROA spectra of DNA and RNA.

We present in Figs. 16.11(a–c), respectively, the Raman and ROA spectra of calf thymus DNA and of phenylalanine-specific transfer RNA ($tRNA^{Phe}$) in the presence and absence of Mg^{2+} ions as examples for RNA and DNA molecules [159]. ROA bands in the region ~900–1150 cm^{-1} originate in vibrations of the sugar rings and phosphate backbone. The region ~1200–1550 cm^{-1} is dominated by normal modes in which the vibrational coordinates of the base and sugar rings are mixed. ROA band patterns in this sugar-base region appear to reflect the mutual orientation of the two rings and possibly the sugar ring conformation. The region ~1550–1750 cm^{-1} contains ROA bands characteristic of the types of bases involved and the particular stacking arrangements. Although the ROA spectra of the DNA and the two RNA molecules are similar there are numerous differences in their details. The main differences originate in the DNA taking up a B-type double helix in which the sugar puckers are mainly C2′-*endo* and the RNAs taking up A-type double helical segments where the sugar puckers are mainly C3′-*endo*. Smaller differences arise between the two RNA spectra and these are most apparent in the sugar-phosphate region ~900–1150 cm^{-1}. It is well known that Mg^{2+} ions are necessary to hold RNAs in their specific tertiary folds. For $tRNA^{Phe}$ in the presence of Mg^{2+} this is a compact L-shaped form; whereas in the absence of Mg^{2+} the $tRNA^{Phe}$ adopts an open cloverleaf structure [160]. The ROA spectrum of the Mg^{2+}-free $tRNA^{Phe}$ shows a strong negative–positive–negative triplet at ~992 cm^{-1}, ~1048 cm^{-1} and ~1091 cm^{-1} which is very similar to that found in A-type polyribonucleotides and has been assigned to the C3′-*endo* sugar pucker. This signature is weaker and more complex in the ROA spectrum of the Mg^{2+}-bound sample, suggesting a wider range of sugar puckers. Switching the sugar pucker conformations from C3′-*endo* to C2′-*endo* would elongate the sugar-phosphate backbone and may assist the formation of the loops and turns that characterize the tertiary structure of the folded form.

Far fewer ROA studies have been reported for RNAs than for proteins, which currently limits analysis of these spectra. In contrast to many proteins, however, the

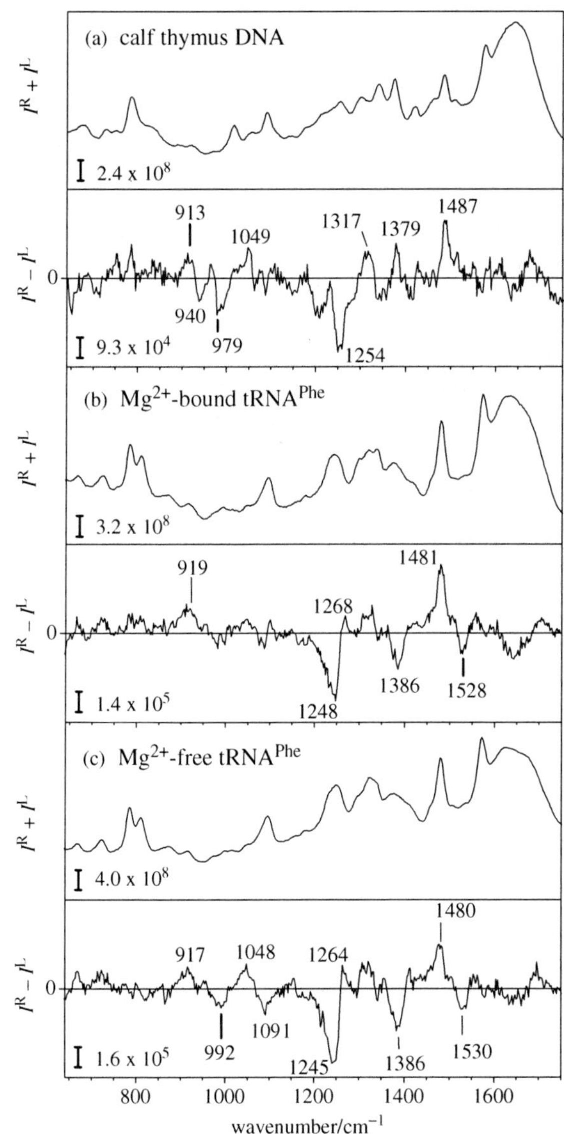

Fig. 16.11. Backscattered ICP Raman ($I^R + I^L$) and ROA ($I^R - I^L$) spectra of (a) calf thymus DNA, (b) Mg^{2+}-bound $tRNA^{Phe}$, and (c) Mg^{2+}-free $tRNA^{Phe}$ at ~pH 6.8 and 20°C.

secondary structure elements of RNA are usually fully stable in the absence of tertiary structures; often complex RNA tertiary structures may thus be created in a hierarchical manner from distinct individual secondary structure motifs. Characterizing these secondary structure motifs and the tertiary structures they create is crucial to understanding RNA structure–function relationships and the principles that govern protein-RNA recognition, which are central to cellular and viral processes ranging from gene expression to cell death. There is considerable scope for the application of ROA spectroscopy in this field.

16.9.5 Studies on viruses

As for proteins and nucleic acids, knowledge of the structures of viruses at the molecular level is essential for understanding their function and hence for the success of enterprises such as structure-guided antiviral drug design [161]. Viruses are also useful models for studying macromolecular assembly processes because viral coat proteins are able to form stable quarternary structures. These structures, called capsids, are made up of a large number of copies of one, or a few, coat proteins. The capsid serves to protect the enclosed nucleic acid (either DNA or RNA). The small coding capacity of viral genomes necessitates the use of multiple copies of the coat proteins, leading to the highly symmetric structures that characterize viruses. However, the widespread application of key techniques in structural biology to intact viruses is often hampered by practical difficulties. High-resolution methods are not applicable to the majority of viruses. X-ray crystal structures have been deposited in the PDB for only a small fraction of the many thousands of viruses as most viruses are very difficult to crystallize. Although the imaging of cryoelectron microscopy data has progressed rapidly, still many problems exist in data collection and sample preparation which need to be overcome and electron microscopy imaging often only probes the exterior surface. In any case, it is preferable to study viruses in aqueous solution, which far more closely resemble the physiological environment than crystalline or vitrified samples do, if at all possible. However, both intact viruses and the precursors of capsid assembly are far too large for determination of their solution structures by high-field NMR spectroscopy. Furthermore, even when a structure at atomic resolution is available for a virus, it rarely contains any information about the encapsidated nucleic acid, as it is often too disordered (and also because of averaging problems related to virus orientation and data processing) to provide useful diffraction data. Although of immense value, X-ray structures at near atomic resolution are only available for several dozen of the thousands of different viruses currently recognized; to study the molecular details of larger and more complex viruses it is necessary to 'dissect' them into well-defined subunits or substructures [162]. Conventional Raman spectroscopy is already a valuable tool in structural virology owing to its ability to provide information about protein and nucleic acid constituents of intact viruses by means of their characteristic vibrational bands [163,164]. Progressive developments, outlined below, promote ROA as a new probe of the structures of viruses and their assembly intermediates.

The first application of ROA to viruses was in studies on filamentous bacteriophages. These form long flexible rods generally of 5–6 nm in width and from several hundred nm to a micron in length and consist of a loop of single-stranded DNA surrounded by the capsid. This capsid consists of several thousand identical copies of a major coat protein containing ~50 amino acids arranged in a helical array. There are also several copies of minor coat proteins. It is known from X-ray fibre diffraction studies that the major coat protein subunits adopt an extended α-helical fold and overlap each other so that the exterior half of each major coat protein is exposed to water while the interior half is 'protected' by neighboring coat protein subunits [165]. Raman and ROA spectra are presented in Fig. 16.12 for three different filamentous bacteriophages, Pf1 [166], fd [120] and PH75 [139]. The ROA spectra are dominated by bands assigned to α-helix in the amide I, extended amide III and backbone skeletal stretch regions and these have already been discussed in Section 16.9.2.1. Furthermore, these spectra provide a good

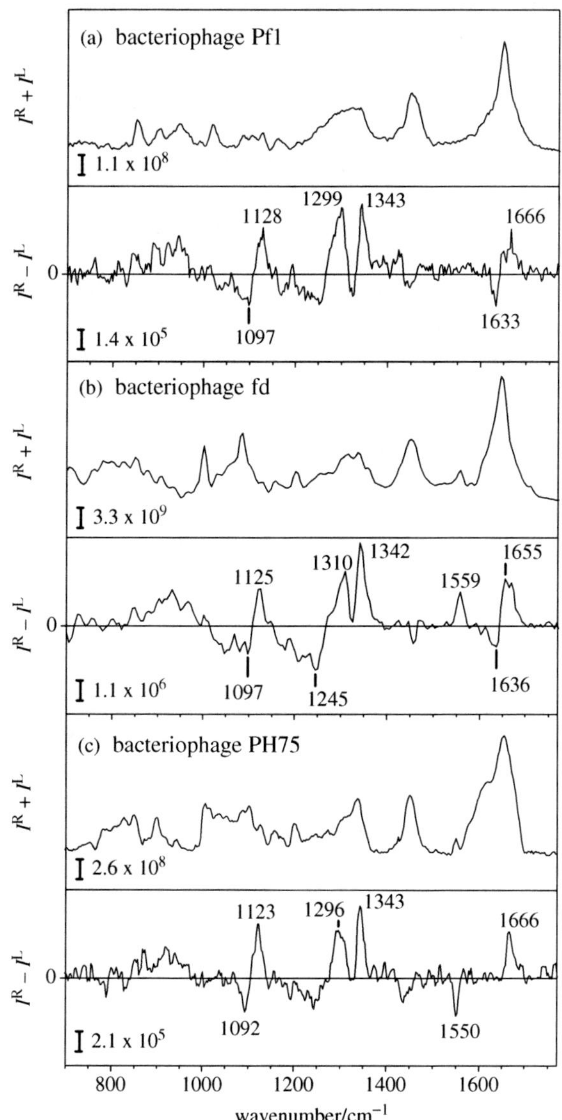

Fig. 16.12. Backscattered ICP Raman ($I^R + I^L$) and ROA ($I^R - I^L$) spectra of the filamentous bacteriophages (a) Pf1, (b) fd, and (c) PH75 at ~pH 7.6 and 20°C.

illustration of the sensitivity to hydration of those ROA α-helical bands in the extended amide III region. The relative intensities of the band at ~1298 ± 2 cm^{-1}, assigned to hydrophobic α-helix, and the band at ~1342 ± 1 cm^{-1}, assigned to hydrated α-helix, correlate well with the hydration of the polypeptide backbone in each major coat protein [166]. For example, the peak intensities of these two bands for Pf1 are virtually identical and the interior protected region of the Pf1 coat protein contains a continuous sequence of 20 residues while the exterior exposed region contains a sequence of 20 mixed hydrophobic and hydrophilic residues. There is a shorter sequence of pure hydrophobic residues in the interior protected region of the fd coat protein and a longer sequence of mixed hydrophobic and hydrophilic residues in the exterior exposed region, which leads to the hydrated α-helix band at ~1342 cm^{-1} in the extended amide III region of the ROA spectrum now being relatively stronger than the band at ~1310 cm^{-1} assigned to α-helix in a more hydrophobic environment. The shift in this latter band to higher wavenumber may also reflect sensitivity to hydration.

Another obvious difference between the ROA spectra of these three filamentous bacteriophages is in the behavior of the W3-type tryptophan vibration observed at ~1550 cm^{-1}. The major coat protein of Pf1 does not contain any tryptophan residues and the ROA spectrum shows consequently no signal at ~1550 cm^{-1}. There is a single tryptophan residue in the major coat protein of the fd bacteriophage, which generates the positive ROA band at ~1559 cm^{-1}, while PH75, which has unusual thermophilic properties, also contains a single tryptophan in its major coat protein, which leads to a negative ROA band at ~1550 cm^{-1}. As has already been mentioned in Section 16.9.3.1, the magnitude and sign of the $\chi^{2,1}$ torsion angle, which relates the orientation of the indole ring to the local peptide backbone, can be determined from ROA spectra due to the fundamental sensitivity of ROA band patterns to chirality [139]. From these spectra, it was determined that $\chi^{2,1}$ is ~+120° in the major coat protein of fd and is –93° in PH75, to within an uncertainty of ~10°. This angle is not usually available from the X-ray fiber diffraction data. Experimental determination of the sign as well as the magnitude of $\chi^{2,1}$ will aid its future application to studies of tryptophan environments in filamentous bacteriophages [96] and towards the development of molecular models [167].

Further studies conducted on the helical array structures of tobacco mosaic virus (TMV), narcissus mosaic virus (NMV) and potato virus X (PVX) [168], and the icosahedral satellite tobacco mosaic virus (STMV) and bacteriophage MS2 [105,120] demonstrated the ability of ROA to probe the folds of the major coat proteins of viruses independent of their morphology. These studies have indicated the potential of ROA to complement high resolution techniques in structural virology. More pertinently, ROA spectra of viruses can probe the structures of the nucleic acid components *in situ*. Determination of the structures of viral nucleic acids has proven difficult in even the best resolved X-ray crystal structures. Several of the ROA spectra of viruses already mentioned also display bands attributed to nucleic acid. However, little analysis was possible in these studies as these bands were generally not sufficiently intense, due to either the low nucleic acid content or its flexible nature. More recently, the nucleic acid content of the cowpea mosaic virus (CPMV) was investigated *in situ* [120]. CPMV is the type member of the comovirus group of plant viruses and has a genome consisting of two molecules of positive-sense RNA (RNA-1 and RNA-2). These are separately encapsidated in icosahedral particles whose structure is known to atomic resolution [169–171]. Virus preparations can be separated into four components, designated top (T), middle (M), bottom-upper (B_U) and bottom-lower (B_L) by centrifugation on density gradients. The four components have identical protein compositions and contain 60 copies each of a large (L) and small (S) coat protein made up of 374 and 213 amino acids, respectively. The T-component contains no RNA, the M-component contains a single molecule of RNA-2 made up of 3481 nucleotides (~24% of the particle mass), and the B_U-component contains a single molecule of RNA-1 made up of 5889 nucleotides (~34% of the particle mass). The X-ray crystal structure of CPMV has already revealed the folds of the L and S proteins to be of the jelly roll β-barrel (sandwich) type [169,170], but no nucleic acid was detectable [170].

Raman and ROA spectra of the T-, M- and B_U-components of CPMV are presented in Figs. 16.13(a–c), respectively. The overall appearance of the ROA spectrum of T-CPMV is very similar to those of proteins with the jelly roll fold such as jack bean concanavalin A, shown in Fig. 16.7(b), as expected. However, the ROA spectra of M- and B_U-CPMV contain a number of additional bands, some of which are quite intense. Subtraction of the ROA spectrum of T-CPMV from those of M-CPMV and B_U-CPMV reveals the spectra of RNA-2 and RNA-1, shown in Fig. 16.14(a) and (b), respectively. The ROA spectrum of Mg^{2+}-free $tRNA^{Phe}$, shown again in Fig. 16.14 (c), displays features, particularly the clear negative–positive–negative triplet from ~996–1094 cm^{-1}, assigned to sugar–phosphate vibrations associated with the C3′-endo sugar ring pucker as found in A-type helices, to those in the two difference spectra for RNA-2 and RNA-1. Similarly, the ROA couplet, negative at ~1669 and positive at ~1702 cm^{-1}, appearing in the two difference spectra may have a similar origin to the weaker couplet in the same region of the ROA spectrum of $tRNA^{Phe}$, which is assigned to ring vibrations of bases involved in the base stacking interactions found in A-type helices [159]. The ROA band pattern in the region ~1200–1500 cm^{-1} in the two difference spectra is also very similar to that in $tRNA^{Phe}$ in the same region: this band pattern was assigned to vibrations involving mixing of vibrational coordinates from both base and sugar rings which reflect the mutual orientation of the sugar and base rings found in A-type helices [159]. Therefore, the authors conclude that much of the viral RNA in both M- and B_U-CPMV adopts an A-type helical conformation. The absence of similar bands to those observed at ~1480 and 1530 cm^{-1} in $tRNA^{Phe}$ may reflect a single-strand helix since these bands possibly originate in vibrations of bases within the base-paired regions of $tRNA^{Phe}$. The close similarity of the two difference spectra revealed that the RNA adopts very similar conformations in the M- and B_U-components.

16.10 SURFACE-ENHANCED ROA

As ROA has a relatively small effect with dimensionless CIDs adopting numerical values of the order of 10^{-3}–10^{-5}, an obvious desire exists for increasing ROA band intensities through the use of potential enhancement mechanisms. Surface enhanced Raman scattering (SERS), which measures Raman scattering of molecules adsorbed onto

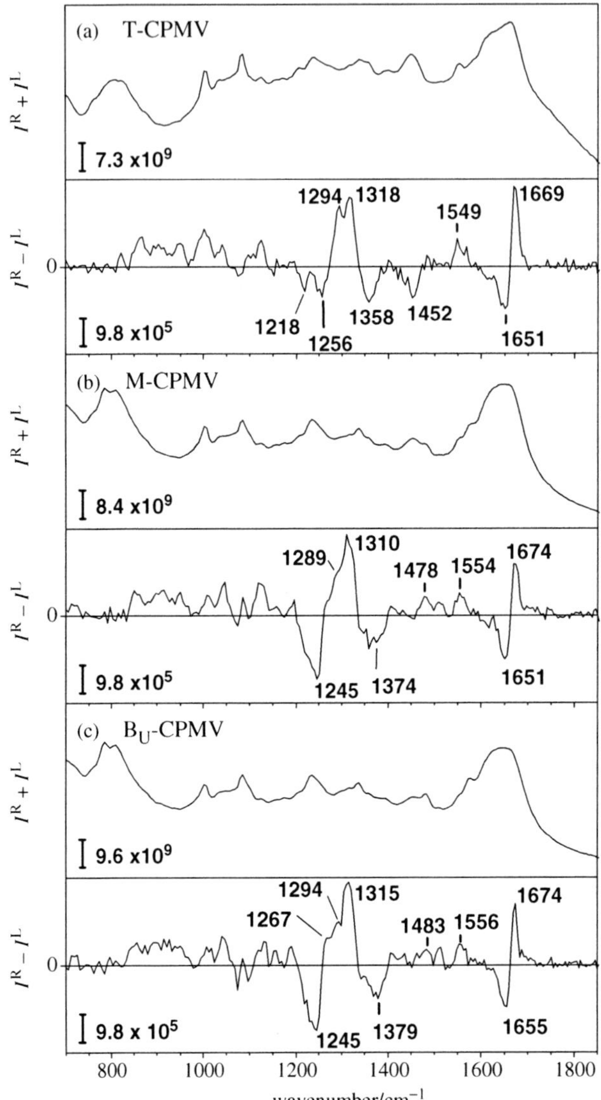

Fig. 16.13. Backscattered ICP Raman ($I^R + I^L$) and ROA ($I^R - I^L$) spectra of the CPMV isolates (a) T-CPMV, (b) M-CPMV, and (c) B_U-CPMV at ~pH 7.0 and 20°C.

specially prepared metal surfaces [172], that could provide such a sought after enhancement mechanism. The SERS technique is characterized by the selective enhancement of certain Raman signals over conventional scattering intensities with typical enhancements of 10^4–10^6 and maximum enhancement of 10^{14} being reported [173]. This high sensitivity, coupled with selectivity to particular surfaces, has established SERS as a widely used and powerful technique for trace molecule detection and structural and surface-interaction studies in recent years [174,175 and references cited therein].

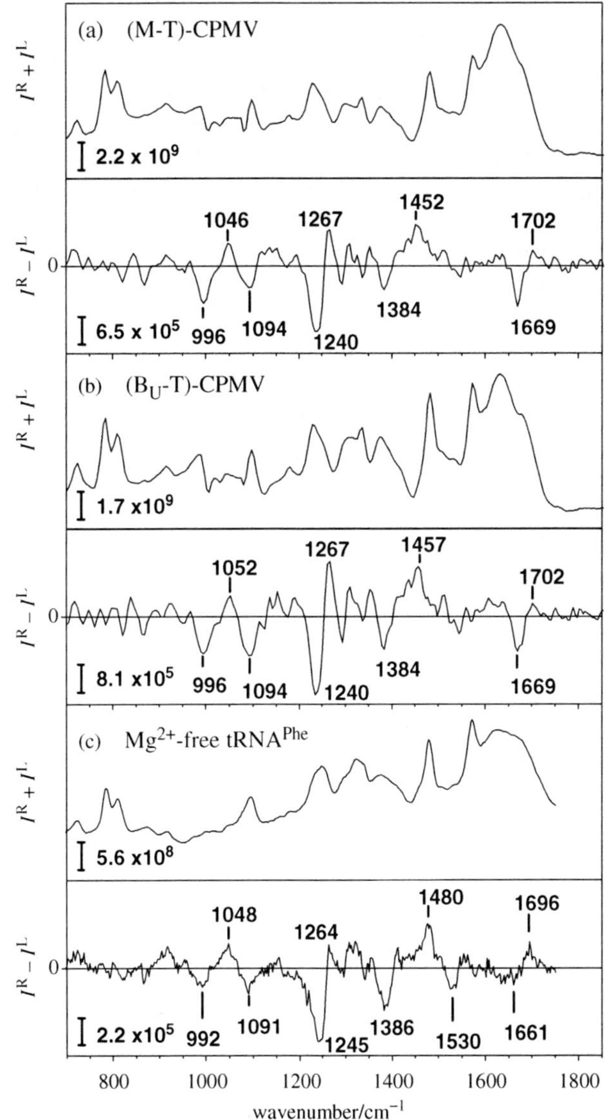

Fig. 16.14. Backscattered ICP Raman ($I^R + I^L$) and ROA ($I^R - I^L$) spectra of the differences (a) (M-CPMV–T-CPMV), and (b) (B_U-CPMV–T-CPMV), and (c) Mg^{2+}-free $tRNA^{Phe}$.

There are two principal mechanisms responsible for generating SERS [174]. The electromagnetic (EM) mechanism involves coupling between photons and electron oscillations at the metal surface, which are also known as surface plasmons. Topological features on a rough metal surface can produce a localized excitation in such surface plasmons, which normally do not couple with photons. If the dimensions of the metal surface features are smaller than the excitation wavelength, the modes of oscillation of surface plasmons can be resonant with both incident and scattered photons producing a large boost in the local electric field. This in turn greatly enhances the intensity of Raman scattering as it is proportional to the square of the electric field strength through the contribution of the induced dipole moment. The frequency of oscillations of surface plasmons is a function of the dielectric constant of the metal and in practice SERS is primarily observed for alkali and coinage metals at, or near, visible excitation wavelengths [176].

Charge transfer complexes between adsorbed molecules and the topographical features on a roughened metal surface give rise to the chemical, or charge transfer (CT), mechanism. Complex formation results in the energy levels of the frontier molecular orbitals of the adsorbed molecule being brought closer to the Fermi level of the metal, the highest occupied molecular level in the valence band at 0 K, and if the difference in energy is close to those of incident photons an enhancement similar to that of the resonance Raman process occurs [177]. Electron-hole pair formation in the vicinity of the topographical features can also lead to the related photon-driven charge transfer (PDCT) mechanism. The observed SERS intensities are proportional to the product of these individual contributions and may be written as [173,178]

$$I_{\mathrm{SERS}}(\omega_{\mathrm{S}}) = N' I(\omega_{\mathrm{L}}) |A(\omega_{\mathrm{L}})|^2 |A(\omega_{\mathrm{S}})|^2 \sigma_{\mathrm{ads}}^{\mathrm{R}}, \tag{18}$$

where N' is the number of molecules involved in the SERS process; $I(\omega_{\mathrm{L}})$ is the excitation laser intensity; $A(\omega_{\mathrm{L}})$ and $A(\omega_{\mathrm{S}})$ are the electromagnetic field enhancement factors at the laser and Stokes frequencies, ω_{L} and ω_{S}, respectively; and $\sigma_{\mathrm{ads}}^{\mathrm{R}}$ describes the chemical enhancement from the increased Raman cross section of the adsorbed molecule. More comprehensive accounts of the mechanisms underlying SERS are available elsewhere [179–182].

Surface enhancement may provide an attractive strategy for increasing the sensitivity of ROA in the form of a new experimental technique, surface enhanced ROA (SEROA). Two instances of this phenomenon have been reported [183,184] but the signal-to-noise level in each case was very low and the results were therefore not entirely convincing. No further development of SEROA had taken place until very recently. Despite difficulties in the reproducibility of SERS and SEROA spectra which relate to well-recognized problems in colloid particle preparation, the advances in instrumentation already described in Section 16.7 make SEROA an interesting possibility for the study of chiral analytes in the micromolar concentration range and at low excitation laser power. Ongoing work in our Glasgow and Manchester laboratories and by Dr. S. Abdali at the Technical University of Denmark is being conducted to develop this novel technique. The adsorption of complex molecules onto colloidal metal particles occurs *via* interaction of electron-rich moieties, such as the carbonyl and amino groups in the polypeptide backbone as well as charged or aromatic side chains, to the metal surface. Biological molecules such as proteins and peptides are likely to be favorable targets due to the large number of potential interaction sites with metal surfaces. Correlation of SEROA spectral features with those of both ROA and SERS spectra are anticipated to extend the application of ROA to samples at what were previously thought to be prohibitively low concentrations. Furthermore, once established, SEROA may also constitute a new chiroptical probe of interactions between biological molecules and metal surfaces.

16.11 APPLICATION OF CHEMOMETRICS TO ROA DATA

The work presented in previous sections clearly illustrates the high information content and sensitivity to stereochemistry of ROA spectra, and that ROA is ideally suited for studying the structure and behavior of biological molecules. More efficient and objective extraction of the information contained in the ROA spectra of biological molecules for future applications in proteomics and related fields will require automated analysis of the spectra using chemometrics and bioinformatics. Pattern recognition methods such as principal component analysis (PCA) are already proving particularly useful for analyzing protein ROA spectra [105,120,122].

PCA is probably the most widely used multivariate chemometric technique and for a set of experimental spectra it calculates a set of subspectra that serve as loadings or basis functions. The algebraic combination of these basis functions with appropriate expansion coefficients may be used to reconstruct any of the original spectra. Fig. 16.15 shows a plot of the expansion coefficients for a set of 75 polypeptide, protein and virus ROA spectra for the two most significant basis functions. The sample positions are color-coded with respect to structural content, with seven different structural classifications being used [105]. For the most significant basis function, large coefficients are associated with proteins and polypeptides containing large amounts of α-helix or β-sheet as proteins and polypeptides containing large amounts of these secondary structures dominate this data set. As the α-helix and β-sheet contents are inversely correlated (the larger the amount of α-helix, the smaller the amount of β-sheet and *vice versa*), the coefficients associated with α-helix and β-sheet have opposite signs. The coefficients of the second basis function reflect disordered structure content. Proteins with small coefficients, either positive or negative, for both basis functions have similar and significant amounts of α-helix and β-sheet. The spectra separate into clusters corresponding to the seven structural classifications, with increasing α-helix content to the left, increasing β-sheet content to the right and increasing disordered structure from bottom to top. Poly(L-lysine) in the α-helical, β-sheet and disordered states appears in the lower left, lower right, and upper centre of the plot, respectively, as expected. The positions of a number of proteins are also indicated to act as illustrative examples of the ability of this analytical approach to discriminate between protein structures.

Different chemometrics and bioinformatics tools display relative advantages and disadvantages and there is tremendous scope for the application of diverse analytical approaches to ROA spectra to investigate problems in structural biology. Two-dimensional (2D) correlation methods appear to be a particularly attractive approach. A number of correlation methods have been developed, the most popular being the generalized 2D-spectroscopic correlation technique [185], which is based on linear relationships between spectral data obtained under a perturbating influence on the system under investigation. This perturbation can involve any change in the physical or chemical environment of the sample. Generalized 2D correlation spectroscopy enhances

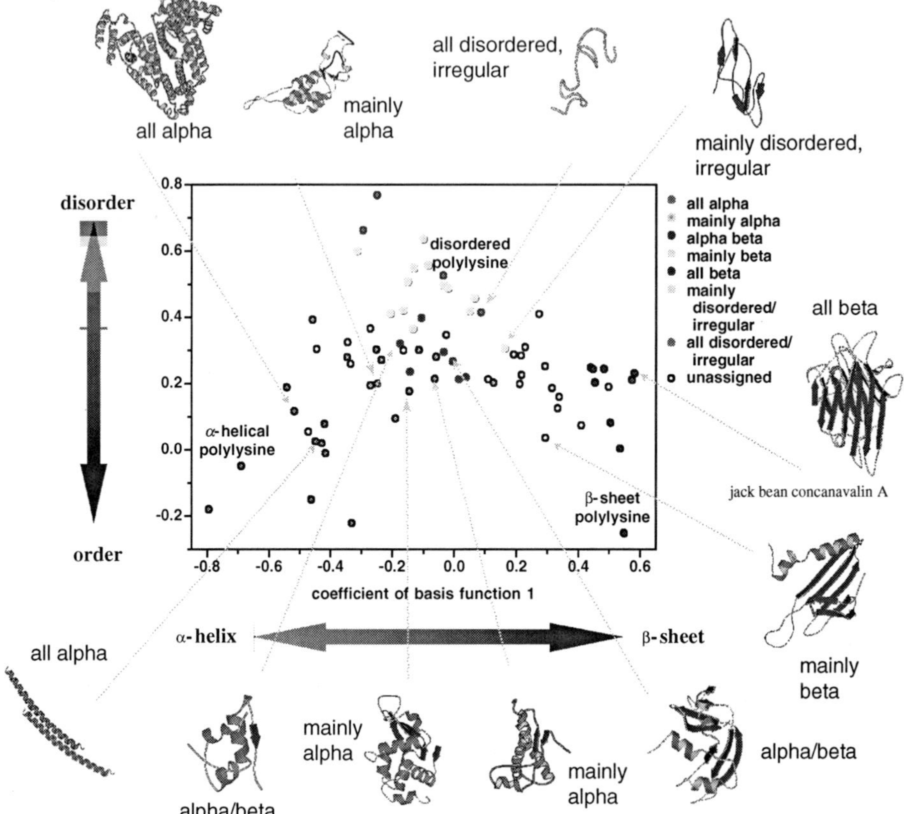

Fig. 16.15. Plot of the PCA coefficients of the two most significant basis functions for a set of 75 polypeptide, protein and virus ROA spectra. The data points are based on structural content as follows: all alpha, >~60% α-helix with little or no other secondary structure; mainly alpha, >~35% α-helix with a small amount (~5–15%) of β-sheet; alpha beta, similar significant amounts of both α-helix and β-sheet; mainly beta, >~35% β-sheet with a small amount (~5–15%) of α-helix; all beta, >~45% β-sheet with little or no other secondary structure; mainly disordered/irregular, little secondary structure; all disordered, no secondary structure; open circles, unassigned. MOLSCRIPT diagrams of several proteins and the positions of α-helical, β-sheet and disordered poly(L-lysine) are shown to illustrate these structural classifications.

spectral resolution by spreading bands over a second dimension, can provide unambiguous band assignments by selectively coupling bands through various interaction mechanisms and can probe the order of spectral changes using the so-called synchronous and asynchronous correlations. The generation of cross peaks in the synchronous and asynchronous 2D maps indicate whether or not spectral changes occur in the same or opposite directions with respect to the externally applied perturbation and also identify their order sequence. A variant of this approach is heterospectral 2D correlation, which is based on the pairwise comparison of spectra obtained from two different techniques, thus allowing the extraction of complementary information [186,187]. Heterospectral 2D correlation of data obtained by two different techniques, such as Raman and

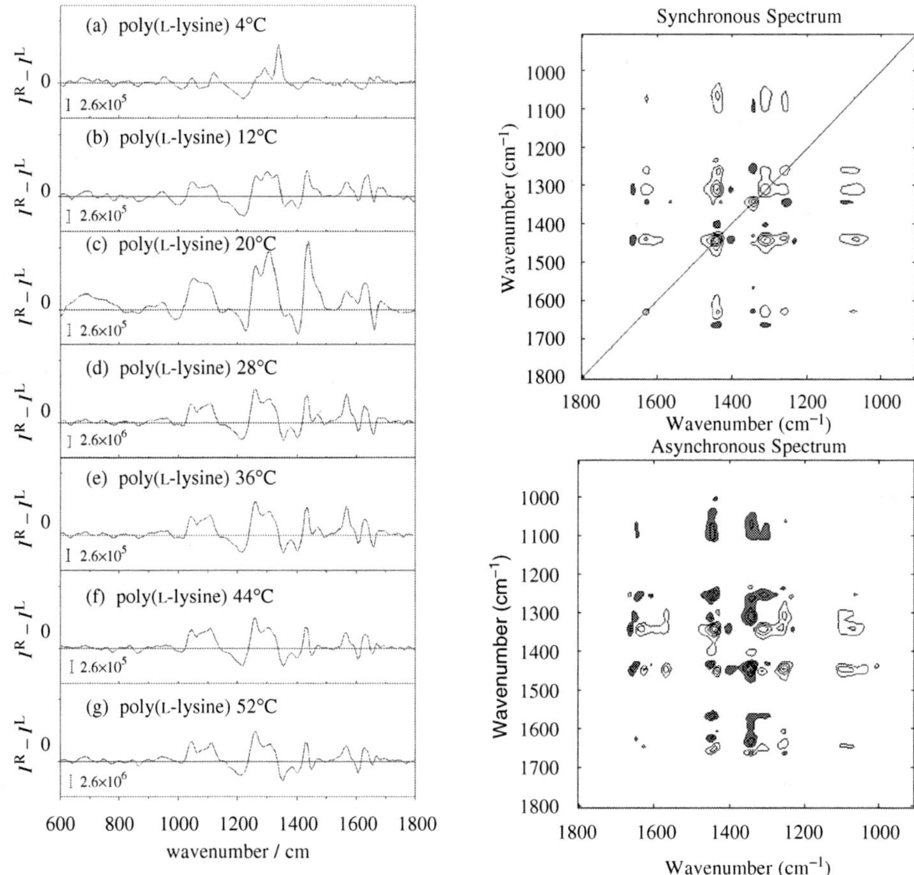

Fig. 16.16. A set of seven ROA spectra (left) recorded as a function of temperature spanning the transition of poly(L-lysine) at ~pH 12 from the α-helical conformation at 4°C to the β-sheet conformation at 52°C. The changes in ROA bands are analyzed as a function of temperature in the 2D synchronous (top right) and asynchronous (bottom right) contour plots. Positive correlations are shown as clear peaks and negative correlations as gray peaks. Band assignments are not shown in this figure but correspond to band assignments already described in this work.

ROA spectroscopies, has great potential for providing new insight into biomolecular conformational dynamics.

We present here an example of this novel approach in Fig. 16.16. A set of seven ROA spectra of poly(L-lysine) spanning the conformational transition from α-helix to β-sheet, generated by heating of the α-helical form at pH 12 in sequential 8°C steps, was analyzed using the generalized 2D-method [188]. The synchronous and asynchronous 2D ROA correlation spectra, respectively, generated from the temperature-dependent spectral intensity variations are also shown. Autopeaks (those correlations situated on the diagonal in the synchronous spectrum) are apparent for regions where the increase in temperature has induced significant changes in intensity, particularly in extended amide III, amide I, backbone skeletal stretch and methylene deformation mode (~1400–1480 cm^{-1}) regions. The numerous synchronous cross peaks between these regions (those correlations on the off-diagonal) indicate where relative similarities in behavior exist, while the numerous asynchronous cross peaks indicate relative dissimilarity in intensity variations between bands. Numerous other cross peaks exist, but as all intensities are autoscaled with respect to the largest intensity variation observed at ~1440 cm^{-1}, they are not apparent in these two plots but are discernable in magnified views (not shown here). Such contour plots allow the relationships between different ROA and parent Raman bands to be elucidated providing a novel approach for the analysis of complementary ROA and Raman spectra of a dynamic system. We are currently exploiting these relationships to obtain several new ROA band assignments for β-sheet and the lysine side chains and to better characterize the complexity of β-sheet formation by correlating the order sequence of intensity variations to changes in backbone Φ,ψ orientations and side chain interactions.

The ability of 2D correlation methods to resolve the order of structural changes should also further extend the application of ROA spectroscopy of biophysical phenomena such as conformational transitions and structural rearrangements governing macromolecular assembly processes and other molecular interactions. As an illustration of the current appeal of ROA spectroscopy in structural biology, Dr. Young Mee Jung at the Pohang University of Science and Technology in South Korea and colleagues are instituting a research program to combine ROA with other spectroscopic data and chemometrics techniques such as PCA through 2D correlation spectroscopy to study protein structure. Such multi-technique approaches will add an additional dimension to the application of ROA spectroscopy in structural biology as they will facilitate new spectra-structure assignments and so enable more efficient extraction and objective mining of the high amount of information existing in ROA spectra of biological molecules.

16.12 BRIGHT FUTURE

Chirality is central to the study and application of the molecular sciences, from the fundamentals of organic chemistry to its apotheosis in the life sciences. The seemingly ever-increasing interest in biological molecules makes this an exciting time for investigating the stereochemistry of complex molecules such as proteins, DNA/RNA and their manifold macromolecular assemblies. This contribution has outlined some of the great advances achieved in recent years in instrumentation and the interpretation of ROA spectra, particularly for complex biological molecules.

Many of the initial difficulties of routine measurement and analysis of ROA spectra have now been overcome; an enormous range of chiral molecular structures is accessible with ROA providing a wealth of new and incisive information not available from other spectroscopic techniques. The recent availability of a high performance commercial instrument has already increased the number of members of the 'ROA community.' Although this expansion of interest is influenced by the great potential of ROA spectroscopy for studies in structural genomics, an open field of more exotic manifestations of ROA spectroscopy is waiting to be explored yet including, but not limited to: resonant ROA in the UV; linear polarization ROA; ROA from chiral interfaces; magnetic ROA; SEROA; variable scattering geometry experiments; and the use of *ab initio*, chemometrics and bioinformatics methods for more detailed analyses of ROA spectra. Chiral systems present an almost infinite number of questions to be explored and ROA spectroscopy has unique advantages for addressing many of these.

16.13 ACKNOWLEDGMENTS

The authors would like to thank the Engineering and Physical Sciences Research Council (EPSRC), Biotechnology and Biological Sciences Research Council (BBSRC) and Royal Society of Chemistry (RSC) for research funding. We are also grateful to the many students and collaborators who have contributed to the ROA work in Glasgow over the years, and to Ms. L. Ashton for performing the 2D ROA study. We are especially grateful to Dr. K. Nielsen for developing the PCA program, Professor V. Baumruk and Mr. J. Kapitan for kindly supplying spectra, and Dr. S. Abdali and Dr. Y. Mee Jung for helpful discussions during the preparation of this chapter.

REFERENCES

1 M. Diem, Modern Vibrational Spectroscopy, Wiley, New York, 1993.
2 L.A. Nafie, G.S. Yu, X. Qu, and T.B. Freedman, Faraday Discuss., 99 (1994) 13–24.
3 G.G. Hoffmann, Infrared and Raman Spectroscopy, VCH Publishers, Weinheim, Germany, 1995, pp. 543–572.
4 T.A. Keiderling, in: G.D. Fasman (Ed.), Circular Dichroism and the Conformational Analysis of Biomolecules, Plenum Press, New York, 1996, pp. 555–598.
5 L.A. Nafie, Appl. Spectrosc., 50 (1996) 14A–26A.
6 L.A. Nafie, Ann. Rev. Phys. Chem., 48 (1997) 357–386.
7 P.L. Polavarapu, Vibrational Spectra: Principles and Applications with Emphasis on Optical Activity, Elsevier, Amsterdam, 1998.
8 W. Hug, in: J.M. Chalmers and P.R. Griffiths (Eds.), Handbook of Vibrational Spectroscopy, John Wiley and Sons Ltd., Chichester, UK, 2002, pp. 745–758.
9 L.D. Barron, Molecular Light Scattering and Optical Activity, 2nd Ed. Cambridge University Press, Cambridge, 2004.
10 L.D. Barron, M.P. Bogaard and A.D. Buckingham, J. Am. Chem. Soc., 95 (1973) 603–605.
11 L.A. Nafie and D. Che, Adv. Chem. Phys., 85 (1994) 105–222.
12 L. Hecht and L.D. Barron, in: J.J. Laserna (Ed.), Modern Techniques in Raman Spectroscopy, Wiley, Chichester, UK, 1996, pp. 265–304.
13 W. Hug, S. Kint, G.F. Bailey and J.R. Scherer, J. Am. Chem. Soc., 97 (1975) 5589–5590.
14 G. Holzwarth, E.C. Hsu, H.S. Mosher, T.R. Faulkner and A. Moscowitz, J. Am. Chem. Soc., 96 (1974) 251–252.
15 L.D. Barron and A.D. Buckingham, J. Am. Chem. Soc., 96 (1974) 4769–4773.
16 A.D. Buckingham, Adv. Chem. Phys., 12 (1967) 107–143.
17 L. Rosenfeld, Z. Phys., 52 (1928), 161.
18 A.D. Buckingham and M.B. Dunn, J. Chem. Soc. A, (1971) 1988–1991.
19 L.D. Barron, Molec. Phys., 21 (1974) 241–246.
20 L. Alagna, T. Prosperi, S. Turchini, J. Goulon, A. Rogalev, C. Goulon-Ginet, C.R. Natoli, R.D. Peacock, and B. Stewart, Phys. Rev. Lett., 80 (1998) 4799–4802.

21 R.D. Peacock and B. Stewart, J. Phys. Chem. B, 105 (2001) 351–360.
22 J. Goulon, A. Rogalev, F. Wilhelm, N. Jaouen, C. Goulon-Ginet and C. Brouder, J. Phys. Condens. Matter, 15 (2003) S633–S645.
23 S. Turchini, N. Zema, S. Zennaro, L. Alagna, B. Stewart, R.D. Peacock and T. Prosperi, J. Am. Chem. Soc., 126 (2004) 4532–4533.
24 P.W. Atkins and L.D. Barron, Molec. Phys., 16 (1969) 453–466.
25 L.D. Barron and A.D. Buckingham, Molec. Phys., 20 (1971) 1111–1119.
26 L. Hecht and L.D. Barron, Appl. Spectrosc., 44 (1990) 483–491.
27 L. Hecht, L.D. Barron, Spectrochim. Acta A, 45 (1989) 671–674.
28 W. Hug, Chem. Phys., 264 (2001) 53–69.
29 L.D. Barron and A.D. Buckingham, Ann. Rev. Phys. Chem., 26 (1975) 381–396.
30 L.A. Nafie and T.B. Freedman, Chem. Phys. Lett., 154 (1989) 260–266.
31 D. Che, L. Hecht and L.A. Nafie, Chem. Phys. Lett., 180 (1991) 182–190.
32 D.A. Long, The Raman Effect, Wiley, Chichester, UK, 2002.
33 L. Hecht, L.D. Barron and W. Hug, Chem. Phys. Lett., 158 (1989) 341–344.
34 L.D. Barron and J.R. Escribano, Chem. Phys., 98 (1985) 437–446.
35 L. Hecht and L.D. Barron, Molec. Phys., 79 (1993) 887–897.
36 L. Hecht and L.D. Barron, Molec. Phys., 80 (1993) 601–606.
37 L.A. Nafie, Chem. Phys., 205 (1996) 309–322.
38 M. Vargeck, T.B. Freedman, E. Lee and L.A. Nafie, Chem. Phys. Lett., 287 (1998) 359–364.
39 L. Hecht and L.A. Nafie, Chem. Phys. Lett., 174 (1990) 575–582.
40 L. Hecht and L.D. Barron, Chem. Phys. Lett., 225 (1994) 525–530.
41 L. Hecht and L.D. Barron, Molec. Phys., 89 (1996) 61–80.
42 T. Petralli-Mallow, T.M. Wong, J.D. Byers, H.I. Lee and J.M. Hicks, J. Phys. Chem., 97 (1993) 1383–1388.
43 A.D. Buckingham and P.J. Stephens, Ann. Rev. Phys. Chem., 17 (1966) 399–432.
44 L.D. Barron and A.D. Buckingham, Molec. Phys., 23 (1972) 145.
45 L.D. Barron and J. Vrbancich, Magnetic Raman Optical Activity, in: R.J.H. Clark, R.E. Hester (Eds.), Advances in Infrared and Raman Spectroscopy, Vol. 12, Wiley, Heyden, 1985, pp. 215–272.
46 L. Hecht and L.D. Barron, J. Raman Spectrosc., 25 (1994) 443–451.
47 L.D. Barron, Nature, 257 (1975) 372–374.
48 L.D. Barron, Chem. Phys. Lett., 46 (1977) 579–581.
49 L.D. Barron, C. Meehan and J. Vrbancich, J. Raman Spectrosc., 12 (1982) 251–261.
50 L.D. Barron and C. Meehan, Chem. Phys. Lett., 66 (1979) 444–448.
51 L.D. Barron, J. Vrbancich and R.S. Watts, Chem. Phys. Lett., 89 (1982) 71–74.
52 L.D. Barron, C. Meehan and J. Vrbancich, Molec. Phys., 41 (1980) 945–947.
53 L.D. Barron and J. Vrbancich, J. Raman Spectrosc., 14 (1983) 118–125.
54 K.R. Hoffman, W. Jia and W.M. Yen, Opt. Lett., 15 (1990) 332–334.
55 K.R. Hoffman, W.M. Yen, D.J. Lockwood and P.E. Sulewski, Phys. Rev. B, 49 (1994) 182–188.
56 D.J. Lockwood, K.R. Hoffman and W.M. Yen, J. Lumin., 100 (2002) 145–154.
57 M.G. Cottam and D.J. Lockwood, Light Scattering in Magnetic Solids, Wiley, New York, 1986.
58 R.D. Amos, Chem. Phys. Lett., 87 (1982) 23–26.
59 L.A. Nafie and T.B. Freedman, in: N. Berova, K. Nakanishi and R.W. Woody (Eds.), Circular Dichroism. Principles and Applications, 2nd Ed. Wiley-VCH, New York, 2000, pp. 97–132.
60 A.D. Buckingham, Faraday Discuss., 99 (1994) 1–12.
61 P.L. Prasad and L.A. Nafie, J. Chem. Phys., 70 (1979) 5582–5588.
62 J.R. Escribano and L.D. Barron, Molec. Phys., 65 (1988) 327–344.
63 W. Hug, G. Zuber, A. de Meijere, A. Khlebnikov and H.-J. Hansen, Helv. Chim. Acta, 84 (2001) 1–21.
64 L.D. Barron and A.D. Buckingham, Acc. Chem. Res., 34 (2001) 781–789.
65 P.K. Bose, L.D. Barron and P.L. Polavarapu, Chem. Phys. Lett., 155 (1989) 423–429.

66 P.L. Polavarapu, J. Phys. Chem., 94 (1990) 8106–8112.
67 J. Costante, L. Hecht, P.L. Polavarapu, A. Collet and L.D. Barron, Angew. Chem. Int. Ed. Engl., 36 (1997) 885–887.
68 T. Helgaker, K. Ruud, K.L. Bak, P. Jorgenson and J. Olsen, Faraday Discuss., 99 (1994) 165–180.
69 P. Bouř, Chem. Phys. Lett., 288 (1998) 363–370.
70 P. Bouř, J. Comput. Chem., 22 (2001) 426–435.
71 K. Ruud, T. Helgaker and P. Bouř, J. Phys. Chem., 106 (2002) 7448–7455.
72 M. Pecul, A. Rizzo and J. Leszczynski, J. Phys. Chem. A, 106 (2002) 11008–11016.
73 P. Bouř, V. Baumruk and J. Hanzlikova, Collect. Czech. Chem. Commun., 62 (1997) 1384–1395.
74 P. Bouř, V. Sychrovský, P. Maloň, J. Hanzliková, V. Baumruk, J. Pospišek and M. Buděšinský, J. Phys. Chem. A, 106 (2002) 7321–7327.
75 K.J. Jalkannen, R.M. Nieminen, M. Knapp-Mohammady and S. Suhai, Int. J. Quant. Chem. 92 (2003) 239–259.
76 G. Zuber and W. Hug, J. Phys. Chem. A, 108 (2004) 2108–2118.
77 L. Hecht, L.D. Barron, E.W. Blanch, A.F. Bell and L.A. Day, J. Raman Spectrosc., 30 (1999) 815–825.
78 L.D. Barron, L. Hecht, E.W. Blanch and A.F. Bell, Prog. Biophys. Mol. Biol., 73 (2000) 1–49.
79 W. Hug and G. Hangartner, J. Raman Spectrosc., 30 (1999) 841–852.
80 W. Hug, in: J. Mink, G. Jalsovsky and G. Keretzury (Eds.), Proceedings of the Eighteenth International Conference on Raman Spectroscopy, John Wiley and Sons Ltd., Chichester, U.K., 2004, pp. 419–420.
81 W. Hug, Appl. Spectrosc., 57 (2003) 1–13.
82 P.L. Polavarapu, Chem. Phys. Lett., 148 (1988) 21–25.
83 P.L. Polavarapu, Spectrochim. Acta A, 46 (1990) 171–175.
84 L.D. Barron, Adv. Infrared Raman Spectrosc., 4 (1977) 271.
85 L.D. Barron and A.D. Buckingham, J. Am. Chem. Soc., 101 (1979) 1979–1987.
86 L.D. Barron and J. Vrbancich, J. Chem. Soc. Chem. Comm., (1981) 771–772.
87 L. Hecht, A.L. Phillips and L.D. Barron, J. Raman Spectrosc., 26 (1995) 727–732.
88 K.M. Spencer, R.B. Edwards, R.D. Rauh and M.M. Carruba, Anal. Chem., 66 (1994) 1269–1273.
89 K.M. Spencer, R.B. Edmonds and R.D. Rauh, Appl. Spectrosc., 50 (1996) 681–685.
90 L.D. Barron and B.P. Clark, J. Chem. Soc. Perk. T, 2 (1979) 1164–1170.
91 L.D. Barron and B.P. Clark, J. Chem. Soc. Perk. T, 2 (1979) 1171–1175.
92 L.D. Barron, L. Hecht and S.M. Blyth, Spectrochim. Acta A, 45 (1989) 375–379.
93 L.D. Barron and L. Hecht, in: N. Berova, K. Nakanishi and R.W. Woody (Eds.), Circular Dichroism: Principles and Applications, Wiley-VCH, New York, 2000, pp. 667–701.
94 L.D. Barron, L. Hecht, A.R. Gargaro and W. Hug, J. Raman Spectrosc., 21 (1990) 375–379.
95 A.T. Tu, Adv. Spectrosc., 13 (1986) 47–112.
96 T. Miura and G.J. Thomas Jr., Subcellular Biochemistry, Vol. 24: Proteins: Structure, Function and Engineering, Plenum Press, New York, 1995, pp. 55–99.
97 R. Schweitzer-Stenner, J. Raman Spectrosc., 32 (2001) 711–732.
98 R. Schweitzer-Stenner, F. Eker, Q. Huang, K. Griebenow, P.A. Mroz and P.M. Kozlowski, J. Phys. Chem. B, 106 (2002) 4294–4304.
99 S.A. Overman and G.J. Thomas Jr., Biochemistry, 38 (1999) 4018–4027.
100 A.J. Adler, N.J. Greenfield and G.D. Fasman, Methods Enzymol., 27 (1973) 675–735.
101 T.-J. Yu, J.L. Lippert and W.L. Peticolas, Biopolymers, 12 (1973) 2161–2176.
102 N.T. Yu, Crit. Rev. Biochem., 4 (1977) 229–280.
103 R.W. Woody, Adv. Biophys. Chem., 2 (1992) 37–79.
104 G. Wilson, L. Hecht and L.D. Barron, J. Chem. Soc. Faraday Trans., 92 (1996) 1503–1510.
105 L.D. Barron, E.W. Blanch, I.H. McColl, C.D. Syme, L. Hecht and K. Nielsen, Spectrosc. Int. J., 17 (2003) 101–126.
106 E.W. Blanch, L. Hecht and L.D. Barron, Methods, 29 (2003) 196–209.
107 I.H. McColl, E.W. Blanch, A.C. Gill, A.G.O. Rhie, M.A. Ritchie, L. Hecht, K. Nielsen and L.D. Barron, J. Am. Chem. Soc., 125 (2003) 10019–10026.

108 S.-H. Lee and S. Krimm, J. Raman Spectrosc., 29 (1998) 73–80.
109 P. Hanson, D.J. Anderson, G. Martinez, G. Millhauser, F. Formaggio, M. Crisma, C. Toniolo and C. Vita, Molec. Phys., 95 (1998) 957–966.
110 K.A. Bolin and G.L. Millhauser, Accs. Chem. Res., 32 (1999) 1027–1033.
111 I.H. McColl, E.W. Blanch, L. Hecht and L.D. Barron, J. Am. Chem. Soc., 126 (2004) 8181–8188.
112 B. Ram Singh (Ed.), Infrared Analysis of Peptides and Proteins: Principles and Applications, ACS Symposium Series, American Chemical Society, Washington DC, 2000.
113 P.R. Carey, Biochemical Applications of Raman and Resonance Raman Spectroscopies, Academic Press, New York, 1982.
114 Z. Chi, X.G. Chen, J.S.W. Holtz and S.A. Asher, Biochemistry, 37 (1998) 2854–2864.
115 N. Sreerama and R.W. Woody, in: N. Berova, K. Nakanishi and R.W. Woody (Eds.), Circular Dichroism: Principles and Applications, Wiley-VCH, New York, 2000, pp. 601–620.
116 T.A. Keiderling, in: N. Berova, K. Nakanishi and R.W. Woody (Eds.), Circular Dichroism: Principles and Applications, Wiley-VCH, New York, 2000, pp. 621–666.
117 J. Kubelka and T.A. Keiderling, J. Am. Chem. Soc., 123 (2001) 12048–12058.
118 J. Kubelka and T.A. Keiderling, J. Am. Chem. Soc., 123 (2001) 6142–6150.
119 S. Krimm and J. Bandekar, Adv. Protein Chem., 38 (1986) 181–364.
120 E.W. Blanch, L. Hecht, C.D. Syme, V. Volpetti, G.P. Lomonossoff, K. Nielsen and L.D. Barron, J. Gen Virol., 83 (2002) 2593–2600.
121 C.D. Syme, E.W. Blanch, C. Holt, R. Jakes, M. Goedert, L. Hecht and L.D. Barron, Eur. J. Biochem., 269 (2002) 148–156.
122 L.D. Barron, L. Hecht and E.W. Blanch, Adv. Protein Chem., 62 (2002) 51–90.
123 E.W. Blanch, D.D. Kasarda, L. Hecht, K. Nielsen and L.D. Barron, Biochemistry, 42 (2003) 5665–5673.
124 M.L. Tiffany and S. Krimm, Biopolymers, 6 (1968) 1379–1382.
125 T.A. Keiderling, R.A.G.D. Silva, G. Yoder and R.K. Dukor, Bioorg. Med. Chem., 7 (1999) 133–141.
126 I.H. McColl, E.W. Blanch, L. Hecht, N.R. Kallenbach and L.D. Barron, J. Am. Chem. Soc., 126 (2004) 5076–5077.
127 B. Bochicchio and A.M. Tamburro, Chirality, 14 (2002) 782–792.
128 Z.S. Shi, R.W. Woody and N.R. Kallenbach, Adv. Protein Chem., 62 (2002) 163–240.
129 K. Ma and K.A. Wang, Biopolymers, 70 (2003) 297–309.
130 N. Sreerama and R.W. Woody, Protein Sci., 12 (2003) 384–388.
131 J.C. Ferreon and V.J. Hilser, Protein Sci., 12 (2003) 447–457.
132 B. Bochicchio, A. Ait-Ali, A.M. Tamburro and A.J.P. Alix, Biopolymers, 73 (2004) 484–493.
133 J.M. Hicks and V.L. Hsu, Proteins, 55 (2004) 330–338.
134 E. Smyth, C.D. Syme, E.W. Blanch, L. Hecht, M. Vasak and L.D. Barron, Biopolymers, 58 (2001) 138–151.
135 A.A. Adzhubei and M.J.E. Sternberg, J. Mol. Biol., 229 (1993) 472–493.
136 B.T. Stapley and T.P. Creamer, Protein Sci., 8 (1999) 587–595.
137 B.E. Raumann, B.M. Brown and R.T. Sauer, Curr. Opin. Struct. Biol., 4 (1994) 36–43.
138 T. Miura, H. Takeuchi and I. Harada, J. Raman Spectrosc., 20 (1989) 667–671.
139 E.W. Blanch, L. Hecht, L.A. Day, D.M. Pederson and L.D. Barron, J. Am. Chem. Soc., 123 (2001) 4863–4864.
140 E.W. Blanch, L.A. Morozova-Roche, D.A.E. Cochran, A.J. Doig, L. Hecht and L.D. Barron, J. Mol. Biol., 301 (2001) 553–563.
141 E.W. Blanch, L.A. Morozova-Roche, L. Hecht, W. Noppe and L.D. Barron, Biopolymers, 57 (2000) 235–248.
142 O.B. Ptitsyn, Adv. Protein Chem., 47 (1995) 83–229.
143 M. Arai and K. Kuwajima, Adv. Protein Chem., 53 (2000) 209–282.
144 P. Haezebrouk, M. Joniau, H. van Dael, S.D. Hooke, N.D. Woodruff and C.M. Dobson, J. Mol. Biol., 246 (1995) 382–387.

145 L.A. Morozova-Roche, J. Zurdo, A. Spencer, W. Noppe, V. Receveur, D.B. Archer, M. Joniau and C.M. Dobson, J. Struct. Biol., 130 (2000) 339–351.
146 M.R.H. Krebs, D.K. Wilkins, E.W. Chung, M.C. Pitkeathly A.K. Chamberlain, J. Zurdo, C.V. Robinson and C.M. Dobson, J. Mol. Biol., 300 (2000) 541–549.
147 C. Holt and L. Sawyer, J. Chem. Soc. Faraday Trans., 89 (1993) 2683–2692.
148 R.W. Carrell and D.A. Lomas, Lancet, 350 (1997) 134–138.
149 C.M. Dobson, R.J. Ellis and A.R. Fersht, Philos. Trans. R. Soc. Lond. B, 356 (2001) 127–227.
150 V.N. Uversky, J. Biomol. Struct. Dyn., 21 (2003) 211–234.
151 Y.C. Sekharudu and M. Sundaralingham, in: E. Westhof (Ed.), Water and Biological Macromolecules, CRC Press, Boca Raton, FL, 1993, pp. 148–162.
152 G. Siligardi and A.F. Drake, Biopolymers, 37 (1995) 281–292.
153 R.K. Dukor and T.A. Keiderling, Biopolymers, 31 (1991) 1747–1761.
154 N. Sreerama and R.W. Woody, Biochemistry, 33 (1994) 10022–10025.
155 A.F. Bell, L. Hecht and L.D. Barron, J. Chem. Soc. Faraday Trans., 93 (1997) 553–562.
156 A.F. Bell, L. Hecht and L.D. Barron, J. Am. Chem. Soc., 119 (1997) 6006–6013.
157 A.F. Bell, L. Hecht and L.D. Barron, Biospectroscopy, 4 (1998) 107–111.
158 A.F. Bell, L. Hecht and L.D. Barron, J. Raman Spectrosc., 30 (1999) 651–656.
159 A.F. Bell, L. Hecht and L.D. Barron, J. Am. Chem. Soc., 120 (1998) 5820–5821.
160 W. Saenger, Principles of Nucleic Acid Structure, Springer, New York, 1984.
161 W. Chiu, R.M. Burnett and R.L. Garcia (Eds.), Structural Biology of Viruses, Oxford University Press, Oxford, UK, 1997.
162 S.C. Harrison, Principles of Virus Structure, in: D.M. Knippe and P.M. Howley (Eds.), Fields Virology, 4th Ed. Lippincott Williams and Watkins, Philadelphia, 2001, pp. 53–85.
163 G.J. Thomas Jr., in: T.G. Spiro (Ed.), Biological Applications of Raman Spectroscopy, Vol. 1, John Wiley and Sons, New York, 1987, pp. 135–201.
164 G.J. Thomas Jr., Annu. Rev. Biophys. Biomol. Struct., 28 (1999) 1–27.
165 D.A. Marvin, Curr. Opin. Struct. Biol., 8 (1998) 150–158.
166 E.W. Blanch, A.F. Bell, L. Hecht, L.A. Day and L.D. Barron, J. Mol. Biol., 290 (1999) 1–7.
167 M. Tsuboi, S.A. Overman and G.J. Thomas Jr., Biochemistry, 35 (1996) 10403–10410.
168 E.W. Blanch, D.J. Robinson, L. Hecht, C.D. Syme, K. Nielsen and L.D. Barron, J. Gen. Virol., 83 (2002) 241–246.
169 G.P. Lomonossoff and J.E. Johnson, Prog. Biophys. Mol. Biol., 55 (1991) 107–137.
170 T. Lin, Z. Chen, R. Usha, C.V. Stauffacher, J.-B. Dai, T. Schmidt and J.E. Johnson, Virology, 265 (1999) 20–34.
171 T. Lin, A.J. Clark, Z. Chen, M. Shanks, J.-B. Dai, Y. Li, T. Schmidt, P. Oxelfelt, G.P. Lomonossoff and J.E. Johnson, J. Virol., 74 (2000) 493–504.
172 M. Fleischmann, P.J. Hendra and A.J. McQuillan, Chem. Phys. Lett., 26 (1974) 163–166.
173 K. Kneipp, H. Kneipp, I. Itzkan, R.R. Dasar and M.S. Feld, Chem. Rev., 99 (1999) 2957–2975.
174 A.G. Brolo, D.E. Irish and B.D. Smith, J. Mol. Struct., 405 (1997) 29–44.
175 A. Campion and P. Khambhampati, Chem. Soc. Rev., 27 (1998) 241–250.
176 R.L. Garrell, Anal. Chem., 61 (1989) 401A–411A.
177 R.J.H. Clark and T.J. Dines, Angew. Chem. Int. Ed., 25 (1986), 131–158.
178 R. Aroca and B. Price, J. Phys. Chem. B, 101 (1997) 6537–6540.
179 M. Moskovits, Rev. Mod. Phys., 57 (1985) 783–826.
180 J.C. Rubim, P. Corio, M.C.C. Ribeiro and M. Matz, J. Phys. Chem., 99 (1995) 15765–15774.
181 A. Otto, Phys. Stat. Sol. A, 188 (2001) 1455–1470.
182 Z.-Q. Tian, B. Ren and D.-Y. Wu, J. Phys. Chem. B, 106 (2002) 9463–9483.
183 S. Higuchi, N. Ikoma, T. Honma and Y. Gohshi, in: S.R. Durig and J.F. Sullivan (Eds.), Proceedings of the Twelwth International Conference on Raman Spectroscopy, Wiley, Chichester, U.K., 1990.
184 G.-S. Yu, Development and Application of Raman Optical Activity. Ph.D. Thesis, Syracuse University, USA, 1994, pp. 134–141.

185 I. Noda, Appl. Spectrosc., 47 (1993) 1329–1336.
186 I. Noda, Appl. Spectrosc., 44 (1990), 550–561.
187 I. Noda and Y. Ozaki, Two-Dimensional Correlation Spectroscopy: Applications in Vibrational and Optical Spectroscopy, John Wiley and Sons Ltd., Chichester, UK, 2004.
188 L.A. Ashton, L.D. Barron, B. Czarnik-Matusewicz, L. Hecht, J. Hyde and E.W. Blanch, Mol. Phys., (2006) accepted.

© 2006 Elsevier B.V. All rights reserved.

CHAPTER 17

Mass spectral methods of chiral analysis

Brandy L. Young, Lianming Wu and R. Graham Cooks

Department of Chemistry, Purdue University, 560 Oval Drive, West Lafayette, IN 47907, USA

17.1 BACKGROUND TO CHIRAL ANALYSIS BY MASS SPECTROMETRY

Life has been under the constant influence of chiral forces, from the initial chemical processes that led to homochirality in the molecules that were subsequently incorporated into living systems, right through the evolutionary processes that led to the present diversity of biological forms [1]. Chirality is an intrinsic property of the biomolecular building blocks of life, being incorporated into the essential components that are necessary for life: amino acids, sugars, proteins, nucleic acids, lipids and complex carbohydrates. For this among many other reasons, chirality continues to attract interest in the chemical, biological, and pharmaceutical sciences.

The chiral nature of living systems makes chiral drugs essential in order to maximize therapeutic efficacy [2,3]. Enantiomeric forms of drugs can be metabolized via different biological pathways and can produce different physiological effects [4]. The recent surge in the chiral drug market follows US FDA guidelines that recognize the differential biological actions of optical isomers. These guidelines mandate that analytical methods determine the contributions of individual enantiomers as well as determining their individual pharmacological and toxicological activity [5]. These rules have encouraged pharmaceutical companies to explore new technologies for chiral analysis [6,7].

Mass spectrometry (MS), was first demonstrated as a method capable of chiral discrimination by Fales and Wright [7] in experiments that used amino acids. In other early work, negatively charged cluster ions were generated and their dissociation kinetics were studied. The experiments showed that the fragmentation rates of clusters, such as the deprotonated complex of D/L proline and camphor, could be used to distinguish enantiomers [8]. After slow development, the last decade has seen a strong interest in improving MS methods for routine and reliable chiral identification and especially for chiral quantification. The focus of this chapter is on MS as a tool for chiral analysis, especially chiral quantitation.

Interest in all aspects of MS is flourishing [9], especially in its applications to the biological sciences. The power and versatility of this analytical method has led to a natural synergy connecting novel instrumentation to creative applications, a conjunction that has nourished this field. With the advent of sophisticated ionization techniques of electrospray ionization (ESI) [10] and matrix-assisted laser desorption ionization (MALDI) [11], and the newer methods of sonic spray ionization (SSI) [12,13] and desorption electrospray ionization (DESI) [14], essentially any type of compound can be ionized from either solution or the condensed phase in a gentle way. These techniques are complemented by the exceptional sensitivity and specificity of mass analyzers. These and other developments have broadened the applicability of MS to the point where its value as a research and development tool is both firmly established and is still rapidly growing [15,16].

Many applications of mass spectrometry, including chiral analysis, often depend on tandem mass spectrometry (MS/MS), a two-stage method used *inter alia* to characterize individual components or to recognize groups of functionally related compounds in mixtures [17,18]. The first stage of a tandem mass spectrometer is often employed to select specific ions of interest from the mixture of ions generated from a complex sample and the second mass analyzer (in ion traps, the same mass analyzer is used later in time) is utilized to characterize the product ions resulting from the selected ion [19,20]. The conversion of the selected precursor ion to a set of product ions can be based on collision-induced dissociation (CID) which yields characteristic fragment ions or ion/molecule (I/M) reactions, which yield characteristic product ions. Various alternative scan modes (precursor ion scan, product ion scan, neutral loss/gain scan), allow one to efficiently examine complex mixtures for groups of compounds with given functionality. The use of MS/MS provides several practical advantages most notably the fact that it can lower detection limits by excluding undesired chemical components ("chemical noise") and so improve signal-to-noise ratios.

The instrumentation used for MS/MS experiments falls into two main categories. The first, the tandem-in-space or beam-type instruments use two or more mass spectrometers assembled in sequence. Two mass-analyzing quadrupoles, two sector analyzers, or hybrid combinations represent common tandem-in-space mass spectrometers. Specific tandem-in-space mass spectrometers include quadrupole/time-of-flight (Q-TOF) [21] instruments, TOF-TOF [22] instruments, and combinations of magnetic sectors and quadrupole mass filters. The second type of MS/MS instrument, the tandem-in-time type, provides an alternative way to perform MS/MS experiments. Quadrupole (or Paul ion traps) [23], Fourier transform ion cyclotron resonance (FT-ICR) instruments [24], and the very recent Orbitrap [25] mass analyzer are examples of trapping devices which provides mass measurement capabilities. These types of mass analyzers can be used to store ions of interest and eject unwanted ions. The selected ions can then be activated to cause dissociation, and the fragment ions can be mass analyzed by reusing the original mass analyzer. This process can be repeated to perform multistage mass spectrometry (MS^n) experiments which are useful in elucidating the structures of compounds.

The triple quadrupole (QqQ) deserves special attention since it is the most widely used tandem-in-space mass spectrometer. The QqQ has a linear assembly of three quadrupoles [26] although only the first and the third quadrupoles are mass analyzers. The mass analyzing quadrupoles are operated using a combination of radio frequency (r.f.)

and d.c. voltages that are necessary for mass selection. The second quadrupole is supplied with only an r.f. voltage and transmits ions of a wide range of m/z ratios. This quadrupole also serves as a collision cell for CID or as a reaction cell for ion/molecule (I/M) reactions, into which argon or other collision gases, or gaseous reagents, can be introduced and from which the products of the respective CID or I/M reactions are transmitted into the subsequent mass analyzing quadrupole for mass analysis. The offset in the potentials at which the ion source and the collision cell are held can be adjusted to vary the collision energy, the range being from a nominal zero to several hundred volts. Adjusting this voltage affects the types of fragmentation processes and the rates of I/M reactions.

Mass spectrometry is still commonly thought of as a "chirally blind" technique since each enantiomer of a pair gives an identical mass spectrum under most operating conditions. Enantiomeric identification, whether done using chemical reactions or physical methods, requires that the enantiomer be subjected to an asymmetric environment. For example, chiral high performance liquid chromatography (HPLC) employs a chiral stationary phase [27,28], while capillary electrophoresis (CE) [29–31], and gas chromatography (GC) [32] are useful complementary methods. Circular dichroism (CD) [33] uses chiral radiation to discriminate between enantiomers, while nuclear magnetic resonance spectroscopy (NMR) [34] and enzymatic technologies [35], can be used for chiral analysis in the presence of chiral compounds.

Turning to quantitative chiral analysis, one can note that internal standards or external calibration methods can be used with linear, quadratic, or curve fitting procedures in chiral chromatography [36]. Enantiomers are separated as a result of diastereomeric interactions with the chiral stationary phase. Optimizing the system to facilitate chiral recognition is largely dependent on selecting the appropriate combination of chiral stationary phase and internal standard. Chiral chromatography is already a mature technique and many chiral stationary phases are available. However, sample preparation and derivatization procedures [37] can be labor-intensive and a successful chiral separation is often only achieved by trial and error, using many combinations of stationary phase and internal or external standards. This problem is often magnified when complex matrices are analyzed.

Parallel procedures are used in quantitative chiral analysis by MS but interest in MS for quantitative chiral analysis lies partly in its potential to provide more rapid analysis. Although particular instances of successful distinction of enantiomers have been reported over the past three decades, general methods of quantitative chiral analysis by mass spectrometry have only been demonstrated quite recently. This chapter describes the current state of chiral MS techniques. It also emphasizes that mass spectrometry, especially tandem mass spectrometry [18], provides several unique analytical advantages for quantitative chiral analysis. These include intrinsically high sensitivity, molecular specificity, and tolerance to impurities, as well as the simplicity and speed of the mass spectrometric measurement [38]. We also note the major limitations of mass spectrometric chiral analysis, which are poor in applicability to samples $<1\%$ *ee*. In addition, empirical methods are used to select chiral references to use in chiral analysis.

17.2 FUNDAMENTAL INTERACTIONS LEADING TO CHIRAL RECOGNITION

The fundamental chemical interactions needed to achieve chiral recognition are often dealt with empirically on a case-by-case basis and the collections of experimental findings whose results are often rationalized by qualitative explanations. Despite the diversity of the interactions that can occur and the diversity of explanations offered, there are some commonalities that can be summarized in the concept of the three-point interaction [39], which inherently involves the use of a chiral selector to facilitate chiral recognition. A chiral selector is an object that can be used to distinguish two enantiomers. Intrinsically, chiral recognition implies and includes a three-point interaction with a chiral selector.

The three-point rule, an idea first formulated by Ogston [40], suggests that if at least three active positions of a chiral selector simultaneously interact with appropriate positions of at least one enantiomer of the analyte, and at least one of these interactions is stereochemically dependent, then the enantiomers can in principle be resolved, Fig. 17.1 [41]. This interaction will be significantly affected by replacing one enantiomer with the other. For example, if the *SS* (*S*-selector and *S*-enantiomer) interaction is different to the *SR* (*S*-selector and *R*-enantiomer) interaction, then the free energy difference ($\Delta\Delta G$) of the two associations will be non-zero and separation of the enantiomers is possible based on thermodynamic enantioselectivity. When there is no three-point interaction for either enantiomer of the analyte, the particular selector cannot be used to distinguish between the *S* or *R* enantiomers [41]. In such associations, the forms *SS* and *SR*, are essentially equivalent; there is no free energy change difference and no chiral recognition. The three-point model has been successfully used in explaining chiral discrimination in solution and in chromatography, and it forms the implicit basis for the current mass spectrometric analysis methods.

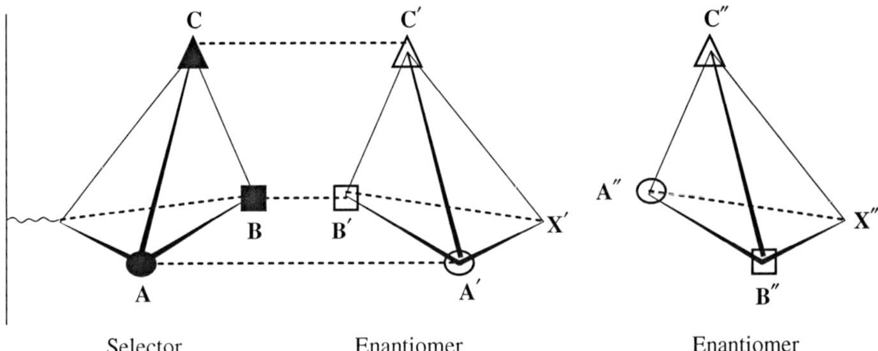

Fig. 17.1. Three-point model of chiral recognition. Group A on the chiral selector can interact with A′ or A″ but not with B′, C″, or X-type sites of the analyte enantiomer. With one enantiomer of the analyte three interactions are possible but only two are possible with the other enantiomer. The selector may be part of a larger structure, e.g. a protein or a silica particle used as stationary phase in liquid chromatography [170]. (Note: This figure is an adaptation of the original)

Chiral selectors are common in nature (e.g. enzymes) or can be prepared artificially (e.g. chiral stationary phases, chiral chemical reagents, etc.). Chiral selector/enantiomer interactions have been categorized into a variety of types—rigid interactions [42], repulsive interactions [41], flexible interactions [41], multiple receptor interactions [43], pseudo interactions [44,45], etc. For molecules with a single chiral center and free rotation of substituents around covalent single bonds to the central atom, the requirement is that the configurational interactions be along three different bonds axes [44–46].

The three-point model takes into consideration the interactions between the chiral selector and analyte only. However, it has been shown that chiral recognition can be assisted by the local environment. This assistance can come from an achiral mediator, where there are interactions between the chiral selector, analyte, and mediator [33,34]. Thermochemically, $\Delta\Delta G$ can become larger as a result of a more rigid two point chiral selector/enantiomer interaction when the third interaction involves a mediator. This magnifies the unequal reactivities of the enantiomers of the analyte as a result of increased steric hindrance with the mediator. The mediator is essentially a multipoint interaction site for the selector and the enantiomer. This mediation can occur *via*: (1) the addition of an achiral molecule, such as a achiral solvent or a metal cation [47,48], (2) interaction with a sorbent surface [47], or (3) interactions with a combination of mediators (achiral molecules and solvents) [46]. Mediation-assisted chiral recognition does not follow the common implementation of the three-point rule—two points of attractive interaction and one point of non-bonding interaction—but this type of mediated chiral recognition is not an exception to the three-point rule [46].

Although, the three-point rule is commonly invoked, a fundamental understanding of the nature of chiral recognition has been pursued on a phenomenological basis [39,49]. Fortunately, MS methods have been developed that borrow from the framework of the three-point rule despite our limited understanding of the nature of the interactions in particular cases. Many chiral selector/analyte combinations are being explored in MS method development and the potential to explore many others is very promising as chiral MS develops.

The simplicity of the solvent-free environment in the mass spectrometer allows the effects of the chiral environment to be probed without interference from solvent or other achiral mediators. This might, in due course, provide a better understating of the nature of chiral recognition as well as yield improved analytical methods for chiral recognition. Furthermore, comparisons of gas-phase and condensed-phase data provide a powerful means for rationalizing the role of solvents and counter-ions in determining the outcomes of many interactions.

17.3 MASS SPECTROMETRIC METHODS FOR CHIRAL ANALYSIS

Current interest in MS for chiral analysis is evident from the variety of chiral MS procedures that have recently been developed. Several reviews on chiral analysis using MS have been published [38,50–53]. The most commonly used methods fall into five main categories: (1) solution-phase kinetic resolution [54,55] followed by MS product analysis, (2) gas-phase host–guest (H–G) diastereomeric adduct formation [56],

(3) I/M (equilibrium) reaction efficiency methods [57], (4) CID of diastereomeric adducts [55,58] with (5) data treatment by the kinetic method [59,60]. These mass spectrometric methods are based on characteristic methods of discrimination and recognition of discrimination in particular chemical systems.

One important feature of each of these techniques is quantitative evaluation of the enantioselectivity of the method. Enantioselectivity is a measure of the change induced when the behavior of pure enantiomers is compared. Typically it is determined from ion abundance ratios in MS and MS/MS experiments and it can be evaluated in the mass spectra when using standard mixtures. Several of the mass spectrometric methods of chiral analysis described in what follows has its own nomenclature to describe enantioselectivity. To emphasize the underlying connections between individual methods, the term R_{chiral}, which in the original literature represents the ratios (ratio of ratios) of fragment ion abundances for the pure enantiomers in a MS/MS experiment, will be adopted to represent enantioselectivity for all methods hereafter (the original notation can be easily re-substituted if desired). In addition to enantioselectivity, the measured response, which is also an ion abundance ratio measurement for various enantiomeric compositions including standard and unknown mixtures, will be designated as R.

Solution-phase kinetic resolution This method for chiral analysis is characterized by the fact that the chirally-selective reactions occur in solution and MS is used merely to analyze the reaction products. These products are generated from chiral reagents that are "mass tagged" and so traceable through a reaction just like an isotopic label would be [54]. While not a purely mass spectrometric method, the experiment is designed for MS recognition of the products and serves as a good connection between solution phase and gas phase chiral analysis. This method requires that a reaction occur with each enantiomer, such that new products of differing enantiomeric content (and mass) are formed. The relative ion abundances of the newly formed mass-tagged products are correlated with the *ee* of the analyte using single-stage MS.

Host-guest (H–G) diastereomeric adduct formation This procedure also uses single stage rather than tandem mass spectrometry but it is entirely a gas phase MS experiment. It starts with the formation of diastereomers from the analyte and chiral reference, their formation being monitored simply by recording their relative abundances in the mass spectrometer. The formation of chiral inclusion complexes occurs as a result of H–G interactions [56]. Like any single-stage MS method, it requires that the enantiomeric guest (analyte) be isotopically (or otherwise) labeled so that the H–G diasteriomeric adducts can be mass-resolved and the ion abundance ratios can be recorded. This method allows quantitative chiral analysis and is readily implemented using any of a variety of ionization techniques, including chemical ionization (CI) [61–64], fast atom bombardment (FAB) [65–67], MALDI [68,69], and ESI [15,70,71]. The method is best used when the examined samples are pure, since MS/MS is not used.

Ion/molecule (equilibrium) reactions As in most of the other methods, diastereomeric complex ions are generated in the mass spectrometer from a chiral analyte and a chiral reference molecule. However, in this MS/MS experiment, the product ions, such as crown ether or β-cyclodextrin complexes, are mass-selected and allowed to react further with a neutral reagent (which can be either chiral or non-chiral) [57,72–74]. Chiral differentiation is achieved by investigating the difference in reaction rates e.g. displacement of the chiral analyte *versus* the chiral reference by the new reagent. This experiment is readily performed in an ion trap such as a quadrupole ion trap (QIT) [74] or a FT-ICR [57,72,73].

CID of diasteriomers In this simple MS/MS method, diastereomeric ions are generated in the ion source of the mass spectrometer, mass-selected, and differences in their fragmentation patterns are recorded [58,75,76]. No isotopic or other labeling is required. In principle this procedure is directly analogous to the use of solution-phase reactions to generate diastereomers, but the method of interrogation of the products is different. Different types of diastereomers can be generated; a very early example was the formation of a proton-bound anion [8] of proline and camphor. This complex was isolated and dissociated and the fragment ion abundance ratio relative to that of the precursor ion (dissociation efficiency) was measured in a CID experiment and this ratio was found to depend on the enantiomeric content of the analyte.

Kinetic method This MS/MS method is based on a particular quantitative relationship between %*ee* and relative ion abundance in the mass spectrometer. It is used to recognize and quantify mixtures of chiral molecules by studying the dissociation kinetics of trimeric cluster ions, which includes the chiral analyte and a chiral selector (the chiral reference). Analytes of known chiral purity serve as calibration samples. The technique builds on earlier work that showed the fragmentation rates of metal-bound trimeric cluster ions could be used to distinguish enantiomers [8,77]. In the most thoroughly studied cases, separation of enantiomers is achieved by forming metal-bound complexes involving the chiral analyte and a chiral reference in the ion source of a mass spectrometer and observing the relative kinetics of competitive dissociations of the cluster ions. The ratio of fragment ions varies with the chirality of the analyte studied. The method has been optimized such that in principle any chiral species can be quantitatively resolved. It has been used with a variety of ionization techniques, including CI and ESI, and various tandem mass spectrometers, particularly the QIT.

17.3.1 Kinetic resolution in solution with mass-tagged reagents

This hybrid solution phase–gas phase method uses MS to quantify diastereomers that are generated in solution (there are complications in specifying the location of ion formation in any spray ionization method since clustering processes are driven by the evaporating solution in the source of the mass spectrometer allowing for the possibility that cluster formation occurs in the solution phase, the gas phase or somewhere between). Chiral quantification by this procedure has been demonstrated using alcohols and amines by means of diastereoselective derivatization and automated quantitative ESI MS.

The methodology was established by using an equimolar mixture of pseudo-enantiomeric "mass-tagged" chiral acylating agents that differ in configuration and in the presence/absence of a mass-tag substituent that is remote to the chiral center. The mass of the reagent molecule therefore correlates with its known configuration. Reactions of the analyte (alcohols and amines, Chart 17.1) with any pair of chiral reagents produce esters and amides, depending on the starting reagent. This reaction, done in the presence of 1,3 dicyclohexylcarbodiimide (DCC), proceeds with unequal rate

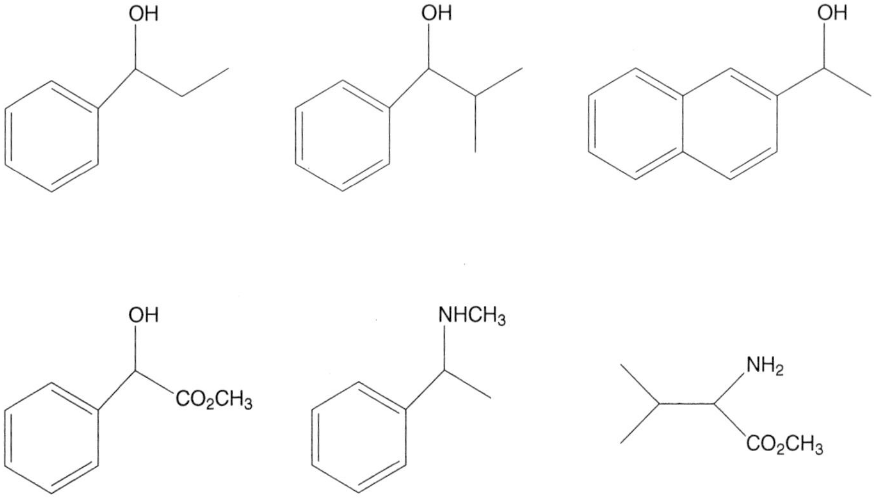

Chart 17.1. Chiral compounds selected for application of the kinetic resolution method.

constants for the two enantiomers of the analyte ($k_f > k_s$; f = fast, s = slow). This is shown in the scheme below for reactions of alcohols with chiral mass-tagged acids A–CO_2H and B–CO_2H [54].

$$\begin{array}{ccccc} & \text{A–CO}_2\text{R} \xleftarrow[\text{fast},\, k_{fA}]{\text{A–CO}_2\text{H, DCC, base}} & \text{R–OH} & \xrightarrow[\text{slow},\, k_{sB}]{\text{B–CO}_2\text{H, DCC, base}} \text{B–CO}_2\text{R} & \\ \text{Mass 1} & & + & & \text{Mass 2} \\ & \text{A–CO}_2\text{S} \xleftarrow[\text{slow},\, k_{sA}]{\text{A–CO}_2\text{H, DCC, base}} & \text{S–OH} & \xrightarrow[\text{fast},\, k_{fB}]{\text{B–CO}_2\text{H, DCC, base}} \text{B–CO}_2\text{S} & \end{array}$$

While the differences seen are largely due to the chirality of the reaction partners (as intended) there is a secondary effect due to differences in reactivity associated with mass-tagging. In the absence of a correction that accounts for the reactivity differences, the relative amounts of the product ester that are formed in the reaction are linearly proportional to the concentration of each enantiomer and to the enantiomeric composition. A linear relationship can be derived simply by assuming that the rates of the esterification reactions are first order in acid and alcohol, Eqs. (1) and (2)

$$\frac{d(\text{mass}_1)}{dt} = k_{fA}[\text{ROH}][\text{A–CO}_2\text{H}] + k_{sA}[\text{SOH}][\text{A–CO}_2\text{H}] \qquad (1)$$

$$\frac{d(\text{mass}_2)}{dt} = k_{sB}[\text{ROH}][\text{B–CO}_2\text{H}] + k_{fB}[\text{SOH}][\text{B–CO}_2\text{H}] \qquad (2)$$

Since the concentrations of the acids are present in large excess the following definitions can be used.

$$k_{fA}[\text{A–CO}_2\text{H}] = k_{FA} \qquad k_{fB}[\text{B–CO}_2\text{H}] = k_{FB}$$
$$k_{sA}[\text{A–CO}_2\text{H}] = k_{SA} \qquad k_{sB}[\text{B–CO}_2\text{H}] = k_{SB}$$

Thus, Eqs. (1) and (2) reduce to the expressions shown as Eqs. (3) and (4), respectively.

$$\frac{\mathrm{d(mass_1)}}{\mathrm{d}t} = k_{\mathrm{FA}}[\mathrm{ROH}] + k_{\mathrm{SA}}[\mathrm{SOH}] \tag{3}$$

$$\frac{\mathrm{d(mass_2)}}{\mathrm{d}t} = k_{\mathrm{SB}}[\mathrm{ROH}] + k_{\mathrm{FB}}[\mathrm{SOH}] \tag{4}$$

Hence the relative ratio (Eq. (3) *versus* Eq. (4)) gives the following relationship, Eq. (5),

$$\mathrm{R} = \text{mass ratio} = \frac{k_{\mathrm{FA}}[\mathrm{ROH}] + k_{\mathrm{SA}}[\mathrm{SOH}]}{k_{\mathrm{FB}}[\mathrm{SOH}] + k_{\mathrm{SB}}[\mathrm{ROH}]} \tag{5}$$

Now using the definition, $y = \mathrm{R}/\mathrm{R}_{\mathrm{chiral}}^{\mathrm{hybrid}}$, (where $\mathrm{R}_{\mathrm{chiral}}^{\mathrm{hybrid}}$ is the observed mass ratio for the racemate and accounts for the secondary effect that results from differences in the reactivity of the non-identical mass-tagged chiral acylating agents; note the original literature refers to $\mathrm{R}_{\mathrm{chiral}}^{\mathrm{hybrid}}$ as q and R as s), we get Eq. (6) (a rearranged form of Eq. (5)).

$$y = \frac{k_{\mathrm{FA}}\left(\frac{[\mathrm{ROH}]}{[\mathrm{SOH}]}\right) + k_{\mathrm{SA}}}{k_{\mathrm{FB}} + k_{\mathrm{SB}}\left(\frac{[\mathrm{ROH}]}{[\mathrm{SOH}]}\right)} \tag{6}$$

Alternatively, an expression for the enantiomeric ratio can be obtained, if the following assumptions ($k_{\mathrm{fA}} = k_{\mathrm{fB}}$) and ($k_{\mathrm{sA}} = k_{\mathrm{sB}}$) are made, then ($k_{\mathrm{FA}} + k_{\mathrm{FB}} = k_{\mathrm{F}}$) and ($k_{\mathrm{SA}} + k_{\mathrm{SB}} = k_{\mathrm{S}}$), and the above equation simplifies further to Eq. (7) (where Eq. (8) is an expanded view of Eq. (7)) with the assumptions already mentioned.

$$\frac{[\mathrm{ROH}]}{[\mathrm{SOH}]} = \frac{yk_{\mathrm{FB}} - k_{\mathrm{SA}}}{k_{\mathrm{FA}} - yk_{\mathrm{SB}}} \tag{7}$$

$$\frac{[\mathrm{ROH}]}{[\mathrm{SOH}]} = \frac{yk_{\mathrm{F}} - k_{\mathrm{S}}}{k_{\mathrm{F}} - yk_{\mathrm{S}}} = \frac{y\left(\frac{k_{\mathrm{F}}}{k_{\mathrm{S}}}\right) - 1}{\left(\frac{k_{\mathrm{F}}}{k_{\mathrm{S}}}\right) - y} = \frac{y\mathrm{R} - 1}{\mathrm{R} - y} \tag{8}$$

From the definition of *ee*, one obtains the expression given in Eq. (9) and the simplified form Eq. (10) (simplified using the expression derived in Eq. (7)).

$$\frac{\%ee}{100} = \left(\frac{[\mathrm{ROH}] - [\mathrm{SOH}]}{[\mathrm{ROH}] + [\mathrm{SOH}]}\right) = \frac{\left(\frac{[\mathrm{ROH}]}{[\mathrm{SOH}]} - 1\right)}{\left(\frac{[\mathrm{ROH}]}{[\mathrm{SOH}]} + 1\right)} \tag{9}$$

$$= \left(\frac{\frac{y\mathrm{R} - 1}{\mathrm{R} - y} - 1}{\frac{y\mathrm{R} - 1}{\mathrm{R} - y} + 1}\right) = \frac{(y\mathrm{R} - 1 - \mathrm{R} + y)}{(y\mathrm{R} - 1 + \mathrm{R} - y)} \tag{10}$$

The enantiomeric excess of an unknown sample is then determined directly from the observed MS intensity ratio, Eq. (11).

$$\%ee = \left[\frac{(y-1)(\mathrm{R}+1)}{(y+1)(\mathrm{R}-1)}\right] \times 100, \quad \text{where} \quad \begin{array}{l} \mathrm{R} = k_\mathrm{f}/k_\mathrm{s} \\ y = \text{corrected intensity ratio.} \end{array} \tag{11}$$

Fig. 17.2 shows the results of applying this method to four secondary alcohols, and one primary and one secondary amine of varying structures, including aliphatic and aromatic substituents. Samples of 20, 50, 70, and 90%*ee* were used. Each point in this figure is the averaged result of two independent acylation reactions per sample, and each reaction was analyzed by three averaged ESI-MS injections. Repeat injections were found to give reproducible intensity ratios with a standard deviation of ~3%. In almost every case, the measured value falls within +/−10%*ee* of the true value. The results showed that the greater the enantiomeric discrimination in the acylation process, the better the *ee* measurement.

The methodology has also been optimized to include *ee* determinations with solid-phase chiral acylating agents, where mass-tagged chiral activated esters were bound to polystyrene resins and used for the *ee* measurement of amines [78]. This procedure made the *ee* determination processes more convenient. Also, this methodology was used to

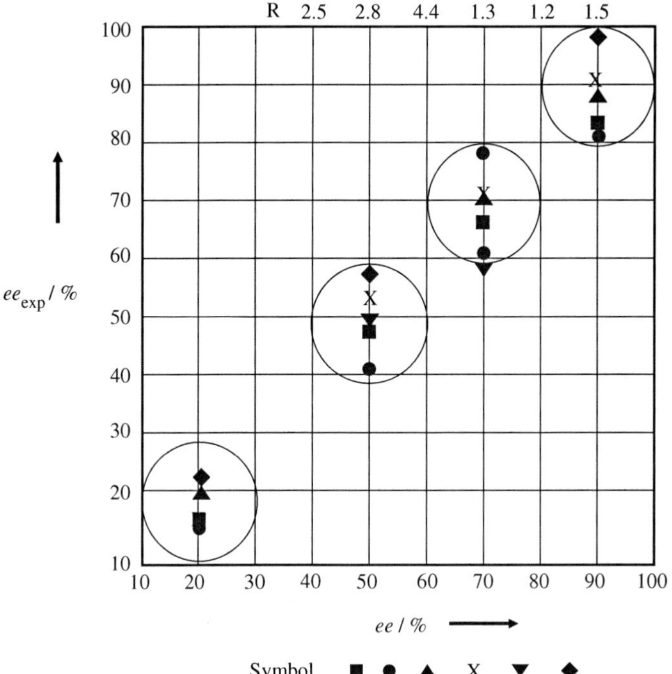

Fig. 17.2. Plots of the actual *versus* measured enantiomeric excesses and R ($k_\mathrm{f}/k_\mathrm{s}$) values for samples of 20–90%*ee* enriched in the *S*-enantiomer [171]. (Note: This figure is an adaptation of the original)

screen a family of chiral phosphate P,N-ligands for activity in rhodium-catalyzed asymmetric hydrosilylation of ketones [79], demonstrating its potential as a combinatorial catalytic screening tool.

As an example of a recent study that falls into this general class of solution-phase *ee* MS analysis experiments but differs in the acylation step, derivatives of *N*-(3,5-dinitrobenzoyl)-leucine, a commonly used chiral stationary phase, were used as chiral selectors in an ESI-MS chiral analysis experiment [55]. Chiral selectors were prepared and used for the purpose of discriminating enantiomers in a single-stage ESI-MS experiment. The selectors were designed so that the ionization site was well-removed from the site required for effective chiral recognition. This minimized any interference in the formation of the selector-analyte complex, which could potentially affect the chiral selectivity, and promoted ionization in which the charge site was far from the chiral centers. The addition of a chiral analyte to a solution that contained two psuedo-enantiomeric chiral selectors, which differ in stereochemistry and in the length of the *N*-alkyl chain, produced selector-analyte complexes that were electrosprayed into a triple quadrupole mass spectrometer.

The enantiomeric composition of the analyte was determined from the natural log of the relative intensities of the selector-analyte complex in the full-scan single stage mass spectrum. A linear variation in the relative intensity of the product ions for the quasi-racemic mixture of mass-labeled pseudo-enantiomeric chiral selectors, with %*ee* of the analyte was observable from the ESI mass spectra. This relationship provides a measure of the extent of enantioselectivity and allows quantitative *ee* determinations to be made. This procedure was successfully demonstrated for several derivatized selectors and analytes [80,81]. Also, the results showed that the method is not dependent on the analyte concentration, provided the selector concentration is in excess. This is important for practical use. The scope and limitations of the method were determined, plus comparisons of the methods enantioselectivities, precision, and accuracy were evaluated side by side with chiral HPLC using a *N*-(3,5-dinitrobenzoyl)-leucine-derived chiral stationary phase.

17.3.2 Single-stage mass spectrometry

Chiral recognition can be achieved in single-stage MS, for example, by H–G complexation that results in enantioselective intermolecular interactions, where MS is used to measure the extent of product formation which differs for the enantiomeric analytes due to the free energy differences in the interactions associated with each enantiomeric configuration. Either the host or the guest must be isotopically labeled in these experiments so that the H–G diastereomeric complexes can be distinguished in the single-stage mass spectrum [56]. Except for the requirement of isotopic labeling, the method is readily implemented and it can be used with various ionization techniques including CI [64], FAB [67], MALDI [82], and ESI [70]. The two versions of the experiment are referred to as the Enantiomer Labeled (EL)-Host (Scheme 17.1a) or the Enantiomer Labeled (EL)-Guest method (Scheme 17.1b), since either the host or guest can be isotopically labeled. Crown ethers [67,83], carbohydrates [84,85] and

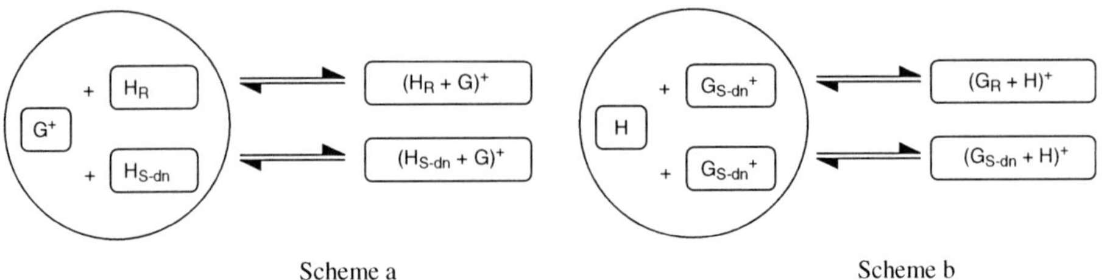

Scheme 17.1. Guest-host diastereomeric ion formation for chiral recognition: (a) host-labeled (EL-host) method; (b) guest-labeled (EL-guest) method [56].

cyclodextrins [86] represent typical host compounds chosen to complex with the chiral guest (analyte).

17.3.2.1 Enantiomer labeled (EL)-host method

Cyclic crown ethers with structures of the types shown in Chart 17.2 are commonly used as host compounds in the EL-Host method. An equimolar mixture of H_R and H_S-*d6* (the chiral *R* and *S* forms of the host, H) is used as the host-pair reagent for *ee* determination of the chiral guest.

The host-pair reagent (H_R/H_S-d6 = 1/1) is mixed with the guest compound (G, which represents the pure guest enantiomer or various compositions) and is ionized using FAB-MS or ESI-MS. Two diastereomeric H–G complex ions simultaneously appear in the single-stage MS spectrum. The ratio of their intensities is the chiral selectivity for the experiment (termed the IRIS value in the original literature, but called R^{IRIS}_{chiral} here to connect these experiments to others discussed in the review) and reflects the degree of chiral discrimination when the pure *R*- and *S*-enantiomers are used, Eq. (12).

$$R^{IRIS}_{chiral} = IRIS = \frac{I[(H_R + G)^+]}{I[(H_{S-d_6} + G)^+]} = \frac{I_R}{I_{S-d6}} \quad (12)$$

This method uses the relative peak intensities (R) of the diastereomeric H–G complex ions as a quantitative measure of chiral composition. The fundamental concept of the EL-Host method is illustrated by the simplified data shown in Fig. 17.3. The R values change with the enantiomeric content of the guest. Fig. 17.3, shows that the optically pure (*R*)-guest forms a complex with the (*R*)-host which is more abundant by an arbitrary factor, in this case a factor of two (Run 1, R = 2), than its complex with the isotopically labeled *S*-enantiomer of the host (H_S-d*n*). Therefore, the optically pure (*S*)-guest should complex the (*S*)-host twice as strongly as the (*R*)-host (Run 2, R = 0.5) because of the mirror image relationship during the G-H complexation. Furthermore, the racemic (*RS*)-guest (0%*ee*) should provide a pair of peaks of equal intensities (Run 3, R = 1) due to the net compensation of a racemic guest/racemic host combination in such a systems. Accordingly, one can expect to be able to determine the %*ee* of an unknown analyte from the measured R value. In practice, the R values obtained by the EL-host method are very close to the calculated $[(H_R \cdot G)^+]/[(H_S \cdot G)^+]$ values. This is illustrated in Fig. 17.4 in the case of the analyte (guest, G) tryptophan 2-propyl ester hydrochloride but in this case the guest is isotopically labeled. The data indicate that the

1

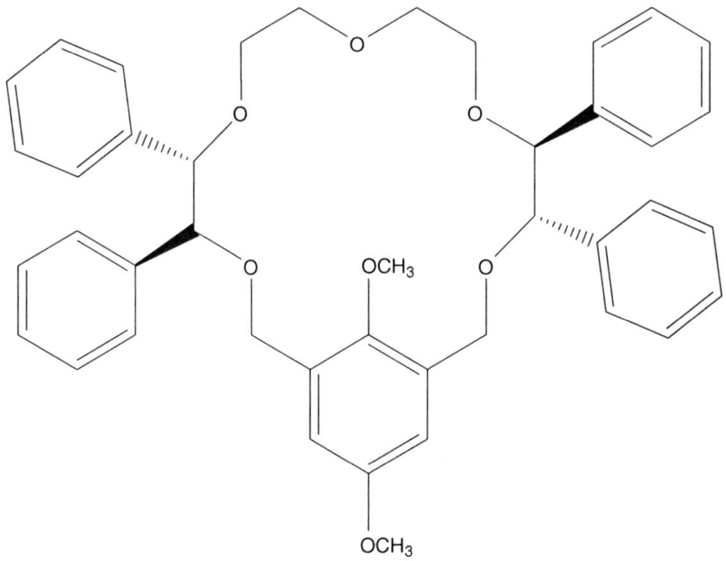

2

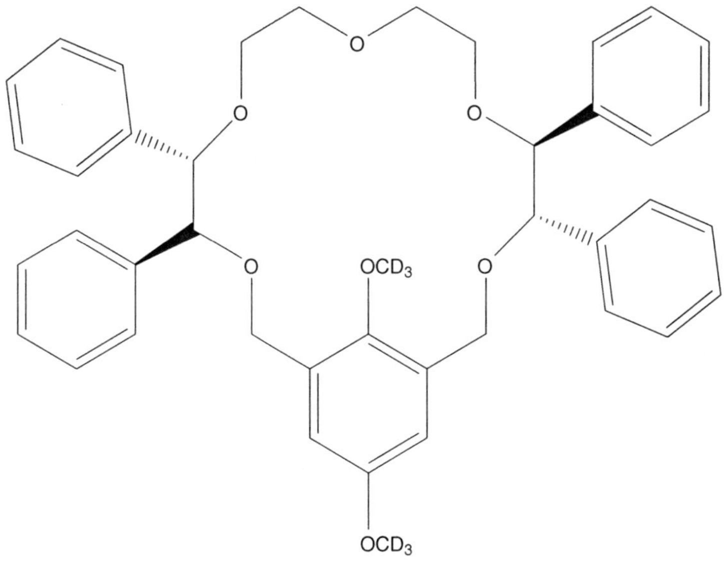

Chart 17.2. Typical cyclic crown ether hosts in EL-host method [56,169].

concentration ratio of the pre-formed diastereomeric ions in the matrix are quantitatively reflected by the corresponding relative ion abundances, at least as far as the diastereomeric host–guest complex ions are concerned. Although this data supports the assumption that the relative ion abundance ratio reflects the relative concentration of the diastereomeric H–G ions in the matrix, discrimination due to isotope effects during

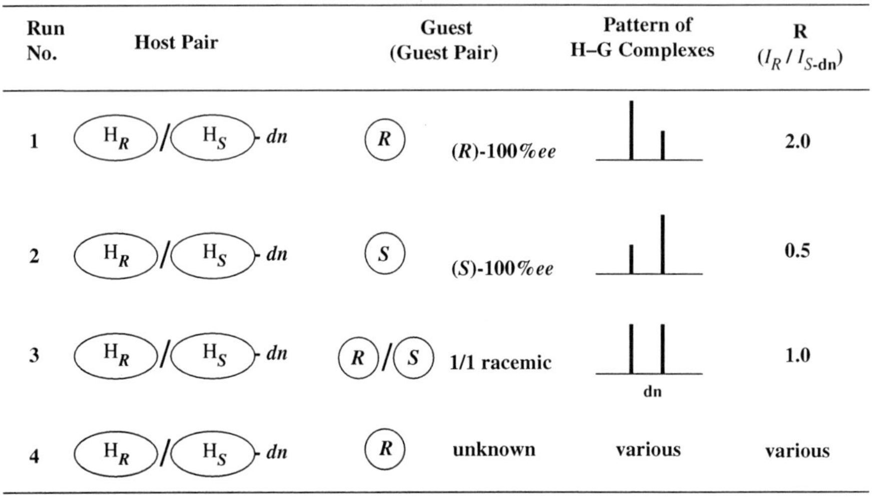

Run No.	Host Pair	Guest (Guest Pair)		Pattern of H–G Complexes	R ($I_R / I_{S\text{-dn}}$)
1	H_R/H_S-*dn*	*R*	(*R*)-100%*ee*		2.0
2	H_R/H_S-*dn*	*S*	(*S*)-100%*ee*		0.5
3	H_R/H_S-*dn*	*R*/*S*	1/1 racemic	dn	1.0
4	H_R/H_S-*dn*	*R*	unknown	various	various

Fig. 17.3. Fundamental concept of the enantiomer-labeled (EL) host method, where *R* and *S* represent the pure enantiomers and H_R and H_S represent the host compounds with opposite enantiomeric configurations [83]. (Reprinted from [83] with permission from Elsevier.)

the ionization process might need to be considered for accurate *ee* determination. A detailed discussion on isotope effects on quantitative *ee* measurement is beyond the scope of this chapter [87].

Since chiral recognition is strongly dependent on the interactions between a given host and a given guest, designer host compounds have been synthesized for the purpose of differentiating different types of chiral guests. For example, linear and flexible gluco- and fructo-oligosaccharides (Chart 17.3) have been successfully used for chiral discrimination of amino acid derivatives by induced-fitting chiral recognition [84,85,88]. Cyclic antibiotics have also been used as host compounds for the analysis of amino acid derivatives [56].

17.3.2.2 Calibration

When making quantitative determinations by the H–G method one must first construct a calibration curve using appropriate standards of known *ee* from optically pure *R*- and *S*-guest forms. The R value determined by the E–L method can then be used to deduce the unknown *ee* value(s). It is worth noting that a single host reagent will not be suitable for all chiral guests. The size of the analyte and the host cavity size must be appropriately matched during method development.

The method was further improved by using double labeling with the corresponding deuterium-labeled *S*- (or *R*-) enantiomer guest compound (for example, $G_{S\text{–d}m}$, *m*: number of deuterium atoms, $m < n$ or $m > n$) as an internal standard. Chiral linear hosts used for double-labeling experiments are shown in Chart 17.4. In such a case, four complex ion peaks $(H_{DD}+G)^+$, $(H_{DD}+G_{S\text{–d}m})^+$, $(H_{LL\text{–d}n}+G)^+$, and $(H_{LL\text{–d}n}+G_{S\text{–d}m})^+$ are independently observed in a *single* mass spectrum (Fig. 17.5) [89]. If the number of deuterium atoms in the labeled host and the labeled internal standard are too similar, it will become difficult to separate the peaks for the complex ions.

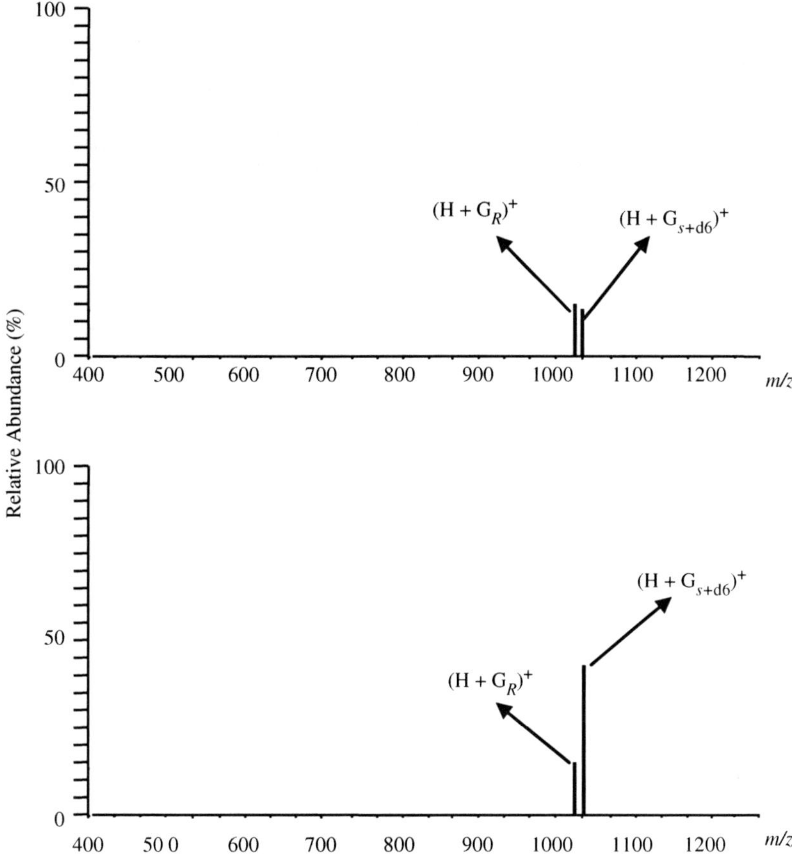

Fig. 17.4. Typical mass spectra in the FAB-MS/EL-guest method. (a) Host: **3A**, guest: Trp-*O*-iPro$^+$ (tryptophan 2-propyl ester hydrochloride), $(\mathrm{H} + \mathrm{G}_R)^+$ *m/z* 1013, $(\mathrm{H} + \mathrm{G}_{S-\mathrm{d}6})^+$ *m/z* 1019; (b) host: **3B**, guest: Trp-*O*-iPro$^+$, $(\mathrm{H}+\mathrm{G}_R)^+$ *m/z* 789, $(\mathrm{H}+\mathrm{G}_{S-\mathrm{d}6})^+$ *m/z* 795 [88]. (Reprinted from [88] with permission from Elsevier.)

Thus, $n > m$ (or $n < m$) is recommended for the labeled enantiomeric host and the internal standard labeled guest.

To make an enantiomeric determination, a given *ee*-unknown amine salt was mixed with (1) the host pair (H_{DD} and $\mathrm{H}_{\mathrm{LL}\text{-d}n}$) and (2) the corresponding labeled *S*-enantiomer guest salt (guest, $\mathrm{G}_{S\text{-d}m}$). The FAB mass spectra were measured in a 3-nitrobenzyl alcohol (NBA) matrix. The peak intensity excess (*I*e) value resulting from the *ee*-unknown guest is defined in Eq. (13).

$$
\begin{aligned}
I\mathrm{e} &= \frac{[I(\mathrm{H}_{\mathrm{DD}} + \mathrm{G})^+ - I(\mathrm{H}_{\mathrm{LL-d}n} + \mathrm{G})^+]}{[I(\mathrm{H}_{\mathrm{DD}} + \mathrm{G})^+ + I(\mathrm{H}_{\mathrm{LL-d}n} + \mathrm{G})^+]} \\
I\mathrm{e} &= \frac{(\mathrm{IRIS} - 1)}{(\mathrm{IRIS} + 1)}
\end{aligned}
\tag{13}
$$

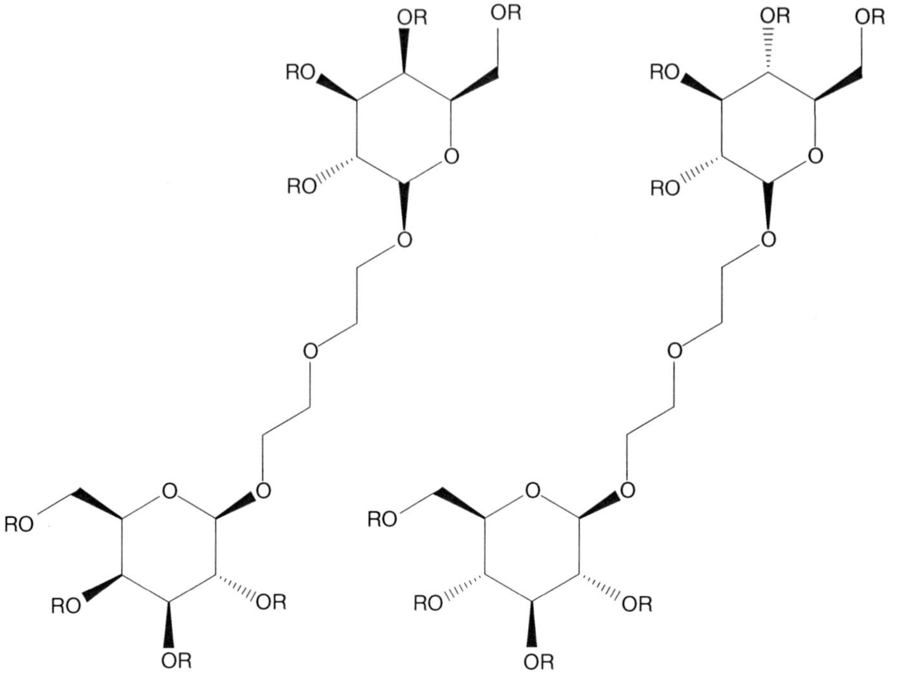

Chart 17.3. Chiral linear hosts using peracetylated derivatives: (a) $R = COCH_3$; (b) $R = CH_3$ [88].

Similarly, the Ie_{std} value resulting from the *ee*-100% standard (labeled) guest ($G_{S\text{-d}m}$) is defined in Eq. (14a) and (14b)

$$Ie_S = \frac{[I(H_{DD} + G_{S-dm})^+ - I(H_{LL-dn} + G_{S-dm})^+]}{[I(H_{DD} + G_{S-dm})^+ + I(H_{LL-dn} + G_{S-dm})^+]}, \tag{14a}$$

$$Ie_R = \frac{[I(H_{DD} + G_{R-dm})^+ - I(H_{LL-dn} + G_{R-dm})^+]}{[I(H_{DD} + G_{R-dm})^+ + I(H_{LL-dn} + G_{R-dm})^+]}. \tag{14b}$$

Therefore, the *ee* value can be obtained as the ratio of I to $I_{S \text{ or } R}$ Eq. (15)

$$\text{ee}(\%) = \frac{Ie}{Ie_{R \text{ or } S}} \times 100. \tag{15}$$

This value was found to correlate linearly with the *ee* value of the guest (Fig. 17.6). Accuracy was within ±1%*ee*, achieved by increasing the accumulating scan number ($n = 50 \sim 70$) and by using the double labeling method, which requires another labeled reference compound. Since the R^{IRIS}_{chiral} values, based on the H–G complexation, are influenced by the initial concentration of host and guest, preparing equal guest concentrations is important when measuring standard mixtures.

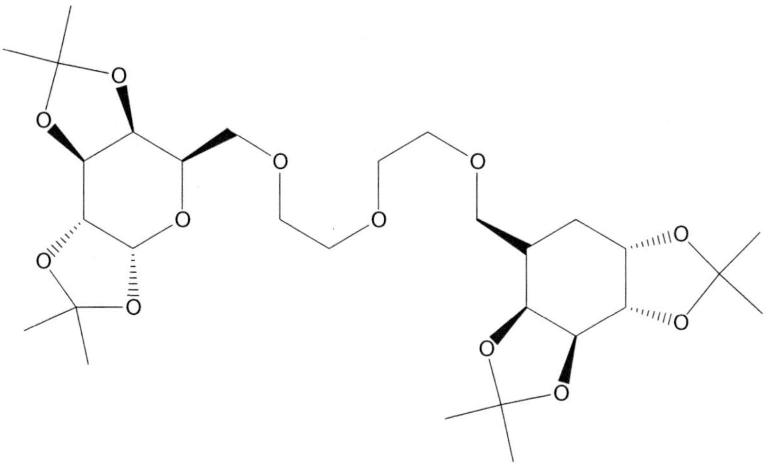

(a) **DD-Gal2deg**

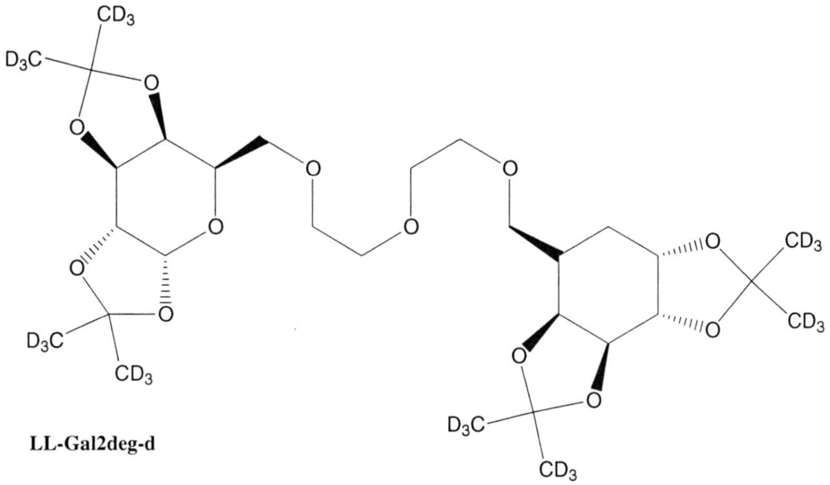

(b) **LL-Gal2deg-d**

Chart 17.4. Chiral linear hosts used for double-labeling experiments [52].

17.3.2.3 Without isotope labeling

Recently, the H–G complexation method has been modified to differentiate chiral amino acids without using isotope labeling [90]. Cyclodextrins have been used as the host molecules. For example, the complexes of α-cyclodextrin and amino acids D/L-Tyrosine (D/L-Tyr) can be easily generated using electrospray ionization and differentiated by comparing the relative peak intensity ratio of the complex to the host ($R = I_{complex}/I_{host}$, Fig. 17.7). The quantitative determination of enantiomeric excess of D-Tyr was then demonstrated by plotting R values obtained from the mixtures against %*ee* of D-Tyr, Fig. 17.8. Although the correlation coefficient ($r^2 = 0.9707$) is not adequate for highly accurate quantitation, further improvement in accuracy might be achieved by optimizing

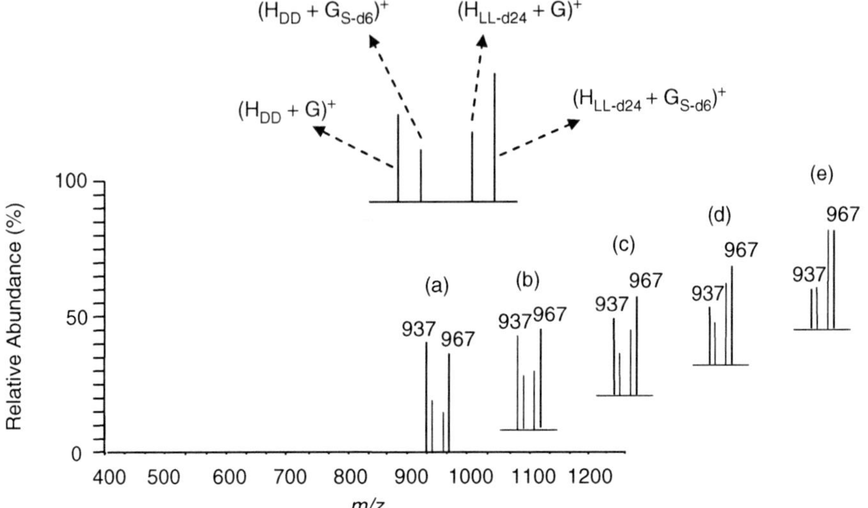

Fig. 17.5. FAB mass spectra of the deuterium-labeled/unlabeled enantiomeric host pair (H_{DD}, $H_{LL\text{-}d24}$) with various *ee* guests [G^+: Trp-*O*-iPr$^+$(Cl^-)] including the deuterium-labeled standard guest [$G_{S\text{-}d6}{}^+$: *S*-Trp-*O*-iPr-$d_6{}^+$(Cl^-)] using NBA as the matrix. (a) *R* 100% *ee*, (b) *R* 50% *ee*, (c) 0% *ee*, (d) *S* 50% *ee*, (e) *S* 100% *ee*. In the case of the spectrum (e), the two pairs of the peaks differing from six mass units show equal intensities because the concentrations of the *S* 100% guest and the deuterium-labeled *S* 100% guest (the standard) are equal [89]. (Reprinted from [89] with permission from Elsevier.)

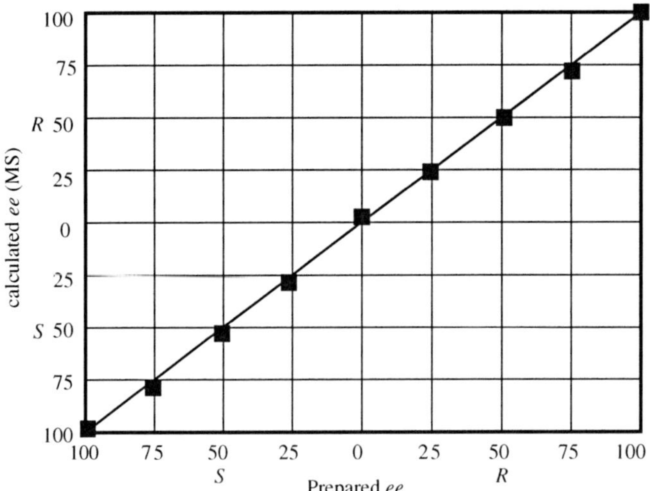

Fig. 17.6. Correlation between the experimental results and actual *ee* values of amine salts [89]. (Reprinted from [89] with permission from Elsevier.)

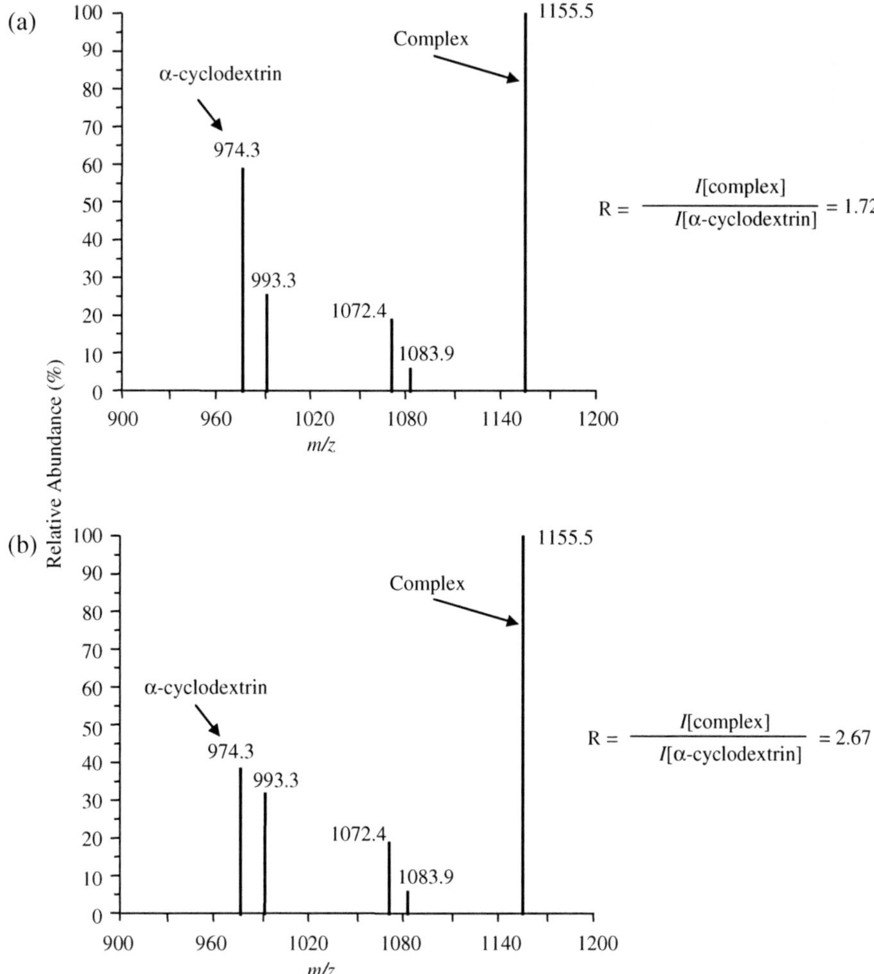

Fig. 17.7. ESI mass spectra of solutions containing α-cyclodextrin (α-CD) and (a) L-Tyr and (b) D-Tyr [90]. (Note: This figure is an adaptation of the original)

the ionization conditions to significantly suppress non-specific complex formation. As a general point, one should note that the use of single-stage mass spectra without isotopic labeling has the advantage of simplicity but the lack of reference parameters, especially in the absence of isotopic labeling, makes accurate quantitation difficult.

17.3.3 Ion/molecule reactions

I/M reactions have had broad impact in both fundamental and applied science. All chiral MS methodologies involve some form of I/M reaction, whether in the formation of

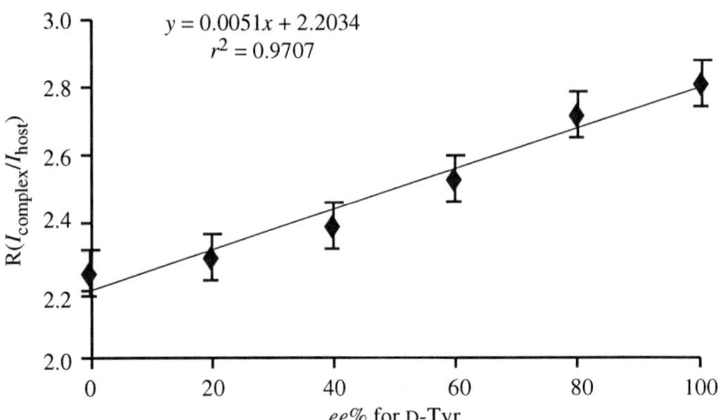

Fig. 17.8. The quantitative determination of enantiomeric excess of D-Tyr determined from plotting experimental R ($I_{complex}/I_{host}$) values *versus* %*ee* of D-Tyr [90], using α-cyclodextrin (α-CD) as the host. (Note: This figure is an adaptation of the original)

a diastereomer (in solution or in the gas phase) or in its analysis. Methods discussed in this section involve I/M reactions performed in MS/MS experiments and they include (1) guest-host (G-H) I/M reactions, where exchange rates between a guest and host in a CID experiment are used to determine the chirality of the guest analyte, (2) I/M gas-phase dependence reaction, where the relative abundance of the analytes fragment ions are used to determine chirality, and (3) I/M metal-bound diastereomeric dissociation reactions similar to the kinetic method (Section 17.3) where chirality is determined from the fragmentation patterns of metal-bound diastereomeric ions. Although single-stage experiments can be imagined (using isotopic labeling), MS/MS has been used in all reported experiments which can be conducted using any commercial mass spectrometer.

17.3.3.1 Guest–host tandem mass spectrometry method

Host molecules such as cyclodextrins (CD) [91–94] and crown ethers [73] can form G-H complexes with analytes in the course of electrospray or other soft ionization methods. These complex ions could in principle be mass-selected and characterized by CID. Alternatively, I/M reactions, e.g. those which result from ligand-exchange with a neutral amine can be used to characterize diastereomeric complexes in MS. This particular amine exchange process is effectively a proton transfer process mediated by a host molecule, where the exchange rate strongly depends on the chirality of the guest molecule (Scheme 17.2). G–H, I/M methods have been successfully used for chiral recognition, and enantiomeric quantification of amino acids and drugs including amphetamine, ephedrine and penicillamine [91].

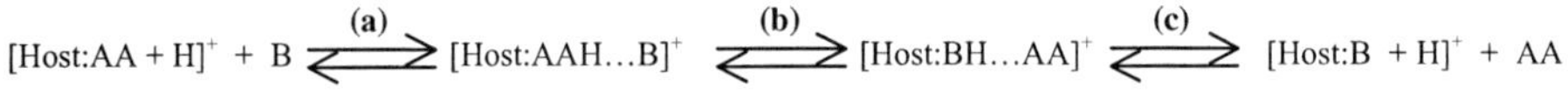

$$[\text{Host:AA} + \text{H}]^+ + \text{B} \underset{}{\overset{\textbf{(a)}}{\rightleftarrows}} [\text{Host:AAH...B}]^+ \overset{\textbf{(b)}}{\rightleftarrows} [\text{Host:BH...AA}]^+ \overset{\textbf{(c)}}{\rightleftarrows} [\text{Host:B} + \text{H}]^+ + \text{AA}$$

Scheme 17.2. Ion/molecule reactions for chiral recognition [91].

Enantioselectivity, defined by the ratio of the rate constants k_L and k_D (k_L and k_D are the rate constants for the pure L and D enantiomers respectively) is given by, Eq. (16) (the term S was used in the original literature to describe enantioselectivity).

$$R^{I/M}_{chiral} = \frac{k_L}{k_D} \tag{16}$$

An $R^{I/M}_{chiral}$ value of 1.0, achieved when using the pure enantiomers, indicates no enantioselectivity for the G–H system in question. For amino acids, $R^{I/M}_{chiral}$ values as large as 5 have been observed [92]. Results from chiral recognition experiments using two hosts and natural and non-natural amino acids, as well as pharmaceutical compounds, are summarized in Table 17.1. An application of this I/M reaction based method for

Table 17.1 Chiral recognition by I/M reactions using MS/MS

	Slectivity (R_{chiral})	
Analytes	Host-permethylated β-CD	Host-maltoheptaose
Natural amino acids		
Alanine	1.6	1.1
Asparagine	0.93	N/A[a]
Aspartic acid	2.2	N/A[a]
Cysteine	2.2	N/A[a]
Glutamic acid	1.9	N/A[a]
Histidine	2.3	N/A[a]
Isoleucine	3.8	2.3
Leucine	3.6	1.9
Methionine	0.37	N/A[a]
Phenylalanine	0.82	4.6
Proline	1.5	N/A[a]
Serine	1.2	N/A[a]
Threonine	0.63	N/A[a]
Tryptophan	2.1	N/A[a]
Tyrosine	0.67	4.9
Valine	3.1	2.1
Non-natural amino acids		
Homoserine	2.2	N/A[a]
cis-4-Hydroxyproline	1.4	N/A[a]
Allo-threonine	22	N/A[a]
Allo-isoleucine	4.1	N/A[a]
Drugs		
Dopamine	0.93 (fast), 2.19 (slow) 1,3-Diamino-propane	1.22 ethylene-diamine
Amphetamine	1.46	0.90
Ephedrine	0.83 (1-amino-2-propanol)	0.78 (1-amino-2-propanol)
Penicillamine	1.9 (fast), 6.18 (slow)	1.07 (fast), 1.85 (slow)

[a]Data taken from Refs. [57,91,93,94] N/A means data not available.

recognizing chiral drugs is demonstrated for amphetamine (AMP) in Fig. 17.9 at various reaction times (CD as the host molecule and *n*-propylamine (NPA) as the reagent) [91].

The I/M reaction method requires that the host and the guest form reasonably stable complexes. It is the cooperative interaction of several weak forces—dipole–dipole, hydrophobic, electrostatic, van der Waals, and hydrogen bonding—that leads to molecular recognition and chiral differentiation. Three-point interactions, as illustrated

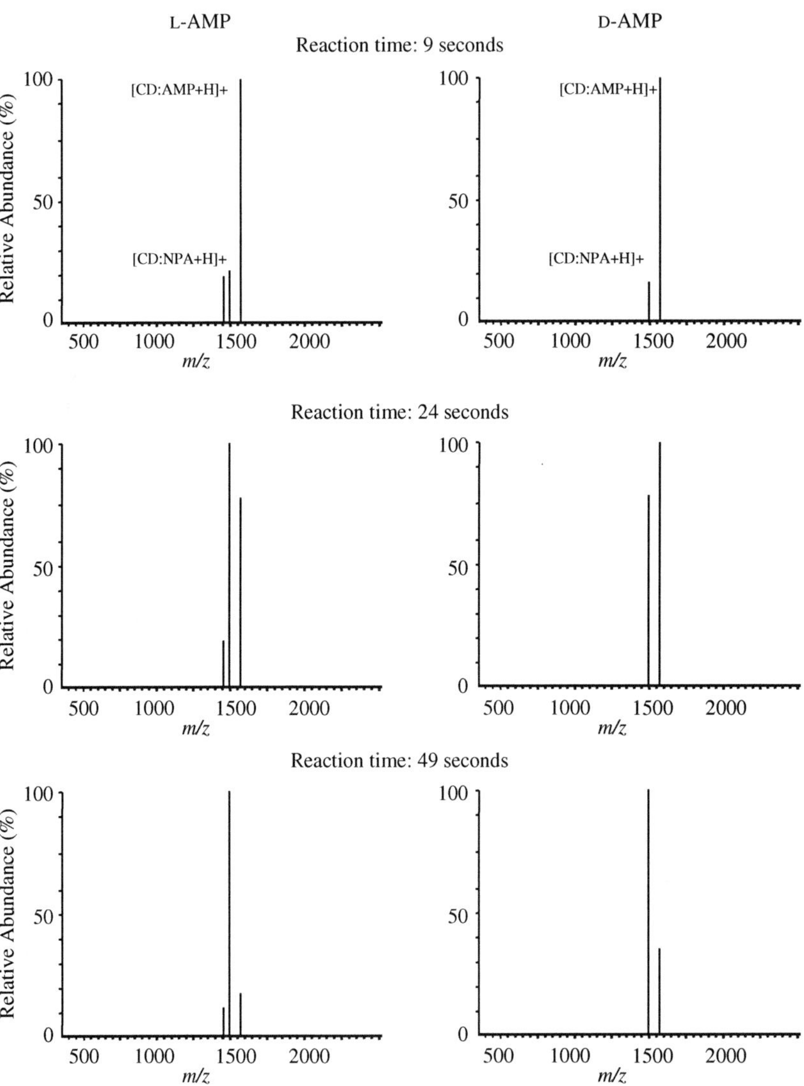

Fig. 17.9. ESI-FTMS spectra of a solution containing β-cyclodextrin (β-CD) and amphetamines (AMP) at various reaction times. The complex $[CD{:}Amp{+}H]^+$ is isolated and allowed to react with *n*-propylamine (NPA) of 3.2×10^{-7} Torr. This reaction has a relatively low selectivity corresponding to $R^{I/M}_{chiral} = 1.46$ [91]. (Note: This figure is an adaptation of the original)

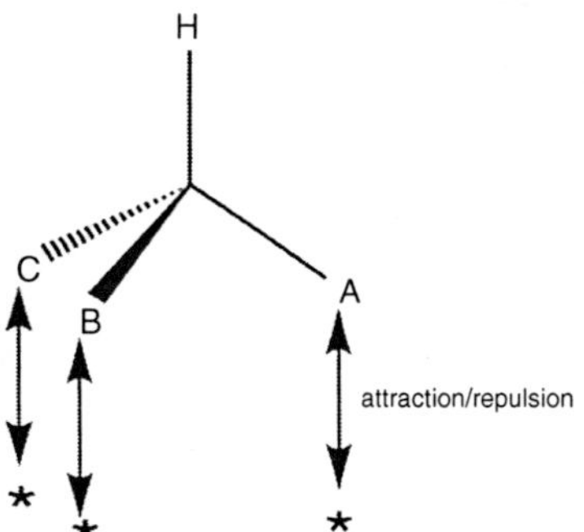

Scheme 17.3. Three-point interaction involving the analyte and the host* [91].

in Scheme 17.3, can either be attractive or repulsive depending on the functional groups [93]. In addition, enantioselectivity is enhanced by congruence of the overall size of the guest and the cavity size of the host. For example, as the alkyl groups of amino acids become larger, the repulsive interactions increase. The attractive interactions of the amino and carboxylic termini have to be reconciled with the increasing steric repulsion. This type of offsetting effect is often seen when enantiomers are forced to adopt specific conformations to promote enantioselectivity.

Calibration curves are generated by directly comparing the relative ion abundance ratios using standard samples of the known *ee*. The construction of a calibration plot normally involves determining the best conditions for the I/M exchange reaction, which includes selecting a reagent, optimizing the pressure in the analyzer chamber, and determining the appropriate reaction time. FT-ICR instruments or ion traps are typically used in this type of experiment. To construct a calibration curve, six or seven analyte solutions are made, normally including the pure D-form (0:100) and the pure L-form (100:0). A calibration curve for phenylalanine is shown in Fig. 17.10. On the ordinate is the ratio of the intensity of the reactant complex (I) to the sum of the intensities of the product and reactant complexes (I_0), while on the abscissa is the mole fraction of the D-Phe isomer (D/[D+L]) is plotted.

The quality of the calibration curve is dependent on the magnitude of the selectivity value ($R^{I/M}_{chiral}$), which in turn depends on both the chiral host in the complex and on the I/M reagent. Both compounds interact selectively with the pure enantiomer. In a particular example, a large selectivity was obtained with a cyclodextrin host and 1,3-diaminopropane as the reagent gas. The corresponding correlation coefficient obtained was $r^2 = 0.996$ [91]. To produce this calibration curve, a reaction time of 88 s was used. This method also showed that with just two reacting species (guest and host) it was possible to construct a valid calibration curve. Further comparison of quantitative accuracy for chiral analysis of amino acids with different chiral selectivities is shown in Table 17.1. In addition, systems with larger selectivity gave better accuracy, as shown in Table 17.2, where phenylalanine gave the best quantitative accuracy.

In a similar application, gas-phase deprotonation reactions of cytochrome-c ions by chiral amines (2-butylamine and 1-amino-2-propanol) exhibited strong chiral specificity [95]. The *R*-isomer of 2-butylamine was shown to be almost ten times more reactive

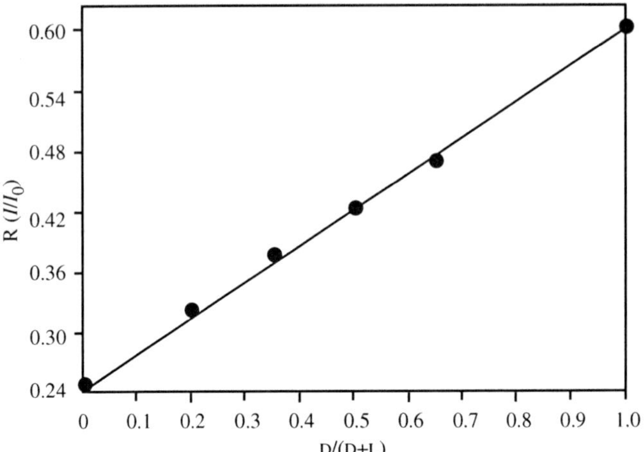

Fig. 17.10. Calibration curve for mixtures of L- and D-phenylalanine complexes reacting with n-propylamine (3.4×10^{-7} Torr). The R (I/I_0) value is the ratio of the intensity of the reactant complex and the sum of the intensities of the product and reactant complexes. The large selectivity ($R_{chiral}^{I/M} = 4.6$) results in high values for r^2 of 0.999 [57]. (Note: This figure is an adaptation of the original)

Table 17.2 Enantiomeric excess determination by I/M reactions[a,b,c]

Analyte	Chiral selectivity (R_{chiral}) and linearity of calibration curve (r^2)	Actual (%)	Experimental (%)	Accuracy (%)[d]
Ala				
		12	22	183
	$R_{chiral} = 1.6$, $r^2 = 0.961$	57	48	84.2
		74	71	96
Leu				
		17	19	112
	$R_{chiral} = 3.6$, $r^2 = 0.975$	58	52	89.7
		78	75	96.2
Phe				
		28	28	100
	$R_{chiral} = 4.6$, $r^2 = 0.998$	58	64	110
		90	89	98.9

[a]Experimental values were obtained directly from the calibration curves.
[b]Using β-CD as the host and n-propylamine as the reagent.
[c]Pressure used to determine the test samples were corrected using the equation:
$\ln R_1 = P_1/P_2$, where R represents the ratio of the intensities of the reactant complex and the sum of reactant and the product complex at a given pressure (where pressure is P_1 or P_2).
[d]Accuracy is calculated based on the original data in Ref. [57].

with cytochrome c than was the S-isomer. With 1-amino-2-propanol, the R-isomer was twice as reactive as the S-isomer and the specificity decreases with increasing proton charge state of cytochrome c. For the charge state 12+, the (2R)-2-butylamine is only 50% more reactive than the S-isomer, compared to 10 times for the 9+ state. These experiments involve simple proton transfer mechanisms. This method was the first to report on proton transfer as a means of deducing chirality, albeit in a complex chemical system. The ability to make stereochemically relevant measurements on systems involving biopolymers is encouraging in regard to more detailed exploration of stereochemistry in such molecules using simple MS methods.

17.3.3.2 Gas-phase ion/molecule reactions

Chiral molecules can be distinguished using simple proton-transfer reactions in MS/MS experiments. For example, chiral differentiation has been reported for sec-butylamines, sec-butanols, and D/L camphor [96]. These experiments are based on the dependence of I/M reaction efficiency as it relates to analyte chirality. Model experiments have been conducted using a triple quadrupole mass spectrometer, equipped with a chemical ionization source. Selection and transfer of the precursor ion (i.e. protonated camphor, *m/z* 153) into quadrupole Q_1, is followed by reaction with a neutral reagent (e.g. either (*R*)-(−) or (*S*)-(+)-sec-butylamine) in Q_2 and the product ions are then mass-selected using Q_3 and monitored

Proton transfer from protonated (D)- and (L)-camphor (*m/z* 153) to (*R*)-(−) or (*S*)-(+)-sec-butylamine ($m/z = 74$) gives fragment ion abundances that directly correlate to the chirality of camphor. The ratio of the relative abundances (*m/z* (74, 109)) varies with the chiral composition as shown in Fig. 17.11. Since the major ions are due to camphor fragmentation and the fragment intensities are similar when comparing L- and D-camphor spectra, the relative abundances of the camphor fragments can be used to deduce chirality. Proton transfer is the most common reaction in the gas phase, making this approach a promising one for simple, rapid, and sensitive chiral recognition.

This type of experiment has also been demonstrated in the negative ion mode, where the steric dependence is expected to be greater. The reason for the increased enantioselectivity is that I/M reactions in the negative ion mode are generally expected to produce closer interactions between the chiral centers in the I/M complexes. The method is also promising because of the simple and rapid procedures employed and the high chiral selectivity demonstrated by the negative-ion mode experiments.

17.3.4 Dissociation of cluster ions

17.3.4.1 Dissociation of diastereomeric ions

CID of diastereomeric adducts gives differences in fragmentation patterns which can be used to differentiate chiral molecules. Using metal-coordinated chiral auxiliaries (either pre-attached or added directly to the sample), diastereomeric adducts can be formed. Analysis of the MS/MS spectra of different diastereomers can be used for quantitative analysis and many also yield trends that allow the assignment of the stereochemistry of certain functional groups in an unknown compound [75,97]. As a result of the different stabilities of diastereomers associated with different analyte chirality, CID of the diastereomers can result in different fragmentation patterns that can be used for molecular recognition [98]. For example, when the complex ion $[Co(DAP)_2(ManNAc)]^+$ (DAP represents diaminopropane) undergoes CID, the MS/MS spectrum shows that the

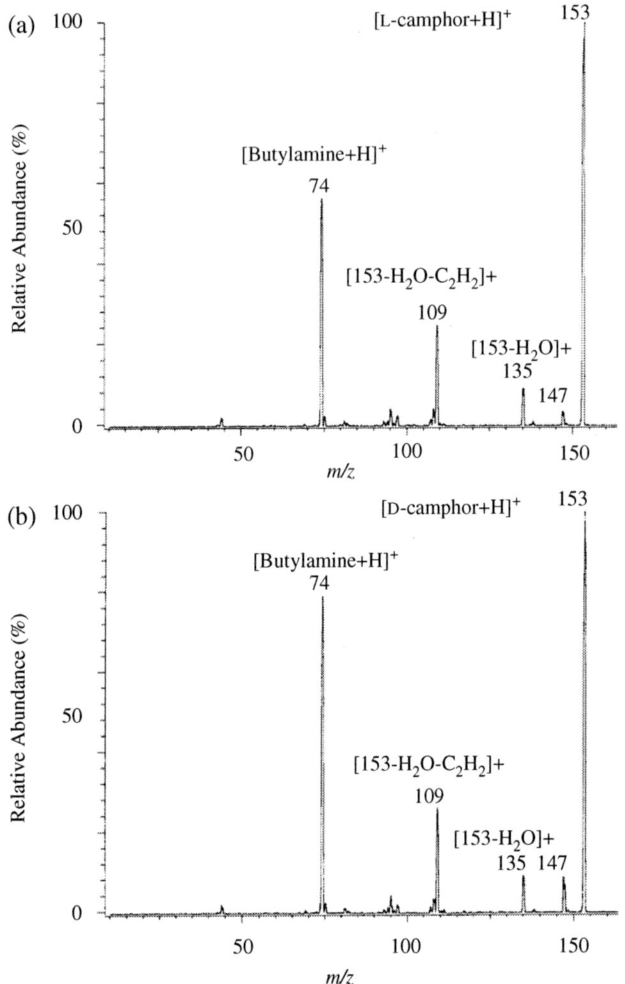

Fig. 17.11. Product ion MS/MS spectra (1 eV, 1 mTorr) showing products of proton transfer from protonated (a) L-camphor and (b) D-camphor to (*R*)-(−)-sec-butylamine [96]. (Reproduced by permission of The Royal Society of Chemistry.)

major product is due to loss of 74, which corresponds to a DAP ligand (Fig. 17.12(a)). There is also a signal corresponding to loss of neutral fragment of mass 120 in the MS/MS spectrum with a relative abundance of about 20%. The product ion spectrum of the complex of the isomers $[Co(DAP)_2(GlcNAc)]^+$ displays this loss at a relative abundance of 4% (Fig. 17.12(b)) and also exhibits a unique loss of 131, while $[Co(DAP)_2(GalNAc)]^+$ loses only the DAP ligand (Fig. 17.12(c)). These MS/MS spectra may be used to distinguish the three diastereomers on the basis of the presence or absence of the two ions, *m*/*z* 306 and 295. Quantitative chiral analysis is also possible in principle.

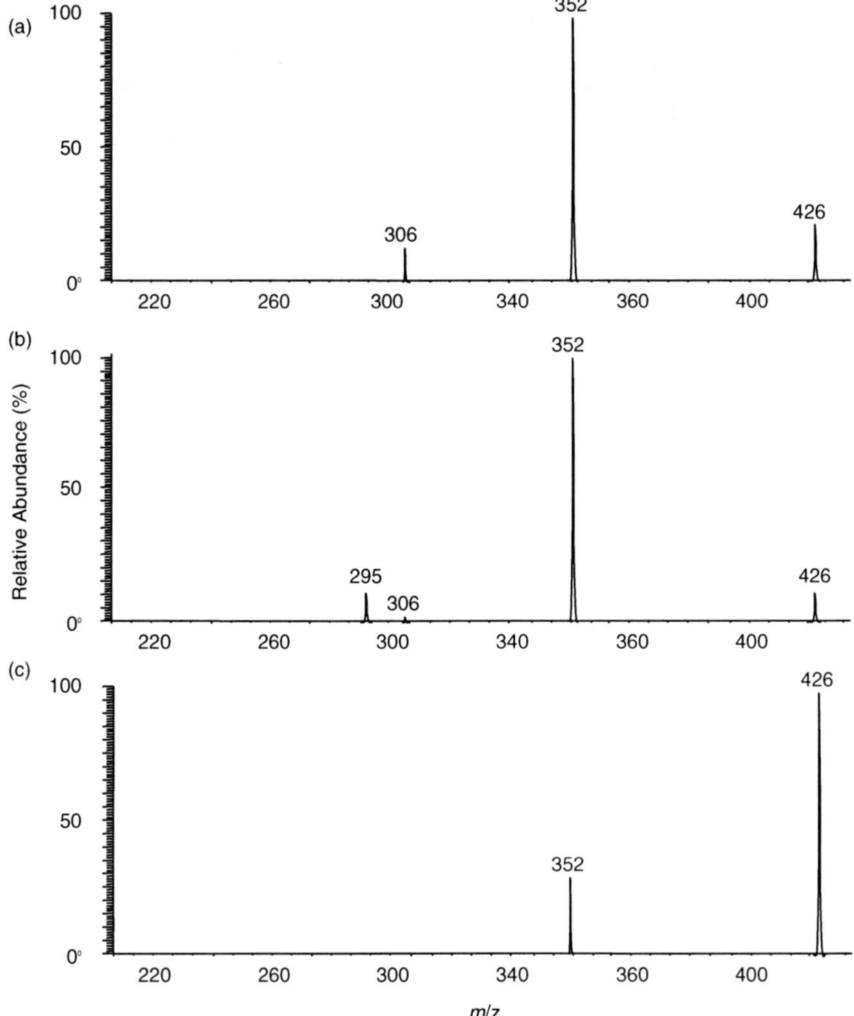

Fig. 17.12. MS/MS of Co(DAP)-bound sugar complexes: (a) ManNAc; (b) GlcNAc; (c) GalNAc [98]. (Note: This figure is an adaptation of the original)

17.3.4.2 Dissociation of metal-bound diastereomeric ions

One method for quantitative chiral analysis (*ee*-determination) of α-amino acids uses a mixture of a chiral analyte (A) and a chiral reference (ref*) compound. The protonated trimers ($[(A)(ref^*)_2H]^+$) are easily generated *via* ESI and are mass selected and dissociated *via* CID. This produces the protonated dimeric ion, $[(A)(ref^*)H]^+$, in the MS/MS spectrum. The basis of this method is that the peak intensity ratio of the product ion to the precursor ion (R, in Eq. (17)) depends on the enantiomeric content of the analyte.

$$R = \frac{I[(A)(ref^*) + H]^+}{I[(A)(ref^*)_2 + H]^+} \tag{17}$$

When the pure *R*-enantiomer is used (A_R), then the ratio becomes $R = R_R$, and when the pure *S*-enantiomer is used (A_S), then the ratio becomes $R = R_S$. The resulting ratio of ratios then becomes R^{CID}_{chiral} (called CR in the original literature but renamed here for internal consistency) and this is to be regarded as a numerical indication of the degree of chiral selectivity, Eq. (18).

$$R^{CID}_{chiral} = \frac{R_S}{R_R} \tag{18}$$

When the racemic sample is used then R becomes R_0 and if the three ratios R_R, R_S, and R_0 are measured, then the equation below can be used for *ee*-determination, Eq. (19) [58,99,100].

$$\%ee = \frac{[1/(R_R - R_S) - 1/(R_S - R_0)]}{[2/(R - R_0) - 1/(R_R - R_0) - 1/(R_S - R_0)]} \times 100 \tag{19}$$

As another example of the application of CID to chiral analysis, consider the ternary complexes of copper(II)-bipyridine with amino acids [101]. A second chiral center in the complex influences the binding preference through its interaction with the chiral amino acids. However, since the fragmentation patterns for different types of complexes are not always predictable, it is not easy to generalize this approach for quantitative chiral analysis.

17.3.4.3 Dissociation of cluster ions using the kinetic method

The use of competitive gas-phase fragmentations of cluster ion to infer thermochemical information has come to be known as the kinetic method [59,60]. In the original form of this experiment, cluster ions bound *via* protons (Eq. (20)) (later other atomic or polyatomic anions or cations) are isolated and the products of their CID are recorded in a MS/MS experiment.

$$[B_1\text{–H–}B_2] \begin{cases} \xrightarrow{k_1} B_1H^+ + B_2 \\ \xrightarrow{k_2} B_1 + B_2H^+ \end{cases} \tag{20}$$

In this equation, k_1 and k_2 represent the rate constants for two competitive dissociation channels of the proton-bound dimer to yield B_1H^+ and B_2H^+, respectively. The ratio of rate constants, *viz.* the branching ratio of two fragment ion abundances, is related logarithmically to the gas-phase basicities of B_1 and B_2, in the standard form of the kinetic method, as shown in Eq. (21).

$$\ln\left(\frac{k_1}{k_2}\right) = \frac{\Delta(\Delta G)}{\mathbf{R}T_{eff}} \tag{21}$$

Here, $\Delta(\Delta G)$ is the difference in gas-phase basicities of B_1 and B_2, **R** is the gas constant, and T_{eff} is the effective temperature (the parameter in the kinetic method which connects the fragmentation of isolated ions to the mean internal energy of the fragmenting population [102]). Note that this relationship only applies when a number of simplifying conditions are fulfilled and a more detailed discussion is given elsewhere, [59,60,103–105].

When trimeric rather than dimeric clusters are generated (trimers create a more sterically hindered complex) from a chiral analyte, chiral reference, and metal cation, dimeric cluster ions are formed as fragmentation products in a CID experiment with the trimeric ion. Fragmentation of the cluster ions leads to different fragmentation patterns that result from differences in the dissociation energies of the two dimeric ions. However, *a reference channel* also exists just as it does in the simple case shown in Eq. (20). This reference channel allows the fragment ion abundances in a spectrum to be used as a thermochemical measure and it is from these two concepts that the kinetic method for chiral analysis is derived.

An early application of these concepts used 2,3-butanediol for stereoisomeric identification [77]. This system was examined by dissociating, in two separate experiments, both positively and negatively charged proton-bound dimeric cluster ions. The $\Delta(\Delta G)$ between the set of diastereomeric complexes can be obtained through the competitive dissociation in the non-reference channel of the cluster ion, and this represents the fundamental basis for chiral distinction [59,60]. Thus, energy differences of < 1 kJ/mol for fragmentation result in large changes in the respective rate constants. This change can be followed through the fragment ion abundance ratio in a MS/MS experiment. It is just one of the features of this experiment that it is highly sensitive to small differences in chiral composition.

As mentioned, the mass-selected diastereomeric cluster ions (the precursor trimer ions) fragment to give diastereomeric products (dimers). However, the clusters are chosen so that the comparative fragments generated by both diastereomeric ions contain the reference ligand. This provides an internal reference so that the systems is calibrated and the measured ion abundance ratios have a predictable (log abundance *versus* enantiomeric excess) relationship, which is illustrated in the next section.

Since trimeric ions as opposed to dimeric ions form more sterically hindered clusters and provide reference fragmentation channels which are free of chiral effects (Fig. 17.13), the interactions have the potential to magnify $\Delta(\Delta G)$ thus improving the enantioselectivity. The kinetic method is related to the product ion branching ratio, Eq. (22), as a result of the analyte (An) and reference (ref*) ligand loss from the cluster ion in an MS/MS experiment [99,106–109].

$$\mathrm{An} \xrightarrow[\mathrm{ESI}]{\mathrm{M^{II}, ref*}} [(\mathrm{M^{II}})(\mathrm{ref^*})_2(\mathrm{An})-\mathrm{H}]^+ \begin{cases} \xrightarrow{k_1} [(\mathrm{M^{II}})(\mathrm{An})(\mathrm{ref^*}) - \mathrm{H}]^+ + \mathrm{ref^*} \\ \xrightarrow{k_2} [(\mathrm{M^{II}})(\mathrm{ref^*})_2 - \mathrm{H}]^+ + \mathrm{An} \end{cases} \tag{22}$$

In this equation, An represents the chiral analyte, $\mathrm{M^{II}}$ represent the transition metal ion, and ref* represents the chiral reference compound. The branching ratio leading to the competitive formation of the two fragment ions (Eq. (23)) is given by the following expression.

$$\mathrm{R} = \frac{[\mathrm{M^{II}}(\mathrm{An})(\mathrm{ref}*) - \mathrm{H}]^+}{[\mathrm{M^{II}}(\mathrm{ref}*)_2 - \mathrm{H}]^+} \tag{23}$$

When the analyte is enantiomerically pure, R becomes R_R or R_S. Enantioselectivity, which is defined as $\mathrm{R_{chiral}}$, serves as a numerical indicator for the degree of chiral

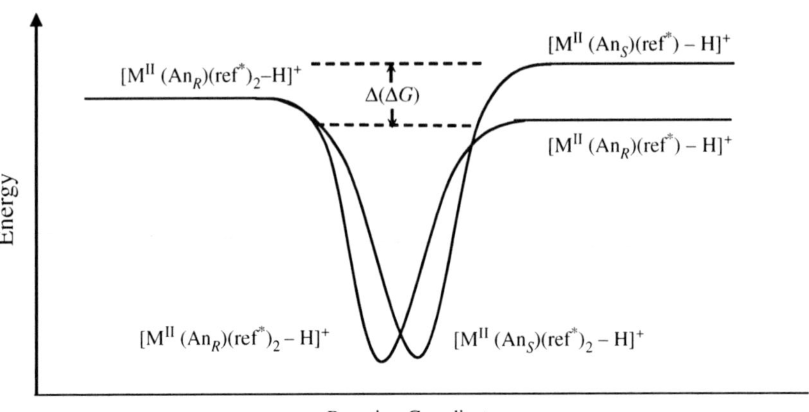

Fig. 17.13. Potential energy diagram showing competitive dissociation for chiral analysis by the kinetic method [51]. (Note: This figure is an adaptation of the original)

sensitivity. When R_{chiral} equals one, this means that the combination of metal, reference, and analyte fails to create a stereochemically dependent interaction. Systems that have R_{chiral} values greater than or less than 1.0 are indicative of chirally selective interactions, Eq. (24).

$$R_{chiral} = \frac{R_R}{R_S} = \frac{[M^{II}(An_R)(ref^*) - H]^+/[M^{II}(ref^*)_2 - H]^+}{[M^{II}(An_S)(ref^*) - H]^+/[M^{II}(ref^*)_2 - H]^+} \tag{24}$$

When choosing the central metal ion it should be tetra-coordinated by a reference and the analyte simultaneously. Transition metal cluster ions have the advantage of easy trimeric cluster ion formation and large chiral recognition, especially when aromatic reference compounds are used to promote π–cation interactions [106]. Table 17.3 compares chiral differentiation of amino acids using different transition metal ions in the case of binding by a proton.

In the kinetic method expression for chiral analysis, $\Delta(\Delta G)$ is defined as the difference in free energies between the reactions in Eqs. (25) and (26) whose reverse barriers are considered to be negligible or equal, as illustrated by Fig. 17.13.

$$[M^{II}(An)(ref^*)_2 - H]^+ \rightarrow [M^{II}(An)(ref^*) - H]^+ \tag{25}$$

$$[M^{II}(An)(ref^*)_2 - H]^+ \rightarrow [M^{II}(ref^*)_2 - H]^+ \tag{26}$$

When the analyte consists of a pure *R*- or *S*-enantiomer, $\Delta(\Delta G)$ becomes $\Delta(\Delta G)_R$ or $\Delta(\Delta G)_S$, respectively, and Eq. (21) takes the form of Eqs. (27) and (28).

$$\ln R_R = \frac{\Delta(\Delta G)_R}{\mathbf{R}T_{eff}} \tag{27}$$

$$\ln R_S = \frac{\Delta(\Delta G)_S}{\mathbf{R}T_{eff}} \tag{28}$$

Table 17.3 Comparison of chiral recognition using Cu(II), Ni(II), and H as the central ions

	Chiral selectivity		
Analyte	Ni[a]	Cu[b]	H[c]
Ala	1.2 (Phe)	2.0 (Phe)	1.1 (BBSer)
Arg	0.68 (Lys)	0.33 (His)	2.0 (Bpro)
Asp	2.4 (Phe)	2.7 (Phe)	0.81 (Bphe)
Asn	2.6 (Gln)	1.8 (Trp)	1.2 (Bpro)
Cys	0.78 (Asn)	Cys oxidized to cystine	1.3 (BBSer)
Glu	2.7 (Phe)	3.1 (Phe)	1.8 (BPRo)
Gln	2.6 (Asn)	6.8 (Trp)	3.0 (BBSer)
His	0.73 (Arg)	0.47 (Arg)	4.3 (BBSer)
Ile	1.7 (Phe)	4.8 (Phe)	1.6 (BBSer)
Leu	1.2 (Phe)	2.3 (Phe)	1.5 (BBSer)
Lys	0.85 (His)	0.56 (His)	2.6 (BBSer)
Met	1.4 (Gln)	7.6 (Trp)	1.5 (Bpro)
Pro	1.8 (Phe)	5.3 (Phe)	2.3 (Bpro)
Phe	3.0 (Trp)	8.3 (Trp)	0.81 (BPhe)
Ser	1.1 (Phe)	1.5 (Phe)	1.3 (Bpro)
Thr	1.2 (Phe)	1.8 (Phe)	1.2 (Bpro)
Trp	7.9 (Asn)	1.8 (Asn)	0.49 (BBSer)
Tyr	2.4 (Phe)	11 (Trp)	1.1 (Bpro)
Val	1.6 (Phe)	4.5 (Phe)	1.5 (BBSer)

[a,b,c]Using Ni [165], Cu [106], and H [99] as the central ions, respectively.

For an enantiomeric mixture with an enantiomeric excess of the *R*-enantiomer given by *ee*, one can derive the expression, Eq. (29).

$$\begin{aligned}\Delta(\Delta G) &= \Delta(\Delta G)_R \frac{1+ee}{2} + \Delta(\Delta G)_S \frac{1-ee}{2} \\ &= \frac{[\Delta(\Delta G)_R + \Delta(\Delta G)_S]}{2} + \frac{[\Delta(\Delta G)_R - \Delta(\Delta G)_S]}{2} ee\end{aligned} \tag{29}$$

Therefore the relationship between R and *ee* can be expressed by combining Eqs. (24), (27), (28), and (29) to obtain Eqs. (30) and (31).

$$\ln \mathrm{R} = \left[\frac{\Delta(\Delta G)_R + \Delta(\Delta G)_S}{2\mathbf{R}T_{\mathrm{eff}}}\right] + \left[\frac{\Delta(\Delta G)_R - \Delta(\Delta G)_S}{2\mathbf{R}T_{\mathrm{eff}}}\right] ee \tag{30}$$

$$\ln \mathrm{R} = \left[\frac{\ln(\mathrm{R}_R) + \ln(\mathrm{R}_S)}{2}\right] + \left[\frac{\ln(\mathrm{R}_{\mathrm{chiral}})}{2}\right] ee \tag{31}$$

A linear relationship between ln R and *ee*, predicted by Eqs. (30) and (31), is observed when using the kinetic method. Equation (31) gives physical meaning to the calibration curve of ln R against *ee*, in which the slope is equal to one half the natural logarithm of the chiral selectivity and the intercept is one half of the sum of the natural logarithm of the branching ratios when the analyte is pure *R* and *S*. It is clear that the larger the chiral

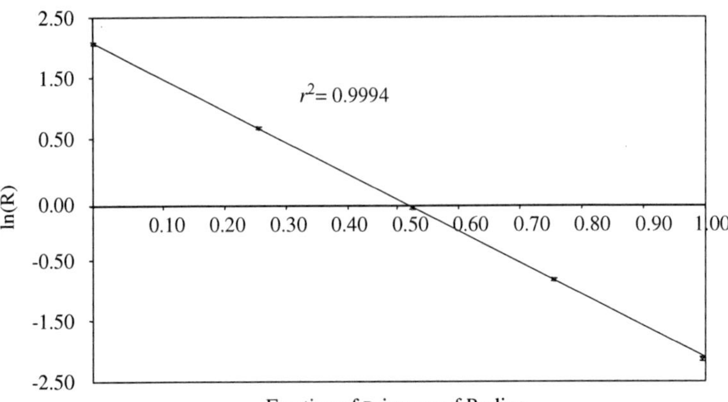

Fig. 17.14. Linear relationship of ln (R) *versus* the fraction of D-proline. The analytes are D- and L-proline [172]. (Note: This figure is an adaptation of the original)

selectivity, the larger the change in the measured ratio R with unit change in *ee*; this will often correspond to a more accurate measurement of %*ee*, unless the peak abundance ratio is so large that the less abundant peak is not easily measured accurately. This relationship has been confirmed for different classes of compounds including amino acids [100,106], α-hydroxy acids [107], sugars [110], and chiral drugs [111–113], as well as chiral [114] and isomeric [115] dipeptides. Although most chiral analysis by the kinetic method has been performed using a quadrupole ion trap, the method has also been applied to *O*-phosphoserine [108], α-aminophosphonic acids [116], and dipeptide derivatives [117] using a triple quadrupole mass spectrometer.

A typical calibration for enantiomeric determination of the amino acid proline shows good linearity with a correlation coefficient (r^2) of 0.9994 (Fig. 17.14) [118]. Furthermore, the kinetic method, as shown by Fig. 17.15, has been applied to enantiomeric quantification of important chiral drug compounds such as pseudoephedrine and DOPA [112]. The results show an excellent correlation between the average measured *ee* and the actual *ee* values. Chiral recognition and quantification of a novel anti-viral drug agent FMAU (2′-fluoro-5-methyl-β-D, L-arabinofuranosyl uracil) (from Triangle Pharmaceuticals, Chart 17.5) has also been successfully demonstrated [111]. The quantification of L-FMAU shows that an average accuracy of 0.6% *ee* is obtained. A summary of applications using the kinetic method to quantitatively determine the %*ee* of different types of chiral drugs is tabulated in Table 17.4. For some cases, as little as 1% *ee* could be measured. The details of these experiments will be addressed later in the Section 17.4.

17.3.4.3.1 Fixed-ligand method A significant improvement in the kinetic method results when one replaces one of the two identical reference ligands by an easily deprotonated compound having high metal affinity, hereafter called the fixed ligand, (L^{fixed}) [119]. Fragmentation still occurs *via* two channels, loss of the reference ligand and the analyte, but the strongly chelated fixed ligand will not be lost, as illustrated in Scheme 17.4.

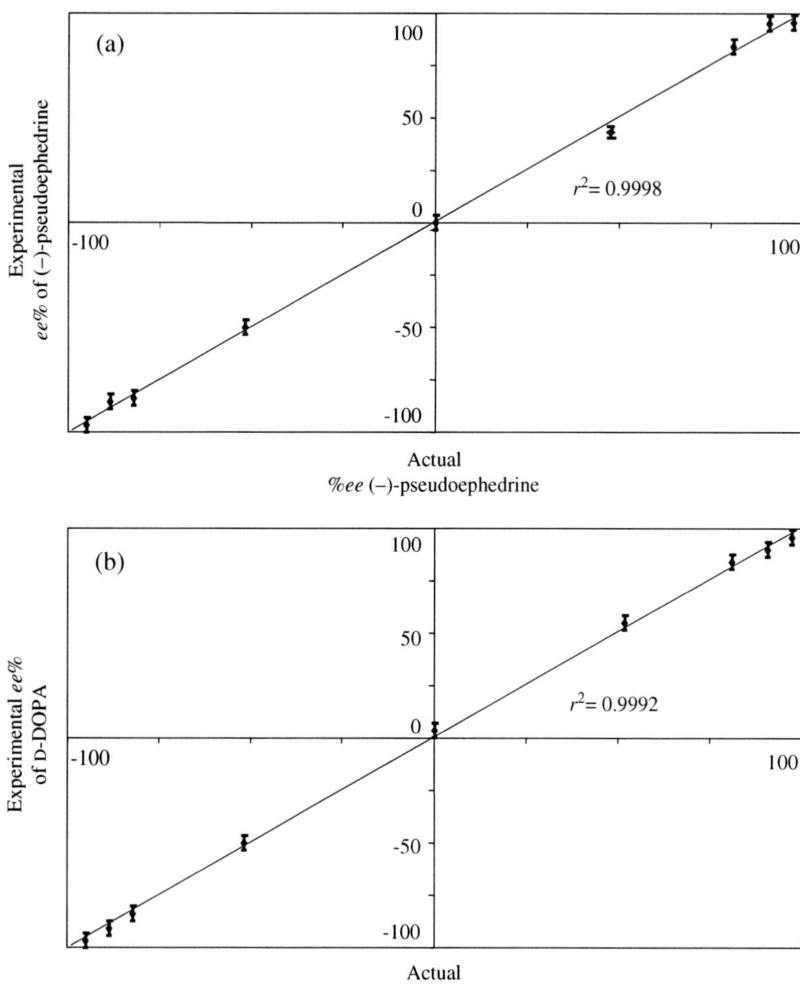

Fig. 17.15. Plots of actual *versus* the measured *ee* for (a) pseudoephedrine and (b) DOPA [167]. (Note: This figure is an adaptation of the original)

Under the fixed ligand conditions, Eq. (22) takes the form of Eq. (32) and accordingly, Eqs. (23) and (24) become Eqs. (33) and (34), respectively.

$$\mathrm{An} \xrightarrow[\mathrm{ESI}]{\mathrm{M^{II}, ref^*, L^{fixed}}} [(\mathrm{M^{II}} + \mathrm{L^{fixed}} - \mathrm{H})(\mathrm{ref^*})(\mathrm{An})]^+ \begin{cases} \xrightarrow{k_1} [(\mathrm{M^{II}} + \mathrm{L^{fixed}} - \mathrm{H})(\mathrm{An})(\mathrm{ref^*})] + \mathrm{ref^*} \\ \xrightarrow{k_2} [(\mathrm{M^{II}} + \mathrm{L^{fixed}} - \mathrm{H})(\mathrm{ref^*})]^+ + \mathrm{An} \end{cases} \tag{32}$$

$$\mathrm{R^{fixed}} = \frac{[(\mathrm{M^{II}} + \mathrm{L^{fixed}} - \mathrm{H})(\mathrm{An})(\mathrm{ref^*})]^+}{[(\mathrm{M^{II}} + \mathrm{L^{fixed}} - \mathrm{H})(\mathrm{ref^*})]^+} \tag{33}$$

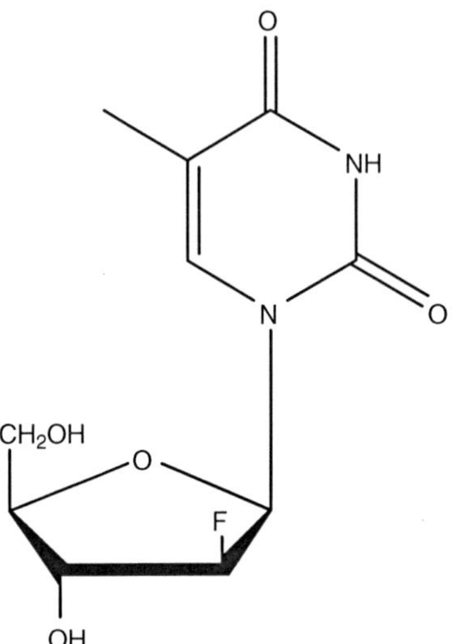

Chart 17.5. *Clevudine* (L-FMAU, 2′-fluoro-5-methyl-β-L-arabinofuranosyluracil) [111].

Table 17.4 Typical applications of chiral quantitation by the kinetic method[a]

System		Actual (%)	Experimental (%)	Absolute error (%)	Reference
Analyte	Central ion and reference				
FMAU	Co (II), *N*-acetyl-L-Pro	30	30.6 ± 0.5	0.6	Ref. [166]
ψ-Ephedrine	Cu (II), L-Trp	−50	−50.7 ± 0.2[b]	0.7	Ref. [167]
DOPA	Cu (II), L-Tyr	90	89.0 ± 1.8[b]	1.0	Ref. [167]
Leu	H, D-BBSer	76	75.1 ± 0.8	0.9	Ref. [100]
Mannose	Cu (II), *N*-FMOC-L-Pro	−52	−51.6 ± 0.5	0.4	Ref. [110]
		−92	−94.0 ± 3.0	2	Ref. [110]
Thalidomide	Zn (II), D-galactose	77	77 ± 3	0	Ref. [113]
		25	26 ± 1	1	Ref. [113]

[a]Chosen from previously published results, but not a complete list.
[b]Standard deviations are estimated based on the original data in Refs. [167] and [100].

$$\mathrm{R}_{\mathrm{chiral}}^{\mathrm{fixed}} = \frac{\mathrm{R}_R^{\mathrm{fixed}}}{\mathrm{R}_S^{\mathrm{fixed}}} = \frac{[(\mathrm{M}^{\mathrm{II}} + \mathrm{L}^{\mathrm{fixed}} - \mathrm{H})(\mathrm{An}_R)]^+/[(\mathrm{M}^{\mathrm{II}} + \mathrm{L}^{\mathrm{fixed}} - \mathrm{H})(\mathrm{ref}^*)]^+}{[(\mathrm{M}^{\mathrm{II}} + \mathrm{L}^{\mathrm{fixed}} - \mathrm{H})(\mathrm{An}_S)]^+/[(\mathrm{M}^{\mathrm{II}} + \mathrm{L}^{\mathrm{fixed}} - \mathrm{H})(\mathrm{ref}^*)]^+} \tag{34}$$

Here $\mathrm{R}_R^{\mathrm{fixed}}$ and $\mathrm{R}_R^{\mathrm{fixed}}$ are the branching ratios for the enantiomerically pure analytes, An_R and An_S, when the fixed ligand is used. Note that the equations are written showing the fixed ligand as the only deprotonation site, which is desired but may not always be achieved. Based on the kinetic method [120], a linear relationship (Eq. (35)) is expected to exist between the free energy $\Delta(\Delta G^{\mathrm{fixed}})$ difference and the natural logarithm of the chiral selectivity (ln $\mathrm{R}^{\mathrm{fixed}}$).

$$\ln \mathrm{R}^{\mathrm{fixed}} = \frac{\Delta(\Delta G^{\mathrm{fixed}})}{\mathbf{R}T_{\mathrm{eff}}} \tag{35}$$

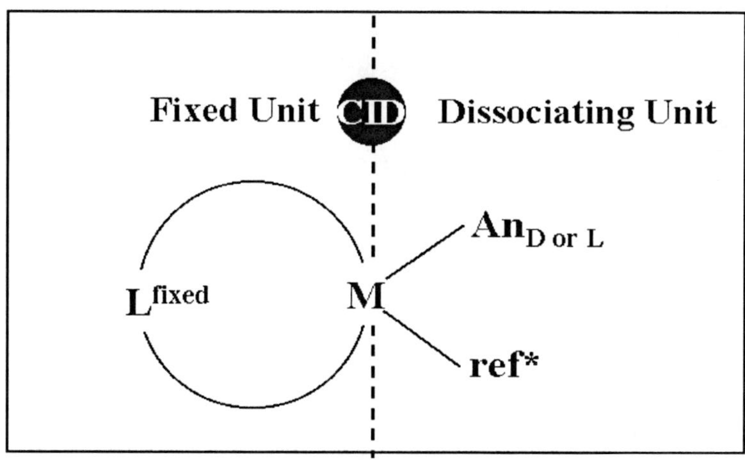

Scheme 17.4. The fixed-version kinetic method for chiral analysis [119].

Here $\Delta(\Delta G^{\text{fixed}})$ is defined as the difference in free energies between reactions Eqs. (36) and (37), the reverse barriers of which are considered to be negligible or equal.

$$[(\mathrm{M}^{\mathrm{II}} + \mathrm{L}^{\text{fixed}} - \mathrm{H})(\mathrm{An})(\mathrm{ref}^*)]^+ \rightarrow [(\mathrm{M}^{\mathrm{II}}+\mathrm{L}^{\text{fixed}} - \mathrm{H})(\mathrm{An})]^+ + \mathrm{ref}^*. \tag{36}$$

$$[(\mathrm{M}^{\mathrm{II}}+\mathrm{L}^{\text{fixed}} - \mathrm{H})(\mathrm{An})(\mathrm{ref}^*)]^+ \rightarrow [(\mathrm{M}^{\mathrm{II}}+\mathrm{L}^{\text{fixed}} - \mathrm{H})(\mathrm{ref}^*)]^+ + \mathrm{An}. \tag{37}$$

When the analyte consists of a pure *R*- or *S*-enantiomer, $\Delta(\Delta G^{\text{fixed}})$ becomes $\Delta(\Delta G^{\text{fixed}})_R$ or $\Delta(\Delta G^{\text{fixed}})_S$, respectively, and Eq. (35) takes the forms of Eqs. (38) and (39).

$$\ln \mathrm{R}_R^{\text{fixed}} = \frac{\Delta(\Delta G^{\text{fixed}})_R}{\mathbf{R}T_{\text{eff}}} \tag{38}$$

$$\ln \mathrm{R}_S^{\text{fixed}} = \frac{\Delta(\Delta G^{\text{fixed}})_S}{\mathbf{R}T_{\text{eff}}} \tag{39}$$

For an enantiomeric mixture with an enantiomeric excess of the *R*-enantiomer given by *ee*, one can write.

$$\begin{aligned} \Delta(\Delta G^{\text{fixed}}) &= \Delta(\Delta G^{\text{fixed}})_R \frac{1+ee}{2} + \Delta(\Delta G^{\text{fixed}})_S \frac{1-ee}{2} \\ &= \frac{[\Delta(\Delta G)_R^{\text{fixed}} + \Delta(\Delta G)_S^{\text{fixed}}]}{2} + \frac{[\Delta(\Delta G)_R^{\text{fixed}} - \Delta(\Delta G)_S^{\text{fixed}}]}{2} ee \end{aligned} \tag{40}$$

Therefore the relationship between R and *ee* can be expressed by combining Eqs. (33), (34), (35), and (40) to obtain Eq. (41).

$$\ln \mathrm{R}^{\text{fixed}} = \left[\frac{\ln(\mathrm{R}_R^{\text{fixed}}) + \ln(\mathrm{R}_S^{\text{fixed}})}{2}\right] + \left[\frac{\ln(\mathrm{R}_{\text{chiral}}^{\text{fixed}})}{2}\right] ee \tag{41}$$

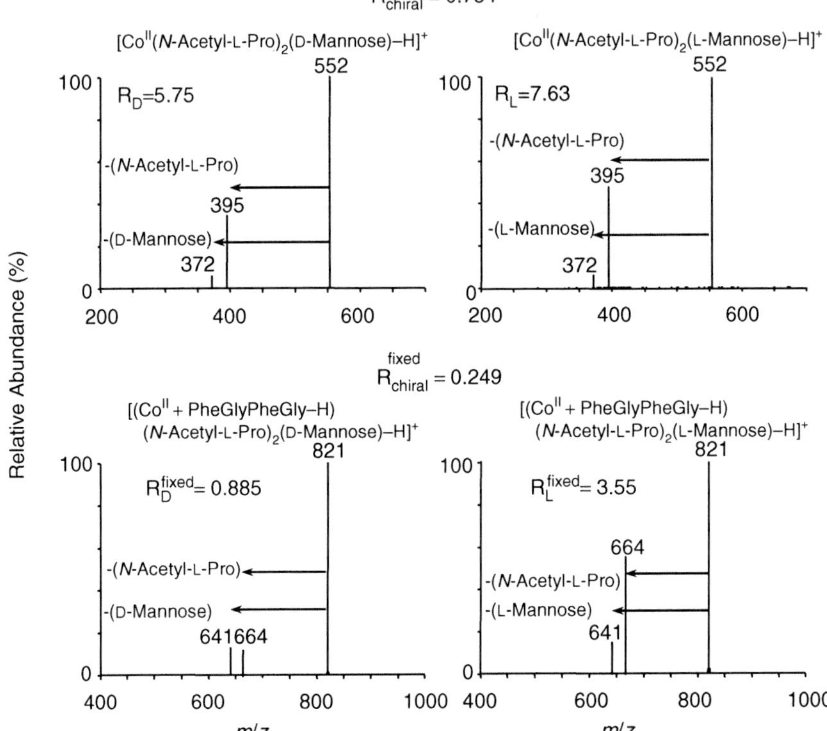

Fig. 17.16. MS/MS product ion mass spectra showing improved chiral recognition of D/L-mannose (An) using the fixed-ligand kinetic method by studying the dissociation kinetics of the trimeric cluster ion $[(Co^{II} + \text{L-Phe-Gly-L-Phe-Gly}-H)(N\text{-Acetyl-L-Pro})(An)]^+$, in comparison with $[Co^{II}(N\text{-Acetyl-L-Pro})_2(An)-H]^+$ [173]. (Note: This figure is an adaptation of the original)

Equation (41) predicts a linear relationship between ln R^{fixed} and *ee* using the fixed-ligand kinetic method based on trimeric cluster ions of the type $[(M^{II} + L^{fixed}-H)(ref^*)(An)]^+$.

The fixed-ligand version of the kinetic method makes it easy to change the properties of the fixed ligand (i.e. its size and functionality), to optimize the chiral interactions resulting from metal–ligand and ligand–ligand contact, and to maximize chiral recognition. Fig. 17.16 shows a factor of three improvement for the chiral recognition of the sugar mannose, as a result of observing the analyte and non-fixed ligand loss [121]. Another advantage of the fixed-ligand kinetic method is that it simplifies the dissociation kinetics. This makes it easy to construct calibration curves with good correlation coefficients, as illustrated by Fig. 17.17 for chiral analysis of the mixture D-Ala-D-Ala/L-Ala-L-Ala [121]. This method was extended to the analysis of drug compounds [119] as has been used in conjunction with other forms of ionization [122].

17.3.4.3.2 Quotient ratio method This variant on the kinetic method addresses the common situation that only one standard with known optical purity is available to

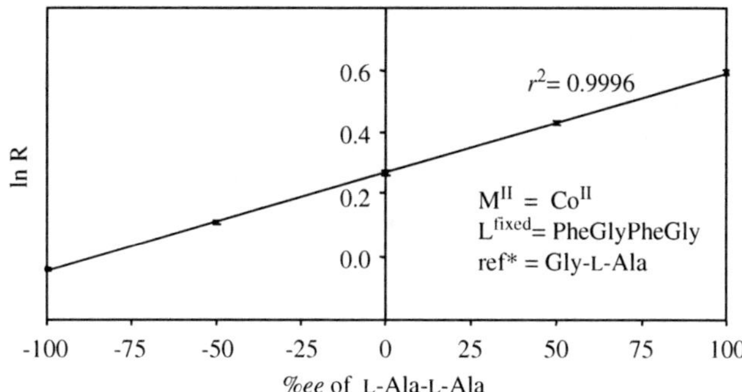

Fig. 17.17. Quantification of D-Ala-D-Ala/L-Ala-L-Ala using L-Phe-Gly-L-Phe-Gly as the fixed ligand, Gly-L-Ala as the reference ligand, and Co^{II} as the central metal ion. The error bars corresponding to each point are standard deviation based on triplicate measurements at 95% confidence level [173]. (Note: This figure is an adaptation of the original)

develop a chiral analysis method. The kinetic method is still applicable, provided that two independent measurements are made for every unknown sample. This is achieved by altering the chirality of the reference ligand in the standard method (precursor ion: $[M^{II}(ref^*)(An)_2-H]^+]$ or that of both the reference ligand and the fixed ligand, precursor ion: $[(M^{II}+L^{fixed}-H)(ref^*)(An)]^+$). This approach is called the quotient ratio method (QR) [123]. The fixed-ligand method can also be used as a variant to the quotient ratio (QR^{fixed}) [124], where singly-charged trimeric cluster ions, $[(M^{II}+L^{fixed}-H)(ref^*)(An)]^+$, are readily generated by soft ionization methods such as ESI. In particular for the case of the fixed-ligand version of the kinetic method, there are reciprocal relationships when inverting the chirality of the fixed/reference ligands, so it is easy to construct two or more single-point calibration curves. This allows data to be cross-checked. For example, by using the dipeptide D-Ala-D-Ala as the fixed ligand and L-Tyr as the reference ligand for chiral recognition of the D/L-DOPA, there are four possible calibration curves, as shown by Fig. 17.18 [124]. There are 8 and 16 calibration curves which can be constructed using tripeptides and tetrapeptides as the fixed ligand. It is easy to see that the numbers of combinations will be doubled when using reference ligands with two chiral centers. Furthermore, by changing the chirality of the fixed ligand, the chiral interactions in the cluster ion are refined, allowing one to maximize chiral differentiation. Note that any alteration in the reference or fixed ligand will result in chirality switching and this produces the largest effect and is often most conveniently used.

17.3.4.4 Enantioselectivity

When analyzing chiral tetramide macrocycles, larger gas-phase enantioselectivity was observed with two different MS chiral methods. Diastereomeric proton-bound complexes were generated in the gas phase, where an I/M reaction method and the kinetic method were used to distinguish the enantiomers. The aim of this work was to determine the factors surrounding the large enantioselectivity and from the combined experimental approaches (I/M reaction method and the kinetic method) as well as calculations, develop correlations between the two methods.

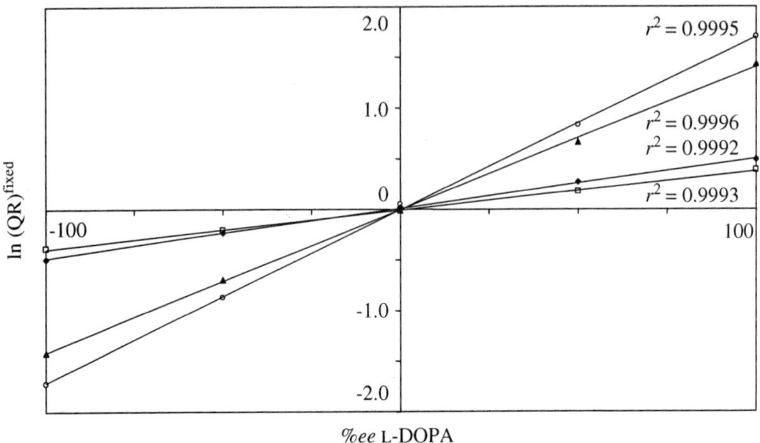

Fig. 17.18. Calibration curves for chiral analysis of D/L-DOPA using the fixed-ligand quotient-ratio (QR^{fixed}) method with four combinations of chirality of the fixed ligands and reference ligands [174]. (Note: This figure is an adaptation of the original)

The I/M reaction method was conducted by measuring exchange rates between a proton bound dimer, $[MHA]^+$, and a chiral reagent ion (an ion that easily displaces the analyte ion) in an FT-ICR that was equipped with an ESI source. The chiral reagents, (*R*)-(−) and (*S*)-(+)-2-butylamine, participated in a substitution reaction with the chiral analyte described by, Eq. (42).

$$[MHA]^+ + B \rightarrow [MHB]^+ + A \tag{42}$$

The rate constant, k', of this reaction was obtained from the slope of a pseudo-first-order rate plot (shown as ln R, which is the ratio of I/I_0 where I is the intensity of complex $[MHA]^+$ at the delay time t and I_0 is the sum of the intensities of $[MHA]^+$ and $[MHB]^+$) *versus* t (time). This rate plot is similar to that seen in Fig. 17.10 and discussed in detail in Section 17.3.3. This experiment generated selectivity values as large as 0.05 (determined from the slope of a calibration plot) and this represents the largest enantioselectivity measured with this type of MS method. Also, the reagent ion was varied so that enantioselectivity effects could be observed. In all the systems investigated there was no significant effect on the selectivity as a result of the varying functionality of the chiral reagent.

Kinetic method experiments were typically carried out using an ESI-QIT, where the diastereomeric ion formed consisted of a proton-bound trimer, $[M_2HA]^+$ (M represents the tetramide macrocycle, H represents the proton and A represents the analytes). The stability gap between the respective enantiomers was as large as 10 (R_{chiral}), as determined from the relative ion abundances in CID experiments. This value varied with respect to the varying functionality of the macrocycles and the analytes substituents. The results for both experiments were compared and confirmed the superior sensitivity of the kinetic method in comparison to the I/M reaction method.

In additional experiments, the free energy ($\Delta\Delta G_{I/M}$) evaluated from the difference in the activation barriers for the displacement reaction of both enantiomers ($\Delta\Delta G_{I/M} = \Delta\Delta G_R - \Delta\Delta G_S$) in the I/M experiment was derived. This value was compared with the free energy term of the corresponding kinetic method experiment ($\Delta\Delta G_{KM}$). In addition calculations confirmed the experimental results. The results suggest that the large enantioselectivity was due to the thermodynamic stability of the diastereomeric complexes, where the more thermodynamically stable system gave the larger selectivity [125]. More systems with improved thermodynamic stability are being explored to corroborate this finding.

17.4 APPLICATIONS

17.4.1 Chiral analysis of ternary mixtures

One of the attractive features of the kinetic method is its possible extension to optical isomer quantification in ternary and higher order mixtures, (compounds with more than two optical isomers) [126,127]. Tartaric acid provides a simple illustration. Based on the preceding discussion the linear relationship between ln R and *ee* using the kinetic method, and a two-point calibration plot permits quantitative analysis in binary mixtures such as D/L-tartaric acid. However, if a third isomer such as *meso*-tartaric acid is present, it will also contribute to the measured branching ratio. Based on the extensive property of free energy change, for a ternary mixture with molar fractions α_D (D-tartaric acid), α_L (L-tartaric acid), and $(1-\alpha_D-\alpha_L)$ (*meso*-tartaric acid), respectively, one can write Eq. (43). This equation explains the difference in the free energy changes associated with the competitive loss of the neutral reference compound and tartaric acid from the cluster ion $[M^{II}(An)(ref^*)_2 - H]^+$. The cluster ion was generated from a mixture containing a chiral reference and any combination of D-, L-, or *meso*-tartaric acid. Eq. (44) describes the measured branching ratio for the ternary dissociation process.

$$\Delta(\Delta G) = \alpha_{D*}\Delta(\Delta G)_D + \alpha_{L*}\Delta(\Delta G)_L + (1-\alpha_D-\alpha_L)*\Delta(\Delta G)_{meso} \tag{43}$$

$$\ln R = \alpha_{D*}\ln R_D + \alpha_L * \ln R_L + (1-\alpha_D-\alpha_L) * \ln R_{meso} \tag{44}$$

If the same analyte is measured under a second set of conditions, a second set of equations is obtained and the system of two equations and unknowns can be solved. For example, one can select: (i) Co^{II} as the central metal ion and L-DOPA as the reference ligand and (ii) Ni^{II} as the central metal ion and *N*-acetyl-L-Phe as the reference ligand. To illustrate the additional contribution from *meso*-tartaric acid, three-point calibration curves can be constructed using the measured branching ratios and displayed *versus* the molar fraction of D-tartaric acid (α_D) (Fig. 17.19).

Using cartesian coordinates, the measured R values can be converted into two groups of points: D1(100, −1.61), L1(0, −2.27), *meso*1(0, −1.90) for system (i) shown as open-diamond symbols and D2(100, 0.0770), L2(0, −0.728), *meso*2(0, −0.442) for system (ii), shown as filled-triangle symbols. Note that the points D, L, and *meso* correspond to cases

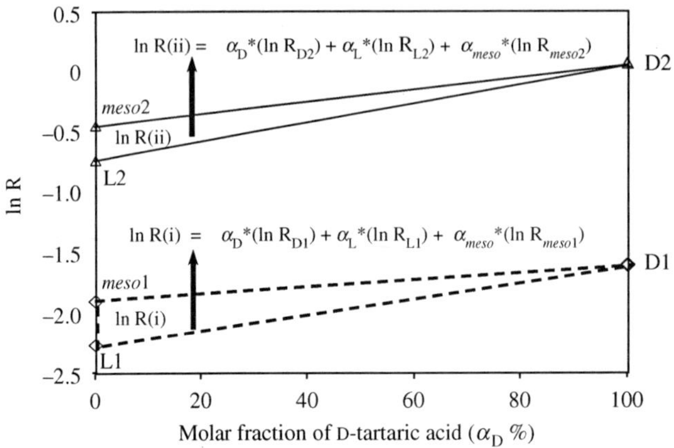

Fig. 17.19. Three-point calibration diagram for chiral quantification of a ternary mixture of tartaric acid using two separate systems: (i) Co^{II} as the central metal ion and L-DOPA as the reference ligand; (ii) Ni^{II} as the central metal ion and *N*-acetyl-L-Phe as the reference ligand. Values corresponding to each point are averages based on triplicate measurements made on separate occasions with less than 2% error [126]. (Reproduced by permission of The Royal Society of Chemistry.)

Table 17.5 Chiral quantification of a ternary mixture of tartaric acids[a] (Reproduced by permission of The Royal Society of Chemistry.)

	Measured R		Actual			Experimental			Relative error(%)		
Entry	Co/L-DOPA	Ni/*N*-Acetyl-L-Phe	%D	%L	%*meso*	%D	%L	%*meso*	D	L	*meso*
1	0.177	0.851	65	30	5	65.3	29.5	5.2	0.5	1.7	4.0
2	0.191	1.01	90	5	5	89.5	5.3	5.2	0.6	6.0	4.0
3	0.143	0.639	10	20	70	9.6	19.7	70.7	4.0	1.5	1.0
4	0.109	0.509	5	90	5	4.9	90.3	4.9	2.0	0.3	2.0
								Average:	1.8	2.4	2.8

[a]CID activation time and energy is optimized and kept constant when the same transition metal and reference ligand are used. (Data taken from Ref. [126].)

in which the analyte is pure D-, pure L-, and pure meso form, respectively. Chiral quantification of ternary mixtures of the optical isomer of tartaric acids was performed using samples with several representative compositions (Table 17.5). If the same analyte is measured under a second set of conditions, a second set of equations is obtained and then a system with three unknowns can be chirally resolved. The experimentally measured molar fractions are consistent with the actual compositions and they show that less than 1% mole fraction of any form of tartaric acid can be determined with relative errors that range from 0.5 to 6.0%. The average relative errors for D-tartaric acid, L-tartaric acid, and *meso*-tartaric acid are 1.8, 2.4, and 2.8%, respectively. This set of data shows that the kinetic method allows quantitative chiral analysis of a ternary mixture of compounds with multiple chiral centers.

17.4.2 Chiral analysis of complex mixtures

The kinetic method can also be extended for chiral analysis of mixtures of different analytes [128]. In this case, Eq. (22) takes the form shown in Eq. (45)

$$\begin{pmatrix} An_1 \\ An_2 \\ \bullet \\ \bullet \\ \bullet \\ An_m \end{pmatrix} \xrightarrow[\text{ESI}]{M^{II},\,\text{ref}^*} [Cu^{II}(\text{ref}^*)_2(An_m) - H]^+ \begin{cases} \nearrow [M^{II}(\text{ref}^*)(An_m) - H]^*(\text{ref}^*) \\ \searrow [M^{II}(\text{ref}^*)_2 - H]^+ + An_m \end{cases} \quad (45)$$

where m is the number of amino acids in a mixture and ref* is an amino acid reference ligand chosen. Note that the quantitative accuracy decreases with increasing the number of analytes in a mixtures, partially due to ion suppression during ESI. Each desired trimeric cluster ion is individually mass-selected and collisionally activated to cause dissociation. For the analyte An_m, the branching ratio $R_{D(m)}$ or $R_{L(m)}$ (related to the analyte in D or L form), and the individual chiral selectivity $R_{\text{chiral}\,(m)}$ are measured based on the procedures described in Eqs. (46) and (47).

$$R_m = \frac{[M^{II}(\text{ref}^*)(An_m - H)]^+}{[M^{II}(\text{ref}^*)_2 - H]^+} \quad (46)$$

$$R_{\text{chiral}(m)} = \frac{R_{D(m)}}{R_{L(m)}} = \frac{[M^{II}(\text{ref}^*)(An_{D(m)} - H)]^+/[M^{II}(\text{ref}^*)_2 - H]^+}{[M^{II}(\text{ref}^*)(An_{L(m)} - H)]^+/[M^{II}(\text{ref}^*)_2 - H]^+} \quad (47)$$

A calibration curve for the analyte An_m is constructed based on Eq. (48)

$$\ln R_{\text{chiral}(m)} = \left[\frac{\ln(R_{D(m)}) + \ln(R_{L(m)})}{2}\right] + \left[\frac{\ln(R_{\text{chiral}(m)})}{2}\right] ee \quad (48)$$

Mixtures of amino acids containing unknown chiral compositions are analyzed by measuring the branching ratio of the two characteristic fragment ions in each of a set of tandem mass spectra recorded for mass-selected cluster ions generated from the mixture. The chiral selectivity is intrinsically related to the energetics of dissociation from trimer to a dimer, and it is expected that chiral selectivity $R_{\text{chiral}(m)}$ will be equal to R_{chiral}.

A series of measurements on a mixture of leucine/proline was made to test this concept. The analysis of each sample only required one measurement in a MS/MS experiment, although triplicate measurements were made on separate occasions to improve precision. The results (Table 17.6) show that less than 2%*ee* could be measured with relative errors ranging from 3.0 to 4.0%. Even though this approach was successful in this particular case, it is recognized that the situation will be much more complicated in a real biological matrix containing a larger variety of compounds. The formation of the target trimeric complex may be increasingly difficult; especially if the analyte concentrations are normally very low for direct quantification. To overcome this problem, a "chiral titration" method could be developed by adding high concentration analytes with known enantiomeric excess.

Table 17.6 Enantiomeric excess (*ee*) of D-amino acids in a binary mixture[a]

		Enantiomeric excess (*ee*) of D-amino acids[b,c,d]			
Analyte	Actual (%)	Experimental (%)			Average (%) (SD[e]), (Relative error (%))
Leu	5	3.6[b]/4.9[c]	6.2/3.4[c]	5.8/6.1[c]	5.2 (1.4)[b]/4.8 (1.4)[c], (4.0[b]/4.0[c])
Pro	10	11[b]/11[d]	8.2[b]/9.8[d]	10[b]/8.7[d]	9.7 (1.4)[b]/9.8 (1.2)[d], (3.0[b]/2.0[d])

[a]Using Cu^{II} as the central metal ion and L-Phe as the reference ligand.
[b]Values obtained from measurements of D-amino acids in a solution of Leu/Pro.
[c]Values obtained from measurements of D-amino acids in a solution of Leu.
[d]Values obtained from measurements of D-amino acids in a solution of Pro.
[e]Standard deviation of the mean based on triplicate measurements on separate occasions.

Table 17.7 Chiral recognition of DOPA measured with the fixed and reference ligand of different chirality[a,b,c] (Data taken from [121])

			$[Cu^{II}(L^{fixed} - H)(An)]^+ / [Cu^{II}(L^{fixed} - H)(ref^*)]^+$				
Entry	L^{fixed}	ref*	D-DOPA	L-DOPA	$(RR)^{fixed}$	$1/(RR)^{fixed}$	$(RR)^{fixed}_{chiral}$
1	D-Ala-D-Ala	D-Tyr	4.47	7.27	0.615	1.63	
	L-Ala-L-Ala	L-Tyr	7.29	4.46	1.63	0.612	0.377
2	D-Ala-D-Ala	L-Tyr	3.42	5.02	0.681	1.47	
	L-Ala-L-Ala	D-Tyr	5.08	3.45	1.47	0.679	0.463
3	D-Ala-L-Ala	D-Tyr	3.98	22.2	0.179	5.59	
	L-Ala-D-Ala	L-Tyr	21.9	3.99	5.49	0.182	0.0326
4	D-Ala-L-Ala	L-Tyr	1.31	5.38	0.243	4.11	
	L-Ala-D-Ala	D-Tyr	5.39	1.29	4.18	0.239	0.0581

[a]Cu^{II} as the central metal ion.
[b]$(RR)^{fixed}_{chiral}$ is defined in Eq. (2).
[c]CID activation level is optimized and kept constant for all measurements of enantiomers.

17.4.3 Chiral morphing

Chiral morphing is a novel technique that allows systematic changes in the chirality of a given ligand for the purpose of optimizing the properties of the compound [124]. For a drug agent, this allows utilization of the spatial diversity of multiple chiral centers to produce drug candidates with improved efficiency, stability, membrane permeability, and oral availability, as well as decreased toxicity and side effects [129].

The fixed-ligand version of the kinetic method, has been utilized with the chiral morphing concept (changing the chirality of the fixed ligand) to further refine the chiral interactions and hence to maximize chiral recognition. Table 17.7 shows how the chirality is affected by the addition of the fixed ligand Ala–Ala. A further advantage is the use of single-point calibration with the ability to obtain multiple calibration curves by chirality-switching of the fixed/reference ligands. The value of this option is that it allows one to cross-check data and prevents errors as well as allowing one to conduct an analysis using just one standard of known %*ee*.

17.4.4 Chiral drug analysis

MS methods, including gas-phase I/M reactions and especially the kinetic method, have been successfully applied to chiral analysis of various types of drugs [119,130,131]. This section deals specifically with kinetic method applications, since the general trend has been to use the kinetic method to develop chiral MS methods. High-throughput *ee* MS analyses for various chiral drugs have been performed, the details of which will be explained.

An increasing number of unnatural L-nucleoside analogs are being investigated as potent chemotherapeutic agents against human immunodeficiency virus (HIV) [132], hepatitis B virus (HBV) [133] and certain forms of cancer [134]. Their enantiomeric counterparts, the D-nucleoside analogs, usually show substantial toxicity likely due to their incorporation into cellular chromosomal and/or mitochondrial DNA [135]. *Clevudine* (L-FMAU, 2′-fluoro-5-methyl-β-L-arabinofuranosyluracil, Chart 17.5), for instance, is a potent anti-HBV agent [136], with EC_{50} (exposure concentration in 50% of the tested population) as low as 0.13 μM and no indication of any apparent clinical toxicity, including interference with mitochondrial function. Its enantiomer, D-FMAU, exhibited unacceptable toxicity in preclinical trials [135].

Fig. 17.20 demonstrates that when the *ee* of L-FMAU is varied, the measured ion branching ratio (R) is affected. A relatively simple relationship between R and *ee* was established, thereby allowing chiral mixtures to be rapidly determined in a MS/MS experiment. A linear relationship between ln (R) *versus ee* was observed (Fig. 17.21) with a correlation coefficient (r^2) of 0.9995, using *N*-Ac-L-Pro as the ref* and Co^{2+} as the central ion. The calibration curve was then used to measure the percent *ee* of unknown samples. An average accuracy of 0.6%*ee* was obtained for this particular case over the wide %*ee* range shown, Fig. 17.21. Besides the speed and high sensitivity—characteristics of most mass spectrometric approaches—this method of quantitative chiral analysis displays several unique features, including accurate chiral quantification without requiring isotopic labeling or wet chemical analysis. The method is tolerant to matrix interference and was implemented using a commercial instrument.

For the generic drugs DOPA, norepinephrine, ephedrine, ψ-ephedrine, propranolol, isoproterenol, and atenolol (Chart 17.6), copper (II)-bound complexes with complementing L-amino acid references were generated *via* ESI [112]. The corresponding trimeric cluster ions that were generated were collisionally activated and underwent dissociation resulting in competitive loss of either the neutral reference or the neutral drug. Given that the product ion branching ratios are related *via* the kinetic method a two-point calibration curve was constructed for quantitative *ee* determination of the generic chiral drugs. The method was optimized to allow *ee* determinations within a few percent.

17.4.5 Method development and validation for chiral purity determination

The single ratio method [51] was used for the quantitative chiral analysis of MaxiPost, a novel potassium channel opener synthesized at Bristol-Myers Squibb Pharmaceutical

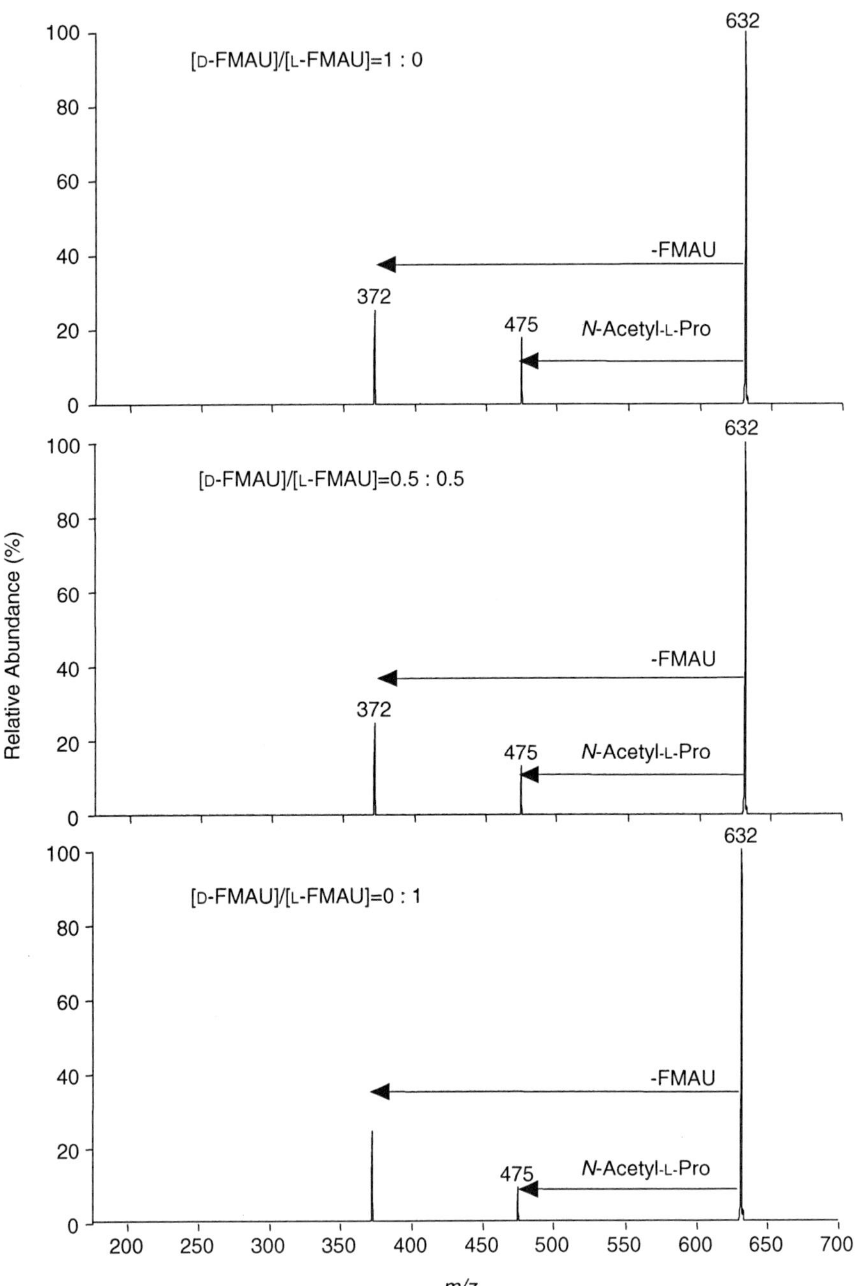

Fig. 17.20. MS/MS product ion spectra of $[Co^{II}(N\text{-Acetyl-L-Pro})_2(FMAU) - H]^+$ ($m/z = 632$) for mixtures of D-and L-FMAU in different proportions. The CID activation level chosen (12%) corresponds to approximately 300 m VAC excitation [166]. (Note: This figure is an adaptation of the original)

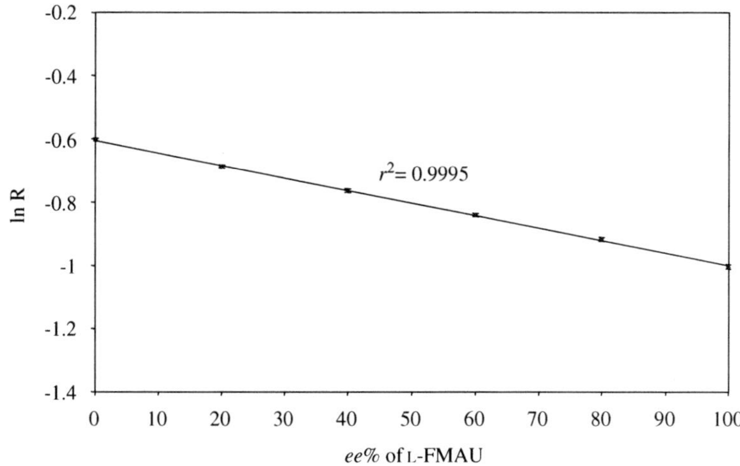

Fig. 17.21. Calibration curve for chiral analysis of enantiomeric excess of L-FMAU using *N*-acetyl-L-Pro as the chiral reference ligand and Co(II) as the central metal ion. Each calibration point shows error bars and represents the average of three individual measurements [166]. (Note: This figure is an adaptation of the original)

Research Institute [137]. Previously, HPLC with ultraviolet detection (LC/UV) and/or circular dichroism detection (LC/CD) was used for the chiral quantification of MaxiPost [138]. The motivation of this work was to benchmark the newly developed chiral MS method against the existing LC/UV and LC/CD methods, using side-by-side comparisons of the method linearity, accuracy, and precision and to demonstrate the industrial applicability of chiral MS in early pharmaceutical development [139].

For chiral LC/UV, the chiral purity of MaxiPost was determined by evaluating the UV area percent (AP) of the individually separated enantiomers (MaxiPost is the *S*-enantiomer with 99.8% chiral purity and its enantiomer BMS-204353 is the *R*-enantiomer that is 100% pure). The chiral purity was determined by dividing the integrated UV area for MaxiPost ($AP_{MaxiPost}$) by the total integrated UV area for both enantiomers ($AP_{MaxiPost} + AP_{BMS\text{-}204353}$). For achiral LC/CD, the chiral purity of MaxiPost was determined by generating a response curve of g-factor [140] (CD response/ UV response) *versus* a set of standards of known chiral purity. The chiral purity of the standards and unknown samples was then evaluated using the least squares fit of the response curve.

Fig. 17.22 shows an ESI mass spectrum of a mixture containing LiCl, MaxiPost, and (+)-5-fluorodeoxyuridine. This figure illustrates how efficiently the trimeric cluster ion of interest at *m/z* 858 $[(Li)(A)(ref^*)_2]^+$, where A represents MaxiPost, BMS-204353, or a mixture of the two enantiomers and ref* represents (+)-5-fluorodeoxyuridine, was generated. Note that chiral analysis *via* the kinetic method has been mainly performed using a quadrupole ion trap but this method was developed on a triple quadrupole mass spectrometer. Also, samples were introduced into the mass spectrometer *via* loop injection, allowing full automation (resulting in reduced analysis time) and improvements in quantitative analysis (resulting from the use of the triple quadrupole

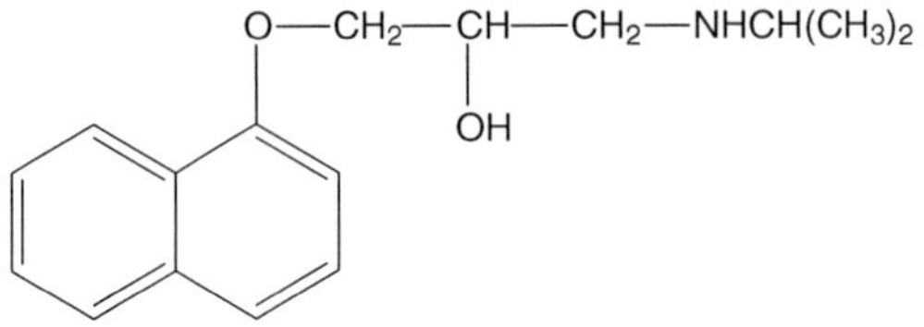

Propranolol

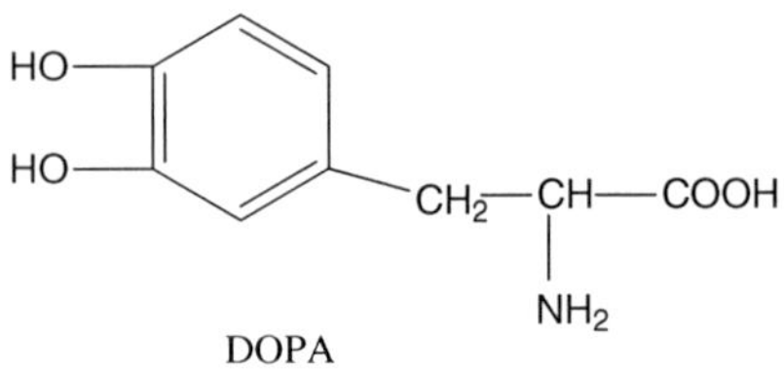

DOPA

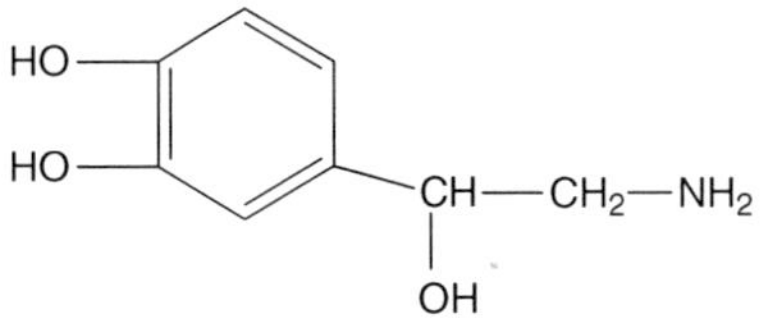

Norepinephrine

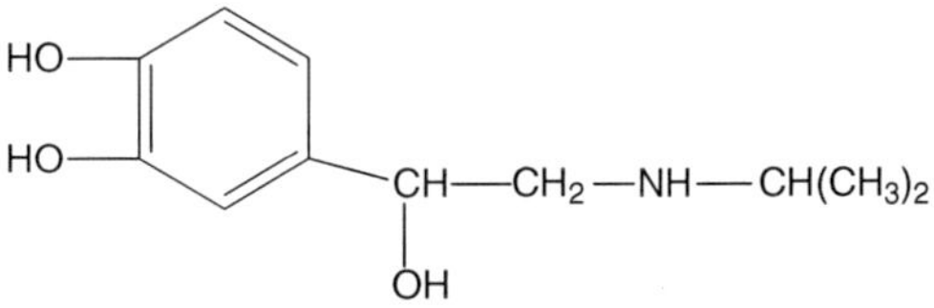

Isoproterenol

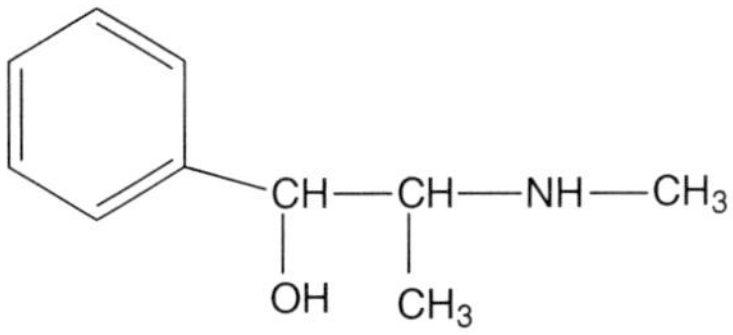

Ephedrine (1*S*, 2*R* & 1*R*, 2*S*)
(ψ)-ephedrine (1*S*, 2*S* & 1*R*, 2*R*)

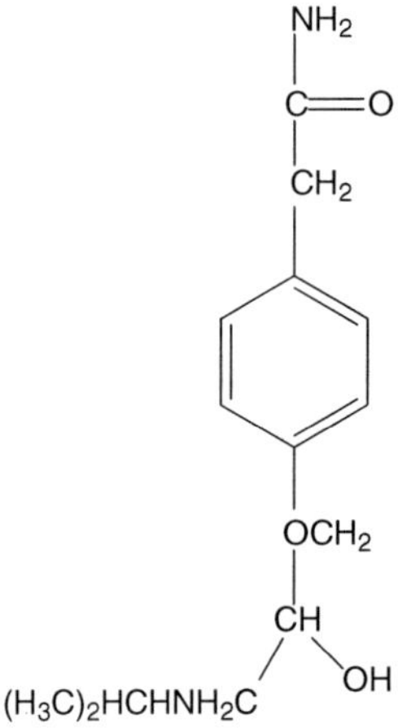

Atenolol

Chart 17.6. Chiral drug compounds [167].

mass spectrometer). Fig. 17.23, illustrates the MS/MS spectrum, where multiple reaction monitor scanning (MRM, an alternative MS/MS scan mode that consists of mass selection of the trimeric cluster ion from Q_1, followed by transfer to the second quadrupole for collisional dissociation, and then subsequent fragment ion mass selection in Q_3) was used. The ratios of the fragment ions $[(Li)(A)(ref^*)]^+$ and $[(Li)(ref^*)_2]^+$ were determined by integrating the areas of their respective MRM transitions, 1.43 and 0.76 for BMS-204353 and MaxiPost, respectively. Linear changes in the chiral purity of MaxiPost were observed from the ratio when standard mixtures were analyzed.

Chiral purity determination using MS/MS was shown to be a fast method of analysis with accuracy and precision that was comparable to the LC/UV and LC/CD methods. Comparisons in accuracy, precision and analysis time obtained from the three methods are shown in Tables 17.8 and 17.9. Significant improvements in the overall analysis time, especially with a large number of samples, were achieved as a result of using MS.

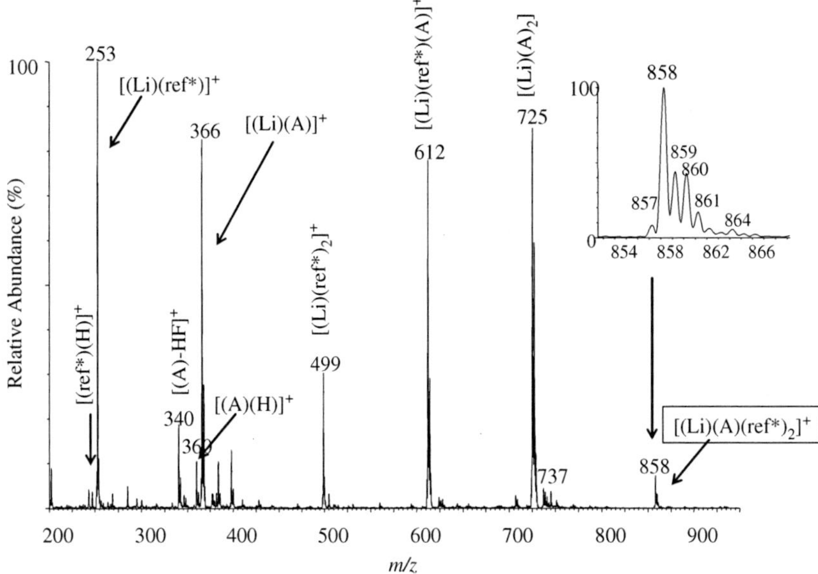

Fig. 17.22. ESI mass spectra of LiCl, (+)-5-fluorodeoxyuridine (ref*) and MaxiPost (A) dissolved in 50/50 methanol/water solution. The spectrum was recorded using a triple quadrupole mass spectrometer. The trimeric cluster is at *m*/*z* 858 [168]. (Note: This figure is an adaptation of the original)

Chiral MS appears to represent a good alternative to chiral chromatography in early pharmaceutical development because it is a high-throughput method with accuracy and precision that compares with the current chiral methodologies with the added advantage of faster analysis time.

17.4.6 Online chiral analysis

In practice, enantiomeric quantification based solely on MS can be challenging from an experimental view point. For instance, when analyzing "real" chiral samples the analyte of interest is generally in a complex matrix and the complexity of the matrix can vary depending on the type of sample being analyzed. "Dirty" samples can increase the difficulty of any analytical method, by reducing the methods sensitivity and selectivity. So, many analytical procedures require some sample clean up prior to the analysis. In the case of MS, very complex samples can suppress ion formation and make peak identification impossible [141].

A new online chiral analysis method that uses HPLC, involves preliminary separation of the chiral analyte from its matrix, then application of the kinetic method for the purpose of chiral discrimination [142]. This has been demonstrated using a commercial liner ion trap (LTQ) mass spectrometer, equipped with an electrospray ionization source, where trimeric cluster ions were generated (fixed and non-fixed trimeric cluster ions) by post-column addition of the reference ligand, metal cation and fixed ligand. The chiral analyte was separated from its matrix using a stationary phase prior to the

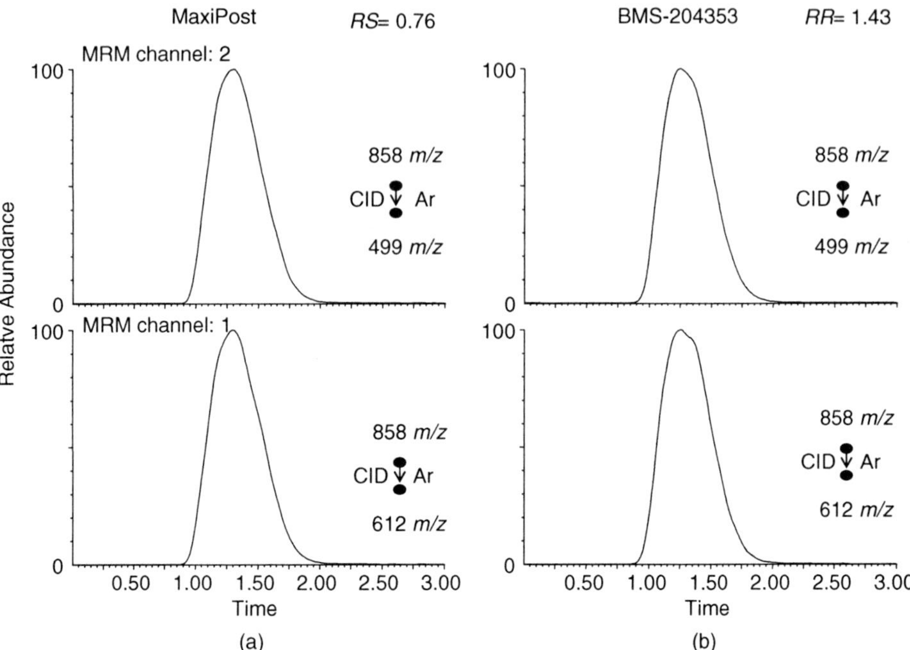

Fig. 17.23. MRM responses for m/z 858 to m/z 499 and m/z 858 to m/z 612 for (a) a solution containing MaxiPost, LiCl, and (+)-5-fluorodeoxyuridine and (b) a solution containing BMS-204353, LiCl, and (+)-5-fluorodeoxyuridine. The ratio for the pure enantiomer MaxiPost and BMS-204353 were calculated using the respective areas from the MRM transition providing an R_{chiral} (R_R/R_S) of 1.88 using a collision energy of 5 eV and a collision gas pressure of 0.6 mTorr [168]. (Note: This figure is an adaptation of the original)

Table 17.8 Comparison of results obtained from the MS/MS, chiral LC/UV, and achiral LC/CD methods [168]

Sample	Chiral purity	Measured chiral purity MS (UV) [CD]	% Bias MS (UV) [CD]	%RSD MS (UV) [CD]
QC 1	89.9	90.0 (90.2) [90.2]	−0.8 (0.2) [−0.3]	4.2 (0.1) [3.4]
QC 2	92.8	94.6 (94.9) [93.6]	−1.0 (0.0) [−0.8]	2.0 (0.1) [0.7]
QC 3	96.8	96.4 (97.0) [96.6]	0.0 (0.0) [−0.5]	1.4 (0.1) [1.5]
QC 4	98.8	98.4 (98.9) [98.3]	−2.9 (0.0) [−0.3]	2.3 (0.0) [0.6]
QC 5	99.3	99.2 (99.4) [98.5]	0.9 (0.0) [−1.3]	1.2 (0.1) [0.2]
QC 6	99.6	99.5 (99.7) [99.3]	1.6 (0.0) [0.3]	1.1 (0.2) [0.4]
Sample 1	Unknown	99.5 (99.9) [99.2]	N/A	1.3 (0.0) [0.2]
Sample 2	Unknown	99.8 (99.9) [99.3]	N/A	0.6 (0.0) [0.6]
Sample 3	Unknown	100.7 (99.9) [99.9]	N/A	1.7 (0.0) [1.2]

Table 17.9 Method development time, number of injections required to determine chiral purity and analysis time for chiral LC/UV, achiral LC/CD, and chiral MS/MS MaxiPost methods [168]

Method development time	Chiral LC/UV <(week)	Achiral LC/CD <(week)	Chiral MS/MS <(week)
Number of injections[a]	3[b]	25[c]	25[d]
Sample-to-sample analysis time (min)	20[e]	9	3
Total analysis time (min) for 1 sample	60	225	75
Total analysis time (min) for 6 samples	160	360	120
Total analysis time (min) for 25 samples	540	873	291

[a]Number of injections required to determine the chiral purity of a single sample for the given method.
[b]Single injection of blank, system suitability, and sample.
[c]Single injection of blank with triplicate injection of seven standards and sample.
[d]Single injection of blank with triplicate injection of seven standards and sample.
[e]Note that the average run time of 20 random chiral LC/UV methods of pharmaceutical compounds found in the literature gave a mean analysis time of ~30 min. with a range of 12–90 min.

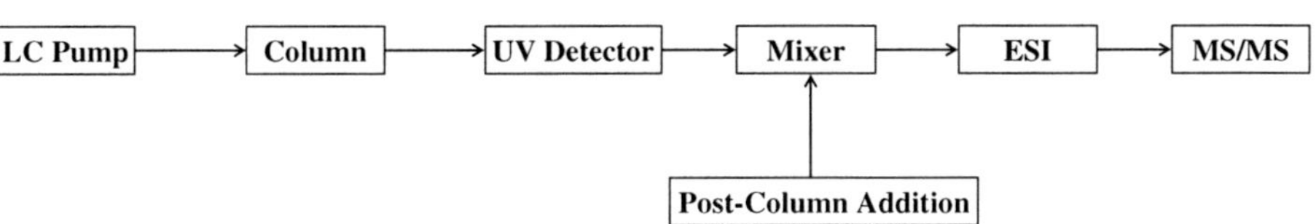

Fig. 17.24. Illustration of post-column addition for on-the-fly chiral analysis by the kinetic method [142]. (Note: This figure is an adaptation of the original)

postcolumn addition, which then included direct infusion of the reference compounds (fixed or non-fixed ligand), and metal cation. The trimeric cluster ion was generated within the source of the mass spectrometer, Fig. 17.24.

The chiral selectivity and reproducibility obtained using the postcolumn addition is similar to results reported using direct infusion of the pure enantiomer pairs. Preliminary tests used Cu^{II}, D/L-DOPA, and L-tryosine, Fig. 17.25. This type of method can be used for chiral mixture analysis as well as for high-throughput screening. These data show the power of online LC/MS chiral analysis using data-dependent MS/MS.

17.5 CONCLUDING REMARKS

17.5.1 Emerging methods

17.5.1.1 Ion mobility

It has been shown that ion mobility spectrometry-mass spectrometry (IMS-MS) can distinguish isomeric compounds that have identical m/z ratios [143]. The mobility of an ion is a measure of how rapidly an ion moves through a buffer (drift) gas under the influence of an electric field. For macromolecules, the mobility depends on the charge

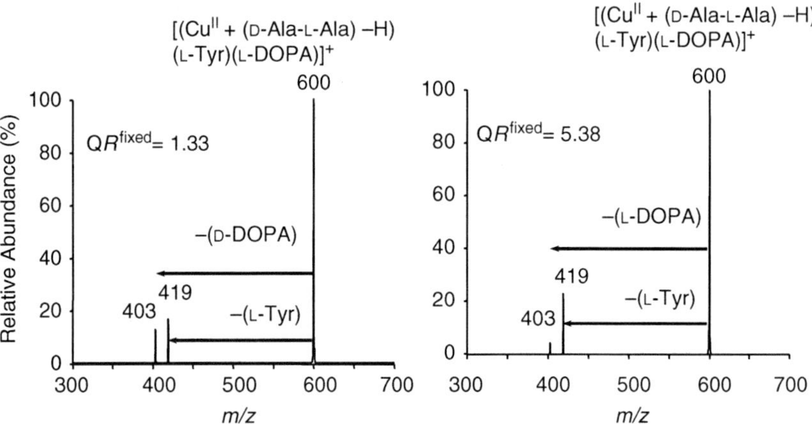

Fig. 17.25. MS/MS product ion spectra shows chiral selectivity (4.05) of D/L-DOPA using Cu^{II} as the central metal ion, D-Ala-L-Ala as the fixed ligand, and L-Tyr as the reference ligand [121]. (Note: This figure is an adaptation of the original)

state and conformation of the ion (referred to as the average collision cross section) rather than mass [144].

A recent demonstration of IMS-MS, applied to isomeric analysis showed that various positional isomeric peptides, electrosprayed into a mass spectrometer, could be separated on the basis of their different collision cross sections [145,146]. With the same mass and charge state, the mobility differences of the peptides were qualitatively related to the collision cross sections of the peptide, which in turn were directly related to the peptide conformations in the gas phase. The analysis of mixtures demonstrated the capability of IMS-MS for separating isomeric ions due to small differences in the structure. Baseline separation that differed by as little as 2.5% in the measured collision cross section was achieved.

High resolution IMS can be implemented using an asymmetric transverse field waveform (high-field asymmetric waveform ion mobility spectrometry (FAIMS)) applied to ions passed through a drift region at atmospheric pressure. The resulting high resolution measurements [147] could be used to separate enantiomeric forms of ephedra alkaloids in natural health products [148]. Isotopic dilution, combined with FAIMS separation, ESI, and MS detection was applied to the separation and quantitation of three pairs of ephedrine-type diastereomers, ephedrine (E), pseudoephedrine (PE) and their metabolites norephedrine (NE), norpseudoephedrine (NPE), methylephedrine (ME), and methylpseudoephedrine (MPE). The alkaloids were all quantitatively resolved (Fig. 17.26) with this IMS-MS methodology by analyzing standard solutions of the six target analytes followed by determining the linear range and limits of detection for the analysis. The precision of the IMS based method was evaluated against a conventional LC/UV method, and the MS method proved to be approximately an order of magnitude higher than the LC/UV method (based on standard deviation) [148]. In addition, with the current configuration of FAIMS, samples were quickly analyzed (for instance, 2 min/sample runtimes were achieved). This is significantly faster than current chromatographic approaches.

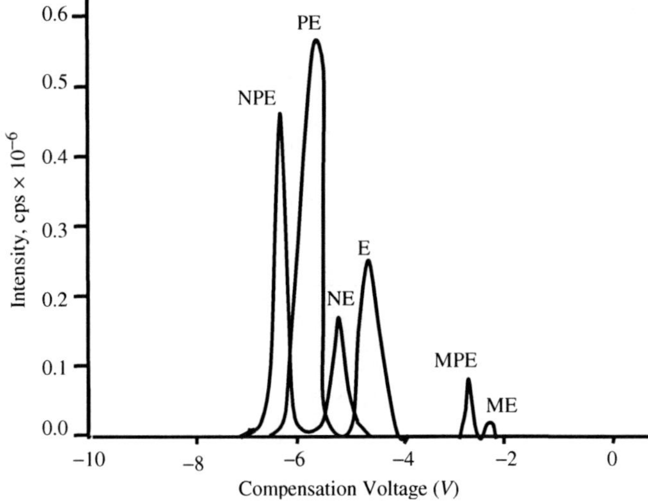

Fig. 17.26. ESI-FAIMS spectrum for a solution of 1 μg/ml E, PE, NE, and NPE and 3 μg/ml ME and MPE in methanol/DDW (90/10) with 0.2 mM ammonium acetate added (P1 mode, DV = 4000 V, and carrier gas is 2.5 l/min nitrogen) [91]. (Note: This figure is an adaptation of the original)

17.5.1.2 High-precision "Vernier" method

The preferred technique for chiral analytical applications is currently chromatography because of its maturity and ease of use combined with exceptional accuracy and precision [149]. High accuracy chiral measurements have been limited with MS because of varying ionization potentials that can result from isotopic labeling [87], concentration effects [55,56], limited selectivity, lack of linearity over the entire calibration range [150] and/or lack of instrumental precision. In an attempt to improve the precision and accuracy of chiral MS a "Vernier" system has been developed, that uses the kinetic method, with the intention of making accurate measurements within a narrow range of *ee* values. The range of concentrations are confined to very high or low *ee* [151] and the system that demonstrated this methodology consists of simple amino acid mixture, e.g. $[(Ni^{II})(Trp)(Asn)_2{-}H]^+$ (Ni represents the metal cation, Trp represents the D/L-tryptophan forms, and Asn represents the D/L-asparagine forms) [152].

One requirement for this methodology is a chemical system that has a large stereochemically dependent interaction, i.e. a large R_{chiral}. A large R_{chiral} can be generated in two ways: first, by obtaining ratios for the pure enantiomers that are switched in selectivity for the analyte and reference (in the case discussed, the L enantiomer has a ratio of analyte-to-reference that equals 15 and for the D enantiomer the ratio equals 0.3, Fig. 17.27(a)); or secondly, by obtaining ratios for the pure enantiomers that differ by two orders of magnitude or more, where one of the ratios is equal to unity (e.g. the L enantiomer can have a ratio of analyte-to-reference that equals 1 and for the D enantiomer the ratio will equal 0.5 or less, Fig. 17.27(b)). The latter case represents the second requirement for the methodology (since one ratio must equal unity) and is applicable for high *ee* measurements using the kinetic method. Values in the

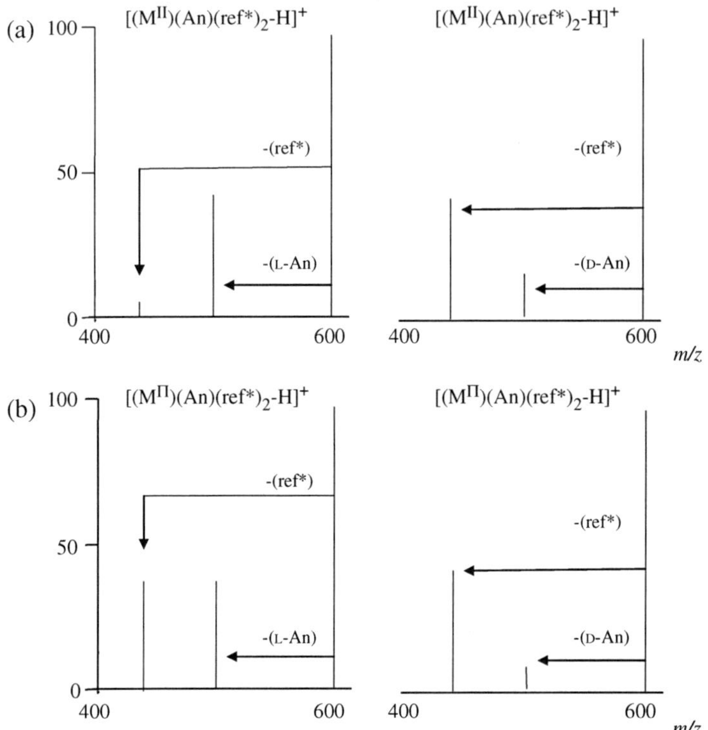

Fig. 17.27. A large R_{chiral} can be generated in two ways, (a) by obtaining ratios for the pure enantiomers that are switched in selectivity for the analyte and reference (b) by obtaining ratios for the pure enantiomers that differ by two orders of magnitude or more, where one of the ratios is equal to unity [151]. (Note: This figure is an adaptation of the original)

extreme regions, 0–10 or 90–100%*ee*, can be accurately determined by measuring a ratio that is closer to unity, since small concentration changes can be easily judged, Fig. 17.28. This results in increased precision over a more limited range of concentrations. This method is useful in instances where the entire calibration curve is not necessary, and the extremes in the calibration curve give high accuracy measurements, Fig. 17.29.

17.5.1.3 Protein analysis by electron capture dissociation

Chiral recognition *via* MS has been traditionally done with the aid of soft ionization techniques such as ESI or FAB, were solution phase components are transferred into the gas phase and solvent effects are absent [64]. A new approach to chiral identification involves the use of electron capture dissociation (ECD), which is inherently a harder ionization technique as compared to ESI or FAB.

ECD, a hard ionization technique, is not an obvious candidate for chiral recognition because of the extensive fragmentation that can occur when dealing with delicate chiral

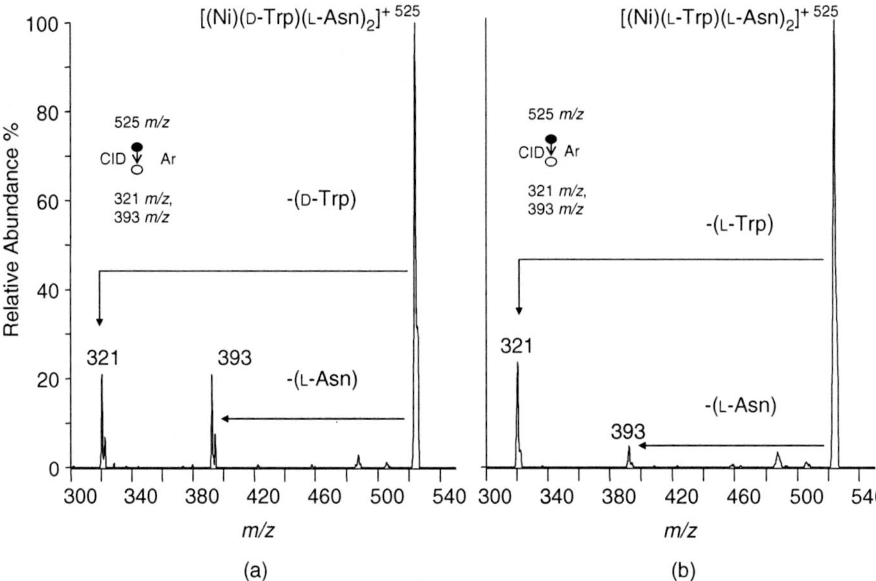

Fig. 17.28. MS/MS of m/z 525 to produce m/z 321 and m/z 393 for: (a) a solution containing D-Trp, $NiCl_2$, and L-Asn and (b) a solution containing L-Trp, $NiCl_2$, and L-Asn. The R_L and R_D values were calculated using the respective relative abundances from the daughter scans providing an R_{chiral} (R_D/R_L) of 4.8, using a collision energy of 10 eV and a collision gas pressure of 2.4 m Torr [151]. (Note: This figure is an adaptation of the original)

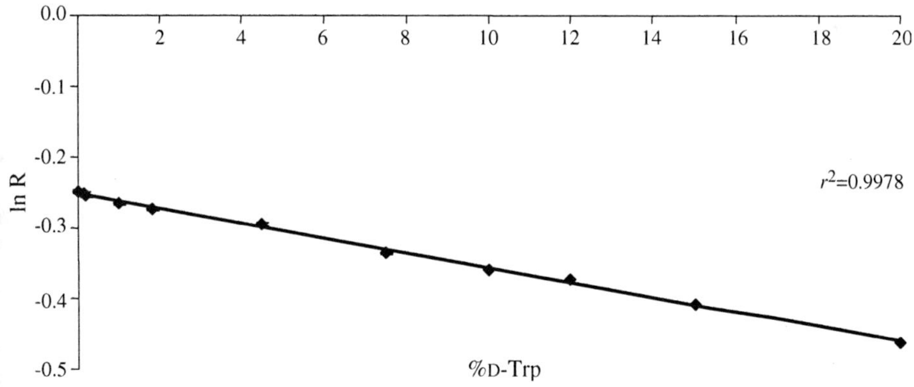

Fig. 17.29. Blow up of the improved linear region on the calibration plot: 0–20% D-Trp using L-Asn as the ref* compound [151]. (Note: This figure is an adaptation of the original)

compounds. However, Zubarev, et. al. discovered that ECD is not only good for resolving enantiomers [153] but it can also be used to resolve small proteins with chiral changes in as little as one residue within the protein. Currently, the largest molecules studied for quantitative chiral MS procedures using ESI have been peptides where D-amino acid detection was demonstrated in peptides up to four and five residues long [150]. Additionally, ESI-IMS-MS has been used to distinguish chiral molecules of similar size and larger [145].

ECD causes N-C_α backbone cleavage without rupture of the intramolecular or intermolecular forces within a macromolecule and this is advantageous for chiral MS analysis [154]. The technique can be used as a probe for secondary and tertiary gas-phase structures and this was demonstrated with ubiquitin and cytochrome c, for which unique but identifiable fragmentation patterns exist, and in general for any compound ionized *via* ECD [155,156]. This potentially represents a good approach for studying large molecules with small chiral differences.

ECD has been used to distinguish stereoisomers in the 2+ Trp-cage ion (NLYIQWLKDGGPSSGRPPPS), which happens to be the smallest and most rapidly folding protein [157]. The fragmentation patterns were used to distinguish a single D-amino-acid substitute on the 20 residue Trp-cage. The method compares fragmentation patterns to differentiate gas-phase tertiary structures, thus a correlation between the native Trp-cage and a substituted Trp-cage (where a D-amino acid was substituted) was observed. By simply measuring the z-ion abundances (z_{n+1}, C-terminal fragments) [158] and determining the sites of protonation, the chiral effect was noted.

In addition to ECD, hot electron capture dissociation (HECD) [159] and sustained off resonance irradiation collision activated dissociation (SORI CAD) [157] have been employed with the intent to identify larger chiral proteins. It seems likely that the recently developed electron transfer dissociation method (ETD) [160,161] which uses an anion/cation reaction to effect electron transfer will prove even more chirally selective with the appropriate choice of chiral anionic reagents.

17.5.1.4 Chiral analysis of solid samples

A new method of ionization, DESI (Section 17.1), generates ions from a condensed phase analyte confined to a surface; charged droplets interact with the surface and produce gaseous ions in the atmospheric interface of the mass spectrometer [14]. These ions are transferred into the mass spectrometer.

DESI was used as the ionization method in a kinetic method chiral assay designed to identify drug chirality [162]. The methodology used was based on published procedures [112] which where optimized for DESI ionization. The trimeric cluster ion, $[(Cu^{2+})(\text{L-His})_2(A)]^+$, where L-His represents the ref* compound histidine, A represents the analyte, *R*- or *S*-propranolol, and Cu^{2+} represents the metal cation mediator), was generated. The results showed that this system responded fairly well to this form of ionization, since the relative abundance of the trimeric cluster ion (in the full MS spectrum) was approximately 15% of the base peak, Fig. 17.30.

The surface on which the sample was deposited was a cotton swab. The R_{chiral} for this system ($R_{chiral} = 2.4$) was comparable to the published ESI value, demonstrating some of the similarities between DESI and ESI. In addition to generating the trimeric cluster ion, the linearity of this system was evaluated. The results show that there is a linear response and that this method can be used to differentiate enantiomers. The DESI and ESI experiments, even though they measure compounds in different

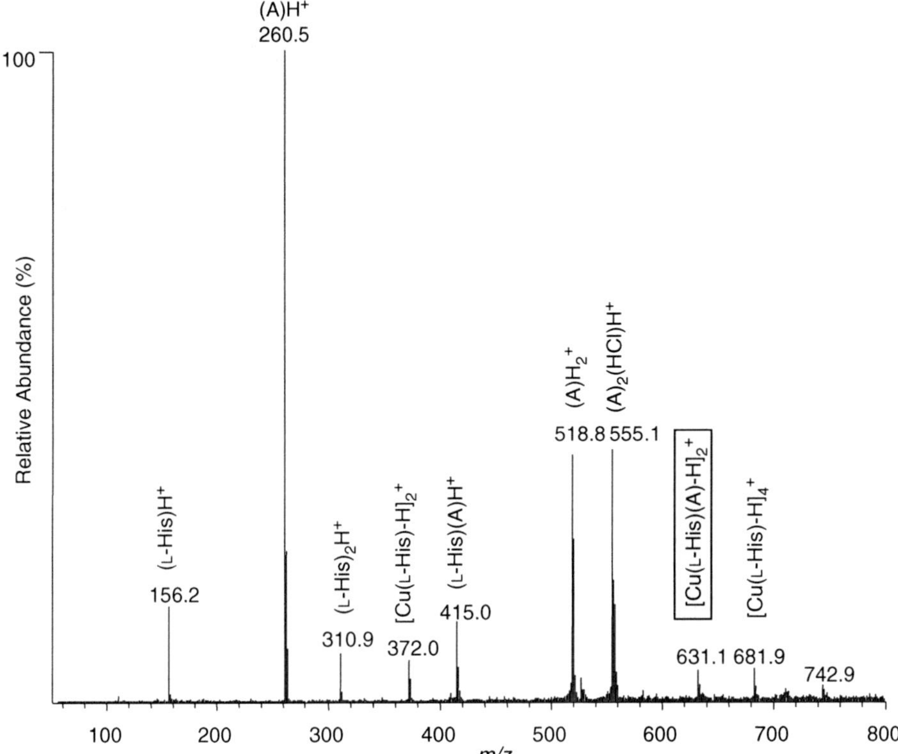

Fig. 17.30. DESI mass spectrum of $CuCl_2(M^{II})$ and L-histamine (ref*) with *S*-propranolol (A) on a cotton swab dissolved in 50/50 $MEOH/H_2O$. The spectrum was recorded on LCQ DECA XP. The trimeric cluster ion is at *m/z* 631 [162]. (Note: This figure is an adaptation of the original)

phases (condensed and solution, respectively) when compared side-by-side, showed comparable linearity responses, Fig. 17.31.

Prior to this analysis a demonstration of this method was conducted using simple amino acids, where the chiral purity of phenylalanine was deduced [14] in the first example of DESI being used for a chiral MS analysis. DESI is advantageous because it is an ambient ionization method, and one which often requires no sample preparation.

17.5.1.5 Sonic spray ionization chiral mass spectrometry analysis

The standard ion trap, ESI methodology for chiral quantitative analysis using the kinetic method was extended [142] to allow even gentler ionization. This was done using sonic spray ionization (SSI) (a relatively new atmospheric pressure ionization technique that uses a very high gas flow rate during spray ionization) which is particularly useful for thermally labile compounds. When coupled with a quadrupole ion-trap-time-of-flight (Q-TOF) or LCQ mass spectrometer accurate, chiral and isomeric determinations (down to 2%*ee*) were demonstrated in parallel experiments [122].

The trimeric cluster ions, which consisted of a fixed (non-dissociating ligand) and a non-fixed ligand, were generated using all possible ionization and instrumentation

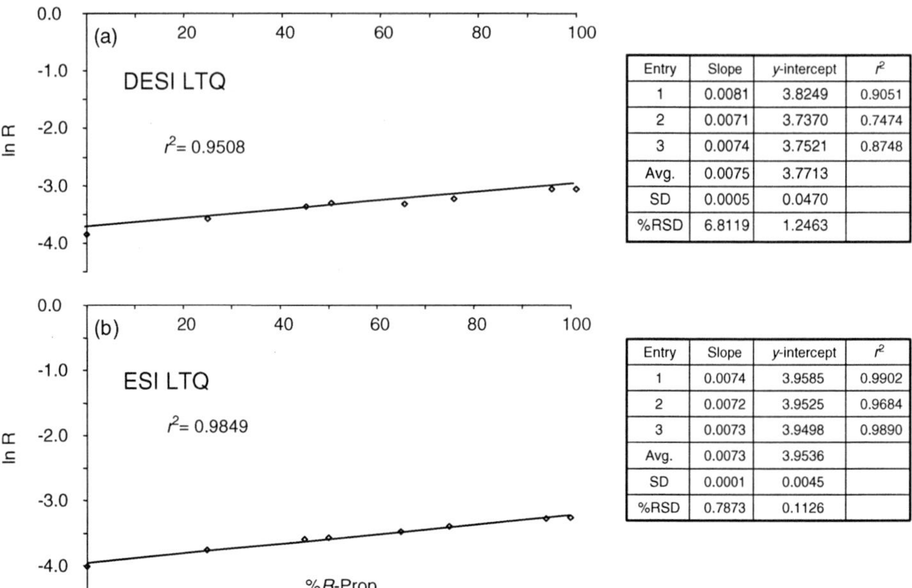

Entry	Slope	*y*-intercept	r^2
1	0.0081	3.8249	0.9051
2	0.0071	3.7370	0.7474
3	0.0074	3.7521	0.8748
Avg.	0.0075	3.7713	
SD	0.0005	0.0470	
%RSD	6.8119	1.2463	

Entry	Slope	*y*-intercept	r^2
1	0.0074	3.9585	0.9902
2	0.0072	3.9525	0.9684
3	0.0073	3.9498	0.9890
Avg.	0.0073	3.9536	
SD	0.0001	0.0045	
%RSD	0.7873	0.1126	

Fig. 17.31. Linearity comparisons of *R*-propranolol in (a) DESI and (b) ESI-LTQ experiment [162]. (Note: This figure is an adaptation of the original)

combinations (ESI-LCQ, ESI-Q-ToF, SSI-LCQ and SSI-Q-ToF). The obtained results gave comparable quantitative accuracies for both the ESI and SSI experiments using the LCQ and Q-ToF. The method showed that the fixed ligand procedure could be used with the SSI source as well as the ESI source in a Q-ToF and the results corroborate early findings that suggest the fixed-ligand promotes improved chiral selectivity, as a result of geometrical constrains [119,121,124].

In addition to evaluating the effect of ionization and variation in instrumentation, experiments have been reported in which concentration effects on the performance of the kinetic method chiral analysis were monitored. It was necessary to investigate whether mixtures of various concentrations will affect the method's chiral quantification ability. This is important especially in situations where limited information is given for an unknown sample. It was determined in a number of cases, that the concentration is independent of the chiral selectivity, an essential feature of a practical chiral MS analysis. This assay demonstrated that not only can an ESI quadrupole ion trap deduce chirality but SSI experiments that include Q-ToF and the LCQ provide an alternative technique that can perform this assay while maintaining the same enantiomeric selectivity. The versatility of this method is still being explored.

17.5.2 Advantages and challenges ahead

Methods of quantitative *ee* determination using MS have been summarized. Single-stage MS methods (kinetic resolution in the solution phase and H–G single-stage complexation) and MS/MS methods (I/M reactions and kinetic method formalism) have both been discussed. These methods encompass results of a large group of scientists from different laboratories who are interested in expanding the versatility of MS as a tool for chiral quantification. Despite the initial perception of MS as a "chirally blind" technique, this chapter shows that MS has much to contribute to chiral analysis.

In particular, mass spectrometry provides a rapid and sensitive method for chiral analysis. In a clean gas-phase environment, enantio-discrimination mechanisms are relatively easy to understand, the experimental conditions are readily optimized, and chiral recognition can be easily obtained. With the rapid development of ionization methods and novel instrumentation (offering high mass resolution and mass measurement accuracy), mass spectrometry may be blossoming into an important tool for chiral analysis, especially in fast combinatorial screening and clinical diagnostics [163].

Although mass spectrometry has shown promising features for rapid determination of enantiomeric excess, several aspects still need improvement in order for MS to become a reliable method for routine chiral analysis. Lower limits of quantification, higher accuracy, increased reproducibility, and reduced ion suppression of matrix effects during the ionization process are all desirable. Some experiments have demonstrated that high accuracy and precision can be obtained but further improvements must be made. Since all mass spectrometric methods share the same feature of using the reference ligand to create a chiral environment, the selection and/or optimization of reference ligands, even the development of new reference ligands, becomes more important when doing precision work and method development.

Lastly, this methodology has the potential to be applied into many fields. The three-point interaction used for gas-phase chiral recognition is useful in the exploration of stereoselective ion channel-chiral drug interactions [164]. Such interactions might give rise to quantitative differences in stereoselectivity which can be utilized to better understand different pharmacological effects of chiral drugs in different forms of enantiomers.

17.6 ACKNOWLEDGMENTS

This work was supported by the National Science Foundation, 0412782-CHE, and the US Department of Energy, Office of Energy Research. L. Wu expresses appreciation for the 2003 American Chemical Society-Division of Analytical Chemistry Graduate Fellowship Award sponsored by Johnson & Johnson Pharmaceutical Research and Development. We are grateful to the colleagues that have served as reviewers for this chapter. Their thoughtful comments, suggestions, and encouragement have been immensely helpful.

REFERENCES

1 M. Quack, How important is parity violation for molecular and biomolecular chirality. Angew. Chem. Int. Ed., 41 (2002) 4618–4630.

2 G.B. Baker and T.I. Prior, Stereochemistry and drug efficacy and development: relevance of chirality to antidepressant and antipsychotic drugs. Ann Med., 34 (2002) 537–543.

3 S.C. Stinson, Chiral chemistry. Chem. Eng. News, 80 (2002) 45–57.

4 S.C. Stinson, Chiral drug interactions. Chem. Eng. News, 77 (1999) 101–120.

5 S.C. Stinson, Chiral drugs. Chem. Eng. News, 78 (2000) 55–79.

6 L.C.R. Pasteur, Louis Pasteur (1822–1895). Hebd. Seance Acad. Sci. Paris, 26 (1848) 535–536.

7 H.M. Fales and G.J. Wright, Detection of chirality with the chemical ionization mass spectrometer. "Meso" ions in the gas phase. J. Am. Chem. Soc., 99 (1977) 2339–2340.

8 M.L. Sigsby, Refinements of Mass Spectrometry and Applications to Organic Analysis. PhD Thesis, Purdue University, West Lafayette, 1980.

9 S.A. McLuckey and J.M. Wells, Mass analysis at the advent of the 21st Century. Chem. Rev., 101 (2001) 571–606.

10 J.B. Fenn, M. Mann, C.K. Meng, S.F. Wong and C.M. Whitehouse, Electrospray ionization for mass-spectrometry of large biomolecules. Science, 246 (1989) 64–71.

11 F. Hillenkamp, M. Karas, R.C. Beavis and B.T. Chait, Matrix-assisted laser desorption/ionization mass spectrometry of biopolymers. Anal. Chem., 63 (1991) 1193A–1203A.

12 A. Hirabayashi, M. Sakairi and H. Koizumi, Sonic spray ionization method for atmospheric pressure ionization mass spectrometry. Anal. Chem., 66 (1994) 4557–4559.

13 A. Hirabayashi, M. Sakairi and H. Koizumi, Sonic spray mass spectrometry. Anal. Chem., 67 (1995) 2878–2882.

14 Zoltán Takáts, Justin M. Wiseman, Bogdan Gologan and R.G. Cooks, Mass spectrometry sampling under ambient conditions with desorption electrospray ionization. Science, 306 (2004) 471–473.

15 J.B. Fenn, N. Mann, C.K. Meng and S.F. Wong, Electrospray ionization-principles and practice. Mass Spectrom. Rev., 9 (1990) 37–70.

16 H. Nonami, K. Tanaka, Y. Fukuyama and R Erra-Balsells, Beta-carboline alkaloids as matrixes for UV-matrix-assisted laser desorption/ionization time-of-flight mass spectrometry in positive and negative ion modes. Analysis of proteins of high molecular mass, and of cyclic and acyclic oligosaccharides. Rapid Commun. Mass Spectrom., 12 (1998) 285–296.

17 J.C. Schwartz, A.P. Wade, C.G. Enke and R.G. Cooks, Systematic delineation of scan modes in multidimentional mass spectrometry. Anal. Chem., 62 (1990) 1809–1818.

18 K.L. Busch, G.L. Glish and S.A. McLuckey, Mass Spectrometry/Mass Spectrometry: Techniques and Applications of Tandem Mass Spectrometry, VCH Publishers, New York, 1988.

19 E.de. Hoffmann, Tandem mass spectrometry: a primer. J. Mass Spectrom., 31 (1996) 129–137.

20 A. Shukla and J.H. Futrell, Tandem mass spectrometry: dissociation of ions by collisional activation. J. Mass Spectrom., 35 1069–1090 (2000).

21 I.V. Chernushevich, A.V. Loboda and B.A. Thomson, An introduction to quadrupole-time-of-flight mass spectrometry. J. Mass Spectrom., 36 (2001) 849–865.

22 A.E. Giannakopulos, B. Thomas, A.W. Colburn, D.J. Reynolds, E.N. Raptakis, A.A. Makarov and P.J. Derrick, Tandem time-of-flight mass spectrometer (TOF-TOF) with a quadratic-field ion mirror. Rev. Sci. Inst., 73 (2002) 2115–2123.

23 R.E. March, An introduction to quadrupole ion trap mass spectrometry. J. Mass Spectrom., 32 (1997) 351–369.

24 A.G. Marshall, C.L. Hendrickson and G.S. Jackson, Fourier transform ion cyclotron resonance mass spectrometry: a primer. Mass Spectrom. Rev., 17 (1998) 1–35.

25 Q. Hu, R. Noll, H. Li, A. Makarov, M. Hardman and R.G. Cooks, The orbitrap: a new mass spectrometer. J. Mass Spectrom., 40 (2005) 430–443.

26 R.A. Yost and D.D. Fetterolf, Tandem mass spectrometry (MS/MS) instrumentation. Mass Spectrom. Rev., 2 (1983) 1–45.

27 T.E. Beesley and R.P.W. Scott, Chiral Chromatography, R.P.W. Scott, C. Simpson and E.D. Katz (Eds.), John Wiley and Sons Inc., New York, 1998.

28 Y. Tang, J. Zukowski and D.W.I. Armstrong, Investigation on enantiomeric separations of fluorenylmethoxycarbonyl amino acids and peptides by HPLC using native cyclodextrins as chiral stationary phases. J. Chromatogr. A, 743 (1996) 261–271.

29 X. Zhu, B. Lin, A. Jakob, S. Wuerthner and B. Koppenhoefer, Separation of drugs by cappillary electrophoresis, part 10. Permethyl-alpha-cyclodextrin as chiral solvation agent. Electrophoresis, 20 (1999) 1878–1889.
30 G. Blaschke and B. Chankvetadze, Enantiomer separation of drugs by capillary electromigration techniques. J. Chromatogr. A, 875 (2000) 3–25.
31 K.D. Altria, M.A. Kelly and B.J. Clark, Current applications in the analysis of pharmaceutical by capillary electrophoresis. II. Trends Anal. Chem., 17 (1998) 214–226.
32 Timothy J. Ward and D.-M. Hamburg, Chiral separations. Anal. Chem., 76 (2004) 4635–4644.
33 T. Hattori, Y. Minato, S. Yao, M.G. Finn and S. Miyano, Use of a racemic derivatizing agent for measurement of enantiomeric excess by circular dichroism spectroscopy. Tetrahedron Lett., 42 (2001) 8015–8018.
34 M.A. Evans and J.P. Morken, Isotopically chiral probes for in situ high-throughput asymmetric reaction analysis. J. Am. Chem. Soc., 124 (2002) 9020–9021.
35 P. Abato and C.T. Seto, EMDee: an enzymatic method for determining enantiomeric excess. J. Am. Chem. Soc., 123 (2001) 9206–9207.
36 T.E. Beesley, B. Buglio and R.P.W. Scott, Quantitative Chromatographic Analysis, Marcel Dekker, New York, 2001.
37 L.G. Blomberg and H. Wan, Determination of enantiomeric excess by capillary electrophoresis. Electrophoresis, 21 (2000) 1940–1952.
38 M.G. Finn, Emerging methods for the rapid determination of enantiomeric excess. Chirality, 14 (2002) 534–540.
39 V.A. Davankov, The nature of chiral recognition: is it a three-point interaction? Chirality, 9 (1997) 99–102.
40 A.G. Ogston, Interpretation of experiments on metabolic processes, using isotopic tracer elements. Nature, 162 (1948) 963.
41 V.R. Meyers and M. Rais, A vivid model of chiral recognition. Chirality, 1 (1989) 167–169.
42 V.A. Davankov, J.D. Navratil and H.F. Walton, Ligand-Exchange Chromatography, CRC Press, Boca Raton, 1988, pp. 446–475.
43 W.H. Pirkle and M.H. Hyun, Effects of interstand distance upon chiral recognition by a chiral stationary phase. J. Chromatogr., 328 (1985) 1–9.
44 S. Topiol and M. Sabio, Computational chemical studies of chiral stationary phase model complexes of methyl N-(2-naphthyl)alaninate with N-(3,5-dinitrobenzoyl)leucine n-propylamide. J. Chromatogr., 461 (1989) 129–137.
45 W.H. Pirkle and T.C. Pochapsky, Intermolecular 1H{1H} nuclear overhauser effects in diastereomeric complexes: support for a chromatographically derived chiral recognition model. J. Amer. Chem. Soc., 108 (1986) 5627–5628.
46 V.A. Davankov, V. Meyer and M. Rais, A vivid model illustrating chiral recognition induced by achiral structures. Chirality, 2 (1990) 208–210.
47 V.A. Davankov and A.A. Kurganov, The role of achiral sorbent matrix in chiral recognition of amino acid enantiomers in ligang exchange chromatography. Chromatographia, 17 (1983) 686–690.
48 A.A. Kurganov, Ya. Zhuchkova and V.A. Davankov, Stereoselectivity in bis(-amino acid) copper(II)complexes—VII. Thermodynamics of N-benzylproline coordination to copper(II). J. Inorg. Nucl. Chem., 40 (1978) 1081–1083.
49 V.A. Davankov, A.A. Kurganov, L. Ya. Zhuchkova and T.M. Ponomareva, Chiral discrimination phenomena in mixed-ligand copper(II) complexes with amino acids and 1,3-dicarbonyl compounds. Chirality, 5 (1993) 303–309.
50 A. Filippi, A. Giardini, S. Piccirillo and M. Speranza, Gas-phase enantioselectivity. Int. J. Mass Spectrom., 198 (2000) 137–163.
51 W.A. Tao and R.G. Cooks, Chiral analysis by MS. Anal. Chem., 75 (2003) 25A–31A.
52 M.Y.T. Sawada, H. Yamada. Depression of the apparent chiral recognition ability obtained in the host–guest complexation systems by electrospray and nano-electrospray ionization mass spectrometry. Eur. J. Mass Spectrom., 1 (2004) 27–37.

53 M. Sawada, Chiral compounds, in: N.M.M. Nibbering (Ed.), The Encyclopedia of Mass Spectrometry, Vol. 4, Elsevier, Amsterdam, 2004.

54 J. Guo, J. Wu, G. Siuzdak and M.G. Finn, Measurement of enantiomeric excess by kinetic resolution and mass spectrometry. Angew. Chem. Int. Ed., 38 (1999) 1755–1758.

55 ChengLi Zu, Bobby N. Brewer, Beibei Wang and M.E. Koscho, Tertiary amine appended derivatives of N-(3,5-dintrobenzoyl)leucine as chiral selectors for enantiomer assays by electrospray ionization mass spectrometry. Anal. Chem., 77 (2005) 5019–5027.

56 M. Sawada, Chiral recognition detected by fast atom bombardment mass spectrometry. Mass Spectrom. Rev., 16 (1997) 73–90.

57 G. Grigorean, J. Ramirez, S.H. Ahn and C.B. Lebrilla, A mass spectrometry method for the determination of enantiomeric excess in mixtures of D,L-amino acids. Anal. Chem., 72 (2000) 4275–4281.

58 Z. Yao, T. Wan, K. Kwong and C. Che, Chiral recognition of amino acids by electrospray ionization mass spectrometry/mass spectrometry. Chem. Commun., 20 (1999) 2119–2120.

59 R.G. Cooks, J.S. Patrick, T. Kotiaho and S.A. McLuckey, Thermochemical determinations by the kinetic method. Mass Spectrom. Rev., 13 (1994) 287–339.

60 R.G. Cooks and P.S.H. Wong, Kinetic method of making thermochemical determinations. Acc. Chem. Res., 31 (1998) 379–386.

61 B. Munson, Chemical ionization mass spectrometry. Int. J. Mass spectrom., 200 (2000) 234–251.

62 W.J. Richter and H. Schwarz, Chemical ionization – a highly important productive mass spectrometric analysis method. Angew. Chem., 90 (1978) 449–469.

63 A.G. Harrison, Chemical Ionization Mass Spectrometry, CRC Press, Boca Raton, 1992.

64 E.N. Nikolaev, E.V. Denisov, V.S. Rakov and J.H. Futrell, Investigation of dialkyl tartrate molecular recognition in cluster ions by Fourier transform mass spectrometry: a comparison of chirality effects in gas and liquid phases. Int. J. Mass Spectrom., 183 (1999) 357–368.

65 K.L. Busch, Desorption ionization mass spectrometry. J. Mass Spectrom., 30 (1995) 233–240.

66 F.M. Devienne and J.C. Roustan, "Fast Atom Bombardment" – a rediscovered method for mass spectrometry. Org. Mass Spectrom., 17 (1982) 173–181.

67 M. Sawada, Y. Takai, H. Yamada, S. Hiriyama, T. Kaneda, T. Tanaka, K. Kamada, T. Mizooku, S. Takeuchi, K. Ueno, K. Hirose, Y. Tobe and K. Naemura, Chiral recognition in host-guest complexation determined by the enantiomer-labeled guest method using fast-atom bombardment mass spectrometry. J. Am. Chem. Soc., 117 (1995) 7726–7736.

68 K. Dreisewerd, The desorption process in MALDI. Chem. Rev., 103 (2003) 395–425.

69 K. Dreisewerd, S. Berkenkamp, A. Leisner, A. Rohlfing and C. Menzel, Fundamental of MALDI-MS with pulsed infrared lasers. Int. J. Mass Spectrom., 226 (2003) 189–209.

70 M. Sawada, Y. Takai, H. Yamada, J. Nishida, T. Kaneda, R. Arakawa, M. Okamoto, K. Hirose, T. Tanaka and K. Naemura, Chiral amino acid recognition detected by electrospray ionization (ESI) and fast atom bombardment (FAB) mass spectrometry (MS) coupled with the enantiomer-labelled (EL) guest method. J. Chem. Soc. Perkin Trans., 23 (1998) 701–710.

71 J.B. Fenn, M. Mann, C.K. Meng, S.F. Wong and C.M. Whitehouse, Electrospray ionization for mass-spectrometry of large biomolecules. Science, 63 (1989) 1193A–1203A .

72 D.V. Dearden, Y. Liang, J.B. Nicoll and K.A. Kellersberger, Study of gas phase molecular recognition using Fourier transform ion cyclotron resonance mass spectrometry (FTICR-MS). J. Mass Spectrom., 36 (2001) 989–997.

73 Y.J. Liang, J.S. Bradshaw, R.M. Izatt, R.M. Pope and D.V. Dearden, Analysis of enantiomeric excess using mass spectrometry: fast atom bombardment/sector and electrospray ionization Fourier transform mass spectrometric approaches. Int. J. Mass Spectrom., 187 (1999) 977–988.

74 G. Grigorean, S. Gronert and C.B. Lebrilla, Enantioselective gas-phase ion-molecule reactions in a quadrupole ion trap. Int. J. Mass Spectrom., 219 (2002) 79–87.

75 T.T. Dang, S.F. Pedersen and J.A. Leary, Chiral recognition in the gas phase: mass spectrometric studies of diastereomeric cobalt complexes. J. Am. Soc. Mass Spectrom., 5 (1994) 452–459.

76 J.C. Tabet, Ion-molecule reactions in the gas-phase-IX-differentiation of enantiomeric menthols using stereospecific SN_2 process induced by a chiral reagent. Tetrahedron, 43 (1987) 3413–3420.

77 W.Y. Shen, P.S.H. Wong and R.G. Cooks, Stereoisomeric distinction by the kinetic method 2,3-butanediol. Rapid Commun. Mass Spectrom., 11 (1997) 71–74.

78 David D. Diaz, Sulan Yao and M.G. Finn, Measurment of enantiomeric excess of amines by mass spectrometry following kinetic resolution with solid-phase chiral acylating agents. Tetrahedron Lett., 42 (2001) 2617–2619.

79 Sulan Yao, Jun-Cai Meng, Gary Siuzdak and M.G. Finn, New catalysts for the asymmetric hydrosilylation of ketones discovered by mass spectrometry screening. J. Org. Chem., 68 (2003) 2540–2546.

80 B.N. Brewer, C. Zu and M.E. Koscho, Determination of enantiomeric composition by negative-ion electrospray ionization-mass spectrometry using deprotonated N-(3,5-dinitrobenzoyl)amino acids as chiral selectors. Chirality, 17 (2005) 456–463.

81 M.E. Koscho, C. Zu and B.N. Brewer, Extension of chromatographically derived chiral recognition systems to chiral recognition and enantiomer analysis by electrospray ionization mass spectrometry. Tetrahedron Asymmetry, 16 (2005) 801–807.

82 M. P. So, T.S.M. Wan and T.W.D. Chan, Differentiation of enantiomers using matrix-assisted laser desorption/ionization mass spectrometry. Rapid Commun. Mass Spectrom., 14 (2000) 692.

83 M. Sawada, H. Yamaoka, Y. Takai, Y. Kawai, H. Yamada, T. Azuma, T. Fujioka and T. Tanaka, Determination of enantiomeric excess for organic primary amine compounds by chiral recognition fast-atom bombardment mass spectrometry. Int. J. Mass Spectrom., 193 (1999) 123–130.

84 M. Shizuma, H. Adachi, A. Amemura, Y. Takai, T. Takeda and M. Sawada, Chiral discrimination of permethylated gluco-oligosaccharide toward amino acid ester salts. Tetrahedron, 57 (2001) 4567–4578.

85 M. Shizuma, Y. Kadoya, Y. Takai, H. Imamura, H. Yamada, T. Takeda, R. Arakawa, S. Takahashi and M. Sawada, New artificial host compounds containing galactose end groups for binding chiral organic amine guests: chiral discrimination and their complex structures. J. Org. Chem., 67 (2002) 4795–4807.

86 M. P. So, T.S. Wan and T.W. Chan, Differentiation of enantiomers using matrix-assisted laser desorption/ionization mass spectrometry. Rapid Commun. Mass Spectrom., 14 (2000) 692–695.

87 K.A. Schug, N.M. Maier and W. Lindner, Deuterium isotope effects observed during competitive binding chiral recognition electrospray ionization – mass spectrometry of cinchona alkaloid-based systems. J. Mass Spectrom., 41(2) (2006) 157–161.

88 M. Shizuma, M. Ohta, H. Yamada, Y. Takai, T. Nakaoki, T. Takeda and M. Sawada, Enantioselective complexation of chiral linear hosts containing monosaccharide moieties with chiral organic amines. Tetrahedron, 58 (2002) 4319–4330.

89 M. Shizuma, H. Imamura, Y. Takai, H. Yamada, T. Takeda, S. Takahashi and M. Sawada, Facile ee-determination from a single measurement by fast atom bombardment mass spectrometry: a double labeling method. Int. J. Mass Spectrom., 210–211 (2001) 585–590.

90 Y. Cheng and D.M. Hercules, Measurement of chiral complexes of cyclodextrins and amino acids by electrospray ionization time-of-flight mass spectrometry. J. Mass Spectrom., 36 (2001) 834–836.

91 G. Grigorean and C.B. Lebrilla, Enantiomeric analysis of pharmaceutical compounds by ion/molecule reactions. Anal. Chem., 73 (2001) 1684–1691.

92 J. Ramirez, F. He and C.B. Lebrilla, Gas-phase chiral differentiation of amino-acid guests in cyclodextrin hosts. J. Am. Chem. Soc., 120 (1998) 7387–7388.

93 S. Ahn, J. Ramirez, G. Grigorean and C.B. Lebrilla, Chiral recognition in gas-phase cyclodextrin: amino acid complexes – is the three point interaction still valid in the gas phase? J. Am. Soc. Mass Spetrom., 12 (2001) 278–287.

94 J.F. Gal, M. Stone and C.B. Lebrilla, Chiral recognition of non-natural a-amino acids. Int. J. Mass Spectrom., 222 (2003) 259–267.

95 E. Camara, M.K. Green, S.G. Penn and C.B. Lebrilla, Chiral recognition is observed in the deprotonation reaction of cytochrome c by (2R)- and (2S)-2-butylamine. J. Am. Chem. Soc., 118 (1996) 8751–8752.

96 Habib Bagheri, Hao Chen and R.G. Cooks, Chiral recognition by proton transfer reactions with optically active amines and alchols. Chem. Commun., 23 (2004) 2740–2741.

97 V. Carlesso, F. Fournier and J.-C. Tabet, Stereochemical differentiation of four mono-saccharides using transition metal complexes by electrospray ionization/ion trap mass spectrometry. Eur. Mass Spectrom., 6 (2000) 421–428.
98 H. Desaire and J.A. Leary, Differentiation of diastereomeric N-acetylhexosamine monosaccharides using ion trap tandem mass spectrometry. Anal. Chem., 71 (1999) 1997–2002.
99 Z.P. Yao, T.S. Wan, K.P. Kwong and C.T. Che, Chiral analysis by electrospray ionization mass spectrometry/mass spectrometry. 1. Chiral recognition of 19 common amino acids. Anal. Chem., 72 (2000) 5383–5393.
100 Z.P. Yao, T.S. Wan, K.P. Kwong and C.T. Che, Chiral analysis by electrospray ionization mass spectrometry/mass spectrometry. 2. Determination of enantiomeric excess of amino acids. Anal. Chem., 72 (2000) 5394–5401.
101 J.L. Seymour and F. Turrecek, Competitive binding studies of D,L-aromatic amino acids to copper chiragen complexes using electrospray ionization mass spectrometry. 49th ASMS Conference on Mass Spectrometry and Applied Topics, Chicago, Illinois, 2001.
102 L. Drahos and K. Vekey, Special feature: commentary – how closely related are the effective and the real temperature. J. Mass Spectrom., 34 (1999) 79–84.
103 P.B. Armentrout, Entropy measurements and the kinetic method: a statistically meaningful approach. J. Am. Soc. Mass Spectrom., 11, (2000) 371.
104 P.B. Armentrout, Special feature: commentary – is the kinetic method a thermodynamic method? J. Mass Spectrom., 34 (1999) 74–78.
105 L. Drahos and K. Vekey, Entropy evaluation using the kinetic method: is it feasible? J. Mass Spectrom., 38 (2003) 1025–42.
106 W.A. Tao, D. Zhang, E.N. Nikolaev and R.G. Cooks, Copper(II)-assisted enantiomeric analysis of D,L-amino acids using the kinetic method: chiral recognition and quantification in the gas phase. J. Am. Chem. Soc., 122 (2000) 10598–10609.
107 L. Wu, W.A. Tao and R.G. Cooks, Ligand and metal ion effects in metal ion clusters used for chiral analysis of alpha-hydroxy acids by the kinetic method. Anal. and Bioanal. Chem., 373 (2002) 618–627.
108 G. Fago, A. Filippi, A. Giardini, A. Lagana, A. Paladini and M. Speranza, Chiral recognition of O-phosphoserine. Angew. Chem. Int. Ed., 40 (2001) 4051–4054.
109 A. Paladini, C. Calcagni, T. Di Palma, M. Speranza, A. Lagana, G. Fago, A. Filippi, M. Satta and A.G. Guidoni, Enantiodiscrimination of chiral alpha-aminophosphonic acids by mass spectrometry. Chirality, 13 (2001) 707–711.
110 D.V. Augusti, F. Carazza, R. Augusti, W.A. Tao and R.G. Cooks, Quantitative chiral analysis of sugars by electrospray ionization tandem mass spectrometry using modified amino acids as chiral reference compounds. Anal. Chem., 74 (2002) 3458–3462.
111 W.A. Tao, L. Wu and R.G. Cooks, Rapid enantiomeric quantitation of an antiviral nucleoside agent (D,L-FMAU, 2′-fluoro-5-methyl-beta, D,L-arabinofuranosyluracil) by mass spectrometry. J. Med. Chem., 44 (2001) 3541–3544.
112 W.A. Tao, F.C. Gozzo and R.G. Cooks, Mass spectrometric quantitation of chiral drugs by the kinetic method. Anal. Chem., 73 (2001) 1692–1698.
113 D.V. Augusti, R. Augusti, F. Carazza and R.G. Cooks, Quantitative determination of enantiomeric composition of thalidomide solutions by electrospray ionization tandem mass spectrometry. Chem. Commun., 19 (2002) 2242–2243.
114 W.A. Tao and R.G. Cooks, Parallel reactions for enantiomeric quantification of peptides by mass spectrometry. Angew. Chem. Int. Ed., 40 (2001) 757–760.
115 W.A. Tao, L. Wu and R.G. Cooks, Differentiation and quantitation of isomeric dipeptides by low-energy dissociation of copper(II)-bound complexes. J. Am. Soc. Mass Spectrom., 12 (2001) 490–496.
116 A. Paladini, D. Scuderi, A. Lagana, A. Giardini, A. Filippi and M. Speranza, Enantiodiscrimination of chiral α-aminophosphonic acids by mass spectrometry. Chirality, 13 (2001) 707–711.
117 J. Chen, C. Zhu and Y. Zhao, Enantiomeric quantification of the bioactive peptide seryl-histidine methyl ester by electrospray ionization mass spectrometry and the kinetic method. Rapid Commun. Mass Spectrom., 16 (2002) 1251–1253.

118 W.A. Tao, D. Zhang, F. Wang, F. Thomas and R.G. Cooks, Kinetic resolution of D,L-amino acids based on gas-phase dissociation of copper(II) complexes. Anal. Chem., 71 (1999) 4427–4429.
119 L. Wu and R.G. Cooks, Chiral analysis using the kinetic method with optimized fixed ligands: application to some antibiotics. Anal. Chem., 75 (2003) 678–684.
120 R.G. Cooks and T.L. Kruger, Intrinsic basicity determination using metastable ions. J. Am. Chem. Soc., 99 (1977) 1279–1281.
121 L. Wu and R.G. Cooks, Improved chiral and isomeric analysis by mass spectrometry using fixed ligands. Eur. J. Mass Spectrom., 11 (2005) 231–242.
122 L. Wu and R.G. Cooks, Chiral and isomeric analysis by electrospray ionization and sonic spray ionization using the fixed-ligand kinetic method. Eur. J. Mass Spectrom., 11 (2005) 231–242.
123 W.A. Tao, R.L. Clark and R.G. Cooks, Quotient ratio method for quantitative enantiomeric determination by mass spectrometry. Anal. Chem., 74 (2002) 3783–3789.
124 L. Wu, M. Eduardo and R.G. Cooks, Chiral morphing and enantiomeric quantification in mixtures by mass spectrometry. Anal. Chem., 76 (2004) 663–671.
125 J. Lah, N.M. Maier, W. Lindner and G. Vesnaver, Thermodynamics of binding of (R)- and (S)-dinitrobenzoyl leucine to cinchona alkaloids and their tert-butylcarbamate derivatives in methanol: evaluation of enantioselectivity by spectroscopic (CD, UV) and microcalorimetric (ITC) titrations. J. Phys. Chem. B, 105 (2001) 1670–1678.
126 L. Wu, R.L. Clark and R.G. Cooks, Chiral quantification of D-, D-, and *meso*-tartaric acid mixtures using a mass spectrometric kinetic method. Chem. Commun., 137 (2003) 136–137.
127 L. Wu, K. Lemr, T. Aggerholm and R.G. Cooks, Recognition and quantification of binary and ternary mixtures of isomeric peptides by the kinetic method: metal ion and ligand effects on the dissociation of metal-bound complexes. J. Am. Soc. Mass Spectrom., 14 (2003) 152–160.
128 L. Wu, A.W. Tao and R.G. Cooks, Kinetic method for the simultaneous chiral analysis of different amino acids in mixtures. J. Mass Spectrom., 38 (2003) 386–393.
129 J.C. Fromme and G.L. Verdine, Structural insights into lesion recognition and repair by the baterial 8-oxaguanine DNA glycosylase MutM. Nature Struct. Bio., 9 (2002) 544–552.
130 C.T. Yu, Y.L. Guo, G.Q. Chen and Y.W. Zhong, Chiral recognition of zinc(II) ion complexes composed of bicyclo[3.3.0] octane-2,6-diol and s-naproxen probed by collisional-induced dissociation. J. Am. Soc. Mass Spectrom., 15 (2004) 795–802.
131 G. Grigorean and C.B. Lebrilla, Quantitative enantiomeric analysis of drugs via FT-ICR MS. ACS National Meeting, Washington DC, United States, 2000.
132 H. Uchida, E.N. Kodama, K. Yoshimura, Y. Maeda, P. Kosalaraksa, V. Maroun, Y.-L. Qiu, J. Zemlicka and H. Mitsuya, In vitro anti-human immunodeficiency virus activities of Z- and E-methylenecyclopropane nucleoside analogues and their phosphoro-L-alaninate diesters. Antimicrob. Agents Chemother., 43 (1999) 1487–1490.
133 S.K. Ono, N. Kato, Y. Shiratori, J. Kato, T. Goto, R.F. Schinazi, F.J. Carrilho and M. Omata, The polymerase L528M mutation cooperates with nucleotide binding-site mutations, increasing hepatitis B virus replication and drug resistance. J. Clin. Invest., 107 (2001) 449–455.
134 H.K. Han, R.L.A. De Vrueh, J.K. Rhie, K.-M.Y. Covitz, P.L. Smith, C.-P. Lee, D.M. Oh, W. Sadee and G.L. Amidon, 5′-Amino acid esters of antiviral nucleosides, acyclovir, and AZT are absorbed by the intestinal PEPT1 peptide transporter. Pharm. Res., 15 (1998) 1154–1159.
135 G.Q. Yao, S. Liu, E. Chou, M. Kukhanova, C.K. Chu and Y. Cheng, Inhibition of EBV replication by a novel L-nucleoside, 2′-fluoro-5-methyl-beta-L-arabinofuranosyluracil. Biochem. Pharmacol., 51 (1996) 941–947.
136 M.R. Blum, G.E. Chittick, L.H. Wang, L.J. Keilholz, L. Fang, G.M. Szczech and F.S. Rousseau, Clevudine (L-FMAU) a new agent under development for the treatment of hepatitis B virus (HBV): evaluation of the safety, pharmacokinetics and effect of food following single-dose administration in healthy male volunteers. Hepatology, 32 (2000) 1702–1706.
137 P. Hewawasam, V.K. Gribkoff, Y. Pendri, S.I. Dworetzky, N.A. Meanwell, E. Martinez, C.G. Boissard, D.J. Post-Munson, J.T. Trojnacki, K. Yeleswaram, L.M. Pajor, J. Knipe, Q. Gao, R. Perrone and J.E. Starrett, The synthesis and characterization of BMS-204352 (MaxiPost)

and related 3-fluorooxindoles as openers of maxi-K potassium channels. Bioorg. Med. Chem. Lett., 12 (2002) 1023–1026.

138 D.D. Dischino, H.A. Dulac, K.W. Gillman, L.S. Keller, E.S. Kozlowski, L.R. Marcin, J.J. Mongillo and J.E. Starrett Jr., Microwave-assisted synthesis and chiral HPLC separation of 18F-labeled maxipost. An agent for post-stroke neuroprotection. J. Labelled Compd. Radiopharm, 46 (2003) 1161–1171.

139 D. Zhang, R. Krishna, L. Wang, J. Zeng, J. Mitroka, R. Dai, N. Narasimhan, R. A. Reeves, N.R. Srinivas and L.J. Klunk, Metabolism, pharmacokinetics, and protein covalent binding of radiolabeled marcipost (BMS-204352) in humans. Drug Metab. Dispos. 33(1) (2005) 83–93.

140 B.R. Baker and R.L. Garrell, g-Factor analysis of protein secondary structure in solutions and thin films. Faraday Discuss., 126 (2004) 209–222.

141 C.R. Mallet, Ziling Lu and J.R. Mazzeo, A study of ion suppression effects in electrospray ionization from mobile phase additives and solid-phase extracts. Rapid Commun. Mass Spectrom., 18 (2003) 49–58.

142 L. Wu and R.G. Cooks, On-the-fly chiral analysis by the kinetic mehtod using data-dependent LC/MS-MS. In preparation.

143 D.F. Hagen, Characterization of isomeric compounds by gas and plasma chromatography. Anal. Chem., 51 (1979) 870–874.

144 H.H. Hill, W.F. Siems, R.H. Louis and D.G. McMinn, Ion mobility spectrometry. Anal. Chem., 62 (1990) 1200A–1209A.

145 C. Wu, W.F. Siems, J. Klasmeier and H.H. Hill, Separation of isomeric peptides using electrospray ionization/high-resolution ion mobility spectrometry. Anal. Chem., 72 (2000) 391–395.

146 C. Wu, W.F. Siems, U.K. Rasulev, E.G. Nazarov and H.H. Hill, 6th International Workshop on Ion Mobility Spectrometry, Dresden, Germany, 1997.

147 M.A. McCooeye, Z. Mester, B. Ells, D.A. Barnett, R.W. Purves and R. Guevremont, Quantitation of amphetamine, methamphetamine and their methylenedioxy derivatives in urine by solid-phase microextraction coupled with electrospray ionization-high-field asymmetric waveform ion mobility spectrometry-mass spectrometry. Anal. Chem., 74 (2002) 3071–3075.

148 M.A. McCooeye, L. Ding, G.J. Gardner, C.A. Fraser, J. Lam, R.E. Sturgeon and Z. Mester, Separation and quantitation of the stereoisomers of ephedra alkaloids in natural health products using flow injection-electrospray ionization-high field asymmetric waveform ion mobility spectrometry-mass spectrometry. Anal. Chem., 75 (2003) 2538–2542.

149 T.J. Wozniak, R.J. Bopp and E.C. Jensen, Chiral drugs: an industrial analytical perspective. J. Pharm. Biomed. Anal., 9 (1991) 363–382.

150 L. Wu, E.C. Meurer, B. Young, P. Yang, M.N. Eberlin and R.G. Cooks, Isomeric differentiation and quantification of alpha, beta-amino acid-containing tripeptides by the kinetic method: alkali metal-bound dimeric cluster ions. Int. J. Mass Spectrom., 231 (2004) 103–111.

151 B.L. Young and R.G. Cooks, Improved precision and accuracy in chiral determination using the mass spectrometric kinetic method. In preparation.

152 D. Zhang, W.A. Tao and R.G. Cooks, Chiral resolution of D- and L-amino acids by tandem mass spectrometry of Ni(II)-bound trimeric complexes. Int. J. Mass Spectrom., 204 (2001) 159–169.

153 M.L. Nielsen, M.M. Savitski and R.A. Zubarev, Distinguishing and quantifying peptides and proteins containing D-amino acids by tandem mass spectrometry. Anal. Chem., 77 (2005) 4571–4580.

154 K.F. Haselmann, T. Jorgensen, B.A. Budnik, F. Jensen and R.A. Zubarev, Electron capture dissociation of weakly bound polypeptide polycationic complexes. Rapid Commun. Mass Spectrom., 16 (2002) 2260–2265.

155 F.W. McLafferty, in: C.J. McNeal (Ed.), Tandem Mass Spectrometry of Large of Large Molecules, J. Wiley, New York, 1986 pp. 107–120.

156 H.B. Oh, K. Breuker, S.K. Sze, Y. Ge, B.K. Carpenter and F.W. McLafferty, Secondary and tertiary structures of gaseous protein ions characterized by electron capture dissociation mass spectrometry and photofragment spectroscopy. Proc. Natl. Acad. Sci., USA 10 (2002) 15863–15868.

157 C. M. Adams, F. Kejldsen and R.A. Zubarev, Electron capture dissociation distinguishes a single D-amino acid in a protein and probes the tertiary structure. J Am Soc Mass Spectrom., 15 (2004) 1087–1098.

158 H. Steen and M. Mann, The ABC's (and XYZ's) of peptide sequencing. Molecular Cell Biology: Nat. Rev., 5 (2004) 699–711.

159 F. Kjeldsen, K. F. Haselmann, E. S. Sorensen and R.A. Zubarev, Distinguishing of Ile/Leu amino acid residues in the PP3 protein by (hot) electron capture dissociation in Fourier transform ion cyclotron resonance mass spectrometry. Anal. Chem., 75 (2003) 1267–1274.

160 J.J. Coon, B. Ueberheide, J.E.P. Syka, D.D. Dryhurst, J. Ausio, J. Shabanowitz and D.F. Hunt, Protein identification using sequential ion/ion reactions and tandem mass spectrometry. Proc. Natl. Acad. Sci. USA, 102 (2005) 9463–9468.

161 J.J. Coon, J. Shabanowitz, D.F. Hunt and J.E.P. Syka, Electron transfer dissociation of peptide anions. J. Am. Soc. Mass Spectrom., 16 (2005) 880–882.

162 B. L. Young, S. Miller and R.G. Cooks, DESI analysis of chiral drugs. In preparation.

163 P. Chen, Electrospray ionization tandem mass spectrometry in high-throughput screening of homogeneous catalysts. Angew. Chem. Int. Ed., 42 (2003) 2832–2847.

164 Y. Kwon,-W. and D.J. Triggle, Chiral aspects of drug action at ion channels: a commentary on the stereoselectivity of drug actions at volatge-gated ion channels with particular reference to verapamil actions at the Ca^{2+} channel. Chirality, 3 (1991) 393–404.

165 D. Zhang, W.A. Tao and R.G. Cooks, Chiral resolution of D- and L-amino acids by tandem mass spectrometry of Ni(II)-bound trimeric complexes. Int. J. Mass Spectrom., 204 (2001) 159–169.

166 W.A. Tao, L. Wu and R.G. Cooks, Rapid enantiomeric quantitation of an antiviral nucleoside agent (D,L-FMAU, 2′-fluoro-5-methyl-beta, D,L-arabinofuranosyluracil) by mass spectrometry. J. Med. Chem., 44 (2001) 3541–3544.

167 W.A. Tao, F.C. Gozzo and R.G. Cooks, Mass spectrometric quantitation of chiral drugs by the kinetic method. Anal. Chem., 73 (2001) 1692–1698.

168 B.L. Young, S.A. Miller, M.C. Madden, M. Bair, J. Jia and R.G. Cooks, Chiral purity assay for MaxiPost™ using tandem mass spectrometry: method development, validation, and benchmarking. In preparation.

169 M. Sawada, H. Yamada, Y. Takai, Y. Kawai, H. Yamada, T. Azuma, T. Fujioka and T. Tanaka, Determination of enantiomer excess for amino acid ester salts using FAB mass spectrometry. Chem. Commun., 1569–1570 (1998).

170 V. R. Meyers and M. Rais, A vivid model of chiral recognition. Chirality, 1 (1989) 167–169.

171 J. Guo, J. Wu, G. Siuzdak and M.G. Finn, Measurement of enantiomeric excess by kinetic resolution and mass spectrometry. Angew. Chem. Int. Ed., 38 (1999) 1755–1758.

172 W.A. Tao, D. Zhang, F. Wang, P. Thomas and R.G. Cooks, Kinetic resolution of D, L-amino acids based on gas-phase dissociation of copper(II) complexes. Anal. Chem., 71 (1999) 4427–4429.

173 L. Wu and R.G. Cooks, Improved chiral and isomeric analysis by mass spectrometry using fixed ligands. Eur. J. Mass Spectrom., 11 (2005) 231–242.

174 L. Wu, E. Meurer and R.G. Cooks, Chiral morphing and enantiomeric quantification in mixtures by mass spectrometry. Anal. Chem., 76, 663–671 (2004).

© 2006 Elsevier B.V. All rights reserved.

CHAPTER 18

Novel chiral derivatizing agents powerful for enantioresolution and determination of absolute stereochemistry by X-ray crystallographic and ¹H NMR anisotropy methods

Nobuyuki Harada, Masataka Watanabe and Shunsuke Kuwahara

Institute of Multidisciplinary Research for Advanced Materials, Tohoku University, 2-1-1 Katahira, Aoba, Sendai 980-8577, Japan

18.1 INTRODUCTION

It is well recognized that molecular chirality is essential for life processes, and that most biologically active compounds controlling physiological functions of living organisms are chiral. Hence in the structural study of biologically active compounds, including natural products, determination of absolute configuration becomes the first major issue. The second issue is the chiral synthesis of natural products and biologically active compounds that become pharmaceutical targets and how efficiently desired enantiomers can be synthesized with 100% enantiopurity or enantiomeric excess (% ee). Furthermore, studies on chiral functional molecules and molecular machines, such as the light-powered chiral molecular motor developed in our laboratory, has been rapidly progressing in recent years. Therefore, the unambiguous determination of the absolute configuration of chiral compounds as well as their chiral syntheses are of vital importance in the field of material science.

We have recently developed chiral carboxylic acids as novel molecular tools proven to be powerful for enantioresolution and simultaneous determination of absolute configuration of various alcohols. Those chiral molecular tools are very useful for the facile synthesis of enantiomers with 100% ee and also for the absolute configurational assignment. The methods using these chiral tools have been successfully applied to various compounds, and their methodologies and applications are explained throughout this chapter.

18.2 METHODOLOGIES FOR DETERMINING ABSOLUTE CONFIGURATION AND THEIR EVALUATIONS

The methodologies to determine the absolute configurations of chiral compounds are classified into the following two categories.

18.2.1 Nonempirical methods for determining absolute configurations of chiral compounds

As the methods of this category, there are the Bijvoet method of X-ray crystallography [1] and circular dichroism (CD) exciton chirality method [2]. These powerful methods provide nonempirical determination of a target molecule's configuration without the knowledge of the absolute configuration of reference compounds. In X-ray crystallography, since the anomalous dispersion effect of heavy atoms can be measured very accurately under proper conditions, the absolute stereostructure obtained is unambiguous and reliable. In addition, the molecule can be projected as a three-dimensional structure, and therefore the method has been employed extensively. However, the X-ray method needs single crystals of suitable size good for X-ray diffraction, and so the critical problem is how to obtain such single crystals. As a consequence, a study using this method often becomes a lengthy trial and error search for ideal single crystals.

The CD exciton chirality method [2] is also useful because the absolute configuration can be determined in a nonempirical manner, and it does not require crystallization. Furthermore, chiral chemical and biological reactions are traceable by CD, and even the absolute configurations and conformations of unstable compounds can be obtained by this method. However, because some compounds are not ideal targets for this method, the results must be interpreted carefully.

18.2.2 Relative methods for determining absolute configuration using an internal reference with known absolute configuration

Absolute configuration can be obtained by determining the relative configuration at the position of interest against a reference compound or substituent with known absolute configuration. A typical example is the X-ray crystallography taken after the introduction of a chiral auxiliary with known absolute configuration (Fig. 18.1) [3–6]. In this case, the absolute configuration of the point in question can be automatically determined using the chirality of the auxiliary introduced as an internal reference. Consequently, the samples do not need to contain heavy atoms for anomalous dispersion effect. The result obtained is very clear, even when the final *R*-value is not small enough due to poor quality of the single crystal. The absolute configuration can be

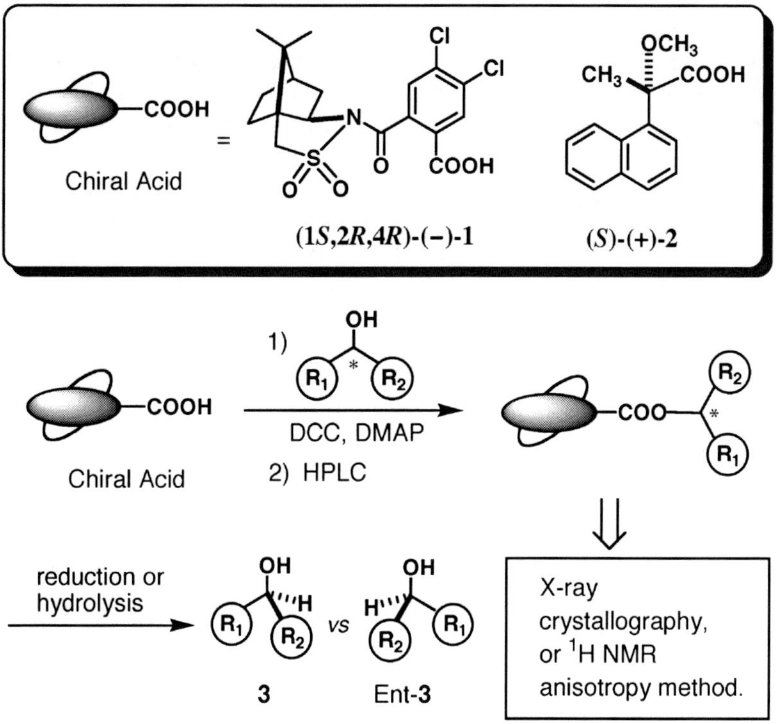

Fig. 18.1. Enantioresolution and determination of absolute configuration of alcohols using chiral carboxylic acids.

determined with certainty, even if only the relative configuration is obtained. A variety of methods to link an internal reference to the target molecule have been developed. For example, there are ionic bonding such as conventional acid-base salts, covalent bonding such as esters or amides, and the use of recently developed inclusion complexes [7–9]. These relative X-ray methods are expected to find widespread application.

Recently, the proton nuclear magnetic resonance (^{1}H NMR) anisotropy method has often been employed as the relative method, and it is useful for the study of the absolute configuration of natural products. In particular, the absolute configurations of secondary alcohols are frequently determined using the advanced Mosher method developed by Kusumi et al. [10–13]. In this case, the absolute configurations of chiral auxiliaries, such as Mosher's reagent [α-methoxy-α-(trifluoromethyl)phenylacetic acid (MTPA)] and Trost's reagent [α-methoxyphenylacetic acid (MPA)], are known, and the preferred conformation of the esters formed with chiral secondary alcohols and MTPA or MPA acid is rationalized. In addition, the aromatic substituent (phenyl group) generates a magnetic anisotropy effect due to the ring current induced under the external magnetic field, and so the proton NMR signals of the alcohol moiety facing to the phenyl group in the preferred conformation are moved to a higher magnetic field (high field shift). By observing the ^{1}H NMR anisotropy effect, the absolute configuration of alcohol part can be determined. This method is very convenient, since it does not require crystallization of compounds and NMR machines are daily used. One problem of this method is that it is based on the assumption of preferred conformation of molecules in solution. However, it is highly reliable since the method itself has a self-diagnostic function. Although the method has been widely applied to secondary alcohols, it is expected to be extended to other kinds of compounds as well.

The absolute configuration can be determined relatively by chemical correlation or comparison of optical rotation, $[\alpha]_D$, and/or CD spectrum with that of reference compounds with known absolute configuration. Although this method is also frequently employed, a careful selection of reference compounds is necessary for reliable analysis.

18.3 METHODOLOGIES FOR CHIRAL SYNTHESIS AND THEIR EVALUATIONS

The task after determination of absolute configuration is the synthesis of chiral compounds. The practical methods to synthesize chiral compounds are roughly divided into two categories, each of which is further divided and has advantages and disadvantages as described below. In this chapter, "chiral synthesis" includes not only the so-called asymmetric synthesis but also enantioresolution. In addition, the method in which covalently bonded diastereomers are formed using a chiral auxiliary, followed by HPLC separation and recovery of the target compound, is also defined as enantioresolution.

18.3.1 Enantioresolution of racemates

Type (*a*) In this method, a chiral auxiliary is ionically bonded to racemates as seen in the conventional cases of acid-base combination, and a mixture of diastereomers formed is subjected to fractional recrystallization to obtain enantiopure compounds. This method is also applicable to the inclusion complexes formed by, for example hydrogen bonding [7–9]. The critical point is whether or not the diastereomer can be obtained with 100% enantiopurity through fractional recrystallization. It should be noted that recrystallization does not always afford 100% enantiopure diastereomer. If this method is successful, it is suitable for mass preparation of the chiral compounds.

Type (*b*) In this method, a chiral auxiliary is covalently bonded to racemates to produce a diastereomeric mixture, which is separated by conventional HPLC on silica gel or other methods to enantiopure diastereomers, and then the chiral auxiliary is cleaved off (Fig. 18.1). This method can yield an enantiopure compound. The point is whether or not diastereomers can clearly be separated by HPLC. If a clear separation is achieved, each diastereomer obtained is enantiopure, and the target compound after cleavage of the chiral auxiliary is also 100% enantiopure. It is advisable to use a chiral auxiliary that can be cleaved off easily.

Type (*c*) This is an excellent method where the racemates are directly enantioseparated by HPLC or GC using columns made of chiral stationary phases, and a number of reports have been published [14]. The question is again whether racemates are clearly separated into two enantiomers or not. If a clear separation is achieved, 100% pure enantiomers are obtained by this method as well. The method is convenient and suitable for analytical separation, as it does not require derivatization. In general, chiral columns are expensive and are, therefore, mostly used for analytical purposes. However in some cases, mass separation is conducted on an industrial scale to obtain chiral compound such as pharmaceutical materials. Careful analysis is required when determining absolute configuration by the elution order, as there are many exceptions.

Type (*d*) This is a unique method where racemates undergo an enzymatic or asymmetric reaction to yield enantiomers by the kinetic resolution effect. In particular, high stereoselectivity of the enzymatic reaction leads to high enantiopurity [15]. However, care should be taken, since the method does not always yield 100% enantiopurity.

18.3.2 Asymmetric syntheses

Type (*a*) This is a highly efficient and powerful method to obtain chiral products by the action of a chiral reagent or chiral catalyst on achiral compounds. Being a well-known method, many eminent reviews have been published for these asymmetric syntheses, and so no further explanation is required here. The problem with this method is that the products obtained are not always enantiopure. Furthermore, it is generally difficult to determine the absolute configuration of the products based on the reaction mechanism. Accordingly, an independent determination of the absolute configuration by the methods described above is suggested.

Type (*b*) There is also another method to obtain chiral compounds such as by enzymatic reaction on achiral or meso compounds. The asymmetric reaction of a meso compound by an enzyme is particularly interesting and is defined as the desymmetrization reaction. In this case too, the enantiopurity is not always 100%, and the absolute configuration must be determined separately.

18.4 CAMPHORSULTAM DICHLOROPHTHALIC ACID (CSDP ACID (−)-1) AND CAMPHORSULTAM PHTHALIC ACID (CSP ACID (−)-4) USEFUL FOR ENANTIORESOLUTION OF ALCOHOLS BY HPLC AND DETERMINATION OF THEIR ABSOLUTE CONFIGURATIONS BY X-RAY CRYSTALLOGRAPHY

The authors consider that the most reliable and powerful method for determining the absolute configuration is the X-ray crystallography of compounds containing a chiral auxiliary with known absolute configuration as the internal reference, as described above. Namely, the absolute configuration of the point in question can be unambiguously determined from the ORTEP drawing showing a relative stereochemistry, because the absolute configuration of the chiral auxiliary is already known. Therefore

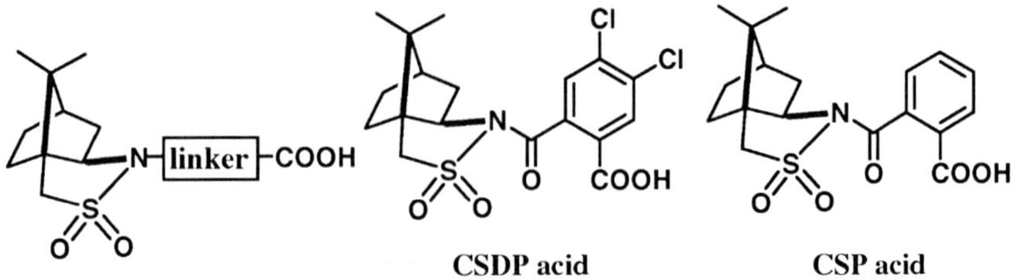

N—linker—COOH

CSDP acid
(1*S*,2*R*,4*R*)-(−)-1

CSP acid
(1*S*,2*R*,4*R*)-(−)-4

Fig. 18.2. Design of chiral molecular tools, CSDP and CSP acids containing 2,10-camphorsultam moiety.

it is easy to determine the absolute configuration, and there is no possibility to make a mistake in the assignment.

We also consider that the highly efficient method for preparing an appropriate amount of various chiral compounds with 100% enantiopurity in a laboratory scale is the enantioresolution of type (b) in Section 18.3.1, as illustrated in Figs. 18.1 and 18.2. In this method, a chiral auxiliary is covalently bonded to racemates, and the obtained diastereomeric mixture can be separated by conventional HPLC on silica gel. If the chromatogram shows a base-line separation, the diastereomers obtained are enantiopure. This method is characterized by a clear and efficient separation even with a small amount of sample, compared to the fractional recrystallization method described in type (a) in Section 18.3.1.

As chiral auxiliaries satisfying these two requirements, we have designed and prepared the chiral molecular tools, camphorsultam dichlorophthalic acid (CSDP acid) (−)-**1** and camphorsultam phthalic acid (CSP acid) (−)-**4**, connecting (1*S*,2*R*,4*R*)-2,10-camphorsultam and 4,5-dichlorophthalic or phthalic acid and have applied those chiral tools to various compounds (Fig. 18.2) [16–18, 20–33,42,46]. The 2,10-camphorsultam was selected because of its good affinity with silica gel used in HPLC, allowing good separation of two diastereomers. In addition, the sultam amide moiety is effective for providing prismatic single crystals suitable for X-ray diffraction experiment. Furthermore, the (1*S*,2*R*,4*R*) absolute stereochemistry of 2,10-camphorsultam established is useful as the internal reference of absolute configuration. To connect the alcohols, an ester bond was chosen, because it could readily be formed and cleaved off. Accordingly, phthalic acids, especially 4,5-dichlorophthalic acid, were selected as a linker [16,17]. In telephthalic acid and succinic acid, the two chiral moieties are separated spatially. However, in phthalic acid, they are close enough to result in a stronger interaction. So we have expected its diastereomeric recognition would be effective in HPLC (Fig. 18.2).

The desired molecular tool, CSDP acid (−)-**1**, was synthesized by reacting (1*S*,2*R*,4*R*)-(−)-2,10-camphorsultam anion with 4,5-dichlorophthalic anhydride [(−)-**1**, mp 221°C from EtOH; $[\alpha]_D^{20}$ −101.1 (*c* 1.375, MeOH); Fig. 18.2]. In a similar way, CSP acid (−)-**4**, was also prepared from (1*S*,2*R*,4*R*)-(−)-2,10-camphorsultam and phthalic anhydride [(−)-**4**, mp 184–187°C from $CHCl_3$; $[\alpha]_D^{20}$ −134.7 (*c* 2.218, MeOH); Fig. 18.2]. The compounds **1** and **4** should be formally called phthalic acid amides. However, here we adopted its common names, CSDP and CSP acids. These carboxylic acids were

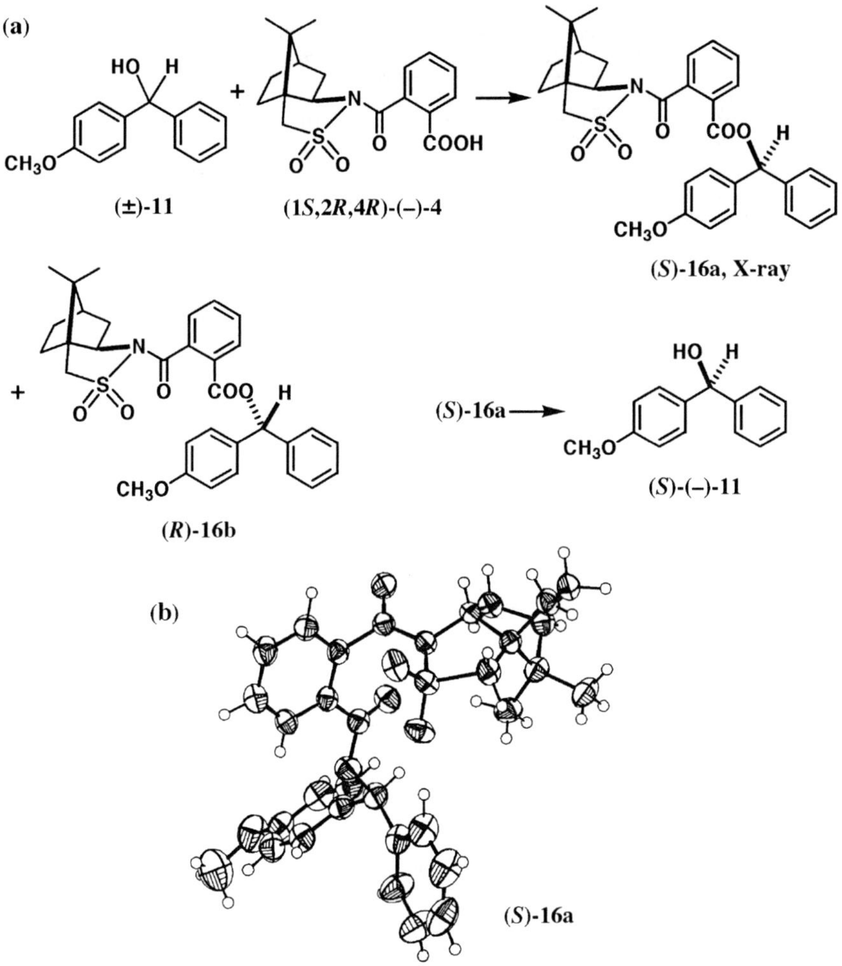

Fig. 18.3. (a) Enantioresolution of alcohol (**11**) using CSP acid (–)-**4** and determination of absolute configuration by X-ray crystallography. (b) ORTEP drawing of ester (*S*)-**16a**.

condensed with alcohol under the conditions of 1,3-dicyclohexylcarbodiimide (DCC) and 4-dimethylaminopyridine (DMAP) [16,17].

The following exemplifies a general procedure of this method. The CSP acid (−)-**4** was allowed to react with (±)-(4-methoxyphenyl)phenylmethanol (**11**) using DCC and DMAP in CH_2Cl_2 yielding diastereomeric esters, which were separated by HPLC on silica gel: hexane/EtOAc = 4:1; $\alpha = 1.1$, $R_s = 1.13$ (Fig. 18.3 and Table 18.1). The first-eluted ester **16a** obtained was recrystallized from EtOAc giving large colorless prisms: mp 172°C. A single crystal of **16a** was subjected to X-ray analysis affording the ORTEP drawing as shown in Fig 18.3(b), from which the absolute configuration of the alcohol part was determined as *S* based on the absolute configuration of camphorsultam moiety used as an internal reference. The *S* absolute configuration of **16a** was also

Table 18.1 Enantioresolution of alcohols by HPLC on silica gel using (1*S*,2*R*,4*R*)-(−)-CSP acid, and determination of their absolute configurations by X-ray crystallography

Alcohol	Solvent[a]	α	R_s	X-ray	Abs. Config. First Fr.	Ref.
5	H/EA = 3/1	1.1	1.3	2nd Fr.	*R*	[16]
6	H/EA = 5/1	1.1	1.3	1st Fr.	*R*	[16]
7	H/EA = 4/1	–	0.73	–	*R*	[4]
8	H/EA = 7/1	1.1	0.8	–	3*R*,4*R*	[17,18]
9	CH_2Cl_2/EA = 50/1	–	–	1st Fr.	*Msc*	[19]
10	H/EA = 2/1	1.2	1.3	–	a*R*,a*R*	[20,21]
11	H/EA = 4/1	1.1	1.3	1st Fr.	*S*	[22]
12	H/EA = 5/1	1.1	1.6	1st Fr.	R	[22]
13	H/EA = 3/1	1.3	2.8	1st Fr.	*R*	[23]
14	H/EA = 2/1	1.1	1.0	1st Fr.	S	[23]
15	H/EA = 3/1	1.1	1.6	2nd Fr.	*R*	[16]

[a]H—*n*-hexane, EA—ethyl acetate.

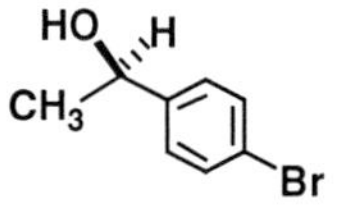

(*R*)-[CD(+)260.2]-**5** 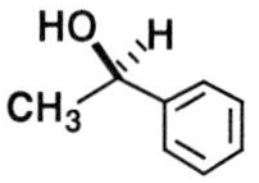 (*R*)-(+)-**6** 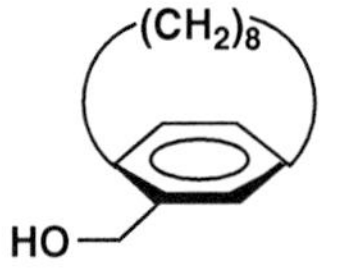 (*R*)-**7** 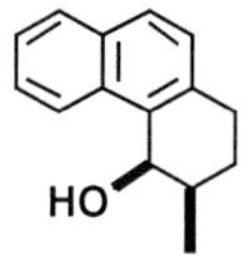 (3*R*,4*R*)-(+)-**8**

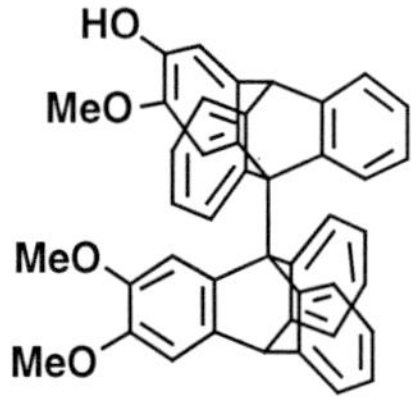

(*Msc*)-**9** 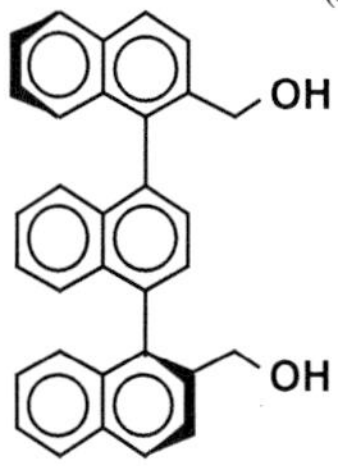 (a*R*,a*R*)-(−)-**10** 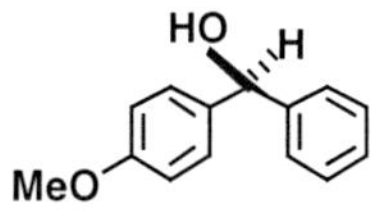 (*S*)-(−)-**11** 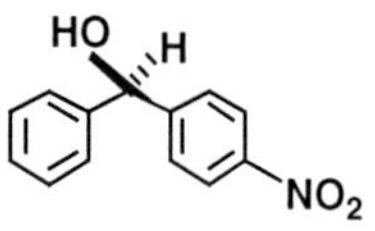 (*R*)-(−)-**12**

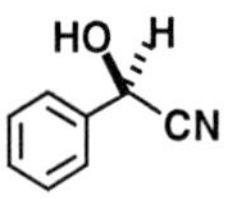

(*R*)-**13** 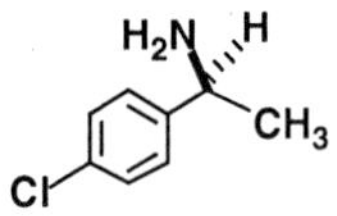 (*S*)-(−)-**14** 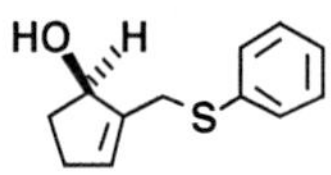 (*R*)-[CD(−)282.3]-**15**

confirmed by the heavy atom effect of a sulfur atom contained. The reduction of the first-eluted ester (*S*)-**16a** with $LiAlH_4$ yielded enantiopure alcohol (*S*)-(−)-**11**.

Other examples are listed in Table 18.1. The successful enantioresolution of various alcohols, determination of their absolute configurations by X-ray analysis, and recovery of enantiopure alcohols listed in the table proved the effectiveness of this method.

The compound **9** is an interesting example of atropisomeric and chiral substance studied by Toyota et al. In general, it is very difficult to determine the absolute configuration of chiral compounds of atropisomerism based on steric hindrance, however, Toyota et al. successfully solved this problem by application of the method described above [19].

In cyanohydrin **13** and amine **14**, the diastereomeric separation and determination of their absolute configurations were possible; however, there remains a problem to recover enantiopure compounds **13** and **14**, because the amide bond is not easily hydrolyzed in general. Amine **14** can be enantioresolved as the salt of (2*R*,3*R*)-(+)-tartaric acid [23], and its absolute configuration was established as (*S*)-(−) by this method. For compound **15**, its absolute configuration was unambiguously determined by this internal reference method, in spite of the large *R*-value, owing to its poor crystallinity.

We have then found that another chiral molecular tool, CSDP acid (−)-**1**, is much more powerful for enantioresolution of alcohols by HPLC on silica gel and also for providing prismatic single crystals suitable for X-ray analysis. As in the case of CSP acid (−)-**4**, this CSDP acid (−)-**1** is also useful as an internal reference in determining absolute configuration by X-ray analysis. Moreover, CSDP acid (−)-**1** contains two chlorine atoms as heavy atoms in addition to a sulfur atom, which leads to efficient determination of absolute configuration by the anomalous dispersion effect of heavy atoms.

An example is illustrated in Fig. 18.4. Alcohol (±)-**8** was condensed with CSDP acid (−)-**1** in the presence of DCC and DMAP. The diastereomeric mixture of the two esters obtained was subjected to HPLC on silica gel: hexane/EtOAc = 7:1, $\alpha = 1.18$, $R_s = 1.06$. While ester **38a** obtained as the first eluted fraction afforded silky and fine needle-like crystals which were unsuitable for X-ray analysis when recrystallized from MeOH, the second eluted fraction **38b** gave larger prisms good for X-ray analysis when recrystallized from EtOAc. The absolute configuration of **38b** was unambiguously determined to be (3*S*,4*S*) using the 2,10-camphorsultam part as an internal reference and also the heavy atom effect. Ester **38b** was reduced with $LiAlH_4$ to remove the chiral auxiliary, yielding enantiopure alcohol (3*S*,4*S*)-(−)-**8** [17,18]. This absolute configuration was consistent with the result obtained by the CD exciton chirality method applied to the corresponding 4-bromobenzoate of alcohol **8** [17].

Recently we have found that the solvolysis with K_2CO_3/MeOH was very effective to recover enantiopure alcohols from CSDP acid esters in a high yield.

Table 18.2 lists other application examples. CSDP acid esters of *para*-substituted diphenylmethanols **11**, **12**, and **19**–**23** were clearly separated by HPLC on silica gel, although the *para*-substituents governing the chirality of the molecule are apart from the stereogenic center, i.e. the carbon atom with hydroxyl group [22,24,26]. These results indicate that the CSDP acid recognizes well the molecular chirality, i.e. the difference between a hydrogen atom and a *para*-substituted functional group.

It was impossible to separate the diastereomeric CSDP acid esters of (4-methylphenyl) phenylmethanol (**21**) by HPLC on silica gel, as listed in Table 18.2. Namely, the chirality recognition of **21** as CSDP acid ester was difficult. The difference between hydrogen and methyl group constituting the molecular chirality of **21** is small, and so it is very hard to recognize such a trivial difference [22]. The following strategy was then adopted. First, (4-bromophenyl)(4′-methylphenyl)methanol (**22**) was selected as a precursor, which was well enantioresolved as CSDP acid esters, and their absolute configurations were determined by X-ray crystallography. The reduction of the bromine

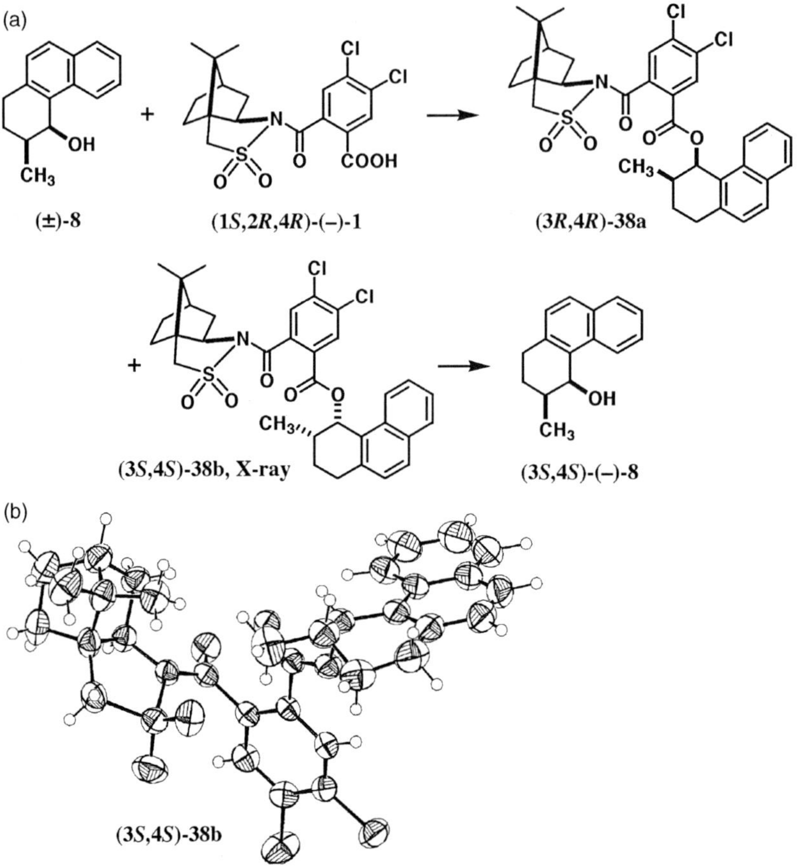

(a)

(±)-8 (1*S*,2*R*,4*R*)-(–)-1 (3*R*,4*R*)-38a

(3*S*,4*S*)-38b, X-ray (3*S*,4*S*)-(–)-8

(b)

(3*S*,4*S*)-38b

Fig. 18.4. (a) Enantioresolution of alcohol (**8**) using CSDP acid (–)-**1** and determination of its absolute configuration by X-ray crystallography. (b) ORTEP drawing of ester (3*S*,4*S*)-**38b**.

atom led to the desired and enantiopure alcohol (*S*)-(–)-**21** [22]. This strategy was also useful for the synthesis determination of absolute configuration of isotope-substituted chiral diphenylmethanols as described below.

The molecular chirality can also be generated by the substitution with isotopes as shown in ^{2}H-substituted diphenylmethanol **24** and ^{13}C-substituted diphenylmethanol **26** [26,27]. It is almost impossible to directly enantioresolve these isotope-substituted chiral compounds by usual HPLC with chiral stationary phase or by HPLC on silica gel using chiral auxiliary. In such a case, the next strategy is suggested: (a) selection of (±)-(4-bromophenyl)(phenyl-2,3,4,5,6-d_5)methanol (**25**) as a precursor; (b) enantioresolution as CSDP acid esters; (c) determination of its absolute configuration; and (d) subsequent removal of the bromine atom providing the desired and enantiopure (phenyl-2,3,4,5,6-d_5)phenylmethanol (**24**) [26].

The preparation of (phenyl-2,3,4,5,6-d_5)phenylmethanol (**24**) was carried out as follows. Racemic alcohol (±)-**25** was condensed with CSDP acid and then the esters

Table 18.2 Enantioresolution of alcohols by HPLC on silica gel using (1*S*,2*R*,4*R*)-(−)-CSDP acid, and determination of their absolute configurations by X-ray crystallography

Alcohol	Solvent[a]	α	R_s	X-ray	Abs. Config. First Fr.	Ref.
8	H/EA = 7/1	1.18	1.06	2nd Fr.	3*R*,4*R*	[17,18]
17	H/EA = 7/1	1.23	1.27	1st Fr. 2nd Fr.	1*R*,2*S*	[25]
18	H/EA = 10/1	1.30	1.74	1st Fr.	1*S*,4*R*	[17]
11	H/EA = 4/1	1.20	0.91	–	*S*	[22]
12	H/EA = 5/1	1.26	1.37	–	*R*	[22]
19	H/EA = 8/1	1.17	0.80	Yes[b]	*R*	[26]
20	H/EA = 6/1	1.17	0.95	–	*R*	[24]
21	H/EA = 7/1	1.00	–	–	–	[22]
22	H/EA = 8/1	1.18	0.83	1st Fr.	*R*	[22]
23	H/EA = 4/1	1.1	1.0	–	*R*	[24]
25	H/EA = 8/1	1.21	1.07	Yes[b]	*S*	[26]
27	H/EA = 4/1	1.27	1.20	Yes[b]	*S*	[27]
28	H/EA = 5/1	1.12	1.01	1st Fr.	*S*	[28]
30	H/EA = 4/1	1.14	0.91	2nd Fr.	*R*	[28,29]
31	H/EA = 10/1	1.26	1.03	–	*R*	[28]
32	H/EA = 6/1	1.26	1.29	–	*R*	[30]
33	H/EA = 5/1	1.16	1.11	1st Fr.	*S*	[31]
34	H/EA = 5/1	1.12	0.87	1st Fr.	*S*	[31]
35	H/EA = 2/1	1.11	0.88	–	*R*	[31]
36	H/EA = 2/1	1.38	1.19	1st Fr.	*R*	[31]
10	H/EA = 3/1	1.2	1.6	2nd Fr.	a*R*,a*R*	[20,21]
37	H/EA = 4/1	1.27	1.14	Yes[c]	*S*	[33]

[a]H—*n*-hexane, EA—ethyl acetate.
[b]X-ray analysis of camphanate ester.
[c]X-ray analysis of 4-bromobenzoate.

formed were separated by HPLC on silica gel. Both diastereomeric esters separated gave only fine needle-like crystals even after a series of recyrstallizations. Accordingly, after recovering the enantiopure alcohol (−)-**25** from the first-eluted fraction **39a**, a part of (−)-**25** was converted to ester **40** using (−)-camphanic acid chloride (Fig. 18.5). Ester **40** showed good crystallinity, providing prismatic crystals suitable for X-ray analysis, and the absolute configuration of the alcohol moiety was determined as *S*, using the absolute configuration of (−)-camphanic acid part as an internal reference. Subsequently, alcohol (*S*)-(−)-**25** was reduced with H_2NNH_2/H_2O in the presence of Pd-C to yield the isotope-substituted and enantiopure (phenyl-2,3,4,5,6-d_5)phenylmethanol [CD(−)270.4]-(*S*)-**24**, making possible the unambiguous determination of the chirality generated by the substitution with isotopes. The specific rotation $[\alpha]_D$ measured at the wavelength of the sodium D-line (589 nm) is usually used to distinguish enantiomers. However it is difficult to measure $[\alpha]_D$ value of compounds with isotope-substitution chirality. We have proposed the new definition method of enantiomers by the use of CD data, because CD is not only more sensitive than $[\alpha]_D$, but it is also accurately measurable even with a small amount of samples. For example, [CD(−)270.4]-(*S*)-**24** stands for the enantiomer with negative CD at 270.4 nm and *S* absolute configuration [26].

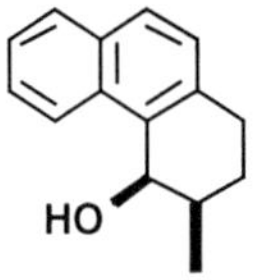

(3*R*,4*R*)-(+)-**8** 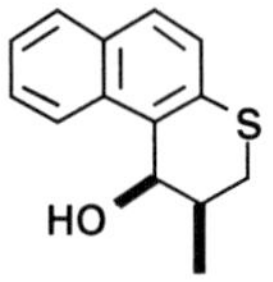 (1*R*,2*S*)-(+)-**17** 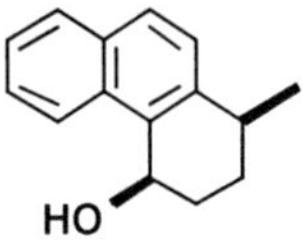 (1*S*,4*R*)-**18**

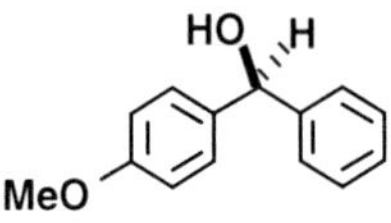

(*S*)-(−)-**11** 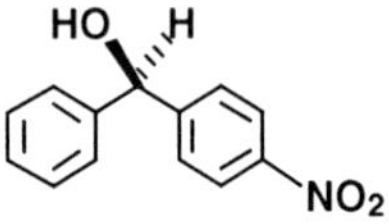 (*R*)-(−)-**12** 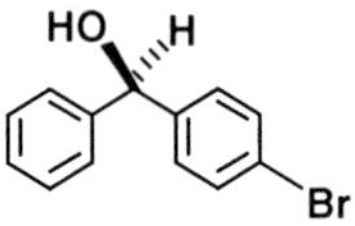 (*R*)-(−)-**19** 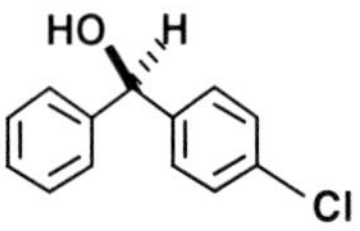 (*R*)-(−)-**20**

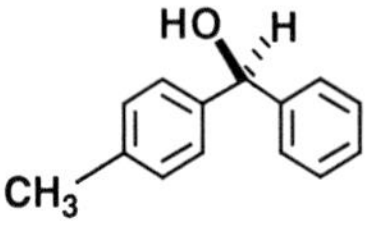

(*S*)-(−)-**21** 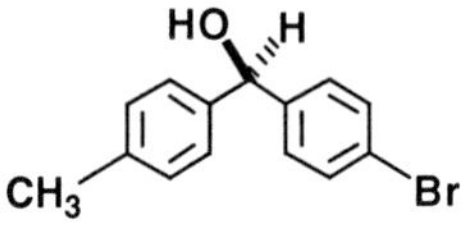 (*R*)-(−)-**22** 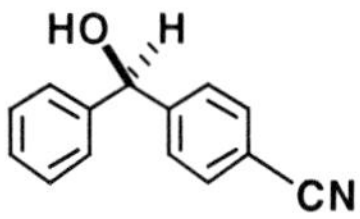 (*R*)-(−)-**23**

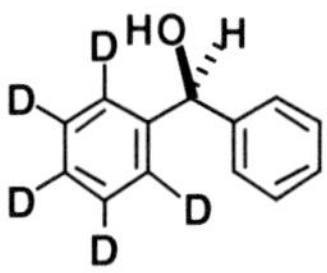

[CD(−)270.4]-(*S*)-**24** 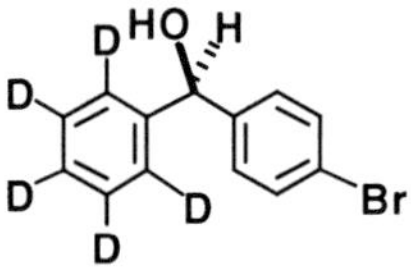 (*S*)-(−)-**25** 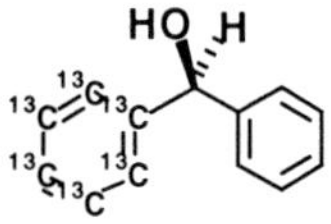 [CD(−)270]-(*S*)-**26** 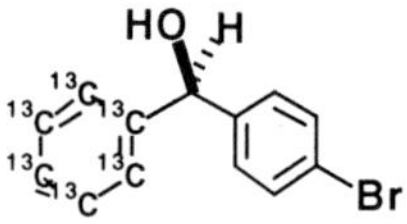 (*S*)-(−)-**27**

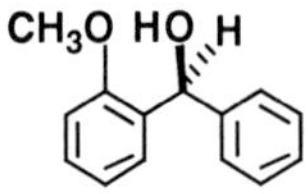

(*S*)-(−)-**28** 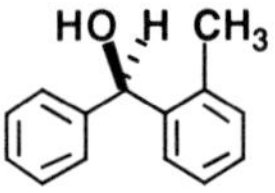 (*R*)-(−)-**29** 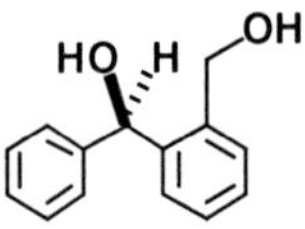 (*R*)-(+)-**30** 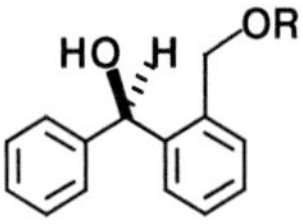 (*R*)-**31**: R = TBDMS (*R*)-**32**: R = TBDPS

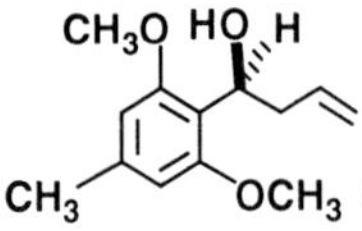

(*S*)-(−)-**33** 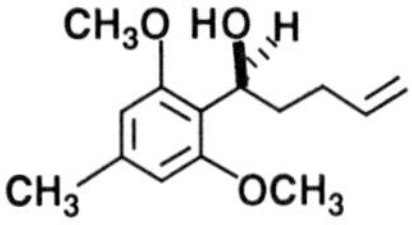 (*S*)-(−)-**34** 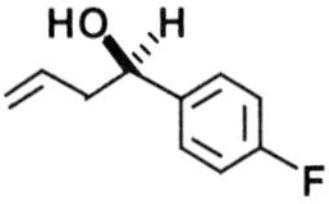 (*R*)-(+)-**35** 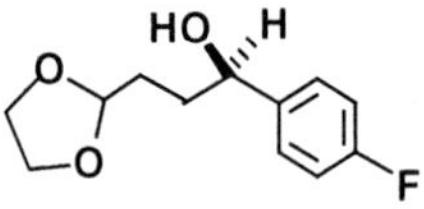 (*R*)-(+)-**36**

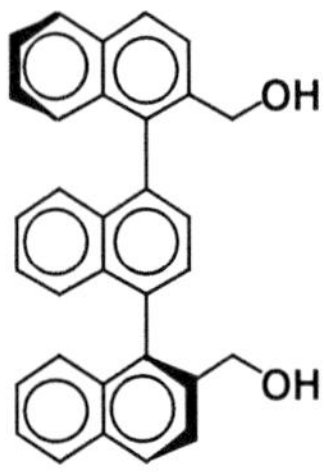

(a*R*,a*R*)-(−)-**10** 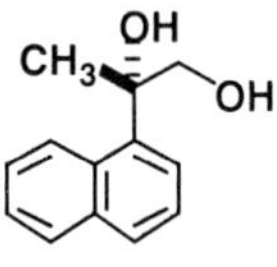 (*S*)-(−)-**37** 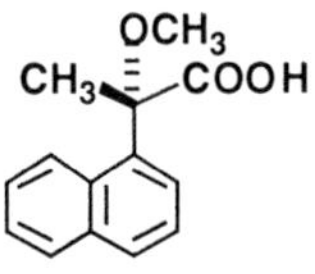 (*S*)-(+)-**2**

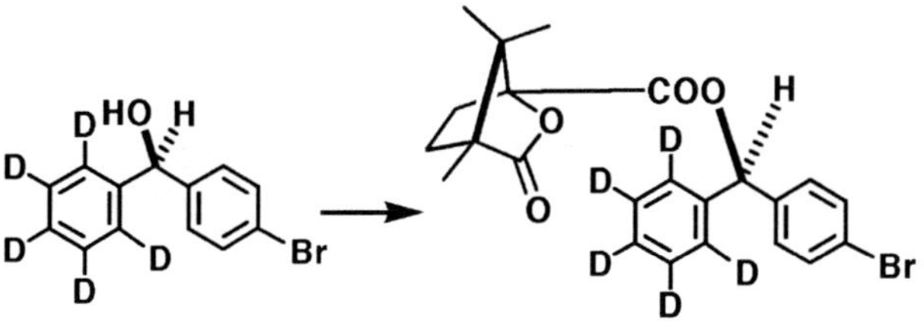

(*S*)-(−)-**25**

(*S*)-**40, X-ray**

Fig. 18.5. Preparation of camphanate ester (*S*)-**40**, whose absolute configuration was determined by X-ray crystallography.

It would be even more advantageous, if (−)-camphanic acid could be used from the beginning as the chiral auxiliary for enantioresolution. However, the enantioresolution power of (−)-camphanic acid is generally weak. For example, camphanic acid esters **40** prepared from racemic alcohol **25** could not be separated by HPLC on silica gel. Thus it is sometimes necessary to select two chiral auxiliaries depending on the situation.

A similar scheme could be applied to synthesize ^{13}C-substituted chiral diphenylmethanol (**26**) [27]. Namely, (4-bromophenyl)(phenyl-1,2,3,4,5,6-$^{13}C_6$)methanol (±)-(**27**) was chosen and then enantioresolution and determination of absolute configuration were carried out in a similar way as described above. Subsequently, the bromine atom was reduced to yield the ^{13}C-substituted chiral (phenyl-1,2,3,4,5,6-$^{13}C_6$)phenylmethanol [CD(−)270]-(*S*)-**26**. Despite the very small molecular chirality due to the slight difference between isotopes ^{12}C and ^{13}C, the CD spectrum of ^{13}C-substituted chiral diphenylmethanol (*S*)-**26** is clearly observable [27].

Interesting results were also obtained in the case of *ortho*-substituted diphenylmethanols. Using the method described above, enantiopure (2-methoxyphenyl)phenylmethanol (−)-(**28**) was prepared and its absolute configuration was determined as *S* [28]. Chiral (2-methylphenyl)phenylmethanol (**29**) had previously been synthesized by asymmetric catalytic reaction, and its absolute configuration had been estimated based on the chiral reaction mechanism. Is the absolute configurational assignment based on the reaction mechanism reliable? In such a case, the independent and unambiguous determination of the absolute configuration is necessary. To solve this problem, we have carried out the following experiments. The direct enantioresolution of **29** as CSDP acid esters was unsuccessful, as in the case of *para*-substituted alcohol **21**. Namely, it was difficult to discriminate hydrogen atom and the methyl group in the *ortho* position. We then attempted the enantioresolution of (4-bromophenyl)(2′-methylphenyl)methanol (**41**), because *para*-methyl analogue **22** was separable as described above. However, the HPLC analysis exhibited only a single peak.

Next we adopted the following indirect method [28–30]. The strategy consisted of enantioresolution of (2-hydroxymethylphenyl)phenylmethanol (**30**) as CSDP acid ester, determination of absolute configuration by X-ray analysis, and conversion of the enantiopure derivative of **30** obtained to the desired alcohol **29**. CSDP acid (−)-**1** (1 eq.) was allowed to react diol (±)-**30** to yield a diastereomeric mixture of esters, in which the primary alcohol group was selectively esterified. In this case, the chiral auxiliary group bonding to the primary alcohol moiety was remote from the stereogenic center of **30**, but the diastereomeric esters were clearly separated. Single crystals were obtained from the second-eluted fraction (−)-**42b**, leading to the determination of its absolute configuration as *S* by X-ray analysis. The first-eluted fraction (*R*)–(−)-**42a** was converted to the desired enantiopure alcohol (*R*)–(−)-**29** *via* several reaction steps. The above result indicated that the absolute configuration of **29** previously assigned on the basis of an asymmetric reaction mechanism was wrong [29]. Since the absolute configurational assignment based on the reaction mechanism sometimes may result in an error, it is advisable to determine the absolute configuration of products independently.

The CSDP acid method is also effective for the preparation of a variety of benzyl alcohols **33**–**36** [31], which would be useful as chiral synthons for the total synthesis of natural products because of their unambiguous absolute configurations and 100% enantiopurity.

Atropisomer **10** is a unique chiral compound containing three naphthalene chromophores. Its enantioresolution and determination of absolute configuration were carried out by the following method [20,21]. Racemic diol, (±)-1,1′:4′,1″-ternaphthalene-2,2″-dimethanol (**10**), was combined with CSDP acid **1** forming diesters, which were separated by HPLC on silica gel: hexane/EtOAc = 2:1, $\alpha = 1.18$. While the first-eluted fraction (−)-**43a** yielded fine needle-like crystals by recrystallization from hexane/EtOAc, the second-eluted fraction (+)-**43b** yielded large crystals.

In general, single crystals suitable for X-ray analysis have prismatic or columnar forms or thick plate-like forms with definite surfaces and edges. The second-eluted ester (+)-**43b** gave odd crystals resembling airplanes with triangular wings upon recrystallization, and they did not look like single crystals. However, after removal of the wings, the body part was subjected to X-ray analysis, which revealed that it was a single crystal. Interestingly, the formula weight of an asymmetric unit estimated from the preliminary lattice constant did not agree with the molecular weight of (+)-**43b**. So, we had initially thought that molecular structure might be incorrect, assuming that the asymmetric unit of **43b** contained one molecule because of the asymmetric structure of CSDP acid moiety. Careful investigation of the data obtained, however, revealed that a half of the molecule **43b** was equivalent to one asymmetric unit. Namely, ester (+)-**43b** has a C_2 symmetric structure even in crystals, despite the fact that the molecule contains complex CSDP acid moieties. The absolute configuration, i.e. torsion among three naphthalene chromophores, was unambiguously determined as (a*S*,a*S*) on the basis of internal reference. The chiral auxiliaries were removed from ester (a*S*,a*S*)-(+)-**43b** to yield enantiopure diol (a*S*,a*S*)-(+)-**10**. The absolute configuration obtained from this X-ray analysis was consistent with that obtained by the application of the CD exciton chirality method to (+)-**10** [20,21].

The compound, 2-(1-naphthyl)propane-1,2-diol (**37**), was isolated as a chiral metabolite of 1-isopropylnaphthalene in rabbits (Fig. 18.6). The metabolite, however, was not enantiopure and its absolute configuration had been only empirically estimated based on the reaction mechanism. To obtain the enantiopure diol **37** and to determine its absolute configuration in an unambiguous way, the method of CSDP acid was applied to (±)-**37** [33]. In this case, only the primary alcohol part was esterified, and a diastereomeric mixture obtained was clearly separated by HPLC on silica

(a)

OH
H_3C
OH
1) CSDP acid
(−)-**1**
2) HPLC
(±)-**37**
(*S*)-(−)-**37**
(*S*)-(−)-**45, X-ray**

(b)

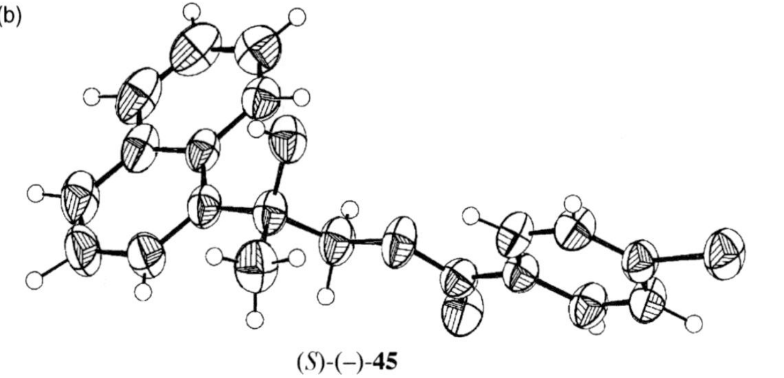

(*S*)-(−)-**45**

Fig. 18.6. (a) Enantioresolution and determination of the absolute configuration of 2-(1-naphthyl)propane-1,2-diol (**37**). (b) ORTEP drawing of ester (*S*)-(−)-**45**.

gel: hexane/EtOAc = 4:1, $\alpha = 1.3$, $R_s = 1.1$. In this HPLC, the presence of free tertiary hydroxyl group is important, because the protection of the tertiary alcohol group led to poor separation.

Despite the repeated recrystallizations, both diastereomers were obtained only as amorphous solids. Therefore the first-eluted fraction (−)-**44a** was reduced with $LiAlH_4$ to yield enantiopure glycol (−)-**37**, which was further converted to 4-bromobenzoate (−)-**45** (Fig. 18.6). By recrystallization from EtOH, (−)-**45** gave good single crystals suitable for X-ray analysis, and consequently its absolute configuration was explicitly determined as *S* by the Bijvoet pair measurement of the anomalous dispersion effect of the bromine atom contained (Table 18.3) [33]. As listed in Table 18.3, the ratio of structural factors |Fc(hkl)|/|Fc(hk–l)| calculated for the *S* absolute configuration agreed with those obtained from the observed data, establishing the *S* configuration of ester (−)-**45**.

Furthermore, we have obtained enantiopure 2-methoxy-2-(1-naphthyl)propionic acid (MαNP acid) (*S*)-(+)-(**2**) *via* several reactions from diol (*S*)-(−)-**37** (Fig. 18.7) [33]. We have discovered that this novel carboxylic acid, MαNP acid (**2**), was also powerful for enantioresolution and simultaneous determination of absolute configuration of various secondary alcohols by the ^{1}H NMR anisotropy method [34–46]. The results obtained by the ^{1}H NMR anisotropy method are of course consistent with those by the X-ray method. Therefore, the methods of CSDP and MαNP acids are useful as complementary molecular tools.

References pp. 691–692

Table 18.3 The Bijvoet pairs of (*S*)-(–)-2-(1-naphthyl)propane-1,2-diol 1-*p*-bromobenzoate (**45**): observed and calculated values of the structural factors for (h,k,l) and (h,k,–l) reflections, and their ratios[a,b]

h	k	l	|Fo(hkl)| [|Fc(hkl)|]	|Fo(hk–l)| [|Fc(hk–l)|]	|Fo(hkl)/Fo(hk–l)| [|Fc(hkl)|/Fc(hk–l)]
1	4	1	39.4 [35.4]	32.1 [27.9]	1.23 [1.26]
1	5	1	39.3 [37.7]	46.2 [42.2]	0.85 [0.89]
2	8	1	78.4 [74.1]	73.8 [68.4]	1.06 [1.08]
4	1	1	102.6 [91.1]	92.8 [84.6]	1.11 [1.08]
5	5	1	10.1 [11.3]	20.2 [19.0]	0.50 [0.59]
2	1	2	162.2 [154.3]	149.3 [143.6]	1.09 [1.07]
4	4	2	83.0 [81.0]	90.7 [87.6]	0.92 [0.92]
5	6	2	71.0 [68.1]	66.4 [62.6]	1.07 [1.09]
1	3	3	76.0 [74.9]	83.6 [79.6]	0.91 [0.94]
2	1	3	75.8 [72.8]	69.5 [66.6]	1.09 [1.09]
2	3	3	89.7 [86.5]	99.6 [94.5]	0.90 [0.90]
2	5	3	80.9 [77.3]	73.8 [69.4]	1.10 [1.11]
3	7	3	66.8 [63.6]	73.2 [69.1]	0.91 [0.92]
5	4	3	40.0 [40.1]	46.4 [45.6]	0.86 [0.88]
2	1	4	104.6 [99.5]	98.0 [92.6]	1.07 [1.07]
2	10	3	49.4 [49.7]	45.0 [43.7]	1.10 [1.07]
7	5	3	42.2 [40.9]	36.3 [35.3]	1.16 [1.16]
4	4	4	80.9 [75.5]	87.0 [80.7]	0.93 [0.94]

[a]In this crystal, |Fo(–h–k–l)| = |Fo(hk–l)|.
[b]Reflections satisfying | |Fo(hkl)| – |Fo(hk–l)| | > 10 σ(Fo) were selected,
where $\sigma(\mathrm{Fo}) = [\sigma \mathrm{count}^2 + (0.007\ |\mathrm{Fo}|)^2]^{0.5}$.

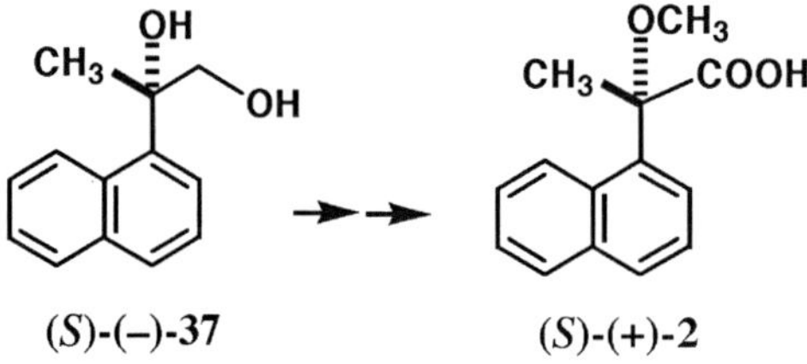

Fig. 18.7. Synthesis of a novel chiral molecular tool, (*S*)-(+)-2-methoxy-2-(1-naphthyl)propionic acid (MαNP acid) (**2**) from glycol (*S*)-(–)-**37**, and determination of its absolute configuration.

18.5 A NOVEL CHIRAL MOLECULAR TOOL, 2-METHOXY-2-(1-NAPHTHYL) PROPIONIC ACID (MαNP ACID (*S*)-(+)-2), USEFUL FOR ENANTIORESOLUTION OF ALCOHOLS AND DETERMINATION OF THEIR ABSOLUTE CONFIGURATIONS BY THE ^{1}H NMR ANISOTROPY METHOD

We have discussed the design and applications of CSP and CSDP acids useful for both the synthesis of enantiopure compounds and the unambiguous determination of their absolute configurations by X-ray analysis. The X-ray crystallographic method using the internal reference of absolute configuration thus leads to the unambiguous and reliable determination of absolute configuration. However, the drawback of X-ray crystallography is that the method needs single crystals, and therefore it is not applicable to non-crystalline materials. However, in routine experiments, prismatic single crystals suitable for X-ray analysis are not always obtainable. So is there any other method applicable to non-crystalline materials? In addition, most of the applications listed in Tables 18.1 and 18.2 are limited to aromatic compounds. So, a powerful method applicable to aliphatic compounds is required.

We have recently discovered that 2-methoxy-2-(1-naphthyl)propionic acid (MαNP acid (**2**), Fig. 18.8), is remarkably effective on enantioresolution of aliphatic alcohols, especially acyclic aliphatic alcohols [34–46]. In the ^{1}H NMR spectra of the esters formed from MαNP acid **2** and alcohols, the chemical shifts of the protons in the alcohol moiety are strongly affected by the magnetic anisotropy effect induced by the naphthyl group. Therefore this MαNP acid **2** can be used as the chiral auxiliary of the advanced Mosher method [10–13] useful for determining the absolute configuration of secondary alcohols. Another advantage of the MαNP acid **2** is that it does not racemize, because the α-position of **2** is fully substituted, and therefore it is easy to prepare the enantiopure acid **2**. As discussed below, MαNP acid **2** is a very powerful chiral derivatizing agent, which simultaneously enables both enantioresolution of secondary alcohols and determination of their absolute configurations. Namely, the MαNP acid method explained here is very useful for enantioresolution of racemic alcohols and also for the determination of the absolute configurations of natural products and biologically active synthetic chiral compounds, e.g. chiral drugs. In this sense, the chiral MαNP acid **2** is hence superior to the conventional chiral acids, Mosher's MTPA acid [10], Trost's MPA acid [12], 1- and 2-NMA acids developed by Riguera [11] and Kusumi [10] groups.

The following sections describe in detail the principle and applications of this chiral MαNP acid method: (a) synthesis of chiral MαNP acid **2**, (b) determination of its absolute configuration by X-ray analysis and chemical correlation, (c) enantioresolution of racemic acid **2** with chiral alcohols, (d) absolute configurational and conformational analyses of MαNP acid esters by NMR and CD spectroscopic methods, (e) enantioresolution of racemic alcohols and determination of their absolute configuration using chiral MαNP acid **2**, and (f) recovery of chiral alcohols with 100% enantiopurity from the separated diastereomeric esters.

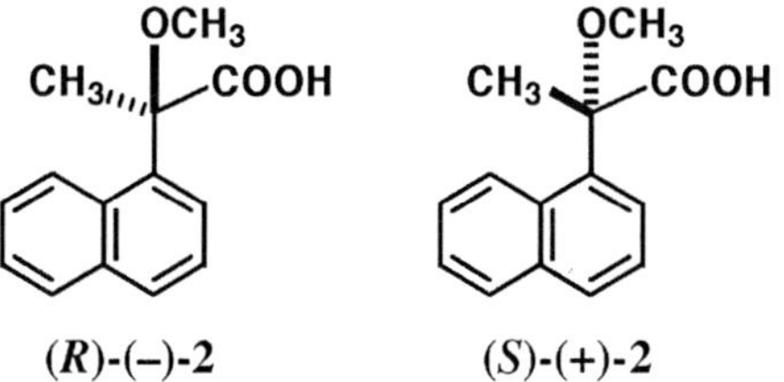

Fig. 18.8. Novel chiral MαNP acids with powerful ability to enantioresolve alcohols and strong NMR anisotropy effect.

18.5.1 Facile synthesis of MαNP acid (2) and its extraordinary enantioresolution with natural (–)-menthol [33,35,45]

To synthesize a large amount of enantiopure chiral MαNP acid (**2**), the facile synthesis and enantioresolution of racemic acid **2** were carried out as shown in Fig. 18.9. In general, chiral synthetic amines or alkaloids have been used for enantioresolution of carboxylic acids. However, we have adopted the following novel strategy to use chiral alcohols. In this method, chiral alcohols are condensed with racemic acid **2** and the diastereomeric esters formed are separated by HPLC on silica gel. The separated esters are then hydrolyzed to yield both enantiomers of the desired carboxylic acids.

As a chiral alcohol, naturally occurring (–)-menthol was selected and esterified with racemic acid **2**. It was much surprising that the diastereomeric esters **48a** and **48b** formed were very easily separated by HPLC on silica gel (hexane/EtOAc =10:1) as illustrated

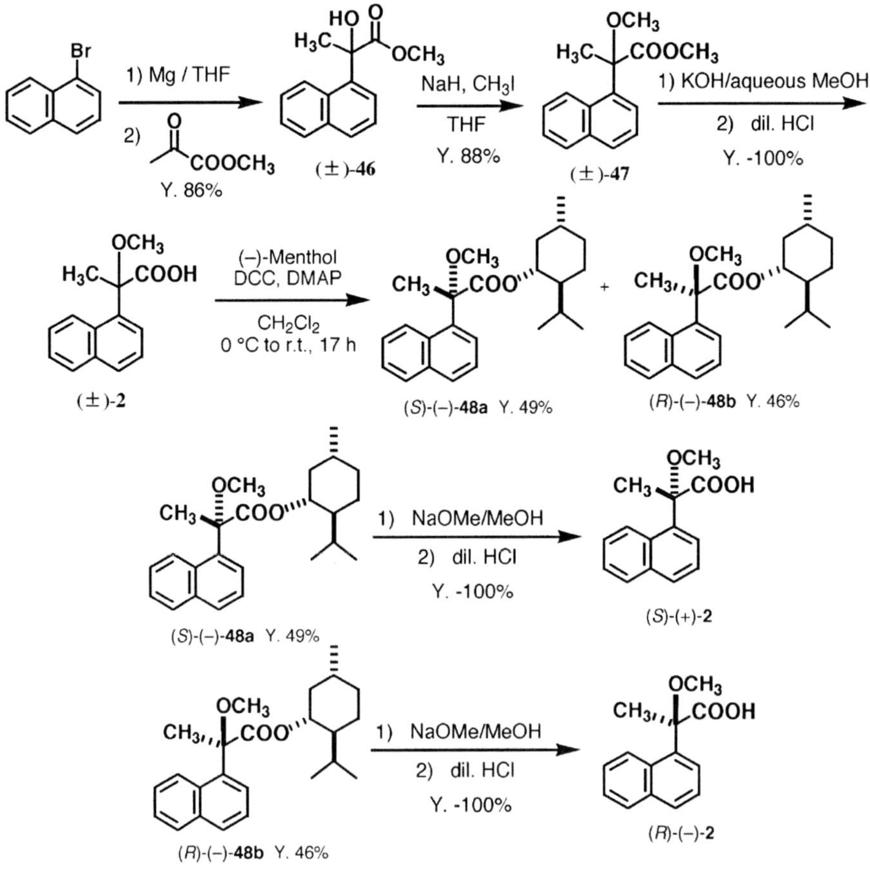

Fig. 18.9. Facile synthesis and enantioresolution of novel chiral MαNP acid.

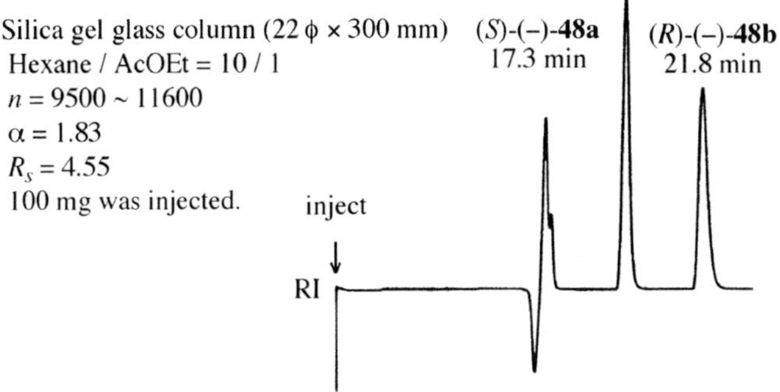

Fig. 18.10. HPLC separation of MαNP acid menthol esters.

in Fig. 18.10. The separation and resolution factors were extraordinarily high ($\alpha = 1.83$, $R_s = 4.55$), indicating that acid **2** has a great ability to recognize the chirality of the alcohols. The efficiency in separation enabled the HPLC of a preparative scale: esters **48a**/**48b** (1.0–1.8 g) were separable in one run using a glass column of silica gel (25 ϕ × 400 mm). The first-eluted ester **48a** was subjected to solvolysis to yield chiral acid (+)-**2**, while the second-eluted ester **48b** gave acid (−)-**2**. To determine the absolute configurations of chiral acids **2** obtained, those were converted to methyl esters, the CD spectra of which were measured. By comparison of those CD spectra with that of the authentic sample with known absolute configuration established by X-ray analysis and chemical correlation, the absolute configurations of chiral acids **2** were determined as (*S*)-(+) and (*R*)-(−), respectively, leading to the assignment of (*S*)-(−)-**48a** and (*R*)-(−)-**48b** (Fig. 18.9).

18.5.2 The ^{1}H NMR anisotropy method for determining the absolute configuration of secondary alcohols: the sector rule and applications [35,42]

As described above, the ^{1}H NMR anisotropy method has been frequently used as a relative and empirical method for determining the absolute configurations of chiral organic compounds [10–13]. In particular, the advanced Mosher method for chiral secondary alcohols has been successfully employed in the field of natural products. In the cases of Mosher's MTPA and Trost's MPA acids, the phenyl group exhibits the magnetic anisotropy effect induced by the aromatic ring current, affecting the chemical shift (δ) of protons in the alcohol part. Therefore, the absolute configuration of chiral alcohol can be determined by the difference ($\Delta\delta$) of the chemical shifts of esters formed with (*R*) and (*S*) carboxylic acids: $\Delta\delta = \delta(R) - \delta(S)$ or $\Delta\delta = \delta(S) - \delta(R)$. We have found that MαNP acid **2** is superior to the Mosher's MTPA and Trost's MPA acids, because the magnetic anisotropy effect of naphthyl group is much larger than that of the phenyl group and therefore larger $\Delta\delta$ values are obtained. So, the absolute configuration of chiral alcohols can be unambiguously determined, while using MαNP acid **2** as a chiral NMR anisotropy reagent. Moreover, MαNP acid has another advantage that it does not racemize, because the α-position of **2** is fully substituted. From these reasons, it is advisable to use MαNP acid **2**, rather than other conventional chiral acids, for determining the absolute configuration of chiral alcohols including natural products.

All NMR proton peaks of diastereomeric MαNP esters **48a** and **48b** were fully assigned by various methods including two dimensional ones (^{1}H, ^{1}H-^{1}H COSY, ^{13}C, ^{1}H-^{13}C COSY, HMBC, Fig. 18.11a). The protons of the isopropyl group in ester **48b** appeared at much higher fields than in ester **48a**. On the other hand, the protons in the 2-position in **48a** appeared at higher fields than in ester **48b**. Those high-field shifts are obviously due to the magnetic anisotropy effect induced by the naphthyl group of MαNP acid moiety.

To determine the absolute configuration from the NMR anisotropy effect, it is required to determine the preferred conformation of each diastereomer. In esters **48a** and **48b**, the absolute configurations of MαNP acid and menthol moieties are established as described above, and so the following stable conformations are proposed to satisfy the anisotropy effects observed in the NMR spectra (Fig. 18.11). Namely, the two oxygen atoms of the methoxyl and ester carbonyl groups are synperiplanar (*syn*) to each other in their stable conformations. In addition, the ester carbonyl oxygen atom is also *syn* to the alcohol methine proton. Therefore, the methoxyl group, ester group, and alcohol methine proton lie in the same plane, which is called the MαNP plane (Figs. 18.11 and 18.12). These *syn* conformations are similar to those proposed for MPA esters. In ester **48a**, the naphthyl group and H-2 protons are on the same front side of the MαNP plane, and the H-2 protons are located above the naphthyl plane. Therefore the H-2 protons feel the magnetic anisotropy effect of high-field shift, and so they appear at higher-field. In ester **48b**, the naphthyl group is close to the isopropyl group, and the high-field shifts of isopropyl protons are observable.

The predominance of the *syn* conformations in esters **48a** and **48b** are also supported by the comparison of the NMR data with those of 2-hydroxy-2-(1-naphthyl)propionic acid (HαNP) menthol esters shown in Fig. 18.11(b). From the NMR chemical shift and IR data, it is obvious that the tertiary hydroxyl group of HαNP esters is intramolecularly hydrogen bonded to the oxygen atom of the ester carbonyl group. Namely, the hydroxyl group and the ester carbonyl oxygen atom take a *syn* conformation. We have found a very interesting fact that the NMR chemical shift data of MαNP acid menthol ester (*S*;1*R*,3*R*,4*S*)-(–)-**48a**, especially those of the menthol part, are very similar to those of HαNP acid menthol ester (*S*;1*R*,3*R*,4*S*)-(–) as shown in Fig. 18.11(a) and (b). The same is true for the pairs of other diastereomers, (*R*;1*R*,3*R*,4*S*)-(−)-**48b** and HαNP acid menthol ester (*R*;1*R*,3*R*,4*S*)-(−) (Fig. 18.11). These facts indicate that MαNP acid menthol esters take the *syn* conformation, as HαNP acid menthol esters usually do. This fully explains the observed magnetic anisotropy effects.

By using the NMR anisotropy effect of MαNP esters, the sector rule for determining the absolute configuration of secondary alcohols can be deduced (Fig. 18.12). The basic procedure is as follows; (*R*)-MαNP and (*S*)-MαNP acids are separately allowed to react with a chiral alcohol, the absolute configuration of which is defined as *X*. So, the ester prepared from (*R*)-MαNP acid has the (*R*,*X*) absolute configuration, while the other ester from (*S*)-MαNP acid has the (*S*,*X*) absolute configuration. All NMR proton signals of (*R*,*X*)- and (*S*,*X*)-esters are fully assigned by careful analysis. If necessary, the use of

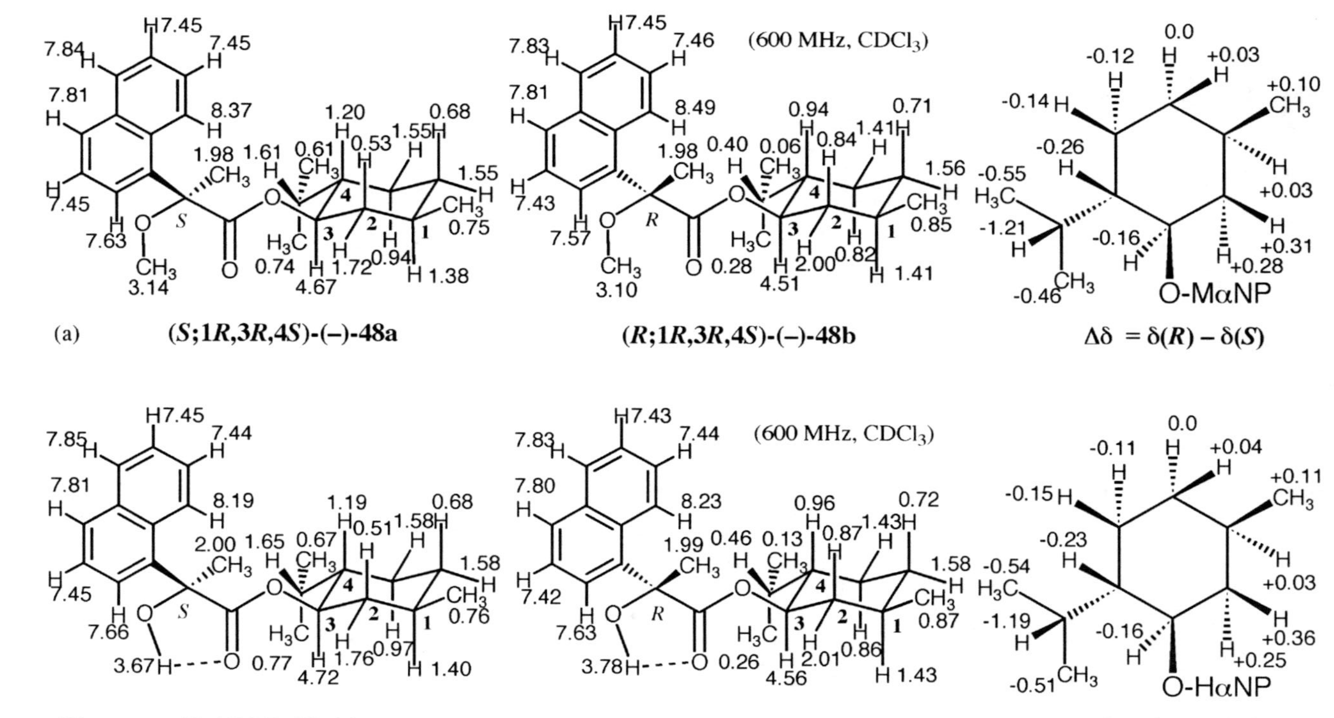

Fig. 18.11. NMR data of MαNP and HαNP acid menthol esters.

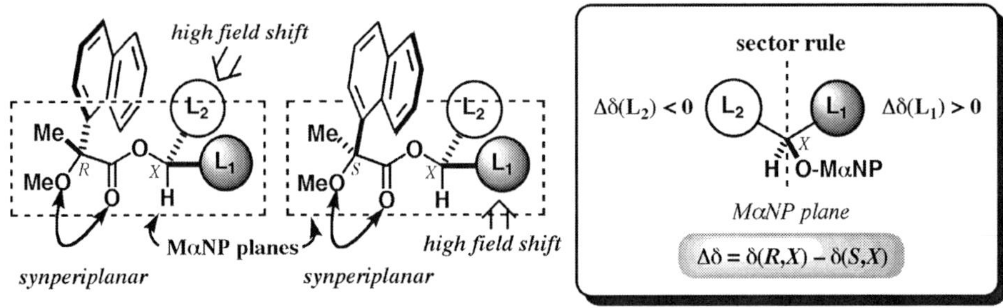

Fig. 18.12. The preferred conformation of MαNP esters, and the sector rule for determining the absolute configuration of chiral alcohols by use of NMR $\Delta\delta$ values.

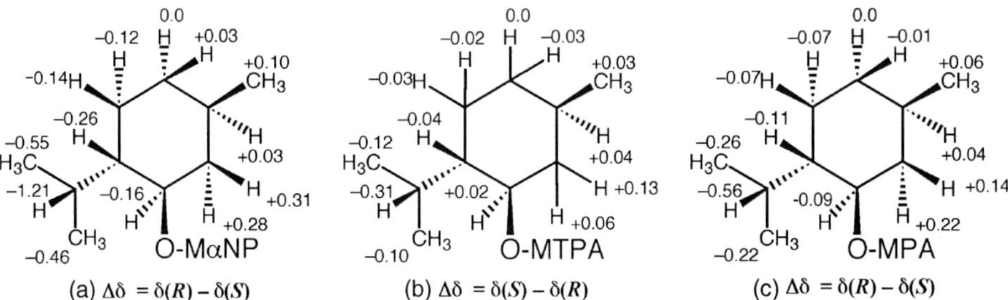

Fig. 18.13. Comparison of the NMR $\Delta\delta$ values of menthol esters formed with chiral carboxylic acids.

two-dimensional spectra is suggested. The $\Delta\delta$ values ($\Delta\delta = \delta(R,X) - \delta(S,X)$) are calculated for all protons in the alcohol moiety. Fig. 18.12 shows the sector rule for MαNP ester, where the MαNP group is placed in the down and front side, while the methine proton of the secondary alcohol in the down and rear side. The group L_1 with protons exhibiting positive $\Delta\delta$ values is placed in the right side, while the group L_2 with protons showing negative $\Delta\delta$ values in the left side. From this projection, the absolute configuration X of chiral alcohol can be determined.

The magnetic anisotropy effect of the chiral MαNP acid is much stronger than that of the conventional chiral carboxylic acid (Fig. 18.13). For instance, the $\Delta\delta$ values of MαNP-menthol ester are ca. four times larger than those of Mosher's MTPA ester [10] (Fig. 18.13(b)); twice for Trost's MPA ester [12] (Fig. 18.13(c)); comparable to 1-NMA and 2-NMA esters reported by Riguera [11] and Kusumi [10] et al. MαNP acid is thus effective for determining the absolute configuration of natural products.

Some application examples of this MαNP acid method to chiral alcohols are shown in Fig. 18.14.

18.5.3 Enantioresolution of various alcohols using MαNP acid and simultaneous determination of their absolute configurations [37,45]

Another extraordinary quality of MαNP acid is its excellent ability in chiral recognition. For example, as discussed above, racemic MαNP acid could be successfully

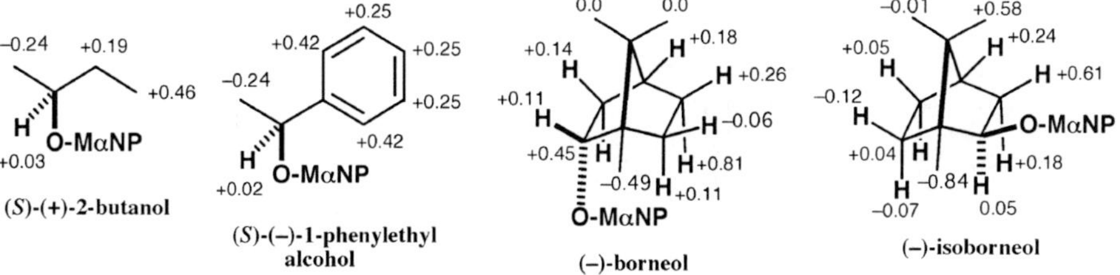

Fig. 18.14. The NMR Δδ values and absolute configurations determined by the MαNP acid method.

enantioresolved as the esters of natural (−)-menthol; the diastereomeric esters formed were clearly separated by HPLC on silica gel. MαNP acid could be also enantioresolved with other chiral alcohols as listed in Fig. 18.14. These facts logically indicate that if enantiopure MαNP acid is used, racemic alcohols can be enantioresolved. In fact, we have succeeded in enantioresolution of various alcohols using enantiopure MαNP acid (*S*)-(+)-**2** as exemplified in Fig. 18.15.

This novel chiral MαNP acid (*S*)-(+)-**2** has thus a remarkable enantioresolving power for alcohols, especially for aliphatic alcohols. For instance, in the case of 2-butanol, the diastereomeric esters can be baseline separated with the separation factor $\alpha = 1.15$ and resolution factor $R_s = 1.18$. In this case, it is obvious that the chiral carboxylic acid **2** recognizes well the slight difference between the methyl and the ethyl groups. This is an excellent and practical method since the chiral acid **2** exhibits a high resolving power to aliphatic alcohols, to which in general asymmetric syntheses are hardly applicable.

The next question is then how the absolute configuration of alcohol moiety is determined. The absolute configurations of separated diastereomers can be determined by applying the NMR anisotropy method using chiral MαNP acid described above. A general scheme is illustrated in Fig. 18.16. Racemic alcohol is esterified with MαNP acid (*S*)-(+)-**2** yielding a mixture of diastereomeric esters, which is separated by HPLC on silica gel. The absolute configuration of the first-eluted ester is defined as (*S*,*X*), where *S* denotes the absolute configuration of MαNP acid part, while *X* denotes that of the alcohol part. So, the absolute configuration of the second-eluted ester is expressed as (*S*,–*X*), where –*X* indicates the opposite absolute configuration of *X*. The original definition of Δδ value is $\Delta\delta = \delta(R,X) - \delta(S,X)$, and so the value of $\delta(R,X)$ is required to calculate the Δδ value. However, the enantiomer (*R*,*X*) does not exist in this scheme, and so the original equation of Δδ is not useful here.

To solve the above problem, the following conversion of the equation was performed. Since the ester (*S*,–*X*) is the enantiomer of ester (*R*,*X*), their NMR data should be identical with each other i.e. $\delta(R,X) = \delta(S,-X)$. Therefore, $\Delta\delta = \delta(R,X) - \delta(S,X) = \delta(S,-X) - \delta(S,X) = \delta(\text{2nd fr.}) - \delta(\text{1st fr.})$. So, the absolute configuration *X* of the first-eluted fraction can be determined from the Δδ value which is obtained by subtracting the chemical shift of the first-eluted fraction from that of the second-eluted fraction (Fig. 18.16). This method has been applied to the esters shown in Fig. 18.15, giving Δδ values and the absolute configurations of the first-eluted esters (Fig. 18.17). The Δδ values are reasonably distributed, positive values at the right, and negative value

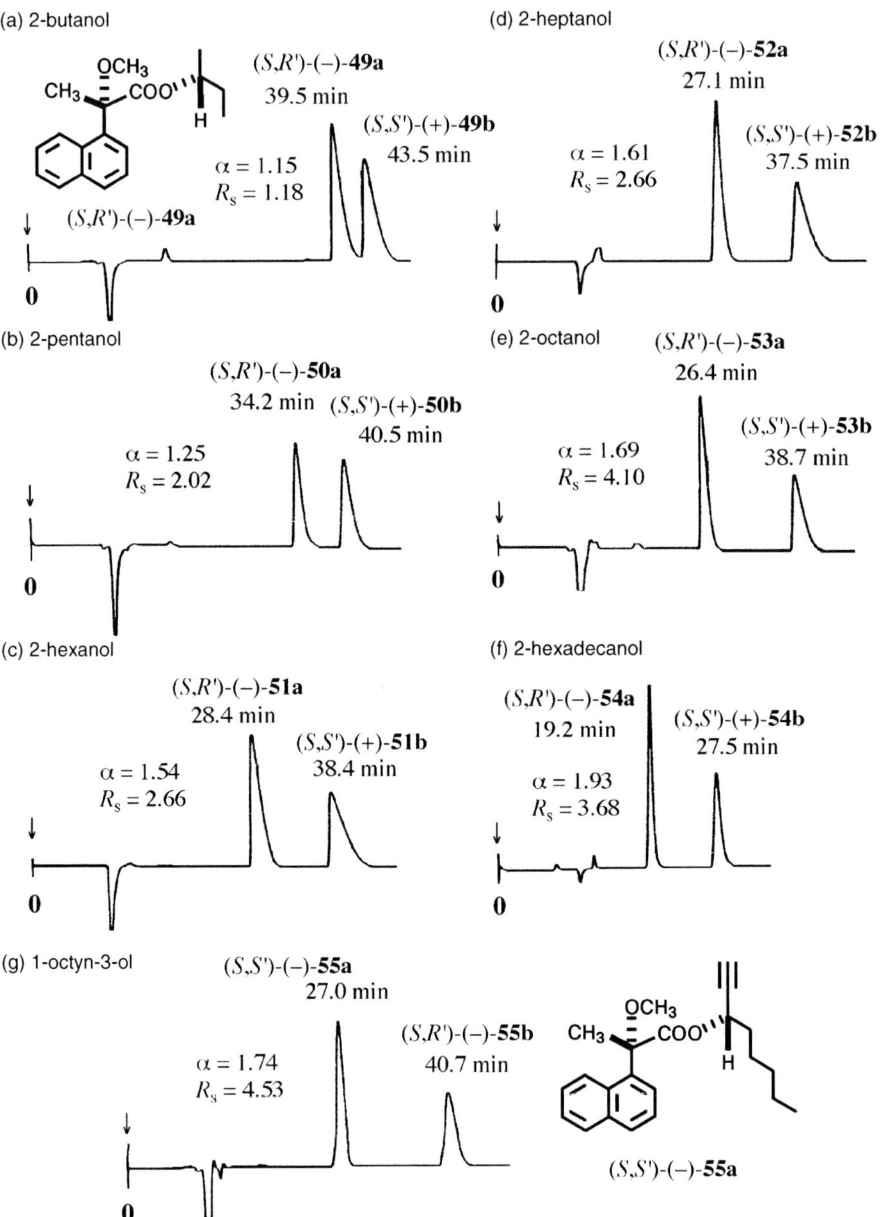

Fig. 18.15. HPLC separation of diastereomeric esters formed from aliphatic alcohols and (*S*)-(+)-MαNP acid (silica gel, 22 ϕ × 300 mm, hexane/EtOAc = 20:1).

at the left. The absolute configuration of the first-eluted ester can thus be determined, and the opposite absolute configuration is of course assigned to the second-eluted ester. It should be noted that when MαNP acid (*R*)-(−)-**2** is used, the $\Delta\delta$ value is defined as $\Delta\delta = \delta(R,X) - \delta(S,X) = \delta(R,X) - \delta(R,-X) = \delta(\text{1st fr.}) - \delta(\text{2nd fr.})$.

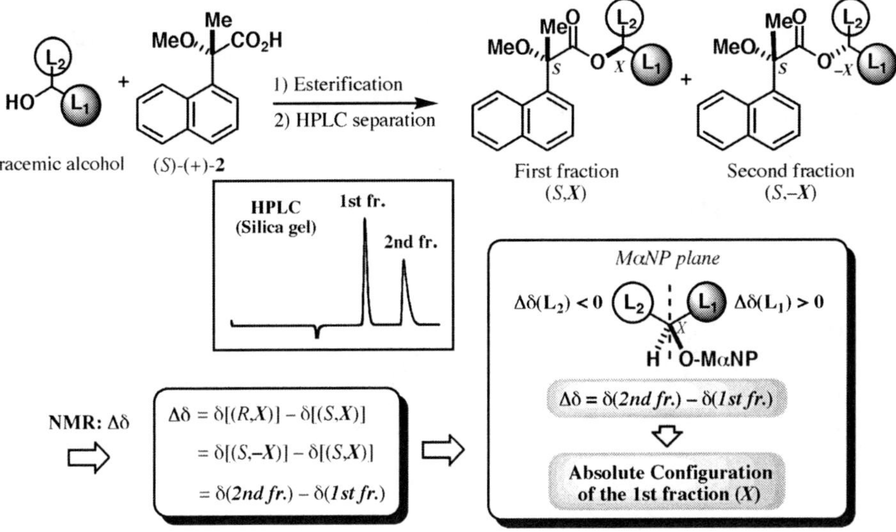

Fig. 18.16. Enantioresolution of racemic alcohol as (*S*)-MαNP esters, and determination of the absolute configuration of the first-eluted fraction by the NMR anisotropy method.

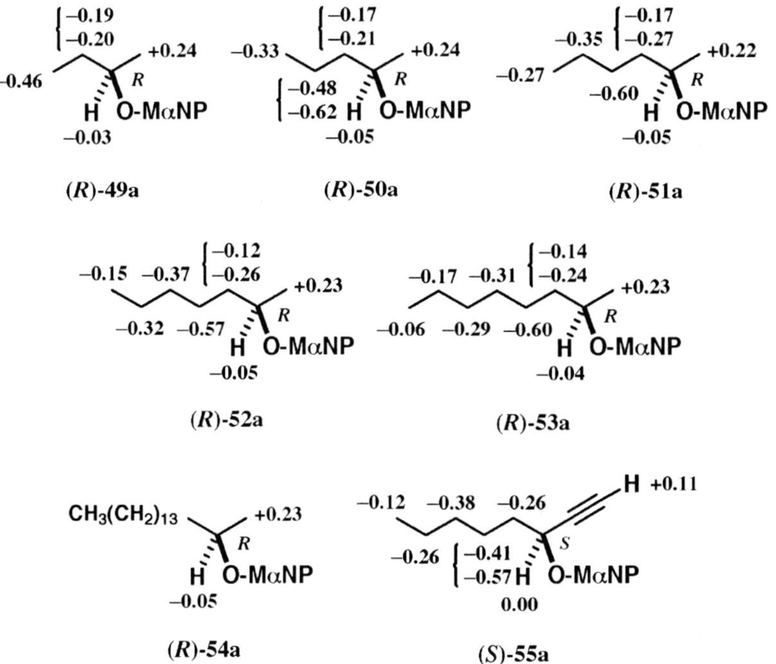

Fig. 18.17. Determination of the absolute configurations of the alcoholic part of the first-eluted esters by the NMR anisotropy method using (*S*)-(+)-MαNP acid, and the observed $\Delta\delta$ values.

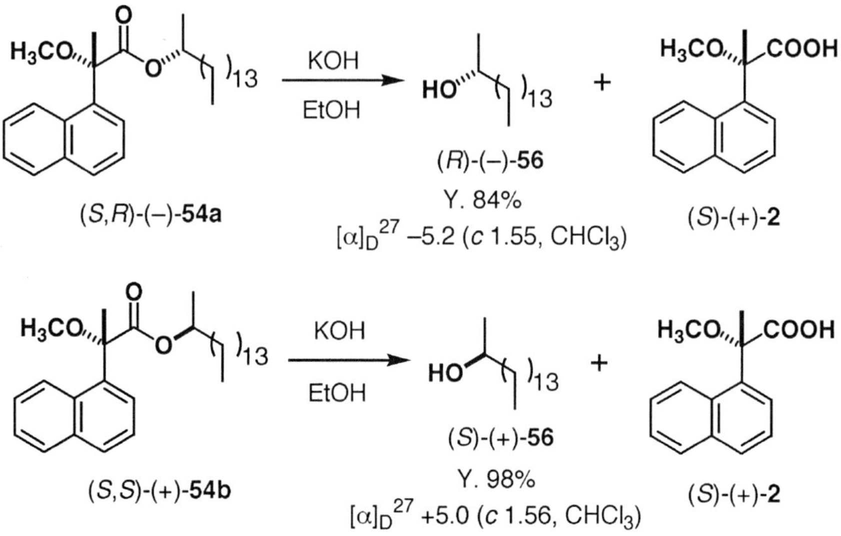

(*S*,*R*)-(−)-**54a** → KOH, EtOH → (*R*)-(−)-**56** Y. 84% $[\alpha]_D^{27}$ −5.2 (*c* 1.55, $CHCl_3$) + (*S*)-(+)-**2**

(*S*,*S*)-(+)-**54b** → KOH, EtOH → (*S*)-(+)-**56** Y. 98% $[\alpha]_D^{27}$ +5.0 (*c* 1.56, $CHCl_3$) + (*S*)-(+)-**2**

Fig. 18.18. Recovery of enantiopure alcohol and MαNP acid.

The next step is the recovery of enantiopure alcohol and chiral MαNP acid **2**. As exemplified in Fig. 18.18, both enantiopure alcohols were readily obtained by the solvolysis of the separated esters [40]. The chiral MαNP acid **2** was also recovered and could be recycled.

How good is the enantiopurity of the recovered alcohols? In our method, both diastereomeric esters obtained are enantiopure, if MαNP acid **2** used is enantiopure, because they are fully separated in HPLC. The MαNP acid **2** was enantioresolved with natural (−)-menthol, the enantiopurity of which was confirmed as 100% by the gas chromatography using the chiral stationary phase [45].

As described above, MαNP acid has excellent enantioresolving power regardless of its simple molecular structure and the absence of hetero atoms. Besides, the chiral acid **2** is superior to the Mosher's MTPA and Trost's MPA acids in the magnetic anisotropy effect, and therefore further development is expected.

18.6 COMPLEMENTARY USE OF CSDP ACID (−)-1 AND MαNP ACID (*S*)-(+)-2 FOR ENANTIORESOLUTION OF ALCOHOLS AND DETERMINATION OF THEIR ABSOLUTE CONFIGURATIONS BY X-RAY CRYSTALLOGRAPHIC AND ^{1}H NMR ANISOTROPY METHODS [42,46]

As discussed above, the method of MαNP acid is very useful for the preparation of enantiopure secondary alcohols and the simultaneous determination of their absolute configurations. However, those absolute configurations were determined by the empirical sector rule, which is based on the magnetic anisotropy of naphthalene ring and the preferred conformation of MαNP ester. Therefore, the absolute configurations obtained by this method have the empirical nature. How reliable are those absolute configurations? To evaluate the reliability of the MαNP acid method, we have compared the results by the MαNP acid method with those by the X-ray crystallographic analysis, as described below.

Recently, much attention has been focused on chiral fluorinated organic compounds, since some chiral synthetic drugs consist of fluorinated aromatic moieties. To prepare chiral fluorinated diphenylmethanols and to determine their absolute configurations by X-ray crystallography, we have applied the CSDP acid method [46]. For example, (4-trifluoromethylphenyl)phenylmethanol **57** was esterified with CSDP acid (−)-**1** yielding a diastereomeric mixture of esters, which was clearly separated by HPLC on silica gel: separation factor $\alpha = 1.34$; resolution factor $R_s = 2.37$ (Table 18.4). The first-eluted ester **66a** was recrystallized from EtOH giving prisms, one of which was subjected to X-ray crystallography. Although the trifluoromethyl moiety took a disordered structure, the absolute stereochemistry of the first-eluted ester was unambiguously determined as *R* by the internal reference method using the (1*S*,2*R*,4*R*) absolute configuration of the camphor part and also by the heavy atom effect (Fig. 18.19). Enantiopure alcohol (*R*)-(−)-**57** was easily recovered by treating the first-eluted CSDP ester **66a** with K_2CO_3 in MeOH. Although the absolute configuration of (+)-**57** had previously been estimated as *S* by the Horeau's method, the abnormality of its application had been pointed out [47]. Therefore, the direct and unambiguous determination of its absolute configuration has been desired for a long time.

Table 18.4 Silica gel-HPLC separation of diastereomeric esters formed from diphenylmethanols with CSDP acid and/or MαNP acid, determination of their absolute configurations by X-ray crystallography and/or by the ^{1}H NMR anisotropy method

Acid	Alcohol	Solvent[a]	α	R_s	X-ray	^{1}H NMR $\Delta\delta$[b]	Ester (1st Fr)
(−)-**1**	(4-CF_3)-**57**	H/EA = 5/1	1.34	2.37	1st Fr.	–	(*R*)-(−)-**66a**
(−)-**1**	(3-CF_3)-**58**	H/EA = 5/1	1.16	1.22	1st Fr. 2nd Fr.	–	(*R*)-(−)-**67a**
(−)-**1**	(2-CF_3)-**59**	H/EA = 4/1	1.00	–	–	–	**68**[c]
(−)-**1**	(4-F)-**60**	H/EA = 5/1	1.11	1.33	–	–	(−)-**69a**
(−)-**1**	(3-F)-**61**	H/EA = 5/1	1.05	0.77	–	–	**70**[c]
(−)-**1**	(2-F)-**62**	H/EA = 4/1	1.00	–	–	–	**71**[c]
(−)-**1**	(2,6-F_2)-**63**	H/EA = 4/1	1.21	2.50	1st Fr.	–	(*S*)-(−)-**72a**
(*S*)-(+)-**2**	(4-CF_3)-**57**	H/EA = 8/1	1.39	4.84	–	Yes	(*R*)-(−)-**73a**[d]
(*S*)-(+)-**2**	(3-CF_3)-**58**	H/EA = 10/1	1.07	0.84	–	Yes	(*R*)-(−)-**74a**[d]
(*S*)-(+)-**2**	(2-CF_3)-**59**	H/EA = 15/1	1.08	1.28	–	Yes	(*R*)-(+)-**75**
(*S*)-(+)-**2**	(4-F)-**60**	H/EA = 10/1	1.18	2.55	–	Yes	(*R*)-(−)-**76a**
(*S*)-(+)-**2**	(3-F)-**61**	H/EA = 10/1	1.07	1.05	–	Yes	(*R*)-(−)-**77a**
(*S*)-(+)-**2**	(2-F)-**62**	H/EA = 7/1	1.00	–	–	–	**78**[c]
(*S*)-(+)-**2**	(2,6-F_2)-**63**	H/EA = 8/1	1.08	1.28	–	Yes	(*S*)-(−)-**79a**[d]
(−)-**1**	(3-OMe)-**64**	H/EA = 4/1	1.15	1.34	1st Fr.	–	(*S*)-(−)-**80a**
(−)-**1**	(3,5-diOMe)-**65**	H/EA = 5/1	1.16	1.42	1st Fr.	–	(*S*)-(−)-**81a**[d]

[a]H—*n*-hexane, EA—ethyl acetate.
[b]$\Delta\delta$–^{1}H NMR anisotropy effect.
[c]Diastereomeric esters were not separated.
[d]Absolute configuration determined by ^{1}H NMR anisotropy method agreed with that by X-ray crystallography.

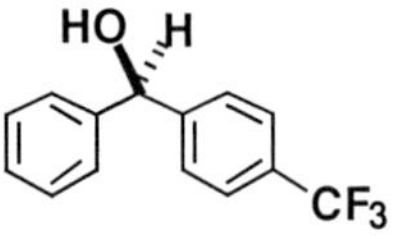

(*R*)-(−)-**57** 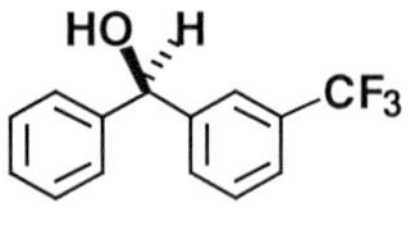 (*R*)-(−)-**58** 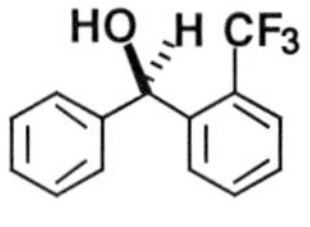 (*R*)-(+)-**59**

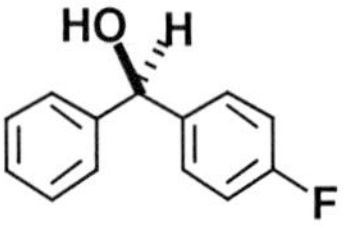

(*R*)-(−)-**60** 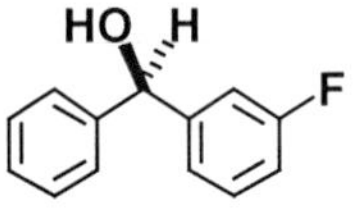 (*R*)-(−)-**61** 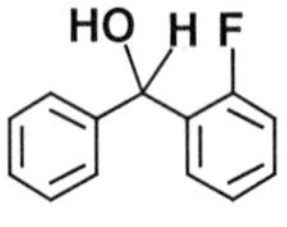 **62**

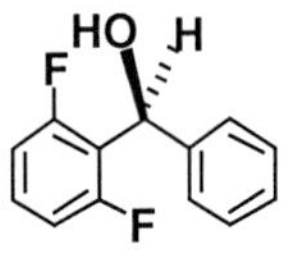

(*S*)-(−)-**63** 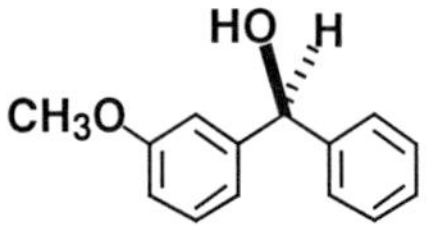 (*S*)-(−)-**64** 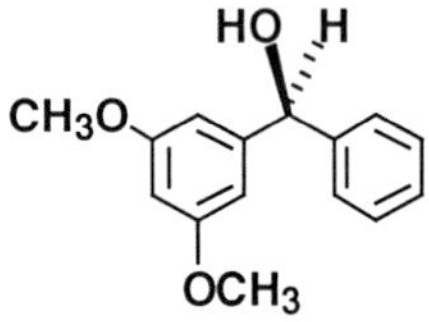 (*S*)-(+)-**65**

The absolute configuration of (4-trifluoromethylphenyl)phenylmethanol (*R*)-(−)-**57** was thus first determined by X-ray analysis.

Other fluorinated diphenylmethanols were similarly esterified with CSDP acid (−)-**1**, and the diastereomeric mixtures obtained were subjected to HPLC on silica gel. As shown in Table 18.4, diastereomeric CSDP esters of (3-trifluoromethylphenyl)phenylmethanol **58**, (4-fluorophenyl)phenylmethanol **60**, and (2,6-difluorophenyl)phenylmethanol **63** were separated well with α values of more than 1.1. In the case of (3-fluorophenyl)phenylmethanol **61**, its CSDP esters were partially separated: $\alpha = 1.05$, so it was unsuitable for the separation of preparative scale. The CSDP esters of the remaining alcohols, (2-trifluoromethylphenyl)phenylmethanol **59** and (2-fluorophenyl)-phenylmethanol **62**, appeared as single peaks in HPLC on silica gel, indicating no separation at all. As listed in Table 18.4, the enantioresolution method using CSDP acid was thus applicable to four fluorinated diphenylmethanols among seven compounds [46].

In the case of (3-trifluoromethylphenyl)phenylmethanol **58**, the final *R*-value of the first-eluted CSDP ester **67a** remained higher because of low crystallinity. However, even in such a case, its absolute configuration was unambiguously determined, based on the internal reference of absolute configuration of the camphor moiety. The second eluted CSDP ester **67b** crystallized as prisms, allowing the X-ray crystallographic determination of absolute configuration (Table 18.4). In the case of (4-fluorophenyl)phenylmethanol **60**, no single crystals suitable for X-ray analysis were obtained from both CSDP esters, and therefore their absolute configurations could not be determined by X-ray crystallography. However, the absolute configuration of (4-fluorophenyl)phenylmethanol **60** was determined by the ^{1}H NMR anisotropy method as discussed below [46]. The first-eluted CSDP ester **72a** of (2,6-difluorophenyl)phenylmethanol **63** was recrystallized

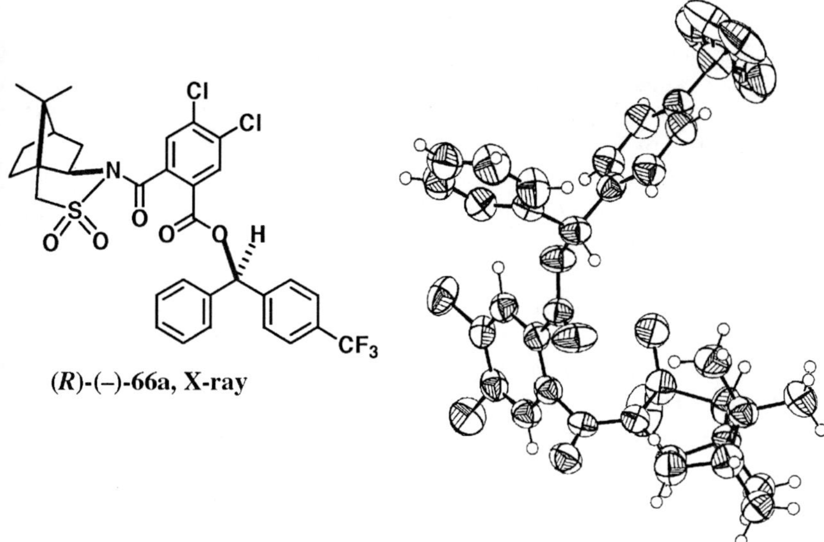

Fig. 18.19. ORTEP drawing of CSDP ester (*R*)-(–)-**66a**.

from EtOH giving prisms; the X-ray analysis led to the unambiguous assignment of an *S* absolute configuration to (–)-**63** (Table 18.4).

To apply the MαNP acid method, racemic fluorinated diphenylmethanols were esterified with MαNP acid (*S*)-(+)-**2**. For instance, (4-trifluoromethylphenyl)phenyl-methanol (±)-**57** was allowed to react with MαNP acid (*S*)-(+)-**2**, giving a diastereomeric mixture of esters. The mixture was separated well by HPLC on silica gel (hexane/EtOAc 8:1): $\alpha = 1.39$; $R_s = 4.84$ (Table 18.4). To determine the absolute configuration by the ^{1}H NMR anisotropy method, all NMR signals were fully assigned by the ^{1}H, ^{1}H-^{1}H COSY, ^{13}C, HMQC, and HMBC methods. In the case of fluorinated diphenylmethanols, the ^{1}H–^{19}F coupling was helpful for assigning ^{1}H NMR signals. The $\Delta\delta$ values of ^{1}H NMR anisotropy effect of (4-trifluoromethylphenyl)phenylmethanol MαNP esters **73a** and **73b** were calculated; the phenyl group showed large positive $\Delta\delta$ values ($+0.11 \sim +0.43$ ppm), while the 4-trifluoromethylphenyl group large negative $\Delta\delta$ values ($-0.58 \sim -0.37$ ppm). By applying the MαNP sector rule shown in Figure 18.16, the *R* absolute configuration was assigned to the first-eluted MαNP ester.

Chiral fluorinated diphenylmethanols were recovered by reduction with $LiAlH_4$ from the corresponding diastereomeric MαNP esters. So, the first-eluted MαNP ester **73a** of (4-trifluoromethylphenyl)phenylmethanol was treated with $LiAlH_4$, yielding enantio-pure alcohol (*R*)-(–)-**57**, which was identical with the authentic sample recovered from the first-eluted CSDP ester of **57**. The *R* absolute configuration of (4-trifluoromethyl-phenyl)phenylmethanol (–)-**57** determined by the ^{1}H NMR anisotropy method was thus confirmed by X-ray crystallography (Table 18.4).

Other fluorinated diphenylmethanols were similarly enantioresolved by the MαNP method. It was relatively difficult to separate diastereomeric MαNP esters of

(3-trifluoromethylphenyl)phenylmethanol **58**, (2-trifluoromethylphenyl)phenylmethanol **59**, (3-fluorophenyl)phenylmethanol **61**, and (2,6-difluorophenyl)phenylmethanol **63**, because their α values were around 1.07–1.08. However, it should be noted that (2-trifluoromethylphenyl)phenylmethanol **59** was enantioresolved by the MαNP method, while its CSDP esters were not separated at all. For (2-fluorophenyl)-phenylmethanol **62**, both chiral CSDP and MαNP acids were useless. The absolute configurations of alcohols **58** and **63** determined by the MαNP method agreed with those obtained by the X-ray method. Therefore it is concluded that the ^{1}H NMR anisotropy method using MαNP acid is reliable.

The CSDP acid method was applicable to *meta*-substituted diphenylmethanols **64** and **65** (Table 18.4). For example, racemic (3,5-dimethoxyphenyl)phenylmethanol **65** was esterified with CSDP acid (−)-**1** yielding a diastereomeric mixture of esters, which was separated well by HPLC on silica gel. The first-eluted ester (−)-**81a** was obtained as single crystals, when recrystallized from EtOH. The X-ray crystallography of ester **81a** led to the unambiguous determination of its absolute configuration as *S*. Ester (*S*)-(−)-**81a** was treated with K_2CO_3 in MeOH giving enantiopure alcohol (*S*)-(+)-**65**. The absolute configuration of (3,5-dimethoxyphenyl)phenylmethanol **65** was thus unambiguously determined for the first time.

The MαNP acid method was next applied to alcohol (*S*)-(+)-**65**, to check the reliability of the NMR anisotropy method. Namely, alcohol (*S*)-(+)-**65** was esterified with MαNP acids (*R*)-(−)-**2** and (*S*)-(+)-**2**, respectively, yielding esters (*R*,*X*) and (*S*,*X*), where *X* denotes the absolute configuration of alcohol moiety. From the ^{1}H NMR data of both the esters, the $\Delta\delta$ values ($\Delta\delta = \delta(R) - \delta(S)$) were calculated. Since the 3,5-dimethoxyphenyl ring showed positive $\Delta\delta$ values and the remaining phenyl group negative $\Delta\delta$ values, the *S* absolute configuration was assigned to alcohol (+)-**65**. This result, of course, agrees with the assignment by X-ray crystallography. The theory that the results obtained by the ^{1}H NMR anisotropy method using MαNP acid are consistent with those by X-ray crystallography, was again corroborated. The MαNP acid method is thus reliable and powerful as a complementary molecular tool for studying the chemistry of chirality.

18.7 CONCLUSIONS

We have developed several novel chiral molecular tools, in particular, chiral carboxylic acids, and successfully applied those CDAs (chiral derivatizing agents) to the enantioresolution of alcohols by HPLC separation, and determination of absolute configuration by X-ray crystallography and/or by ^{1}H NMR anisotropy method. The X-ray crystallographic method using an internal reference is, of course, the best for determining absolute configuration. However, ideal single crystals are not always obtained. In such a case, the ^{1}H NMR method using MαNP acid, which requires no crystallization, is effective. In enantioresolution, chiral CSDP acid and MαNP acid are thus useful as the complementary molecular tools. If the resolution with one CDA is unsuccessful, the use of the other is suggested. The methods described above are very powerful for the preparation of enantiomeric alcohols with 100% enantiopurity and also for the simultaneous determination of their absolute configurations. Further application of those methods to various compounds and studies toward the mass production of enantiomers are now in progress.

18.8 ACKNOWLEDGMENTS

The authors thank the graduate students and colleagues in the laboratory for their contributions and all the collaborators for their efforts and cooperation, whose names are listed in references cited. This project has been supported by grants from the Ministry of Education, Science, Sports, Culture, and Technology, Japan and the Japan Society for the Promotion of Science.

REFERENCES

1 J.M. Bijvoet, A.F. Peerdeman and A.J. Van Bommel, Nature, 168 (1951) 271.
2 N. Harada and K. Nakanishi, Circular Dichroic Spectroscopy—Exciton Coupling in Organic Stereochemistry, University Science Books, Mill Valley, California, and Oxford University Press, Oxford, 1983.
3 N. Harada, T. Soutome, S. Murai and H. Uda, Tetrahedron: Asymmetry, 4 (1993) 1755.
4 N. Harada, T. Soutome, T. Nehira, H. Uda, S. Oi, A. Okamura and S. Miyano, J. Am. Chem. Soc., 115 (1993) 7547.
5 N. Harada, T. Hattori, T. Suzuki, A. Okamura, H. Ono, S. Miyano and H. Uda, Tetrahedron: Asymmetry, 4 (1993) 1789.
6 T. Hattori, N. Harada, S. Oi, H. Abe and S. Miyano, Tetrahedron: Asymmetry, 6 (1995) 1043.
7 F. Toda, Top. Curr. Chem., 140 (1987) 43; F. Toda, in: J.L. Atwood, J.E. Davis and D.D. MacNicol (Eds.), Inclusion Compounds, Oxford University Press, Oxford, 4 (1991) 126–187; F. Toda, in: G.W. Gokel (Ed.), Advances in Supramolecular Chemistry, JAI Press, London, 2 (1992) 141–191.
8 F. Toda, K. Tanaka, I. Miyahara, S. Akutsu and K. Hirotsu, J. Chem. Soc., Chem. Commun., (1994) 1795; F. Toda, K. Tanaka, C.W. Leung, A. Meetsma, B.L. Feringa, J. Chem. Soc., Chem. Commun., (1994) 2371; F. Toda, K. Tanaka, M. Watanabe, T. Abe and N. Harada, Tetrahedron: Asymmetry, 6 (1995) 1495–1498.
9 F. Toda, J. Synth. Org. Chem. Jpn, 47 (1990) 1118.
10 I. Ohtani, T. Kusumi, Y. Kashman and H. Kakisawa, J. Am. Chem. Soc., 113 (1991) 4092. T. Kusumi, H. Takahashi, X. Ping, T. Fukushima, Y. Asakawa, T. Hashimoto, Y. Kan and Y. Inouye, Tetrahedron Lett., 35 (1994) 4397; T. Kusumi, H. Takahashi, T. Hashimoto, Y. Kan and Y. Asakawa, Chem. Lett., (1994) 1093; H. Yamase, T. Ooi and T. Kusumi, Tetrahedron Lett., 39 (1998) 8113; S. Arita, T. Yabuuchi, T. Kusumi, Chirality, 15 (2003) 609.
11 J.M. Seco, Sh.K. Latypov, E. Quinoa and R. Riguera, Tetrahedron Lett., 35 (1994) 2921; J.M. Seco, Sh. Latypov, E. Quinoa and R. Riguera, Tetrahedron: Asymmetry, 6 (1995) 107; Sh.K. Latypov, J.M. Seco, E. Quinoa and R. Riguera, J. Org. Chem., 60 (1995) 504; J.M. Seco, Sh.K. Latypov, E. Quinoa and R. Riguera, Tetrahedron, 53 (1997) 8541; Sh.K. Latypov, J.M. Seco, E. Quinoa and R. Riguera, J. Am. Chem. Soc., 120 (1998) 877; J.M. Seco, E. Quinoa and R. Riguera, Tetrahedron, 55 (1999) 569; J.M. Seco, E. Quinoa and R. Riguera, Tetrahedron: Asymmetry, 11 (2000) 2781; J.M. Seco, L.H. Tseng, M. Godejohann, E. Quinoa and R. Riguera, Tetrahedron: Asymmetry, 13 (2002) 2149; J.M. Seco, E. Quinoa and R. Riguera, Chem. Rev., 104 (2004) 17.
12 B.M. Trost, J.L. Belletire, S. Godleski, P.G. McDougal, J.M. Balkovec, J.J. Baldwin, M.E. Christy, G.S. Ponticello, S.L. Varga and J.P. Springer, J. Org. Chem., 51 (1986) 2370.

13 H. Ohruij, J. Synth. Org. Chem. Jpn, 56 (1998) 591. K. Fukushi, Nippon Nogeikagaku Kaishi, 72 (1998) 1345.
14 Review articles: J. Chromatogr. A, 906 (2001) 1–482. C. Yamamoto and Y. Okamoto, Maku, 25 (2000) 277. Y. Kobayashi, A. Matsuyama and A. Onishi, Fine Chemicals, 29 (2000) 59.
15 R.N. Patel (Ed.), Stereoselective Biocatalysis, Marcel Dekker, New York, 2000.
16 N. Harada, T. Nehira, T. Soutome, N. Hiyoshi and F. Kido, Enantiomer, 1 (1996) 35.
17 N. Harada, N. Koumura and M. Robillard, Enantiomer, 2 (1997) 303.
18 N. Harada, N. Koumura and B.L. Feringa, J. Am. Chem. Soc., 119 (1997) 7256.
19 S. Toyota, A. Yasutomi, H. Kojima, Y. Igarashi, M. Asakura and M. Oki, Bull. Chem. Soc. Jpn., 71 (1998) 2715.
20 N. Harada, V.P. Vassilev and N. Hiyoshi, Enantiomer, 2 (1997) 123.
21 N. Harada, N. Hiyoshi, V.P. Vassilev and T. Hayashi, Chirality, 9 (1997) 623.
22 N. Harada, K. Fujita and M. Watanabe, Enantiomer, 2 (1997) 359.
23 T. Nehira and N. Harada, to be published.
24 K. Fujita and N. Harada, to be published.
25 N. Koumura and N. Harada, to be published.
26 N. Harada, K. Fujita and M. Watanabe, Enantiomer, 3 (1998) 64.
27 N. Harada, K. Fujita and M. Watanabe, J. Phys. Org. Chem., 13 (2000) 422.
28 S. Kuwahara, M. Watanabe, N. Harada, M. Koizumi and T. Ohkuma, Enantiomer, 5 (2000) 109.
29 M. Watanabe, S. Kuwahara, N. Harada, M. Koizumi and T. Ohkuma, Tetrahedron: Asymmetry, 10 (1999) 2075.
30 H. Taji and N. Harada, to be published.
31 M. Kosaka, M. Watanabe and N. Harada, Chirality, 12 (2000) 362.
32 N. Harada, M. Watanabe, S. Kuwahara and M. Kosaka, J. Synth. Org. Chem. Jpn., 59 (2001) 985.
33 S. Kuwahara, K. Fujita, M. Watanabe, N. Harada and T. Ishida, Enantiomer, 4 (1999) 141.
34 A. Ichikawa, S. Hiradate, A. Sugio, S. Kuwahara, M. Watanabe and N. Harada, Tetrahedron: Asymmetry, 10 (1999) 4075.
35 N. Harada, M. Watanabe, S. Kuwahara, A. Sugio, Y. Kasai and A. Ichikawa, Tetrahedron: Asymmetry, 11 (2000) 1249.
36 A. Ichikawa, S. Hiradate, A. Sugio, S. Kuwahara, M. Watanabe and N. Harada, Tetrahedron: Asymmetry, 11 (2000) 2669.
37 H. Taji, Y. Kasai, A. Sugio, S. Kuwahara, M. Watanabe, N. Harada and A. Ichikawa, Chirality, 14 (2002) 81.
38 T. Fujita, S. Kuwahara, M. Watanabe and N. Harada, Enantiomer, 7 (2002) 219.
39 A. Ichikawa, H. Ono, S. Hiradate, M. Watanabe and N. Harada, Tetrahedron: Asymmetry, 13 (2002) 1167.
40 H. Taji, M. Watanabe, N. Harada, N. Naoki and Y. Ueda, Org. Lett., 4 (2002) 2699.
41 Y. Kasai, M. Watanabe and N. Harada, Chirality, 15 (2003) 295.
42 M. Kosaka, T. Sugito, Y. Kasai, S. Kuwahara, M. Watanabe, N. Harada, G.E. Job, A. Shvet and W.H. Pirkle, Chirality, 15 (2003) 324.
43 A. Ichikawa, H. Ono and N. Harada, Tetrahedron: Asymmetry, 14 (2003) 1593.
44 T. Nishimura, H. Taji and N. Harada, Chirality, 16 (2004) 13.
45 Y. Kasai, H. Taji, T. Fujita, Y. Yamamoto, M. Akagi, A. Sugio, S. Kuwahara, M. Watanabe, N. Harada, A. Ichikawa and V. Schurig, Chirality, 16 (2004) 569.
46 J. Naito, M. Kosaka, T. Sugito, M. Watanabe, N. Harada, and W. H. Pirkle, Chirality, 16 (2004) 22.
47 B. Wu and H.S. Mosher, J. Org. Chem., 51 (1986) 1904.

INDEX

A

B

K

L

M

Q

R

S

T

U

V

W

X

Z

CPSIA information can be obtained at www.ICGtesting.com
Printed in the USA
LVOW101959150512

281913LV00010B/60/P